[redacted]

The Biology of People

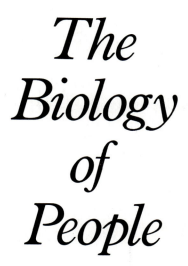

A Series
of Books
in Biology

CEDRIC I. DAVERN
Editor

The Biology of People

SAM SINGER

HENRY R. HILGARD

University of California, Santa Cruz

W. H. Freeman and Company
San Francisco

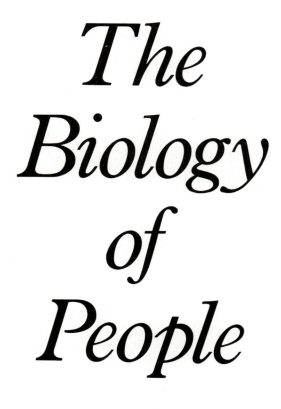

Cover: A drawing by Auguste Rodin of the face of Mme. Séverine (courtesy of Szépmüvészeti Muzeum, Budapest).

Library of Congress Cataloging in Publication Data

Singer, Sam, 1927–
The biology of people.

(A Series of books in biology)
Includes index.
1. Human physiology. 2. Human biology.
3. Human genetics. I. Hilgard, Henry R., joint author.
II. Title.
QP34.5.S56 612 77-17893
ISBN 0–7167–0026–3

Printed in the United States of America

9 8 7 6 5 4 3 2 1

Contents

Preface

In his clear and enthusiastic way, George Gaylord Simpson has recently proclaimed that biology and people are the two most interesting subjects in the world. We agree that the two are at least challenging and inviting subjects and have written this book in hopes of sharing with you what we think are the fundamentals—and some of the highlights—of the biology of people.

For whom is this book intended? Most of our readers will probably be college students enrolled in introductory, relatively nontechnical, courses in biology or human biology. But we warmly welcome the attention of general readers, too, and have therefore tried to make our discussion of the biology of the human species intelligible to anyone who has enough interest in the subject to pick up this book and browse through it.

Human beings are surely the most perplexing and intensively studied creatures that have yet evolved on earth. In fact, so much is already known about human biology that any general discussion of the subject runs the risk of being haphazard. In an effort to make this book both comprehensive and understandable to those who have little or no background in biology, we decided to divide our discussion into three main parts.

Part I is "The Human Species," in which we first establish the context for the rest of the book by discussing the theory of evolution and the concept of biological species and then turn our attention to the evolution, behavior, and ecology of the human species.

Part II is "The Human Machine." Here we first discuss the structure and function of cells and tissues, as well as how the activities of the myriad parts of the human organism are interconnected and coordinated. We then provide a system-by-system description of the structure and function of the human machine and an overview of the major bodily changes that take place in the course of a human lifetime.

Part III is "Human Genetics," a relatively nontechnical introduction to the principles of genetics in general, and of human genetics in particular. (This part of the text is also available as a separate paperback.)

It took four years to write and produce this book, and both authors would like to express thanks to some of the people whose comments and artistic talents were particularly helpful to us as the manuscript progressed. First, we would like to thank each other. Our collaboration was productive and gratifying in that our relative strengths and deficiencies happen to offset one another rather nicely. Thanks are also due to professors J. Z. Young, Kenneth S. Norris, Arthur J. Vander, C. Ladd Prosser, Cedric I. Davern, Robert S. Edgar, and Ursula W. Goodenough, whose critical comments surely helped to improve our rough drafts. Finally, we happily express our appreciation for the expert editorial guidance of Gunder Hefta, John Painter, and especially Linda Chaput; the finely wrought drawings of Eric Heiber and Associates; the sense of design of Jack Nye, who styled the pages; the vigilance of Ruth Allen, who helped with the permissions; and the careful efforts of Betsy Wootten, who typed many, and proofread all, of the chapters.

January 1978

SAM SINGER

HENRY R. HILGARD

The Biology of People

PART I

The Human Species

These people have just unearthed the fossil remains of a giant bony fish, Portheus. *About 70 million years before this photograph was taken, the creature was entombed by sediments in an ancient sea that then covered Western Kansas. (Courtesy of The American Museum of Natural History.)*

CHAPTER 1

Evolution, Species, and People

Our planet is home to an astounding array of living things. A leisurely walk through a green canopied forest, along a rocky ocean shore, or even across an abandoned lot or a well-manicured city park is all that it takes for any of us to encounter several kinds of familiar creatures. But no one person can ever experience the total diversity of life on earth, even in a lifetime of observing nature closely. This is because of the enormous diversity of living creatures; about one and a half *million* kinds have already been described and named by persons engaged in the science that studies living things—biology. Nonetheless, the task of describing the diversity of living things remains largely undone: biologists estimate that between 5 and 10 million kinds of creatures actually exist.

Environmental conditions on the surface of the earth are far from uniform. Yet living things are found virtually everywhere—in the sea, in the desert, on mountaintops, in tropical lowlands, and even in sunless caves deep within the earth. Wherever they may be found, all organisms are especially suited to life in the particular environment in which they normally live. In other words, there is a *good fit* between an organism and its particular surroundings. This common observation is an extremely important one. Living things cannot be understood in isolation from the environment in which they are normally found. Whatever else they may be, organisms are things that do something somewhere.

1–1
Our modern understanding of how evolution works dates from 1859, at which time Charles Darwin (right) and Alfred R. Wallace (opposite page) published the same theory based on independent observations of plants and animals in different parts of the world. Darwin usually gets most of the credit. He had been working at documenting his theory for about twenty years before 1859. (Photo courtesy of The Bettmann Archive.)

Although it is easy to be impressed by the diversity and the environmental good fit of the creatures around us, it is not so easy to explain how the present situation may have come about. But it is worthwhile to try to find an explanation. After all, human beings are living things, and like other creatures, we are well suited to our environments. In trying to explain why living things are diverse and well fitted to the environment, we find ourselves addressing a question well known to all of us: "Where did I come from?"

Biologists explain both the diversity and the goodness of fit by the theory of evolution. Indeed, evolution is at the core of all biological explanations of the natural world. The theory that underlies the evolutionary explanation, at least as it relates to the goodness of fit, was first proposed in 1859 by Charles Darwin and Alfred R. Wallace, independently (Figures 1-1 and 1-2). The theory was formulated to explain the appearance, distribution and activities of those organisms with which naturalists of about a century ago were most familiar—plants, and especially animals. But since that time, year by year and decade by decade, the evolutionary explanation has been successfully applied to newly discovered creatures whose existence was not even known of when the theory was first thought up. That is a sign of great strength for any scientific explanation.

But evolution is not the *only* explanation for the existence and order of the natural world. How one explains nature depends on what one sees there, and it is well known that not all people perceive the world in the same way. What people see in nature depends upon their world view, which is one aspect of a uniquely human activity called culture (see Chapter 3). In some nontechnological societies, people look at the natural world and see, not isolated creatures in different environments, but rather a continuum, a blending of living and nonliving things into a harmonious whole. It is not surprising that their explanations of how things got that way may differ from ours. The theory of evolution is a product of the world view held by twentieth-century science. It is a fitting explanation for the state of the natural world insofar as it makes sense to those of us who share the same world view.

The theory of evolution maintains that all living things evolve. There are no exceptions. This means that human beings have evolved and are evolving, a fact that has never been taken lightly. When the wife of the Bishop of Worcester first heard in 1860 of the idea that people evolved from some kind of nonhuman ancestors, she supposedly exclaimed to her husband, "My dear, let us hope it is not true, but if it is, let us pray that it will not become generally known!" But although it has, in spite of the wishes of the Bishop's wife, become generally known that people have evolved, it is not generally appreciated how clearly the theory of evolution explains characteristics observed in human populations and in individual persons. This book describes people primarily in evolutionary terms. In chapters to come, we will try to explain many human characteristics as the products of evolution. In so doing we will learn more about other living things and more about ourselves. But before we can apply evolutionary concepts specifically to people, we must first discuss the theory of evolution itself. That is the purpose of the present chapter. Let us begin our discussion of evolution at the beginning, with the origin of life on earth.

The Origin of Life on Earth

Living things have inhabited the earth for about 3.5 *billion* years. We are able to make this extraordinary statement with assurance because a record exists from which we can infer the time of the first appearance of living things and in which are found some of the remains of the organisms themselves. These remains are called *fossils,* and the record is in the form of sedimentary rocks. Sedimentary

1–2
Alfred Russel Wallace (left) and Charles Darwin (opposite page) reached the same conclusions concerning evolution at about the same time in the mid-nineteenth century. The two men presented a paper before the Linnaean Society of London on July 1, 1858. (Photo courtesy of The Granger Collection, New York.)

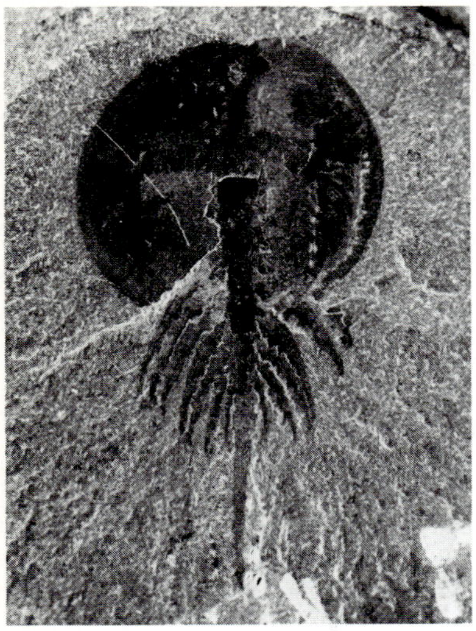
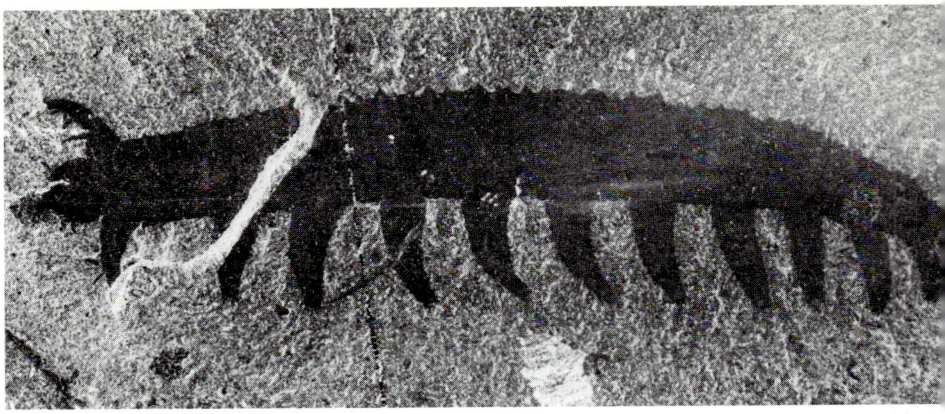

1–3
Fossils from the Burgess Shale in British Columbia. These animals were entrapped by sediments about 550 million years ago. All photos are slightly larger than the original creatures. (Courtesy of the Smithsonian Institution Press.)

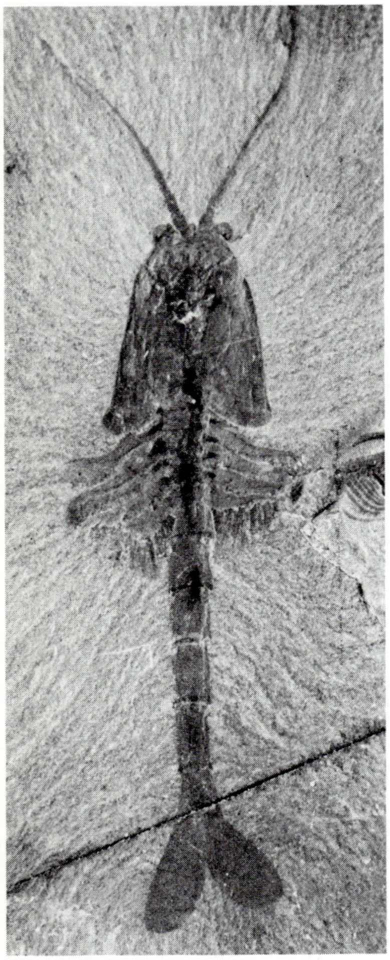

rocks are formed by the accumulation and consolidation of sediments over long periods of time, and some of these sediments provide direct evidence for the existence of ancient organisms (Figure 1-3).

When life began, the envelope of gases surrounding the earth—the atmosphere—had a composition very different from its present one. It contained mostly hydrogen, ammonia, carbon dioxide, and various other gases that are generally toxic to most modern organisms. These simple molecules were converted into more complicated ones by means of energy supplied from various sources such as ultraviolet rays from the sun and lightning during ancient storms. (Figure 1-4 illustrates an artificial means of achieving the same end.) This generation of complicated molecules from the simple compounds present in the atmosphere and oceans of the primitive earth is called *chemical evolution.* Well over 3.5 billion years ago, it resulted in the formation of a variety of chemical compounds that were dissolved in the ancient oceans, thus forming a soup of nonliving chemicals from which the first living things were born.

We can only guess how the transition from nonliving to living things occurred, and we have no way of knowing exactly what the first living things were like. But most biologists agree that the first living creatures showed at least the following characteristics which are common to all living things today: (1) they must have been composed of complicated molecules, (2) they must have been separated from the environment by some kind of selective barrier (a membrane of sorts), and (3) they must have reproduced themselves in nearly exact copies.

All living things consist of complicated molecules that are constantly being broken down and replaced by new ones. In order for this to happen, living things require a source of energy. The first organisms almost certainly found their source in energy stored in the chemical bonds of some of the compounds that surrounded them in the soup of the ancient oceans. Organisms that depend on obtaining preformed energy-rich compounds from somewhere in the environment are called *heterotrophs* (a word whose literal meaning is *other feeders).* We can well imagine that as living creatures reproduced themselves and became more numerous, they probably tended to use up these environmental compounds faster than the compounds were generated by chemical evolution. Unless living things were to remain few in number or perhaps to live only in widely scattered places, they could not continue to survive on such a limited source of energy. And, in fact, they did not have to for long because evolution occurred.

What happened was that some organisms acquired the capacity to manufacture their own energy-rich compounds from very simple substances that were readily available in the environment. Creatures that synthesize energy-rich compounds (mostly sugars) from simpler molecules obtained in the environment are called *autotrophs* (self-feeders). Today virtually all autotrophs engage in a chemical reaction called *photosynthesis.* In photosynthesis, light energy is used to convert molecules of carbon dioxide and water into molecules of sugar and oxygen. The sugar then serves as an energy source for the organism that manufactured it (see Chapter 4).

We have no way of telling how long it took living things to develop the ability to carry out photosynthesis, but we do know that photosynthesizing organisms existed by about 3 billion years ago. We presume that they evolved from nonphotosynthesizing ancestors. This presumption is a reasonable one because among the most primitive kinds of organisms now living those that do not photosynthesize are generally less complicated biochemically than those that do. Thus, photosynthesis can be thought of as a biochemical process that some organisms added to and integrated with the basic chemical machinery of life that had already evolved earlier.

With the evolution of photosynthetic autotrophs, living things probably became more abundant, perhaps because they were able to colonize parts of the seas where heterotrophs could not find enough large preformed molecules to sustain them.

As unlikely as it may seem, we have a good idea of what some of these ancient photosynthesizing organisms were like. We find unmistakable evidence of their presence in rocks that are nearly 3 billion years old. Their remains are in the form of structures called *stromatolites,* which are shown in Figure 1-5. Stromatolites are produced when sediments accumulate on top of colonies of simple organisms called *blue-green algae.* We know this because stromatolites are still being formed this way today, and they are indistinguishable from those that appear in the fossil record. This raises an important question. Why is it that the process of evolution, which has resulted in the fantastic diversity of living things we see around us, has not appreciably changed the appearance of blue-green algae over the course of 3 billion years? Although the matter is far from settled, the answer is probably found in the fact that blue-green algae consist of basic units (called *cells*) that are very different from those which make up more familiar creatures.

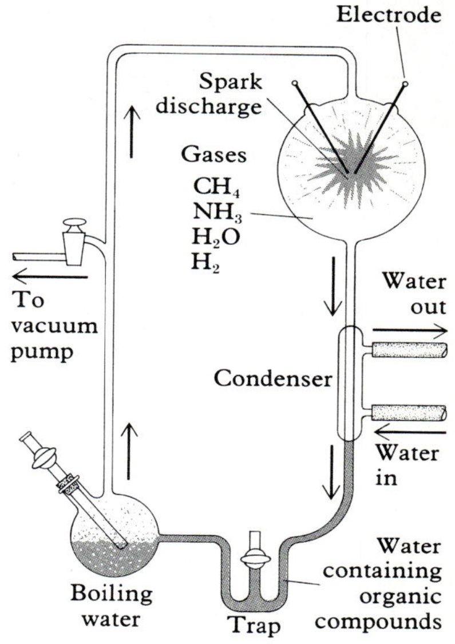

1–4
In this apparatus, complicated chemical compounds can be generated from simpler gases by means of an electrical discharge. Lightning during ancient storms probably had a similar effect on the gases of the primitive atmosphere. (After Sidney W. Fox and Klaus Dose, Molecular Evolution and the Origin of Life. *Copyright © 1972, W. H. Freeman and Company.)*

1–5
Cabbagelike stromatolites are among the oldest fossils known, yet they are still being formed today. The ruler in the photo at the right is 1.5 meters (about 5 feet) long. (Courtesy of Paul Hoffman.)

The Two Basic Types of Cells

Cells are the fundamental structural and functional units of living organisms. This has been known since about 1830. But only relatively recently, with the invention of the electron microscope and the development of precise biochemical techniques, have we come to the important realization that all cells can be classified into one of two basic types. The first are called *prokaryotic* cells, and, though they are enormously complicated in themselves, they are by comparison much simpler than the second type, *eukaryotic* cells.

Figure 1-6 is a diagram that compares the structure of prokaryotic and eukaryotic cells. Some of the obvious differences are: (1) prokaryotes are 10 to 10,000 times smaller than eukaryotes, (2) prokaryotes lack a distinct membrane-bound nucleus, a structure present in all eukaryotic cells in which the genetic material (DNA) is combined with protein to form complex structures called chromosomes that generally are visible only in dividing cells, (3) eukaryotes contain various elaborate structures called *organelles* that are not found in prokaryotes (Figure 1-6).

Most modern organisms are composed of eukaryotic cells. The structure and function of these basic units, which are the building blocks of the human body, are discussed in Chapter 5. But blue-green algae, the creatures that have persisted apparently unchanged for nearly 3 billion years, are prokaryotes.

The only other prokaryotic organisms are the ubiquitous *bacteria*. Traces of objects that look like modern bacteria have been preserved in ancient sedimentary rocks. But because bacteria do not produce characteristic structures such as stromatolites when they interact with the environment, it is more difficult to be certain of their presence than of that of blue-green algae. Nonetheless, most biologists agree that bacteria have more ancient evolutionary origins than any eukaryotes. There is also general agreement that eukaryotes were derived from prokaryotes, because despite the differences we have mentioned both cell types are quite similar in basic biochemistry.

1–6
Top, a typical eukaryotic cell as seen through the electron microscope. Bottom left, a typical prokaryotic cell (bacterium) as seen through the electron microscope. Bottom right, a group of cells drawn to illustrate the relative sizes of eukaryotic and prokaryotic cells.

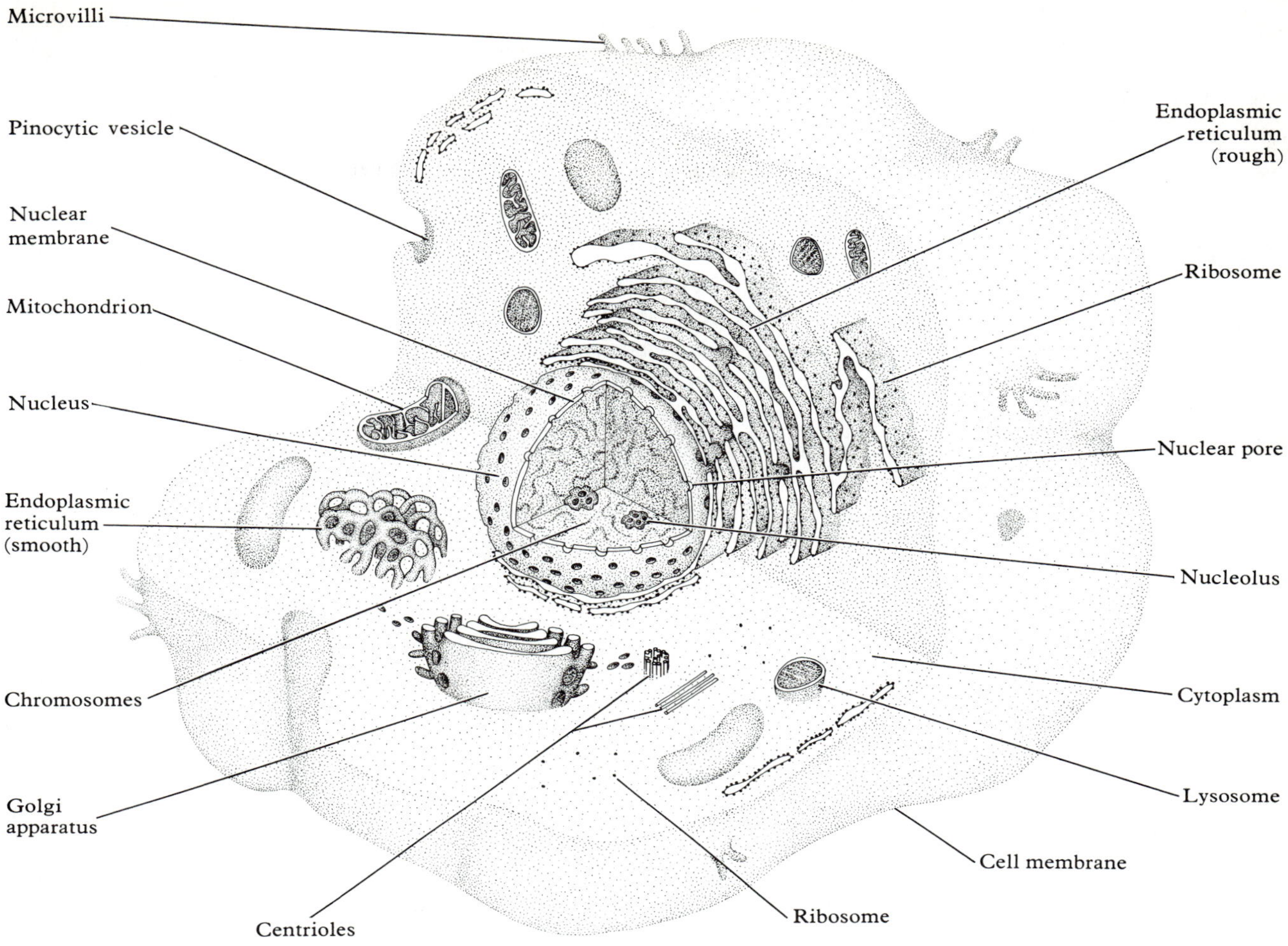

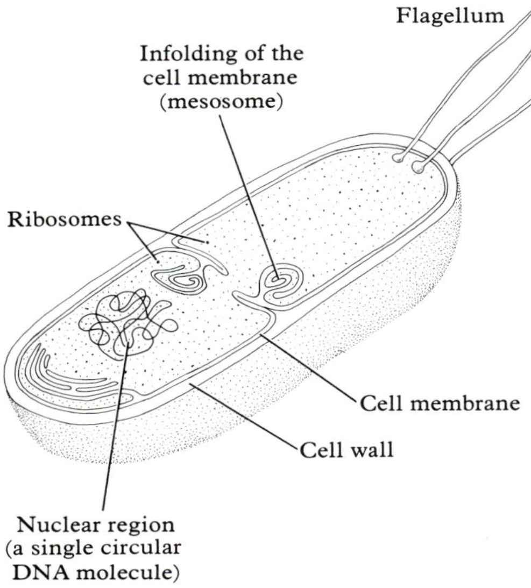

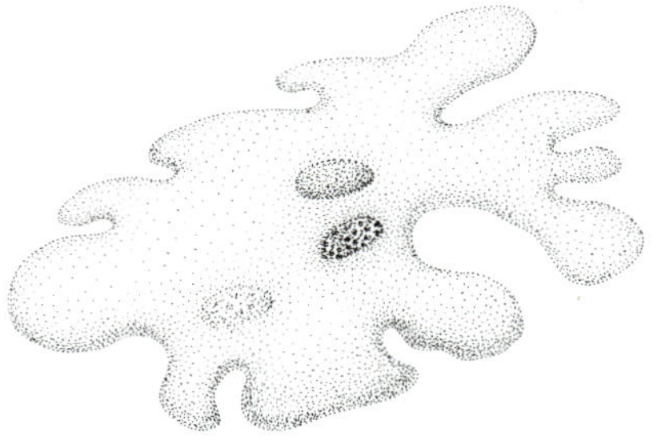

Ameba, a creature that consists of a single unusually large eukaroytic cell that has a diameter of about 0.01 centimeter, is just visible to the naked eye.

A typical eukaryotic cell (mammalian tissue), is about ten times smaller than the *Ameba* and too small to be seen without the aid of a microscope.

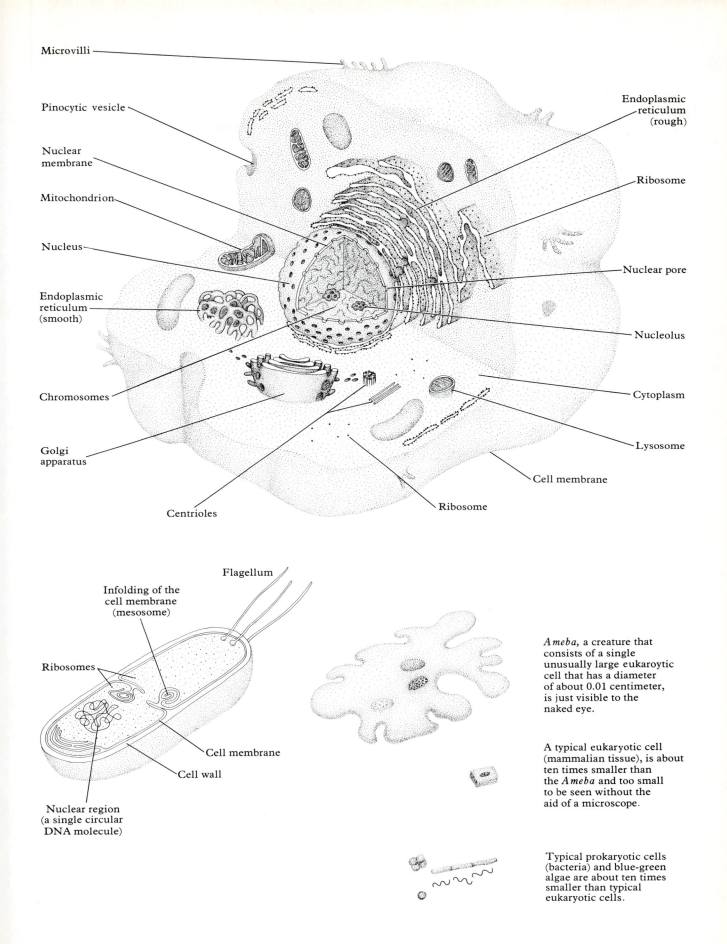

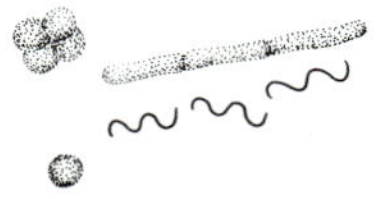

Typical prokaryotic cells (bacteria) and blue-green algae are about ten times smaller than typical eukaryotic cells.

It has been suggested that the first eukaryotic cells resulted from the fusion of prokaryotic cells. As we shall see in Chapter 5, this is probably true of at least some of the components of eukaryotic cells. Nonetheless, the origin of eukaryotic cells, like the origin of life itself, is a subject about which we can only speculate.

Because virtually all of the multicellular organisms we know of, from the microscopic to the leviathan, are composed of eukaryotic cells only, the evolution of eukaryotes from prokaryotes was one of the most significant events in the evolution of living things. Eukaryotic cells contain more DNA than prokaryotic cells, and most organisms composed of eukaryotic cells regularly indulge in a mechanism called *sex,* which results in the constant reshuffling of the composition of DNA in their descendants. Prokaryotes engage in sexual reproduction only sporadically, and this is discussed further in the chapters that follow. The end result of sexual reproduction is that eukaryotes, which have more DNA that is constantly being reshuffled by sex as it is passed on to offspring, are *more variable* than prokaryotes. The enormous capacity for variability in eukaryotic cells as compared to prokaryotic cells probably accounts for the evolution of multicellular organisms from eukaryotic cells only.

Multicellular animals are called *metazoans.* There is no doubt that metazoans originated from some kind of single-celled eukaryotes, but the events surrounding their origins are far from clear. We know that eukaryotic cells had evolved from prokaryotes by about 1 *billion* years ago, because we can distinguish their first traces in the fossil record of that time. From then until about 600 *million* years ago, the record shows only a few scattered traces of single-celled eukaryotes. Then, beginning just before 600 million years ago, an abrupt change took place. Suddenly, over the course of a few million years, the record shows the unmistakable presence of many kinds of metazoans, and it shows that they were found nearly world wide. This sudden explosion of diversity in metazoans remains largely unexplained, and it is as vexing to modern biologists as it was to Darwin over a century ago.

The Cambrian Explosion of Life

Metazoans first appear in the fossil record about 600 million years ago, just before the beginning of the stage of earth history known as the *Cambrian Period.* Figure 1-7 shows the fossil remains of some typical animals of the Cambrian Period, and Figure 1-8 is an overview of geological time and a recap of important events in the history of living things. The record shows that invertebrates, or animals without backbones, were abundant in Cambrian times but that there were no backboned animals (vertebrates). Some simple kinds of plants existed, and they and the animals of that period lived in water: none were land dwellers. How can we explain this Cambrian explosion of metazoan diversity?

People have been trying to solve this problem from Darwin's time on. Darwin himself was well aware of the problem. Indeed, the Cambrian explosion was more of a mystery to him than it is to us because he knew nothing of the Precambrian fossils we have already discussed. In fact, in the last edition of his famous *Origin of Species* he wrote that the Cambrian explosion could be considered as a "valid argument against the views here entertained."

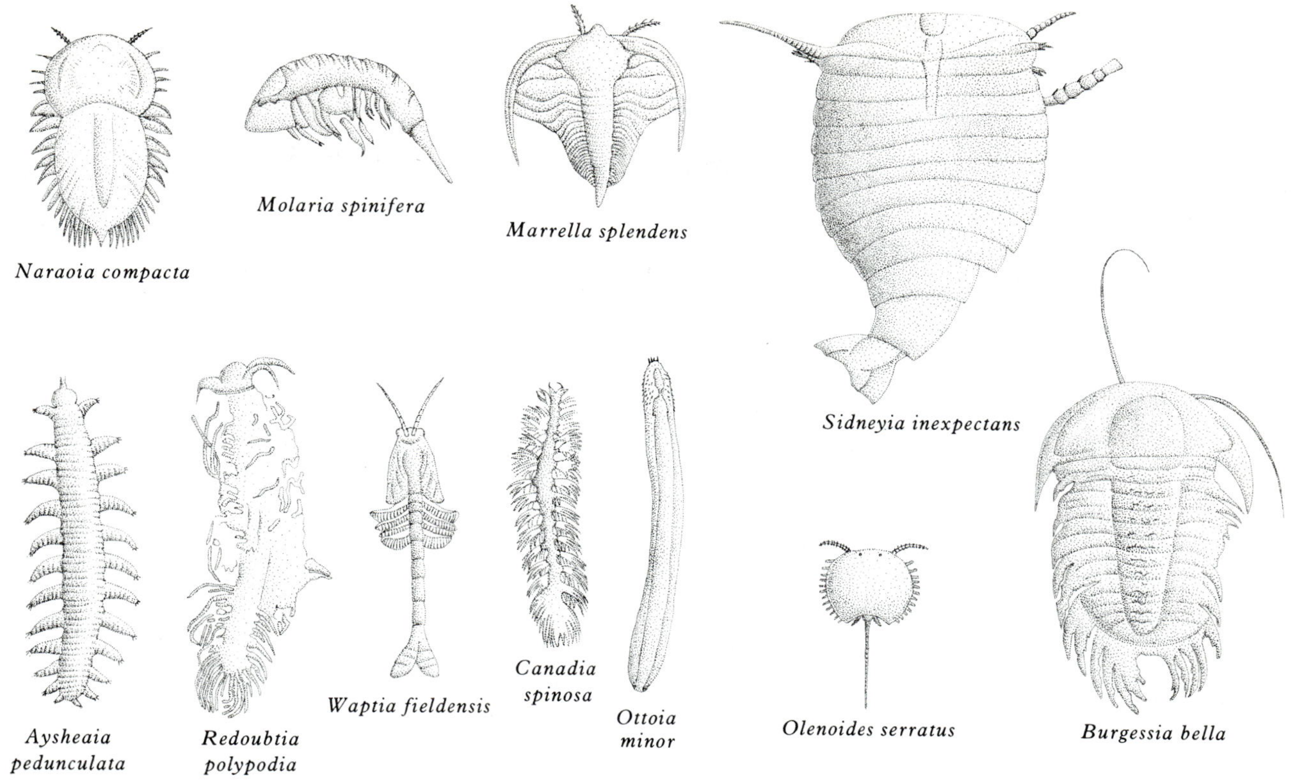

1–7
Some typical Cambrian fossils. (From R. A. Stirton, Time Life and Man. *Copyright © 1959, John Wiley & Sons, Inc.)*

Many explanations have been offered to account for the sudden origin of metazoans. But most biologists agree on several points. First, metazoans did evolve rather quickly by evolutionary standards. And second, metazoan origins must have been related to changes in either the physical or the biological environment that occurred roughly 600 million years ago.

Changes in the physical environment are thought to have taken place mostly in the amount of oxygen present in the atmosphere. It is argued that metazoans did not evolve until atmospheric oxygen reached a certain critical concentration and that this concentration was reached just before the beginning of the Cambrian Period.

One change in the biological environment that may have contributed to the Cambrian explosion is this: perhaps it took until just before the Cambrian Period for certain eukaryotic cells to evolve a way of feeding on the abundant eukaryotic algae that had probably evolved by that time. Among modern organisms, the introduction of a creature that eats a standing "crop" of some other organisms (like the Precambrian crop of algae) results, rather paradoxically, in an *increase* in the diversity of the organisms that are eaten. If this occurred in the Precambrian seas, it could have led to specialization in the creatures that did the eating, which means that they, like the creatures they ate, could also become more diversified.

It seems likely that changes in both the physical and biological environment acted together to result in the Cambrian explosion. In the end, we really do not know what happened. But we probably should not be so surprised that

ERA	PERIOD	EPOCH	(years ago)	MAJOR EVENTS IN EARTH HISTORY	
CENOZOIC	Quaternary	Holocene	10,000		Ice Ages
		Pleistocene	2,500,000	Earliest humans	Modern horse evolves in North America, then dies out; Grand Canyon carved; Cascade Range and Pacific Coast Ranges formed
	Tertiary	Pliocene	7,000,000		
		Miocene	26,000,000	Rapid spread and evolution of grazing mammals	
		Oligocene	38,000,000	Earliest elephants	
		Eocene	54,000,000	First primitive horses, rhinoceroses, and camels	
		Paleocene	65,000,000	First primates	Initial uplift and folding of Rocky Mountains
MESOZOIC	Cretaceous		136,000,000	Extinction of dinosaurs Great evolution and spread of flowering plants	Half of North America covered by seas; Initial uplift of Sierra Nevada
	Jurassic		190,000,000	First birds and mammals Dinosaurs at their peak	Dinosaurs
	Triassic		225,000,000		Arid climates in much of western North America
PALEOZOIC	Permian		280,000,000	Mammal-like reptiles	Ice Ages in Southern Hemisphere World climate much like today Deserts in western United States
	Pennsylvanian		325,000,000	First reptiles	Large insects; Widespread swamps, coal source Tropical climate in United States; Uplift and folding of Appalachian Mountains
	Mississippian		345,000,000		Sharks abundant; Trilobites; Widespread flooding of North America, limestone deposited
	Devonian		395,000,000	First amphibians First forests	
	Silurian		430,000,000	First air-breathing animals (scorpions) First land plants	Deserts in eastern and central U.S.
	Ordovician		500,000,000	Trilobites at peak First vertebrates (fish)	Widespread flooding of North America by seas
	Cambrian		570,000,000	Marine shelled invertebrates common First abundant animal fossils	
PRECAMBRIAN	PROTEROZOIC		2,500,000,000	Marine invertebrates probably common, few with shells (1,200,000,000)	Glaciation—probably worldwide; Many mountain systems uplifted and eroded
	ARCHEOZOIC		4,500,000,000	Earliest plants (3,200,000,000)	

Origin of the Earth

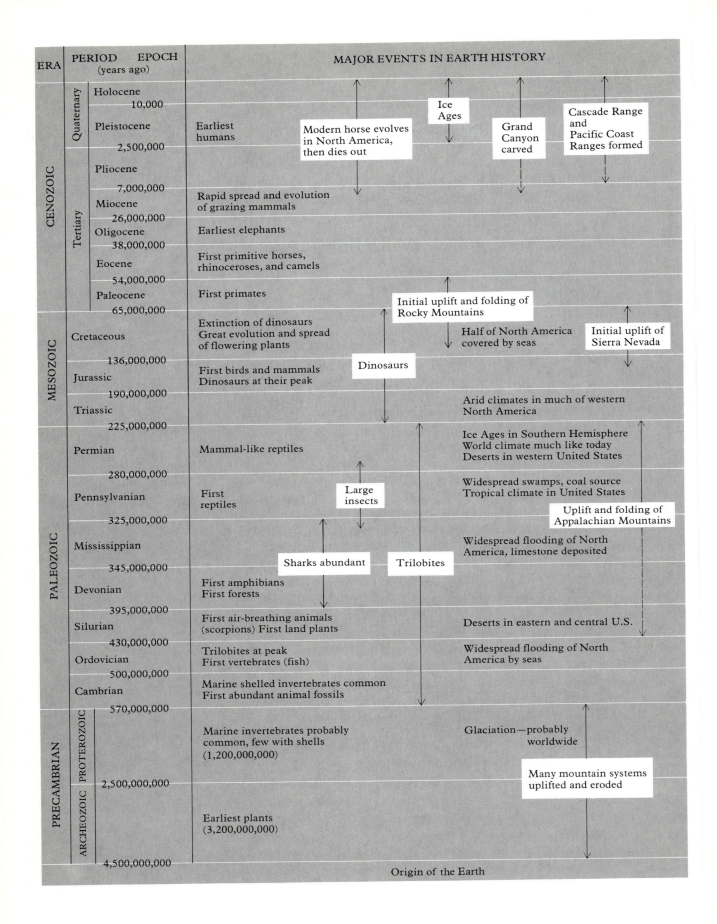

metazoans appeared suddenly. After all, "suddenly" means *millions of years* in this context, and a million years is a very long time. The fossil record from the Cambrian to the present shows consistently that evolution can occur at very different rates. The rate is fastest when creatures with some new adaptation invade a new environment, as we shall discuss in the section that follows. Perhaps the fastest rate of all was attained at the time of the evolution of multicellularity.

Classifying Living Things—The Five Kingdoms

The fossil record is woefully incomplete as an inventory of the diversity of living things at any given time in earth history. Nonetheless, even a superficial examination of the record from the Cambrian to the present leaves us with the overwhelming impression that during this time diversity has increased. Before we discuss how the diversity of modern organisms came about, we should first say something about how biologists classify modern organisms into one of several major groups called *kingdoms.* This will be helpful for several reasons. First, it will provide a better idea of the range of variability among modern organisms. Second, it will set the scene for the upcoming discussion of the evolution of the vertebrates. And third, it will help you to understand where we humans fit into a classification of living things.

All living things can be classified into one of five kingdoms. In the past, biologists were satisfied with only two great divisions—plants and animals. But this is no longer adequate, and never really was, because some organisms meet the criteria for being both plants *and* animals. To eliminate such difficulties in classification, and to emphasize differences in both the organization and the means of nutrition among modern organisms, most biologists agree that a classification that consists of five major divisions is necessary. (Some even insist that there should be six—with an extra kingdom to include viruses, which some people say are living things and some say are not.) The five great kingdoms of living things are as follows:

1. *Kingdom Monera.* Members of the kingdom Monera are the prokaryotes, the bacteria and blue-green algae that we have already discussed. They are among the smallest and simplest living things.
2. *Kingdom Protista.* Protistans are composed of eukaryotic cells. Most are one-celled, but some form multicellular colonies. Though most consist of but a single cell, they may be very elaborate (Figure 1-9). Protistans have been around for at least a billion years, and they have evolved many different ways of life. Some are photosynthetic, some eat other organisms and some are specialized for a parasitic existence in which they obtain nourishment at the direct expense of another organism. (An example of the latter is the organism that causes malaria, which is discussed in Chapter 4.)

1–8
The geological time scale, which shows important events in the history of living things. (After Investigating the Earth, *Revised ed., by the American Geological Institute. Copyright © 1976, 1973 by Houghton Mifflin Company. Used by permission of the publisher.)*

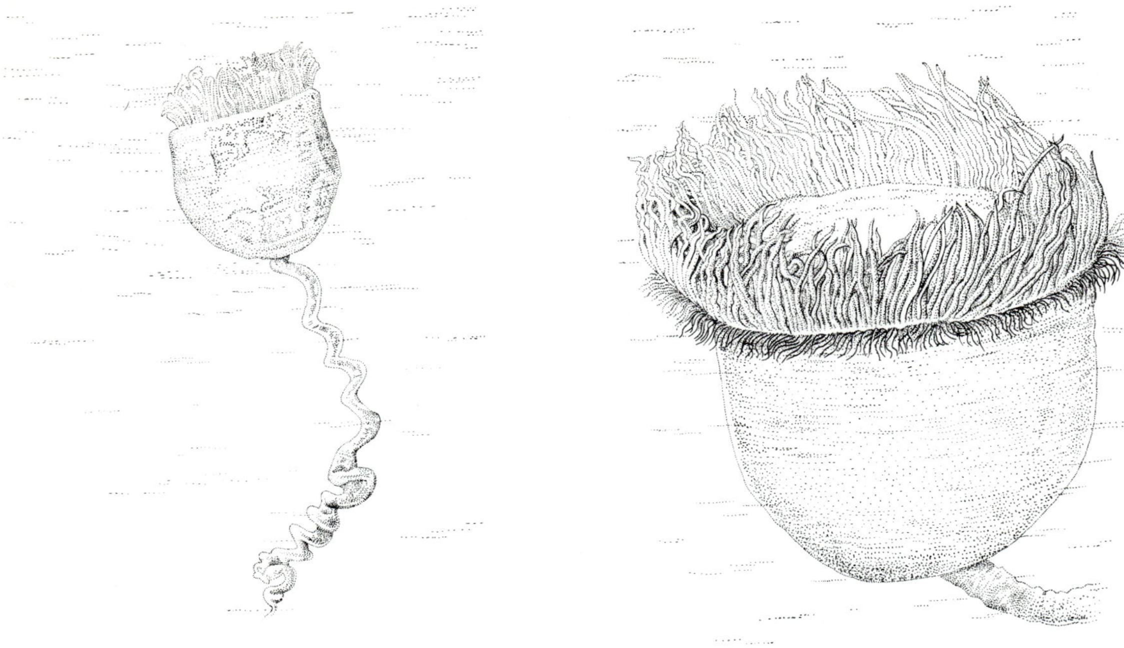

1–9
The single-celled bodies of protistans are sometimes very elaborate. This creature is a vorticella, as seen through the scanning electron microscope.

3. *Kingdom Fungi.* Fungi consist of eukaryotic cells that are usually organized into tissues and organs (see Chapter 5) for at least part of their lives. They are all specialized for absorbing nutrients directly from the environment, and thus are heterotrophs. Familiar examples are mushrooms and molds (Figure 1-10). There are nearly as many different kinds of fungi as there are different kinds of plants.
4. *Kingdom Plantae.* Members of the kingdom Plantae are the so-called true plants, some of which are familiar to all of us. They are made of eukaryotic cells. The great majority, from grasses to redwood trees, are photosynthetic autotrophs, but some are parasites. They are largely responsible for the generation of atmospheric oxygen, and all animals ultimately depend on them as a source of food (Figure 1-11).
5. *Kingdom Animalia (Metazoa).* Animals, in all their shapes and forms, make up the kingdom Animalia. They are multicellular and composed of eukaryotic cells. All are heterotrophs and some are parasites. Characteristically animals ingest food. This means they eat the tissues of other organisms and digest them inside their bodies.

Within the five kingdoms about 60 to 70 distinct subgroups, or phyla, of organisms have been identified, mostly by morphological distinctness (differences in appearance). The animal kingdom comprises about 33 of these distinct groups. Most of the animal phyla date from Cambrian times. (In fact, the only

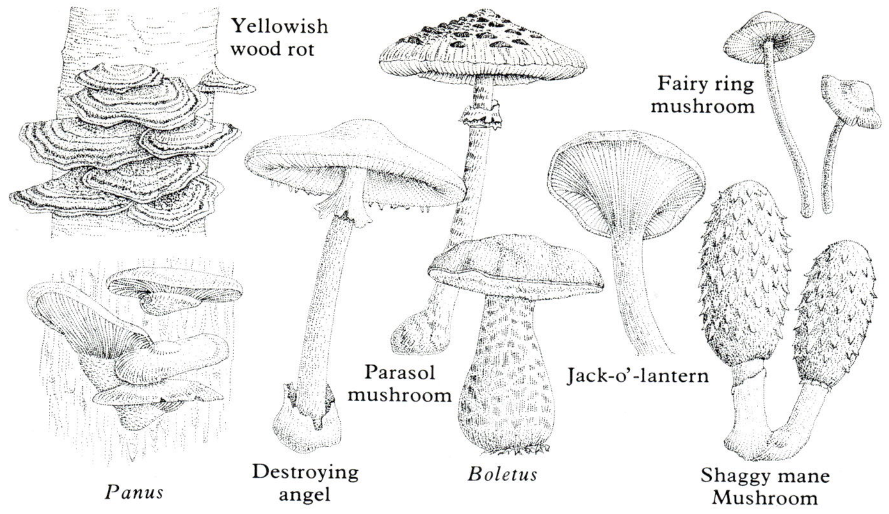

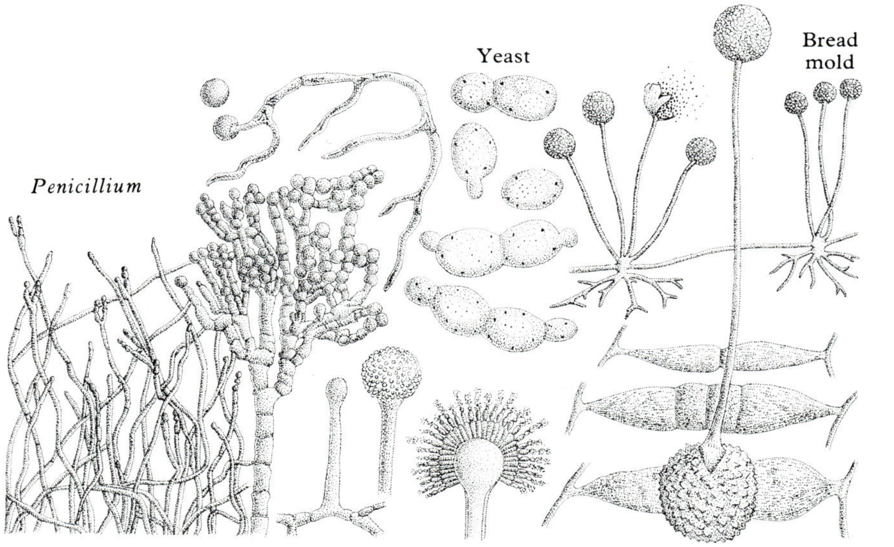

1–10
Some representative mushrooms and molds, the most familiar members of the kingdom Fungi. The molds (bottom) are shown in microscopic detail. (From "Molds and Men," by Ralph Emerson. Copyright © 1951 by Scientific American, Inc. All rights reserved.)

major group that had not appeared in the Cambrian explosion of diversity is the one to which human beings and other animals with backbones belong, the phylum Chordata.)

Now let us consider how the process of evolution resulted in the rich diversity of living things we have just discussed. Most of what is said in the pages that follow applies to all living things, but our main concern will be with the evolution of animals. There are several reasons for our special interest in animal evolution. First, animals, especially vertebrates, have left a fairly complete outline of their evolution in the fossil record, and second, we ourselves are animals.

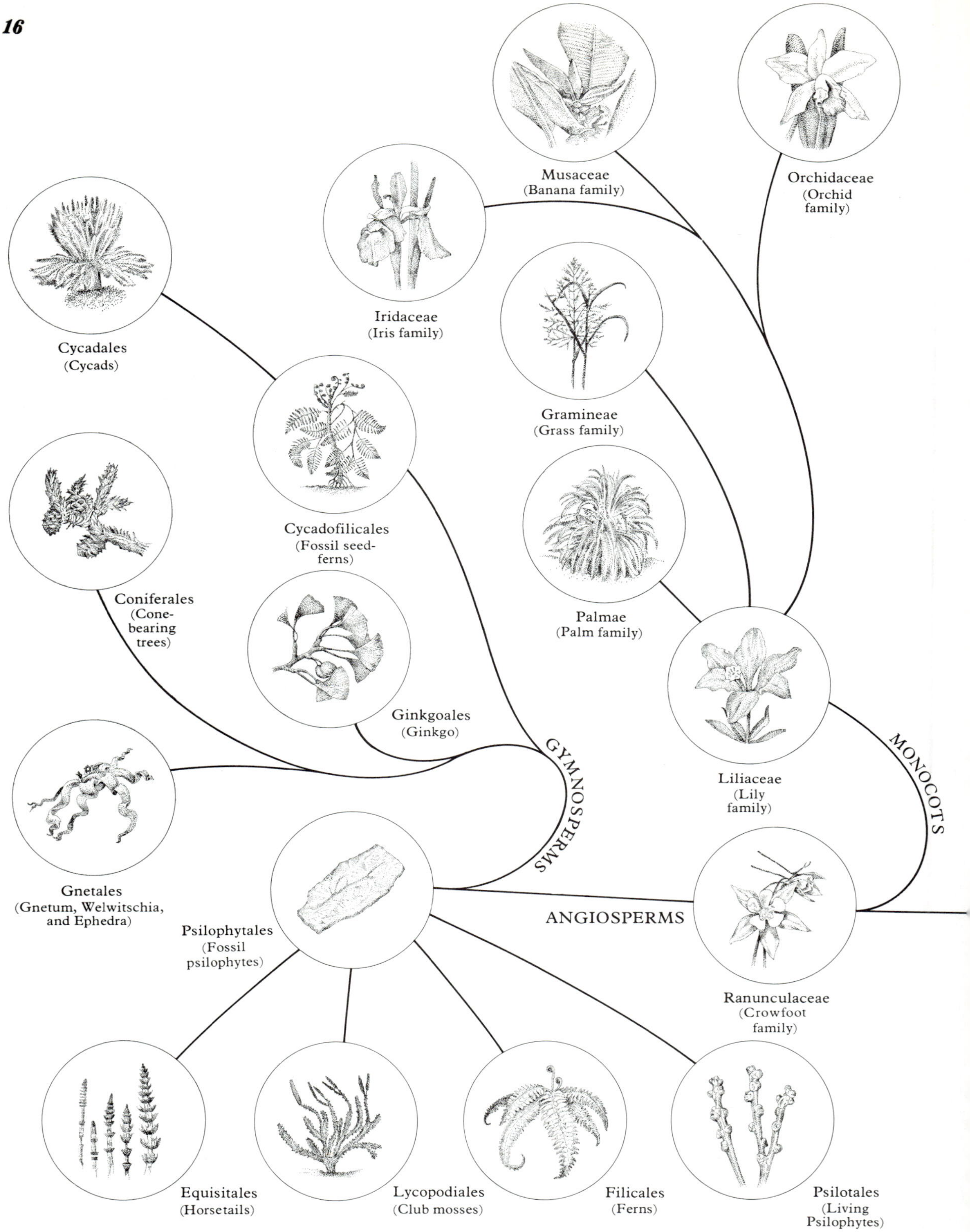

1–11
The evolution and classification of vascular plants. Vascular plants have tubular conducting systems for nutrients and water, and almost all of them are characterized by the presence of roots, stems, and leaves. Vascular plants are classified into three main groups: the primitive spore-bearing ferns and their allies, the gymnosperms

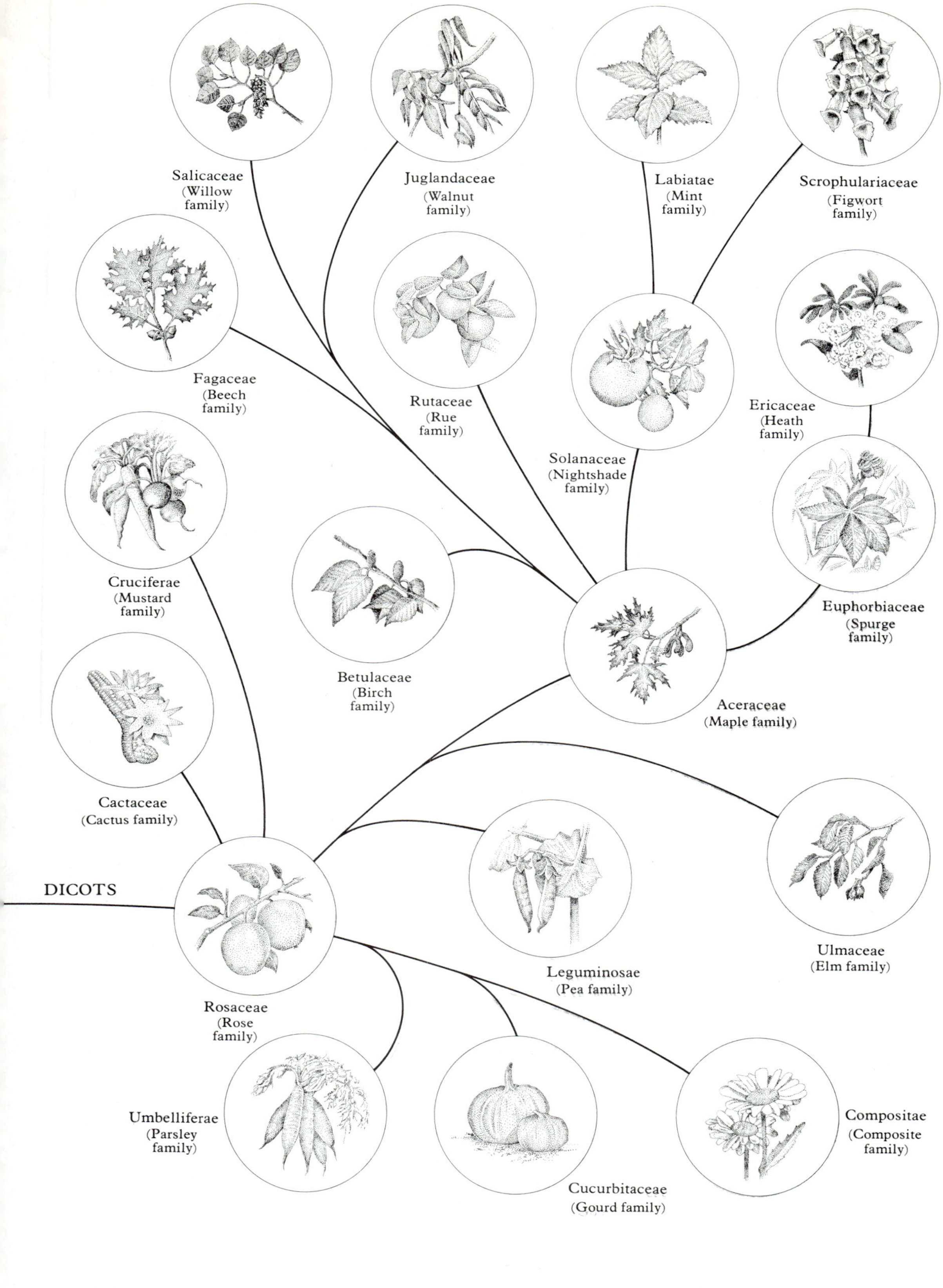

(naked seed-bearers), and the angiosperms (flowering plants.) Angiosperms in turn belong to one of two groups, monocots and dicots, depending on the number of seedling leaves and other characteristics. Some of the most familiar families of monocots and dicots are shown here.

Adaptation and Natural Selection

As we have mentioned, most of the major animal phyla had appeared in the fossil record by Cambrian times. Nonetheless, the record shows an undeniable, almost bewildering, increase in animal diversity from the Cambrian to the present. We must now try to explain how this increase in diversity came about.

We can begin by making a very simple observation. Within a given population of animals of a particular kind (that is, within a given *species),* no two individuals are exactly alike. This is true not only of visible external characteristics such as size and color, but also of less apparent traits like differences in body chemistry and in behavior. This naturally occurring variability among animals in any population serves as the raw material for evolution.

Ultimately, variability between individual animals depends on the composition of the genetic material within the nuclei of the eukaryotic cells that make up their bodies. But this is not to say that the particular morphological, biochemical or behavioral traits of a given animal are therefore fixed and never subject to change during its lifetime. It is well known that the environment has a powerful effect on how an animal's intrinsic genetic makeup is expressed. In other words, individuality among animals is a result of the interaction of genetic and environmental factors. Although the genetic material within the cells of a given animal allows for the expression of a range of characteristics, the animal's environment determines exactly which characters are expressed and to what degree. But, in the final analysis, new raw materials for evolution are generated only by heritable changes that occur within the genetic material. So although the environment may affect the expression of these heritable changes, it cannot account for the origin of the changes themselves.

We need not concern ourselves here with the details of how changes in the genetic material originate. That subject is considered in later chapters, in which we discuss the genetic basis of the variability we all observe among our fellow human beings. For now, all we need to know is that variations in the composition of the genetic material may occur in basically two ways. The first of these is a spontaneous, random but heritable, change called *mutation.* The second is the production of new combinations of DNA by means of *sexual reproduction.* It is through mutation and through sexual reproduction that animals generate and pass on to their offspring the variability that serves as the raw material for evolution.

We are all aware that animals generally have characteristics that are advantageous to them in the environment in which they normally live. Such characteristics are called adaptations, and the process of acquiring them is called *adaptation.* For example, animals that live in the desert, whether they are insects, lizards, birds or rodents (among others) all have physiological and behavioral characteristics that help them to conserve body water. In other words, they are adapted to life in the desert. Similarly, animals that live in mountain streams, in open fields, in saltwater marshes, in the ever-dark abyss of the deep ocean, or anywhere else on earth are well adapted to life in that particular environment. The theory of evolution offers a rather simple explanation of how living things became so well adapted. The explanation is called *natural selection.* Simply put, natural selection results in the maintenance of adaptation. How does this occur?

As you know, mutation and sexual reproduction interact to produce animals that are variable in almost every characteristic. Some of these characteristics may by chance be advantageous to the survival of a given animal in a given environment. If this is so, then those animals that are best suited to survival there will, on the average, *leave more descendants* than those that are not so well suited. The result of this differential reproduction is that eventually only those animals that are best suited to their environment survive to reproduce. And because offspring inherit the adaptive characters that favored the survival of their parents, adpative traits come to predominate in the population over the course of generations.

In chapters to come, we will discuss natural selection specifically in relation to human populations and identify some ponts in the lives of human beings at which natural selection exerts a particularly strong effect. But for now, let us return to the fossil record and attempt to explain some of the fascinating events in the history of animal evolution as recorded in sedimentary rocks in light of what we have just discussed about adaptation and natural selection.

Adaptive Radiation and Extinction

The fossil record shows that the evolution of animal diversity has by no means proceeded at a constant rate. We have already mentioned that metazoan evolution began with an explosion of diversity about 600 million years ago and that virtually all of the major phyla suddenly made their appearance in the fossil record at that time. We can use what we have just discussed about adaptation and natural selection to help explain the occurrence of this remarkable event.

One of the reasons why the phyla appeared to quickly must be that the original metazoans found themselves in an environment that offered many possibilities for adaptation. When they first originated, each of the major animal phyla probably represented distinctive ways metazoans had adapted to life in different environments. Once the evolution of multicellularity had occurred, some metazoans became specialized for a free-floating existence in the seas in which they had evolved. Others became adapted to life in soft mud, on hard rocky bottoms, or in those areas of the earth where the seas lapped up on the then lifeless surface of the land. These ancient environments were not already populated by animals with which the original metazoans had to compete. So natural selection could operate very quickly on naturally occurring variability and result in rather explosive diversification.

Once the major phyla were established, animal evolution proceeded largely by means of modifications and elaborations of the basic body plans that characterize each phylum. Further evolution proceeded in many directions as members of various phyla invaded new environments. Eventually, animals became so diverse and evolved in so many different directions that a given environment became occupied by animals belonging to many phyla, as is true today.

Following the appearance of some new adaptation that enables animals to invade a previously unoccupied environment (or to utilize a given environment more successfully than the animals that already live there), a burst of diversification occurs within the new environment. This "elaboration on a theme" following the first appearance of some major new adaptation is called *adaptive radiation.* Excluding the initial radiation of metazoans we have just discussed, the most outstanding example of adaptive radiation in the history of living things is found among those animals that are by far the most diverse of all—the insects.

There are more kinds, or *species,* of insects than of all other living things combined. Of the well over one million kinds of animals that are known, about *three out of every four* are insects! To what can we attribute the phenomenal diversity of these six-legged creatures?

Insect diversity depends on several factors. But among them, the most important is probably this: insects were the first flying animals. Although they probably first evolved in the sea and now are rather widely distributed in fresh water, insects are primarily land animals. And from the time they first appear in the fossil record about 300 million years ago, many have had well-developed wings. On these wings they soared over the surface of the land, which was then occupied by simple kinds of plants but devoid of other animals. The combination of the evolution of the capacity for flight and the presence of many unoccupied environments into which they could fly was probably in large part responsible for the unequaled adaptive radiation of insects that ensued shortly after they first appeared.

Although adaptive radiation is observed fairly frequently in the fossil record, an even more frequent occurrence is the absolute disappearance of a particular kind of animal from among the multitude of creatures populating the earth at any one time. This dying off is called *extinction,* and the fossil record shows that throughout the long course of the history of evolution, extinction is the rule, not the exception. Charles Darwin was so impressed by the universality of extinction that he wrote: ". . . certainly no fact in the long history of the world is so startling as the wide and repeated exterminations of its inhabitants." Why is it that most populations of animals eventually go extinct?

There is no simple answer to this question. Indeed, most often we can only guess why a given species of animal became extinct. But we can make some generalizations. To begin with, recall that bursts of diversification occur when any environment is invaded for the first time. So it is reasonable to expect that

1–12
The heath hen, one of many creatures now extinct because of human activities. (From James C. Greenway, Jr., Extinct and Vanishing Birds of the World, *Copyright © 1967, Dover Publications, Inc.)*

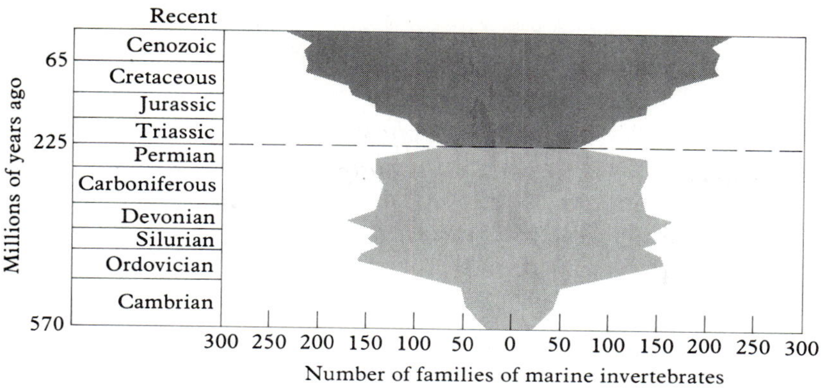

1–13
The great Permian die-off of marine invertebrates. The graph shows how the number of families (and thus the overall diversity) of well-skeletonized invertebrates has varied over the last 570 million years. (From "Plate Tectonics and the History of Life in the Oceans," by James W. Valentine and Eldridge M. Moores. Copyright © 1974 by Scientific American, Inc. All rights reserved.)

extinctions, too, may have something to do with the environment. In particular, if any environment changes to such a degree that animals especially adapted to life there can no longer survive and reproduce, then they may become extinct. We know that this is true because we can observe it happening today. For example, the last heath hen, a lone male, died on Martha's Vineyard, Massachusetts, in 1932 (Figure 1-12). Heath hens became extinct because people overran the land in which they normally lived and, in so doing, changed the environment to such a degree that it was difficult for the birds to survive and reproduce there. (Many were also shot for market because they were good to eat.)

There is also evidence in the fossil record that major changes in the global environment can result in the widespread simultaneous extinction of many kinds of animals. The best example of this wholesale dying-off occurred about 225 million years ago at the end of the Permian Period. The record shows that at about that time nearly one-half of the thousands of kinds of animals living in shallow seas suddenly died off within a few million years (Figure 1-13). What could have happened about 225 million years ago to result in the simultaneous extinction of so many kinds of marine animals?

Many explanations have been offered. However, in the last decade it has become obvious that the major cause of the great Permian die-off was a unique change in the environment. About 225 million years ago a rather startling event occurred: the continents all came together and formed a great supercontinent called *Pangaea* (Figure 1-14). When this happened the amount of living space available for animals in shallow seas was greatly reduced, and global weather must have changed drastically. The net result was that about half of the different kinds of animals living in shallow seas found themselves either in inhospitable environments or unable to compete successfully with other animals for the reduced living space.

Changes in the environment and the failure to compete successfully with other animals for the use of a particular resource are probably important in most extinctions. But how much of a change is required to make an animal become extinct varies enormously and it depends in large part on how specialized the animal is in extracting from the environment the resources it requires to survive and reproduce. Animals that are adapted to a highly specific set of environmental conditions are called *specialists,* and they are more susceptible to the effects of fluctuations of these conditions than *generalists,* which can survive within a fairly wide range of circumstances. For example, the ivory-billed woodpecker (Figure 1-15) is so highly specialized in its eating habits that several square miles of

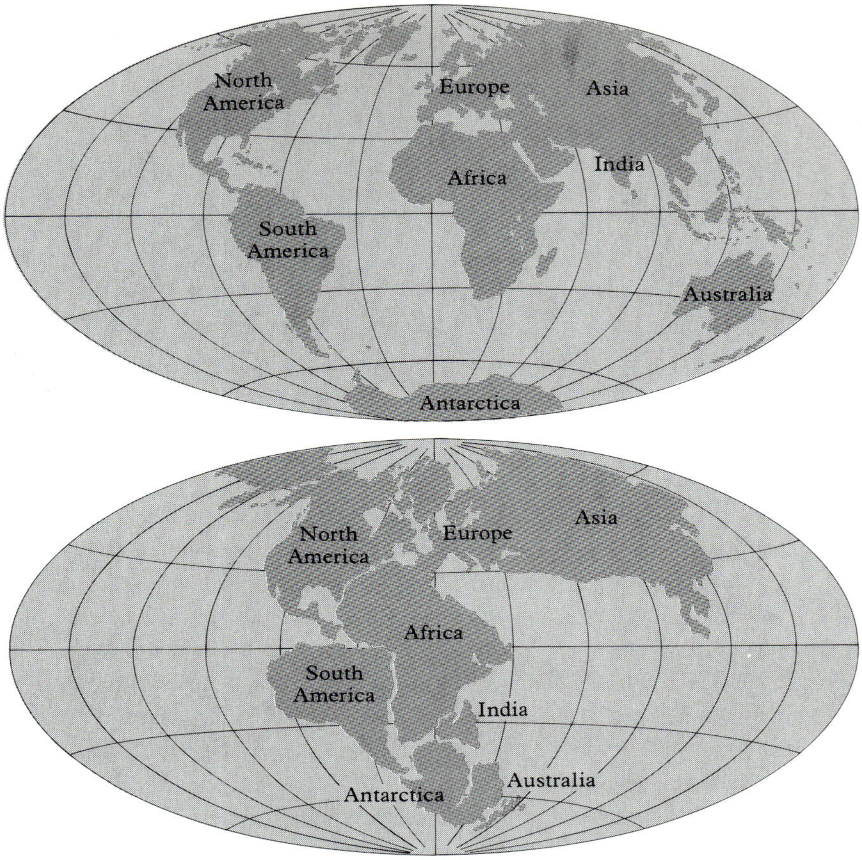

1–14
Top, the continents today. Bottom, the supercontinent of Pangaea, which began to break up about 200 million years ago. (From "Alfred Wegener and the Hypothesis of Continental Drift," by A. Hallam. Copyright © 1975 by Scientific American, Inc. All rights reserved.)

swamp are needed to feed a single pair of birds. The primeval swamps are nearly gone because of human activities, so it is not surprising that the ivory-billed woodpecker is North America's rarest bird. The common robin, on the other hand, is a generalist that can survive on insects, worms, snails, berries, fruits, and other foods, whether they are to be found in fields, forests, or cities. Accordingly, the robin (Figure 1-16) is one of North America's most common birds, and its numbers have increased with the environmental changes brought about by people.

Additional evidence that generalists are less susceptible to extinction than specialists is found in the existence of some extraordinary animals known as *living fossils.* These animals are very unusual because they have existed nearly unchanged from the time they first made their appearance in the fossil record tens or hundreds of millions of years ago to the present day. Some of these remarkable creatures are shown in Figure 1-17. At least part of the reason that these animals have escaped extinction is that they are generalists, and are adapted to tolerate a rather wide range of environmental conditions.

But living fossils are clearly exceptions to the rule. Most animals become extinct. Whether this is caused by overspecialization, the inability to compete successfully against other animals, fluctuations in environmental conditions, or by other factors, extinction is probably inevitable for all living things. After all, the sun will burn out some day, and when it does there will no longer be an energy source to drive the chemical reactions on which all living things depend.

1–15
The ivory-billed woodpecker. Because it requires such highly specialized living conditions this creature has become a rare bird indeed. (From James. C. Greenway, Jr., Extinct and Vanishing Birds of the World. *Copyright © 1967, Dover Publications, Inc.)*

This means, of course, that all the inhabitants of earth may finally become extinct. But that will not happen for at least another 5 billion years. In the meantime, we can content ourselves with trying to understand how the animals that inhabit our planet evolved from ancestors long since extinguished by whatever means.

Perhaps the best way to gain an understanding of how evolution works is to follow the evolution of a particular group of animals as evidenced in the fossil record. And because we are most interested in the evolutionary sequence that led to the emergence of people, let us now turn our attention to the evolution of vertebrates. A discussion of vertebrate evolution will provide further examples of adaptive radiation and extinction, show that the evolution of animal diversity is opportunistic, and provide the context for the discussion of human evolution in the chapter that follows.

1–16
Unlike many species of animals, the robin has directly profited from human activities.

The Evolution of the Vertebrates

The vertebrate story probably began about 450 million years ago, but its opening chapter has yet to be discovered in the fossil record. Vertebrates must have evolved from animals that lacked true backbones, but few fossils that look as if they might bridge the gap between these two great divisions of the animal kingdom have been reported, and those that have are not very convincing. Nonetheless, we can get a clue about how the transition may have occurred from the appearance of some unusual modern animals that are thought to be primitive relatives of the vertebrates. One such justly famous creature is the lancelet, usually known by its scientific name, amphioxus. This small translucent animal (which is eaten with great delight in the Orient) looks rather fishlike, as shown in

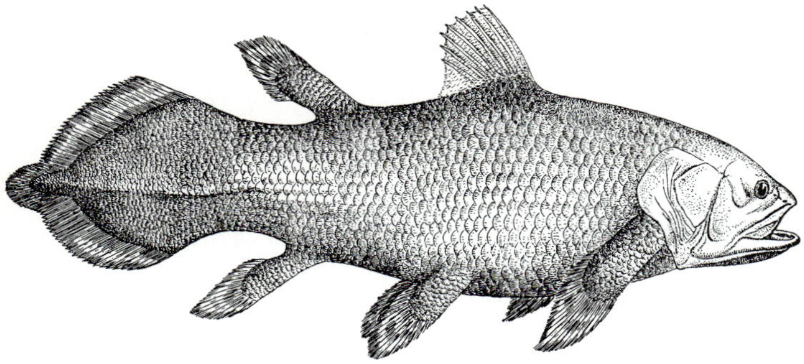

Coelacanth

Tuatara

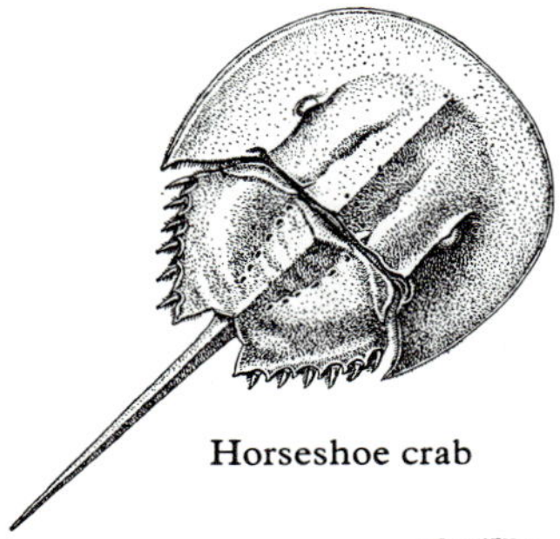

Horseshoe crab

Opossum

1–17
Some living fossils. These animals have not changed appreciably over millions or hundreds of millions of years. (From "Ionizing Radiation and Evolution," by James F. Crow. Copyright © 1959 by Scientific American, Inc. All rights reserved.)

Figure 1-18. Amphioxus is considered a close relative of the vertebrates partly because it has a stiff stabilizing rod (the notochord) running along its back, but it has no true backbone. However, the notochord of amphioxus and the backbone of vertebrates are found in exactly the same anatomical position and it is not hard to envision that the first vertebrates may have arisen from some kind of amphioxuslike animals. But this is not to say that amphioxus itself was an immediate ancestor of the first vertebrates or of anything else. More likely, amphioxus and the first true vertebrates both evolved from a common ancestor long since extinct.

The first animals that had primitive kinds of true backbones were small, fishlike, jawless, heavily armored creatures called ostracoderms (Figure 1-19). They first appear in the fossil record of about 425 million years ago, and they lived in a shallow sea that once covered much of Colorado and several other Western states. A comparison of Figures 1-18 and 1-19 makes obvious how an amphioxuslike animal might have evolved into an ostracoderm. Once bony armor and an internal skeleton appeared, natural selection must have strongly favored the survival and reproduction of animals equipped with it. The external armor probably provided protection against the giant invertebrate water scorpions present at that time, and the internal skeleton provided a sturdy scaffolding for the attachment of muscles. Eventually, the internal skeleton furnished the mechanical support required before any large animal could leave its watery home and take up residence on dry land.

Ostracoderms became extinct. As usual, we can only speculate about what factors were responsible for their demise. But their decline is probably related to the appearance of more efficient kinds of fishes that first evolved when ostracoderms were still flourishing. These new kinds of fish (Figure 1-20) may have been favored by natural selection because, unlike ostracoderms, they had evolved movable jaws. Following the appearance of the jaw, fish began the huge adaptive radiation that has resulted in their being by far the most diverse of the major vertebrate types: fish, amphibians, reptiles, birds, and mammals. And it is from a peculiar kind of jawed fish that lived about 390 million years ago that the first land vertebrates evolved.

These peculiar fishes were known as *lobe-finned fishes* or *crossopterygians* (Figure 1-21). Like modern lungfish (to which they were related), these animals had evolved air bladders that were connected to the back of their throats. This arrangement made it possible for them to gulp air, and to extract some of the oxygen from it. Also, within their fleshy front fins crossopterygians had evolved

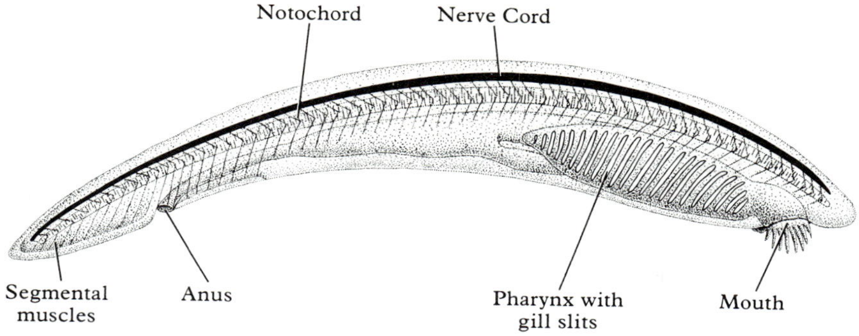

1–18
Amphioxus, a primitive relative of the true vertebrates.

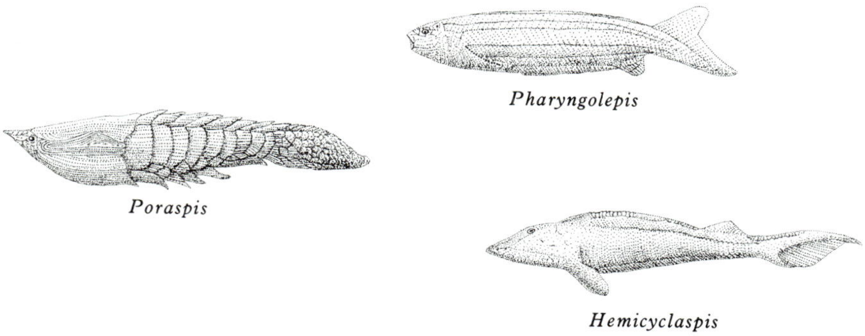

1–19
Some fossil ostracoderms, the oldest animals known to have had primitive backbones. (After Alfred Sherwood Romer, The Vertebrate Body, *4th Ed. Copyright © 1970, W. B. Saunders Company.)*

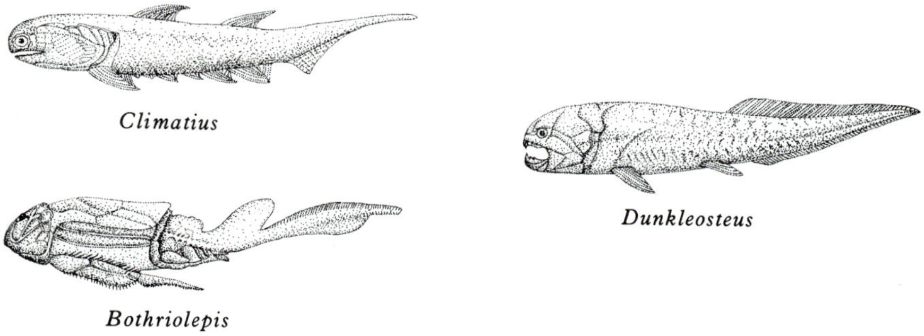

1–20
Some jawed fishes from the Devonian Period. These creatures are not drawn to scale. The creature in the middle was a gigantic animal; many of its kind exceeded 30 feet in length. (After Alfred Sherwood Romer, The Vertebrate Body, *4th Ed. Copyright © 1970, W. B. Saunders Company.)*

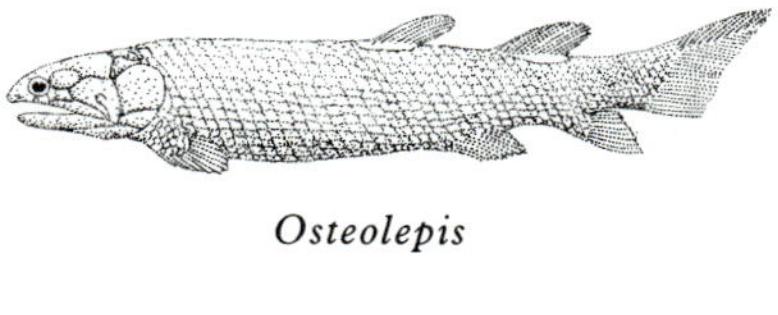

1–21
An extinct crossopterygian from the Devonian Period. (After Alfred Sherwood Romer, The Vertebrate Body, *4th Ed. Copyright. © 1970, W. B. Saunders Company.)*

a rather complicated arrangement of bones. At first, the crossopterygians used their "lungs" to gulp air whenever oxygen was in short supply, and they probably used their front limbs to pull themselves through heavy vegetation on pond bottoms. Later they took fuller advantage, both of their primitive lungs and of their ability to breath air. Some of them simply crawled out of the water and began to take up residence on land. And in time those crossopterygians with the biggest lungs and the sturdiest, most efficient front limbs evolved into another major vertebrate type, the amphibians, a group that includes such modern-day animals as frogs, toads, and salamanders.

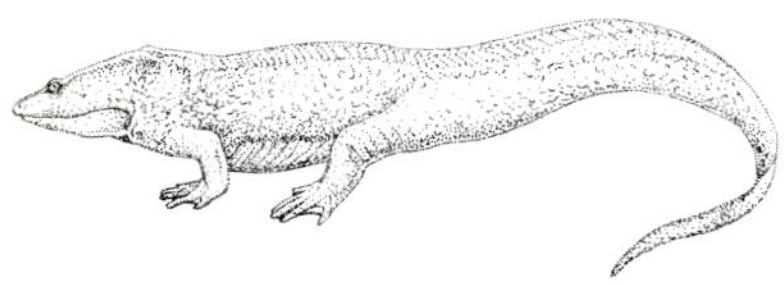

Diplovertebron

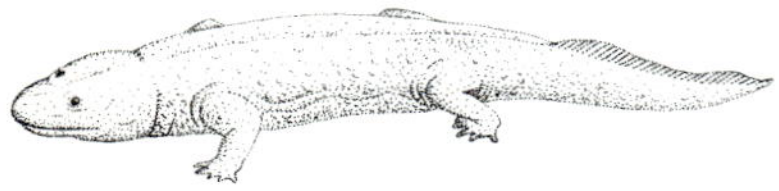

Ichthyostega

1–22
Two of the most ancient known amphibians. Both of these animals were about three feet long.

The first amphibians appeared about 365 million years ago, and they looked like crossopterygians that had developed limbs on which they could pull their low-slung bodies over solid land (Figure 1-22). From what we know happened before, we might expect that amphibians, the first land vertebrates, would undergo adaptive radiation on a grand scale. But although an adaptive radiation of amphibians did occur, it was not a very elaborate one, probably because most amphibians were never totally successful in leaving the water. Even though many spend most of their lives on land, the great majority of modern amphibians must return to water to reproduce. This was probably true of ancient amphibians, too. Also, compared to other land vertebrates, amphibians have a rather inefficient circulatory system. All in all, amphibians (both ancient and modern) have never been an enormously successful vertebrate group. But some amphibians eventually evolved into animals that were the first truly successful vertebrates on land—the reptiles.

In rocks formed about 280 million years ago in what is now the state of Texas, lie the remains of animals that are either reptilian amphibians or amphibious reptiles, usually assumed to be the latter (Figure 1-23). Animals that are clearly reptiles appear in the same rocks of a slightly later period. Presumably these animals, like all modern reptiles, had evolved the real secret of reptilian success—the land egg. The complicated structure of the reptilian egg allowed reptiles to free themselves from the watery environments in which they had evolved. In essence, the land egg provides embryonic reptiles (and birds) with individual watery environments in which to develop. These portable, self-contained bodies of water that do not dry out were largely responsible for the enormous success of reptiles on land.

Reptiles are still fairly abundant. Modern representatives include lizards, snakes, turtles, and crocodilians. But reptiles saw their evolutionary heyday

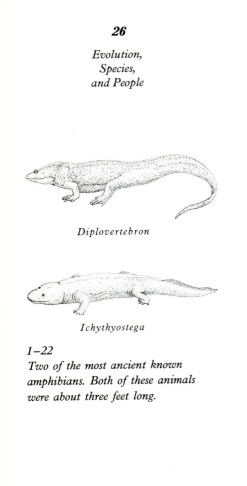

shortly after they first evolved. They underwent extensive adaptive radiation and virtually covered the land because there were no other truly terrestrial vertebrates there to oppose them. Their radiation produced some spectacular and terrifying animals, including a great variety of dinosaurs and ancient flying creatures that had wingspans of over 15 meters (about 50 feet). (Figure 1-24).

But eventually, for reasons that remain largely unknown, the "age of reptiles" came to a rather rapid close. Before this happened, another major vertebrate type, birds, had evolved. Birds eventually replaced flying dinosaurs as the dominant flying vertebrates and in fact probably evolved directly from birdlike dinosaurs. They remain the unquestioned vertebrate masters of the sky today.

When the ruling reptiles became extinct a new vertebrate type, mammals, began to diversify on a grand scale. They now dominate on land. The fossil record shows that small, mammal-like reptiles had already evolved when dinosaurs roamed the earth, but they did not begin their main adaptive radiation until the huge dinosaurs had begun to die off. These early mammals resembled reptiles in their skeletons and they probably still laid land eggs, as do some primitive mammals today (Figure 1-25). But unlike reptiles, which usually lay their eggs and leave never to return, mammals suckle their young and generally take care of them after birth until they can fend for themselves. Other typically mammalian traits include warm-bloodedness, the presence of hair, and the presence in the pregnant female of a well-developed placenta. The placenta is an organ whereby an embryo can be supplied with food and oxygen while it develops within a fluid-filled sac inside its mother's body (see Chapter 15).

Because they spend life before birth within such a protected environment, and because they are nursed and cared for by their mothers from the time they are born, mammals are able to reproduce more efficiently over a wide range of environmental conditions than are reptiles.

The adaptive radiation of mammals produced many of the animals with which we are most familiar (see Figure 2-1). They are wonderfully diverse, ranging in size from tiny shrews to gigantic whales. Yet, as is the rule in evolution, the record shows that in the 100 million years or so that they have been the dominant land animals, more kinds of mammals have become extinct than have survived.

Among those that became extinct are the immediate ancestors of human beings. In the chapter that follows, we will discuss the evolution of the human

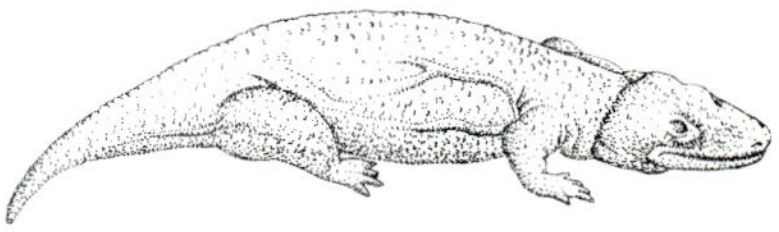

Seymouria

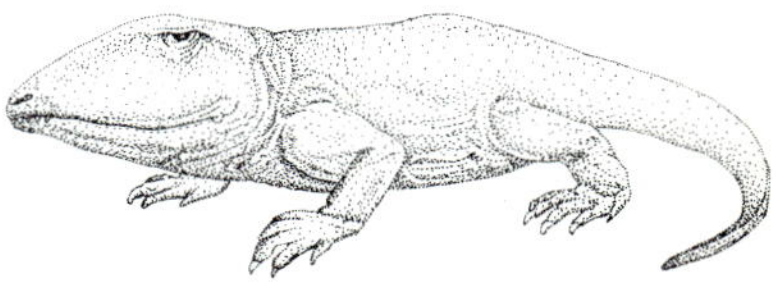

Ophiacodon

1–23
The earliest known reptiles had characteristics intermediate between those of reptiles and amphibians. (After Alfred Sherwood Romer, The Vertebrate Body, *4th Ed. © 1970, W. B. Saunders Company.)*

1–24
A panorama of types of dinosaurs and some of their more primitive relatives. This mural spans about 330 million years of geologic time beginning with the Devonian Period (animals at far right) and ending with the end of the Cretaceous Period (animals at far left). (Courtesy Peabody Museum of Natural History, Yale University, New Haven, Connecticut.)

species and consider the evolutionary relationship between people and those mammals that are our closest relatives—the nonhuman primates. But before we do so, we should first turn our attention to an important concept in evolutionary theory, that of *biological species.*

The Species Concept

The notion of species has been an essential one in evolutionary theory since Darwin first published his history-making *Origin of Species* in 1859. The concept is important because species can be thought of as the natural units of evolution, the things that actually evolve. But before we discuss the special, biological connotation of the word species, we should first say something of how the word is used in everyday speech.

Duckbill platypus

Spiny anteater

1–25
The platypus and the spiny anteater are the only egg-laying mammals.

In popular usage, *species* means *kind, variety,* or *sort.* Biologists use the word in this way, too. For example, each different kind of fossilized or otherwise preserved organism (such as those that are part of a properly named and catalogued museum collection), is referred to as a species. In the classification and naming of organisms, each different species, as judged by how it looks and what it did when it was alive (if this is known), is assigned a two-worded Latin name. We will have more to say about how animals are classified in the following chapter. But when used in discussions of animal evolution, *species* means something special, something in addition to *different kind.*

In the context of the evolution of animals, species are above all populations. That is, species is a collective term that refers to groups of animals within which almost any male can successfully fertilize almost any female to produce fertile offspring. This is so whether the groups are small and highly localized or large and widely scattered at various locations around the world. In nature, species are isolated from one another because animals of different species are not normally attracted to one another sexually; even if they are, they generally cannot produce fertile offspring.

The mating of males of one species with females of a different species generally does not produce offspring, because the entire population that makes up a given species is an isolated genetic unit. That is, the program, in the form of DNA within the nucleus of the cell, that directs the development of a given animal from a tiny fertilized egg is different for each species. But within any species, every individual animal contains essentially the same program. (Of course the program within a given species is not *exactly* the same in any two individuals. The study of variations in the program and of the transmission of various traits within a given species is an important part of the subject matter of the science of Genetics, which we will discuss in Part III.) To sum up, species are populations of interbreeding organisms that are reproductively isolated from one another because of genetic differences between populations.

A major way in which new species evolve is this: some of the groups of animals within a species become so genetically different from other groups of the same species that the two groups can no longer reproduce with one another. This usually occurs when the groups are separated from one another by some kind of natural barrier, like an ocean or a mountain range. In populations of the same species that are separated from one another by long distance or by natural barriers (so that they do not normally come together to reproduce), natural selection may favor different traits in different locations. And in time, these accumulated genetic differences can result in the formation of two different species. Over the past 600 million years, this genetically determined drifting apart of subpopulations with a given species has been responsible in large part for the evolution of the astounding array of living things that now inhabit our planet.

People as a Biological Species

Of the million and a half species of living things already known to science, none has been more intensively studied than the one called *Homo sapiens*—the human species. And none is more difficult to understand. People are such enormously complicated animals that any general discussion of their biology runs the risk of being haphazard. That is why we have just spent time considering evolution and the concept of species. In the rest of this book we will use the notion of *people considered as a biological species* to provide a framework for a discussion of the biology of people.

In the remaining chapters in Part I, we discuss the evolution, behavior and ecology of the human species. Then, in Part II, we will describe the human body as an evolutionarily derived, marvelously complicated, self-regulating machine. Part III is about human genetics. In it we will discuss the multitudinous differences in looks and behavior of members of the human species. We will discuss the biological basis of this variability and its means of transmission from parents to offspring. We will also see how natural selection is presently at work in the human population and discuss the prospects for future human evolution.

What will all this accomplish? Several things, we hope. First, in the following pages we will learn more about ourselves and more about other living things, and that is worthwhile in itself. But perhaps more important, consideration of people as a biological species will provide us with a sense of perspective about ourselves that we can gain in no other way. *We are animals, we have evolved, and we know it.* This makes us wonderfully unique among all our fellow creatures. But, at the same time, the human species is, as we shall see, in many ways a very ordinary one.

Summary

The rich diversity of the living things that now inhabit our planet is the result of evolution. Organisms first evolved from nonliving chemicals in the ancient oceans over 3 billion years ago. Major events that occurred in the very early history of living things include the evolution of autotrophs from heterotrophs, and later, the evolution of eukaryotic cells from prokaryotic cells.

Living things can be classified into five kingdoms, and among these, members of the animal kingdom (metazoans) first appeared in the Cambrian explosion of diversity about 600 million years ago. Since then, animal diversity has increased, mainly by means of variations on certain basic traits that evolved in Cambrian times.

Organisms are closely adapted to the environments in which they live because natural selection favors the survival and reproduction of those creatures that happen to have physiological and behavioral traits that are advantageous in a particular environment. Following the appearance of some major new adaptation, or following the invasion of a previously unoccupied environment, a burst of diversification may occur. Nonetheless, the fossil record shows that most animal populations eventually become extinct.

Vertebrates first evolved about 450 million years ago, and the evolution of the major vertebrate types (fish, amphibians, reptiles, birds, and mammals) can be deduced from their fossil remains. People are clearly mammals, the vertebrate type that is presently dominant on land.

Species are populations of organisms of a particular kind that are reproductively isolated from other such populations because of genetic differences among the groups. Human beings are the most intensively studied of the million and a half species of living things that have been described and named. The human species is in many ways a very ordinary one, but at the same time people are unsurpassingly unique among other living things, mostly because of the extraordinary ways in which they behave. The species concept provides a framework for discussing the biology of people, and a consideration of people as a biological species provides us with a sense of perspective about ourselves that we can gain in no other way.

Suggested Readings

1. *The Meaning of Evolution,* by George Gaylord Simpson.Yale University Press, 1967. An authoritative and enjoyable introduction to the principles of evolution.
2. *Populations, Species, and Evolution,* by Ernst Mayr. Harvard University Press, 1970. A scholarly discussion of animal population biology and of the evolutionary role of animal species.
3. *The Vertebrate Story,* by Alfred Sherwood Romer. University of Chicago Press, 1959. A classic work that describes the major branches of the vertebrate family tree.
4. *The Hot Blooded Dinosaurs,* by Adrian J. Desmond. Dial Press, 1976. A popular discussion of the recent upheavals in the science of paleontology. Provides an update on some of the topics discussed in Simpson's and Romer's books.
5. "An Unsung Single-celled Hero," by Steven Jay Gould. *Natural History,* 83, Nov. 1974. Readings 5 and 6 are about the Cambrian explosion.
6. "The Interpretation of Diagrams," by Steven Jay Gould. *Natural History,* 85, Aug.-Sept. 1976.
7. "Charles Darwin" by Loren C. Eisley. *Scientific American,* Feb. 1956, Offprint 108. How Darwin arrived at his earth-shaking theory.
8. "The Oldest Fossils" by Elso S. Barghoorn. *Scientific American,* May 1971, Offprint 895. A discussion of the oldest known organisms, especially blue green algae and bacteria.
9. "Crises in the History of Life" by Norman D. Newell. *Scientific American,* Feb. 1963, Offprint 867. A discussion of some of the aspects of the problem of mass extinction.

10. "Continental Drift and the Fossil Record," by A. Hallam. *Scientific American,* Nov. 1972, Offprint 903. How the movement of continents can explain similarities between fossils in widely separated areas.

11. "Symbiosis and Evolution," by Lynn Margulis. *Scientific American,* Aug. 1971, Offprint 1230. A discussion of some of the evidence that eukaryotic cells evolved from prokaryotic ones.

12. "Dinosaur Renaissance," by Robert T. Bakker. *Scientific American,* April 1975, Offprint 916. Dinosaurs were "warm-blooded" animals and modern-day birds are their descendants.

Human evolution is not limited to changes in fossilizable body parts. These four renditions of game animals illustrate the evolution of style among artists of the Stone Age. The oldest rendition is at the bottom right and the most recent is at the top left. (From "The Evolution of Paleolithic Art," by André Leroi-Gourhan. Copyright © 1968 by Scientific American, Inc. All rights reserved.)

CHAPTER 2

Human Evolution

In late December of 1857 Charles Darwin answered a letter he had received several months earlier from his fellow naturalist, Alfred Russel Wallace. Like Darwin, but independent of him, Wallace had concluded that evolution by means of natural selection is responsible both for the diversity of living things and for their good fit with the environment. Wallace had written to ask Darwin a question. He wanted to know whether Darwin was going to say anything about the evolution of *people* in his soon-to-be-published manuscript.

Darwin replied that the subject of human evolution was "so surrounded with prejudices" that he chose not to discuss it. But he was quick to say that he thought the question of human origins was the "highest and most interesting problem for the naturalist."

The highest and most interesting problem? Well, at least a very engaging one. From well before the first edition of the *Origin of Species* in 1859 till the present day, people have debated, philosophized, and conjectured about the origin of the human species. Indeed, they have always shown great interest in the subject, sometimes violently so.

In the first edition, Darwin said only that this theory would "throw light" on "the origin of man and his history." (Later, he would publish an entire book on the subject.) But in 1859 that was all he needed to say, for the implication was clear. The theory of evolution would allow for no exceptions. If it was true, human beings, like all other animals, must have evolved.

2–1
The eighteen orders of mammals.

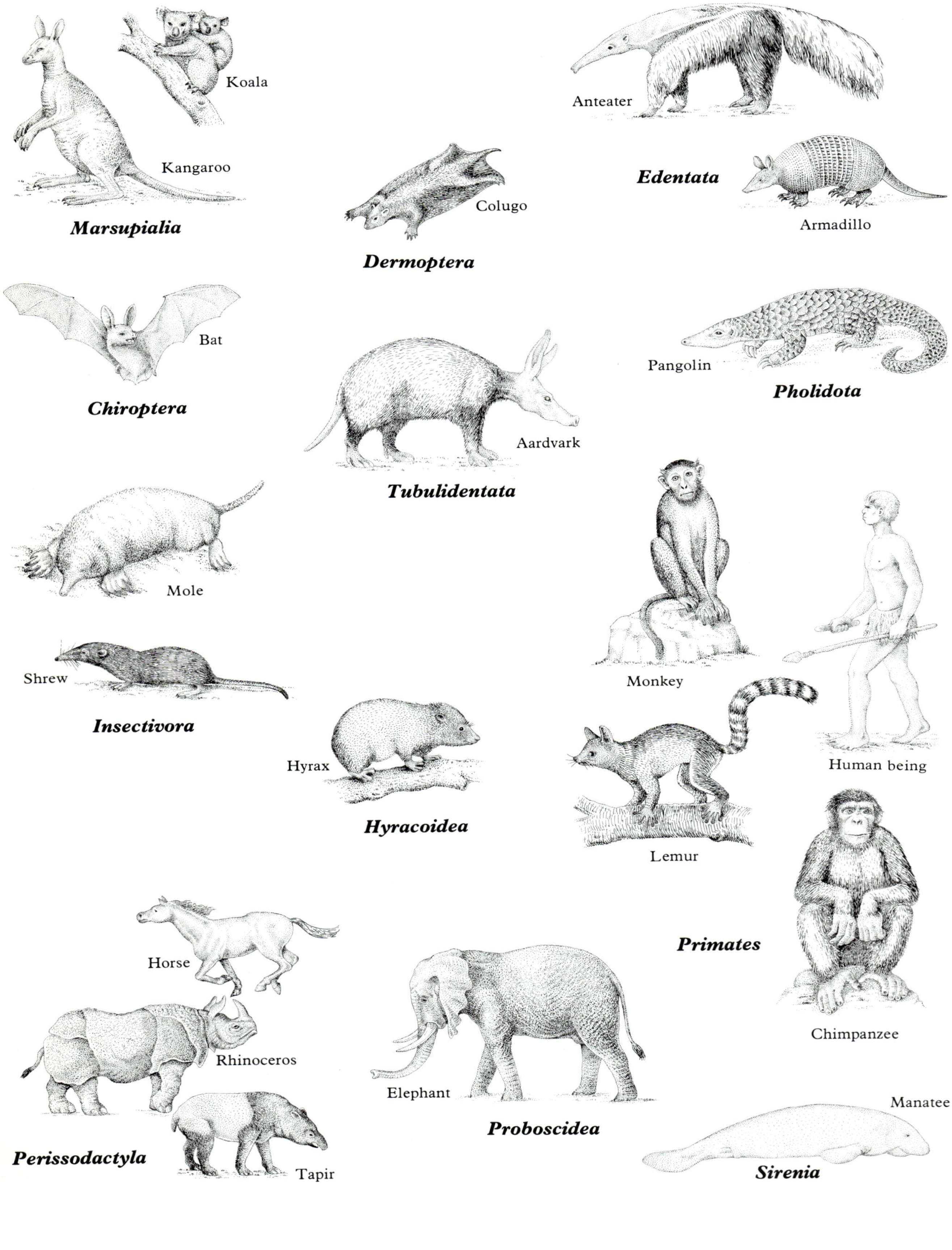
Koala
Kangaroo
Marsupialia
Colugo
Dermoptera
Anteater
Edentata
Armadillo
Bat
Chiroptera
Aardvark
Tubulidentata
Pangolin
Pholidota
Mole
Shrew
Insectivora
Hyrax
Hyracoidea
Monkey
Human being
Lemur
Chimpanzee
Primates
Horse
Rhinoceros
Tapir
Perissodactyla
Elephant
Proboscidea
Manatee
Sirenia

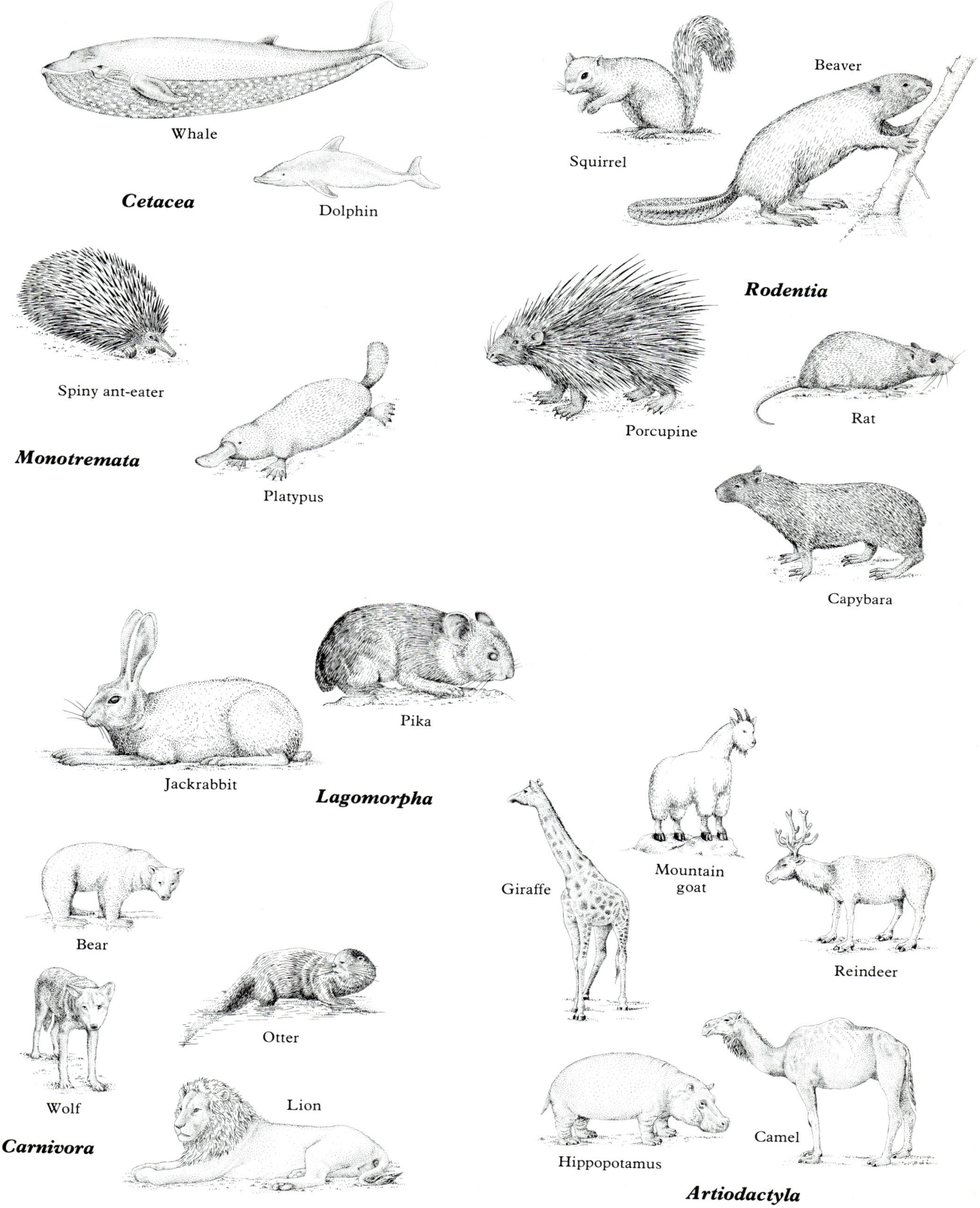
Whale
Dolphin
Cetacea
Squirrel
Beaver
Rodentia
Spiny ant-eater
Platypus
Monotremata
Porcupine
Rat
Capybara
Jackrabbit
Pika
Lagomorpha
Mountain goat
Giraffe
Reindeer
Bear
Otter
Wolf
Lion
Carnivora
Hippopotamus
Camel
Artiodactyla

The very idea of human evolution was earthshaking. From the start it was vigorously opposed by individuals who for various reasons, many of them religious, refused to believe in evolution at all. It was bad enough that Darwin's theory made no exception for human beings. But worse, the theory suggested that people must have evolved from ancestors that looked, and presumably acted, like apes!

Down through the years, the issue has been hotly debated in scientific chambers, brought to trial in Tennessee courtrooms, pronounced upon in papal encyclicals, and discussed on countless street corners. While there are still those who do not believe it, it is now known that human beings did indeed evolve from ancestors that were generally apelike.

In this chapter, we will discuss evidence—both from fossils and from other sources—for human evolution. But first it is helpful to put the human species' place in nature in perspective by discussing the group of mammals to which human beings belong—the primates.

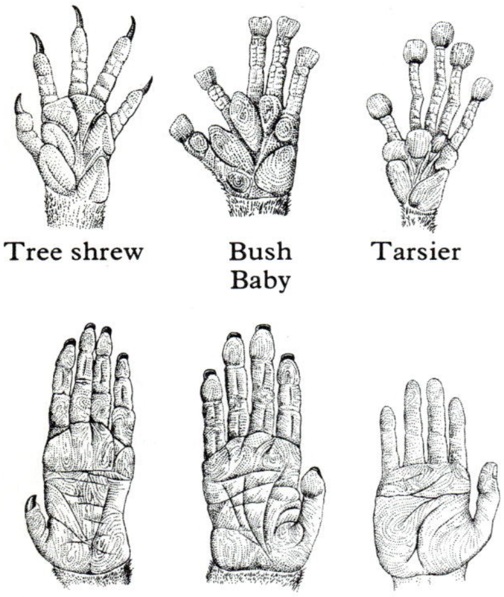

2–2
Some primate hands, palm side up. Illustrations of each of these creatures are found elsewhere in this chapter. (Reprinted by permission from Sherwood L. Washburn, ed. From J. Biegert, Classification and Human Evolution. *Chicago: Aldine Publishing Company. Copyright © 1963 by Wenner Gren Foundation for Anthropological Research, Inc.)*

What Is a Primate?

To answer the deceptively simple question of what a primate is, let us begin by recalling what we already know about animal classification from Chapter 1.

Biologists classify animals into about 33 phyla. People, like all other animals with backbones, belong to the phylum Chordata. This division of the animal kingdom includes about 45,000 species and can be further subdivided into fish, amphibians, reptiles, birds, and mammals.

In all, there are perhaps 4,300 species of mammals, and they can be classified into 18 distinct groups called *orders* (Figure 2-1). Mammals within a given order are considered as related because they evolved from common, or at least very similar, ancestors. People, the great apes, monkeys, and some other, less familiar, relatives make up the order Primates.

Above all, primates are creatures of the trees. They evolved there, and the great majority of them are still arboreal. Nonetheless, a few primates, including baboons, chimps, and people, have forsaken the trees and taken up residence on the ground. But even these animals show unmistakable signs, especially in their skeletons, that they are the descendants of animals adapted for tree life.

Let us consider some characteristic primate adaptations, most of which are related to life in the trees and most of which have been important factors in human evolution. First, primates can climb by grasping objects with their hands and pulling themselves along. Most other tree-dwelling mammals (squirrels, for example) hang onto branches by means of claws. But primates grab branches by their hands and feet and sometimes by their tails, and instead of claws most have evolved flat, protective *nails* at the end of each digit. Figure 2-2 shows some representative primate hands.

Second, primates have a highly developed sense of vision. In order to move quickly through treetops, primates must accurately see the branches they grab, and they must see them in three dimensions. Compared to other mammals, whose eyes are located in the side of the head, most primates have eyes that are rotated forward. This arrangement allows for stereoscopic vision, which means that most primates can see clearly in three dimensions.

A third characteristic of primates, especially those further along the evolutionary series, is the presence of a large and complicated brain. The grasping hands we have just discussed were partly responsible for the evolution of this trait. The increase in brain size and complexity seen in primate evolution probably first occurred because of expansion of those areas of the brain that control the hands and feet. (Later, these expanded areas of the brain came to serve other functions. The meaning of brain size and the relationship between brain and behavior in evolution are discussed in Chapter 3.)

Finally, we should mention briefly two other adaptations. Primates have less highly specialized teeth and they also produce fewer offspring than most other mammals. Both of these factors have been important in primate evolution. The less specialized teeth allowed primates to eat many different kinds of food, and the small number of offspring allowed (or perhaps required) primate parents to devote a relatively long period of time to caring for their young.

Although they all share most of these traits, the primates are still a rather heterogeneous group. There are about 200 living species, including such seemingly unrelated animals as tree shrews and gorillas. We now turn our attention to how biologists classify them.

Classifying Primates

In classifying primates, or any other creatures, one of the major objectives is to arrange the animals into groups that reflect evolutionary history. Most often, this requires at least a little guesswork, so when classifying a given group of animals it is essential to consider all evidence about how they are related. This evidence may come from many sources. First, morphological evidence provides important clues. This includes everything from an animal's outward appearance to the detailed description of its teeth, or of the location of various holes, bumps and ridges on its skeleton. (Bones and teeth are fossilizable, so they receive a lot of attention in comparative studies.) As we would expect, those animals that have the greatest number of anatomical features in common are considered as most closely related.

Second, comparative biochemistry can provide evidence about animal relatedness. The composition of various large molecules, usually proteins, obtained from the blood stream or tissues of the species in question can be compared. Because the composition of these molecules is genetically determined and because it varies from species to species, those animals whose molecules are most alike are considered to be most closely related. (This technique suggests that chimpanzees are the primates most closely related to people.)

Nonetheless, not all biologists agree about the best way to use the evidence in actually devising a scheme of primate classification. This is especially true when it comes to classifying animals known only from fossils, but some living primates are difficult to classify too. For our purposes, we can begin classifying primates by dividing them into two clear-cut categories—*prosimians* and *anthropoids.*

Prosimians

Prosimians are primitive kinds of primates that saw their evolutionary heyday shortly after they first evolved about 65 to 70 million years ago. Since that time, most have become extinct, perhaps because when monkeys evolved they were better adapted for survival in that environment than were prosimians. Among those surviving are lemurs, lorises, tarsiers, and bush babies (Figure 2-3).

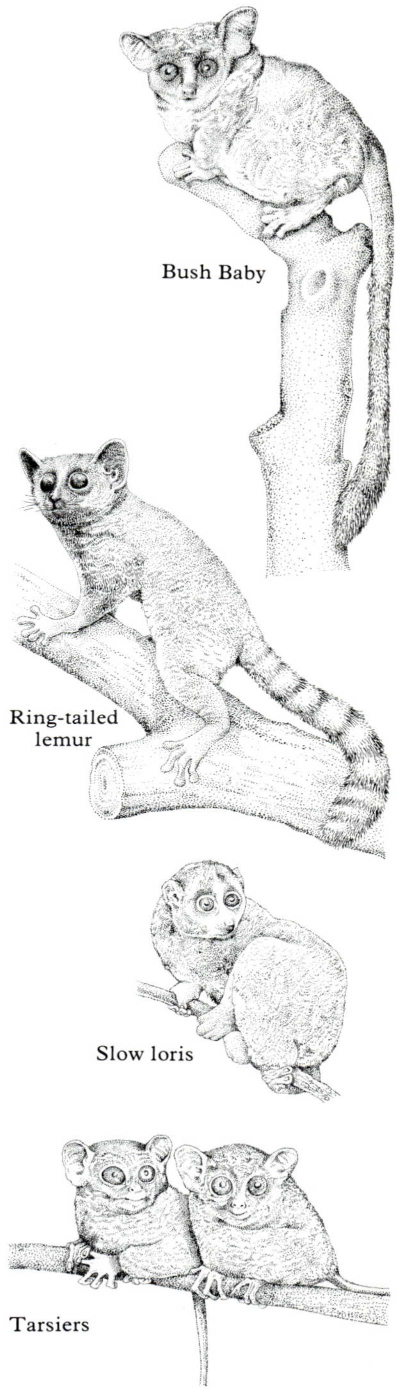

2–3
Some prosimians.

Prosimians probably evolved from ancestors that in some ways resembled tree shrews (Figure 2-4). Modern tree shrews lack many characteristic primate adaptations, but most biologists classify them with the prosimians. This is because anatomical, biochemical, behavioral, and fossil evidence strongly suggests that in spite of their unprimatelike appearance, tree shrews are indeed primates, or at least very close relatives.

2–4
A tree shrew, the most unprimatelike of the primates.

Anthropoids

Most primates are not prosimians, but *anthropoids.* This category includes monkeys, apes, and people. Fossil evidence suggests that monkeys evolved from some kind of extinct prosimian by about 40 million years ago and that apes had split off from the monkey line by perhaps 25 million years ago. (We will discuss the human line later.)

Anthropoids can be subdivided into six groups, or *families:* family *Callithricidae* (marmosets), family *Cebidae* (New World monkeys), and family *Cercopithecidae* (Old World monkeys). The remaining three families of anthropoids consist of two ape families and the human family. Apes and humans are known collectively as *hominoids.* As judged by anatomical and biochemical evidence, apes and humans are more closely related to one another than to any of the three families of monkeys.

2–5
A white-faced gibbon. Siamangs are similarly long armed but bulkier than gibbons. (From R. A. Stirton, Time, Life and Man. *Copyright © 1959, John Wiley & Sons, Inc.)*

The three families of hominoids are: the lesser apes, the great apes, and human beings. To be a little more precise, the superfamily of hominoids consists of three families:

1. Family *Hylobátidae* consists of the "lesser apes"—gibbons and siamangs (Figure 2-5).
2. Family *Póngidae* is made up of three living species known as the "great apes"—the orangutan, the gorilla, and the chimpanzee (Figure 2-6).
3. Family *Homínidae* consists of a single living species, *Homo sapiens,* which includes all human beings. There are also a number of fossil species in the family. Collectively, human beings and the extinct members are called *hominids.*

We now turn our attention to some characteristics we would expect to find in any fossil primate before we classified it as a hominid; that is, as a member of the human family.

A

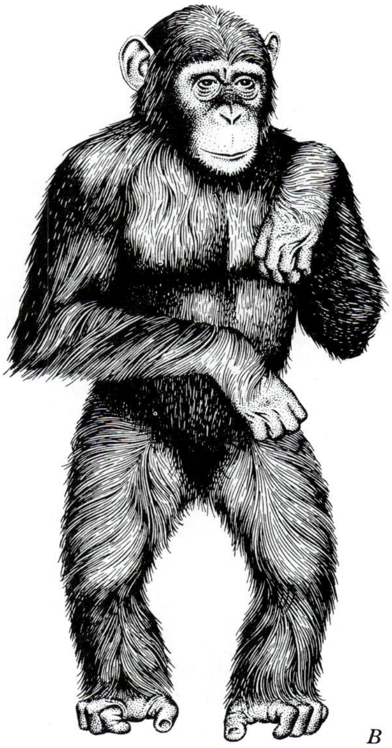

B

C

2–6
A, an orangutan; B, a chimpanzee; C, a mountain gorilla. (A and B from R. A. Stirton, Time, Life and Man. *Copyright © 1959, John Wiley & Sons, Inc.)*

When Is Human?

Because people are the only living hominids, they are the standard of comparison for deciding whether a fossil primate belongs to the human family. Of course, we wouldn't expect all fossil hominids to be equally human in all characteristics. In fact, the fossil record suggests that many of our characteristically human features have evolved at different rates. But we would expect to find at least some evidence that all hominids had evolved, or were in the process of evolving, some fossilizable human characteristics.

For the most part, fossilizable characteristics are part of the skeleton. (However, human excrement—feces—can be fossilized too, and this unlikely source provides us with information about ancient diets in the later stages of human evolution.) Fortunately, human beings have very distinctive skeletons that reflect in solid bone some of our most characteristic adaptations as a species. In particular, our distinctive skulls encase our enormous brains, and throughout our skeletons there is evidence that we stand erect and walk on two legs. These two adaptations are such important indicators of humanness that we must discuss them and their relationship to the skeleton further.

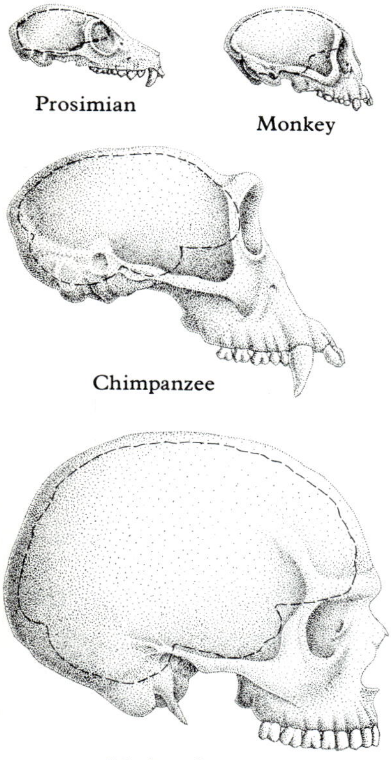

2–7
The skulls of a prosimian, a monkey, a chimpanzee, and a modern human being reflect the relative increase in brain size that occurred during the evolution of the primates. (From "The Casts of Fossil Hominid Brains," by Ralph L. Holloway. Copyright © 1974 by Scientific American, Inc. All rights reserved.)

Figure 2-7 compares the skulls of a prosimian, a monkey, a chimpanzee, and a person. This figure reflects a general trend in primate evolution—the evolution of increasingly massive brains. As you can see, people have enormous brains and correspondingly vaulted and globular skulls compared to other primates. Accordingly, the size of the brain, as reflected by the volume of the skull (*cranial capacity)*, is an important statistic in deciding whether a particular fossil primate is a hominid. But cranial capacity is a very crude measure of an organ whose complexity should be judged through the microscope. You should keep this in mind when we refer to cranial capacities in the discussion of human evolution that follows. (The meaning of brain size is discussed more fully in Chapter 3.)

Besides the presence of a large and complicated brain, another adaptation that is characteristic of people and that leaves fossil traces is upright posture. Standing erect and walking on two legs have had distinctive and widespread effects on the human skeleton. Figure 2-8 shows some of the effects of upright posture on the human pelvis as compared to the pelvis of a gorilla. The bones of the toes, feet, legs, vertebral column, chest, and shoulder also show the effects of bipedal locomotion, and provide us with clues about whether fossilized bones were involved in producing the same kind of locomotion. As we shall see, upright posture apparently is an ancient adaptation in the hominid line.

The parts of the skeleton that are most durable and most frequently fossilized are teeth. This is fortunate, because the patterns of the cusps of the teeth of different primates (among other dental features) are highly characteristic, as shown in Figure 2-9. The teeth of hominids are very distinctive, and it is often possible to infer the presence of fossil hominids in a given location solely on the basis of isolated teeth discovered there. In fact, the skeletons of the Peking Man (a member of the species *Homo erectus)* were located in this way.

There is no denying that human beings have distinctive skeletons and other anatomical features that distinguish them from nonhuman primates. But, in the final analysis, what makes people so special is what they do. The really distinctive thing that human beings do is to engage in an intricate kind of learned behavior called *culture* (this topic is discussed more fully in Chapter 3). Although they are not really fossilizable, cultural activities can nonetheless leave

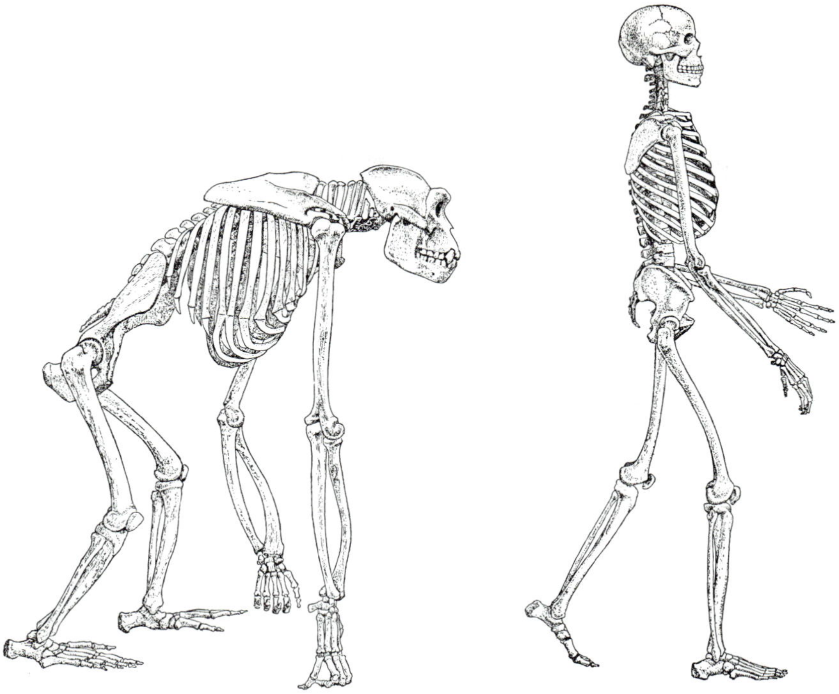

2–8
The pelvis of a gorilla is longer, heavier, and more tilted than that of a human being. These differences are related to posture. (From "The Antiquity of Human Walking," by John Napier. Copyright © 1967 by Scientific American, Inc. All rights reserved.)

traces in the fossil record. These traces of human activity are most often found in tools—that is, in objects that were manufactured to serve a purpose. As we will see, some fossil hominids whose skeletons are regularly found in close proximity to tools are considered probable human ancestors. And during the course of human evolution, tools themselves have evolved; when they were first in use their form was crude and awkward, but the finely made tools of later years were a tribute to the skill of the person who made them.

Evidence of other cultural activities in which humans engaged in later stages of evolution has been found in fossil-like traces. Such activities include the use of fire, burying the dead, drawing and painting on cave walls, and otherwise making works of art. But we find no obvious clues in the earlier stages of hominid evolution about "when is human?" and the decision is often an arbitrary one. Let us begin discussing the evolution of our fellow hominids by proposing a classification of the hominid family.

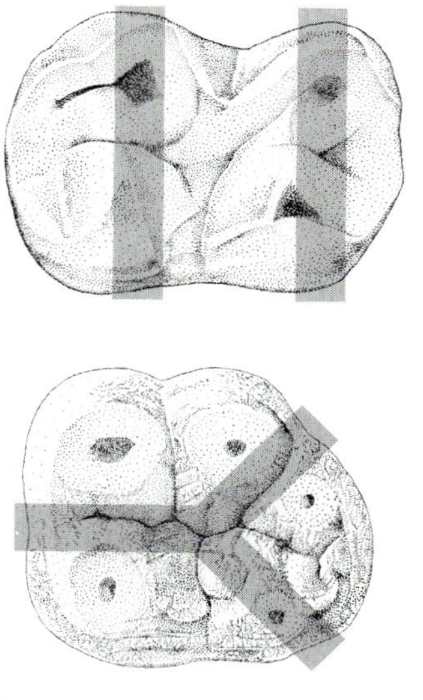

2–9
The teeth of monkeys (top) are shaped differently from those of the great apes and people (bottom).

Classifying Hominids—An Overview

Like other families in schemes of classification, hominids are subdivided first into groups called *genera* (singular, *genus)* and then into individual species. (Table 2-1 provides a recap of the major categories used in classifying animals.) Experts disagree, some of them with surprising enthusiasm, about what is the best way to classify the hominids. The following classification is a reasonable one, though perhaps oversimplified. As with any scheme of classification, not all

TABLE 2-1
The major categories used in formally classifying humans and other members of the animal kingdom.

	HUMAN BEING	ROBIN	LION
Phylum	Chordata	Chordata	Chordata
Class	Mammalia	Aves	Mammalia
Order	Primates	Passeriformes	Carnivora
Family	Hominidae	Turdidae	Felidae
Genus	*Homo*	*Turdus*	*Panthera*
Species	*Homo sapiens*	*Turdus migratorius*	*Panthera leo*

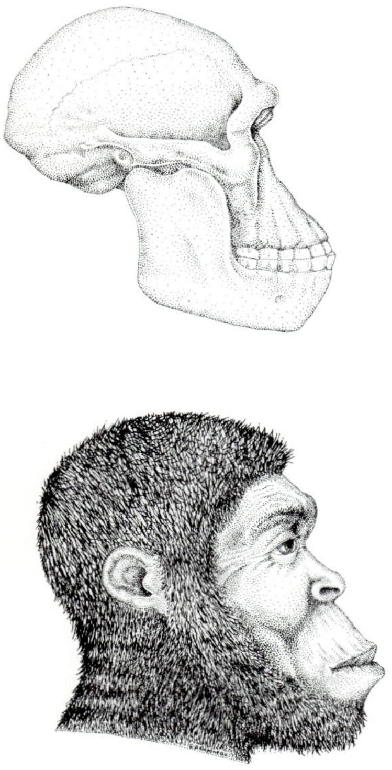

2–10
The gracile australopithecines were slender creatures about 4 feet tall. Top, a skull from South Africa; bottom, a reconstruction of what Australopithecus africanus *may have looked like.*

experts would agree to it, and it will have to be revised as more fossil hominids are discovered.

Family: Hominidae
- Genus: *Ramapithecus*
- Genus: *Australopithecus* (2 species)
- Genus: *Homo* (4 species at least)
  - *Homo habilis* (?)
  - *Homo* "1470"
  - *Homo erectus*
  - *Homo sapiens neanderthalensis* (Neanderthal people)
  - *Homo sapiens sapiens* (Modern people)

We will discuss each of these categories in turn and see how some of these hominids may have evolved from common ancestors, or indeed, from one another.

Ramapithecus

Ramapithecus is the most ancient hominid group discovered so far. We know that the animals of this genus lived in India, East Africa and many other parts of the Old World between 15 and 8 million years ago, and that they were about the size of gibbons. Their teeth and jaws were clearly hominid in type, which distinguished them from some of their contemporaries that had evolved into the ancestors of modern apes.

But we know little else about *Ramapithecus* because the only fossil remains of these creatures that have been found are teeth and jawbones. Nothing at all is known concerning *Ramapithecus*'s cranial capacity or posture. Nonetheless, the fossils that have been found allow us to state with confidence that hominids had evolved by about 15 million years ago. (*Ramapithecus* itself may have evolved from a kind of generalized primate known as *Dryopithecus* that was fairly common in Europe about 20 million years ago. In fact, *Dryopithecus* may well have been ancestral not only to hominids, but to monkeys and apes too.)

The Australopithecines

In 1924, fossil remains of the skull of an ancient hominid child were discovered in a South African quarry. Although part of the skullcap and face were missing, the fossil was complete enough to suggest to the experts who examined it that it had some hominid characteristics. Nonetheless, it had apelike features too, and after a lot of debate it was decided that the creature that had unwittingly contributed this skull to the fossil record should be called *Australopithecus africanus.* The name means "the southern ape from Africa."

In the years since the initial discovery, the remains of hundreds of australopithecines have been found and studied. The fossils leave no doubt that the animals they came from were indeed hominids, which were quite variable as a group. And, though opinions have changed frequently through the years, it now appears almost certain that this group of fossil hominids were *not* direct ancestors of modern people.

The australopithecines were of basically two types—*gracile* and *robust;* but, like most species, the group was a variable one. As the name suggests, the gracile australopithecines were slender and lightly built. They were about four feet tall and probably weighed about 60 or 70 pounds. They had rather protruding jaws (Figure 2-10) and their teeth were the least specialized and most humanlike of all the australopithecines. (The dental evidence suggests that their diet probably included meat, at least when it was available.)

The members of the robust type (*A. robustus)* were much bulkier. They were about five feet tall and weighed perhaps twice as much as their gracile neighbors. Some *A. robustus* skulls show protruding ridges of bone over the eyes called brow ridges (Figure 2-11). Characteristically, the robust australopithecines had massive jaws and large molars, which probably indicates that they were vegetarians, perhaps seed eaters.

Both the gracile and robust varieties stood erect, but they must have had a different kind of bipedal locomotion than we have. They probably waddled from side to side, especially when they ran. However they got around, evidence from East and South Africa shows that both the gracile and robust varieties often existed in the same location (Figure 2-12).

The australopithecines had evolved by about four million years ago. Until very recently, it was widely thought that at least some of them had attained "human" status because their fossils were found near some crudely made stone tools. As it turns out, the tools were probably made not by them but by a different kind of hominid that was a contemporary of the Australopithecines about three million years ago. But before we decide who made them, we should first say more about the tools themselves.

Until 1972, the oldest reported human artifacts were crude stone tools from *Olduvai Gorge* in East Africa. (See Figure 2-12). Figure 2-13 shows what some of these "pebble tools" look like. These artifacts from Olduvai make up what archeologists call the Oldowan cultural tradition, and they date from about two million years ago. Then in 1972, older, but apparently better made, tools were reported from East Rudolf in East Africa. Though these tools are more highly developed than those from Olduvai, in the words of their discoverer (Richard E. Leakey) there are "distinct parallels" between the tools from both locations.

Where do the australopithecines fit in? Originally, they were thought to be the hominids that had manufactured the tools from Olduvai, but lately they have lost support as candidates for bearers of Oldowan culture. There are several

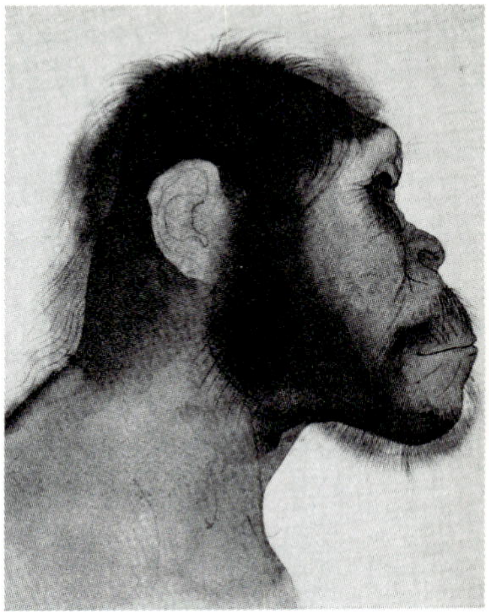

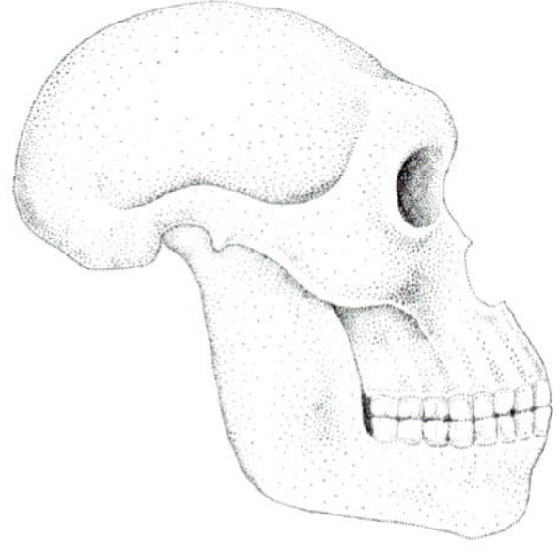

2–11
The robust australopithecines were stocky 5-footers. Bottom, a skull from Olduvai Gorge; top, a reconstruction of what Australopithecus robustus *may have looked like. (Reconstruction by permission of the Trustees of the British Museum, Natural History.)*

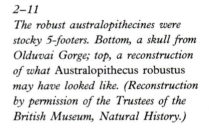

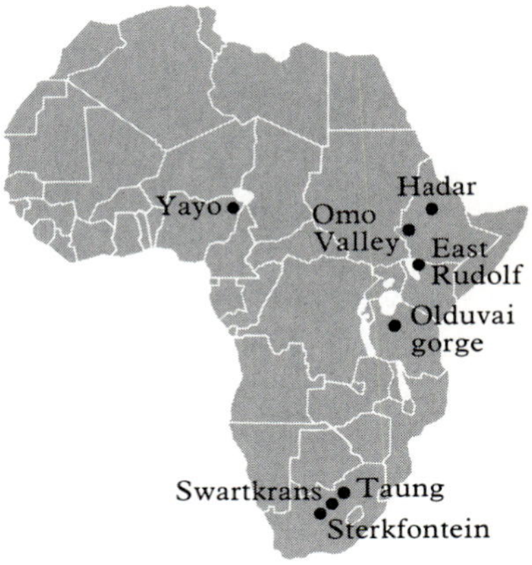

2–12
African sites from which fossils of Australopithecus *have been recovered.*

reasons for this. First, the older artifacts at East Rudolf are associated with a more advanced kind of hominid that is at least as old as the artifacts. There are australopithecines at East Rudolf, and they are similar to those at Olduvai. But we should probably attribute the more advanced tools at East Rudolf to the recently discovered, more advanced hominid. And second, there is evidence that a more advanced form than the australopithecines, perhaps indistinguishable from the hominid from East Rudolf, lived at Olduvai when the Oldowan tools were manufactured about two million years ago.

In the end, we really do not know which hominids manufactured the Oldowan artifacts. But the australopithecines are losing favor, mainly because of the discovery of several bigger-brained and more advanced hominids with whom they were contemporaries. Perhaps a more likely candidate as the first tool maker at Olduvai is *Homo habilis.*

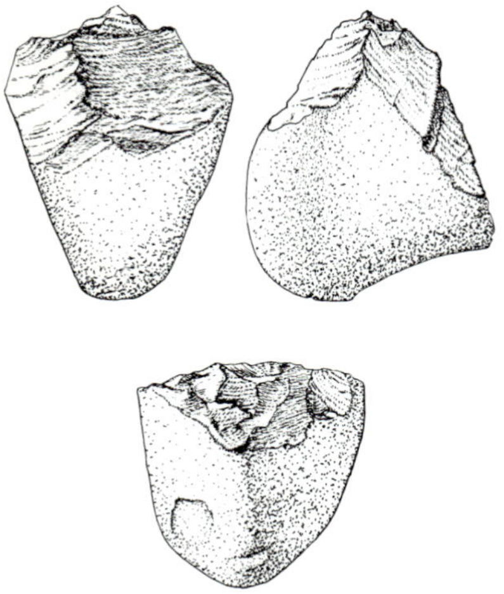

2–13
Two-million year old "pebble tools" from Olduvai Gorge. (From W. F. Bodmer and L. L. Cavalli-Sforza, Genetics, Evolution, and Man. *Copyright © 1976, W. H. Freeman and Company.)*

Homo habilis

Homo habilis was first described from 2-million-year-old fossils found at Olduvai Gorge in the 1960's. Their discoverer, Dr. Louis S. B. Leakey, was impressed that the fossils were far less primitive than australopithecines found in the same deposits, so he placed them in the genus *Homo.* Others were less impressed and thought *Homo habilis* should be considered as an advanced australopithecine. Today, many experts accept *Homo habilis* as a distinct hominid, and this tendency will perhaps continue, because Mary D. and Richard E. Leakey (Louis's wife and son) have recently found evidence, not only that *Homo habilis* may have inhabited East Rudolph, but that this species may have been present at Olduvai 3.5 million years ago. At any rate, the status of *Homo habilis* should become clearer as more discoveries are made at Olduvai, at East Rudolf, and elsewhere.

Homo habilis had a cranial capacity of about 600 cubic centimeters (cc) as compared with roughly 500 for both the robust and gracile australopithecines. That is not much difference, especially when the meaning of brain size is obscure. Nonetheless, the fossils of *Homo habilis* probably represent a group of relatively larger-brained and otherwise more advanced hominids that lived alongside the australopithecines at least 2 million (and perhaps as long as 3.5 million) years ago. As we have seen, some experts, especially the Leakeys, believe that the Oldowan artifacts are the handiwork, not of australopithecines, but of other hominids. This view is becoming very widespread in the wake of some spectacular discoveries at East Rudolf in 1972. Richard E. Leakey discovered not only what he interprets as probably remnants of *Homo habilis,* but also a previously unknown fossil hominid that may be even more advanced than *Homo habilis,* yet older.

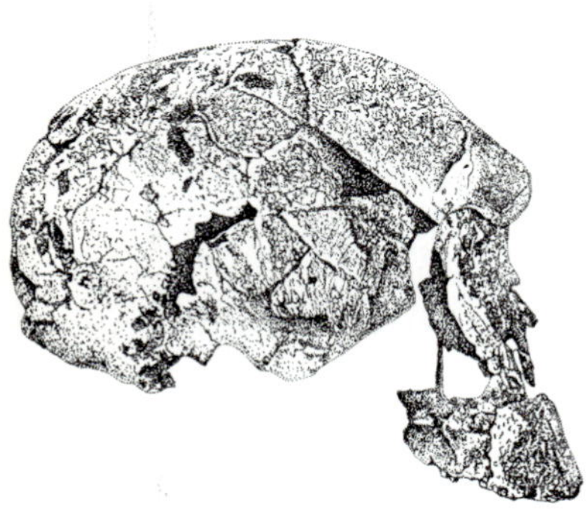

2–14
The skull of Homo *"1470." Note the relative greater size of the braincase as compared with the skulls of the australopithecines. (Published with permission from the National Museum of Kenya.)*

Homo "1470"

The recently discovered fossil hominid *Homo* "1470" has not yet been formally classified or named, but it will surely be ranked at least as "human" as any fossil hominid we have discussed so far. (1470 is the fossil's registration number in the National Museum of Kenya, and the species may eventually be classified as *Homo habilis.*) The major find was that of a nearly complete skull that dates from perhaps 2.8 million years ago. The skull has humanlike contours and a cranial capacity of 800 cc (Figure 2-14). Leg bones that have been recovered from the same site (and are presumably of the same age) are nearly indistinguishable from those of modern people. This hominid is also known from fossils discovered at Hadar in Ethiopia. In close proximity to *Homo* 1470 at East Rudolph are found the stone artifacts just mentioned. As we said, although they are more finely made than those of Olduvai, these tools are older and they date from the same age as the skull and leg bones—perhaps 2.8 million years. We can get some idea of the importance of this remarkable discovery by considering that the next (and more advanced) hominid in line as a probable human ancestor, *Homo erectus,* though found throughout the Old World, has never been dated at older than one million years. (In 1976, Mary D. Leakey discovered, not far from Olduvai Gorge, a fossil hominid that is probably 3.6 million years old and that has a mixture of definite features of the genus *Homo* as well as more primitive traits. Where this creature fits into the human family tree remains to be seen.)

Homo erectus

The group of extinct people called *Homo erectus* is known from fossils from East and South Africa, Europe, and Asia (Figure 2-15). *Homo erectus* people lived from at least one million to about 150,000 years ago. They were "people" in that their fossils are regularly associated with evidence of cultural activities, but they were a primitive form of human being.

As its name implies, *Homo erectus* was fully bipedal. (The species was named before it was known that the australopithecines, *Homo habilis,* and *Homo* 1470 also stood erect.) Most members of this species were about five feet tall, and although the species was scattered throughout the world, all of its members had characteristic features in the architecture of their skulls.

The skulls of *Homo erectus* generally have low foreheads and massive brow ridges. Figure 2-16 shows a reconstruction, based on skeletal features, of what a *Homo erectus* person may have looked like. Perhaps the most characteristic feature of the skulls is their cranial capacity. As usual, there is geographic variation, but most *Homo erectus* skulls have a cranial capacity of from 800 to 1100 cc. (For modern people, the range is from about 1200 to 1800 cc or more, but the generally accepted average is about 1400 cc.) As judged by the contours of the skulls and by studies of plaster casts of their interiors, the brains of *Homo*

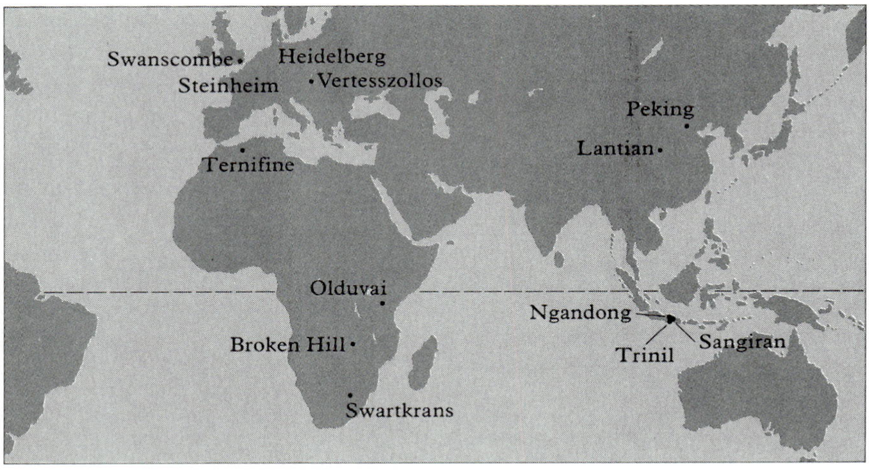

2–15
Places where fossils of Homo erectus *have been unearthed. (From* "Homo erectus," *by William W. Howells. Copyright © 1966 by Scientific American, Inc. All rights reserved.)*

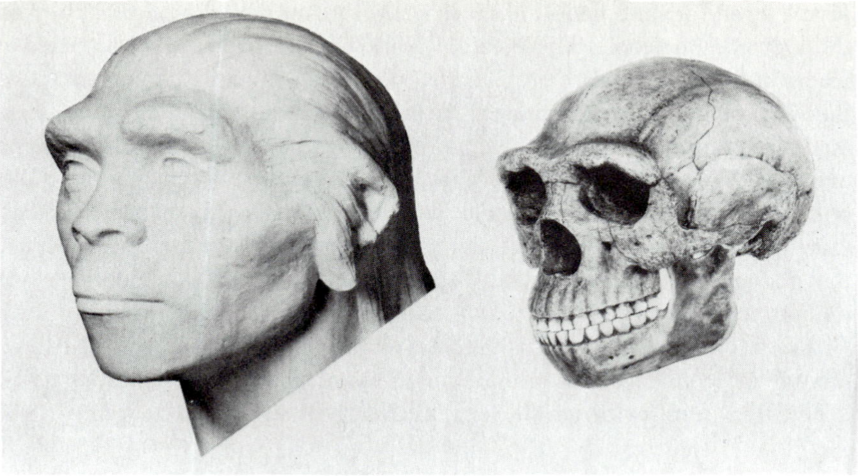

2–16
A restoration of a Homo erectus *person, left, and the skull upon which it was based, right. This person lived near what is now Peking. (Courtesy of the American Museum of Natural History.)*

erectus were surely less highly developed than our own, but nonetheless more human than those of any other fossil hominid we have encountered so far. From this point on in human evolution (and probably in the earlier stages, too) there is a parallel between the evolution of larger, more complicated brains and evidence of cultural activities. *Homo erectus* people with their relatively simple brains clearly engaged in cultural activities. Although there is no doubt that the performance of cultural activities had an effect on the further development of the brain in human evolution, exactly how this happened remains unclear. Nonetheless, by studying the cultural remains of *Homo erectus,* we can get a clue about what some of the influences on the further development of the human brain may have been.

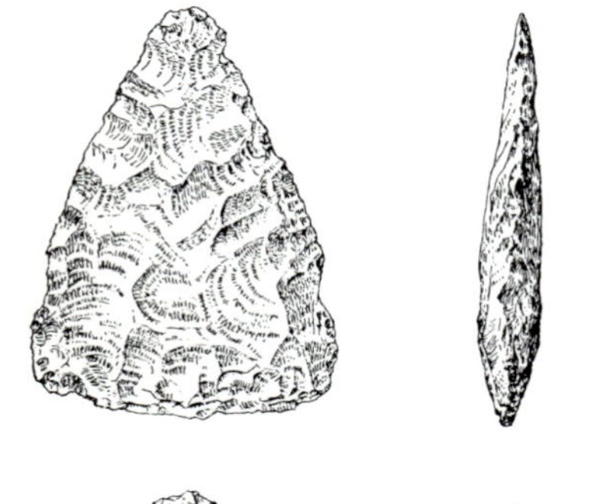

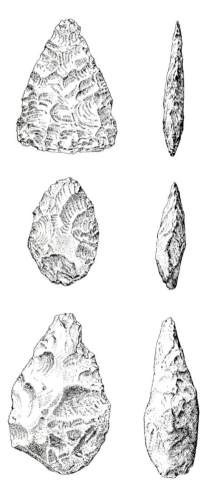

2–17
Some stone tools manufactured by European Homo erectus *people. (From "Tools and Human Evolution," by Sherwood L. Washburn. Copyright © 1960 by Scientific American, Inc. All rights reserved.)*

To begin with, *Homo erectus* people regularly made stone tools, such as those shown in Figure 2-17. (They may have made other objects from materials such as wood and bark, but these are lost to us because they leave no enduring traces.) Some of the tools in Figure 2-17 must have required considerable skill, and it is likely that as tools became more essential to human survival, any kind of improvement in the organization of the brain that could result in the more capable manufacture of tools would be strongly favored by natural selection.

Another factor that must have influenced brain development is social organization. It has been estimated that *Homo erectus* people lived in groups of perhaps 50, and at least some of them lived in caves. They made their living primarily by hunting and gathering, as do many aboriginal people today. Also, *Homo erectus* people were responsible for a cultural milestone in human evolution—the domestication of fire. Human beings have used fire for cooking and warmth (and camaraderie) for at least 500,000 years, and among the first to do so were *Homo erectus* people in both Hungary and China. (Both of the sites where their remains have been found are at about the same rather northerly latitude, and both date from a glacial period of the Ice Age (see below). Although it is difficult to directly relate social behavior to the evolution of the brain, the two must somehow be related. *Homo erectus* people, with their moderate-sized brains and their use of fire, were in many ways already acting like modern people 500,000 years ago.

Of all of the cultural factors that have influenced the evolution of the brain, surely none is so important as the development of spoken *language.* As we will see in Chapter 3, communication by means of the symbol system of language is a uniquely human behavioral trait. And in the opinion of many biologists, the development of language is the most likely of many possible influences to account for the tremendous growth in complexity of the human brain in its later stages of evolution. (In Chapter 3 we will discuss in some detail how language relates to the brain.) *Homo erectus* people, as judged by the contours of their brains and by their cultural development, had probably evolved at least a rudimentary form of language. And the people who follow them in the fossil record, the *Neanderthalers,* almost certainly had.

Homo sapiens neanderthalensis

By about 150,000 years ago another, more advanced species of people had evolved from *Homo erectus* ancestors. These were the Neanderthalers, named after Neanderthal, the German valley in which some of their fossils were discovered. These people were even more human than *Homo erectus* in that they had larger brains and left behind them evidence of a more sophisticated culture. In brain volume, the Neanderthalers were fully as human as ourselves, with an average cranial capacity of about 1400 cc. In fact, some Neanderthal populations may have had relatively larger brains than modern people. But endocasts of skull interiors show that the Neanderthal brain was shaped differently from ours and that it had not yet attained fully modern contours. Nonetheless, these ancient people are considered very close relatives of living human beings. In fact, the extinct Neanderthal population, which was quite variable, is officially known as *Homo sapiens neanderthalensis.* (In naming species, the use of three Latin names instead of two indicates that a particular species is made up of two or more distinct but closely related types, or *subspecies.* It is often difficult to distinguish subspecies from one another by morphological characteristics alone. To emphasize the overlap in skeletal and cultural features between living and extinct people, modern, living humans are officially called *Homo sapiens sapiens.)*

Neanderthal people inhabited both Africa and Southeast Asia, but they are best known from remains in Western Europe and the Mediterranean area. It has been suggested that a Neanderthal man dressed in blue jeans and a sweat shirt would not attract much attention in the checkout line of the local supermarket. This is probably true for some of the more progressive Neanderthalers, but others were downright brutish and would surely cause at least some heads to turn. According to most experts, the Neanderthal population is best divided into classic and progressive types.

Paradoxically, the progressive, more modern-looking, variety is first known from fossils about 150,000 years old, and the oldest remains of the classic, more brutish Neanderthalers date from about 75,000 B.P. (B.P. means before present). The classic Neanderthaler had a receding chin, prominent cheeks, and very obvious brow ridges that curved over each eye and were connected across the bridge of the nose (Figure 2-18). These people were short by modern standards, about five feet tall. They were powerfully built, and had short extremities and stubby hands and fingers. They must have been very strong. The progressive Neanderthalers are so called because their skeletons more closely resemble those of modern people than do those of the classic Neanderthalers. This is especially true of the contours of the progressive Neanderthaler's skull (Figure 2-19).

All Neanderthalers regularly made stone tools of the Mousterian tradition, named after the town in southwestern France near which they were found, Le Moustier. Some of these were quite delicately chipped, as Figure 2-20 illustrates. These tools were used to butcher gigantic animals such as mammoths and woolly rhinoceroses, which the people trapped in excavated pits lined with

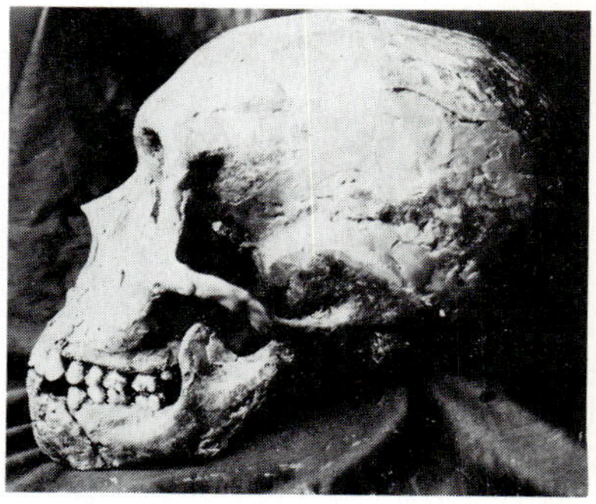

2–18
A restoration of a "classic" Neanderthaler from Europe and the skull on which it was based. (Courtesy of The American Museum of Natural History.)

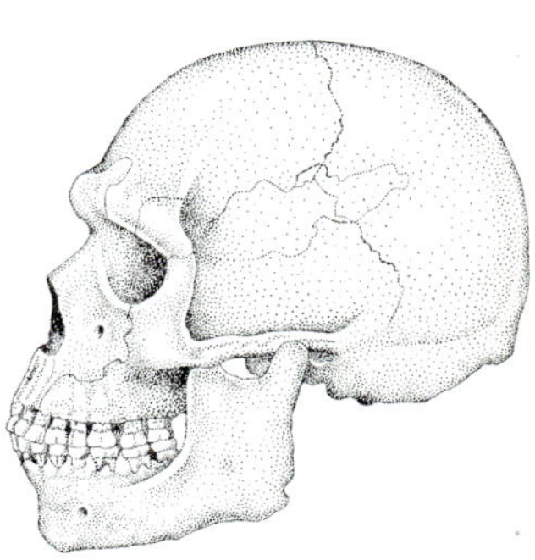

2–19
The skull of a "progressive" Neanderthaler who lived about 45,000 years ago.

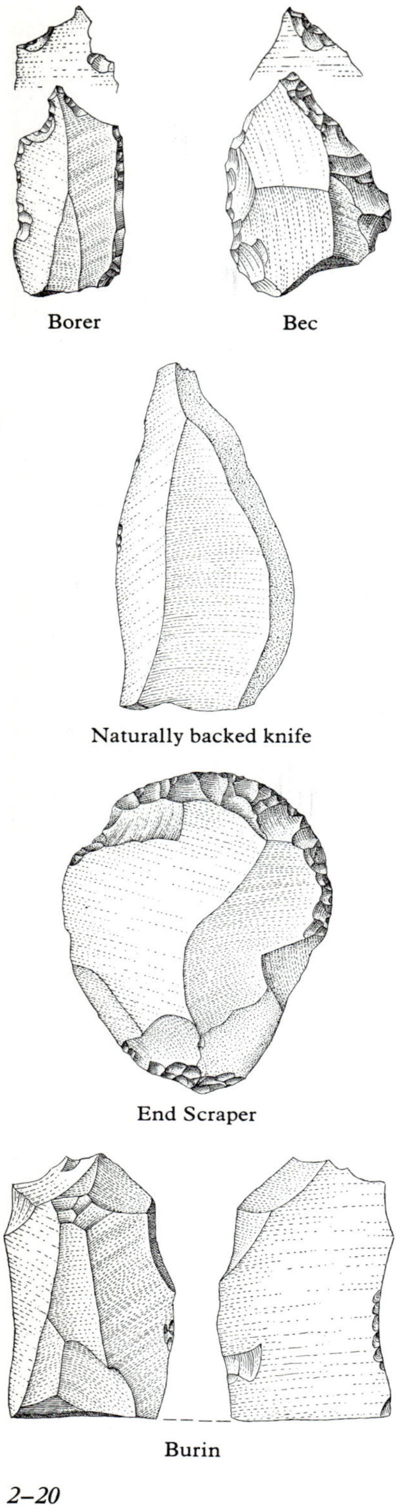

2–20
Stone tools manufactured by European Neanderthalers. (From "Stone Tools and Human Behavior," by Sally R. Binford and Lewis R. Binford. Copyright © 1969 by Scientific American, Inc. All rights reserved.)

spiked stakes. The Neanderthalers regularly made use of fire, but as far as we know, they did not make ornaments for their bodies; nor did they leave behind them any traces of deliberate art work.

Nonetheless, the Neanderthalers were the first people we know of to engage in the distinctly human practice of burying their dead. The oldest burials date from about 500,000 years ago, and they consist of skeletons carefully placed in either natural or excavated recesses. But by 40,000 years ago, burials had become more elaborate, as evidenced by a remarkable Neanderthal "cemetery" discovered in France. The plot contained the remains of two adults and four children, one of whom was probably stillborn. Such burials are evidence that Neanderthal people had amassed some kind of beliefs concerning life and death; that is, that they had evolved a world view.

Additional evidence that Neanderthal people had a system of beliefs and that they engaged in at least some ritualistic activities is found in the remains, not of the people themselves, but of giant cave bears, which they probably hunted or collected. The Neanderthalers may have had special, perhaps ritualistic, regard for these huge animals, because cave bear skulls and leg bones that were carefully and purposefully arranged have been discovered in several caves known to have been inhabited by Neanderthalers. (By the time modern people had evolved, we have evidence that a thriving cave bear "cult" had arisen.)

What happened to the Neanderthalers? Once again, the experts disagree, but what probably happened can be summarized in this way. First, the classic Neanderthalers became extinct because modern people, who were more adept at survival than they, evolved. Second, the progressive Neanderthalers may have directly evolved into an early form of *Homo sapiens sapiens,* or at least have interbred with such people. This idea is supported by fossils found at the Skhul cave in Israel, where skulls that have characteristics that lie between Neanderthal and modern people have been discovered and dated at about 35,000 B.P. (A nearby cave, dated 10,000 years earlier, contained the remains of typical Neanderthalers, but no intermediate fossils were found.)

Neanderthal people, most of whom were of the more classic type, were found throughout Europe till about 35,000 years ago. Then the record shows that they were abruptly replaced by fully modern people, and from then on we never find traces of Neanderthalers in Europe again. *Homo sapiens sapiens* people apparently first invaded Europe from the south (perhaps coming from the Middle East) during an interglacial interval of the Ice Age or *Pleistocene Epoch* (Figure 2-21). When this happened, the Neanderthalers may have been forced to retreat from modern people. Or the Neanderthalers may have chosen to follow the retreating ice sheets northward because they were accustomed to making their living in harsher, more glacial, environments. In fact, Mousterian tools have been discovered in Asiatic Russia, where a tundra-like environment persisted long after it had disappeared from Europe. These tools apparently date from more recent times than any Mousterian artifacts found in Europe, and they may have been manufactured by Neanderthalers who were following the ice sheets northward. At any rate, modern people first moved into Europe during an interglacial period about 35,000 years ago, and the Neanderthalers disappeared shortly thereafter. Where did these modern people come from?

Homo sapiens sapiens—The Transition

Like the Neanderthalers and the *Homo erectus* people who preceded them, modern people are a highly variable species. Nonetheless, as judged by the contours of the skull and by evidence of cultural activities, people indistinguishable from ourselves have existed for at least 40,000 years. The earliest remains come from Borneo, but by 35,000 B.P. traces of modern people are found in Europe, Asia, the Middle East, and Africa. Although the evidence is scanty, it suggests that modern people first evolved in either Africa or Asia. The question is, evolved from whom?

Nobody knows for sure exactly who the immediate ancestors of modern people were, but there are several contenders for the title. Probably first in line are the progressive Neanderthalers. As we have seen, Skhul cave in Israel dates from 35,000 B.P. and it contained skulls that show a blending of Neanderthal and modern features. This suggests that the local Neanderthalers had evolved directly into a more modern type. However, we cannot be sure that the people from Skhul were not actually progressive Neanderthalers in the process of interbreeding with modern people who had evolved elsewhere and were moving into the area. Also, there is evidence from Europe that a very modern type was alive there about 250,000 years ago—100,000 years before the first progressive Neanderthalers appeared.

In the 1930s parts of two very intriguing skulls were discovered in Europe. One came from Swanscombe, England, and the other from Steinheim, Germany. What was so interesting about both of these skulls is that they came from people whose heads were shaped roughly like our own in back, but whose faces, with thick brow ridges and rather low foreheads, were far from modern. These people existed at least 250,000 B.P., and are distinct enough from progressive Neanderthalers that they are usually classified as a separate subspecies, *Homo sapiens steinheimensis*. More evidence is clearly needed, but it may be that these rather advanced "Steinheim people" were widespread in Europe (and perhaps elsewhere); they may even have been direct ancestors of modern people, or at least ancestors of the progressive Neanderthalers.

In the end, we do not know which ancient peoples were our most immediate ancestors. Nor do we know exactly when or where the transition to fully modern people occurred, or over how large a geographic area the nearly modern population was evolving. But we do know that fully modern people with fully modern brains had invaded most of the Old World by 35,000 years ago. What makes us sure that they were very much like us is the way they acted.

Homo sapiens sapiens—*Cultures of the Late Stone Age*

Ancient human beings evolved rich and distinctive cultures in various parts of the Old World. The cultures that are best known are those evolved by Europeans beginning about 35,000 years ago. Figure 2-22 shows some tools from "Late Stone Age" cultures of Europe. The people responsible for one of the best known of the Late Stone Age cultures in Europe are called "*Cro-Magnon* people," after a cave in the south of France that housed some of their more spectacular remains. Cro-Magnon people, as judged by their skeletons, were about as tall as modern Europeans but had slightly wider faces and were perhaps more muscular (Figure 2-23).

By 25,000 years ago, the Cro-Magnons had clearly crossed what Carl Sandberg has called the "margins of animal necessity." And, to paraphrase the poet a little, they had clearly come to the deeper rituals of their bones, to the time for dances, songs and stories, to the time for thinking things over once having so marched. They were as human as we, for they had evolved a world view that led them to create works of art that had symbolic and ritualistic significance.

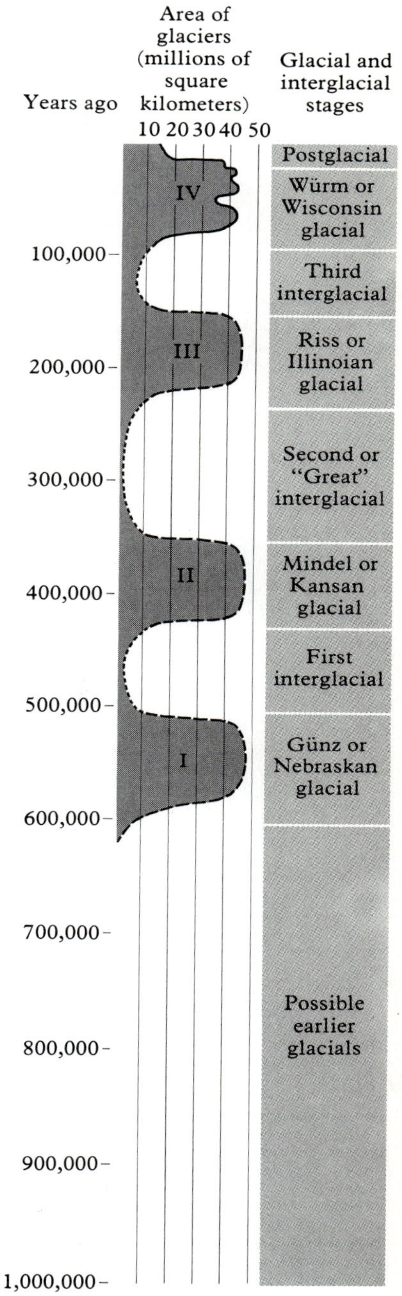

2–21
The chronology of the Pleistocene epoch, or "Ice Age." Glacial advances and retreats are shown by the solid black line. Modern people first invaded Europe at the end of the most recent glacial period, about 35,000 years ago. (From "Tools and Human Evolution," by Sherwood L. Washburn. Copyright © 1960 by Scientific American, Inc. All rights reserved.)

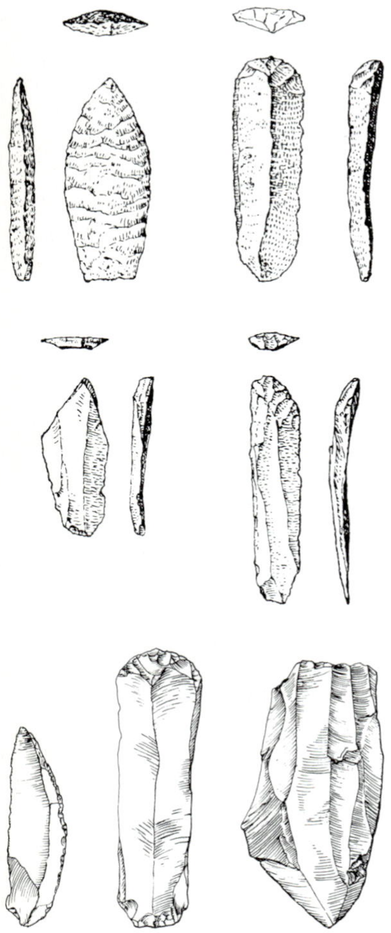

2–22
Some European tools manufactured by modern people during the Late Stone Age. Notice that these tools are more finely chipped and otherwise more refined than those made by earlier members of the human family. (From "Tools and Human Evolution," by Sherwood L. Washburn. Copyright © 1960 by Scientific American, Inc. All rights reserved; and from W. F. Bodmer and L. L. Cavalli-Sforza, Genetics, Evolution and Man. *Copyright © 1976, W. H. Freeman and Company.)*

Perhaps the most spectacular cultural achievement of Cro-Magnon people is their cave paintings (Figure 2-24). There are over 70 known caves in France alone, some of whose walls are literally covered with pictures of the animals most often hunted by Cro-Magnons. These paintings were clearly not decorative, for most of them are in sunless caves deep within the earth, and even there some are painted on the least visible walls. The artists, who made their paints from clays and mineral oxides mixed with fat and charcoal, either dabbed the paints directly on the limestone walls or blew powder onto the walls through hollow bones. Almost certainly, these priceless paintings were produced in connection with rituals by whose performance the Cro-Magnons had hoped to insure themselves of a successful hunt.

Besides making cave paintings, Cro-Magnons carved statuettes from stone, bone, and ivory. For the most part, these figurines are rather small, and they look like women with greatly exaggerated breasts, bellies, and buttocks (Figure 2-25). Though they are usually called "earth goddesses" or "fertility figures," we really do not know what significance they had for the Cro-Magnons. But unlike cave paintings, the figurines were often found in Cro-Magnon dwelling places, and many of them taper to a point at the bottom, as if the artist designed them to be stuck in the earth or placed in a base of some sort.

The Cro-Magnons buried their dead carefully and with great respect. One famous woman of the Gravettian culture who was buried about 25,000 years ago and later disinterred in what is now Czechoslovakia was found with the paws and tail of an arctic fox in one hand and its teeth in the other. She had been painted with red ochre, and her grave was covered over by two huge shoulder blades from a mammoth. We can only guess about what all this meant to the woman, or to the people who buried her. But clearly, the Cro-Magnons had developed a rich and complicated culture that had a world view in which the woman was considered worth burying with appropriate ceremony.

Still further evidence that Cro-Magnon people had a complicated world view comes from the remains of cave bears whose skeletons were arranged in a way that was clearly symbolic to the people who arranged them, although of unknown meaning to us. In one cave in Switzerland, a chest of cave bear skulls and leg bones was discovered buried in the floor, and overwhelming evidence that the Cro-Magnons had a special, probably ritualistic regard for this animal is found throughout Europe.

In all, the Cro-Magnons, with their finely made tools, their statuettes and burials, their paintings, and their reverence for cave bears, were in no way less human than ourselves.

Human Evolution since the Stone Age

Like our Cro-Magnon ancestors, some of us still make paintings, carve statuettes, show reverence for the dead, and have rituals that concern bears. (An example of the latter are the Ainu people of northern Japan.) But, there have been changes in our species since it first evolved 40,000 years ago. By and large, these changes have not been physical, but behavioral.

Physically, the human species is much as it was 40,000 years ago, at least as judged by fossilizable parts. Facial features have perhaps become more delicate, and the size of the molar teeth has definitely decreased a little, but there have

2–23
An artist's conception of what a band of Cro-Magnons may have looked like. (Courtesy of The American Museum of Natural History.)

been no appreciable changes either in cranial capacity or in the contours of the skull. However, like more ancient peoples, modern humans are extremely variable in such physical features as skin color, height, muscle mass, width of the nose, and texture of hair. In following chapters we will discuss the genetic basis of this variability and then consider the prospects for the future evolution of our species. But variability within the species was probably as extensive as it is today by about 12,000 B.P. By that date, the most recent glacial interval of the Ice Age had ended (though another one may be beginning now), and in the milder climates that now prevail, modern people first colonized the entire Old World and then made their ways to America, the islands of the Pacific, and Australia. As populations became isolated from one another by great distance or by cultural rules, genetic differences between them became more pronounced.

Although the human species has not changed much physically, human behavior has changed considerably since the time of the Cro-Magnons. Perhaps the most important change in human behavior was the switchover from a relatively unsettled hunting and gathering way of life to a more sedentary one of raising crops. The first traces of agriculture are from Asia, and they date from at least 9,000 B.P. From Asia, the knowledge of agriculture spread outwards, and it has had an enormous effect on the size of the human population and on human social organizations ever since (see Chapter 4).

Then, between 6,000 and 7,000 B.P., another breakthrough occurred—the practice of metallurgy. At first, copper and lead ornaments were the only objects manufactured from smelted ores. But by 5,000 B.P. bronze had been invented by adding tin to copper, and since that time metals have been widely used to manufacture tools and weapons.

Meanwhile, in several river valleys of the Old World, most notably those of the Tigris-Euphrates and the Nile, agriculture and ever-improving technology were instrumental in producing the rise of the first cities, indeed of the first ancient civilizations of historic times. The complexity of the civilizations of the

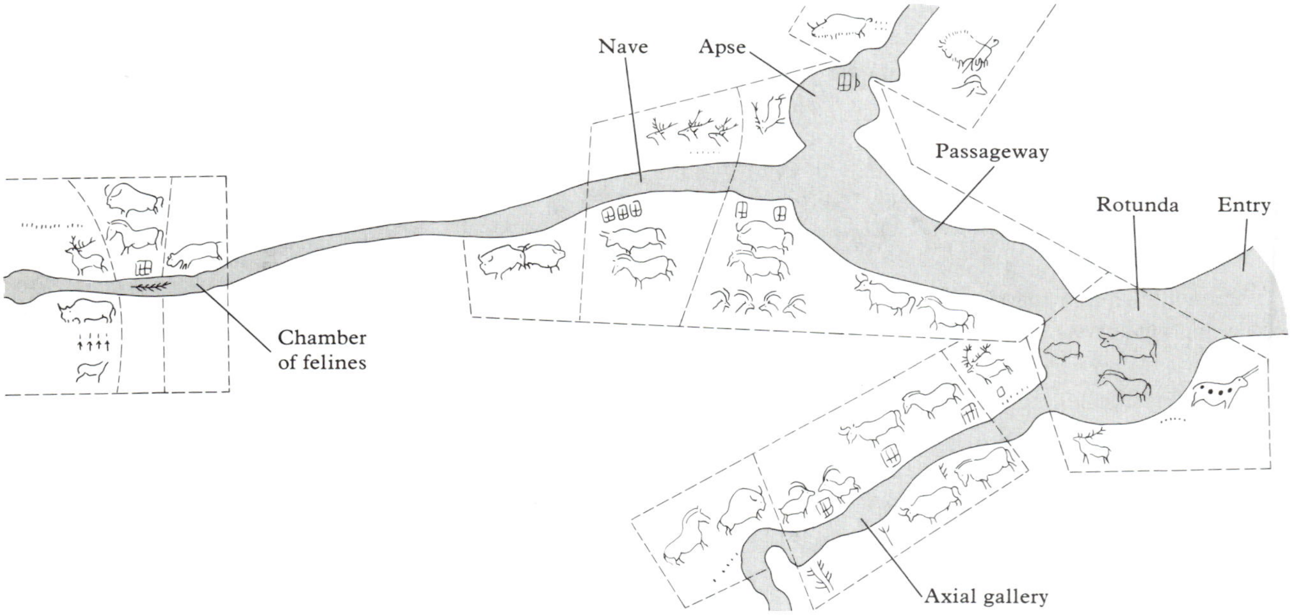

2–24
Top, an artist's conception of Cro-Magnons creating their cave art. Bottom, the layout of Lascaux cave in France. The paintings in the Chamber of Felines are more than 300 feet away from the entrance. (Artist's conception from The American Museum of Natural History. Layout from "The Evolution of Paleolithic Art," by André Leroi-Gourhan. Copyright © 1968 by Scientific American, Inc. All rights reserved.)

ancient Sumerians and Egyptians indicate that an enormous change in social organization had occurred since the time of the Cro-Magnon bands of hunters and gatherers. Yet even today, in spite of the stunning technological achievements of some people, others just as human and just as civilized still make their living much as the Cro-Magnons did (Figure 2-26).

As we have seen, traces of cultural activities have been associated with human, or near-human, beings for at least 2.8 million years. Our species had evolved to its present form 40,000 years ago, and people had invented civilization by about 7,000 years ago. Since then the rate of technological development has been explosive, and technology will surely have some effect on the survival and future evolution of human beings. But as we will discuss in the following chapter, culture is really a constellation of human activities whose rate of evolution is difficult to measure overall. Although our species has changed rapidly in its technology, in many ways we are not culturally very different from the Cro-Magnons. After all, most of us still make fires, cook food, use tools, respect and bury the dead, wear jewelry, and perform rituals. Perhaps more important, some of us still make works of art. And most important, we all still communicate with one another primarily by expressing our thoughts verbally in the symbol system of language.

2–25
A Stone-age fertility goddess.

Summary

Biologists classify people, apes, monkeys, and some other less familiar relatives as members of the order Primates. Most of the characteristic adaptations of primates are related to life in the trees, and many of these adaptations have been important factors in human evolution.

Primates can be subdivided into two groups, the more ancient and primitive prosimians, and the more recently evolved anthropoids. Anthropoids in turn are classified into six families: three families of monkeys, two families of apes, and the living and extinct members of the human family. Members of the human family are called hominids, and in deciding whether a fossil represents a hominid all available evidence about the creature's appearance and behavior must be taken into account.

The most ancient member of the human family is *Ramapithecus,* but very little is known about how these animals looked or behaved. The *Australopithecines,* of which there were two different species, were rather primitive hominids that lived in Africa, and they were probably not direct ancestors of modern people.

The oldest known human artifacts are crude stone tools that are at least two million years old. Such tools have been discovered at East Rudolph and at Olduvai Gorge in East Africa. These tools are probably the handiwork of the hominid known as *Homo habilis,* which inhabited both of these sites. A recently discovered hominid, which is probably at least three million years old and is presently known as "Homo 1470," may have been slightly more advanced than *Homo habilis.*

Homo erectus people were primitive, but undeniably human, beings who inhabited Africa, Europe, and Asia from about one million years ago to roughly 150,000 years ago. *Homo erectus* people domesticated fire and may have evolved a crude form of spoken language.

2–26
Artifacts (at one-seventh their natural size) made by present-day African Pygmies. Most of these objects are perishable and would leave no fossil traces.

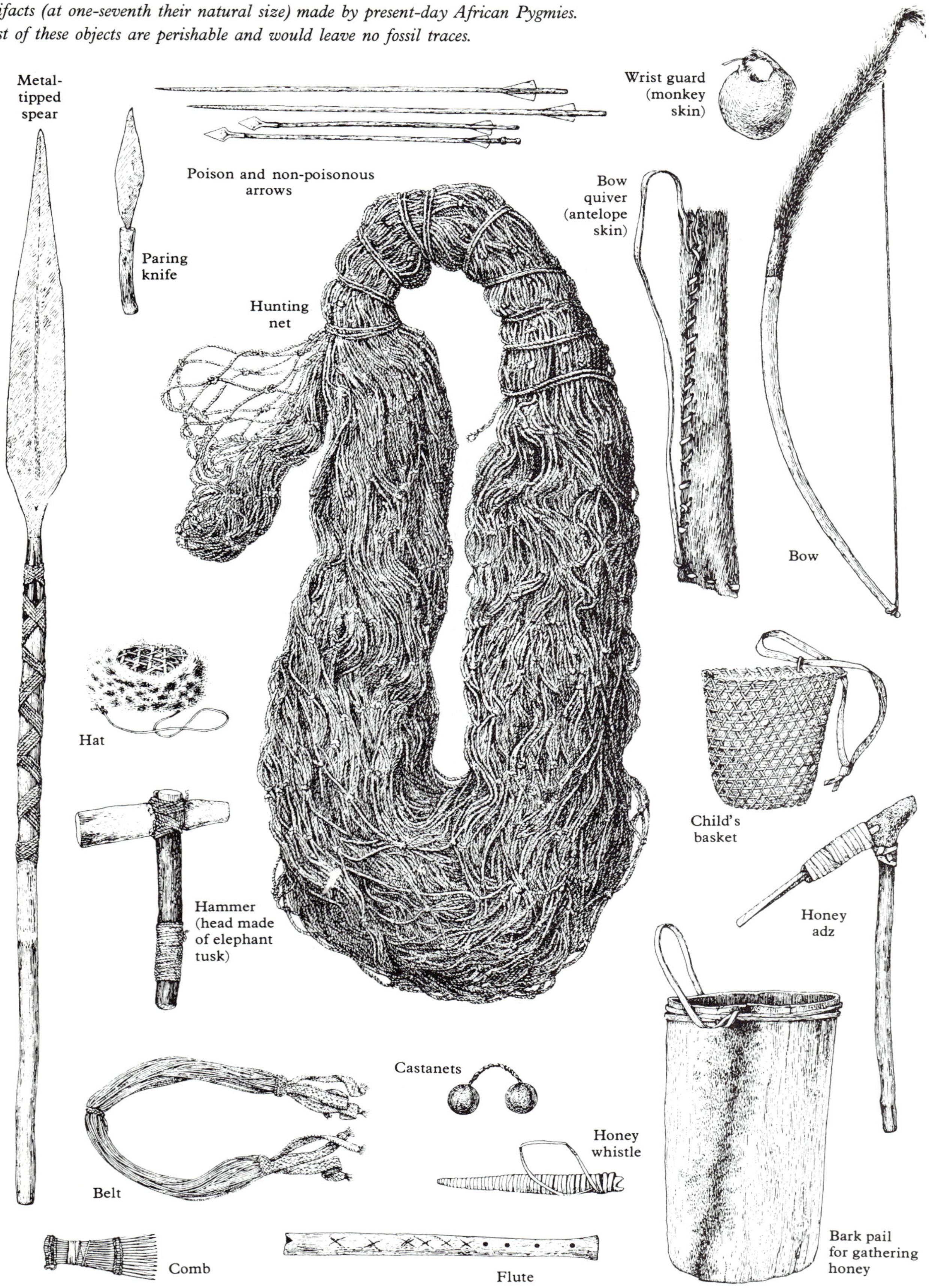

Neanderthal people *(Homo sapiens neanderthalensis)* were of two types, classic and progressive. The classic Neanderthalers became extinct, and the progressive Neanderthalers may have directly evolved into modern people. The Neanderthalers sometimes buried their dead and they almost certainly communicated by means of language, but they did not engage in deliberate artwork.

Modern people *(Homo sapiens sapiens)* first evolved by 40,000 years ago in either Africa or Asia. By 25,000 years ago Cro-Magnon people had evolved a world view that led them to create works of art that had symbolic and ritualistic significance. Physically, modern people have not changed much in the past 40,000 years, yet human behavior has changed considerably since the time of the Cro-Magnons. Among the most important behavioral changes are the practice of food raising and the invention of metallurgy.

Suggested Readings

1. *The Natural History of Man,* by J. S. Weiner. Universe Books, 1971. A broadly based inquiry about the evolutionary history and present diversity of the human species.
2. *Man,* 2d Edition, by Richard H. Harrison and William Montagna. Meredith Corporation, 1973. An overview of human biology with special emphasis on similarities and differences between people and nonhuman primates.
3. *Earliest Man and Environments in the Lake Rudolf Basin: Stratigraphy, Paleoecology, and Evolution,* Y. Coppens, C. Howell, G. L. Isaac, and R. E. F. Leakey, Editors. University of Chicago Press, 1976. This rather technical work discusses some of the relevant facts concerning the hominid known as Homo "1470."
4. *Peking Man,* by Harry L. Shapiro. Simon and Schuster, 1974. An entertaining account of the disappearance of some famous fossils and of the lifestyles of *Homo erectus* people.
5. "Skull 1470" by Richard E. Leakey. *National Geographic, 143,* 6 (June 1973). The story of its discovery and some of its implications for human evolution.
6. "Ethiopia Yields First 'Family' of Early Man" by Donald C. Johanson. *National Geographic, 150,* 6 (Dec. 1976). An account of some spectacular discoveries of fossil hominids at Hadar, Ethiopia.
7. "The Earliest Apes," by Elwyn L. Simons. *Scientific American,* Dec. 1967, Offprint 636. A discussion of the immediate evolutionary ancestors that apes and people have in common.
8. *"Homo Erectus,"* by William W. Howells. *Scientific American,* Nov. 1966, Offprint 630. A discussion of the now-extinct populations of *Homo erectus* people.
9. "Tools and Human Evolution," by Sherwood L. Washburn. *Scientific American,* Sept. 1960, Offprint 601. How the evolution of stone tools reflects cultural changes.
10. "Stone Tools and Human Behavior," by Sally R. Binford and Lewis R. Binford. *Scientific American,* April 1969, Offprint 643. How statistical analysis can be used to identify groups of stone tools that were used for various kinds of jobs.
11. "The Casts of Fossil Hominid Brains," by Ralph L. Holloway. *Scientific American,* July 1974, Offprint 686. The contours of the brain can sometimes be related to cultural changes during human evolution.
12. "The Coprolites of Man," by Vaughn M. Bryant, Jr. and Glenna Williams-Dean. *Scientific American,* Jan. 1975, Offprint 687. Fossilized human excrement can sometimes provide information about ancient diets.

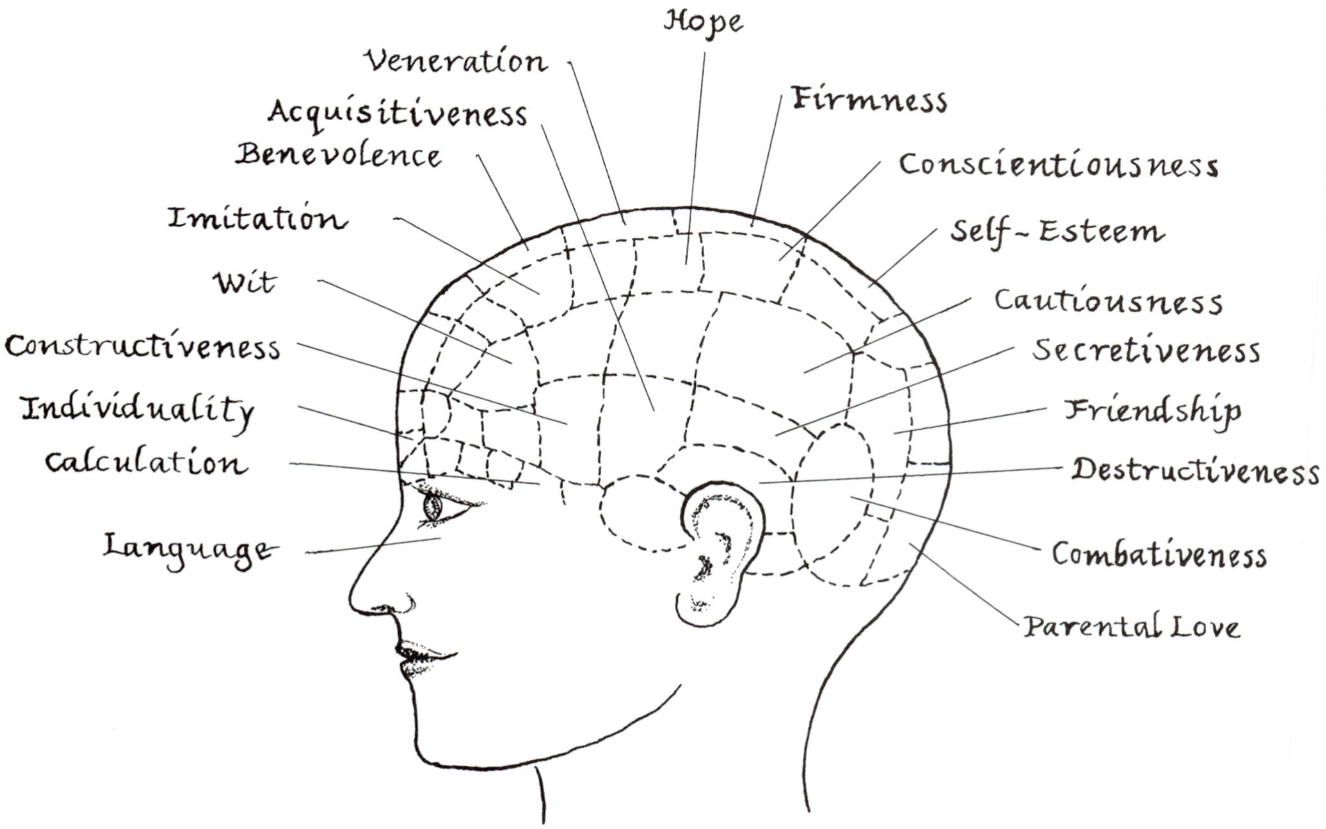

Phrenology, the analysis of character and mental aptitudes as determined by the shape and protuberances of the skull, was most popular in the early nineteenth century. Although there is no basis for the localization of the "moral" qualities shown here, phrenology offered one of the first suggestions that "higher mental functions" are localized in specific areas of the brain. (Courtesy of Galen H. Hilgard.)

CHAPTER 3

Human Behavior, Its Basis in the Brain

Anyone who has watched a spider carefully spinning her web or a flock of geese winging southward as winter approaches is aware that animals seem to know what they must do to survive. Biologists explain this behavior in evolutionary terms. The explanation is a familiar one (see Chapter 1): animals do what they do because natural selection has resulted in the survival of those animals that are best fitted to a particular environment, both where they live and what they do there. What animals *do* is as much a product of evolution as how they look or how they are distributed. Animals cannot be understood simply as things. They are things that do something somewhere.

What an animal does is referred to as its behavior. In its broadest sense, behavior includes *homeostasis*—the ability of living things to maintain their internal environment within the narrow limits compatible with life. Homeostatic mechanisms that regulate the internal environment are discussed in Chapter 6. In this chapter, we are concerned with the biological basis of the day-to-day activities of animals, especially those of people. These activities are referred to as *adaptive behavior* to emphasize their relation to survival in a particular environment.

For all but the simplest animals, adaptive behavior is a product of the *brain;* in general, the kind of adaptive behavior that is characteristic of a given animal depends on how its brain is organized. This is nowhere more obvious than in the brains and behavior of people. The human brain is undoubtedly the most complicated organ that natural selection has produced so far, and people behave in ways more complicated than those of any other animals. It is above all what we people *do* that makes us so different from other living things. In this chapter we discuss some kinds of behavior that are unique to people and try to describe these uniquely human activities as products of the human brain. But before we do so, we should discuss the relationship between brain and behavior among animals in general.

Neurons, Brains, and Body Plans

All brains are made up of building blocks called *neurons,* or nerve cells, and the cells that nourish and support them. Figure 3-1 shows what some neurons from different parts of the human nervous system look like. Neurons are connected to one another by *synapses.*. When neurons transmit information to one another, electrical impulses originate in the area of the cell body then travel down the axons and across the synapses. Because neurons are highly interconnected, the firing of a given nerve cell may cause the many other neurons connected to it to fire, too. (The physiology of neurons is discussed more fully in Chapter 6.) Those neurons that carry information about the environment to the brain are called *sensory neurons.* And those that produce some kind of activity—usually the contraction of muscles—in response to incoming sensory information are called *motor neurons.*

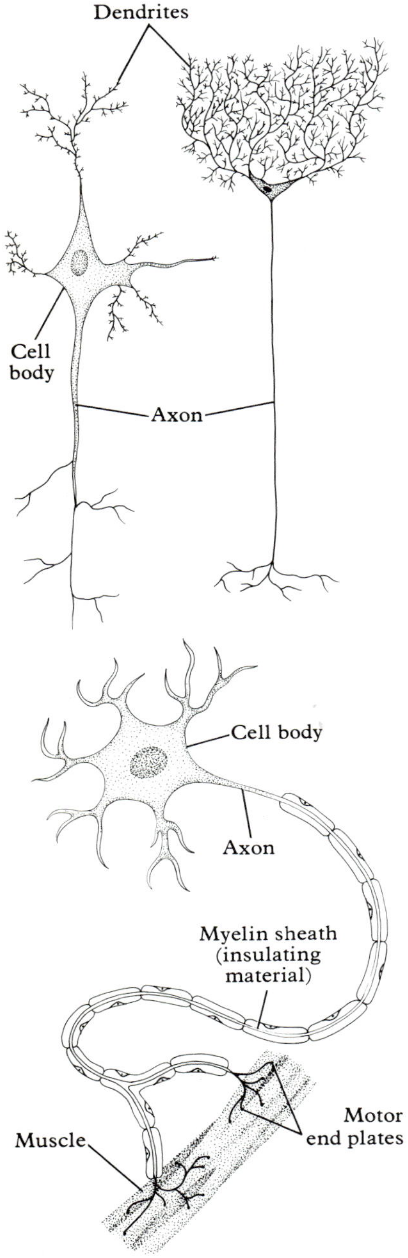

3–1
Some neurons from different parts of the human nervous system. Also see Figure 6–1.

Both sensory and motor neurons tend to be localized in definite areas within the brain, and the brain itself it characteristically located in an animal's head end. It makes sense that sensory and motor neurons became concentrated in this end. In fact, the first brains probably originated in the heads of simple, wormlike animals. These early brains may have served to test the environment in the animal's direction of motion and to coordinate the body movements necessary either to continue in that direction or to change course, as appropriate.

Although it is well known that the arrangement of sensory and motor neurons in the brain is related to the adaptive behavior that characterizes a given animal, the exact nature of this relationship remains largely unexplained. Nonetheless, we can make some generalizations about it, and to do so it is helpful to consider that adaptive behavior comes in basically two kinds—instinctive and learned.

Instinct and Learning

Instinct is a kind of adaptive behavior that is genetically determined in the structure of the brain. If every animal of a given species responds the same way when confronted by a certain stimulus (whether or not the animal has ever encountered it before), then that response can be considered as instinctive. For example, all female digger wasps (which live for only a few weeks) perform a series of complex actions beginning as soon as they emerge from their eggs. During her few weeks in the sun, a female digger wasp must first find and mate with a male. Then she builds a nest that has cell-like chambers within it, deposits

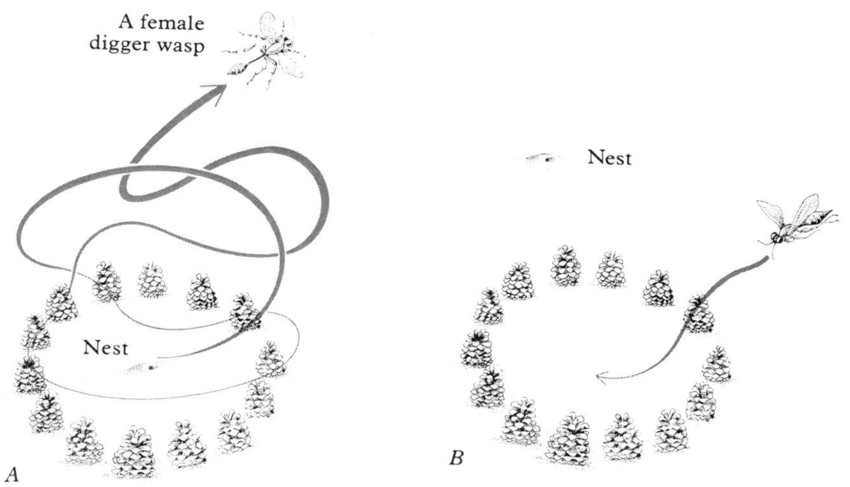

3–2
Although virtually all of the female digger wasp's activities are determined by instinct, she uses learned visual cues in locating her nest. In an experiment illustrated here, a ring of pine cones was placed around the nest entrance while the wasp was inside (A). While the wasp was away, the ring of pine cones was moved (B). The wasp returned to the center of the ring, not to the tree nest entrance. (Redrawn from N. Tinbergen, The Study of Instinct. *Copyright © 1951, Oxford University Press.)*

her eggs, and stores provisions such as caterpillars in the chambers to serve as food for her offspring, whom she will never see. This whole series of adaptive responses is determined by instinct alone because all female digger wasps go through the same routine, even those that never see another female wasp (Figure 3-2).

How does instinct relate to the brain? Simply put, instinctive behavior probably depends on specific, genetically determined circuits connecting sensory and motor neurons. Such circuits result in the production of the same response every time a given stimulus is encountered. So instinct can be considered as a kind of collective memory of activities so essential to the survival and reproduction of a species that by natural selection they have become part of the brains of each member of that species.

In addition to instinct, all animals (except perhaps the very simplest ones) have a second kind of adaptive behavior called *learning.* Learning is the capacity to modify behavior by experience. Learned responses are not precoded in the brain, but in the final analysis learning too depends upon the way in which sensory and motor neurons are interconnected. The learned responses themselves are not genetically determined, but how much an animal can learn and the kinds of things it can learn to do are genetically determined and depend on the structure of the brain.

But although it is relatively easy to envision how pre-established connections between sensory and motor neurons can result in an instinctive response to a given stimulus, the cellular basis of learned behavior is not at all obvious. Learning can be defined as a lasting *change* in behavior that depends on past experience. In order for this to occur, there must exist within the brain some kind of record, or *memory,* of what happened when a given stimulus was encountered in the past. The brain somehow decides what course of action is most appropriate in response to a given stimulus depending on the outcome of past actions taken in response to the same or similar stimuli.

We really do not know much of how memory and learning relate to neurons, but this much we can say: complex learned responses depend on the presence of many highly interconnected neurons that are located between sensory and motor nerve cells (Figure 3-3). These interposed neurons, called *association neurons,* allow incoming sensory information to be processed in complicated ways before the appropriate learned response is produced.

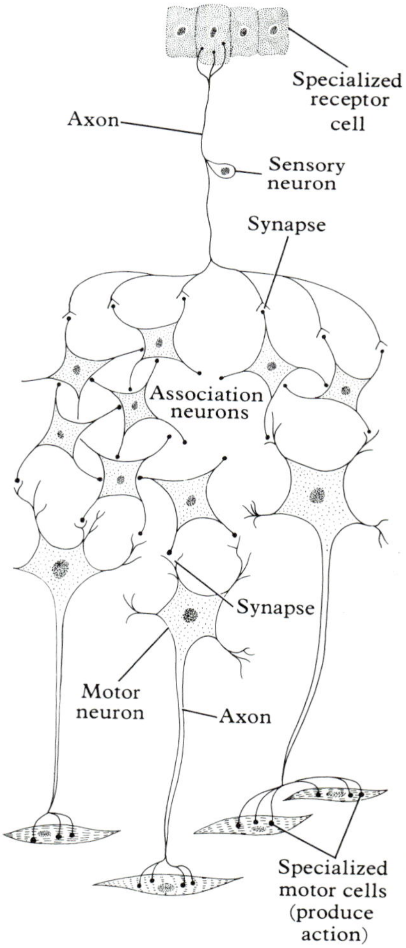

3–3
Association neurons are nerve cells that are interposed between sensory and motor neurons.

Most animals engage in both instinctive and learned behavior, and the distinction between the two is often not clear cut. Both kinds of behavior are essential to the survival of most animals. But with the evolution of the vertebrates (see Chapter 1), learning, with the behavioral flexibility it provided, became more and more important to survival as backboned animals evolved more complicated brains. And, as we shall see, learning is of greatest importance to the animals that have the most complicated brains of all—people.

Brain and Behavior in the Major Vertebrate Groups

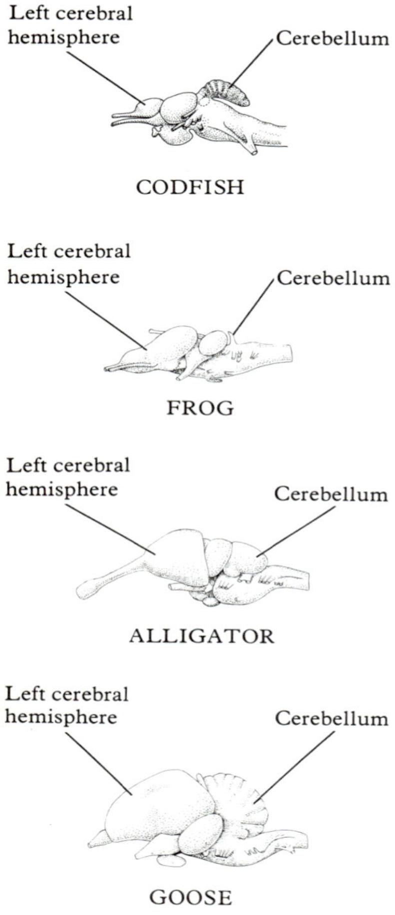

3–4
The brains of a codfish, a frog, an alligator, and a goose.

The major change in the adaptive behavior of vertebrates along the evolutionary sequence from fish to amphibians, to reptiles, and finally to birds and mammals is in the complexity that individual animals can handle in learning. This is best measured by comparing how different animals perform on behavioral tests. But animals differ so widely, both in the kinds of sensory information they receive and in their abilities to manipulate objects employed in testing their learning capacities, that it is often difficult to devise truly comparable test situations. This is especially true for animals that are adapted to life in very different environments, like porpoises and people (see the following section), or for animals that perceive the world differently than people do. For example, the reason it took a long time to find out how homing pigeons find their way home is that the pigeons probably use a magnetic sense that is so far not known to occur in human beings.

Nonetheless, everyday experience and data from behavioral tests make it relatively easy to arrange the major vertebrate types according to increasing ability to learn. Among the more ancient vertebrate types (fish, amphibians, and reptiles), the differences in learning ability are not so obvious—none of these three types are particularly good learners of complex tasks. But behavioral tests show that the sequence for learning ability is generally the same as the evolutionary one. The same tests show that birds are comparatively better learners than any of the three more ancient vertebrate types and that mammals are far and away the best learners of all.

Figure 3-4 is a diagram of the brains of a fish, a frog, an alligator, and a bird. A glance at the figure shows that there are obvious differences between these brains, but even more striking are the similarities in their gross organization. The evolution of the vertebrate brain has consisted of an elaboration on a theme. The plan of concentrating sensory and motor neurons in specific areas (which is characteristic of even the simplest brains) is most apparent in the brains of vertebrates. The various swellings or *lobes* that are visible on the surfaces of all the brains shown in Figure 3-4 represent concentrations of neurons that participate in specific functions. It is, in fact, possible to deduce the kind of adaptive behavior most important to the survival of some vertebrates simply by looking at their brains. Compare, for example, the part of the brain called the *cerebellum* in the brains of the alligator and the bird in Figure 3-4. This part of the brain contains concentrations of neurons that play a part in the production of coordinated body movements. The cerebellum of the alligator is relatively small, because agility and the performance of finely coordinated body movements are not very important in the adaptive behavior of alligators. On the other hand, agility and maneuverability are of extreme importance in the adaptive behavior of birds. Accordingly, birds have a relatively large cerebellum since many neurons connected in complicated ways are necessary to coordinate the intricate maneuvers performed in flight.

The most obvious difference between the four brains diagrammed in Figure 3-4 is in the relative size of that part of the brain called the *cerebral hemispheres.* These symmetrical outgrowths of the front part of the brain are no larger than the other protuberances in the brain of the fish, but they are progressively larger in the brains of amphibians, reptiles, birds, and mammals. As the hemispheres evolved from one group to the next, they pushed backwards, expanding around the areas of the brain underlying them; they eventually became so large that they hide most of the other regions of the brain of species furthest along the evolutionary sequence. They are most highly developed in mammals.

Brain and Behavior in Mammals

Figure 3-5 is a diagram of the brains of a tree shrew, a cat, and a human being. Once again, the most obvious difference between the brains of these mammals is in the relative size of the cerebral hemispheres. In fact, the hemispheres of most mammals are so greatly elaborated that their surfaces are thrown into folds or *convolutions* that increase the total surface area of the hemispheres and allow for more neurons to be concentrated there. We have already seen that certain areas of the brain, like the cerebellum, are larger and more highly developed if the functions they perform are of a particular importance in adaptive behavior. The obvious inference concerning the evolution of the cerebral hemispheres is that whatever their function is, it must become increasingly important to survival further along the vertebrate sequence. Because we know that the ability to learn is of extreme importance in the adaptive behavior of mammals, we might expect that the progressive elaboration of the hemispheres is related to the ability to learn.

But the exact relationship between elaboration of the cerebral hemispheres and the ability to learn is not clear cut. To being with, not all learning depends on the cerebral hemispheres. In fact, the exact place within the brain where learning occurs has never been found, probably because there is no single place where learning occurs. But this much we do know: elaborate cerebral hemispheres with many highly interconnected association neurons allow for a kind of refinement of learned behavior that is not apparent in animals whose hemispheres are poorly developed. We also know that the hemispheres are the seat of the higher mental functions of the brain, and the most complicated kinds of learning do depend upon their presence.

From the appearance of the brains of tree shrews, cats, and people in Figure 3-5, we would probably expect that people, with their very elaborate hemispheres, should be able to learn more than tree shrews or cats, and this is borne out by experience. But now look at Figure 3-6, which compares the brains of a chimpanzee, a porpoise, and a person. All three of these animals have large, highly convoluted cerebral hemispheres, and if we had to predict which of these animals would be able to learn the most simply on the basis of the gross appearance of its brain the decision would not be easy. In fact, all three of these

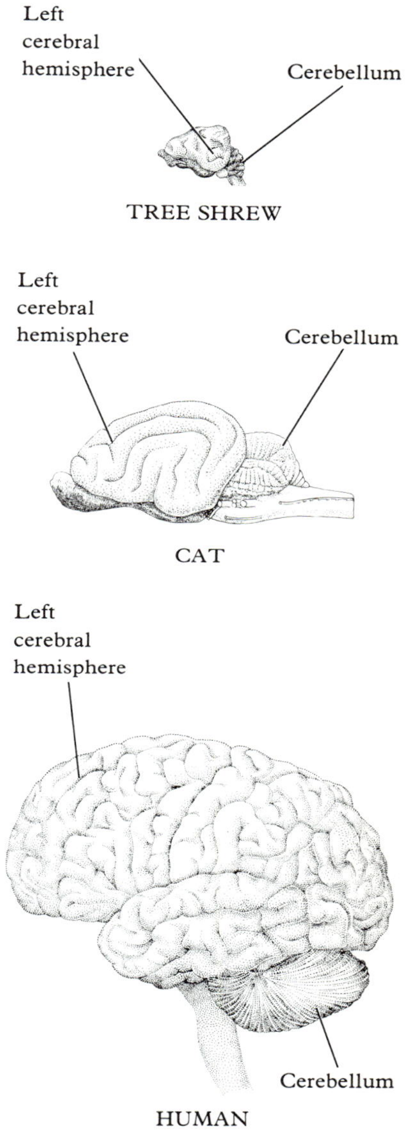

3–5
The brains of a tree-shrew, a cat, and a human being. Notice the relative size of the cerebral hemispheres.

animals are able learners, as we would expect. Yet there are marked differences in the learning abilities of chimpanzees, porpoises, and people. Evidently, there must be something more to the biological basis of learning ability than the mere presence of large, convoluted cerebral hemispheres. That something has to do with features of brain architecture that are not so obvious on first glance. But before we look more closely at the structure of these three brains, we must first decide which of these animals is able to learn the most.

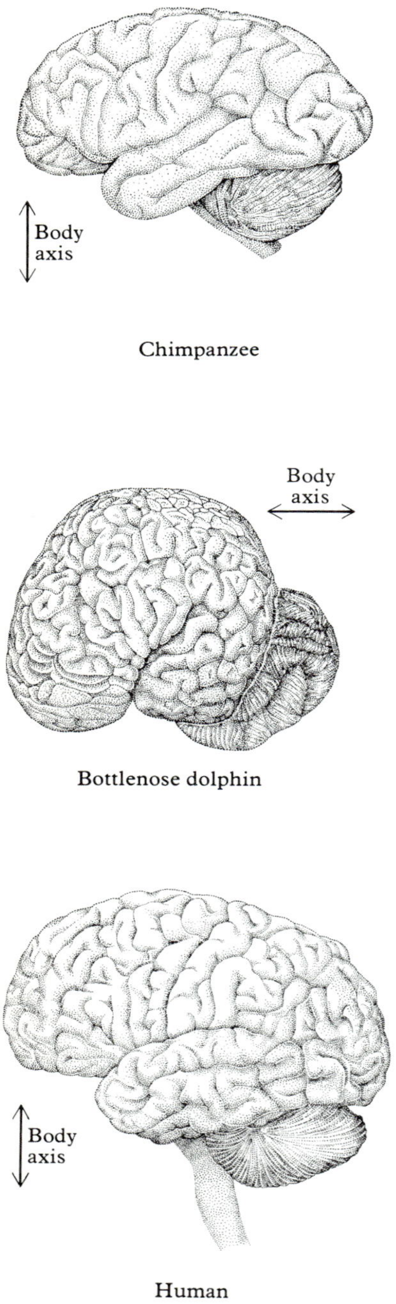

3–6
The brains of a chimpanzee, a porpoise, and a human being.

Some Problems in Measuring the Capacity to Learn

To begin with, there is no doubt that people can, and routinely do, learn more than chimpanzees. People and chimps are both primates and are both built on the same general body plan. This makes it easy to devise tests to compare their learning abilities. And such tests, in combination with observations of the animals as they go about their day-to-day activities in their natural setting, leave no doubt that people readily learn more than chimps or any other primates.

But now consider the difficulties of devising a test to compare the learning abilities of chimps and porpoises. Chimps are visually oriented land animals that can manipulate objects well with their hands. Porpoises live in water, have no hands, and orient themselves in their environment primarily by means of hearing. In devising tests of the learning abilities of porpoises, we may underestimate their intelligence if we measure their ability to learn by standards that apply to land animals that are visually oriented and can manipulate objects well. It is only relatively recently that we have become aware of the remarkable ability of porpoises to learn, primarily because they are adapted to life in a watery environment that is so different from our own. Also, very little is known of how these animals behave and learn in their natural environment. So all we can say in comparing the ability of chimps and porpoises to learn is that both are very capable of learning.

But what about the learning ability of porpoises as compared with that of people? Here again, the same reservations apply—we have no way of accurately measuring differences between these two mammals in learning ability. Nonetheless, in spite of some popular literature to the contrary, there is no reason to believe that porpoises are as intelligent as people. Evidence indicates that they are not. To begin with, porpoises have much more stereotyped kinds of behavioral responses than people do. Also, baby porpoises have short periods of dependency on their parents, as compared to people. There is really little doubt that if an appropriate test could be devised, people would prove to be much more intelligent than porpoises. We already know that people are by far the most intelligent animals on land.

Now recall that we could not predict that people would be more intelligent than chimps or porpoises simply by looking at the three brains in Figure 3-6. But we do know that the most complicated kinds of learned behavior depend upon the cerebral hemispheres. What exactly is it then about the human hemispheres (which do not look much different from those of the chimp or the porpoise) that enables people to learn so much more than any other animals?

If instead of a diagram we had the actual brains of a chimp, a porpoise, and a person in front of us, we would notice that the chimp's brain is smaller than the other two and that the brains of the porpoise and the person are more nearly equal in size. Perhaps it is the relative *size* of the cerebral hemispheres that determines how much an animal can learn. And, since larger brains have larger hemispheres, maybe brain size in itself is an indicator of learning ability. This is generally true, but then large animals have large brains to begin with. Consider, for example, that the blue whale, the largest animal that has ever lived, has a brain that weighs about 9000 grams, as compared with the average 1400 grams for the brains of people. But the blue whale may weigh as much as a hundred *tons,* as compared to the average person's 70 kilograms (160 or so *pounds*). If we compare the *ratio* of brain weight to total body weight, we find that the brain of the blue whale accounts for about one ten-thousandth of the animal's total body weight, whereas the human brain accounts for roughly one forty-fifth of the total weight of an average person. The same kinds of calculations for chimps and porpoises give brain weight to body weight ratios of $\frac{1}{110}$ and $\frac{1}{38}$ respectively. Thus we can conclude that it is not the size of the human brain alone that makes people able to learn so much.

Nonetheless, the ratio of brain weight to body weight provides us with a kind of index of "braininess" that is useful in comparative studies of brain evolution. Comparison of the brains of fish, reptiles, birds, and mammals reveals clear-cut differences in their ratios (Figure 3-7). And, as you may recall from our discussion of human evolution (Chapter 2), cranial capacity is an important statistic in helping to decide how close to human a particular fossil relative is considered to be. But clearly, cranial capacity as an indicator of brain size provides us with only a very crude estimate of intelligence and learning ability

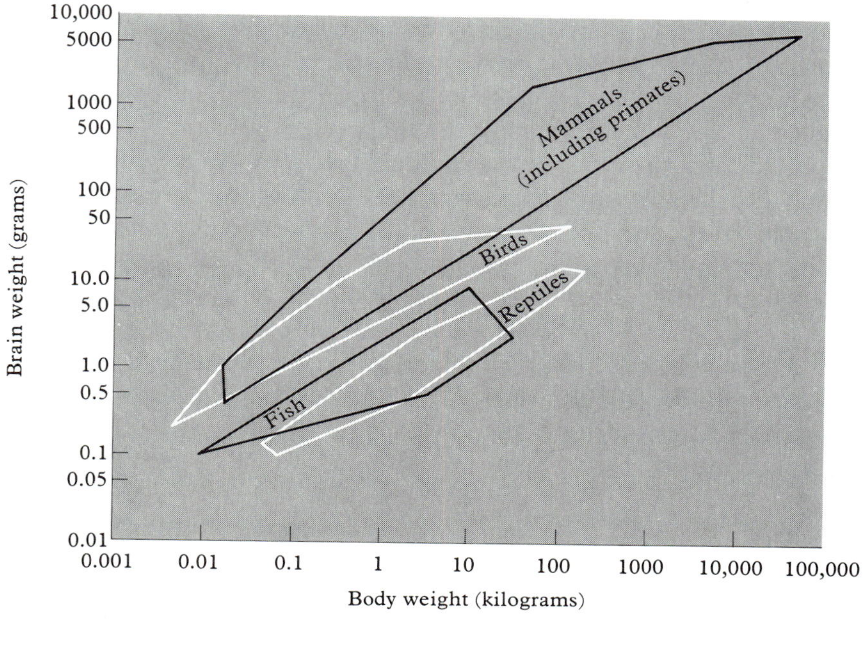

3–7
A graph of brain weight versus body weight for some major vertebrate types. (From Harry J. Jerison, Evolution of the Brain and Intelligence. *Copyright © 1973, Academic Press, Inc.)*

among our distant ancestors. Indeed, some Neanderthal peoples who made only very crude stone tools and who left behind them no works of art when they became extinct 40,000 years ago had relatively *larger* brains than do modern people.

The Cerebral Cortex

Because the size of the cerebral hemispheres does not by itself make people such outstanding learners, there must be something within the hemispheres that enables people to learn so much. Figure 3-8 shows what the human brain looks like if we cut it in two across its width. Notice that the outside surface of the convoluted hemispheres is uniformly covered by a thin mantel of gray tissue. This "gray matter" consists of millions upon millions of neurons that cover the entire surface of the cerebral hemispheres and compose the *cerebral cortex*. Underlying the relatively thin sheet of cortex is the "white matter" consisting of huge numbers of axons (each surrounded by a white insulating material) that originate from the neurons of the cortex and connect by way of synapses with other neurons located elsewhere in the cortex of the opposite side of the brain or with neurons elsewhere in the nervous system. Once again, the cross-sectioned brains of chimps and porpoises are very similar to those of people when looked at with the naked eye. Moreover, if we look at very thin sections of the cortex of these three animals under the microscope it is often difficult or impossible to tell which is which. Thus the cortical basis of the differences in learning ability between porpoises, chimpanzees, and people is not obvious, no matter how closely we look.

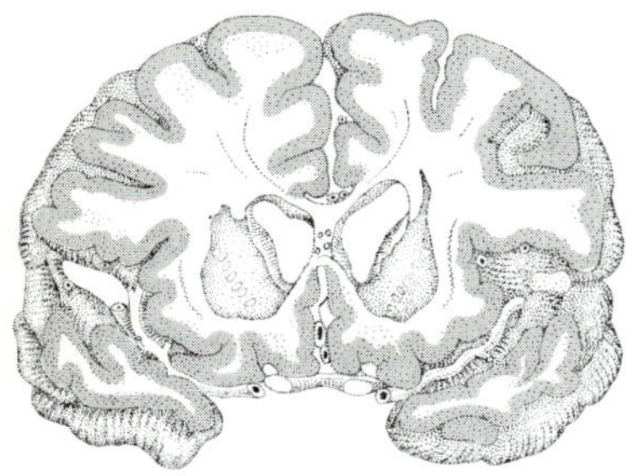

3–8
A cross-section of the human brain. Notice the thin covering of gray matter.

How the cortex affects differences in learning ability from one species to another is only poorly understood. Nonetheless, most biologists would probably agree on the following two points concerning the cortical basis of the human species' extraordinary learning ability. First, the human cortex contains more neurons than any other brain, and this surely has some bearing on the capacity to learn. Second, what makes the human cortex so unique is the way it is organized. As we shall soon discuss, the presence of elaborate, well-localized yet highly interconnected cortical areas that play a direct part in language and other higher mental functions is probably unique to the human species.

It is difficult to directly relate learning ability to either cell numbers or to complicated brain circuitry. But this much we do know: most of the neurons in the human cortex do not directly receive sense data or send out motor impulses to produce action. They are, rather, association neurons, the kind of nerve cells we know are of importance in the learning process. There are large areas of the human cortex that contain nothing but association neurons. These *association areas* enable people to behave in more complicated ways than any other animals.

As the hemispheres evolved and became progressively more elaborate, they took over functions previously served by the well-defined lobes of the brain that we have discussed before. For example, the brain of the tree shrew does not have the discrete, isolated, optic lobe that the brains of more primitive vertebrates have. Instead, the hemispheres themselves have assumed control over making sense out of incoming sensory data from the eyes, and the information is received and processed by the hindmost region of each cerebral hemisphere. This region of the hemisphere is referred to as the optic lobe even though it is less obviously a specific lobe of the brain in the tree shrew than in primitive vertebrates. There are also specific areas of the hemispheres that participate in the reception of information from the other senses, as well as localized areas that exert control over movement of various parts of the body. Figure 3–9 shows the location of the various lobes and of the sensory and motor areas in the human cerebral hemispheres.

Mapping the Cortical Areas

The location of the sensory and motor areas has been discovered primarily by two methods. The first is the electrical stimulation by means of an electrode of different areas of the brains of people undergoing neurosurgery. The people are fully conscious, because the surgery is performed under local anesthesia. The electrode generates a small electric current which stimulates neurons that are close by to discharge. If the electrode is inserted into the area of the brain shown by light shading in the diagram (Figure 3-9), and a small electric current is delivered, the conscious patient is surprised to find (because he did not intend it) that various parts of his body *move,* depending on what part of this *primary motor area* is stimulated. Moving the electrode produces such responses as movement of a finger, the twitching of a leg muscle, the contraction of the muscles of the face, and many others. The control of specific body parts is highly localized within the motor area; this topic is discussed further in Chapter 6. However, you should remember for future reference that stimulation of the motor area of the *right* side of the brain results in movement of the appropriate part of the *left* side of the body. And stimulation of the left side of the brain causes movement on the right side of the body.

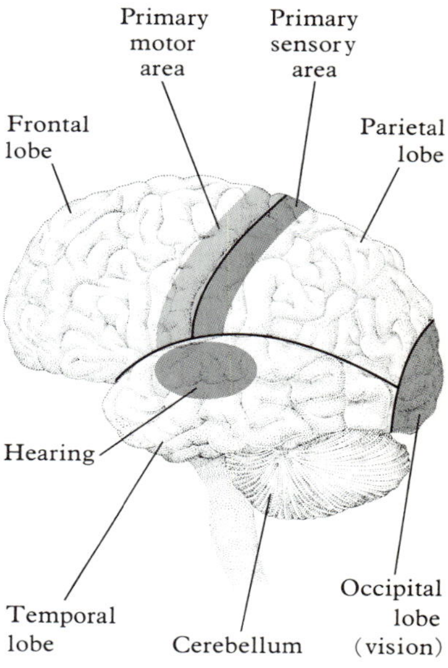

3–9
The lobes of the human cerebral hemispheres and the location of the primary motor and sensory areas.

Stimulation of the darkly shaded area shown in Figure 3-9, the *primary sensory area* (parallel to the motor area), results in a person's experiencing a vague sensation of touch from some definite area of the body on the side opposite from that of the hemisphere that was actually stimulated. There are also sensory areas concerned with vision and hearing, as shown in Figure 3-9. Stimulation of these areas results in the perception of flashes of light or ill-defined sounds. In general, the sensory areas for sight and hearing are not as clearly associated with one side of the body or the other as the primary sensory and motor areas are.

In addition to studies with electrodes, a second method of localizing functions within the brain is to observe the behavior of people who have had a part of their cortex removed by surgery (usually because of the presence of a tumor), or whose cortical areas have been damaged for other reasons. Results of all these studies show that only about 25 percent of the cerebral cortex participates in well-localized sensorimotor processes such as those we have been discussing. The remaining 75 percent or so of the cortex consists of the association areas, and they are larger and more highly developed in people than in any other animals. These relatively huge and poorly understood areas of our brains are of great importance in processes like reasoning, thinking, and abstraction—the higher mental functions that are so important in learning and the adaptive behavior of people. In the rest of this chapter, we will describe some kinds of behavior that are unique to people and try to relate them to the association areas of the human brain. But before we do so, we must first turn our attention to some kinds of behavior that we share with many nonhuman animals, whose biological basis is found in parts of the human brain that—unlike the cerebral cortex—we have inherited relatively unchanged from our nonhuman ancestors.

Emotions and the Limbic System

Internal, highly subjective feelings, or *emotions,* strongly influence the way all people behave. Nonhuman animals also show unmistakable signs of experiencing emotion—for example, they cower in fear, snarl in rage, and purr in contentment (Figure 3-10). Actually, we only infer that other animals experi-

3–10
These drawings first appeared in a book entitled The Expression of the Emotions in Man and Animals, *by Charles Darwin, published in 1872. Drawn from life by Mr. Wood.*

ence emotion, because unlike people they cannot tell us that they do. But emotion has measurable physiological qualities as well as purely subjective ones. And because animals show the same kinds of measurable changes in blood pressure, heart rate, and muscle tone (for example) that people do when they are exposed to frightening situations, we have reason to believe that other animals have subjective feelings under such circumstances that are similar to our own. We must now enquire about the biological basis of emotion and its importance in learning and adaptive behavior.

From personal experience and the observation of nonhuman animals, we know that particular emotional states are often associated with fulfillng specific goals that are of fundamental importance in survival. Such activities, like eating to relieve hunger, fleeing from frightening situations, and finding an appropriate outlet for built-up sexual energy, have importance not only for the survival of individual animals, but for the species to which they belong. And natural selection has insured that they are equipped with the "drive" or *appetite* to perform such activities: otherwise they would not have survived. Once these kinds of goal-directed activities are fulfilled, the animal experiences satisfaction, or the feeling of relief, to signal that a particular need has been (at least temporarily) fulfilled. Though nobody knows for sure, emotions may have first evolved as adaptive mechanisms that reinforced the motivation to perform some kinds of instinctive behavior, or to signal the animal when its needs were fulfilled.

Evidence in the structural organization of the brain supports the idea that emotions first evolved in relation to the performance of instinctive behavioral patterns necessary to survival. Once again, most of the evidence comes from animals whose brains have been stimulated by means of an electrode. We already know that the insertion of an electrode into the cerebral cortex can help us to localize various cortical areas concerned with the reception of sense data and with the production of voluntary movement. But emotional reactions are conspicuously absent when we stimulate various areas of the cortex itself. If instead we insert an electrode deeper into the substance of the brain and lodge it in certain more primitive areas underlying the cortex, the relation between brain and emotion becomes more clear.

The insertion of an electrode into specific areas of the brains of rats, cats, and monkeys has been known for some time to result in behavior that is apparently emotion-charged. Areas producing the most dramatic responses are located within a complicated circuit of nerve tissues just under the hemispheres referred to as the *limbic system.* This area, diagrammed in Figure 3-11 as it appears in the human brain, is evolutionarily older and has undergone much less obvious change throughout the course of evolution than have the cerebral hemispheres. It is possible to localize definite areas within this portion of the brain in which stimulation by an electrode causes a cat to react by hissing, spitting, growling, and, if the animal is cornered and not allowed to flee, by well-directed biting and clawing. In short, such stimulation makes the cat act frightened and ready to fight if it cannot flee. But if we move the electrode and stimulate other areas of the limbic system not far removed from the area that we stimulated first, the cat may become abnormally placid and virtually impossible to frighten or enrage no matter how much we provoke it in other ways.

Even more interesting are the reactions of monkeys whose limbic lobes are stimulated in specific areas that apparently result in their experiencing a feeling of pleasure. In some experiments, electrodes are arranged so that the monkey

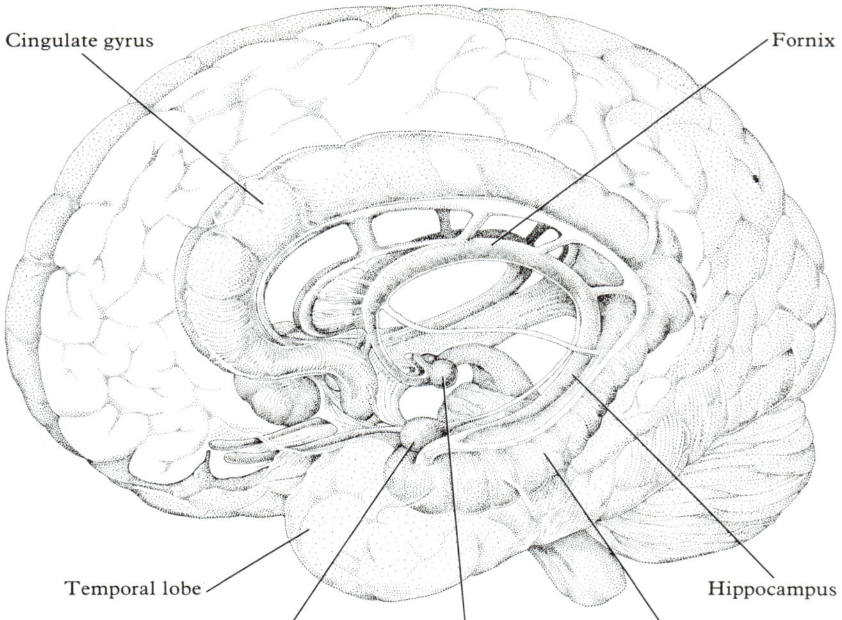

3–11
The main components of the human limbic system. (From "Opiate Receptors and Internal Opiates," by Solomon H. Snyder. Copyright © 1977 by Scientific American, Inc. All rights reserved.)

itself can cause the electrode to discharge by pressing a bar located in the cage with it (Figure 3-12). Once it has discovered that pressing the bar results in a feeling of pleasure, the monkey may begin pushing the bar over and over again. Some monkeys ignore food and drink and continue to press the bar until they fall over exhausted. Certain monkeys have been clocked at 17,000 bar presses per hour. The areas that produce such responses are called pleasure centers, though we really have no way of knowing exactly what inner feelings are experienced by the monkey when it presses the bar. There are other areas nearby that apparently result in very unpleasant (perhaps painful) feelings when stimulated, and monkeys carefully avoid pressing the bar if the electrode is inserted into one of these areas.

Both feelings of fear (like we saw in the cat) and feelings of pleasure have also been reported by conscious people whose limbic systems were stimulated by electrodes during the course of neurosurgery. The sensations produced when different areas are stimulated are described by such patients as ranging from feelings of "ease and relaxation" or "great satisfaction," to feelings of "fright" or "horror." Some patients who have brain damage in the area of the limbic system are subject to violent outbursts of rage in response to rather trivial stimuli arising from everyday interactions with other people.

Emotion and Learning

All of these observations of the behavior of people and other animals leave no doubt that the limbic system plays a part in the production of emotions. And emotions are of extreme importance in normal behavior. We have seen that emotions such as pleasure and rage may have evolved as adaptive mechanisms to reinforce innate behavioral patterns. But although they probably first arose in association with instinctive behavior, emotions are also very important in the learning process. Evidence indicates that the early stages of memory formation may occur in the area of the limbic system called the *hippocampus* (see Figure 3-11), and that from there, memories are transferred to other areas of the brain (including the cortex), where they are stored permanently. The hippocampus may influence the association of emotional states with specific experiences in such a way that, when the experience is recalled, the subjective inner feelings associated with it are recalled as well. And our emotional states have direct bearing on our abilities to learn.

The Frontal Lobes

The least understood portion of the limbic system is the part that probably most influences human behavior. This is the area where the limbic system is directly connected to the cerebral cortex. Although the limbic system has many interconnections with the lower brain centers controlling physiological processes (like blood pressure and heart rate), there are surprisingly few connections between the limbic system and higher cortical centers. Those that do exist connect the limbic system with the frontmost section of the brain, the *frontal lobes* (see Figure 3-9).

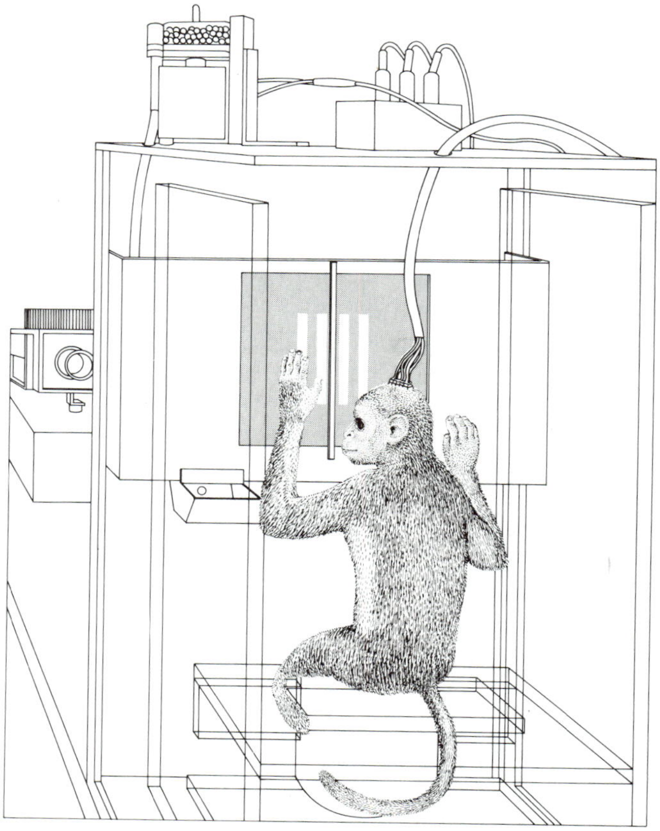

3–12
A monkey stimulating one of its pleasure centers. (From "The Neurophysiology of Remembering," by Karl H. Pribram. Copyright © 1968 by Scientific American, Inc. All rights reserved.)

The frontal lobes are huge association areas known to be involved (though exactly how remains unknown) with the human qualities of self-control and restraint. It is perhaps through the interconnections of the limbic system and the frontal lobes that people can exercise the control over their emotions and associated instinctive drives that is necessary for successful interaction with other people.

The role of the frontal lobes in human behavior is best demonstrated by the behavior of people who have suffered damage either to the frontal lobes themselves or to their connections to the limbic system. A famous case is that of a railroad construction foreman named Phineas Gage, who in 1848 was packing blasting powder into a hole with an iron rod when the powder exploded and drove the rod through his skull, nearly obliterating his frontal lobes as it went (see Figure 3-13). Phineas survived this remarkable injury and lived for another twelve years. But, although there was surprisingly little, if any, change in his intelligence following the accident, Phineas's physician and friends soon became aware that the injury had somehow altered his personality. Thus Phineas's physician, who knew his patient to be an "efficient and capable" foreman before the accident, described the change in his personality in this way:

> He is fitful, irreverent, indulging at times in the grossest profanity (which was not previously his custom), manifesting little deference to his fellows, impatient of restraint or advice when it conflicts with his desires, at times . . . obstinate yet capricious and vacillating. His mind was radically changed, so that his friends and acquaintances said he was no longer Gage.

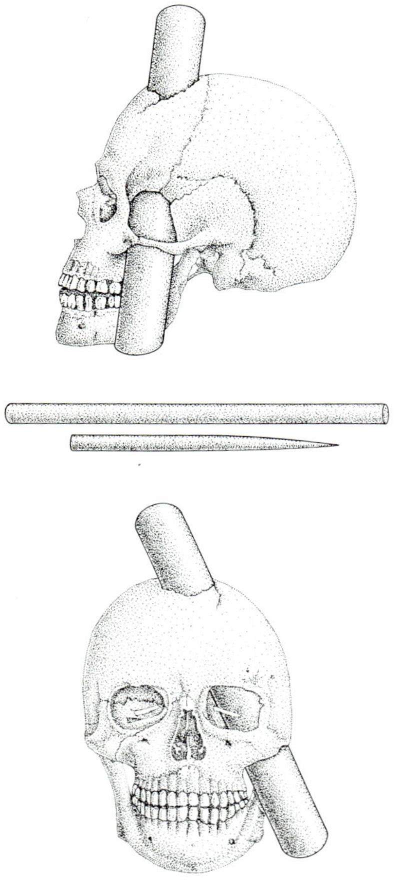

3–13
The skull of Phineas Gage and the iron rod that went through part of it.

Admittedly, an iron rod is a very imprecise instrument for localizing functions carried out by different areas of the cerebral cortex. But many other cases of injury to the frontal lobes have been reported since Phineas's time, and the behavior of the injured people leaves no doubt that these huge association areas are concerned with conscious cortical control over the kinds of behavior that influence social interactions—a form of learned behavior that is absolutely essential to the survival of the human species. People who have frontal lobe injuries frequently make tactless, indiscreet, or unduly frank remarks, and they are prone to quick mood changes from states of joy and exhilaration to those of sadness and rage.

In the past it was considered acceptable to surgically disconnect the frontal lobes from the rest of the brain of persons who had certain severe mental disorders. Classically, this operation, called frontal lobotomy, was performed on severely agitated and supposedly incurable patients, because experimental work on normal monkeys had shown that this procedure often resulted in an apathetic, placid animal, at least shortly after the operation was performed. We might expect that the study of the behavior of people following frontal lobotomy would tell us much about the normal function performed by the association areas of the frontal lobes. But this is *not* true, for two important reasons. First, people upon whom lobotomy is performed have very abnormal behavior to begin with, and second, the results of this procedure are extremely variable. Although frontal lobotomy may result in less violent behavior in some severely disturbed people, it can also produce dangerously aggressive behavior in others; in still others, there may be little if any apparent effect on behavior a few weeks or months following the operation. But because the results of the procedure are so variable, and because there exists a wide range of tranquilizing drugs that can alter violent emotional states as necessary for the protection of the patients and those around them, there is really no place in modern medical practice for the permanent destruction of brain tissue as a means of altering undesirable social behavior.

Human Behavior, Its Special Qualities

The behavior of people who have damaged frontal lobes suggests that natural selection has provided us with cortical association areas by means of which we can exercise control over limbic drives and emotions in our day-to-day interactions with others. Like most other primates, people normally live together in social groups. Mammals who live in social groups must *learn* not only how to care for themselves, but also how to interact with one another. It is only relatively recently that the interactions among individual members of various primate social groups have been well studied in the animals' natural habitats. Such studies show that many members of primate social groups (like those of chimpanzees, gorillas, and baboons) engage in rather complicated interactions. It has been suggested that we can learn much about human behavior (at least insofar as we interact with one another) by observing the behavior and interactions of other primates, our closest animal relatives. In fact, some authors have compared a traffic jam in Rome to the social behavior of a troop of howler monkeys. Such comparisons are simply analogies, and they make the serious mistake of assuming that because two kinds of behavior resemble one another, they must have arisen in response to similar needs in the two social groups.

Some investigators have applied the methods used by field biologists studying primate social behavior to the study of human beings. Most methods consist simply of accurate but tedious observations of human beings as they go about their day-to-day affairs. These observations show that, regardless of where they live or how they explain the world around them, human beings do have some simple kinds of behavior in common. For example, people everywhere, but no other animals, *smile* when they are happy (and sometimes when they are not). It may also be true that all people recognize the raised, outstretched palm of the hand as a symbol of greeting and nonhostile intent. But the really astounding differences between the day-to-day behavior of human beings and other animals are found in very complicated forms of learned behavior. Like other social animals, people learn how to behave primarily from other members of their own species. But what makes people so different from any other animals is that we learn *so much,* not only from our own experiences, but from those of others. Our enormously complicated brains enable us to do this.

Our brains enable us to store accurate memories of past events, even those we may have experienced only a single time and perhaps years ago. This huge storehouse of perceptions, based on past experience, is on file in our brains, and is available for an almost endless variety of uses in our day-to-day activities. Through our associations and communications with other human beings, we can vastly expand our individual memory stores and learn from what has happened to others in circumstances we have not yet encountered, and perhaps never will. And the most important thing that human beings learn from one another is how to cooperate to survive. The necessity for interaction and cooperation is so important to the survival of our species that all human beings have evolved specific rules of conduct that tell them how to behave towards one another. These sets of rules are one aspect of culture, learned behavior that has its biological basis in the elaborate association areas found only in the human brain.

Culture: Its Social, Material, and Mental Components

Culture includes not only rules of social interaction (social culture) but also other behavioral means of adapting to the environment such as making tools. The objects that people manufacture to meet their biological and social needs in adapting to the environment are called material culture. Different cultural groups make different kinds of tools, build different kinds of dwellings, dress differently, and use the environment in very different ways. They do these things by means of objects that they *learned* to make from other people.

The ways in which different groups of people interpret their own experiences, origins, and relationships to other living and inanimate things are also culturally determined and variable. The system of beliefs that provides an explanation for the world around us is called world view or mental culture. And mental culture, like material and social culture, is also a form of learned behavior. The main way we come to learn all culturally determined things is through another kind of learned adaptive behavior, language.

The Uniqueness of Language

Like the cultural information it is used to transmit, language is usually considered to be an exclusively human characteristic, the major means by which *people* communicate with one another. What makes language different from any other forms of animal communication we know of is that language is essentially a symbol system through which arbitrary sounds come to stand for objects or concepts that bear no relation to the sounds themselves. Many animals communicate information about the environment or about their feelings to one another by means of sound. But although this is communication, it is not language. The sounds produced by nonhuman animals are largely affective—they depend on and communicate about the physiological and emotional state of the animal at the moment they are expressed. Such utterances are not symbolic; they do not name, describe, or generalize as language does. Through language, people can refer to things that are not actually present when they are spoken about, to events in the past or to probable events in the future, or even to concepts that cannot be fully understood, like "infinity." It is largely the symbol system of language that enables people to learn so much so rapidly, and it is not surprising that such complicated symbol systems are employed only by people with their fabulously complicated brains.

This is *not* to say, however, that the process of *abstraction,* which must occur before a concept or object can be associated with a given sound, may not occur in animals other than people. In fact, it is known that chimps can learn to use sign language or various electronic devices to communicate with people in ways that show that they understand that certain symbols stand for specific objects, perhaps even concepts. But not all chimps can learn to do this, and there is no evidence that chimps routinely communicate to one another in this way in their natural environment, whereas people everywhere naturally communicate with one another primarily by means of language. There is also a distinct possibility that some kind of abstraction takes place in the clicking and whistling sounds by which porpoises communicate with one another, but their abilities to use true language have yet to be demonstrated.

Language, as the vehicle of culture, is a behavioral adaptation that is essential to the survival of the human species. What then, is the biological basis in the human cortex of our capacity to express our thoughts symbolically in words?

Language and the Brain

Our understanding of how language relates to the brain has come primarily from the study of people who have suffered damage to the cortical areas concerned with the production of language. This most often happens when a person suffers a stroke, which usually is caused by an interruption of the blood supply to the particular cortical area because of a blood clot located in one of the arteries supplying the surface of one of the hemispheres (Figure 3-14).

Disturbances of language resulting from damage to the cortex are collectively called *aphasia,* and there are several types of this condition. The symptoms depend on the area of the cortex that has been damaged.

If, for example, there is localized damage to the area called *Broca's area* (named after its discoverer) in Figure 3-15, the patient usually shows symptoms of *motor aphasia.* Such patients usually show no disturbances of conscious awareness, and they can readily understand what others around them are saying. But in spite of the fact that these people know what they want to say, when they try to speak they either cannot do so at all or their words come out garbled.

Now notice that Broca's area is very close the the primary motor area of the cortex (Figure 3-15). We might suspect that the reason why people who have motor aphasia cannot speak is simply that the adjacent motor areas controlling the muscles of the face, jaw, tongue, palate, and larynx (voice-box) have been damaged along with Broca's area, and that these muscles are paralyzed and no longer under voluntary control. But this explanation is not correct. Studies have shown that direct damage to the portion of the motor area controlling these muscles usually produces only mild weakness of the facial muscles of the opposite side and no permanent weakness of the jaw, tongue, palate, or vocal cords. The control of the motor activity of one side of the body by the opposite side of the brain is much less evident in the head than on the rest of the body. The muscles used in the production of speech can generally be controlled by either side of the brain. But curiously, language itself is usually a function of only one side of the brain, the left. So for now we can simply assume that brain damage resulting in aphasia occurs on the left side of the brain.

Why is it then, if they can still control the muscles used in speech, that people who have motor aphasia cannot use language properly? We can get a clue by listening to the speech of people who have suffered only minimal damage to Broca'a area and are still able to communicate in varying degrees by language.

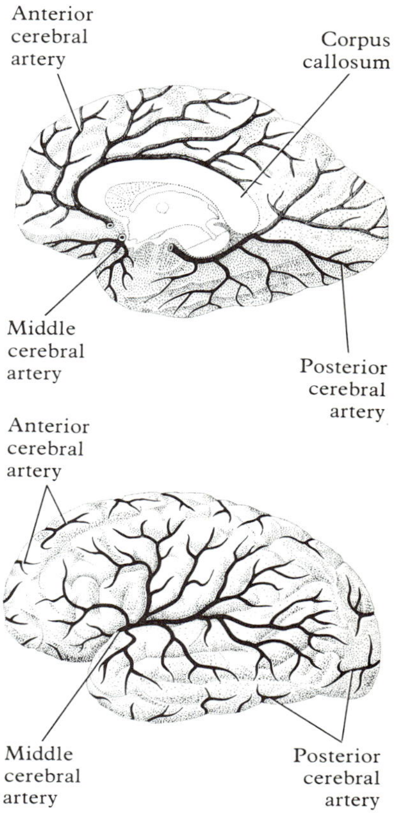

3–14
The cerebral hemispheres are nourished by arteries that supply specific regions. In many patients who suffer from inadequate oxygen supply to the brain, the damage is not within the area of a single blood vessel but rather in "border zones." These are the regions between the areas served by the major arteries where the blood supply is marginal. The brain at the top has been cut in two lengthwise. (From "Language and the Brain," by Norman Geschwind. Copyright © 1972 by Scientific American, Inc. All rights reserved.)

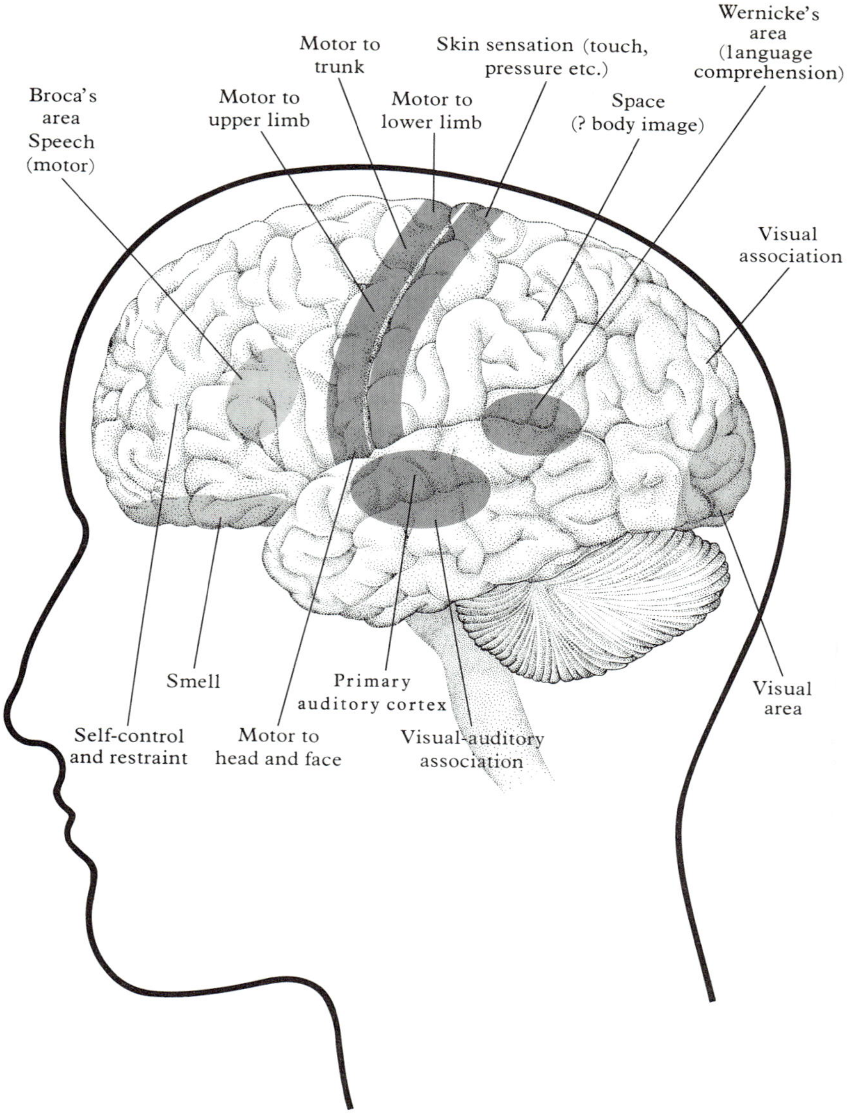

3–15
The areas of the left cerebral hemisphere that participate in language functions, and a recap of the main sensory, motor, and association regions of the human brain.

Such people usually speak only in very simple sentences, so their speech sounds like a message sent by telegram. The ability to speak intelligible words generally varies in proportion to the amount of damage to their Broca's area. People who have motor aphasia are often aware of the mistakes they make when they speak, and they sometimes become very angry and frustrated by their errors; we can conclude that Broca's area has something to do with the translation of thoughts into articulated patterns of words. Apparently, Broca's area controls the complex patterns of muscle coordination that are necessary for the production of articulated words.

People who have suffered damage to the area of the cortex labeled "Wernicke's area" in Figure 3-15 have an entirely different kind of language disorder, *sensory aphasia.* Brain damage in Wernicke's area results in a loss of comprehension of spoken language. To people who have this unfortunate condition, language seems meaningless; though they may still be able to speak words (assuming that Broca's area is still intact), they are meaningless to them.

Now notice that Wernicke's area is very close to the area of the primary auditory cortex (Figure 3-15), that part of the sensory cortex that receives sound. When a word is heard, the sensory neurons in the primary auditory cortex discharge, and they may cause the adjacent association neurons of Wernicke's area to do so, too. The primary auditory cortex is normal in people who have sensory aphasia (they hear words), so we can conclude that Wernicke's area must somehow be involved with attaching meaning or comprehension to the sounds received by the primary auditory cortex. It is also known that Wernicke's area carries out the formation of thoughts into words that can be communicated.

Let us review what we know about Broca's and Wernicke's areas so far. Broca's area, associated with the primary motor cortex, contains the brain circuitry necessary for the production of articulated patterns of words. As such, it is an association area for the transformation of symbols (words) that exist only as thoughts into the complicated motor patterns that compose a spoken message. Wernicke's area is also an association area. It is concerned with attaching meaning to the sounds received by the primary auditory cortex, and with the choice of words with which to express our thoughts. In order for normal language to result, these two association areas must somehow be connected with one another. That they normally are is best illustrated by the language disorder called *conduction aphasia,* which results when the connections between Broca's and Wernicke's areas are interrupted without causing damage to either of the areas themselves.

Not many cases of conduction aphasia have been reported, because the area between Broca's and Wernicke's areas is rarely the only part of the cortex affected by a stroke or brain damage. In fact, most cases of aphasia are said to be mixed aphasia, because most often, brain damage is extensive enough to damage both areas of the cortex, and the resulting language disorder is somewhere between the extremes of sensory and motor aphasia that we have discussed.

But the few cases in which only the cortex interconnecting Broca's and Wernicke's areas was damaged are most informative. Wernicke's area is functioning normally, so comprehension is intact. Therefore, these people can understand language and formulate their thoughts into mental words. Because Broca's area is functioning normally, we would not expect people who have conduction aphasia to have difficulty speaking words, and they do not. But their speech is very abnormal, because they cannot *transfer* the association of words with meaning, brought about by Wernicke's area, to Broca's area, where the motor patterns for transforming thought words into spoken words are found. As we would expect, these people have great difficulty in repeating spoken language. They hear and understand the words but cannot translate their understanding into the spoken words themselves, because the two cortical association areas where these two different functions are performed have been disconnected by brain damage.

Association areas that play a part in the comprehension of *written* language are located between and connected to both the primary visual cortex and Wernicke's area (Figure 3-15). Apparently, visual patterns of written words must first be converted into an auditory form by Wernicke's area before the written word can be comprehended.

The Split Brain

Figure 3-15 shows a recap of all of the areas of the cortex that we have been able to associate with specific functions so far. You may recall that association areas account for about 75 percent of the cerebral cortex, and the figure makes it apparent that those used in language account for only a relatively small proportion of the total. (Remember that association areas for language are usually found in only the left hemisphere.) Now we must ask: what do the remaining association areas do?

This is very difficult to answer and it is even harder to pinpoint where the functions take place. The reason for this is that our higher mental functions (in addition to language), which are the products of the remaining association areas, are probably not processes that result from the isolated activity of separate, highly localized areas of the cortex, but rather from the interaction of many different cortical association areas. Nonetheless, we do have evidence that the human brain is *not* bilaterally symmetrical (the same on both sides) with respect to the kinds of mental activites that it performs. The evidence comes from the study of the behavior of people who have so-called split brains, those whose right and left cerebral hemispheres have been disconnected from one another.

The right and left cerebral hemispheres are normally connected to one another by the *corpus callosum,* as shown in Figure 3-16. This structure consists of millions of axons that originate from cortical neurons of one hemisphere and terminate on cortical neurons of the opposite side. The corpus callosum is not the only bundle of axons that interconnects the hemispheres, but it is by far the largest, and, as we said, it does connect the cortex of one side directly to the other.

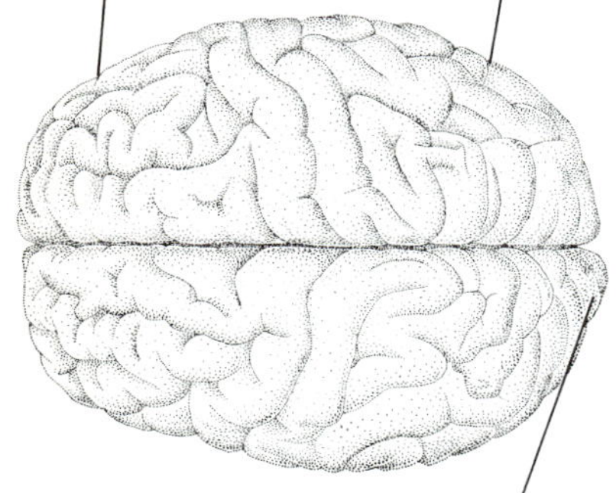

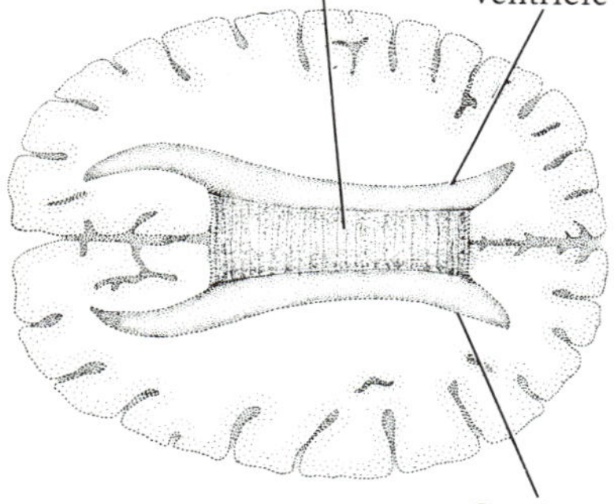

3–16
The corpus callosum directly interconnects the two cerebral hemispheres Above, the top of an intact brain. Below, the top of the brain has been cut away to show the fibers of the corpus callosum. The cerebral ventricles are fluid-filled spaces that are part of a system of ducts and channels through which cerebrospinal fluid circulates. See Chapter 6.

Drastic as it may sound, it has been found that cutting the corpus callosum of persons who have certain unusually severe kinds of epilepsy (a disease that is characterized by losses of consciousness or the sudden onset of involuntary movements) can help to relieve their symptoms. Somewhat surprisingly, after the corpus callosum has been cut, these people seem normal in their abilty to function in everyday life. But if they are tested in some situations that they do not encounter in everyday life, results suggest that not only their cerebral hemispheres but also some of their mental functions have been separated from one another by the surgeon's knife.

As we have seen, language is usually a function of the left cerebral hemisphere only. The localization of language function to the left side of the brain is apparent in people who have split brains, but unlike people who suffer from damage to Broca's or Wernicke's areas of the left hemisphere these people have no obvious impairment of the ability to use language, because all of the association areas are still intact.

If we blindfold a person who has a split brain and place a familiar object, such as a comb, in his *right* hand, nerve impulses that signal the presence of the object are sent to the primary sensory areas of the *left* hemisphere, as usual. This incoming sense data is processed by association areas surrounding the primary sensory cortex of the left side, and the person thereby comes to know exactly what the object in his right hand is. If we ask him to name the object, he does so without hesitation. However, if we had originally placed the comb in the person's *left* hand, so that sensory impulses signaling its presence arrived at the primary sensory cortex of the *right* hemisphere, we would find him unable to name the object, however hard he may try. But this is not to say that he does not know what the object is. In fact, he can show us that he knows that the object is a comb by demonstrating how to use it if we ask him to. He can also remember what the object is, and retrieve it from a tray of dissimilar objects by means of touch. But he cannot verbally communicate that the object in his left hand is a comb unless we remove the blindfold and allow him to see the object. Then, he immediately says "comb," and may even appear embarrassed for not being able to say so sooner.

What can we conclude from such observations? Because the corpus callosum has been cut, there is apparently no way for the right hemisphere (which knows what the object is) to transfer this knowledge to the left hemisphere, where the ability to associate mental concepts with words is found. So the person may know what the object in his left hand is, but be unable to say so, because sensory information from the left hand is received by the right hemisphere, which has been disconnected from the left one. The person can immediately name the object in his left hand if he is allowed to *see* it, because sensory information from the eyes (unlike that from the hands) is received almost equally by *both* hemispheres and does not depend on the corpus callosum for the transfer of information from one side to the other. (Sense data from the right hand arrive directly at the left hemisphere, so the person has no difficulty naming objects placed in his right hand even if he cannot see them.) The reason why people who have split brains usually have no difficulty in everyday life is that vision, not touch, is normally our most important means of recognizing objects. It is only when we devise special tests like those described above that we can demonstrate that cutting the corpus callosum has an effect on the ability to name objects recognized by the right side of the brain.

Unlike adults, children (especially under six years old) who have suffered damage to the speech areas of the left hemisphere can often regain the capacity for language. It has been suggested that children can do so because they can train the right hemisphere to take control over language, a function in which it would not normally participate. Apparently because most adults have used the left hemisphere for language functions since they first learned to speak it is much harder for them to teach the right hemisphere how to communicate by language, and most adults remain aphasic after damage to the left side of the brain. But the fact that children can relocate language function to the right hemisphere following injury suggests that some kind of training plays a part in determining what areas of the brain come to serve at least some of the higher thought processes so important in human behavior. We are not sure exactly how general this property of relocating intellectual functions to undamaged cortical areas following injury is, but it is probably easier for younger brains to do so than it is for more experienced ones.

Many other tests for comparing the kinds of mental activities engaged in by the two cerebral hemispheres have been devised and carried out on people who have split brains. The results suggest that in addition to language, many other mental activities, in which we normally engage and which we collectively call "consciousness," may also be localized primarily in one hemisphere or the other. Thus, the left hemisphere (the side that speaks, reads, and writes) may be thought of as primarily involved in mental activites that are logical and well ordered in time and that can be expressed in words. The right hemisphere, on the other hand, is associated with the appreciation of objects in space (rather than time), with the recognition of faces, with the appreciation of nonsymbolic sounds like music, and perhaps with many other mental processes that are difficult, or even impossible, for any of us to express in words. Whether or not such mental processes can be consistently related to one cerebral hemisphere or the other must await further studies. But there is no doubt that at least some of our mental activities are more characteristic of one cerebral hemisphere than the other. And there is also no doubt that our understanding of our thought processes as they relate to the association areas of either the right or left hemispheres has barely begun.

Summary

Adaptive behavior is of basically two kinds, instinctive and learned. The evolution of large and complicated brains, especially their cerebral hemispheres, generally corresponds to the ability of animals to learn. Human beings have the most complicated brains of all, and people have correspondingly huge learning capacities.

Learned behavior is undoubtedly the human species' most important means of adapting to the environment. People must learn not only how to best adapt to the local environment, but also how to act toward one another according to culturally determined rules that are transmitted from generation to generation primarily by means of the learned symbol system of language.

Observation of the behavior of people who have damaged brains, as well as studies that employ electrodes for stimulation, has helped to localize certain functions to particular areas of the brain. Through such techniques we have come to know that emotions are associated with the limbic system, self-control and restraint are related to the frontal lobes, and language depends on the interaction of Broca's and Wernicke's areas, which are usually found only in the left cerebral hemisphere.

But no matter how curious we may be about how our minds operate and where within our brains particular mental capacities reside, the rules of culture make it clear that acting toward one another as human beings is more important to the survival of our species than understanding exactly how or where within our brains the capacity to do so resides. Our cultural rules strictly prohibit us from experimenting on one another's brains simply to find out how they work. And this is clearly adaptive behavior because it amounts to natural selection working to protect the human brain—that most complicated organ that insures our survival as a species by enabling us to behave toward one another in endlessly complicated ways.

Suggested Readings

1. "The Great Ravelled Knot," by George W. Gray. *Scientific American,* Oct. 1948, Offprint 13. Discusses how certain areas of the human brain are known to be concerned with specific functions.
2. "Pathways in the Brain," by Lennart Helimen. *Scientific American,* July 1971, Offprint 1227. Discusses how recently devised microscopic techniques have added to our knowledge of the brain's circuitry.

3. "The Evolution of Intelligence," by M. E. Bitterman. *Scientific American,* Jan. 1965, Offprint 490. Discusses how some of the different kinds of vertebrates vary in their learning abilities.

4. "Teaching Language to an Ape," by Ann James Premack and David Premack. *Scientific American,* Oct. 1972, Offprint 549. Discusses how a young chimpanzee was taught the meaning of over a hundred different "words."

5. "Brain Changes in Response to Experiences" by Mark R. Rosenweig, Edward L. Bennett, and Marian Cleeves Diamond. *Scientific American,* Feb. 1972, Offprint 541. The brains of rats kept in a "lively" environment show consistent differences compared to those of animals kept in a "dull" environment.

6. "The Control of Short-Term Memory," by Richard Atkinson and Richard M. Shiffrin. *Scientific American,* Aug. 1971, Offprint 538. Memory has two components: short term and long term. Information in short term memory is somehow transferred into the long-term form.

7. "Memory and Protein Synthesis," by Bernard W. Agranoff. *Scientific American,* June, 1967, Offprint 1077. On exposure to certain chemicals that interfere with protein synthesis, goldfish forget what they have been taught.

8. "The Functional Organization of the Brain," by A. R. Luria. *Scientific American,* March 1970, Offprint 526. How the brain's higher mental functions are difficult to localize to specific regions.

9. "The Asymmetry of the Human Brain," by Doreen Kimura. *Scientific American,* March 1973, Offprint 554. How the right and left human hemispheres are concerned with different mental processes.

10. "Language and the Brain," by Norman Geshwind. *Scientific American,* April 1972, Offprint 1246. A discussion of how different kinds of aphasia are related to damage of the cerebral hemispheres.

11. "On Telling Left from Right," by Michael C. Corballis and Ivan L. Beale. *Scientific American,* March 1971, Offprint 535. How the concept of "right" and "left" is related to the symmetry of the nervous system.

12. "The Split Brain in Man," by Michael S. Gazzaniga. *Scientific American,* Aug. 1967. Offprint 508. Discusses some of the consequences of cutting the human corpus callosum.

"The Dream," by Henri Rousseau, 1910. (Oil on canvas, 6′8½″ × 9′9½″. Collection, The Museum of Modern Art, New York. Gift of Nelson A. Rockefeller.)

CHAPTER 4

Human Ecology

The surface of the earth is virtually covered by living things. In the oceans tiny free-floating plants and animals are abundant in surface waters, while below them marine organisms in seemingly endless variety swim through the waters or supply their needs on the sea floor, sometimes at depths that can be measured in miles. On land, except for parts of the hottest, driest deserts and the coldest polar regions, there is hardly a spot that is not inhabited or at least visited by some kind of living thing. In the air, birds and insects make their way, usually over the land, while microscopic spores of bacteria and fungi are wafted everywhere by the movement of global air currents. All of these creatures and all of these places—that is, all living things on earth and all their environments—together make up the *biosphere.*

The biosphere began about three and a half billion years ago when the first living things evolved from nonliving chemicals dissolved in the ancient oceans (see Chapter 1). At that early stage in evolution, the interaction between organisms and environment was relatively simple—the organisms merely utilized certain preformed molecules surrounding them in the oceanic soup. As evolution proceeded, organisms became increasingly diverse and evolved more sophisticated ways of nourishing themselves. As they did so, they regularly invaded previously unoccupied environments, and the living and nonliving components of the biosphere gradually became interwoven in increasingly complicated ways.

The study of the interaction between living things and the environment is known as *ecology*. Ecology is not concerned with individual organisms, but rather with populations of organisms and with how these populations interact with one another and with the nonliving environment. As used by ecologists, the word *environment* is easy to define: anything that is not part of a particular organism is part of its environment.

As you know from reading earlier chapters, populations are groups of organisms that belong to the same species. In fact, biological species are defined as populations of organisms of the same kind that are somehow reproductively isolated from other populations (see Chapter 1). Ecologists call populations of different species living together in the same general area *communities*. This emphasizes that no matter where it is found the survival of a given species is usually highly dependent on the presence of other species. Communities of different species, together with the nonliving environment in which the communities exist, are called *ecosystems*, to emphasize that organisms do not exist in isolation from the nonliving environment any more than they exist in isolation from one another. Finally, all the ecosystems on earth together make up the biosphere, which for all its complexity is, as we shall see, a self-sufficient system, except that it requires an outside source of energy.

Although all species affect and are in turn affected by the living and nonliving environment, some have more pronounced effects than others. In particular, in the past 8,000 years (especially in the last few hundred) the human species has had an enormous and unprecedented effect on the biosphere. Dams made by people interrupt the flow of the earth's mightiest rivers, many of the planet's forests and grasslands have yielded to cultivation or to urban sprawl, and the human species now has the capacity to put an end to the entire biosphere merely by pushing a few strategic buttons. Yet the environmental effects of the human species are not brought about equally by all human populations. Those people who still live much as our ancestors did when the species first evolved about 40,000 years ago do not affect the environment very differently than other large, nonhuman animal species. What makes the human species ecologically unique is that by means of technology some human populations are able to manipulate the living and nonliving environment in ways that are unknown among other living things.

In this chapter we will discuss what it is about people that makes our species ecologically unique, and we will see how this uniqueness has far-reaching effects on the finely geared workings of the biosphere. We begin our discussion of human ecology by considering some important characteristics of ecosystems.

How Ecosystems Are Organized

The biosphere is a mosaic of individual ecosystems, some of whose boundaries meld with one another so gradually that it is impossible to say where one ecosystem stops and a neighboring one begins. But this does not mean that ecosystems are arbitrary divisions of nature that exist only in the minds of ecologists. On the contrary, discrete assemblages of different species that inhabit a given area and interact with one another in order to survive and reproduce are familiar to all of us. For example, hardwood forests, grasslands, alpine meadows, and seemingly desolate expanses of tundra harbor similar collections of organisms that utilize environmental resources similarly whether they exist in North America, Europe, or Asia.

Although ecosystems may be as small as a pond or as large as an ocean, they all have certain features in common. First, all organisms and therefore all ecosystems ultimately depend on radiant energy from the sun.

Green plants, whether microscopic algae or gargantuan redwood trees, form the basis of all ecosystems because their cells contain a green pigment called *chlorophyll.* There are several kinds of chlorophylls, but all of them enable plants that possess them to carry out *photosynthesis,* an important chemical reaction about which we will have more to say in following pages. Simply put, plants that contain chlorophyll can transform the energy of sunlight into chemical bonds that can be used as sources of energy by other living things.

In a given ecosystem those organisms that can directly trap the energy of the sun and store it in chemical bonds are called *producers.* In turn, *consumers* utilize the energy stored in producers by eating them. Some consumers—cows and rabbits, for example—simply eat green plants that have directly stored radiant energy from the sun. Others eat animals that have eaten green plants, and thereby obtain second hand energy to survive and reproduce; indeed, many consumers are several times removed from the original solar energy source.

Most consumers are not very efficient at extracting energy from what they eat. In fact, in all ecosystems nearly 90 percent of the energy contained in the chemical bonds of producers is lost as energy is funneled through each successive consumer in what is generally called a *food chain,* or better, *food web* (Figure 4-1).

When consumers have eaten and extracted as much energy from the producers as they can, *decomposers* take over. Most decomposers are either bacteria or fungi, and these creatures reclaim what is left of the biologically usable energy after the consumers have extracted all the energy they can. They do this first by extracting energy from the waste products of living animals; then later, when producers or consumers die, decomposers transform their remains into materials that can be recycled through the ecosystem.

Only about a tenth of one percent of the total energy reaching the surface of the Earth as sunlight is made available to other living things by photosynthesizing green plants. Ultimately, this energy is reradiated into space as heat that is produced during chemical reactions, in which chemical bonds are made and broken. Thus, all ecosystems (and therefore the entire biosphere) are dependent on a continuing input of solar energy, and are constantly releasing stored chemical energy in the form of heat.

But except for their energy needs, ecosystems are self-contained and self-sufficient units in which essential materials are recycled again and again through the ecosystem. How this recycling works is perhaps best illustrated by the carbon cycle.

The Carbon Cycle

Carbon is unique among the elements in that it can form many chemical bonds both with itself and with other elements: carbon-containing compounds are present in all living things without exception. In general, carbon is cycled from the atmosphere, where it exists in the form of carbon dioxide gas (CO_2), into the

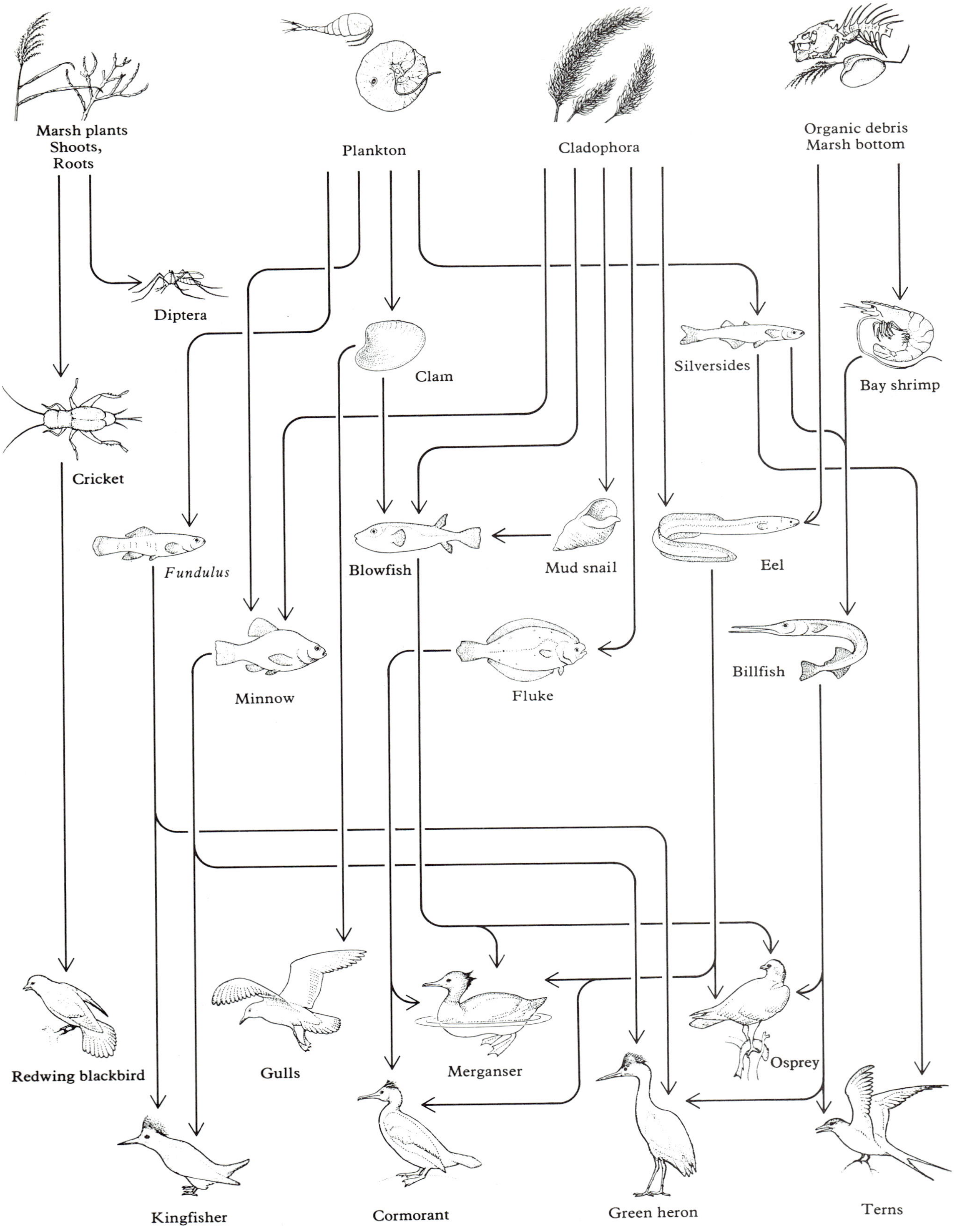

Marsh plants
Shoots,
Roots
Plankton
Cladophora
Organic debris
Marsh bottom
Diptera
Clam
Silversides
Bay shrimp
Cricket
Fundulus
Blowfish
Mud snail
Eel
Minnow
Fluke
Billfish
Redwing blackbird
Gulls
Merganser
Osprey
Kingfisher
Cormorant
Green heron
Terns

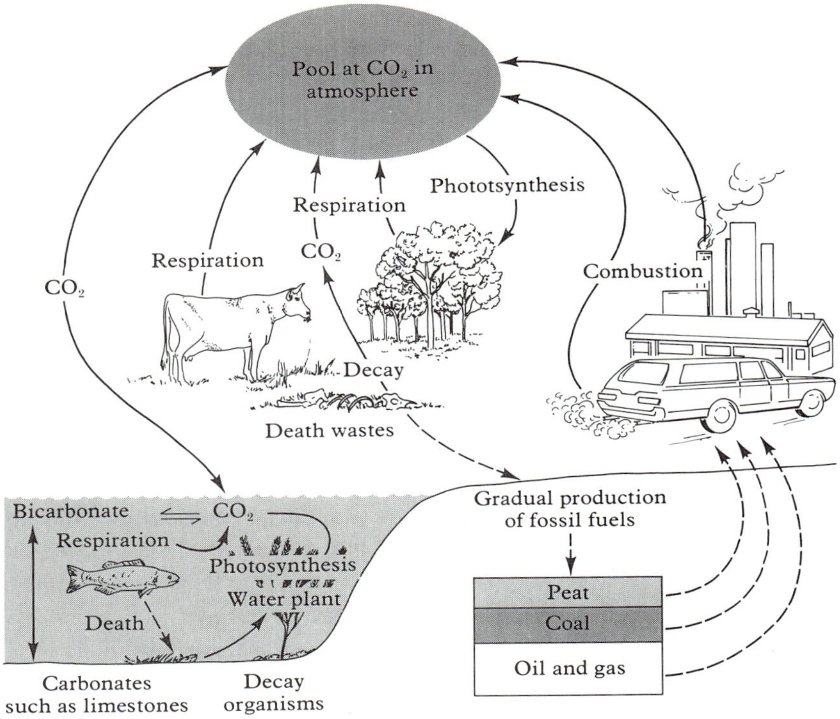

4–2
The carbon cycle. Solid arrows represent the flow of carbon dioxide and interrupted ones depict the subcycles discussed in the text. (From Paul R. Ehrlich and Anne H. Ehrlich, Population, Resources, Environment, *2d ed. Copyright © 1972, W. H. Freeman and Company.)*

chemical components of living tissues and then back to the atmosphere as CO_2 again (Figure 4-2). The chemical basis of this continuous recycling of carbon is found primarily in two chemical reactions—photosynthesis and respiration—that are almost each other's opposite.

Photosynthesis can be summed up by the following word equation:

$$\underset{\text{(sunlight)}}{\text{Energy}} + \text{carbon dioxide} + \text{water} \xrightarrow{\text{chlorophyll}} \text{oxygen} + \text{carbohydrate}$$

The carbon dioxide comes from the atmosphere, and the water comes either from the soil in which land plants grow or directly from the surrounding medium of plants that live in the oceans or in fresh water. The reaction takes place only in the presence of chlorophyll, as we have indicated by placing this ingredient above the arrow in the equation. Carbon-containing carbohydrates that are produced in the reaction store converted solar energy in the form of chemical bonds and serve as the immediate source of energy at the base of every food web. The oxygen gas that is produced by photosynthesis is released into the atmosphere and it is an essential ingredient in the process of respiration.

4–1
A food web is a complex network through which energy passes from producers to consumers. The web pictured on the opposite page shows some of the interactions between the plants and animals in a Long Island estuary and along the nearby shore. (From "Toxic Substances and Ecological Cycles," by George M. Woodwell. Copyright © 1967 by Scientific American, Inc. All rights reserved.)

The word equation for respiration is the opposite of that for photosynthesis, except that chlorophyll and sunlight do not play a part:

Oxygen + carbohydrate ⟶ carbon dioxide + water + energy

Both plants and animals engage in respiration, and the process depends on the presence of oxygen from the atmosphere. Carbohydrates are used up when they combine with oxygen, and the end products of the reaction are water and carbon dioxide gas, which is returned to the atmosphere. As you will learn in Chapter 8, respiration is a controlled "burning" of carbohydrates by means of which organisms extract usable forms of energy from chemical bonds.

The overall result of these two complementary chemical reactions is that carbon is cycled from the atmosphere into the chemical components of living things and then back to the atmosphere again. Actually, the carbon cycle is more complicated than that, because some of the carbon is removed from the main cycle by a much slower cycle-within-a-cycle in which huge amounts of carbon are incorporated either into sedimentary rocks, particularly limestone, or into fossil fuels. The carbon in limestone, as well as that in coal, oil, and natural gas, is derived from the dead bodies of organisms that for various reasons failed to decompose quickly after they died. Carbon in these forms can be removed from the main cycle for very long periods of time, but some carbon eventually returns to the main cycle as CO_2 in two ways. First, the weathering of sedimentary rocks, mostly by means of flowing water, results in the release of some CO_2. Second, CO_2 is returned to the atmosphere when people burn fossil fuels to obtain energy.

Carbon is not the only material recycled through ecosystems. There are cycles for water, nitrogen, phosphorus, sulfur, and many other materials. In all ecosystems, substances necessary for life are taken from the air, soil, or oceans and used by green plants to manufacture carbon-containing compounds that in turn provide food for animals that eat plants or that eat other animals that have eaten plants. Then when organisms die, essential substances are returned to the soil or oceans where they are eventually reused by plants again. The key to survival in all ecosystems is in this recycling of substances necessary for life.

The complexity of ecosystems depends not only on how materials are cycled through them but also on how many species make up the community that interacts with the nonliving environment. In describing how populations of different species interact with one another and with the nonliving environment, it is helpful to use the concept of niche. When we begin to consider the functional role and relative importance of organisms in different ecosystems, it is here that we begin to see what is ecologically so special, and in some ways so ominous, about the human species.

The Niche Concept

Within a given community, each species has evolved a characteristic lifestyle. In other words, in a given environment and under given conditions, different species interact with the environment in different ways. The way in which a species makes its living in nature is called the species' *niche.* More generally, a species' niche is considered to include not only everything the population does to satisfy its needs, but also the physical circumstances within which members of the population can survive and reproduce.

Although each species generally has a unique niche, some niches overlap, and some niches are clearly more variable and complicated than others. For example, the survival of the yucca moth of the American Southwest depends on the presence of yucca plants (Spanish bayonet), in the base of whose flowers the female moth deposits her eggs. The plant in turn cannot survive unless the moth carries yucca pollen from plant to plant. Because yucca plants and yucca moths are so highly specialized that they cannot get along without one another, both are said to have narrow niches (the yucca plant is shown in Figure 4-3).

In contrast, a mountain lion living in the same area (though probably at a higher altitude) can eat any animal it can successfully hunt. Because its means of livelihood is much more broadly based than the Yucca's, the mountain lion is said to have a broad niche.

The broadest niches of all are those of animals called *generalists.* Generalists can eat either plants or animals, and they can usually tolerate rather wide fluctuations in environmental conditions. (Some generalists have escaped extinction for hundreds of millions of years and are truly living fossils, as discussed in Chapter 1.) And among generalists, the broadest niche of all is that of the human species, especially those modern human populations that have evolved space-age technology.

The Human Niche

What exactly is the niche of the human species? There are so many people, and they engage in so many kinds of interactions with one another, with other living things, and with the nonliving environment, that the human niche is all but impossible to specify. Nonetheless, the human niche has at least two distinctive characteristics.

First of all, the human niche covers the largest area, at least on land. No other species is so widely distributed as the human species, with the possible exception of some species that directly benefit from human activities and that have therefore followed people as they migrated over the continents. (This includes houseflies, body lice, and mice, at least.) Second, the human niche is unique in that people, as they go about their day-to-day affairs, are capable of more widespread modification of the living and nonliving environment than any other species. In other words, people are the only animals that dominate many different ecosystems at the same time.

Yet not all human populations do this equally. Some populations of the human species still make their livings much as people did when our species first evolved 40,000 years ago; these people presumably affect the environment in much the same way now as they did then. By studying the ecology of these modern hunting and gathering peoples, we can get an idea of what the human niche was like when our species first evolved.

The Ecology of Hunter-Gatherers

The ecology of modern peoples who live by means of hunting and gathering has been studied in various locations, including Africa, South America, and Australia. In general, hunting and gathering is not an easy way of life, and most

4–3
The beautiful yucca plant and the yucca moth cannot exist without one another. (Photo copyright © 1972 Fritz Henle, courtesy of Photo Researchers, Inc.)

4–4
Groups of hunter-gatherers like the Fuegians pictured here, who live along the Strait of Magellan, generally do not alter the environment more than other large animals in the same ecosystem. (Courtesy of The American Museum of Natural History.)

modern hunter-gatherers devote a large part of their time to extracting food from the ecosystems in which they live. Modern hunter-gatherers live in groups of only 50 people or so, and in general these groups do not alter their environment much more than groups of other large, nonhuman animals living in the same ecosystem (Figure 4-4).

Although modern hunter-gatherers do not affect the environment very much, there is evidence that hunter-gatherers may have had more widespread effects in the past. This is especially true of some hunter-gatherer groups that lived during the last glacial period of the Ice Age (Pleistocene Epoch). Indeed, between 15,000 and 12,000 B.P., practically all species of large, mostly herbivorous, game animals in North America suddenly became extinct, perhaps because they were simply killed off by groups of hunter-gatherers moving southward (Figure 4-5). It has been suggested that the rather abrupt extinction of about 40 percent of large African mammals that occurred about 50,000 B.P. may have had similar causes, because people probably developed the hunting technique of using fire to drive animals into swamps (where they were relatively easy prey) about that long ago. In the end, we do not know whether many of the Pleistocene game animals of Europe and North America became extinct because

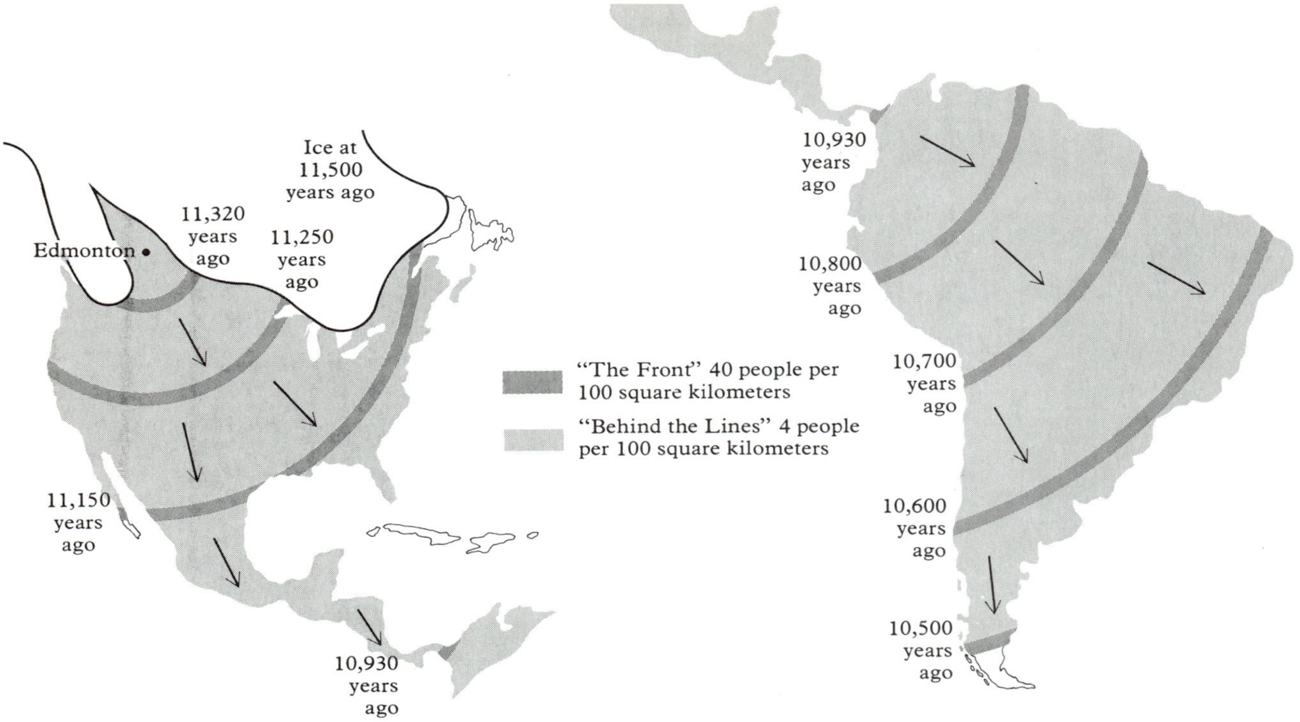

4–5
People were probably directly responsible for the extinction of several kinds of large mammals that occurred toward the close of the last glacial period. As animals became scarce nearby, the hunters followed their prey farther south. (From "The Discovery of America," by Paul S. Martin. Science, 179, *969–974, 9 March 1973. Copyright © 1973 by the American Association for the Advancement of Science.)*

they were killed off by people, or because of natural causes. (We will discuss the role of modern people in causing extinction in following pages.)

But whether or not human beings were responsible for the Pleistocene extinctions of large mammals, they, with their huge brains, dexterous hands, wills to survive, and unmistakably human inventiveness, have in the past 12,000 years come to dominate not only most ecosystems but the entire biosphere. Perhaps the most important human "invention" that made all of this possible occurred over 10,000 years ago. What happened was that some groups of hunter-gatherers settled down and started tending herds and raising crops.

The Invention of Food Raising

Since about 12,000 B.P., human beings not only have harvested grains, but also have fed, sheltered, and cared for certain animals that they later killed and ate. Although the domestication of some animals may have occurred independently from that of plants, there is evidence that these two ways of raising food often were employed together, at least in archaeological sites in Southeast Asia and in the Middle East, the areas where the transition from hunting and gathering to a more settled life of food raising probably first took place. The Natufian people, who lived in Jordan and Israel about 10,000 years ago, had almost invented agriculture in that they did not directly plant crops, but nonetheless left behind them flint sickles that show obvious signs of having been used, perhaps for cutting kinds of wheat that grew wild in the area at that time. This idea is supported by the fact that the Natufians made mortars and pestles in which they probably ground grain.

By 8,000 B.P. farming settlements that relied on crops of planted wheat, barley, and peas, as well as on meat from domesticated sheep and goats, had been formed at various locations in Asia and the Middle East. From there, the farming way of life spread to China, Europe, and Africa. Then, from about 6,000 B.P. on, there is evidence of the practice of agriculture in the New World (perhaps first in Mexico), where it was probably "invented" independently.

In ecological terms, the invention of food raising, particularly the planting and harvesting of crops, is perhaps the human species' greatest accomplishment. There are several reasons for this. First, food raising ushered in an unprecedented, ongoing, and almost overwhelming increase in the size of the human population. Second, food raising allowed people to live together in much larger groups, and from some of these human aggregations emerged cities and eventually the civilizations of ancient history and the industrialized societies of the present day. And third, the practice of food raising, especially plant agriculture, greatly altered the nature of the human food web because food raising greatly simplifies ecosystems.

How Food Raising Simplifies Ecosystems

The first practice of agriculture as we know it occurred about 9,000 B.P. and was perhaps related to the drying up of grasslands as the climate of Western Asia became warmer with the retreat of the ice sheets. Wheat was the first crop to be raised, then barley, and later, peas, oats, rye, and other grains. In order to domesticate plants, people had to invent not only techniques such as planting and cultivation but also the use of fertilizers and, in arid regions, of irrigation. Such activities are likely to simplify relationships within a given ecosystem. Why is this so?

First of all, in clearing land and planting crops, people replace the natural mixture of species of wild plants in a given ecosystem with a few, or even a single, plant species that may be cultivated over an extensive area. Thus the introduction of cultivation directly reduces the diversity of most ecosystems. Diversity is further reduced when people try to minimize or eliminate the effects of competition by other members of the ecosystem (insects, crows, and gophers, for example) that are just as interested in eating crops as the people who planted them. And finally, through irrigation and flood control people sometimes change or stabilize the physical aspects of an ecosystem so that conditions are optimal for a crop plant but not for the members of the community that lived in the area before cultivation was introduced. In sum, when human beings introduce food raising into an ecosystem, the natural variability within that ecosystem decreases.

Diversity and Stability in Ecosystems

The problem with any simple ecosystem is that it is unstable and very susceptible to catastrophe. Field studies have repeatedly shown that the stablest naturally occurring ecosystems are those that have the most stable climates and are most complex. And generally, the most complex ecosystems contain the

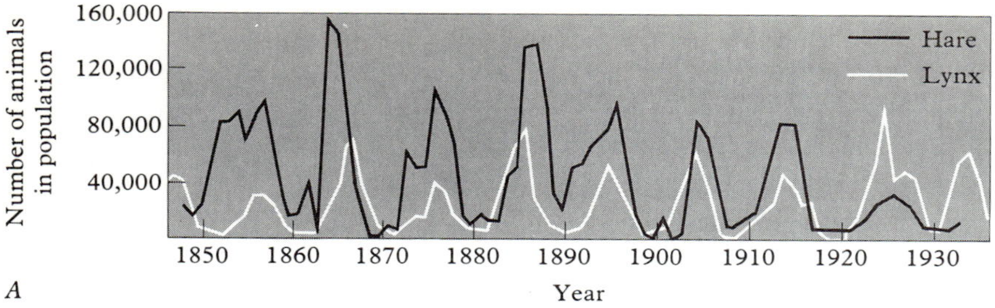

A

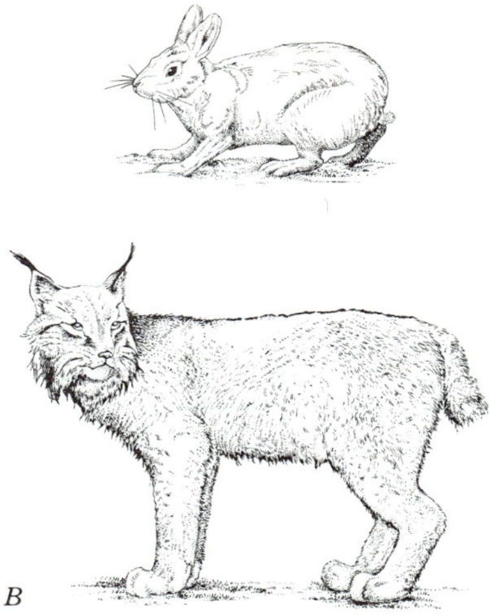

B

4–6
The cyclic variations in the size of the populations of snowshoe hare and lynx depend on each other, as shown by the records of the Hudson Bay Company. In simple ecosystems the number of animals in a population often fluctuates widely.

most species. Field studies have also shown that complex ecosystems, such as that of a coral reef, do not vary much in the number of members of a given species that are present from year to year. Apparently, in complicated ecosystems the complexity of the interactions between the many species in some way dampens fluctuations in population size for all the species of the community. In contrast, ecosystems that are less complex are characterized by rather wide fluctuations in population size. For example, some populations of the arctic tundra vary enormously from year to year, as shown in Figure 4-6.

How all of this relates to people is perhaps best illustrated by what happened in Ireland in 1845, the worst year of the Irish potato blight. Following its introduction a few years earlier, the potato found itself so well suited to the conditions of the Irish countryside that by about 1840 it had been very extensively planted, and as a result, the well-nourished Irish population had grown to more than 8 million and had also grown to depend upon the potato for nutrition. But like all simple ecosystems, this one, which directly converted solar energy into potato starch that was then consumed by people, was especially susceptible to catastrophe. Catastrophe struck in the 1840s when the potato crop failed because it was attacked by a fungus or "blight" (Figure 4-7). The result was that somewhat more than a million people died of starvation and at least a million more emigrated.

Although people have become much more efficient at growing potatoes, at controlling competition from crop-eating organisms, and at modifying the physical environment, in many ways the ecological position of the human species has not changed much since the time of the Irish potato blight. The human species is still almost totally dependent on a relatively limited number of species of domesticated plants and animals for which people regularly sacrifice rich and diversified ecosystems in order to produce monotonous, but nourishing, fields of grains and grasses. The susceptibility of such human-generated ecosystems to catastrophe is still great. Insect pests, viruses, blights, and other organisms still exact weighty tolls, and humans suffer and starve when crops fail because of drought or other vagaries of the weather. Yet in spite of their vulnerability, the simplified food relationships in ecosystems dominated by people who raise crops have been enormously successful. After all, nearly *4 billion* human beings now live on our planet, and in large part, this is because agriculture can feed so many human mouths.

The Human Population: Its Size and Growth

If they are given the opportunity, all living things are capable of increasing their numbers prodigiously. We have all heard of the proverbial pair of houseflies that could, within a few months, cover the earth with their offspring, provided that all of the offspring of every generation survived to reproduce. Charles Darwin once calculated that a single pair of elephants (which are the slowest breeders of all animals) could leave 19 million living descendents over the course of 750 years. In fact, our planet is neither knee-deep in houseflies nor overrun by elephants because in all naturally occurring ecosystems a series of checks and balances effectively limits the size of any population. In a given community these checks and balances include the effects of predators, parasites, and the agents of infectious diseases, as well as the generally limiting effects of food supply.

The human population has probably never been heavily preyed upon by other large animals, but parasites and infectious diseases have had and still have some regulatory effects on population size. (We will discuss the effects of medicine, parasites, and infectious diseases on the human population later.) However, when people invented food raising they removed from the human population the most general constraint to population growth of all—the lack of an adequate food supply. And the results have been just as you might expect. Our planet may not be overrun by houseflies or elephants, but it is rapidly being overrun by human beings as the size of the human population continues to increase explosively.

4–7
Top, the cultivated potato plant, and bottom, the fungus that causes potato blight. The fungus sends out threadlike extensions that permeate the leaf, as shown in this magnified cross section. (From "The Late Blight of Potatoes," by John S. Niederhauser and William C. Cobb. Copyright © 1959 by Scientific American, Inc. All rights reserved.)

Initially, the practice of food raising did not have the profound effects on population growth that it does today because agriculture was much less efficient in the past than it is now. Before agriculture was invented, it probably took about 30 square kilometers of land (about 12 square miles) to support a single person; this is true for modern hunting-gathering groups. Since then, there has been a general tendency for the amount of land required to support one person to decrease in area as agriculture has become increasingly efficient. In other words, the *productivity* of the land has been greatly increased by the development of more refined agricultural techniques.

It has been estimated that when food raising was first invented 10,000 years ago, there were about 5 million people on earth. By 2000 B.P. the number had grown to about 250 million, which means that the size of the human population was doubling about every 1,600 years, largely because of food raising. But since then, the number of years required for doubling has decreased dramatically, to the point where the human population is now doubling about every 35 years! Figure 4-8 shows the growth rate of the human population in the past million years, and Table 4-1 emphasizes the population's ever-accelerating rate of growth by showing the number of people added to it in each decade of the present century. Given the present rate of increase, by the year 2000 there will be four living people for every person alive in 1900. What is so ominous about this accelerating rate of population growth is that modern agriculture has become so highly dependent on technology that it may not be able to support the burgeoning population much longer. The reason is simple—many of the resources on which technological agriculture depends are rapidly running out.

The Technological Basis of Modern Agriculture

The high levels of agricultural output that presently support the human population depend in large part on three products of modern technology: fertilizers, pesticides, and machines that run on fossil fuels.

Like all other living things, crop plants require certain nutrients in order to grow. The opulent productivity of modern agriculture depends on a constant supply of fertilizers to replace nutrients that are taken up by crops or otherwise lost from the soil by erosion. Perhaps the nutrients on which crops most depend for this are compounds of nitrogen and phosphorus. Ultimately, the nitrogen in fertilizers comes from the atmosphere, and because nitrogen accounts for about 80 percent of the air we breathe, it is present in ample supply. But not so phosphorus, which is produced primarily by mining phosphorus-rich rocks. People now add phosphorus to cultivated land at a rate of over 7 million metric tons per year, and what is not used by crops is soon diluted beyond recovery in soils or lost irretrievably as flowing water carries it to the oceans, where it is deposited in sediments. At present rates of population growth and phosphorus consumption, it has been estimated, unless people can develop effective and economical methods of recovering and recycling phosphorus, the world supply will run out in about a hundred years. (At that time there will be a whopping 20,000 million human mouths to feed, more than five times the present number.)

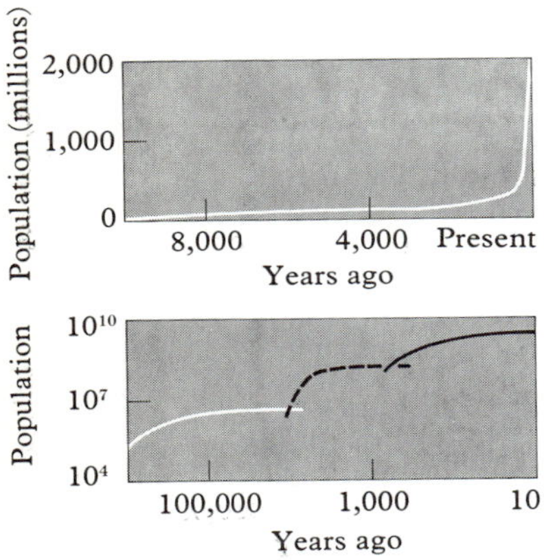

4–8
Top, the growth of the human population in the past 10,000 years (arithmetic scale). Bottom, an estimate of the growth of the human population in the past million years (logarithmic scale). The three surges in population size reflect the invention of toolmaking, the introduction of agriculture, and the industrial revolution. (From "The Human Population," by Edward S. Deevey, Jr., Copyright © 1960 by Scientific American, Inc. All rights reserved.)

TABLE 4-1
The number of people added to *the total world population in each decade of the twentieth century.*

DECADE	PEOPLE ADDED (MILLIONS)
1900–1909	120
1910–1919	132
1920–1929	208
1930–1939	225
1940–1949	222
1950–1959	488
1960–1969	604
1970–1979 (projected)	848

Industrialized agriculture is also highly dependent on manufactured pesticides that reduce or eliminate unwanted weeds and plant-eating insects. In natural ecosystems, especially those that are most diverse, if a particular weed or insect pest rapidly increases its numbers, then the system of checks and balances within the ecosystem operates to return things to normal. But as we discussed earlier, the introduction of agriculture greatly simplifies any ecosystem and virtually eliminates its natural system of checks and balances. So if a weed or insect pest begins to increase its numbers, it must be controlled by chemicals because natural controls were eliminated by people when they introduced agriculture to the area. (Some pests can be controlled if their natural enemies are introduced directly into the simplified agricultural ecosystem, and this method of control is more ecologically sound than the use of chemicals.) There is no shortage of pesticides today. But there is evidence that chemical control of crop pests is getting out of hand. To begin with, a rapidly increasing number of insects have already developed resistance to many pesticides, and will develop resistance to others. But more important, pesticides used to control weeds and insects in artificially simplified ecosystems have had disastrous effects on neighboring natural ecosystems and even on those far removed from the place where chemicals were used. For example, DDT has become widely distributed in many food webs, with disastrous effects in some, and this deadly chemical can even be recovered from Antarctica, where no crop plants are raised (Figure 4-9). (Later in this chapter, we will discuss the ecological effects of the use of insecticides as a means of controlling infectious diseases in the human population.)

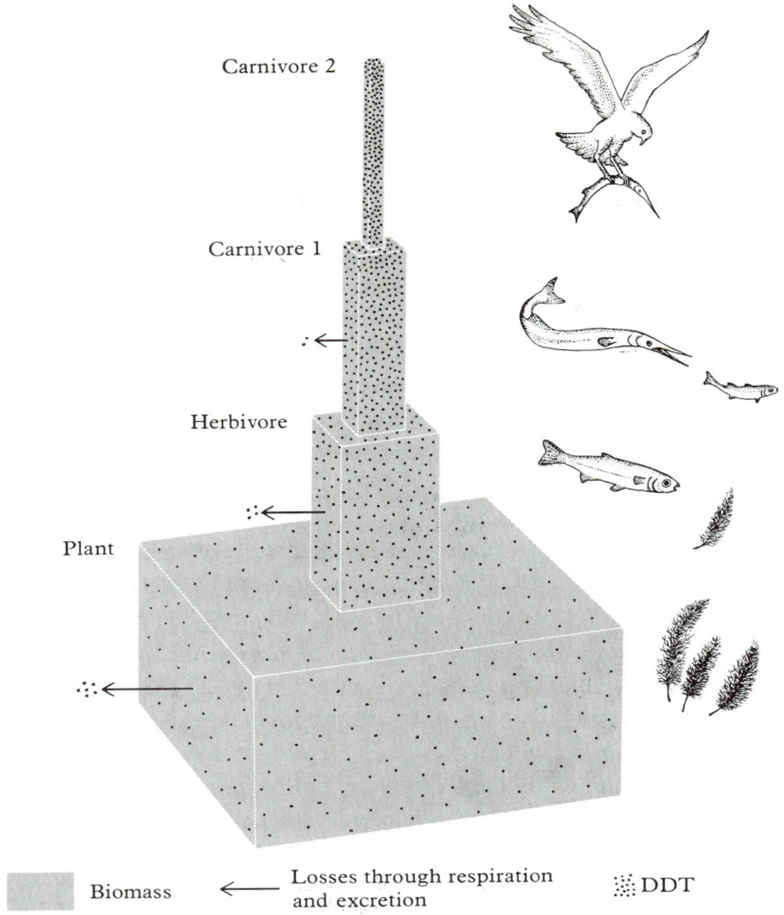

4–9
The concentration of DDT increases with each link in a food web. (From "Toxic Substances and Ecological Cycles," by George M. Woodwell. Copyright © 1967 by Scientific American, Inc. All rights reserved.)

In addition to fertilizers and pesticides, modern agriculture also depends on machines that operate on the energy contained in fossil fuels. Machines are used to plant crops, to deliver water, fertilizer, and pesticides to them, and finally to harvest, process, and distribute them. As we are all aware, the world supply of fossil fuels is rapidly running out, and ultimately some other energy source will have to be devised to turn the wheels of agriculture and thereby feed the human population.

Perhaps the best way to sum up the relationship between agriculture and technology is by estimating what will be required of technology when the human population simply doubles its present size, an event that will probably occur by the year 2000. Analysis of worldwide data shows that to produce twice as much food will require 6.5 times as much fertilizer, 6 times as much pesticide, and 2.8 times as much power as we now have. And as if that were not enough, the same figures show that while worldwide food production is now growing at a rate of 3 percent per year, food-related industries are growing at a rate of 6 percent per year. This means that about 50 percent of the growth of food-related industries is concerned not with producing food, but rather with devising technologial means of delivering produced food to human consumers. Thus, not only is energy-consuming industry growing at a faster rate than the human population can produce food, but in fact, technology is growing two to three times faster than the population itself!

Some Further Effects of Technology

The almost endless array of machines, appliances, buildings, chemicals, vehicles, and other devices that are so characteristic of modern, industrialized societies are ultimately the products of a revolution that began in England in the early eighteenth century, and then spread to other countries. This Industrial Revolution depended on advances in the practice of metallurgy, a skill first discovered about 7,000 years earlier (see Chapter 2). What happened in eighteenth century England was that people first learned how to manufacture iron by a process that used coal. This discovery led not only to an enormous expansion of the iron industry but also to the invention of the steam engine, the machine that for the first time provided people with a means of concentrating enormous amounts of usable energy. This series of events surely ranks with the domestication of fire, the discovery of tool making, and the invention of food raising as one of the human species' greatest accomplishments.

The rise of technology has been a mixed blessing. Technology has contributed to the human population explosion in that it provides us with fertilizers, pesticides, and machines. But technology can also be considered, not as a cause, but rather as a result of population growth, because the development of technology amounts to human beings meeting their needs by being inventive. Technology has undoubtedly raised the standard of living of much of the human population and it has allowed for a great increase in knowledge and in leisure time. But technology has had some undesirable effects too, not only on human beings, but indeed on the entire biosphere. Although technology provides us with goods and conveniences undreamed of by our ancestors, some of them are directly harmful to living things, and the manufacture of any product is always accompanied by the generation of waste materials that must be disposed of.

Of course, living things generate waste materials themselves, but they are usually no problem to dispose of because they can be recycled through the biosphere by the action of decomposers. But for huge aggregations of human beings living together in modern cities, there is a problem. Such concentrations of people generate so much excrement that it cannot be handled by local decomposers and is therefore flushed away into lakes, streams, rivers, and estuaries, often before decomposers have even had a chance to begin recycling it. This tremendous outflow of sewage from heavily populated areas has had a correspondingly enormous effect on aquatic ecosystems. After all, sewage contains fertilizer, and its introduction into a pond, stream, or lake can result in such overstimulation of plant growth that decomposers within the community cannot keep up with the amount of newly generated vegetation.

One should not confuse waste products from living things with waste products from industry. Decomposers generally have no effect on most industrial waste products, and therefore some of the waste products accumulate in the biosphere as *pollutants.* Some pollutants, such as heavy metals and the oxides of nitrogen and sulfur, are produced as by-products of manufacturing, heating, transportation, or the generation of power. Others, namely pesticides, are deliberately manufactured and spread over crops and forests. Many of these pollutants are extremely toxic, and, though they may be released into the environment in relatively small quantities, they can be concentrated in the tissues of living things as materials pass from producers to consumers in a given food web. The exact effects of some of these toxic substances on people and on

4–10
Some of the creatures that have become extinct because of human activities. Top left, South African ungulates. Top right, Steller's sea cow. Center left, passenger pigeon. Center right, great auk. Bottom left, Labrador duck. (Bird drawings from James C. Greenway, Jr., Extinct and Vanishing Birds of the World. *Copyright © 1967, Dover Publications, Inc.)*

other living things remain to be determined, especially when the substances are present in low concentrations over long periods of time. Effects of other pollutants are more obvious, and some of them are terrifying. For example, a killer smog produced as a side effect of British industry killed nearly 4,000 Londoners in 1952, and mercury-contaminated fish and shellfish are known to have caused people to suffer and die in Japan. It is essential, now that we have realized the dangers, that human beings make every possible effort to dispose of these noxious substances without allowing them to produce their deleterious effects on people or other members of the biosphere.

In addition to directly generating industrial pollutants and indirectly allowing for the efflux of huge amounts of sewage from cities, technology (and the population explosion) have affected the biosphere in still another way. As you have read in earlier chapters, the complete disappearance of a given species (extinction) has been the rule, not the exception, throughout the course of evolution. Yet there is no doubt that people have been directly responsible for the eradication of many species that probably would not otherwise have become extinct when they did. Gone forever are the Labrador duck, the great auk, and Steller's sea cow, to mention a few (Figure 4-10). Also, at least 200 species of mammals, including the majestic Indian tiger and the enormous blue whale (the largest creature that ever lived), are now "endangered species," mostly because of human activities. Some creatures have been hunted to near extinction, and the reason for the decline of others is that people have greatly altered the ecosystems to which the species are adapted. The irretrievable loss of these species is regrettable not only to those of us who delight in the wondrous variety of nature, but also to those who do not. In the past, people have greatly benefited from nature's vast array of species, and many of our most valuable drugs (especially antibiotics), and other products are still derived from undomesticated creatures. Because we really do not know what the future has in store for the human species, we owe it to future generations to preserve as many of these threatened creatures, and as many of the ecosystems in which they live, as we possible can.

In sum, technology has given us food, knowledge, leisure time, and myriad conveniences, but it has produced sewage, pollution, and unprecedented rates of extinction. Technology has also affected the practice of medicine, a human activity that, like technology itself, has been a double-edged sword for the human species.

How Medical Practice Influences Human Ecology

The practice of medicine is perhaps best defined as the thoughtful application of all those arts and sciences devoted to the maintenance of health and to the prevention and treatment of human diseases. As such it is a very ancient human endeavor. Nonetheless, only in the past few hundred years has medical practice had an appreciable effect on the human population. Before that time, physicians applied poultices, concocted herbal tonics, dressed wounds, splinted broken limbs, and otherwise tended to their patients' needs. But though they may have sometimes made their patients feel better, any cures rendered probably had no overall effect on the human population. Then, with the discovery of disease-causing microbes and the rise of technology, things began to change. And the outcome has been at once gratifying and disconcerting. Modern medical practice has provided us with ways of combatting infectious diseases, has increased the newborn's chance of survival, and has prolonged our lifetimes. But at the same time, medical practice has helped to produce human miseries, because it has contributed to overpopulation.

Medical science's most outstanding achievement in human ecology has been its victory over many infectious diseases. Maladies of this sort are produced by microscopic organisms, especially bacteria and viruses, that invade the body and reproduce there. These diseases are "infectious" in that the microbes that cause them can be passed from person to person and, if conditions are favorable, they can rapidly spread throughout the population and cause disease in epidemic proportions.

The conditions that favor the spread of infectious diseases vary depending on the causative organism, but the spread of almost all infectious diseases is favored by crowding, inadequate sanitation, and poor personal hygiene. All of these conditions were well met by city dwellers in ancient and medieval times, and still are met by city dwellers in many parts of the world. Accordingly, throughout human history infectious diseases have sometimes had drastic effects on the size of the human population.

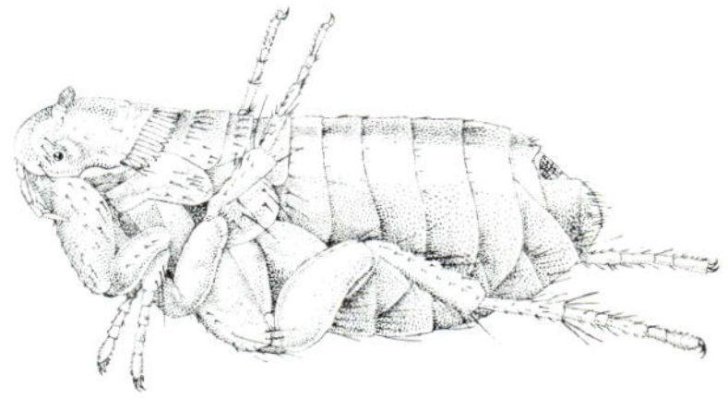

Jumping flea

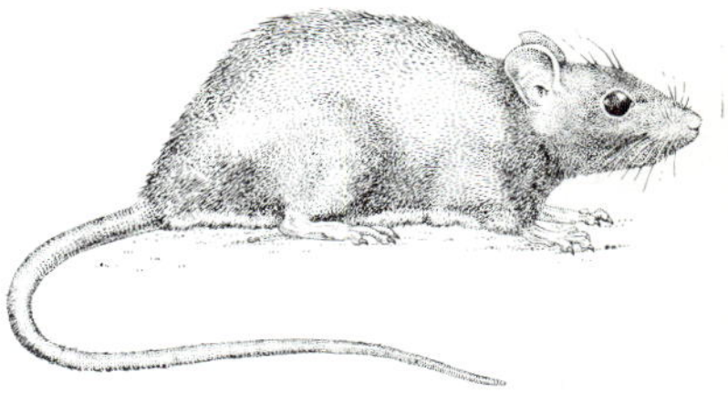

Black rat

4–11
The rat flea and the black rat. The plague bacterium is transmitted from infected rats to people by means of flea bites. (The rat flea from "Fleas," by Miriam Rothschild. Copyright © 1965 by Scientific American, Inc. All rights reserved. The black rat from "Rats," by S. A. Barnett. Copyright © 1966 by Scientific American, Inc. All rights reserved.)

Perhaps the most renowned example is that of the infamous black death that ravaged Europe in mid-1300s. This dreaded disease is passed to people by the bite of fleas that have fed on the blood of rodents infected with the plague bacterium, *Pasturella pestis* (Figure 4-11). In the black death (so called because of the dusky color of its dying victims), the rodent was one that has been a constant companion of human beings since the invention of food raising—the rat. The result of this unfortunate combination of infected rats, plentiful fleas, and crowded, unsanitary conditions was that about one out of every four people died of the plague.

Although minor outbreaks of plague still occur at various places around the world, it is not likely that our species will ever see another black death. In large part this is because technology has given us, not only pesticides that kill rats, but antibiotics that kill bacteria inside the human body without killing the infected person. But this is not to say that epidemics are unknown in modern times or that infectious diseases have been eliminated as causes of human misery. Following World War I an epidemic of influenza spread rapidly through all the cities of the world and killed 10 to 15 million people. Then in 1967 and 1968 another outbreak of influenza, the Asian flu, killed about 9 million people. (The main reason that influenza can still reach such epidemic proportions is that it is caused by a virus. Most antibiotics have no effect on viruses, and technology has yet to provide us with chemical means of stopping them from reproducing once they have gained access to our bodies.) Yet overall, technology has greatly changed the role of infectious diseases in the regulation of the size of the human population.

Nobody laments the demise of such dreaded diseases as plague and typhus fever, but the technological control of infectious diseases has also had an undesirable effect in that it has contributed to the explosive growth of the human population. Basically, any population will increase in size if during a given period of time more members are added to it than die. The control of infectious diseases has greatly increased the number of people added to the population, because it has markedly increased the chances that a newborn person will survive infancy and live to reproductive age, if not longer. The control of infectious diseases (along with other medical practices) has contributed to the population explosion in still another way by lengthening human lifetime and therefore producing a relative decrease in the number of people who die during a given interval (death rate) as compared to the number born (birth rate). Although the double-edged sword of technology has also provided us with effective ways of limiting population size by means of birth control, the worldwide statistics show that the effects of increased infant survival and lengthening lifetime have vastly overridden any effects that birth control has had, at least so far. (The physiological bases of present means of birth control are discussed in Chapter 14.)

The control of infectious diseases has not only profoundly affected the size of the human population, but it has provided us with excellent examples of how complicated human ecology really is. Some of the examples are worth considering further.

4–12
The life cycle of the parasitic worm that causes schistosomiasis.

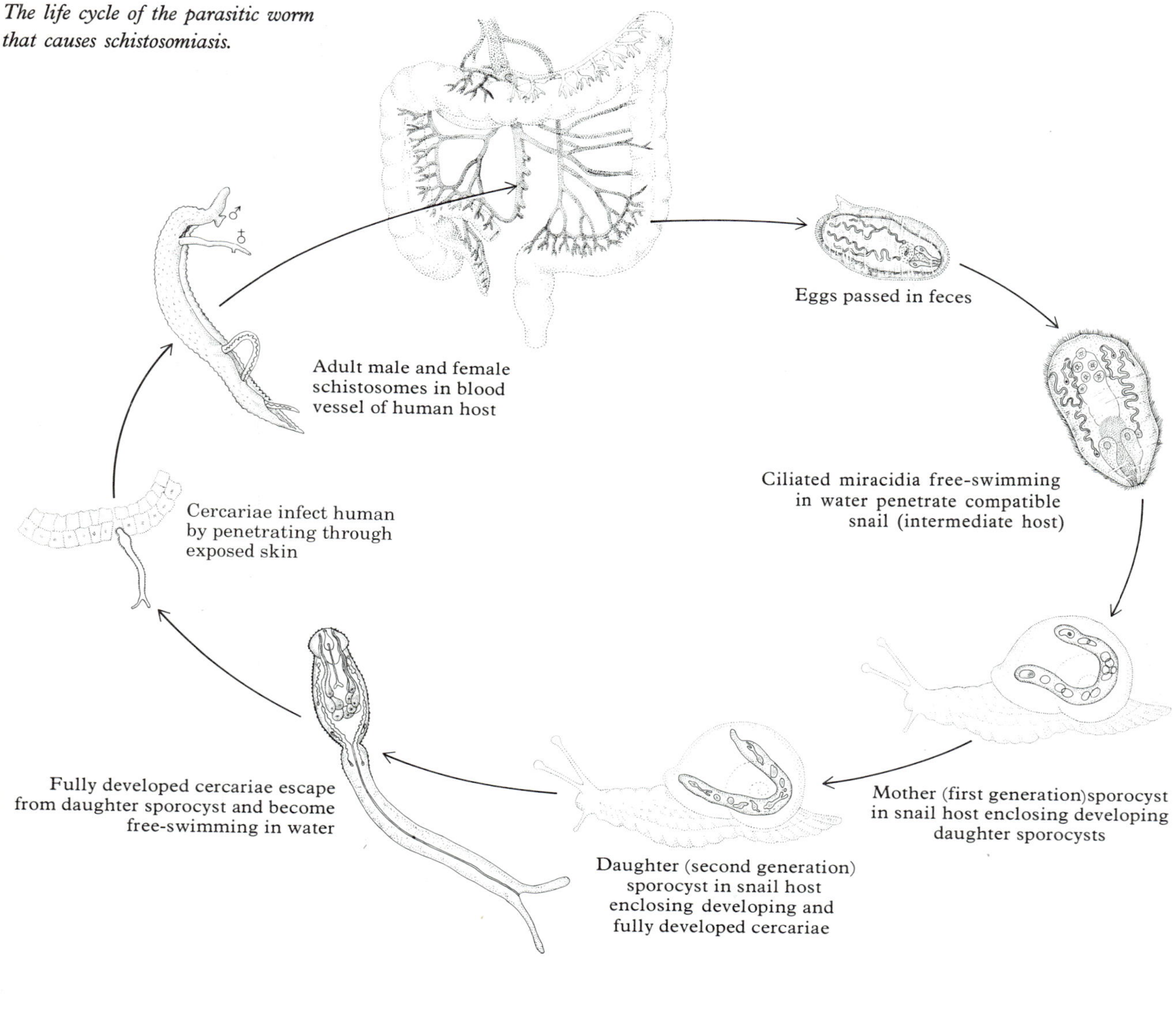

Schistosomiasis is at present the most prevalent of serious infectious diseases. In large part this is because people have altered the environment so that it favors the spread of this "snail-borne curse" in Africa and Asia. Today about 200 million people are afflicted by this infestation of parasitic worms that spend the early phases of their lives inside the bodies of certain species of freshwater snails (Figure 4-12). From there, during a free-swimming stage in their life cycle, the worms contact and penetrate the skin of people who by wading, defecating, or other means come into contact with fresh water. Eventually the worms become established and reproduce within intestinal blood vessels that carry digested nutrients from the intestine to the liver. Chronically, the results are loss of appetite, weight loss, liver dysfunction, and sometimes brain damage and death.

What has favored the spread of schistosomiasis, especially in Africa, is the contruction of mammoth hydroelectric power projects, particularly the Aswan High Dam. Upper Egypt, an area formerly cultivated once a year after annual flooding, is now cultivated year round because the Aswan Dam provides an enormous reservoir of fresh water for irrigation. But with the irrigating waters have come schistosome-laden snails that release their infectious products into irrigation streams and ditches from which the organisms eventually find their way to people and produce their serious pathological effects—not once a year, but continually. The result: schistosomiasis is the most widespread infectious disease our species now faces largely because of efforts to produce more crops by means of irrigation.

Another disease that has environmental repercussions is malaria. Malaria is caused by a microscopic creature that has a complicated life cycle (Figure 4-13). The disease is transmitted to people by the bite of the *Anopheles* mosquito, an almost unavoidable event wherever this six-legged menace is found. In the 12 years from 1959 to 1970 more than one billion people have been freed from the risk of contracting malaria through the use of insecticides (mostly DDT) that can effectively eliminate the mosquitoes from areas to which they once were endemic. The reduction in the incidence of malaria around the world is surely one of technology's crowning victories over infectious diseases.

But some unexpected things happened when the World Health Organization (WHO) sprayed the huts of villagers in Malaysia in an effort to control *Anopheles* mosquitoes in the area and thereby reduce the incidence of malaria, which in some villages was as high as 90 percent. Shortly after their huts were sprayed, most of the villagers found that the roofs of their homes were rotting and caving in. This proved to be a result of a population explosion of moth larvae (immature forms) that normally live in the thatched roofs in only small numbers. The DDT that killed the mosquitoes also killed off the moth's natural enemy, a wasp, and the resulting population explosion of moth larvae consumed so much thatch that the roofs collapsed. Also, although cockroaches that normally inhabited the huts were not killed outright, they did absorb DDT, and when they were eaten by the small house lizards (geckos) that normally prey on them, the DDT accumulated in the tissues of the geckos. When house cats then ate the geckos that had eaten the cockroaches, most of them died because in passing along the food chain from roaches to geckos to cats DDT had been

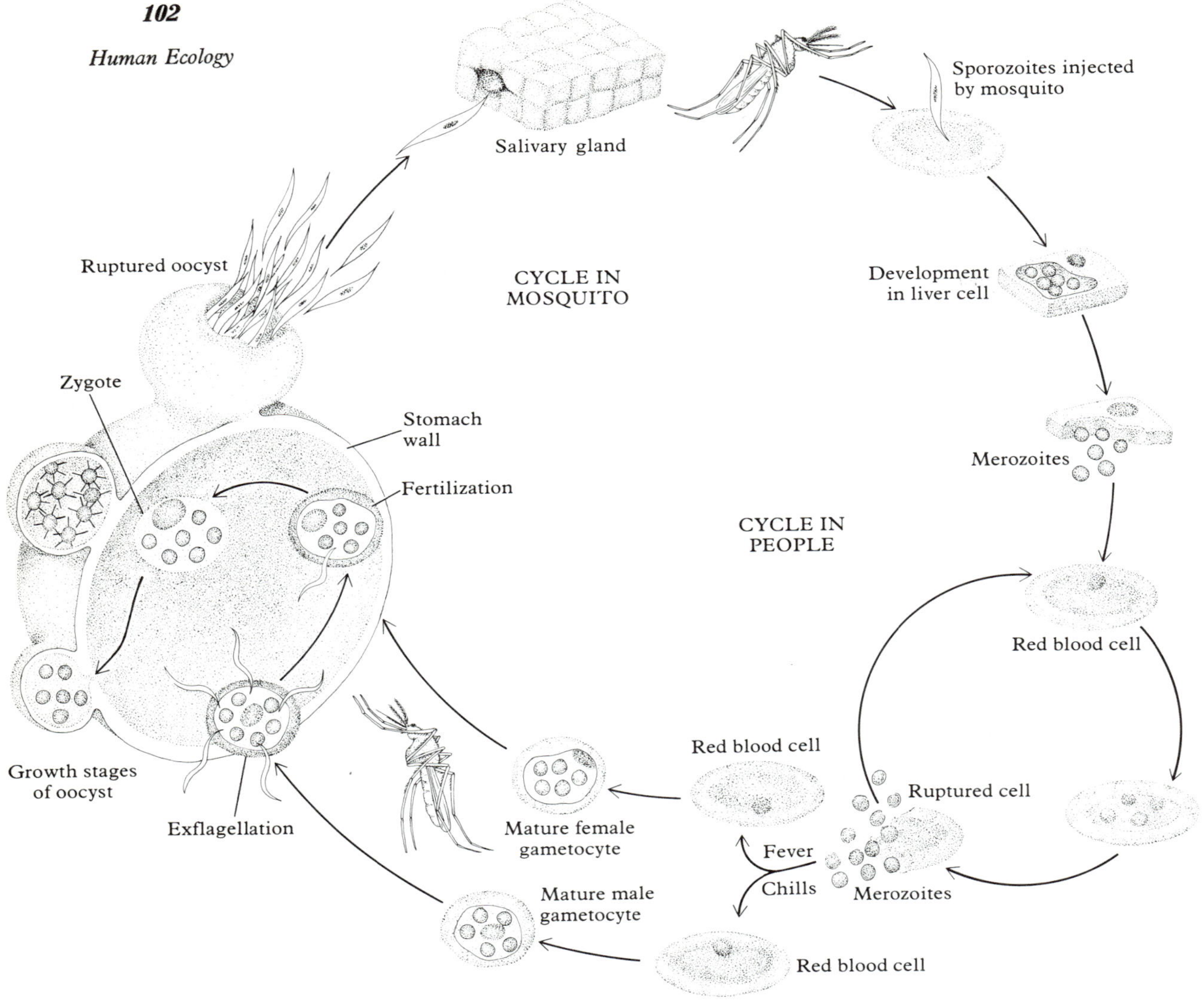

4–13
The life cycle of the microscopic creature that causes malaria.

concentrated at each step until it reached a lethal concentration in the cat's tissues. When all the cats died, the ever-present rats underwent a population explosion, and with their expanded numbers came the threat of typhus and plague. Accordingly, the WHO instituted a remarkable venture known as "Operation Cat Drop," in which cats were parachuted into villages to kill the rats.

In Venezuela, the spraying of huts to eliminate mosquitoes also killed off spiders, centipedes, and other small predators that normally feed on bedbugs. As a result, the population of bedbugs exploded, and their bites so irritated the local human inhabitants that spray teams were persuaded by means of sticks and rocks to leave the area.

Worse yet, in India, mosquito spray teams effectively eliminated malaria in thousands of villages, but produced the most disconcerting effect of all. In two years' time the survival rate of children under three years of age doubled, but many of them starved to death.

Clearly, human ecology is a very complicated subject whose study reveals that the members of the biosphere are linked together in ways that are sometimes not apparent at first glance.

People Are Ecosystems

In the end, the study of human ecology leaves no doubt that people are the most influential organisms in the biosphere, in that they can raise crops, control infectious diseases, and alter any ecosystem to meet their needs. Yet in a very real sense, people are themselves ecosystems because the existence of all human beings depends on the presence of bacteria that live on or in our bodies. Madison Avenue notwithstanding, human well-being is highly dependent on contact with environmental microbes. Within our guts bacteria recycle metabolic products from the liver and thereby enable us to digest fats. Gut bacteria probably also supply us with at least some of our daily requirements of vitamins, and skin microbes prevent the colonization of our body surfaces by potentially pathogenic organisms. From the outmost reaches of the biosphere to the very center of our guts, we people are not alone on Earth, nor are our lives isolated from those of other living things.

Summary

All living things and all of the environments in which they live compose the biosphere. Populations of different species living together in the same area are called communities, and ecosystems consist of both communities and the local nonliving environment. Ecosystems are highly organized units in which complicated interactions take place among producers, consumers, and decomposers.

Except for the input of energy from the sun, ecosystems are self-sufficient because essential materials are recycled through them over and over again. This is illustrated by the carbon cycle, which depends on the nearly opposite chemical processes of photosynthesis and respiration.

The way in which a species makes its living in nature is called the species' niche, and in general the niche of each species is unique.The human niche is the broadest niche of all in that people are widely distributed and are capable of dominating many ecosystems at the same time.

The invention of food raising ushered in an explosive ongoing increase in the size of the human population. Because food raising results in a decrease in natural diversity, its practice greatly simplifies ecosystems and makes them more vulnerable to catastrophe.

Modern agriculture depends on fertilizers, pesticides, and machines that run on fossil fuels. Fertilizers and fossil fuels may soon be in short supply, and some pesticides have widespread effects on neighboring ecosystems—even on the entire biosphere.

Technology has provided us with food, leisure time, and myriad conveniences, but it has produced sewage, pollution, and accelerated rates of extinction and has added to the burgeoning population through modern medical practices that increase the infant survival rate and lengthen the human lifetime. Infectious diseases have been important regulators of the size of the human population in the past, and some infectious diseases, such as schistosomiasis, are now more widespread than ever before, in large part because of environmental alterations brought about by people attempting to raise more food. The unwanted and unpredicted side effects of controlling malaria by means of pesticides have provided some insights into how complicated human ecology really is. In the end, each person is an ecosystem because human existence depends on the presence of organisms that live on or in the human body.

Suggested Readings

1. "The Ecosphere," by La Mont Cole. *Scientific American,* April 1958, Offprint 144. Provides an estimate of the total weight of living matter that our planet could support.
2. "The Biosphere," by G. Evelyn Hutchinson. *Scientific American,* Sept. 1970, Offprint 1188. How living things depend on complex cycles of energy and chemicals.
3. "The Flow of Energy in the Biosphere," by David M. Gates. *Scientific American,* Sept. 1971, Offprint 664. How all living things depend ultimately on the energy of sunshine.
4. "Human Food Production as a Process in the Biosphere," by Lester R. Brown. *Scientific American,* Sept. 1970, Offprint 1196. How many people can the biosphere support without upsetting its overall operation?
5. "Human Energy Production as a Process in the Biosphere," by S. Fred Singer. *Scientific American,* Sept. 1970, Offprint 1197. How the release of energy stored in fossil and nuclear fuels can affect some of the cycles of the biosphere.
6. "Human Materials Production as a Process in the Biosphere," by Harrison Brown. *Scientific American,* Sept. 1970, Offprint 1198. How people must devise ecologically sound cycles in order to conserve unrenewable resources.
7. "The Human Population," by Ronald Freedman and Bernard Berelson. *Scientific American,* Sept. 1974. What are the prospects that the explosively increasing human population will level off because of high death rates or low birth rates?
8. "The History of the Human Population," by Ansely J. Coale. *Scientific American,* Sept. 1974. Some rough estimates of the numbers of people on earth from one million years ago to the present.
9. "Toxic Substances and Ecological Cycles," by George M. Woodwell. *Scientific American,* March 1967, Offprint 1066. How radioactive elements and pesticides that are released because of human activities can reach dangerous concentrations in many different ecosystems.
10. "The Flow of Energy in a Hunting Society," by William B. Kemp. *Scientific American,* Sept. 1971, Offprint 665. How the energy obtained from the natural resources of Baffin Island is channeled through a community of modern Eskimos.

11. "The Flow of Energy in an Agricultural Society," by Roy A. Rappaport. *Scientific American,* Sept. 1971, Offprint 666. The energetics of "slash and burn" agriculture as practiced by a tribe in New Guinea.

12. "The Flow of Energy in an Industrial Society," by Earl Cook. *Scientific American,* Sept. 1971, Offprint 667. What factors ultimately limit the energy consumed by industrial societies?

PART II

The Human Machine

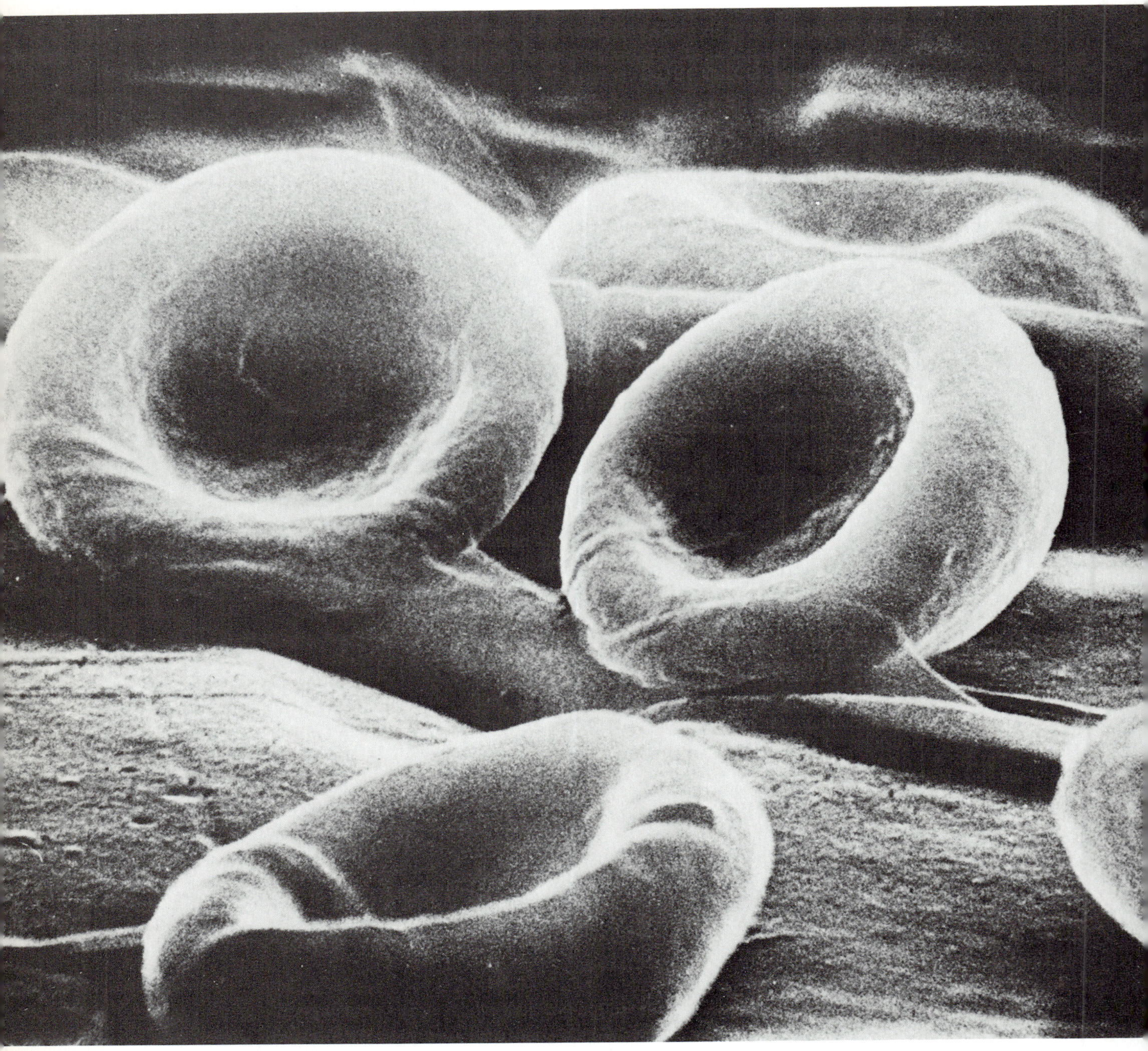

Red blood cells (×20,000), as seen through an electron microscope. Because of their unique shape, these cells have a large surface area for exchange with the environment. Hemoglobin contained within them carries oxygen and gives the blood its red color. (From Lennart Nilsson, Behold Man, *Little, Brown and Company, 1974. Courtesy Albert Bonniers Förlag, Stockholm.)*

CHAPTER

5

Cells and Tissues

The first living things on earth evolved from nonliving chemicals dissolved in the ancient oceans (as you may recall from our discussion in Chapter 1). We have no way of knowing exactly what the first organisms were like, but fossils indicate that one of the greatest milestones in the evolution of living things had occurred by about three and a half billion years ago. That momentous occurrence was the formation of cells.

Cells are the fundamental structural and functional units of all living things. This important generalization, known as the cell theory, was first proposed in the late 1830s, only a few decades before Darwin published his famous theory of evolution. Darwin's theory maintains that evolution by means of natural selection has resulted in the incredible diversity of living and extinct species. The cell theory maintains that underlying the diversity of living things is a fundamental similarity—all organisms are composed of units called cells.

As discussed in Chapter 1, cells can be classified into two basic types. The first are prokaryotic cells, and though they are vastly complicated in themselves, they are by comparison much simpler and much less diverse than the second cell type, eukaryotic cells. In this chapter we are concerned with the structure and function of the eukaryotic cells that are the building blocks of the human body.

The body of an adult human being is made up of about 100 trillion (100×10^{12}) cells, and although they vary enormously in function, most of them have certain features in common. For example, each cell is separated from its environment by an outer membrane, the *cytoplasmic membrane.* Through its cytoplasmic membrane the cell obtains nutrients, secretes products, excretes wastes, and receives signals from other cells. The cytoplasmic membrane encloses the *cytoplasm,* which refers to the cell's contents except for the *nucleus,* a membrane-bound region that is usually located near the center of the cell. The cytoplasm is responsible for many activities of the cell, including energy production and the manufacture of all the body's structural components. The cytoplasm contains a variety of specialized structures, or *organelles,* that carry out specific functions. Among the most studied organelles are the mitochondria, responsible for energy production, and ribosomes, which are centers for the manufacture of large molecules such as the enzymes that facilitate all chemical reactions in the body. As you will learn, the nucleus contains the blueprints that specify all of the activities of the cell. But before we discuss the structural features that most cells have in common, let us first take a look at some of the ways in which cells differ from one another.

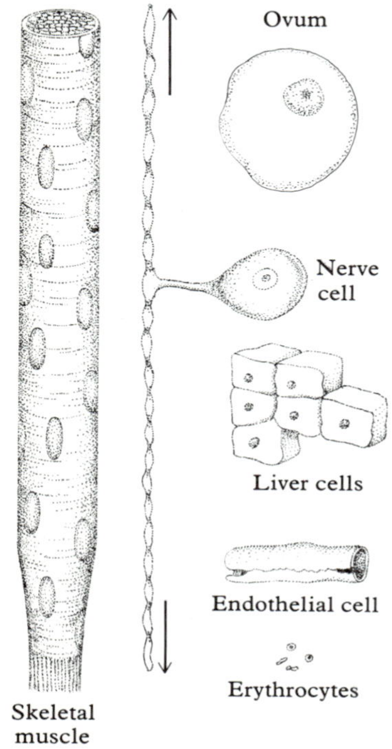

5–1
The body's cells vary considerably in size and shape. They are drawn to scale here, about 300 times their actual size.

The Size and Shape of Human Cells

The cells of the body are a diverse lot, and they vary considerably in size and shape (Figure 5-1). The egg (ovum), about 100 micrometers in diameter, is one of the body's largest cells. (There are 1000 micrometers in a millimeter, so egg cells are $\frac{1}{10}$ millimeter across—just large enough to be barely visible to the naked eye.) This large size allows the egg cell to carry large numbers of energy-rich molecules, which are needed if the egg becomes fertilized and then spends several days in the uterine tube before implantation, when it begins to get its nourishment from its mother. On the other hand, red blood cells, which have no nuclei, are among the smaller human cells, only 7 to 8 micrometers in diameter. Small size suits the main function of the red cell: to deliver oxygen to the tissues as it passes through the capillaries. Small size is an advantage because smaller objects have more surface area in relation to their contents, and the amount of oxygen that can diffuse in and out of the cell is proportional to the surface area. Red blood cells are shaped like discs (Frontispiece) and are indented on each side, a shape that provides even more surface area for oxygen exchange than a spherical cell.

Shape differences also reflect the different functions of cells. Nerve cells, which carry information between widely separated areas, are the most elongated. For example, a nerve cell that picks up sensations from the big toe may extend for several feet from the big toe to the spinal cord; at the magnification shown in Figure 5-1 the cell would be over a quarter of a mile in length. Most skeletal muscle cells are fused to form muscle fibers. These fibers usually extend the entire length of the muscle, which facilitates muscle action. The nuclei of the cells that fuse to form the muscle fiber remain intact, so each muscle fiber contains thousands of nuclei. The epithelial cells that form the lining of the

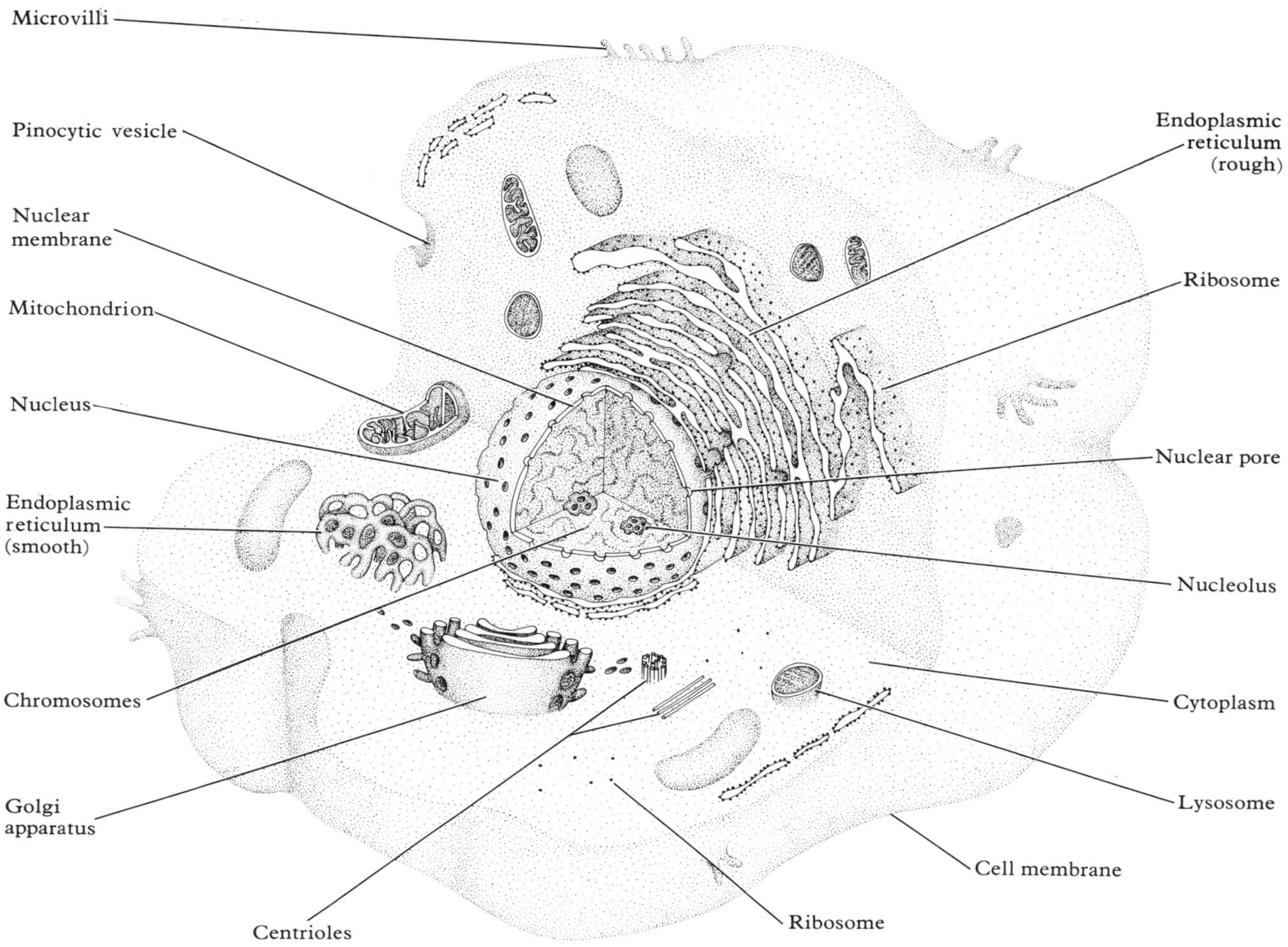

5–2

Diagram of a typical cell based on what is seen through the electron microscope.

smallest vessels of the circulatory system, the capillaries, are flattened and arched into a cylindrical shape. This shape permits the flow of blood through the cylinder, and the thinness of the flattened cell facilitates the passage of small molecules from one side of the cell to the other.

Cells vary not only in size and shape but in internal structure as well. As a result, there is no such thing as "typical" cell contents any more than there is a "typical" cell shape. Nevertheless, certain structural elements are invariably present in the most active cells—those cells capable of division, movement, and the manufacture of molecules used elsewhere in the body. Figure 5-2 is a drawing of the main structural elements of such a cell, and Table 5-1 lists important functions of these structures. You should use Figure 5-2 and Table 5-1 as references as we discuss these structures and their functions further.

TABLE 5-1

Structural elements of eukaryotic cells and some of their functions.

STRUCTURE	FUNCTIONS
Nucleus	Site of all chromosomal materials and chromosomal DNA. Responsible for DNA and RNA synthesis.
Nuclear envelope	A porous double membrane that completely encloses the nucleus. During cell division it breaks into a series of vesicles that fuse around the new nuclei once they have formed. The nuclear envelope allows passage of metabolites in and out of the nucleus but constrains DNA and nuclear proteins. It is continuous in places with the membranes of the endoplasmic reticulum.
Nucleolus	An area of the nucleus, usually spherical, that contains many granules. The site of synthesis of rRNA (ribosomal RNA), a type of RNA that is an important structural component of ribosomes but that does not usually carry genetic information.
Cytoplasm	All material contained within the cell, except what is inside the nucleus. Carries out all functions associated with cytoplasmic structures.
Cell membrane	Completely encloses the contents of the cell. It is responsible for active transport, determines which molecules enter and leave the cell, and has receptors on its outer surface for certain hormones.
Mitochondria	Organelles within the cytoplasm where energy is generated.
Endoplasmic reticulum (ER)	Membranous sacs present in the cytoplasm. ER studded with ribosomes is called *rough* ER, and ER without ribosomes is *smooth* ER. The sacs of the ER transport proteins and other large molecules.
Ribosomes	Small particles, present in the cytoplasm, that are the sites of protein synthesis. If the protein being synthesized is destined for secretion, the ribosomes are associated with ER.
Golgi apparatus	A cluster of parallel membranous sacs involved in the transport of proteins and other large molecules. Secretory vacuoles and lysosomes both grow from the golgi apparatus.
Lysosomes	Membrane-bound organelles that contain degradative enzymes. They fuse with phagosomes within the cell and digest their contents.
Phagosomes	Membrane-bound organelles that contain material taken into the cell by endocytosis. They bring either nutritive materials or scavenged debris into the cell. They fuse with lysosomes and their contents are degraded.
Centrioles	Paired cylindrical structures containing microtubules found in the cytoplasm. They are the origin of the spindle fibers that attach to chromosomes during cell division.

TABLE 5-1 (continued)
Structural elements of eukaryotic cells and some of their functions.

STRUCTURE	FUNCTIONS
Microvilli	Fingerlike projections from the surface of the cell. They increase the area of the cell membrane available for transport, and they may help cells that move actively, such as fibroblasts and lymphocytes, to sense the environment.
Microtubules	Long cylindrical tubules that are structural components of spindle fibers, centrioles, cilia, and sperm tails. They are important in cell division and cell movement.
Microfilaments	Long fibers traversing the cell cytoplasm that participate in cell movement.

Cell Movement

Some cells must keep their shape in order to function well, whereas others change shape and move about. It has recently become known that movement and change in shape are dependent upon structural elements known as *microfilaments* (Figures 5-3 and 5-4), tiny filamentous strands present in most cells. Microfilaments are composed of the same protein molecules that participate directly in muscle contraction (actin, myosin, and others). Some of these microfilaments traverse the cell and are attached at their tips to the outer cell membrane. Because they are attached at each end, their contraction alters cell shape and produces the kind of motion of white blood cells known as amoeboid movement. Microfilaments are also responsible for the shape change in platelets that enables them to participate in blood clotting. Microvilli (Figure 5-5 and Frontispiece, Chapter 10), tiny projections that increase the surface area of specialized cells, are packed with microfilaments, and the action of microfilaments may account for the fact that microvilli have been found to push out from and then melt back into the cell membrane.

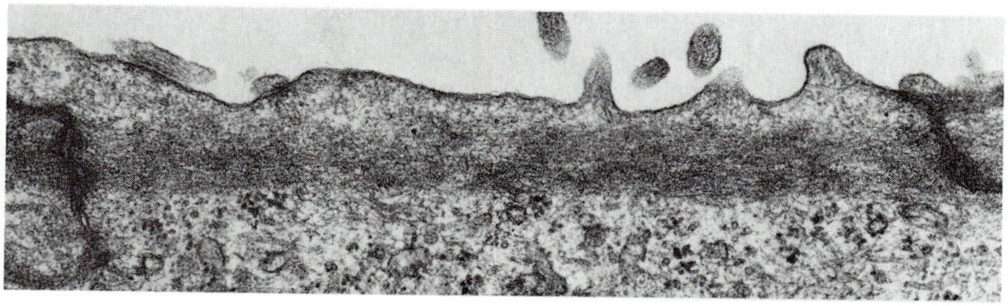

5–3
Cell "muscle," which can produce changes in the shape of cells, is made up of microfilaments arrayed in bundles. Such a bundle is seen in this electron micrograph, lying parallel to the surface of an epithelial cell in the oviduct of a bird. (Photo courtesy of Joan T. Wrenn.)

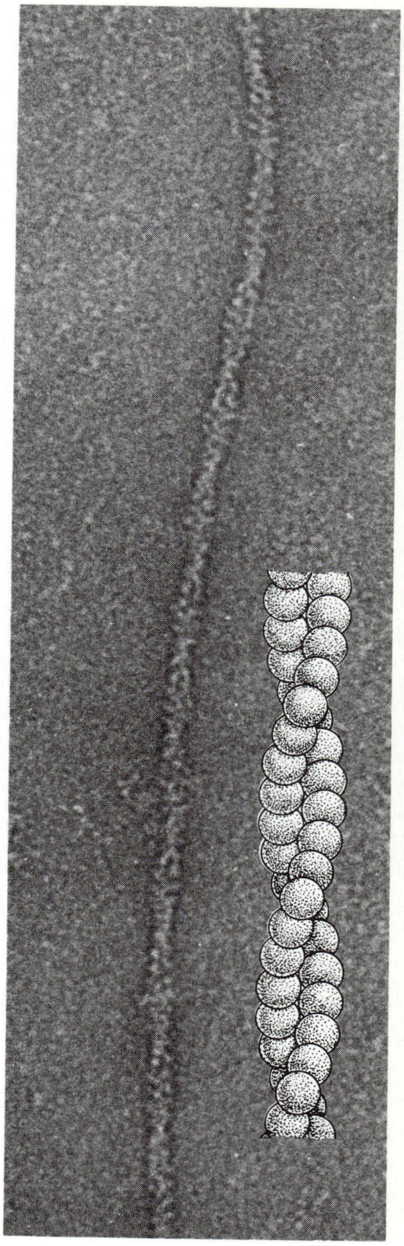

5–4
An electron micrograph of an individual microfilament (×420,000). The inset diagram shows that the microfilament is a double helix, consisting of two strands of actin molecules (circles in the diagram). (The electron micrograph courtesy of Dr. H. E. Huxley. The drawing from "The Mechanism of Muscular Contraction," by H. E. Huxley. Copyright © 1965 by Scientific American, Inc. All rights reserved.)

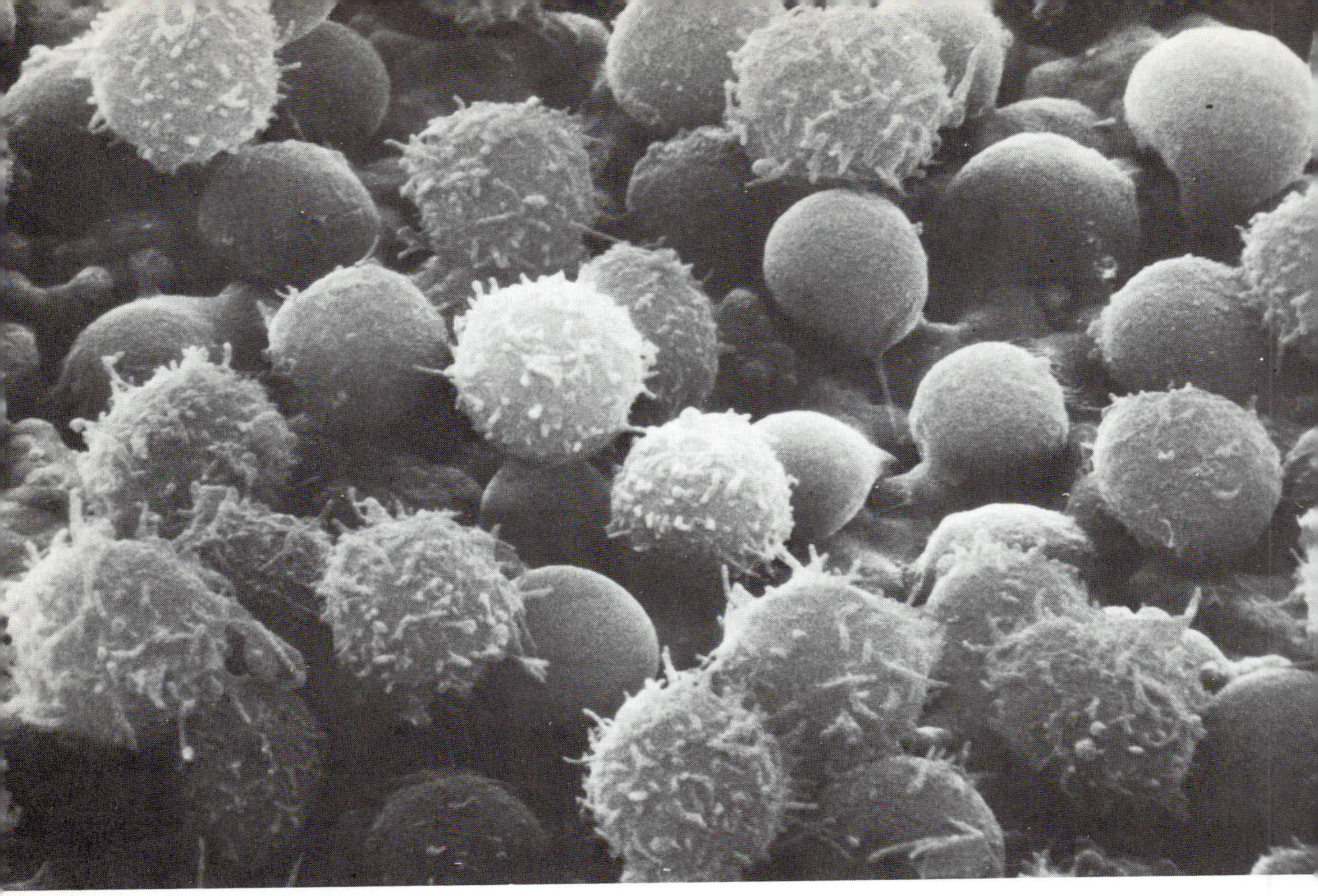

5–5
Human lymphocytes (×5160), many covered with microvilli, viewed through the scanning electron microscope. These lymphocytes were taken from a patient who had lymphocytic leukemia; consequently, large numbers of lymphocytes are present in the blood. (Courtesy of E. de Harven, A. Polliack, and N. Lampen, Sloan-Kettering Institute for Cancer Research. Journal of Experimental Medicine, 138, *607, 1973.)*

The movement of cell parts is also strongly influenced by *microtubules* (Figure 5-6), which are hollow cylindrical structures composed of a special protein called tubulin. Microtubules are the structural elements of *centrioles* and the spindle fibers that emerge from centrioles during cell division (see the following section on mitosis). The importance of microtubules can be shown by treating dividing cells with colchicine, a drug that attaches to microtubules and breaks them down. In colchicine-treated cells the spindle fibers, which are microtubules that become attached to chromosomes during cell division, cannot form; consequently, the cell cannot divide because its chromosomes cannot separate.

Cilia, tiny projections from the surface of specialized cells, and the tails of sperm cells (Figure 5-6) also contain special arrangements of microtubules that further facilitate their motion. The microtubules of both cilia and sperm are arranged precisely: two single microtubules are located in the center and nine doublet microtubules are evenly spaced around the periphery (Figure 5-7). The whiplike motion of cilia and sperm is not well understood, but it seems to depend upon the sliding of doublet microtubules and the integrity of the projecting arms. Recently a condition was described in which the arms of the doublet microtubules were not present in both cilia and sperm of certain males (Figure 5-7). Affected men had frequent colds, bronchitis, and pneumonia, presumably because the cilia of their respiratory tracts were inactive and could not propel inhaled organisms out of the tract. Their sperm cells were immotile as well.

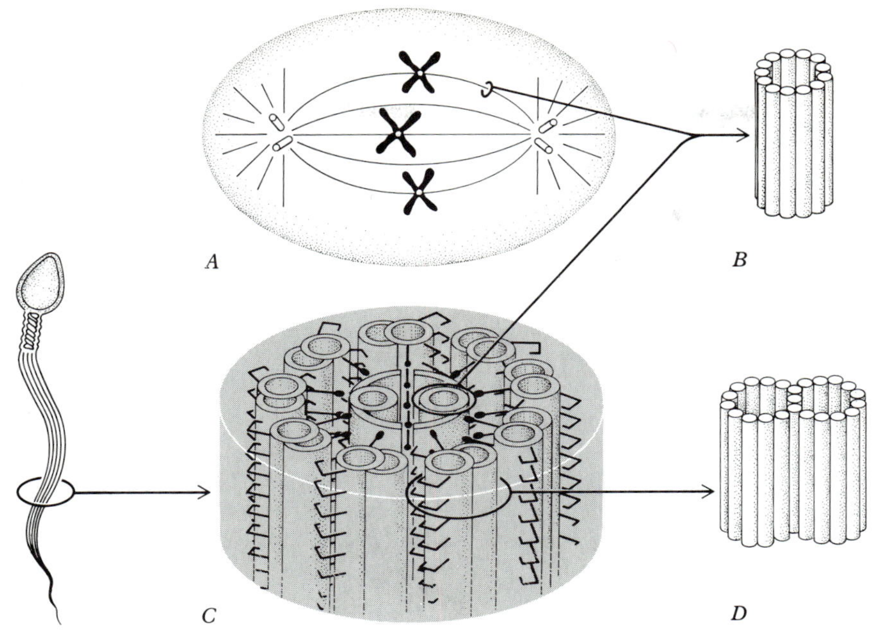

5–6
Microtubules are found in such dissimilar cellular structures as the mitotic spindle (A) that controls the separation of the chromosomes during cell division and the flagellum (C) that propels the sperm cell. Each single microtubule (B) consists of 13 filaments. Doublet microtubules (D) are made up of 23 filaments. Each filament in turn is composed of globular tubulin molecules. (After "How Living Cells Change Shape," by Norman K. Wessells. Copyright © 1971 by Scientific American Inc. All rights reserved.)

Cell Connections

Cells form connections with nearby cells, and these connections help to maintain shape and serve other functions as well. The outer membranes of cells may directly contact each other and form *tight junctions* (Figure 5-8). These junctions, which characterize the boundaries between the epithelial cells that line organs containing body fluids (blood, bile, urine), help to reduce the possibility of molecules diffusing between the cells. Somewhat looser connections, in which there is still some space remaining between the cells, are *desmosomes* (Figures 5-8 and 5-9). A desmosome is a specialized area of cellular adhesion where adjacent cells appear to be spot-welded by fibrous tissue. Finally, there are *gap junctions* (Figure 5-8) which not only connect exterior cell membranes but provide channels, or pipes, between the cells to permit the passage of certain substances. For example, gap junctions are found between smooth muscle cells where contraction of one cell is followed by contraction of an adjacent cell. The result is not the rapid contraction we associate with our skeletal muscles, but a slow wave of contraction passing through the tissue, as in intestinal peristalsis. (See Figure 10-2.)

The spaghetti-like microvilli of cells that wander through tissues much of the time (such as lymphocytes and fibroblasts) form attachments to other structures that are necessary for the propulsion of the cells.

Transport across Cell Membranes

The transport of substances in and out of cells is crucial, both to the cells and to the entire organism. For example, coordination within the nervous and endocrine systems is possible only because chemical substances are released

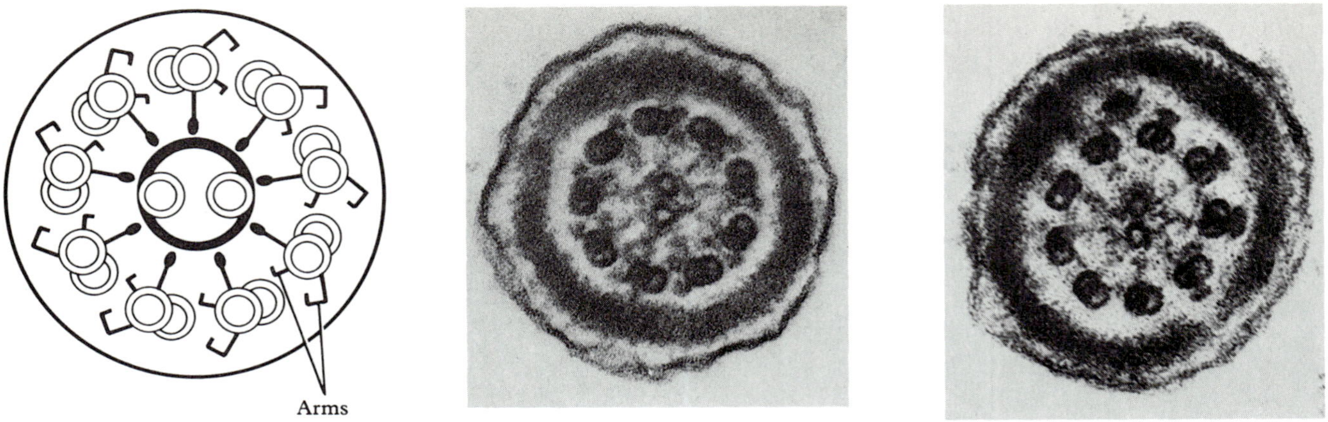

5–7
Electron micrographs of cross sections of sperm tails. Above left, normal sperm tail showing arms on doublet microtubules. Above right, immotile sperm tail, whose doublet microtubules lack arms. Diagram at left shows the usual arrangement of arms found in cilia and sperm tails. (Photos courtesy of Dr. B. A. Afzelius.)

from one cell and received by another. And survival of the individual cell depends upon its receiving nutrients and excreting wastes.

A variety of processes participate in the passage of substances in and out of cells. The particular process being used at any time depends in large part on the size of the substance being transported. For small molecules, diffusion and active transport are employed. *Diffusion,* the tendency of molecules to pass from a region of high concentration to a region of low concentration, is a passive process. The gases O_2 and CO_2 pass across cell membranes through diffusion. *Active transport* is a process in which molecules are transported by means of energy expended by the cell from a region of low to a region of high concentration. You can understand why this movement against a concentration gradient requires energy—it is the opposite of the more "natural" flow that takes place in diffusion. A good example of the process of active transport is provided by the workings of the kidney tubules, which are discussed in Chapter 10.

The structure of the cell membrane plays an important role in determining what passes in and out of the cell, but biologists are only beginning to learn about membranes and how they work. We do know, however, that cell membranes (Figure 5-10) have a framework consisting of two layers of lipid (fatty) molecules that have protein molecules either in them or around them. The inserted protein molecules are responsible for the active transport of certain substances (such as sodium and potassium ions), and for the rapid diffusion of some small molecules (glucose, urea, and amino acids, for example) through the cell membrane. Because of the presence of protein, glucose diffuses through the membrane 100,000 times faster than it would if the cell membrane contained only lipid.

The lipid molecules that are part of the cell membrane are called phospholipids (Figure 10-3). Each phospholipid molecule has a fatty end and a more water-soluble (charged) end bearing a phosphate group. The two layers of phospholipids in the membrane are arranged so that the water-soluble ends of the molecules are on the surfaces of the membrane, allowing interaction with

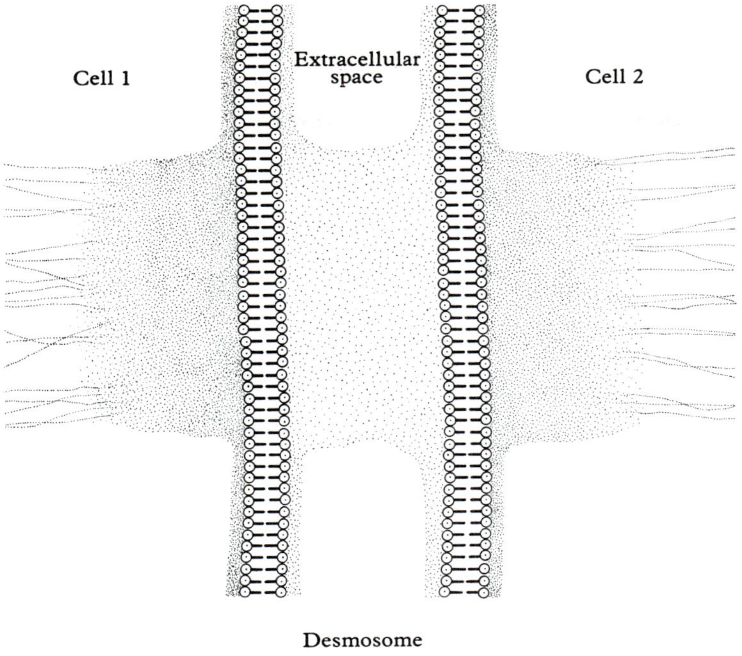

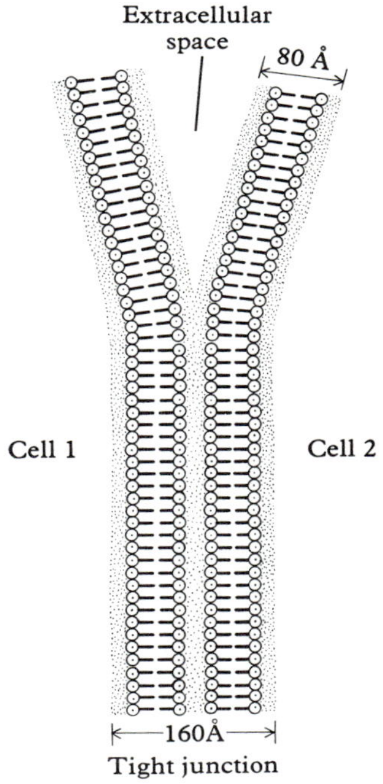

5–8
Three types of cell junctions. Desmosomes are specialized areas of cell adhesion. Tight junctions prevent passage of molecules between the cells and are characteristic of epithelial cells. Gap junctions provide a channel directly connecting the cytoplasms of two adjacent cells.

water inside and outside of the cell (Figure 5-10). The fatty ends of the molecules are on the inside, providing a barrier to indiscriminate passage of water-soluble molecules through the membrane.

Large water-soluble molecules (such as proteins) and small insoluble particles (such as bacteria) enter the cell by acquiring a complete covering of host cell membrane on the way in. When they leave the cell their covering membrane leaves them and fuses with the outer membrane of the host cell. When water-soluble molecules enter the cell by acquiring a covering of host cell membrane, the process is called *pinocytosis,* or cell drinking, and when insoluble particles enter this way the process is called *phagocytosis,* or cell eating. However, because pinocytosis and phagocytosis are essentially the same process (although applied to objects of different sizes), it has been suggested that the term *endocytosis* be used in place of both other terms. The use of the term endocytosis (endo means in) has been joined by the term *exocytosis* (exo means out) to describe exit from the cell by the reverse process.

Phagocytes are cells that specialize in endocytosis as a form of transport; they travel around the body prepared to engulf debris and bacteria as well as large molecules; they help to defend our bodies against some kinds of diseases. When a phagocyte encounters an invading bacterium (Figure 5-11), the membrane of the phagocyte surrounds the bacterium and the reluctant bacterium is drawn inward. The membrane eventually encloses the bacterium. When this happens the bacterium is said to be inside a *phagosome.* Typically, the membrane of the phagosome then merges with the membranes of one or more *lysosomes,* which are

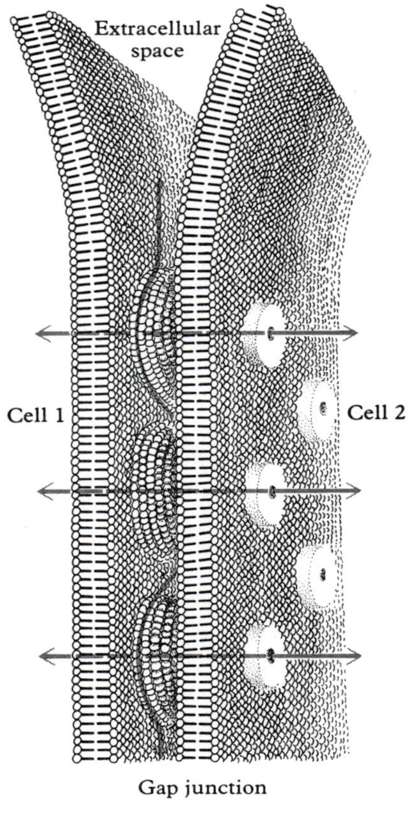

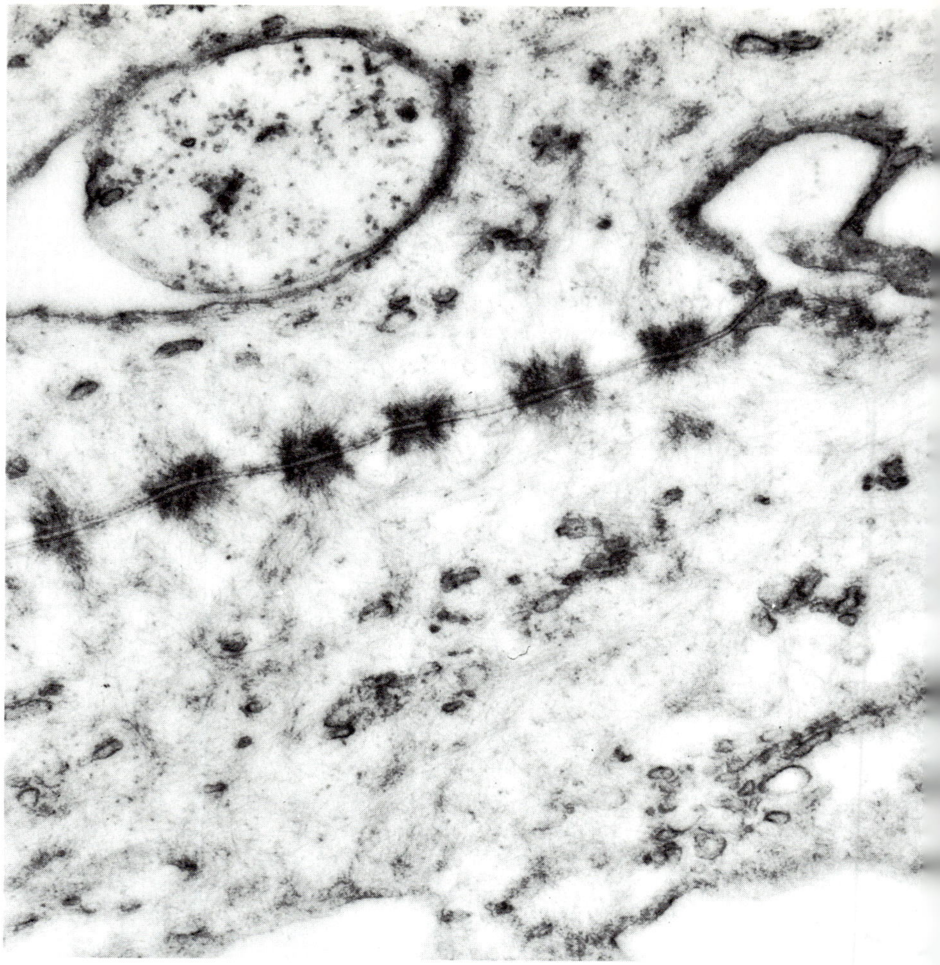

5–9
Desmosomes are apparently special devices for mutual attachment of cells across their membranes. In this electron micrograph (×35,000) more than a dozen such bridges (dark, squarish areas) connect two skin cells from a salamander larva. The cell membranes run horizontally across picture. (Photo courtesy of K. R. Porter.)

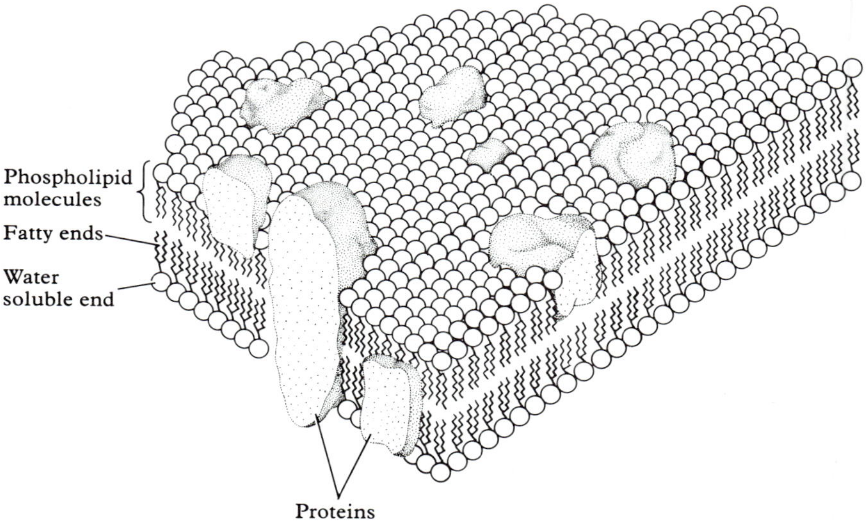

5–10
A cell membrane. Phospholipid molecules provide the basic structure of the membrane. Sometimes several proteins are bound into a single functional complex. Proteins can occupy any position with respect to the phospholipid bilayer: they can be entirely outside or inside, they can penetrate either surface, or they can extend through the membrane.

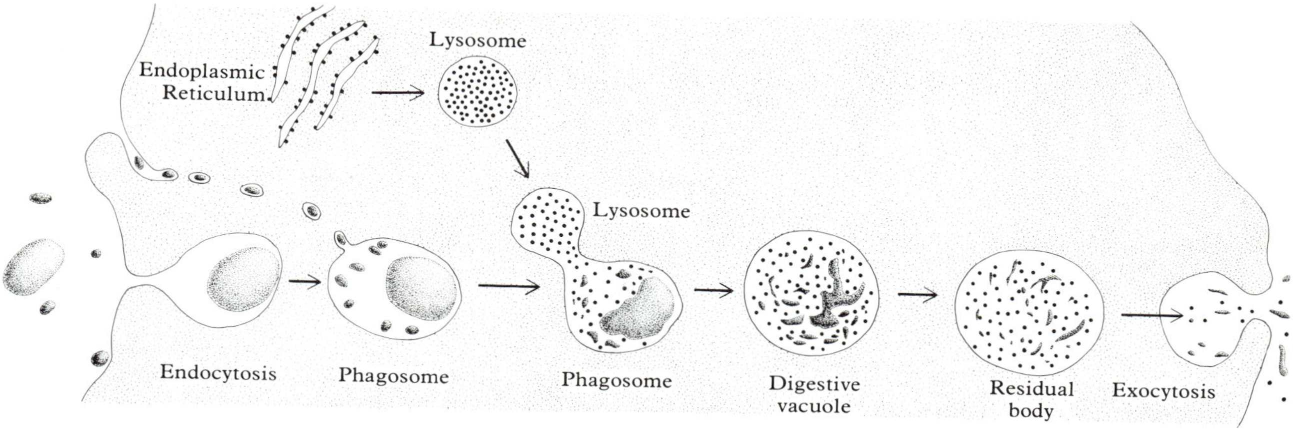

5–11
Large molecules and particulate substances enter the cell by endocytosis, forming a membrane-bound phagosome (at left in diagram). The phagosomes may then merge with a lysosome, which brings together the digestive enzymes of the lysosome and the contents of the phagosome. After digestion has taken place, the membrane of the residual body may fuse with the cell membrane and release its contents outside of the cell. The exit process is known as exocytosis. (After "The Lysosome," by Christian de Duve Copyright © 1963 by Scientific American, Inc. All rights reserved.)

membrane-bound sacs that contain enzymes that facilitate digestion of bacteria. The fusion of the phagosome and lysosomes results in the formation of a *residual body,* which contains the partly digested remains of the bacterium. Finally the membrane surrounding the residual body fuses with the outer cell membrane, and the unusable contents are released. This entire process depends upon two remarkable properties of cell membranes: first, they can fuse with one another (Figure 5-12), and second, they can pinch off into smaller membrane fragments.

Membranes also aid in the export of important products such as plasma proteins and digestive enzymes from the cell. The synthesis of proteins takes place on *ribosomes,* and if a cell manufactures proteins for export, the ribosomes are always associated with a system of membranous intracellular sacs knows as the *endoplasmic reticulum* (ER). (When endoplasmic reticulum is studded with ribosomes, as Figure 5-13 illustrates, it is called *rough* ER as opposed to *smooth* ER, which does not have ribosomes.) The proteins that are synthesized for export enter the sacs of the ER and are moved along with it, first through the smooth ER and then through the *golgi apparatus,* a cluster of closely packed membranous sacs derived from smooth ER, until they are finally discharged by exocytosis (see Figure 5-12). Throughout their journey from ribosome to cell exterior, secreted proteins remain inside membrane-bound structures.

Inside living cells there is a constant turnover of molecules; in fact, entire cells as well as molecules are regularly replaced. Let us now turn our attention to some aspects of the turnover process.

Turnover of Body Cells—Mitosis

Some of the cells of the human body are highly specialized and do not divide, but others reproduce rapidly. Examples of those that do not divide are nerve cells, mature red blood cells that have lost their nuclei, and skeletal muscle cells that have fused to form a muscle fiber. In general, nerve cells, which are characterized by intricate connections with one another, cannot be replaced if destroyed. However, some muscle fibers can be replaced by the division of certain cells (satellite cells) that remain unfused in the muscle fiber.

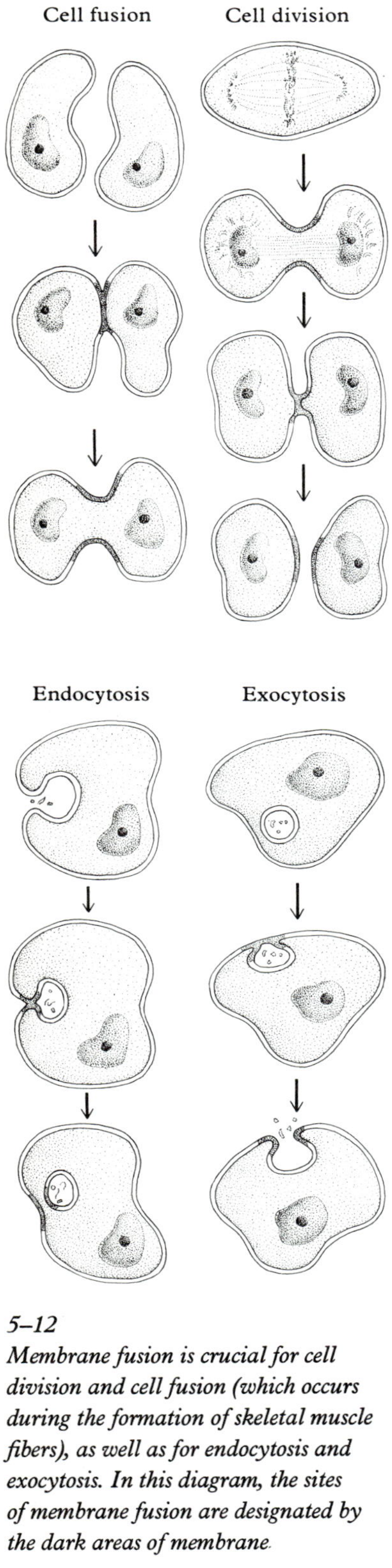

5–12
Membrane fusion is crucial for cell division and cell fusion (which occurs during the formation of skeletal muscle fibers), as well as for endocytosis and exocytosis. In this diagram, the sites of membrane fusion are designated by the dark areas of membrane.

On the other hand, among the most rapidly dividing cell populations of the human body are those that provide replacements for the cells lining the intestines and for worn-out red blood cells. It has been estimated that the entire cellular lining of the intestines is completely replaced by new cells about once every four days. About 0.8 percent of the body's red blood cells are replaced each day, and, although 0.8 is a small percentage, an adult has so many red cells that this amounts to over a million replacements released into the circulation each second!

The process by which body cells or *somatic* cells divide to produce two nearly duplicate copies of themselves is called *mitosis.* As discussed here and in following chapters, the biological basis of individuality among cells and organisms is encoded in the genetic material (*deoxyribonucleic acid,* or *DNA)* that is found inside the nucleus of each cell. Within the nucleus this important DNA is packaged in separate structures called *chromosomes.* Chromosomes consist of DNA and various proteins that provide a scaffold for the lengthy DNA molecules and that help to regulate the activity of the genetic material. In general, chromosomes are coiled, threadlike structures that are visible only when cells are in the act of dividing. At other times the chromosomes are still inside the nucleus, but they are elongated and highly entangled. (It is in this stretched condition that DNA directs the synthesis of proteins and, in dividing cells, the synthesis of new DNA molecules.)

Mitosis is a mechanism that ensures that the genetic material packaged in the chromosomes is distributed in an orderly fashion to each of the two daughter cells that are produced. The strategy is simple: the chromosomes are duplicated before division occurs and a complete set is distributed to each of the two resulting cells. But the details of mitosis, especially of the biochemical events that take place before and during cell division, are complicated and are at present incompletely understood.

As shown in Figure 5-14, mitosis is considered to occur in a series of stages, each designated by a name. For simplicity's sake only six chromosomes are illustrated, but you should bear in mind that human beings actually have 46 chromosomes, as discussed in Chapter 16. The first visible evidence that mitosis is occurring is that the chromosomes become long, thin, intertwined filaments. The chromosomes become progressively shorter and more rodlike, and at this stage each chromosome consists of two separate strands, called *chromatids,* that are joined by a single *centromere.* Each chromatid contains a copy of the DNA, which has already duplicated itself. While these events are taking place inside the nucleus, the surrounding nuclear membrane disintegrates and the spindle fibers, which consist of microtubules that spread out from the centrioles, are laid down. The chromosomes become attached to the spindle fibers by means of their centromeres and then line up around the equator of the spindle fibers. *The centromeres then divide* and when this happens each of the two chromatids of the duplicated chromosomes becomes a separate, single-stranded chromosome. The two sets of single standed chromosomes then begin to move away from one another and eventually arrive at opposite ends of the cell. At this stage the spindle fibers begin to disappear. Meanwhile, a deep furrow has been forming on the surface of the cell, and as it becomes progressively deeper it eventually cuts completely through the cell, producing two new cells. Each new cell contains the same number of chromosomes and approximately half of the structural elements and organelles present in the original cell. After division is completed, each cell grows in size, duplicates some of its organelles, and remains relatively quiescent

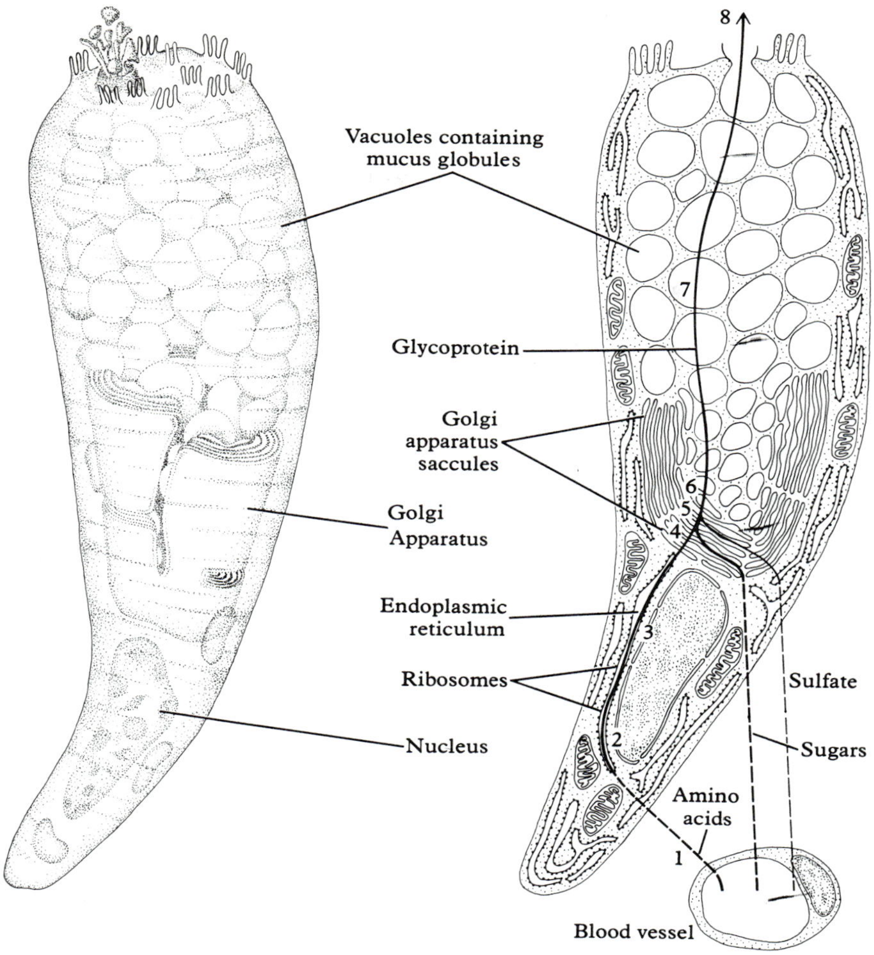

5–13
Far left, a three-dimensional representation of a goblet cell, which secretes mucus into the intestine. Near left, a cross section that shows the processes associated with the formation of mucus. Mucus is a complex glycoprotein, containing protein, carbohydrate, and sulfate components. Precursors of the mucus enter the cell from a blood vessel (1). Amino acids are synthesized on ribosomes (2) into proteins, which move up through the endoplasmic reticulum (3) to enter the Golgi saccules. Meanwhile, simple sugars are taken up into the saccules, there to combine with the incoming protein (4) to form glycoprotein, to which sulfate from the blood is also added (5). The saccules in which the glycoprotein formed are transformed into globules of mucus (6). The globules migrate to the top of the cell (7), ultimately leaving the cell and releasing the mucus to coat the surface of the intestine (8). (After "The Golgi Apparatus," by Marian Neutra and C. P. Leblond. Copyright © 1969 by Scientific American, Inc. All rights reserved.)

for a variable period of time before it duplicates its DNA and mitosis begins again.

From this consideration of how somatic cells are duplicated by the process of mitosis, we now turn our attention to how the body replaces some of the molecules that are constantly turned over inside cells.

Turnover of Molecules—Metabolism

Providing replacements for lost molecules is one of the body's most important tasks. Replacement molecules must exactly match the ones they replace: an individual molecule, whether part of a cell or not, will carry out its function only if properly constructed. Products from dead and dying cells must also be expelled or recycled so that toxic breakdown products do not accumulate and so that their components can be recycled or excreted effectively. Protein molecules known as *enzymes* are responsible for the precision and accuracy associated with the buildup, or *anabolism,* and breakdown, or *catabolism,* of the body's molecules. (The term *metabolism* includes both buildup and breakdown.) Most biochemical reactions are influenced by specific enzymes; a reaction usually will

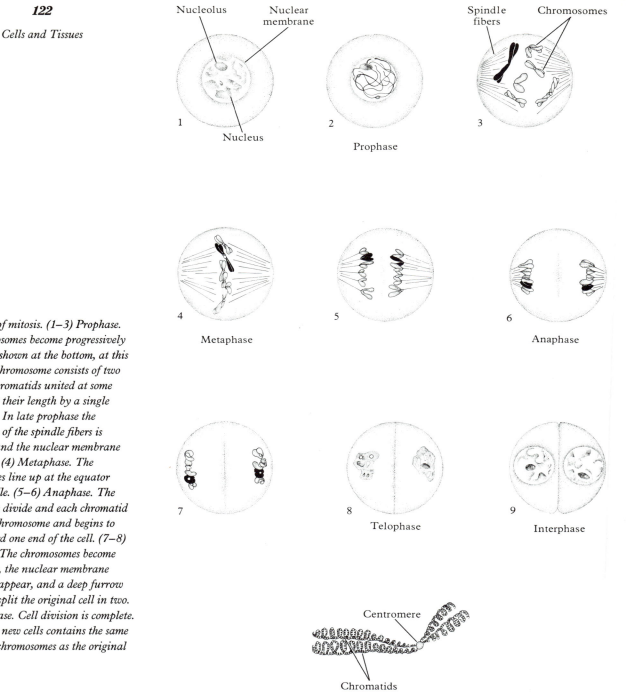

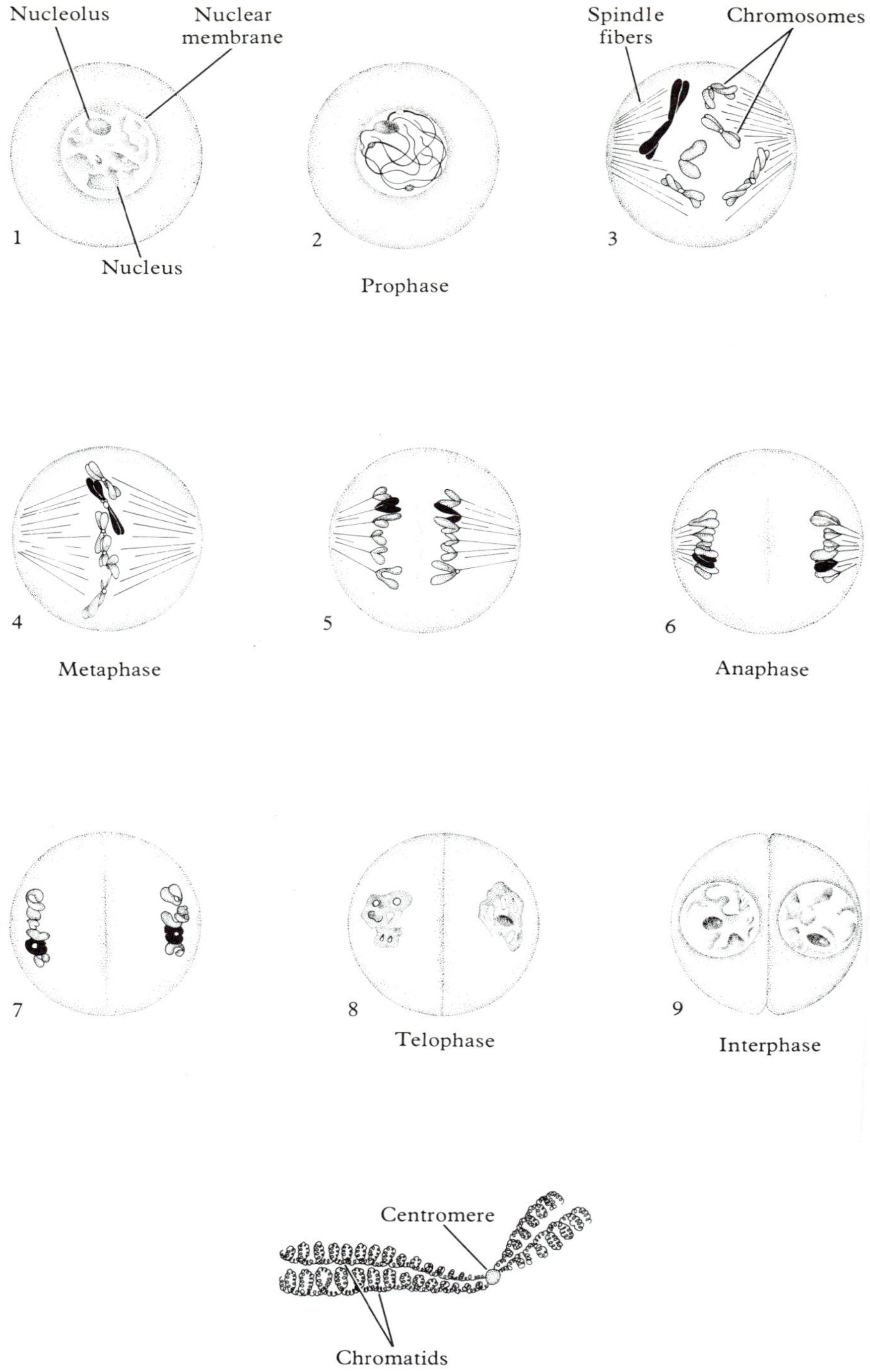

5–14
The stages of mitosis. (1–3) Prophase. The chromosomes become progressively thicker. As shown at the bottom, at this stage each chromosome consists of two identical chromatids united at some point along their length by a single centromere. In late prophase the elaboration of the spindle fibers is completed and the nuclear membrane disappears. (4) Metaphase. The chromosomes line up at the equator of the spindle. (5–6) Anaphase. The centromeres divide and each chromatid becomes a chromosome and begins to move toward one end of the cell. (7–8) Telophase. The chromosomes become less distinct, the nuclear membrane begins to reappear, and a deep furrow has nearly split the original cell in two. (9) Interphase. Cell division is complete. Each of the new cells contains the same number of chromosomes as the original cell.

not proceed if its enzyme is faulty. Enzymes exert this control because they are responsible for bringing together *exactly the right ingredients* involved in a particular reaction. But how do the enzymes themselves come to be "exactly right" so that they can bring together the exact ingredients? The specifications to which each enzyme is constructed ultimately reside in the structure of DNA. The process by which DNA specifies the construction of enzymes and other proteins is so important that we shall consider it in some detail.

The Synthesis of Proteins

Except for a tiny quantity of DNA present in the cytoplasmic organelles known as mitochondria (each mitochondrion contains less than 1/100,000 of the DNA contained in the cell), all the DNA resides on the chromosomes found inside the nucleus of the cell. The backbone of the DNA molecule is made up of alternating phosphate and deoxyribose sugar groups (Figure 5-15). To each deoxyribose along the backbone is attached one of four nucleic acid *bases.* These bases are adenine, cytosine, guanine, and thymine, and they are abbreviated A, C, G, and T. The sequence of these bases in the DNA determines protein structure, because during protein synthesis each *group of three adjacent bases dictates the insertion of a single amino acid into a growing protein molecule.*

Protein synthesis is a two-stage process. The first stage, *transcription,* occurs in the nucleus when the DNA code is transferred (or transcribed) to ribonucleic acid (RNA) molecules. RNA, like DNA, has a backbone of alternating phosphate and sugar molecules, but the sugar molecule in RNA is ribose rather than deoxyribose and the backbone of RNA is single-stranded rather than made up of two strands, as DNA is. Each ribose of RNA is attached to one of four nucleic acid bases. But in RNA, the base uracil (U) replaces the thymine (T) that is found in DNA.

The DNA code can be transferred into an RNA code because each of the four DNA bases attaches to and specifies the insertion of a different (complementary) one of the four RNA bases into a growing strand of RNA. Thus C in DNA codes for G in RNA, G for C, A for U, and T for A (Figure 5-16). The result is that the exact base sequence of DNA dictates an equally exact base sequence in the RNA that is synthesized inside the nucleus.

The newly synthesized RNA, which now carries the code for protein synthesis, is called messenger RNA or mRNA. It is a messenger because it travels from the nucleus of the cell, where it is synthesized, to the cytoplasm, where it attaches to the ribosomes. It passes from nucleus to cytoplasm through small openings known as *nuclear pores.*

The second step of protein synthesis, *translation* (Figure 5-17), occurs at the ribosome, where the base sequence of mRNA is translated into the amino acid sequence of a protein. Each group of three consecutive bases in mRNA selects a transfer RNA (tRNA) that has an amino acid attached to it. Transfer RNA then transfers its amino acid into the growing protein molecule. There are many tRNAs, each carrying both one of the 20 amino acids and a special set of three bases. In the same way that each base in DNA attaches to a specific RNA Base, each mRNA base attaches only to a specific base in tRNA. Therefore each group of three bases in mRNA can attach only to a tRNA carrying a complementary set of three bases. For example, a sequence of GUA in mRNA will attach to the tRNA carrying the complementary sequence CAU and the amino acid valine.

Because of the specific nature of the coding process, a change in only one of the bases in DNA leads to the insertion of a different amino acid into a growing protein chain. At first glance a difference of one amino acid may not seem significant, but in fact only one amino acid (among a total of 574) differs between normal and sickle-cell hemoglobins (see Chapter 18). As discussed in following chapters, people whose DNA codes for the S form of hemoglobin show increased resistance to malaria, and their red cells become sickle shaped when exposed to low oxygen concentrations. Hemoglobin S contains valine in the spot

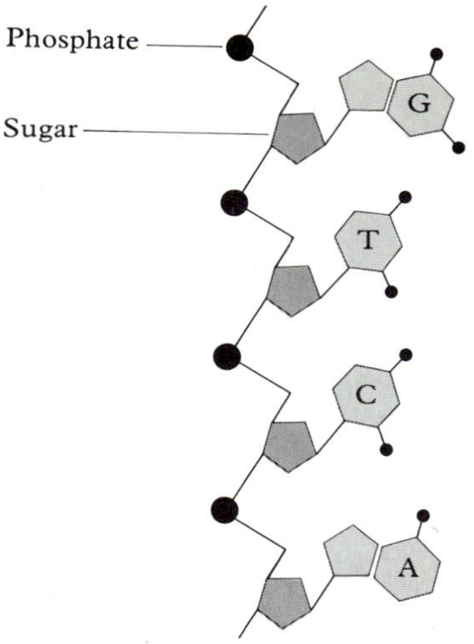

5–15
Portion of a single strand of DNA. Its backbone is composed of alternating sugar and phosphate groups. T, C, G, and A stand for the four nucleic acid bases found in DNA: thymine, cytosine, guanine, and adenine.

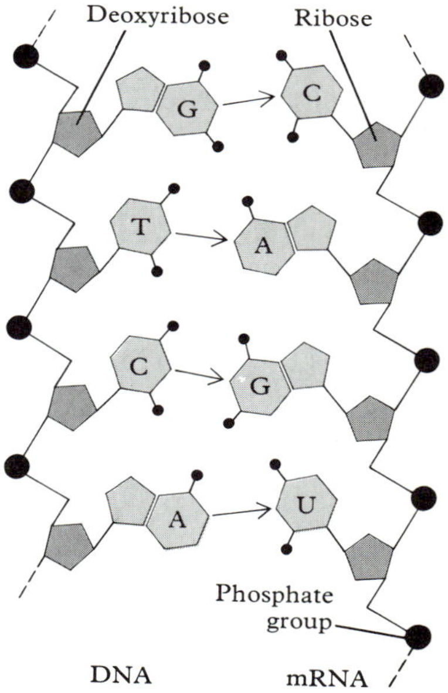

5–16
Portions of DNA and RNA molecules. During transcription, the first state of protein synthesis, each of the DNA bases (A, T, C, G) codes for one of the mRNA bases (U, A, G, C). Therefore, the exact base sequence of the newly synthesized mRNA is determined by the base sequence of the DNA. In RNA the base U (uracil) replaces T (thymine) found in DNA.

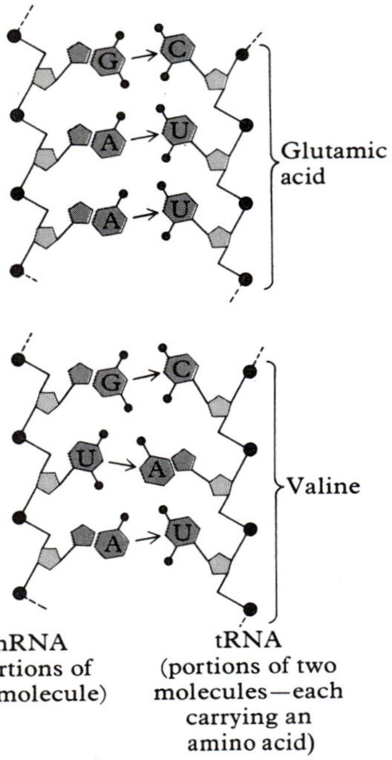

5–17
Portions of an mRNA molecule and two tRNAs. During translation, the second step of protein synthesis, each group of three consecutive bases in mRNA selects a tRNA that has complementary bases. The selected tRNA then transfers its amino acid (glutamic acid or valine) into the growing protein molecules.

where Hemoglobin A contains glutamic acid. This difference is caused by a single base change in the DNA (Figure 5-18).

Both A and S forms of hemoglobin carry oxygen. However, a base change in the DNA often leads to the production of a nonfunctional protein molecule. When a base change leads to synthesis of a nonfunctional *enzyme,* the body becomes unable to carry out the chemical reaction in which the enzyme participates. For example, enzymes participate in the synthesis of the pigment *melanin.* If one of them does not function, the result is an albino. In Chapter 16, we discuss albinism and some other conditions that result from faulty enzymes, which in turn result from heritable alterations in DNA structure.

But even with a complete set of properly functioning enzymes, no synthesis would take place in the body's cells without a supply of energy to drive the reactions. Let us now consider how the body obtains and uses energy.

Obtaining and Using Energy

The synthesis (anabolism) of cell constituents requires energy, and this energy is supplied by the degradation (catabolism) of other cell constituents. These processes take place simultaneously in the cell. Glucose is usually degraded to supply cellular energy, but energy can be obtained from proteins and fats as well. How does the energy released by the degradation of glucose become available to the cell?

In all living things, a single compound regularly functions as a go-between linking energy-giving and energy-requiring reactions. This high-energy compound is adenosine triphosphate (ATP; see Figure 5-19). As glucose is broken down, some of its energy is captured in usable form and stored within the chemical bonds of ATP. Chemically, the formation of ATP is accomplished by the addition of a phosphate group to adenosine diphosphate (ADP). The phosphate group is linked to ADP by means of a chemical bond, which releases a great deal of energy when it is broken.

5–18
Groups of three bases of DNA, mRNA, and tRNA molecules, showing how a single base change from T to A in DNA could result in the insertion of the amino acid valine into a growing protein molecule in place of glutamic acid. The only difference between hemoglobins A and S is that S contains valine, whereas A contains glutamic acid.

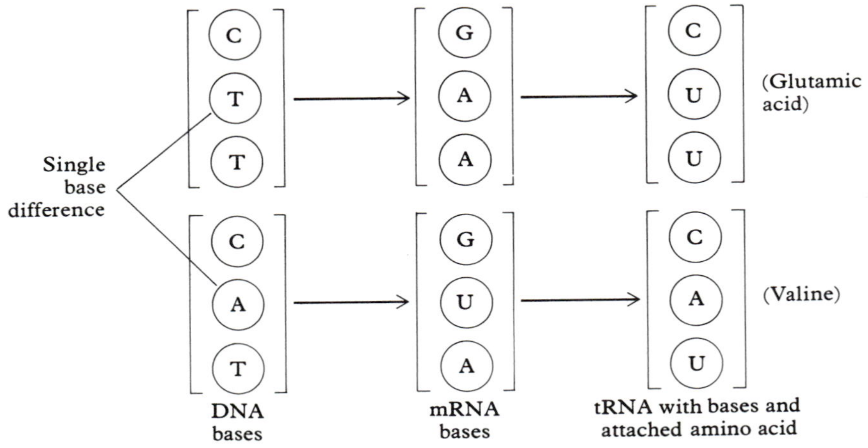

For each glucose molecule that the cell transforms into carbon dioxide and water, 38 ATPs are formed. The following chemical equation shows the link between glucose breakdown and ATP buildup:

$$\underset{\text{Glucose}}{C_6H_{12}O_6} + 6O_2 + 38\ \text{ADP} + 38\ \text{P} \rightarrow 6CO_2 + 6H_2O + 38\ \text{ATP} + 38H_2O$$

Each time an ADP and P combine, a molecule of water (H_2O) is released. This reaction is the product of a complicated series of chemical reactions, most of which occcur inside the specialized cellular organelles known at *mitochondria*

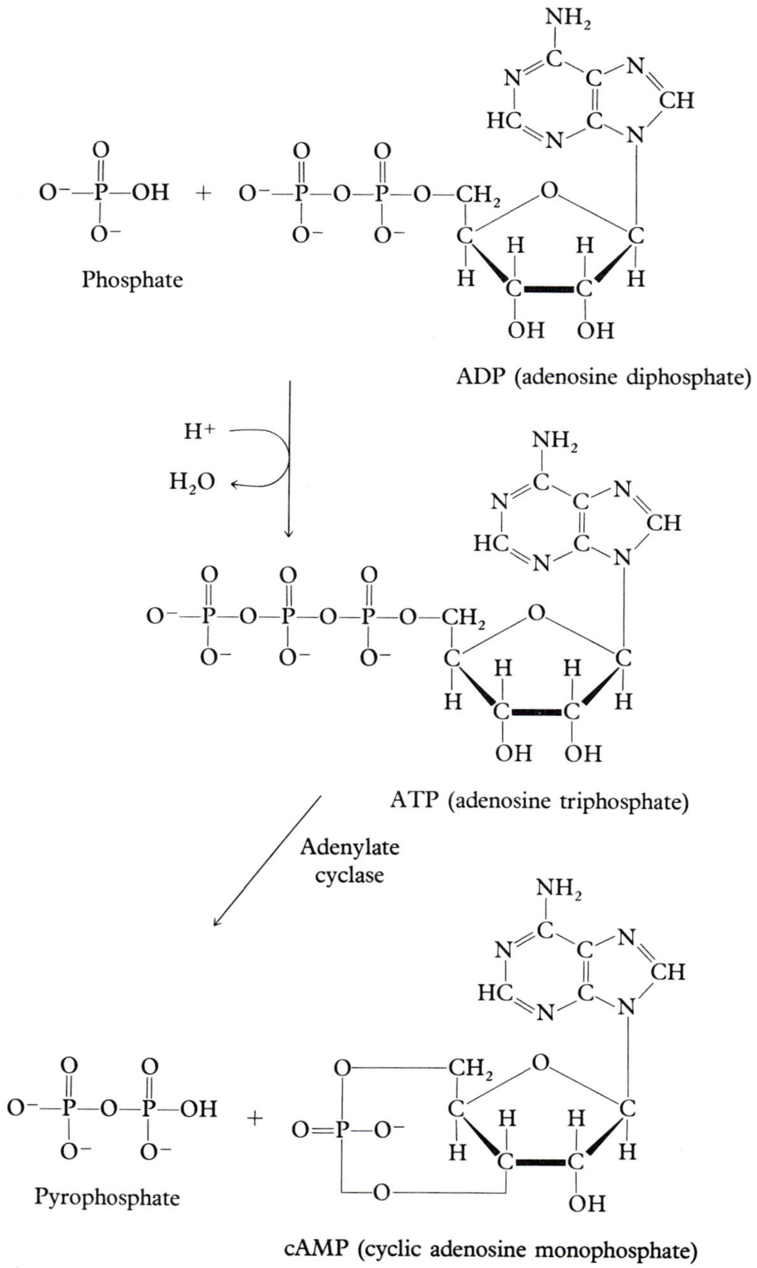

5–19
Two important biological molecules, ATP and cAMP. ATP stores energy, which is released when it is converted to ADP and phosphate. cAMP regulates cell activities. ATP is formed from ADP by the addition of a phosphate group. cAMP is formed from ATP by the action of the enzyme adenylate cyclase.

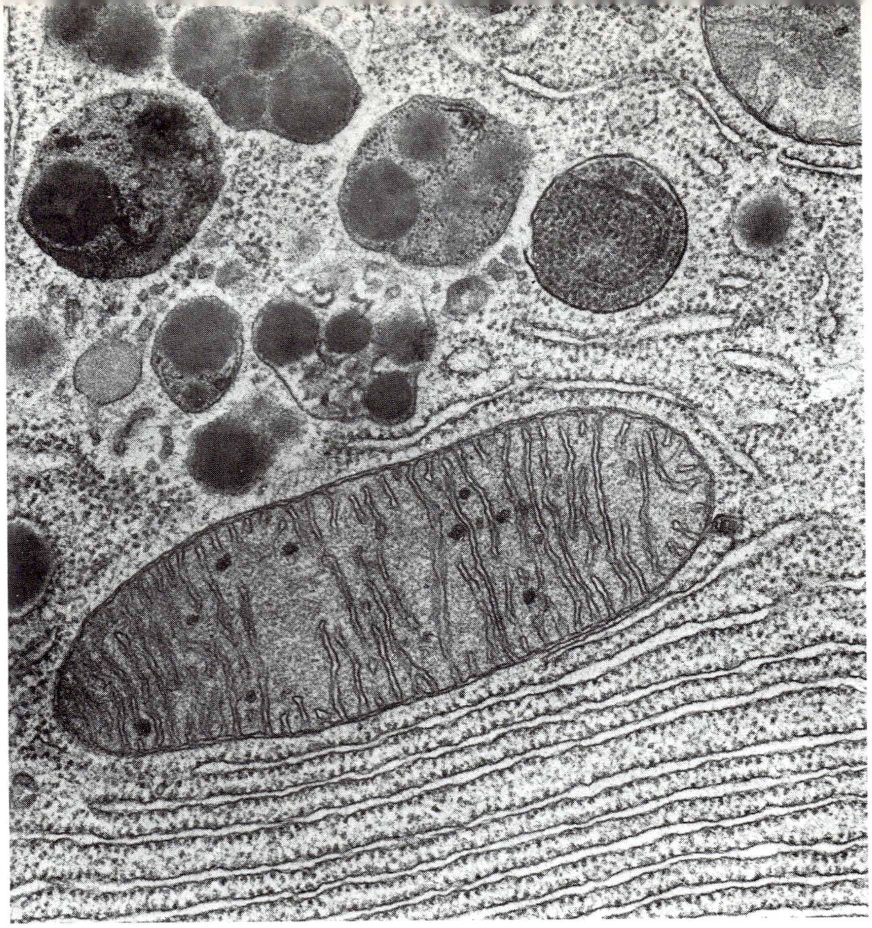

5–20
Electron micrograph of a mitochondrion within cell cytoplasm. (Photo courtesy of K. R. Porter.)

(Figures 5-2 and 5-20). Because they function in this way mitochondria are known as the cell's powerhouses. As you would expect, cells using a great deal of energy have more mitochondria, whereas cells requiring less energy have fewer. For example, liver cells, which require energy to manufacture large molecules such as carbohydrates and proteins, contain around 800 mitochondria, whereas sperm cells, which need energy only for movement, contain about twenty in their middle section. Mitochondria are fascinating structures, particularly because it is now theorized that they were once free-living bacteria that in the distant evolutionary past took up residence inside the more complex (eukaryotic) cells that characterize all animals and plants. Indeed, mitochondria divide like bacteria and contain their own DNA, RNA, and ribosomes. Mitochondrial DNA is very similar to the DNA of modern bacteria (and very different from the DNA of the nucleus) and it codes for enzymes whose amino acid sequence is similar to that of the functionally identical enzyme in bacteria. Even mitochondrial ribosomes are like bacterial ribosomes and are correspondingly smaller than the other ribosomes present in the eukaryotic cell they inhabit. Mitochondria now contain much less DNA than must have been present at the time of their evolutionary origin: they contain about 1000 times less DNA than modern bacteria. It seems likely that nuclear DNA has taken over many of the original functions of the mitochondrial DNA and thus resulted in the elimination of most mitochondrial DNA.

The bacterial origin of mitochondria may account for the toxicity of the antibiotic *chloramphenicol.* Chloramphenicol, like other antibiotics, is useful because it interferes with processes of prokaryotic (bacterial) cells, but not of eukaryotic cells like our own. However, when chloramphenicol is taken to fight bacterial infections it can sometimes mistake mitochondria for bacteria and attack them, too. The occasional side effects of chloramphenicol (including severe anemia) can be best explained by its injurious effect on mitochondrial protein synthesis.

Another interesting aspect of mitochondria is that people inherit them from their *female* parent only. The reason for this is that when the sperm fertilizes the egg, the middle piece of the sperm cell, which contains the mitochondria, apparently remains outside of the egg along with the tail. Mitochondria reproduce themselves inside the cells, so all are descended from the ones present in the egg at the time it was fertilized.

The generation of ATP within mitochondria, like all cell processes, must be regulated according to the demands of the organism at a particular time. We next consider some of the mechanisms that regulate cellular processes.

Regulation of Synthesis and Degradation Within the Cell

Most chemical reactions within the cell are reversible, and this permits the cell to store energy in the form of certain compounds that it synthesizes and then later degrades in order to retrieve the energy. *Glycogen,* which is made up of many glucose molecules strung together, is an energy storage compound. Very active cells that require large amounts of energy, such as liver and muscle cells, almost always have a supply of glycogen on hand. Inside the liver cell, glycogen is broken down into glucose molecules when the cell needs energy by the action of the enzyme *phosphorylase.* Glucose molecules are converted into glycogen for storage by the action of the enzyme *glycogen synthetase,* which has an active form and an inactive form:

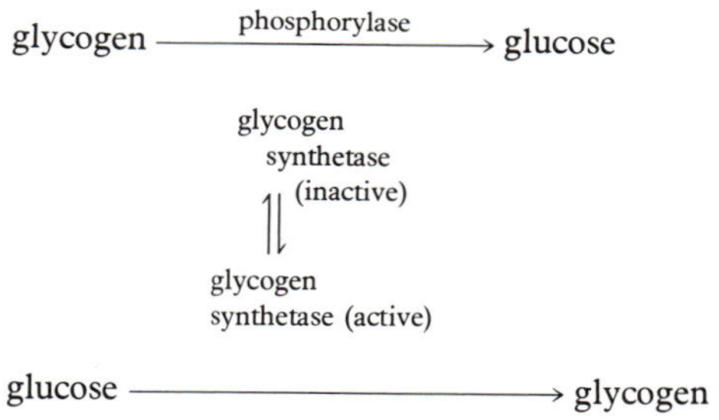

The maintenance of normal concentrations of glycogen inside liver cells depends upon the actions of both phosphorylase and glycogen synthetase. If the phosphorylase enzyme is faulty (because of a defective code in the DNA) the result is *glycogen storage disease,* a serious condition in which glycogen builds up to toxic concentrations in the cell because synthesis is not balanced by breakdown. In the normal cell glycogen cannot reach toxic concentration. High concentrations of glycogen stimulate phosphorylase activity—thereby encouraging the breakdown of glycogen—and convert active synthetase to inactive synthetase—thereby inhibiting the buildup of glycogen. Thus glycogen partly regulates its own concentration inside the liver and muscle cells.

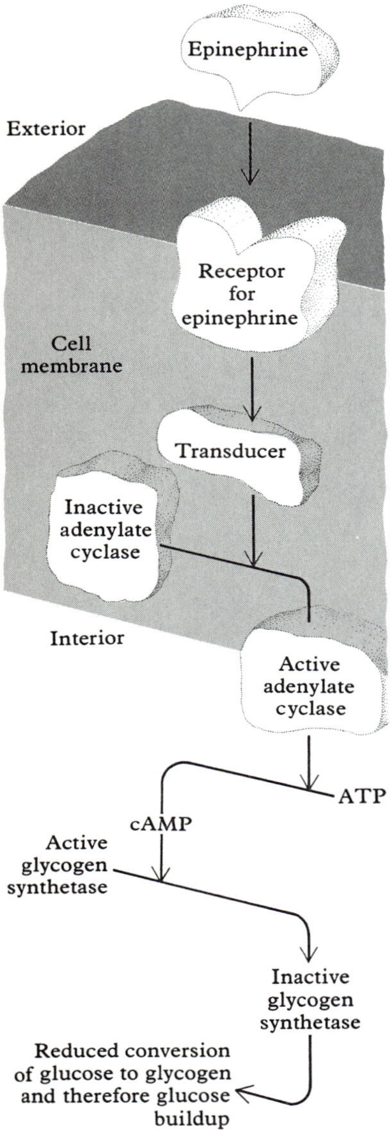

5–21
Epinephrine prepares the body for fleeing or fighting in part by initiating glucose buildup inside cells. In a crucial step, cAMP converts glycogen synthetase to an inactive form, thereby reducing the conversion of glucose to glycogen.

As we might expect, the presence of large amounts of glucose in the cell stimulates the synthesis of glycogen. It accomplishes this by converting inactive synthetase to active synthetase. Thus glucose and glycogen both have an influence on the amount of glycogen present in the cell, both because of their effects on enzymes.

From time to time the activities of an individual cell must be adjusted to serve the needs of the whole organism. Most such adjustments are mediated by hormones (which will be discussed more fully in Chapter 6). For example, when an animal is frightened and is preparing to flee or fight, large quantities of the hormone *epinephrine* (also known as adrenaline) are released from the adrenal glands and distributed almost instantaneously to the cells by means of the circulatory system. Under such stressful circumstances, an animal must be prepared for maximal exertion, and its muscle cells need readily available glucose to form ATP. Epinephrine, which does not enter cells, attaches to receptors on the surface of muscle cells, which causes the formation of the enzyme adenylate cyclase. This enzyme in turn causes the synthesis of cyclic AMP (cAMP), which converts active glycogen synthetase into inactive glycogen synthetase (Figure 5-21). The result is that glucose builds up, because less glucose is converted to glycogen. The overall effect of adrenaline, and indeed of all hormones, is to influence in this way the regulatory mechanisms of individual cells in order to serve the needs of the whole organism.

Environmental Agents and Cell Regulation

Cell regulatory mechanisms are rather delicate, and they may be upset by external agents such as alcohol, nicotine, and certain insecticides (such as DDT). Each of these chemical agents acts on liver cells to stimulate the synthesis of enzymes that degrade the agent, but the enzymes often degrade other substances as well, including some of the body's hormones. Thus the balance between synthesis and degradation for many substances may be altered. We will now discuss the effects of two external agents, alcohol and DDT, on the synthesis and degradation of hormones.

It is well known that many male alcoholics have certain features of hormonal imbalance, including reduced testicle size, impaired sperm production, decreased sexual drive, and abnormalities of metabolism attributable to the sex hormone testosterone. It has recently been shown that a major effect of alcohol is that it reduces the total concentration of testosterone in the body by causing an increase in the concentration of an enzyme that degrades testosterone. This may account for some of the physical and behavioral changes observed in male alcoholics.

People are not the only animals whose hormones have been affected by environmental substances. For example, the decline of certain species of birds (such as the peregrine falcon and the brown pelican) has been attributed to the use of DDT and other insecticides. These birds became endangered because they failed to reproduce. That failure may have been caused at least in part by thin eggshells, which are the result of increased degradation of estrogens, one of the major "female" sex hormones. Adequate concentrations of estrogens are required for the manufacture of strong calcium-containing eggshells because estrogen controls the deposition of calcium into the bones of the bird, and the bones are the storehouse from which calcium is withdrawn when eggshells are being formed. Ring doves fed DDT in tests suffered an increased breakdown of estrogens in the liver and reduced concentrations of the most active estrogens in their blood. These hormone changes may account for the thin eggshells and delay in egg laying that have been observed in DDT-fed birds.

As we have seen, the structures and functions of individual cells are of great importance to the human organism. But most cells do not function in isolation—most are organized into tissues in which all of the cells work together as a single unit. We now turn our attention to how cells are organized into tissues.

Tissues

Not long after a fertilized egg has begun to divide, the newly generated cells become arranged in three distinguishable cell layers: an outer layer of *ectoderm,* a middle layer of *mesoderm,* and an inner layer of *endoderm.* As cell division continues, cells in each layer diversify further and give rise to groups of cells that become specialized in both structure and function. A region dominated by a group of cells that perform a specialized function is called a *tissue.* For example, nervous tissue is dominated by nerve cells (neurons), derived from ectoderm, which conduct impulses. *Connective tissue,* such as bone, cartilage, tendon, and the loose connective tissue that is found beneath the skin, is dominated by fibroblasts, derived from mesoderm, which synthesize fibrous material.

In addition to cells, tissues contain varying amounts of *extracellular material*—material located outside of cells. This extracellular component composes over 50 percent by weight of connective tissues. In fact, the connective tissues, most of which either support or bind together the other tissues of the body, are almost entirely dependent on extracellular material for proper functioning. (The cells that are responsible for the manufacture of connective tissue are embedded within the tissue itself.)

In general, tissues can be divided into five types: blood, muscle, nervous, epithelial, and connective. Each tissue is dependent on others and is usually found interwoven with others to some extent. For example, muscle tissue is always in contact with blood vessels, which are necessary for its proper functioning (Figure 5-22). Similarly, the nervous tissue of the spinal cord contains numerous supporting (glial) cells that do not conduct impulses. We will discuss blood, muscle, and nervous tissue in some detail in following chapters. For now, we turn our attention to a consideration of epithelial and connective tissues.

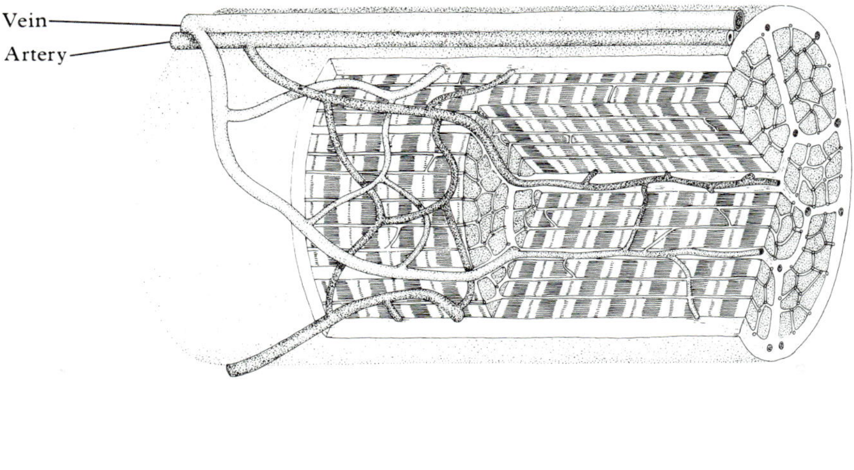

5–22
Muscle fiber is richly supplied with capillaries. Lying atop this dissected muscle fiber are two blood vessels; the smaller, an artery, and the larger, a vein. Most of the capillaries run parallel to the fibrils that make up the fiber. (From "The Microcirculation of the Blood," by Benjamin W. Zweifach. Copyright © 1958 by Scientific American, Inc. All rights reserved.)

Connective Tissues

All connective tissues contain the same basic ingredients: cells, fibers, and ground substance in which the fibers and cells are embedded.

There are two main kinds of connective tissue fibers, *collagen fibers* and *elastic fibers* (Figure 5-23), each composed of a characteristic protein—collagen or elastin. Collagen fibers are usually found in bundles that may equal or exceed the diameter of a red blood cell—one of the smallest cells in the body, as you will recall. Each fiber itself is made up of many microfibrils. Collagen imparts strength and flexibility to connective tissues; the strength of tendons is achieved almost entirely by collagen fibers arranged in parallel bundles.

Elastic fibers are structurally different from collagen, because they are not found in bundles and they do not have an orderly substructure. They give an elastic quality to distensible tissues such as the arteries and the loose connective tissue under the skin.

Connective tissue contains three kinds of cells: fibroblasts, which manufacture fibers of collagen and elastin (Figure 5-24); mast cells, which manufacture large carbohydrate molecules found in the ground substance; and macrophages, phagocytic cells that scavenge for debris. (Both mast cells and macrophages also participate in immune responses—see Chapter 13.)

The carbohydrates found in the ground substance of connective tissue are known as *mucopolysaccharides,* a term that was coined because they were first

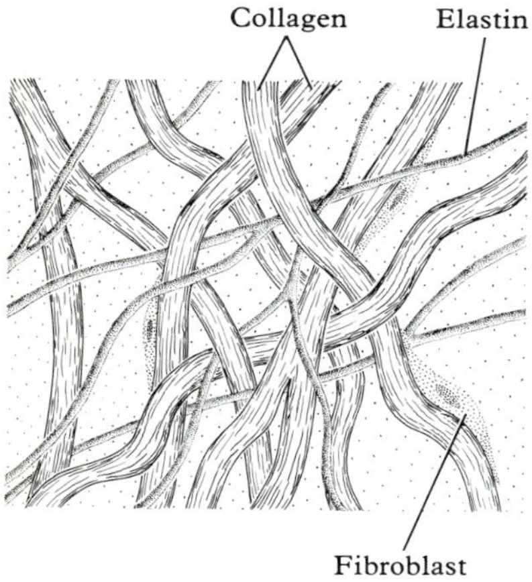

5–23
Collagen and elastic fibers of connective tissue. (After "Giant Molecules in Cells and Tissues," by Francis O. Schmitt. Copyright © 1957 by Scientific American, Inc. All rights reserved.)

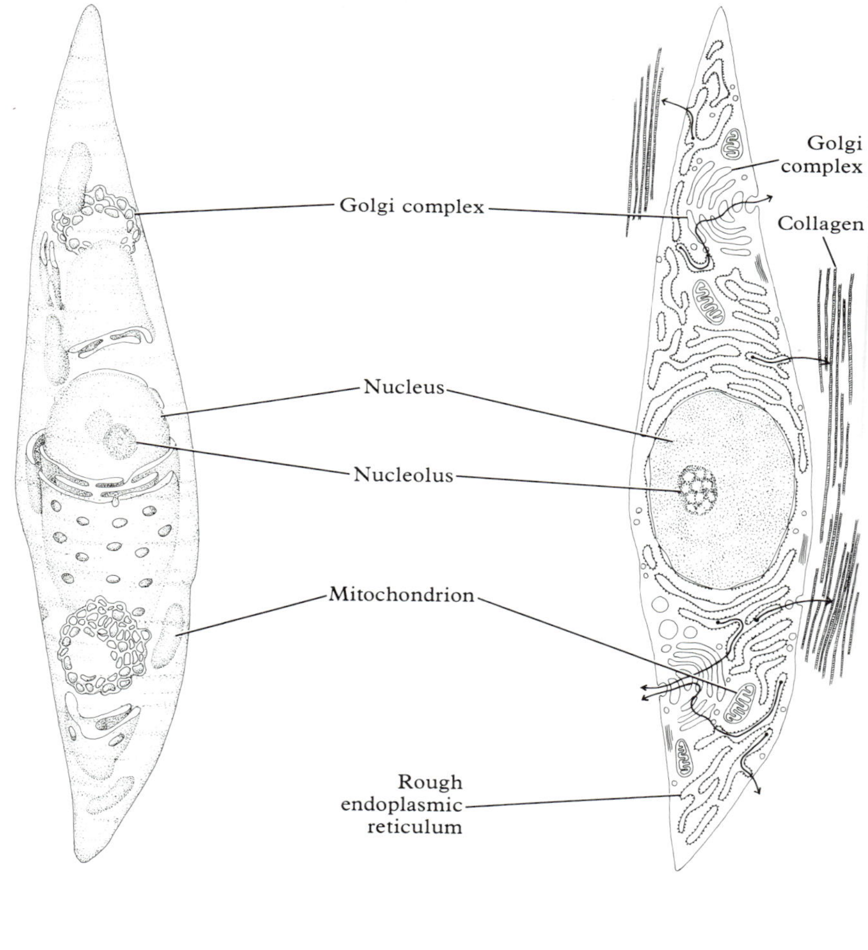

5–24
A fibroblast, found in connective tissue, manufactures collagen and elastin molecules. The pathways of secretion of collagen molecules are shown by arrows. (After "Wound Healing," by Russell Ross. Copyright © 1969 by Scientific American, Inc. All rights reserved.)

described in mucus secretions (hence *muco*) and are composed in part of sugar molecules (*saccharide*). Mucopolysaccharides bind water, and when in contact with collagen and elastin they give the ground substance a semi-firm, gel-like quality in which water does not flow freely, but in which diffusion can nonetheless occur as usual. (The remarkable water-holding properties of mucopolysaccharides are responsible for the characteristic appearance of a cock's comb—in the adult male chicken the comb is 90 percent water and yet is firm and upright because 7 percent of its contents consist of mucopolysaccharides.) Mucopolysaccharides are also present in the joint fluid (synovial fluid), where they serve as excellent lubricants, allowing the linings of joint capsules to slide over one another with a minimum of frictional wear and tear.

Epithelial Tissues

The body's epithelial tissues are found wherever a physical separation of some kind must be maintained. For example, epithelium protects the inner and outer surfaces of the body by forming the outer layer of skin and the inner lining of the intestine. Epithelial tissue also forms the lining of the body's glands, ducts, tubes, and storage sacs, and it serves to isolate the fluids contained within these structures from other body tissues. Because epithelial tissues are responsible for keeping things separate, epithelial cells are usually joined to each other by tight junctions, and very little extracellular material is present. Epithelial tissues are also different from the other tissue types in that most of them are separated from surrounding tissue by a *basement membrane* and they contain no blood vessels.

But the lack of blood vessels does not compromise the functioning of epithelial tissues because their cells are in very thin layers and can receive nutrients directly by diffusion from nearby capillary-rich tissues. In fact, many epithelial tissues known as simple epithelia, are only one cell thick (Figure 5-25). Simple epithelia line the blood vessels; lymphatic vessels; the small intestine; the alveoli of the lungs; and the pleural, pericardial, and peritoneal membranes—all of which surround organs that slide against neighboring tissues.

If epithelial tissue is more than one cell in thickness, it is known as a *stratified epithelium.* Stratified epithelium covers areas that are subjected to considerable surface friction—for example, the skin and the linings of the mouth, esophagus, and vagina. The cells of stratified epithelia are flattened in the plane of the basement membrane so that a single layer of cells may be only 2 micrometers thick. Because the cells are flattened the epithelial tissue stays relatively thin, so outermost layers can receive nutrients by diffusion from surrounding tissue. (A specialized type of stratified epithelium, *transitional epithelium,* lines some of the structures of the urinary system, namely the ureters and the bladder. In transitional epithelium the cells of the layer next to the urine are larger than those underneath. The functional significance of these larger cells is not understood.)

Epithelium lines all of the body's glands and ultimately gives rise to the specialized secretory cells that most glands contain. Those secretory glands that retain a direct connection with the epithelium from which they developed and that discharge their secretions onto that epithelium are known as *exocrine glands* (*exo* means out). The glands of the skin and the intestinal glands that liberate mucus and digestive enzymes are exocrine glands. *Endocrine glands* (*endo* means in) are those that discharge their products directly into the bloodstream rather than onto the epithelium from which they developed, such as the thyroid and

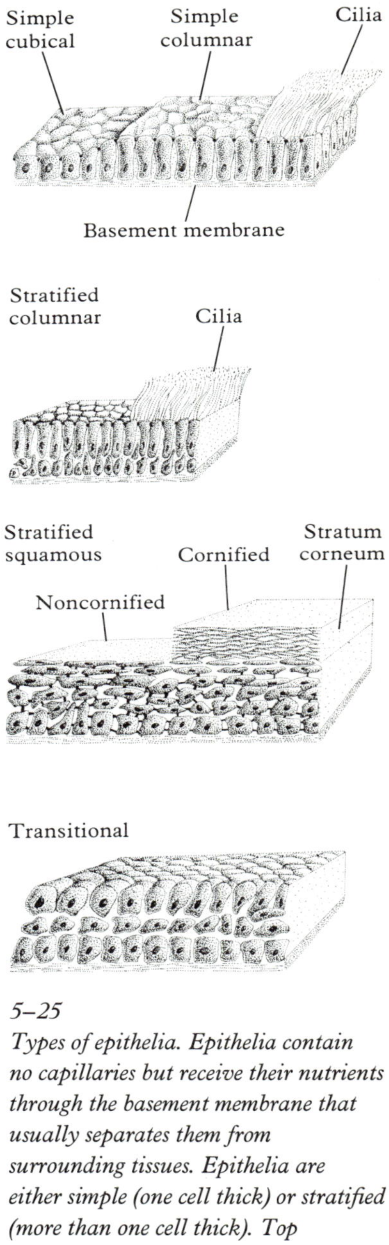

5–25
Types of epithelia. Epithelia contain no capillaries but receive their nutrients through the basement membrane that usually separates them from surrounding tissues. Epithelia are either simple (one cell thick) or stratified (more than one cell thick). Top drawing, simple cubical and columnar epithelia. Other drawings, stratified epithelia.

adrenal glands. We shall have much more to say about the endocrine glands in the following chapter.

Regardless of where they discharge their products, all glands contain not only epithelial tissue but the four other major tissue types as well—blood for nourishment and for the elimination of wastes, nerves for controlling secretion, smooth muscle in vessel walls for controlling the delivery of blood, and connective tissue to hold everything together. The various tissues in a gland work together to accomplish a common task. As we shall discuss in the following chapter, all glands, and in fact all organs, are made up of tissues that collaborate with each other in the interest of regulating the internal environment.

Summary

The body's cells vary considerably in size and shape, depending upon their functions. Cells that remain fixed in place adhere to each other by means of connections between their membranes. There are three kinds of connections: tight junctions, which hold cells together very closely; desmosomes, which are somewhat looser, and gap junctions, which provide channels for the passage of materials between cells. Cell movement and shape change are aided by structures within the cell known as microfilaments and microtubules. The contraction of microfilaments within the cytoplasm causes cell movement. Microtubules are responsible for the beating of cilia, and for the whiplike motion of the tails of sperm cells. They play a major role in cell division because they make up the spindle fibers that draw the duplicated chromosomes apart.

Small molecules enter and leave the cell through its outer membrane where their passage is aided by protein molecules embedded in the membrane's lipid framework. Larger molecules, debris, and structures such as bacterial cells enter the cell by being engulfed by the cell's outer membrane (endocytosis), and leave the cell when the membrane surrounding the engulfed material fuses with the cell's outer membrane (exocytosis).

Cell membranes are important in secretory processes. Protein synthesis, which takes place on cell structures known as ribosomes, occurs on ribosomes bound to membrane sacs if the protein is to be secreted. The newly synthesized proteins enter the sacs, which are known as the endoplasmic reticulum (ER). The proteins are then carried inside the sacs to the outer cell membrane, where they are discharged by exocytosis.

There is a constant turnover of the cells and molecules of the human body, and replacements must be constructed to exact specifications. Somatic cells divide by mitosis to produce two copies of themselves. Mitosis ensures that the genetic material, DNA, is distributed in an orderly fashion to each of the two daughter cells that are produced. The DNA, which is contained within the chromosomes, is duplicated before division occurs, and a complete set of chromosomes is distributed to each of the two resulting cells.

Molecules are constructed according to specifications located in the sequence of nucleic acid bases present in DNA. Through the processes of transcription and translation these specifications take the form of protein enzymes that direct the myriad chemical reactions that occur inside all of us.

The breakdown of glucose supplies most of the energy to run the machinery of the cell. As glucose is broken down, its energy is stored in ATP molecules.

The energy of ATP is released when the molecule is converted to ADP and phosphate. ATP is generated in structures known as mitochondria, which are the powerhouses of the cell. Mitochondria may have originated as bacteria that long ago took up residence inside eukaryotic cells.

Cell processes must be regulated so that synthesis and degradation are kept in balance. To a considerable extent this regulation is mediated by effects on various enzymes that carry out reversible chemical reactions. Hormones influence cells in ways that result in the proper functioning of the entire organism. External agents such as alcohol and DDT can upset the balance of synthesis and degradation, and in particular are known to accelerate the breakdown of hormones associated with reproduction.

Tissues can be defined as regions of the body dominated by groups of cells specializing in a particular function. Tissues contain extracellular material in addition to cells, and this material is especially important for the functioning of connective tissue. Extracellular material consists of fibers and ground substance, both of which contribute to the structural integrity of the tissue.

There are five categories of body tissues: blood, muscle, nervous, epithelial, and connective. Epithelial tissue, which lines the body's tubes, ducts, cavities, and surfaces, has relatively little extracellular material, contains no blood vessels, and is separated from surrounding tissue by a basement membrane. Epithelium is found wherever fluids must be kept compartmentalized, and epithelial cells adhere by tight junctions that prevent fluids from passing between the cells. Epithelial tissue also produces the secretions from endocrine and exocrine glands. Tissues, glands, and other organs work as a team to accomplish the common task of regulating the internal environment.

Suggested Readings

1. "How Living Cells Change Shape," by N. K. Wessells, *Scientific American,* Oct. 1971. Excellent discussion of cell movement and the roles of microtubules and microfilaments.
2. "The Living Cell," by J. Brachet, *Scientific American,* Sept. 1961. A good discussion of the cell and its component parts.
3. *Cell Membranes,* G. Weissmann and R. Claiborne, eds. HP Publishing, 1975. An excellent comprehensive book that contains many chapters about the biochemistry, cell biology, and pathology of cell membranes.
4. *Molecular Biology of the Gene,* J. D. Watson, 3d ed., by W. A. Benjamin, 1976. An outstanding description of the eukaryotic cell is presented in Chapter 16.

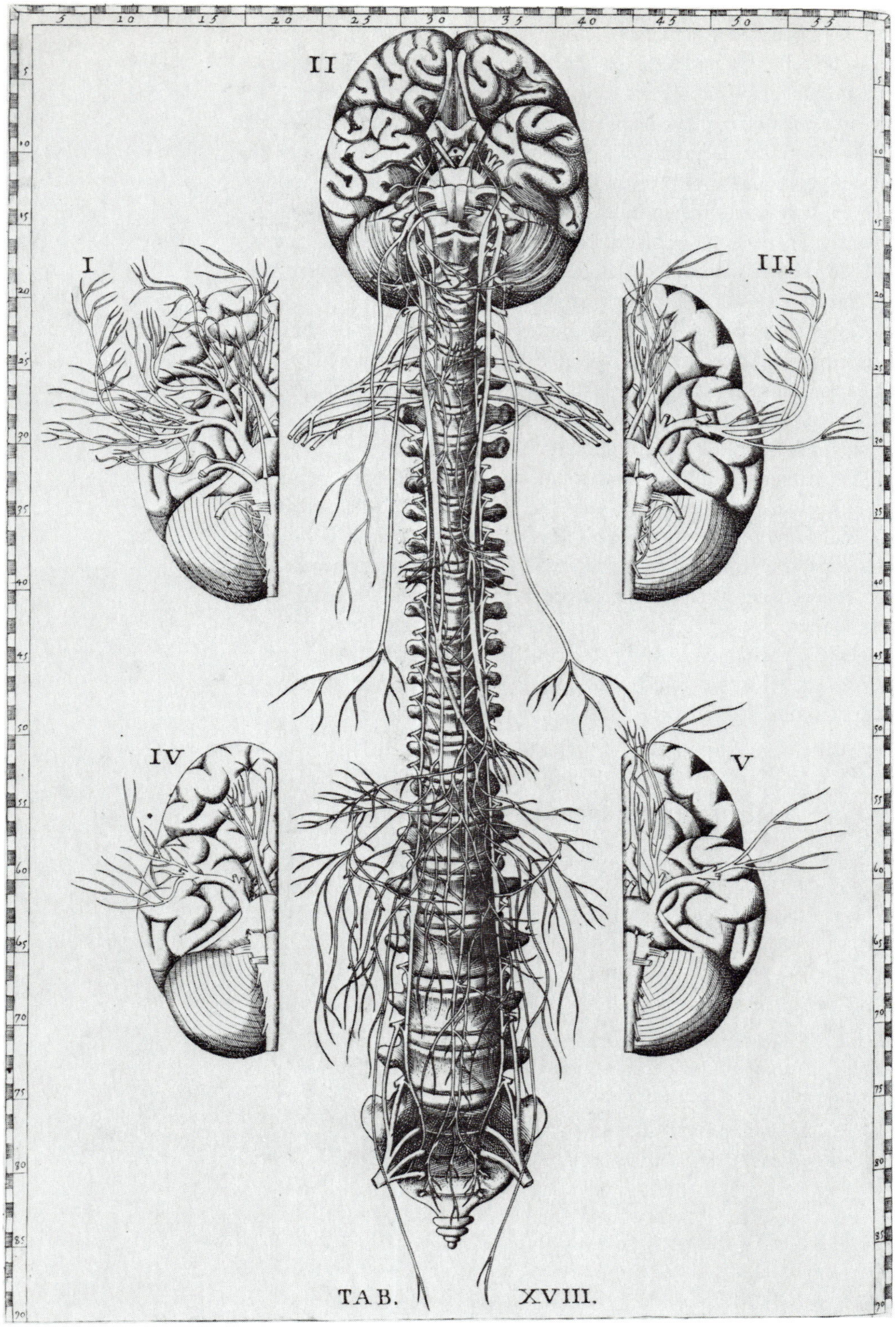

The brain and autonomic nervous system as depicted by Bartolommeo Eustachio (1550–74). (Copyright The British Library, by permission of the Trustees of the British Museum.)

CHAPTER 6

Coordinating the Parts

Although the events surrounding their origins are far from clear, there is no doubt that our planet's first multicellular creatures evolved from single-celled ancestors. These ancestral organisms probably originated in the ancient oceans, and during the course of their further evolution, their bodies became more and more complex. Groups of cells became structurally distinct and began to specialize in performing functions that were of benefit to the entire organism. The trend of specialization among the cells of multicellular creatures in order to benefit the entire organism has continued to the present day, as evidenced by the great variety of complicated tissues and organs of which our own bodies are composed. With this specialization in structure and function has come the necessity for regulating and coordinating the workings of the various parts.

Members of the animal kingdom have evolved two principal systems for regulating and coordinating the functions of their body parts—the nervous system and the endocrine (hormone producing) system. (Plants and fungi, the other two kingdoms of multicellular creatures whose bodies are organized into tissues and organs, regulate their parts by means of hormones, but neither of these groups has evolved nerves.) Nerves can be thought of as information-carrying cables that link and coordinate all parts of the body. Hormones are chemical messengers that are released into the bloodstream, which transports them to their target organs.

It has been known for centuries that the highly branched network of glistening white nerves that extends throughout our bodies brings about its regulatory effects because our nerves ultimately transmit information to and from the most complicated organs that natural selection has produced so far, our brains. But although the descriptions of the body's network of nerves and of the gross features of the human brain were completed hundreds of years ago, the existence of hormones was not suspected and demonstrated until the early 1900s. Since that time a fascinating series of discoveries has brought us to the important realization that nerves, brain, and hormones are functionally interrelated to a degree that was undreamed of only a few decades ago.

In this chapter our main concern is with how the human nervous and endocrine systems interact to coordinate the proper functioning of the rest of the body's tissues, organs, and systems. We begin by turning our attention to the structure and function of the human nervous system, whose basic cellular units are *neurons.* (The special features of the human brain that enable our species to behave in extraordinary and uniquely human ways are discussed in Chapter 3. In this chapter we discuss regulatory functions, not the adaptive behavior of the human species, whose biologic basis in the brain we have discussed before.)

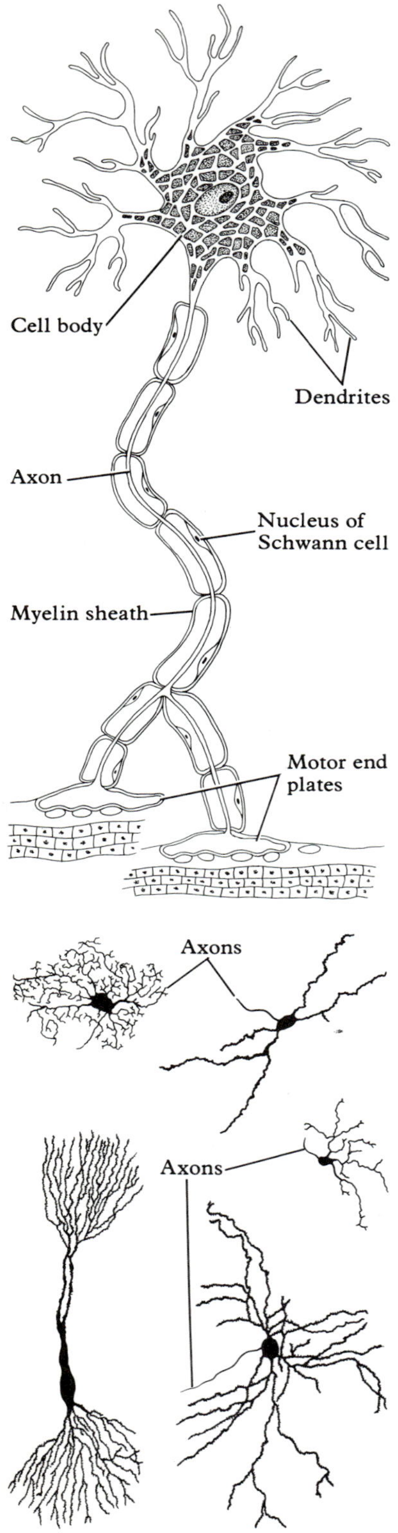

6–1
Top, a typical motor neuron, which is specialized for transmitting nerve impulses to muscle cells. Bottom, some neurons from the brain and spinal cord.

Neurons, Nerve Conduction, and Synapses

Neurons are cells that are specialized for transmitting information in the form of electrochemical impulses. Figure 6-1 compares a typical *motor* neuron, a nerve cell whose stimulation usually results in muscle contraction, to neurons from elsewhere in the human nervous system.

In spite of their apparent differences, all neurons are constructed according to the same general plan. The irregular or star-shaped cell body gives rise to several cell extensions. One of them, the *axon,* has features that distinguish it from the other cell extensions, the *dendrites.*

The axon carries electrochemical impulses away from the cell body when the cell body is stimulated to fire by other nerve or sensory cells with which it is in contact. Compared with the size of the cell body, axons can be enormously long. It has been calculated that if the cell body of a motor neuron that is located in the spinal cord and sends its axon to the muscles of the foot were enlarged to the size of a tennis ball the axon would be nearly a mile long, though only one-half inch in diameter.

Because of their great length, axons in many parts of the nervous system are surrounded by *Schwann cells,* supporting cells that serve two functions. (The brain and spinal cord contain supporting cells of several kinds.) First, Schwann cells help to nourish the elongated axons, parts of which function at great distances from the cell body. Second, Schwann cells invest many axons with a sheath of white insulating material known as *myelin.* The myelin sheath is formed when a Schwann cell wraps itself around an axon several times, thus insulating the axon with layers of its cell membrane and cytoplasm (Figure 6-2). The myelin wrapping serves to increase the speed at which electrochemical impulses are transmitted down the axon, and, as shown in Figure 6-1, the sheath extends almost the entire length of the axon, ending just before the axon of a motor neuron terminates as a motor end plate embedded in muscle fibers. (Some axons in the human nervous system are "unmyelinated" in that they do not have a highly coiled insulating sheath, but even these axons are in contact with Schwann cells. Their axons are simply buried in shallow grooves on the surfaces of the supporting cells.)

Until quite recently it was thought that the shorter, highly branched dendrites that arise from the cell body functioned differently from axons in that dendrites always conducted information toward the cell body, not away from it, as axons do. It turns out that some dendrites may transmit impulses in either direction. This is important because the many dendrites of a given neuron are usually in contact with various parts of many other nerve cells, including a multitude of other dendrites (Figure 6-3). Dendrites may also make direct contact with axons, so, as you can see, nerve cells can transmit impulses to one another by way of a bewildering number of pathways.

Nerve cells communicate with one another by means of electrochemical impulses—that is, by means of electric currents generated because of chemical changes that affect cell membranes. The interior of a neuron, like that of most other cells, contains an excess of negative electric charge as compared with the outside. This excess is due to the fact that cell membranes are not equally permeable to all charged particles, particularly sodium and potassium ions. Normally, the negatively charged interior of a neuron is relatively rich in potassium ions and poor in sodium ions, whereas the fluid surrounding the cell has the reverse composition. What makes the cell membrane of a neuron (and therefore of the axon and dendrites) so distinct is that its permeability to sodium and potassium ions is regulated by the difference in voltage on opposite sides of the membrane. When a nerve impulse arises, the voltage difference is lowered to

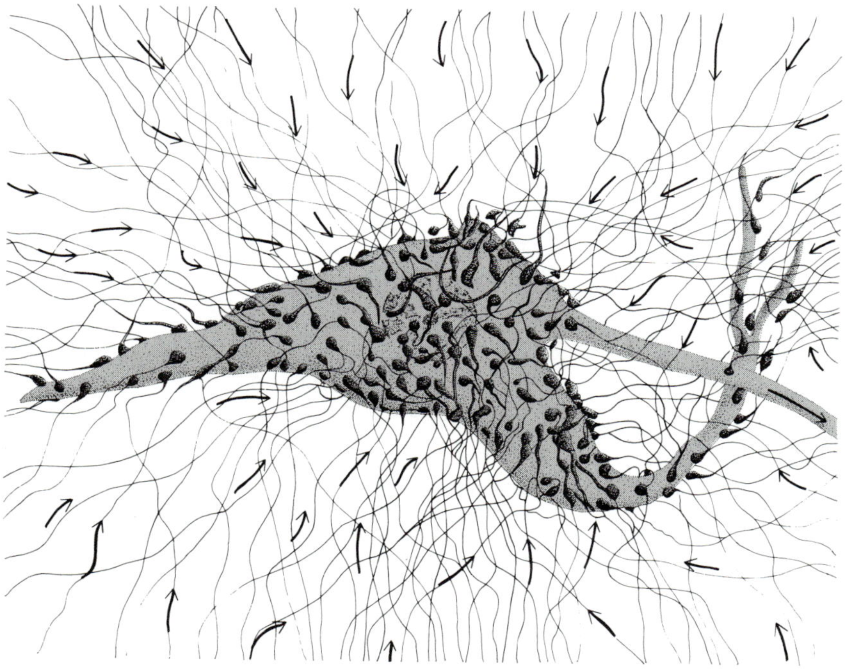

6–3
The cell body of a motor neuron from the spinal cord showing the many dendrites that may make contact with it. Each dendrite swells to form a synaptic knob on the cell surface. (From "How Cells Communicate," by Bernhard Katz. Copyright © 1961 by Scientific American, Inc. All rights reserved.)

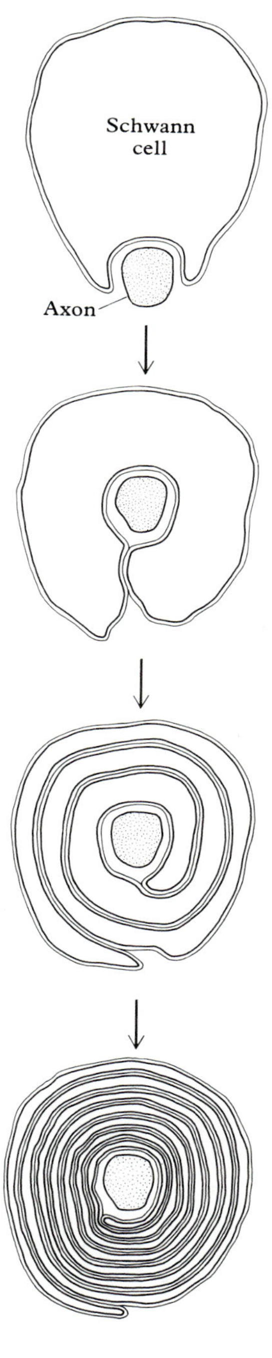

6–2
Schwann cells invest axons with a myelin sheath. The sheath is created when a Schwann cell wraps itself around the nerve axon. After the enfolding is complete, the cytoplasm of the Schwann cell is expelled and the cell's folded membranes fuse into a tough, compact wrapping. (From "How Cells Communicate," by Bernhard Katz. Copyright © 1961 by Scientific American, Inc. All rights reserved.)

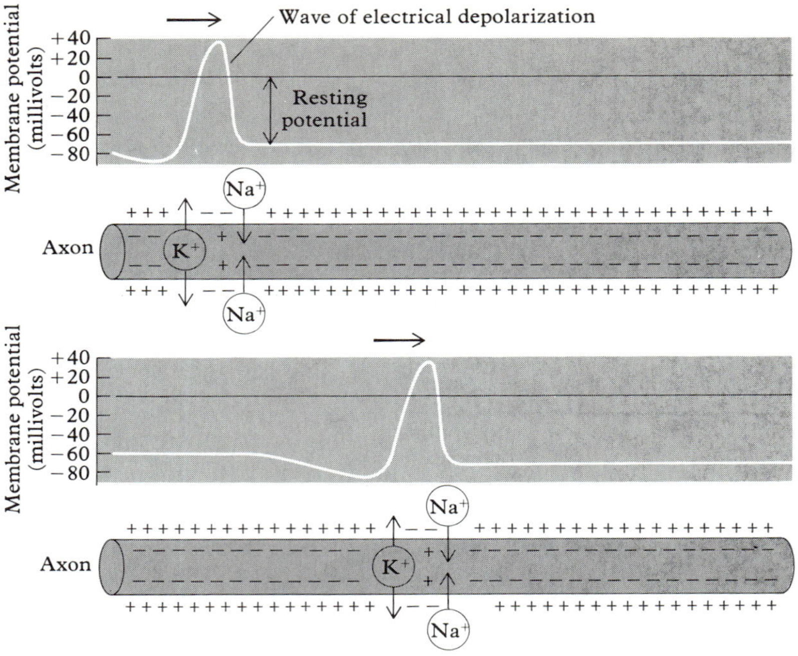

6–4
The propagation of a nerve impulse coincides with changes in the permeability of the cell membrane. As the impulse travels down the axon the area behind it quickly returns to the "resting state" in which the interior of the axon bears a net negative charge. (From "How Cells Communicate," by Bernhard Katz. Copyright © 1961 by Scientific American, Inc. All rights reserved.)

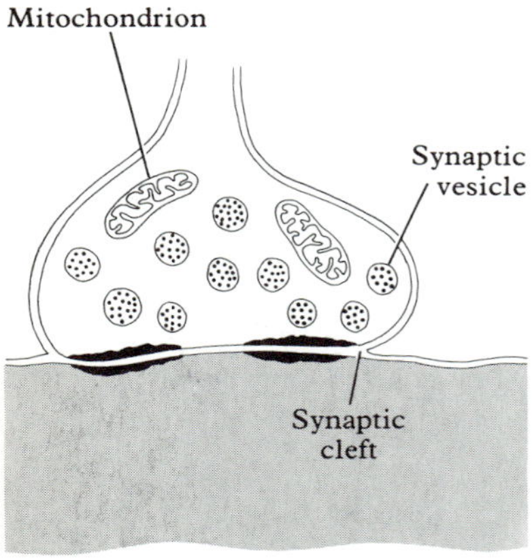

6–5
The structure of a synaptic knob. The synaptic vesicles contain neurotransmitter molecules. Synaptic knobs are designed to deliver short bursts of neurotransmitters into the synaptic cleft, where they can act on the surface of the postsynaptic nerve–cell membrane.

a critical level. The result is that sodium ions from outside the cell enter in such numbers that they momentarily change the relative charge on the inside from negative to positive. This process sets off a wave of depolarization, which then races down the axon and spreads over the surface of the rest of the neuron. When the impulse has passed down the membrane, potassium ions flow *out* of the cell, and the sodium ions that had rushed in shortly before diffuse to the outside once again, thus restoring the original negative charge to the interior of the neuron (Figure 6-4). The inflow and outflow of sodium and potassium ions takes place so quickly and affects so few particles that the overall internal composition of the axon is scarcely affected by its electrical activity.

What causes a neuron to fire in the first place? As we shall discuss later, certain specialized neurons are triggered by light energy, pressure, or other physical means, and some neurons, such as those whose firing results in the rhythmic beating of the heart, discharge spontaneously at a constant rate. But what determines whether most neurons fire is the total of the chemical messages to fire or not fire that are exchanged continually by all nerve cells in contact with one another.

Individual neurons are connected to each other at *synapses.* The synaptic ends of axons and dendrites are enlarged to form *terminal buttons,* or *synaptic knobs,* 5500 of which may encrust the cell body of a single motor neuron in the spinal cord (Figure 6-3). A typical synaptic knob is separated by a tiny synaptic cleft from the part of the other cell with which it appears to be in contact (Figure 6-5). Inside the knob are mitochondria and various membrane-bound packets or vesicles. These vesicles contain one of several kinds of *chemical neurotransmitters,* that is, chemicals whose presence in sufficient quantity in the area of the cell membrane of a nerve (or muscle) cell may cause the cell to fire. When a wave of electrical depolarization reaches the synaptic knob, many of the vesicles fuse with the cell membrane of the synaptic knob and release their contents into the synaptic cleft (Figure 6-6). The liberated neurotransmitter molecules then diffuse to the surface of the membrane of the other cell (the postsynaptic cell), where they usually bring about the firing of the cell. Shortly after they are released, neurotransmitters are inactivated by enzymes that are on or near the postsynaptic cell membrane. The overall function of the synaptic knob is thus to deliver small bursts of chemical neurotransmitters into the synaptic cleft, where the short-lived presence of these molecules usually results in the generation of an electrochemical impulse in the postsynaptic cell.

It has recently been discovered that some neurotransmitters cause the signal "do not fire" to be relayed, probably because such chemicals somehow make postsynaptic cell membranes relatively less permeable to sodium and potassium ions. (Also, some synapses within the brain have neither synaptic knobs nor membrane-bound vesicles, and are thought to communicate with one another directly by means of small electric currents.) Figure 6-7 shows the chemical structures of some of the neurotransmitters known to either stimulate or inhibit the firing of the nerve cells of various mammals. As we shall soon discuss, in some parts of the human nervous system the action of neurotransmitters is well understood, but in other regions, particularly the brain, our understanding of the neurotransmitters is meager.

At any rate, the sum of all the stimulating and inhibitory signals transmitted to it from moment to moment determines whether a particular neuron fires. It was formerly thought that all neurons communicate solely by means of the "all-or-none" waves of depolarization that we have just dicussed. Such currents are all or none because they are generated only when a critical threshold voltage exists on opposite sides of the cell membrane. But we now know that electrical currents of various strengths play a part in communication between neurons,

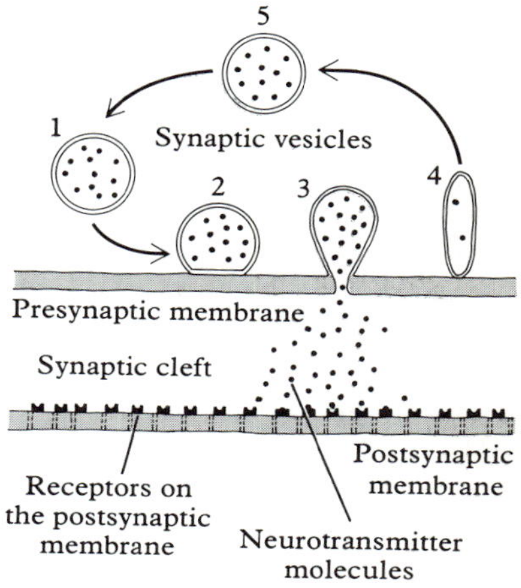

6–6
Synaptic vesicles containing chemical neurotransmitters are distributed throughout the synaptic knob. As shown here, the vesicles usually move up to the synaptic cleft, discharge their contents, and then return to the interior to be filled up again. (From "The Synapse," by Sir John Eccles. Copyright © 1964 by Scientific American, Inc. All rights reserved.)

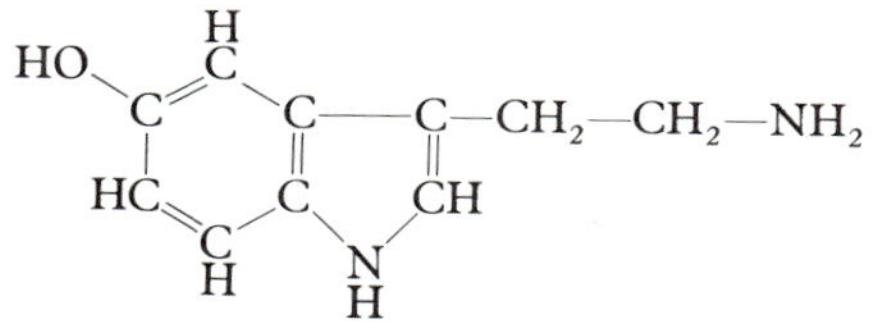

Serotonin

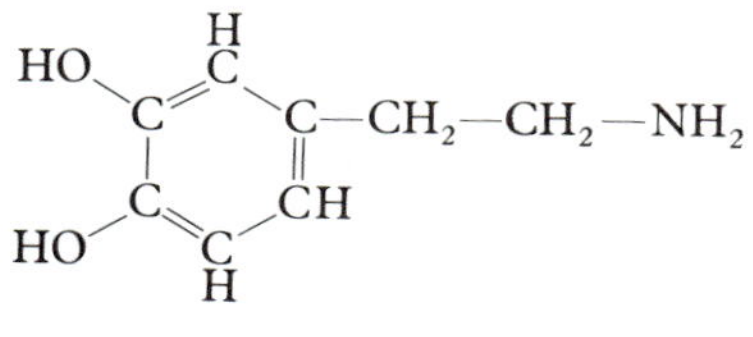

Dopamine

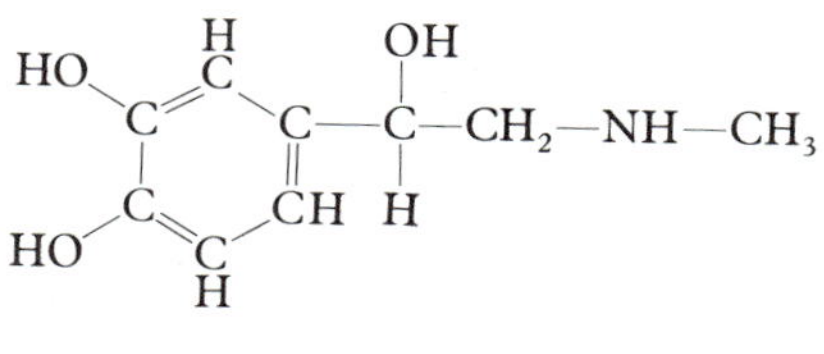

Epinephrine

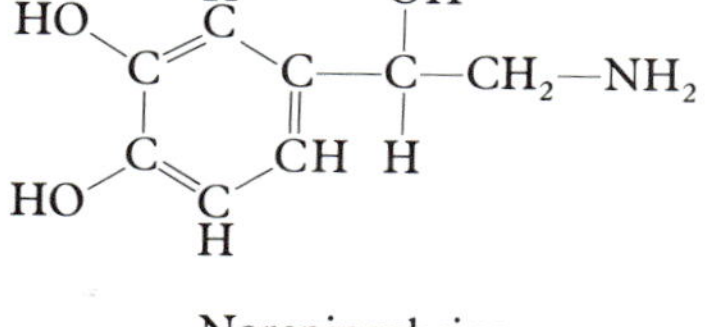

Norepinephrine

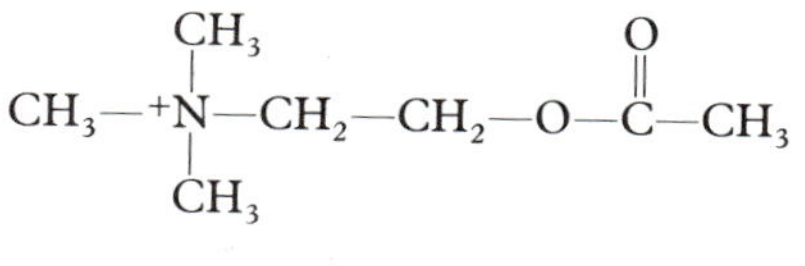

Acetylcholine

6–7
At least four neurotransmitters are found in the brains of mammals: serotonin, dopamine, norepinephrine, and epinephrine. Acetylcholine probably also functions as a neurotransmitter in the brain, and it surely does so elsewhere in the nervous system.

particularly in the complicated circuits connecting the countless dendrites of the human brain. Their generation does not depend on attainment of a critical electrical threshold.

It has recently been suggested that the cause of the mind-boggling effects of certain psychedelic drugs, such as LSD, may be that these chemicals are similar in structure to neurotransmitters employed by the brain or that they somehow interfere with normal synaptic transmission. Thus, LSD and other psychedelics cause sensory dissociation (under the influences of these substances one may "hear" green, "taste" purple, and have other unusual experiences) probably because the drugs temporarily alter the ways in which the billions of nerve cells within our brains communicate with one another.

We now turn our attention to a brief consideration of some highly specialized neurons that are capable of responding directly to physical stimuli such as light, sound waves, pressure, and the presence of various chemicals. These neurons are associated with complicated sense organs and they relay to the brain critical information about the environment.

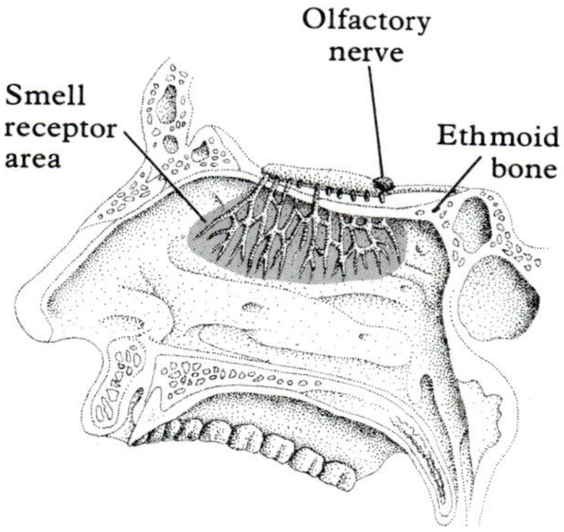

6–8
Smell receptors are located in a spongy region at the top of the nasal cavity, a cross section of which is illustrated here.

The Senses of Smell, Taste, Hearing, and Vision

Various kinds of highly specialized neurons known as receptor cells are found throughout the human body. In general, each kind of receptor cell is specialized to respond to a specific physical or chemical signal, or to a particular form of energy, such as sound waves or light. When receptor cells encounter the particular stimuli that triggers them, they respond by generating electrical impulses that are then routed to the brain for interpretation. At one time, we all learned that there are five senses—smell, taste, vision, hearing, and touch—but in fact human beings have at least *twenty* known senses, and there are undoubtedly more to be discovered. Some of these senses operate below the level of consciousness, which means that we are usually unaware of them. For example, specialized stretch-sensitive neurons located in all of our muscles continually relay electrical impulses to several regions of the brain, especially the *cerebellum,* where the information is processed and from which signals are sent back to the muscles to make fine adjustments in their tone, or to coordinate their movements. All of this goes on without our being conscious of it. Table 6-1 is an inventory of the principal human senses, the stimuli that trigger them, and the locations of the receptor cells that transmit their messages. Many of these sensory mechanisms will be discussed in this or a later chapter; others are listed for the sake of completeness and will not be discussed further. But in the brief discussion of the most familiar senses that follows, you should remember that humans have many more senses than they are aware of.

Smell and Taste

Smell and taste are sometimes called chemical senses because the stimulus that triggers their receptor cells is the presence of chemical substances in the nose and mouth. Smell is among the least understood of all senses. Very little is known of how olfactory impulses originate or of how we come to discriminate such a wide variety of odors, but the shapes of smelled molecules are thought to be important in the process. Smell receptors are located in a spongy area at the top of the nasal cavity, so sniffing is necessary to allow maximal contact between the receptor

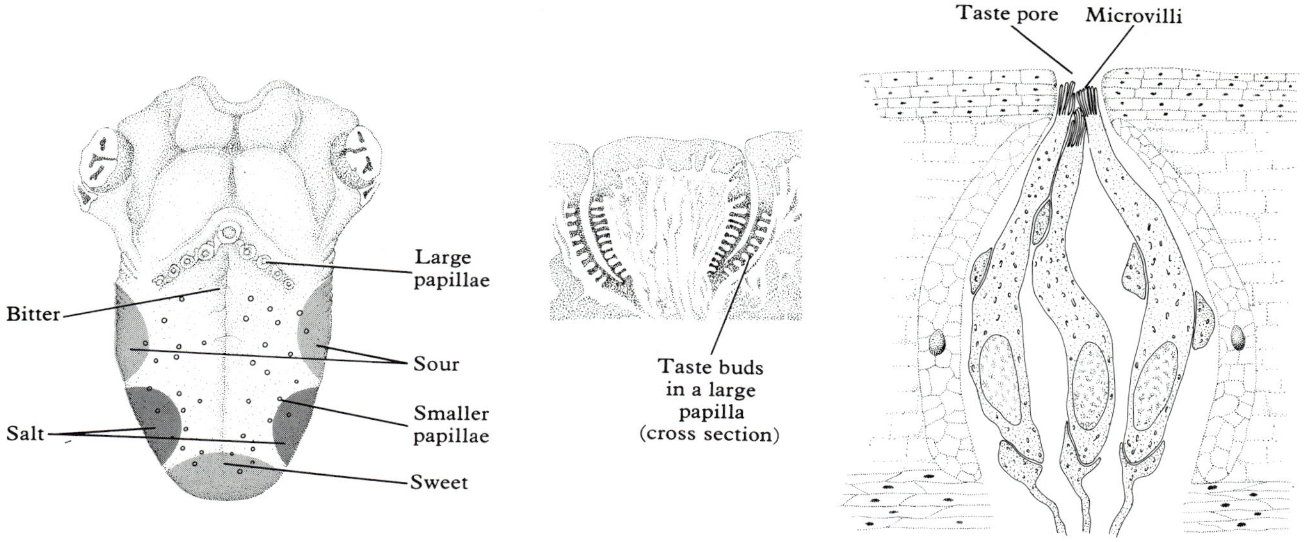

6–9
Taste receptors are located in the tongue, the top of which is shown at left. On the surface of the tongue are numerous round structures called papillae; the larger of these are indicated by the small circles in the drawing. Center, a cross section of one of the large papillae. The small white structures adjoining the fissure around the papilla are the taste buds. Right, a taste bud. Some areas of the tongue are more sensitive to certain tastes than others; these areas are indicated in the drawing on the left.

cells and the airborne chemicals they detect (Figure 6-8). Myelinated axons from the smell-receptor neurons pass through the ethmoid bone, combine to form the *olfactory nerve,* and then relay their impulses to many areas of the brain, including parts of the limbic system (see page 67). This may account for the fact that smells sometimes cause us to recall vividly events that may have occurred years ago.

The sense of taste is closely interwoven with that of smell. Taste is a complex sensation that depends on the temperature, texture, and especially the smell of food, as well as on the responses of taste receptors located on the tongue. Without the sense of smell, the tongue can discriminate only four primary taste sensations—sweet, sour, salty and bitter—and certain areas of the tongue react more strongly to one of these primary tastes than to others (Figure 6-9).Taste receptors are arranged in *taste buds,* from which myelinated axons are sent to the brain by means of two nerves, the *facial* and the *glossopharyngeal.* (Both of these nerves are so-called cranial nerves, of which there are a total of twelve pairs. Cranial nerves are special because they originate directly from the brain rather than from the spinal cord.) Within the brain taste impulses are relayed by way of a complicated pathway to one of the cortical association areas located in the temporal lobe. (Figure 6-10 provides a recap of the various parts of the human brain. The localization of the brain's higher mental functions is discussed in more detail in Chapter 3.)

Hearing

Hearing is the sense that results when receptor cells located in the inner ear generate electrical impulses in response to the vibrations of sound waves. Before they arrive at the inner ear, sound waves are first collected by the outer ear and then funneled through the external ear canal at the end of which they encounter the eardrum (Figure 6-11). This thin fibrous membrane vibrates in proportion to the intensity of the sound waves, and the vibration sets into motion a chain of three tiny bones, or *ossicles,* in the middle ear (Figure 6-12). The last ossicle in line (the stirrup, or *stapes*) has a footplate that fits into the oval window, through which sound waves are finally transmitted to the inner ear.

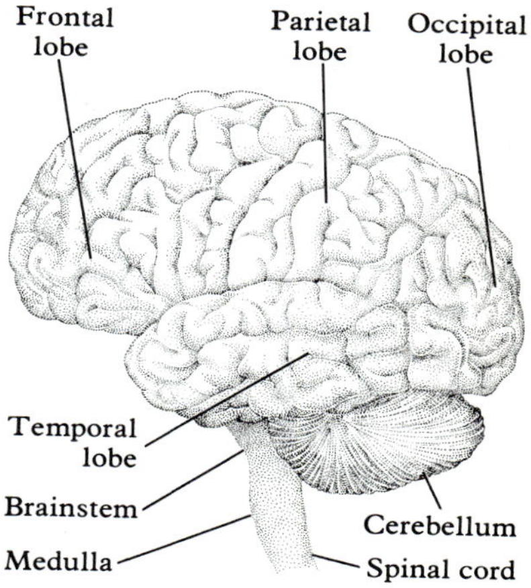

6–10
The lobes of the human cerebral hemispheres and of some of the brain's more readily visible parts. The higher functions of the brain's cortical areas are discussed in Chapter 3. Also see Figure 3–9.

TABLE 6-1
Human senses.

SENSES (SOME OF WHICH ARE NOT EXPERIENCED CONSCIOUSLY)	STIMULI TO WHICH THEY RESPOND	LOCATION OF RECEPTOR CELLS
Vision	Light energy	Retina (the eye)
Hearing	Sound waves	Cochlea (the inner ear)
Smell	Airborne chemicals	Olfactory mucus membrane
Taste	Chemicals in solution	Taste buds
Rotational acceleration	Gravity	Semicircular canals (the inner ear)
Linear acceleration	Gravity	Utricle (the inner ear)
Touch	Contact with the skin	Skin (Meissner's corpuscles and other structures)
Pressure	Pressure	Skin, connective tissues (Pacinian corpuscles)
Warmth	Heat energy	Skin (Ruffini's end organs and other structures)
Cold	Low temperatures	Skin (Krause's end bulbs and other structures)
Muscle stretch	Forces tending to elongate muscles	Skeletal muscles (muscle spindles)
Tendon stretch	Forces exerted on tendons by muscles	Tendons (Golgi tendon organs)
Joint position	Position of the body in space	Nerve endings around joints
Blood pressure in the arteries	Pressure of the circulating blood	Arch of the aorta, the carotid sinus
Blood pressure in the veins near the heart	Pressure of the circulating blood	Walls of the veins, walls of the atria
Lung volume	The degree to which the lung is expanded	Free (unmyelinated) nerve endings in lung tissue
Temperature of the blood in the head	Heat energy	Hypothalamus of the brain
Oxygen concentration	The concentration of oxygen dissolved in the blood	Arch of the aorta and the carotid sinus

TABLE 6-1 (*continued*)
Human senses.

SENSES (SOME OF WHICH ARE NOT EXPERIENCED CONSCIOUSLY)	STIMULI TO WHICH THEY RESPOND	LOCATION OF RECEPTOR CELLS
Acidity of the fluid surrounding the brain	The concentration of hydrogen ions	*Medulla oblongata* of the brainstem
Osmotic pressure of body fluids	The concentration of large molecules, especially proteins, in the blood	Hypothalamus of the brain
Blood sugar concentration	Difference in the concentration of glucose in arteries as compared to veins	Hypothalamus of the brain
Pain	A variety of stimuli	Free (unmyelinated) nerve endings throughout the body

Source: After W. F. Ganong, *Review of Medical Physiology,* 7th ed. Copyright © 1975 by Lange Medical Publications.

The inner ear is composed of two sense organs, one for detecting sound waves and the other for detecting acceleration or deceleration. That part of the apparatus concerned with hearing is a snail-shaped structure known as the *cochlea* (Figure 6-13). The cochlea is filled with a viscous fluid in which pressure waves are created as the footplate of the stapes rocks in and out of the oval window. These pressure waves result in the stimulation of the receptor neurons

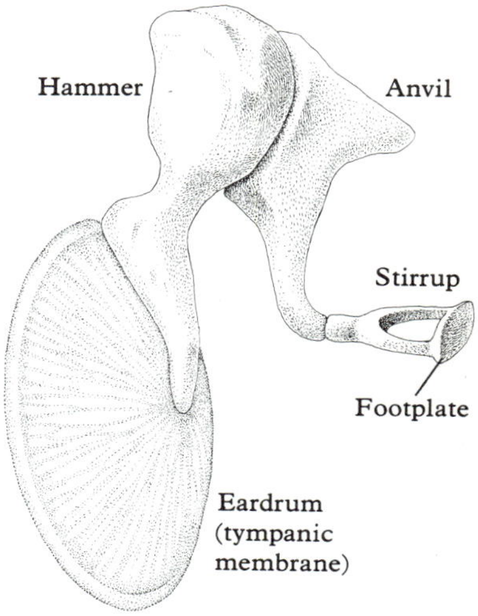

6–12
The three tiny bones, or ossicles, *of the middle ear. The footplate of the stirrup fits into the oval window of the cochlea. (From "The Ear," by Georg von Békèsy. Copyright © 1957 by Scientific American, Inc. All rights reserved.)*

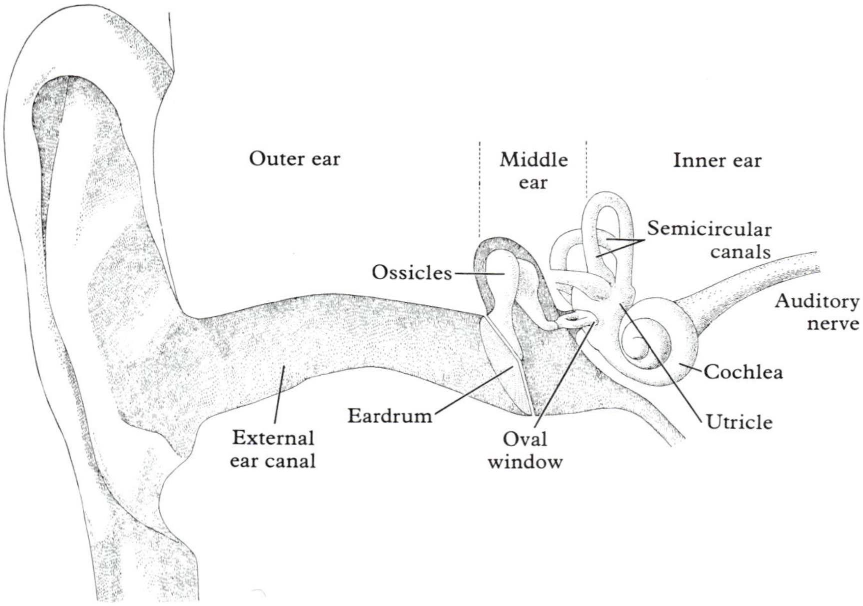

6–11
The anatomy of the human ear. (From "Attention and the Perception of Speech," by Donald E. Broadbent. Copyright © 1962 by Scientific American, Inc. All rights reserved.)

located in the spiraled, pressure-sensitive organ of Corti, which is within the cochlea. From it impulses are eventually relayed to the brain by way of the *auditory nerve.* (The role of the auditory area of the brain in the interpretation of sounds, especially those related to language, is discussed in Chapter 3.)

The remaining part of the inner ear consists mostly of three semicircular canals, each of which lies about at a right angle to the other two. These structures are responsible for the perception of acceleration and deceleration, and for the maintenance of one's sense of balance, orientation, and equilibrium. Movement of the head in any direction stimulates receptor cells in or near the bases of the three semicircular canals, from which information is then relayed to the brain by way of one of the cranial nerves.

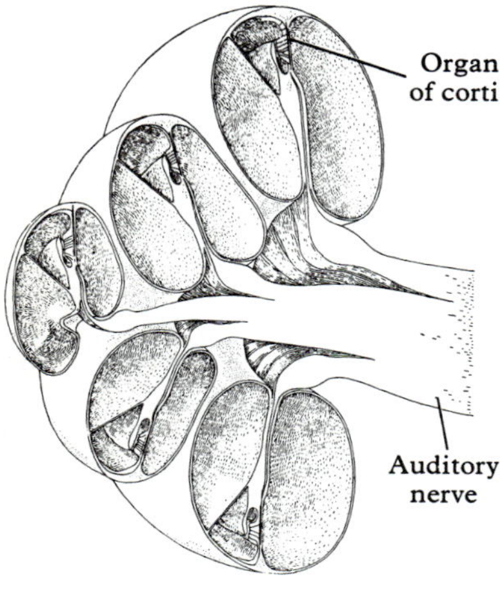

6–13
A cross section of the snail-shaped cochlea, showing the location of the vibration-sensitive organ of Corti. (From "Attention and the Perception of Speech," by Donald E. Broadbent. Copyright © 1962 by Scientific American, Inc. All rights reserved.)

Vision

Vision is the sense whose complicated receptor cells are located in the retina of the eye, where they respond to a particular kind of electromagnetic radiation known as *light.* The delicate and intricate structure of the human eye is shown in Figure 6-14. The eyes of all vertebrates are optically similar to cameras in that both eye and camera use a lens to focus an inverted image on a light-sensitive surface and both have an iris that adjusts the light intensity (Figure 6-15).

When light energy strikes the retina it causes the light-sensitive neurons located there to generate nerve impulses that are ultimately transmitted to the occipital lobes (the hindmost portion of the brain) for interpretation. As shown in Figure 6-16, the human retina is a complex multilayered structure that contains light-sensitive elements of two kinds, *rods* and *cones.* Each retina contains about 125 million light-sensitive nerve cells. Rods and cones differ in their structures and in their reactions to light. Cones are especially sensitive to bright light and to colors, and rods are more sensitive to dim light. (As you might imagine, the retinas of nocturnal animals contain many more rods than cones.)

Although there are about 125 million light-sensitive cells per retina, there are a mere one million myelinated axons in each *optic nerve.* The optic nerves extend from the retina of each eye to the brain, and their pathways partially cross, as Figure 6-17 illustrates. That there are only one million myelinated axons in each optic nerve probably means that the nerve cells of the retina

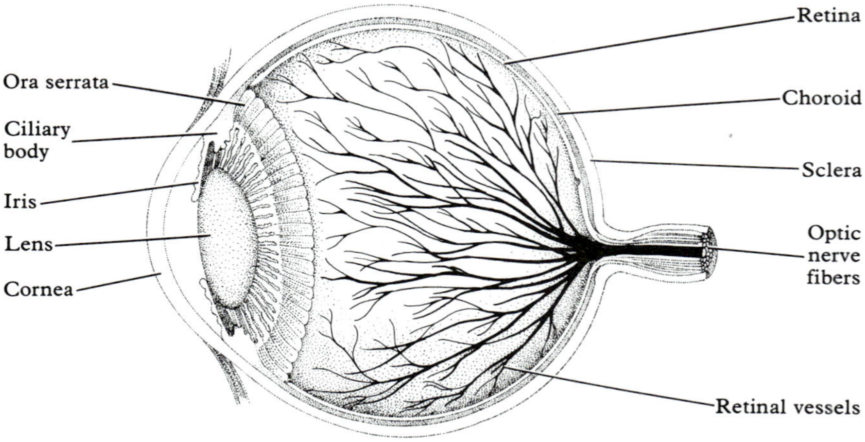

6–14
The anatomy of the human eye.

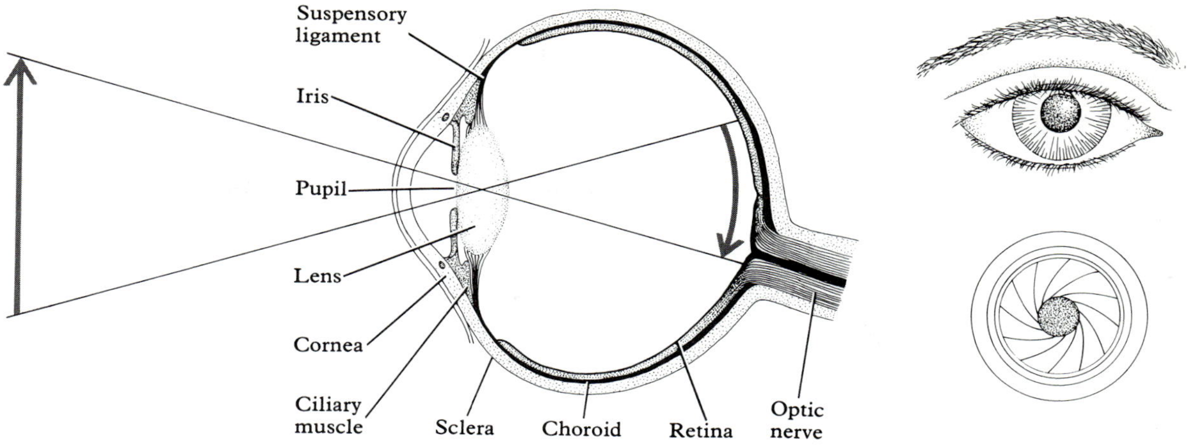

6–15
Both eye and camera use a lens to focus an inverted image on a light-sensitive surface. Both also have an iris to adjust to various light intensities. (From "Eye and Camera," by George Wald. Copyright © 1950 by Scientific American, Inc. All rights reserved.)

somehow encode, or organize, electrical impulses before their axons converge to form the optic nerve. Each fiber in the optic nerve may eventually connect to a specific location in the visual cortex of one of the cerebral hemispheres, but the exact role of the brain's higher centers in the interpretation of incoming visual impulses remains to be discovered.

In sum, the sensory neurons of the nose, tongue, ear, and eye respond to specific stimuli by generating nerve impulses that are promptly relayed to the nearby brain. We now turn our attention to some of the intricate pathways to the brain that are used by the axons of receptor cells that receive stimuli from the body surface and that may therefore be at considerable distances from the brain itself.

The Senses of Body Surfaces—Pathways to and from the Brain

It is not surprising that our body surfaces, which form the interface between ourselves and the external environment, are richly endowed with sense receptors. Within the skin are several kinds of delicate nerve nets and capsulelike structures surrounding nerve endings, all of which respond to various environmental stimuli by generating nerve impulses.

As shown in Figure 6-18, the sense receptors of the skin perceive at least five different sensations experienced on the body surface: touch, pressure, heat, cold, and pain. The sensation of touch results either from the stimulation of *Meissner's corpuscles,* or from the stimulation of an extremely sensitive net of nerve fibers surrounding the bases of individual hairs. *Free* (unmyelinated) nerve endings abound in the skin and are thought to be the peripheral receptors for pain, while the onionlike *Pacinian corpuscles* are known to generate nerve impulses in response to mechanical pressure. Exactly which receptors are concerned with the sensations of warmth and cold is unsettled, but it is usually said that *Ruffini's end-organs* and *Krause's end-bulbs* are involved.

The receptors of the skin are not evenly distributed over the entire body surface. For example, touch receptors are much more numerous in the fingertips than in the skin of the back. But all of the various kinds of sense receptors are present in the skin in sufficient quantities to allow us to experience sensations of touch, pressure, heat, cold, and pain anywhere on our bodies. And the reason that

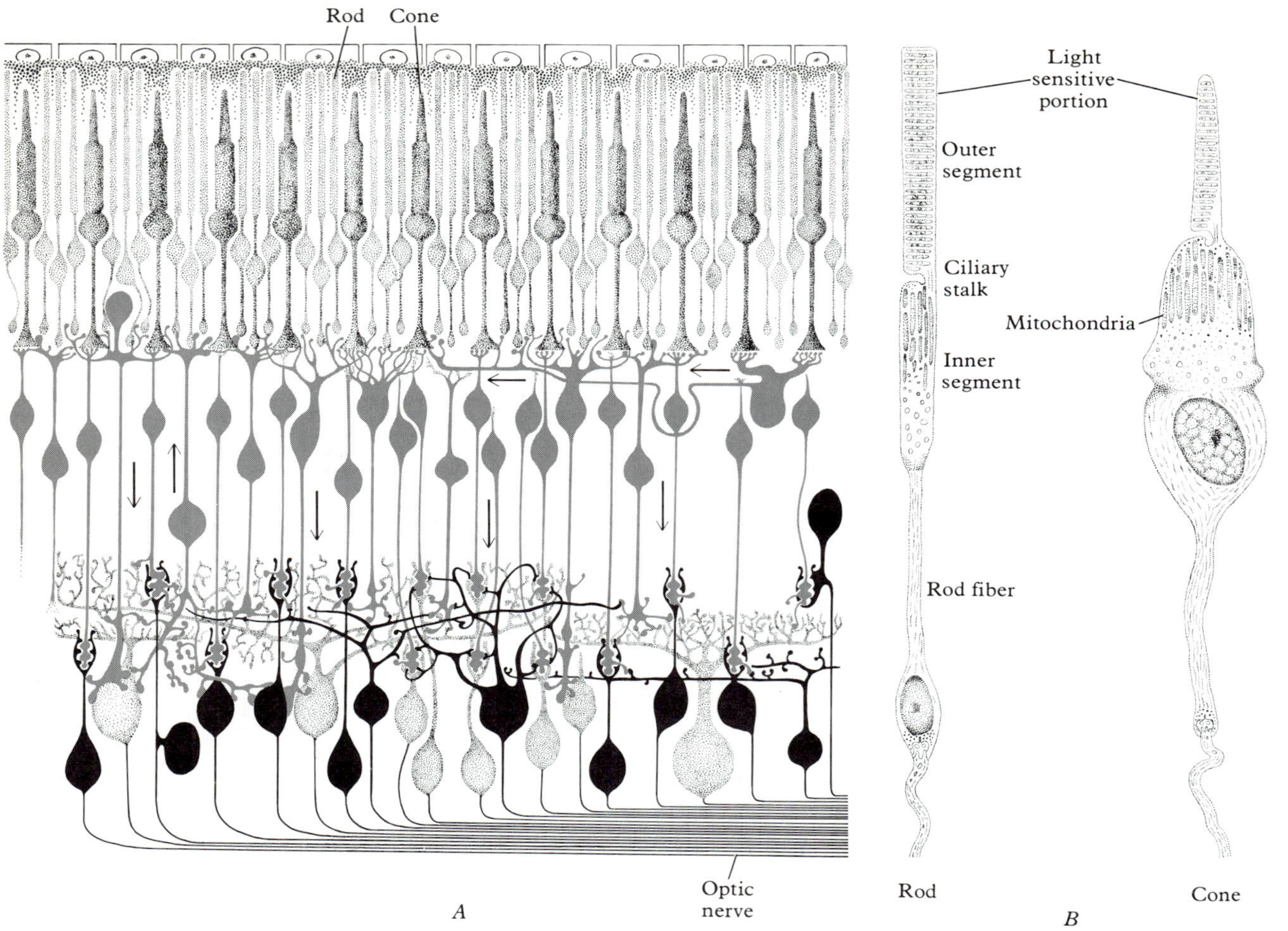

6–16

A, The nerve cell network of the retina. The photoreceptors are the densely packed cells at the top. The thinner cells are rods, the thicker, cones. To reach them the incoming light must traverse a dense but transparent layer of neurons (dark shapes) that have rich interconnections with the photoreceptors and with each other. The output of these neurons finally feeds into the optic nerve. B, The fine structure of a rod, left, and a cone, right. (A from "How Cells Communicate," by Bernhard Katz. Copyright © 1961 by Scientific American, Inc. All rights reserved.)

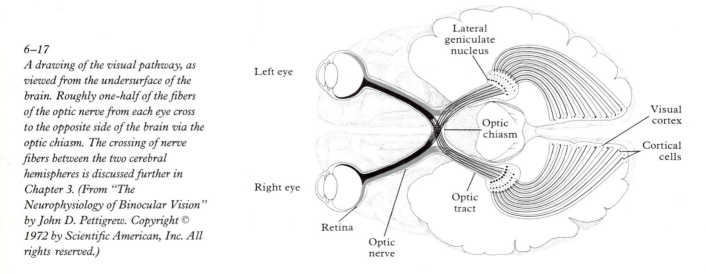

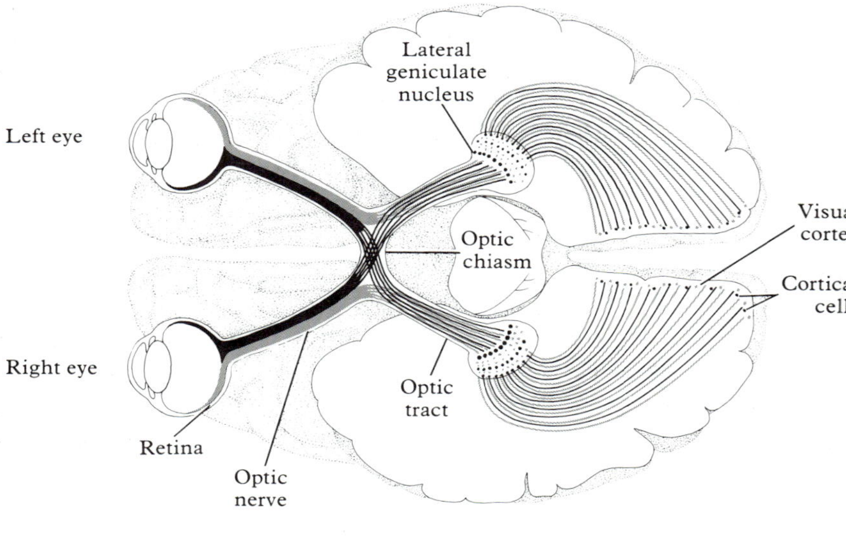

6–17

A drawing of the visual pathway, as viewed from the undersurface of the brain. Roughly one-half of the fibers of the optic nerve from each eye cross to the opposite side of the brain via the optic chiasm. The crossing of nerve fibers between the two cerebral hemispheres is discussed further in Chapter 3. (From "The Neurophysiology of Binocular Vision" by John D. Pettigrew. Copyright © 1972 by Scientific American, Inc. All rights reserved.)

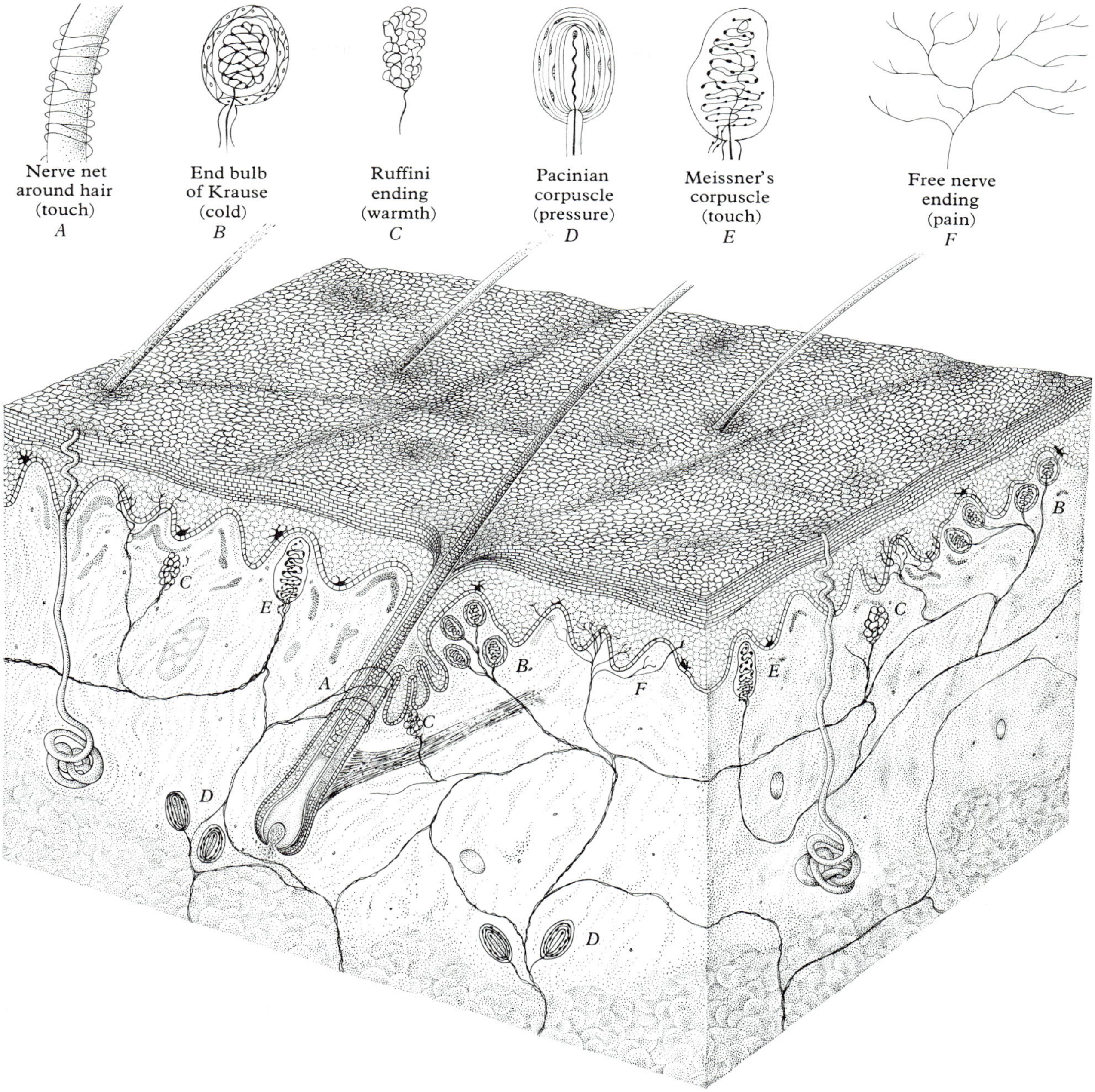

6–18

Top, some receptors of the skin and the senses with which they are thought to be associated. Bottom, a cross section of the skin, showing where some of the receptors are located. (Bottom figure from "What Is Pain?" by W. K. Livingston. Copyright © 1953 by Scientific American Inc. All rights reserved.)

we are able to tell, even blindfolded, when a particular part of the body surface encounters a stimulus is that the nerve impulses generated on specific parts of the body surface are ultimately relayed to certain well-localized areas of the brain. By what pathways do nerve impulses generated on body surfaces eventually make their ways to the brain?

In brief, sensations arising anywhere on the body surface travel by peripheral nerves (the network of tough white cables we mentioned earlier), to the spinal cord, from which they are then relayed to a higher brain center. Consider what happens when someone experiences a painfully hot stimulus, such as when one's right elbow unwittingly comes into contact with a hot radiator. The high temperatures cause both heat and pain receptors in the area of contact to discharge, and the resultant impulses travel first through superficial nerves in the skin, then through larger nerves in the arm, and finally into the spinal cord through the dorsal root of one of the spinal nerves (Figure 6-19).

6–19
Left, the peripheral spinal nerves, as seen from behind. Right, how spinal nerves are related to vertebrate.

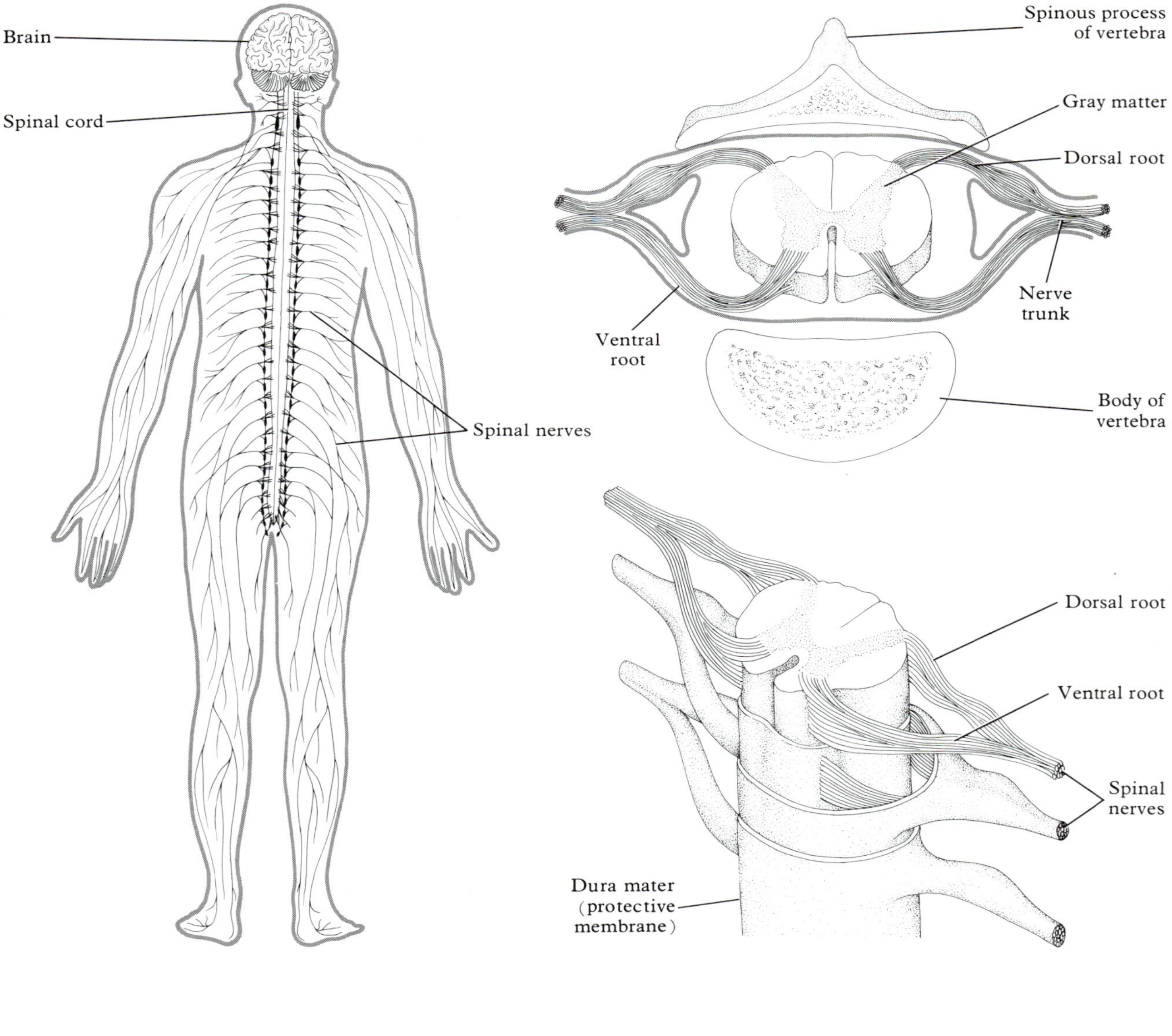

(When the skin of the right elbow becomes hot, sensory fibers feed into the spinal cord through the right dorsal root of the spinal nerve that enters the spinal cord between the sixth and seventh vertebrae.) When sensory impulses signaling "hot" and "painful" enter the spinal cord at this location, two things happen. First, motor neurons located within the butterfly-shaped area of gray matter at that place in the spinal cord are immediately stimulated to send out impulses. These impulses leave the spinal cord by the corresponding ventral root, travel back through the spinal nerve and the nerves of the arm, and finally bring about the contraction of those arm muscles whose activation results in the automatic reflex withdrawal of the affected area from the harmful stimulus. (Because the message doesn't go to the brain, the response is not arrived at consciously.) Second, "hot" and "painful" signals entering the *right side* of the spinal cord at this (or any other) location are relayed to neurons on the *left side* of the cord, and these neurons in turn send out myelinated axons that ascend to the left side of the brain through one of many long tracts of fibers that extend up and down the entire length of the white matter of the spinal cord. (White matter is white because of the presence of huge numbers of myelinated axons side by side, and gray matter is gray because of a relative abundance of cell bodies and dendrites rather than myelinated axons.)

Figure 6-20 shows the pathway to the brain followed by most of the impulses that signal the presence of something hot and painful on the body surface. To continue with our example, after crossing to the opposite side, fibers ascending in a localized region of the spinal cord first pass through the brainstem. They then make synaptic connections with neurons in other parts of the brain from which axons are finally sent out to specific neurons in the cerebral cortex of the primary sensory area of the left hemisphere. Thus, impulses finally arrive at and stimulate that portion of the primary sensory area of the left hemisphere in which electrical activity results in conscious awareness that the right elbow is in contact with a painfully hot object.

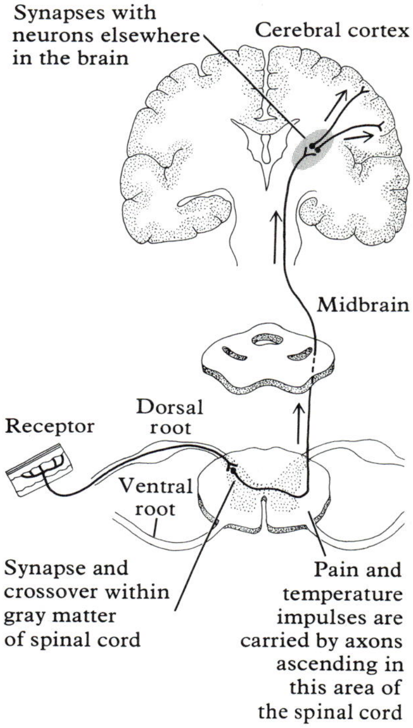

6–20
The pathway to the brain followed by impulses signaling pain and temperature from the body surface.

As shown in Figure 6-21, the *primary sensory area* of the cortex consists of a rather narrow strip of tissue. It is possible to pinpoint which areas of the primary sensory cortex are associated with various parts of the body. Stimulation of a body part results in electrical activity in the part of the sensory cortex with which it corresponds. (Remember that because they follow pathways that cross within the spinal cord, impulses signaling sensations of pain and temperature from one side of the body are generally received and brought to conscious awareness by the cerebral hemisphere of the opposite side.)

The *primary motor area* of the brain (see Figure 6-21) occupies a strip of cortex in front of and parallel to the primary sensory area. Electrical activity in this part of the cortex results in the *movement* of the corresponding body part on the opposite side from the stimulated hemisphere. In general, the map of the motor area is similar to that of the sensory area, and strong stimulation of the sensory cortex often activates the corresponding region of the adjacent motor cortex. In our example, this is experienced as a desire to move away from the hot radiator, an act that presumably would follow the conscious perception of the noxious stimulus almost immediately. How do neurons in the primary motor cortex, in which electrical activity can somehow be voluntarily initiated, bring about the muscle contraction associated with movements we choose to make?

Neurons in all areas of the primary motor cortex, as well as in certain surrounding areas, send out axons that synapse with various other neurons in the brain. Axons from these secondary neurons cross to the opposite side and then

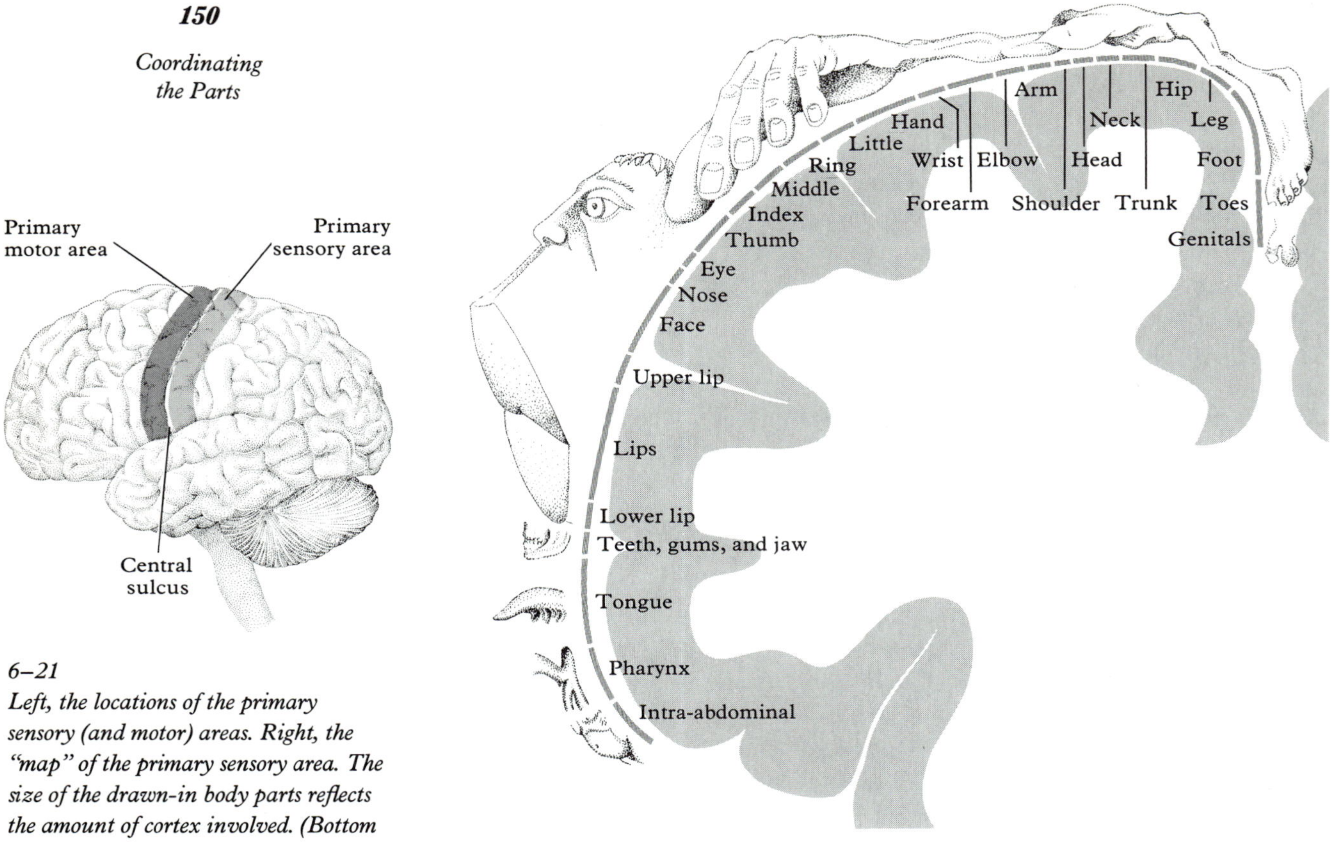

6–21
Left, the locations of the primary sensory (and motor) areas. Right, the "map" of the primary sensory area. The size of the drawn-in body parts reflects the amount of cortex involved. (Bottom drawing from "Tools and Human Evolution," by Sherwood L. Washburn. Copyright © 1960 by Scientific American, Inc. All rights reserved.)

descend in the spinal cord. Figure 6-22 shows the approximate location within the brain where the crossover of fibers in the voluntary motor pathway occurs, as well as the fibers' position as they descend in the white matter of spinal cord. Once the fibers have crossed they descend until some of them synapse with neurons at various levels in the gray matter of the spinal cord. From there impulses that cause muscle contraction are sent out, just as they are during the reflex withdrawal of the scorched elbow. The difference is that this time the impulses to produce action resulted from an act of will, not from the direct processing of incoming sensory impulses by the relatively simple neural circuits located entirely within the spinal cord.

The fiber tracts that enable us to feel pain and temperature from body surfaces and to make voluntary motions are but two of many ascending and descending pathways that make up the white matter of the spinal cord. Fiber tracts also influence conscious awareness of the positions of the joints, the sense of touch, posture, the production of coordinated body movements (many fibers that affect muscle coordination issue from the *cerebellum,* a part of the brain that is richly interconnected with the cortex), and the regulation of the body's internal environment. But a further explanation of these complex fiber tracts within the spinal cord would go beyond the scope of our discussion. From this consideration of pathways to and from the brain we now turn our attention to the functions carried out by the brainstem, which plays a part in processes that are so crucial to sustaining life that they are regulated automatically, without requiring conscious awareness.

The Brainstem

For the purpose of this discussion, you can consider the brainstem to be that portion of the brain between the higher centers and the place where the spinal cord leaves the base of the skull to enter the vertebral column (Figure 6-23). Despite its relatively small size the brainstem is a favorite object of study among neuroanatomists, and rightly so. This inconspicuous portion of the brain is of great importance for at least three reasons: first, all but one of the twelve pairs of cranial nerves arise from the brainstem; second, it houses the reticular formation; and third, within the brainstem are found special groups of neurons directly involved in regulating many aspects of the internal environment.

As mentioned briefly before, the twelve pairs of cranial nerves differ from spinal nerves in that they do not issue from the spinal cord, but rather originate directly within the substance of the brainstem. Each pair of cranial nerves has its own special characteristics. Some are purely motor (for example, three cranial nerves supply the muscles that move the eyes), others are purely sensory (for example, the *auditory nerve* carries sense data from the inner ear), and still others carry both sensory and motor impulses. An example of this is the *facial nerve,* which supplies most of the muscles of the face and at the same time carries taste impulses from the front part of the tongue. Also, as we shall soon discuss, the vagus nerve (the longest cranial nerve) sends fibers to many internal organs and thereby plays a major role in regulating their functions.

Extending the length of the core of the brainstem is the *reticular formation,* as shown in Figure 6-23. In most adult brains the reticular formation is about the size of a little finger, and it is made of millions of small neurons whose axons and dendrites are arranged in a complex highly interconnected network. The

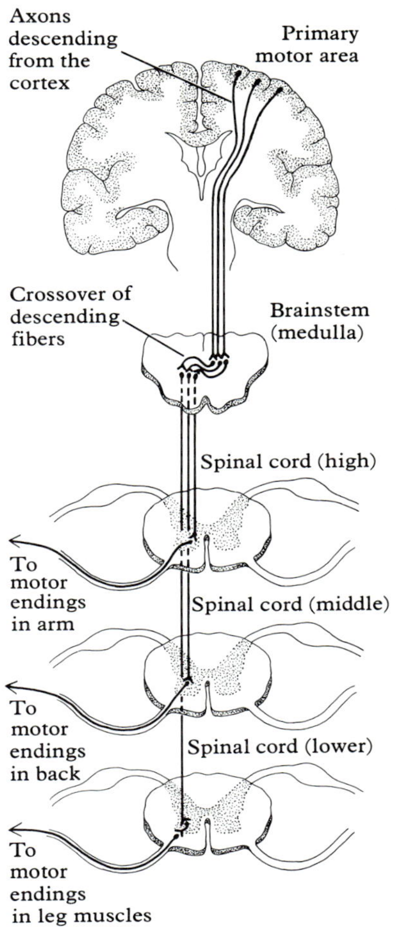

6–22
The voluntary motor pathway.

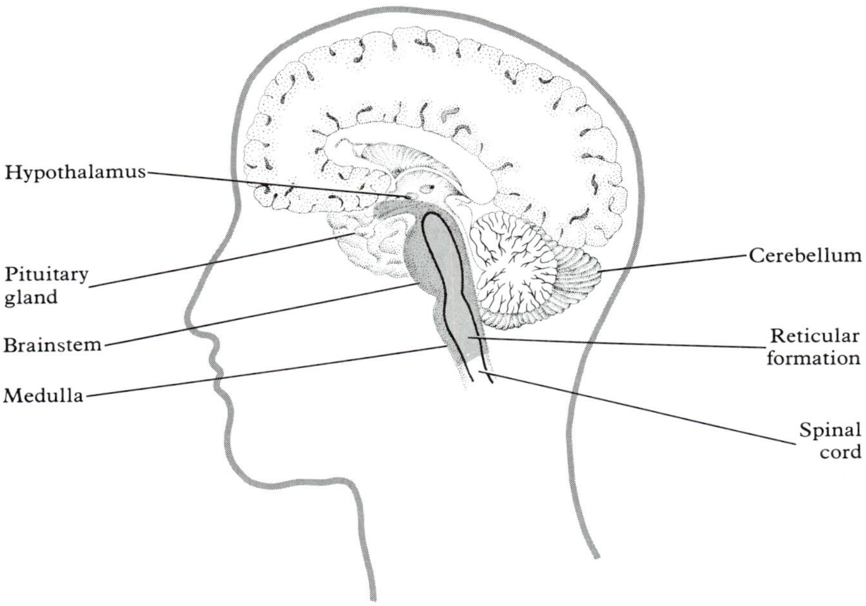

6–23
The brainstem, as seen in a brain that has been cut in two down the middle. The medulla is one of several parts of the brainstem.

reticular formation of the brainstem has several important functions. First, it awakens the cortex of the brain to consciousness and keeps it alert—that is, the electrical activity of the reticular formation maintains wakefulness and intersperses it with periods of sleep. Second, the reticular formation acts to alert, or sensitize, the cortex to incoming sense data in such a way that some of the incoming sensory messages reach consciousness but others do not. Thus, the reticular formation can be thought of as a filter, or a modulator, that sifts through the countless sensory impulses arriving in the brainstem and relays to the cortex only those impulses that past experience has proven to be most relevant. This filtering effect accounts for such familiar experiences as not feeling the presence of the jewelry we wear, or suddenly becoming aware of a faucet that has dripped in the background for hours. Third, the reticular formation is of crucial importance in maintaining muscle tone, and it exerts some control over muscle reflexes and voluntary movements.

In addition to its diffuse core of small, highly interconnected neurons, the brainstem also contains well-localized clusters of larger nerve cells that are concerned with the production of certain protective reflexes and with regulating the internal environment. For the most part, these cell clusters are found in the hindmost portion of the brainstem (not far from where the spinal cord originates), a region of the brain known as the *medulla oblongata,* or *medulla* for short (see Figure 6-23). The medulla contains clusters of neurons concerned with the rather complex reflexes that regulate respiration, blood pressure, and heart rate. We will discuss these reflexes when we consider respiration and circulation in following chapters.

Other reflex responses that are integrated by clusters of neurons within the medulla include swallowing, coughing, sneezing, gagging, and vomiting. Let us use the vomiting reflex as an example for further discussion. The "vomiting center" in the medulla responds to a variety of stimuli, including nauseating smells, sickening sights, irritation of the lining of the upper gastrointestinal tract, and probably certain chemicals occasionally found in the bloodstream. How is the presence of something irritating or nauseating in the stomach relayed to the vomiting center so that the vomiting center can coordinate the reflex muscle contractions that expel the contents of the stomach? Impulses signaling irritation in the stomach follow two main routes to the vomiting center: first, the *vagus nerve,* which is directly connected to the brainstem, and second, sympathetic nerves that feed into the spinal cord before relaying their impulses up to the medulla. The stomach is not unique in being served by two types of nerves. In fact, almost all internal organs have similar setups. In other words, most internal organs are supplied with two sets of nerves through which sense data are sent to the brain and through which regulating impulses from the brainstem and spinal cord are conducted. These different sets of nerves correspond with the two divisions of the *autonomic nervous system,* to which we now turn our attention.

The Autonomic Nervous System

Anatomists divide the human nervous system into two main parts: the central nervous system (CNS) and the peripheral nervous system. The CNS consists of the brain and spinal cord. This part of the nervous system is surrounded by several protective membranes and is bathed by a clear liquid known as *cerebrospinal fluid* (see Figure 6-24). The peripheral nervous system has two subdivisions—the spinal nerves, which we have already discussed (sometimes collectively called the "somatic nervous system"), and the autonomic nervous system (ANS). "Autonomic" means "automatic," and this emphasizes that the motor workings of this division of the nervous system usually occur without conscious awareness. See Figure 6-25 for a diagram of the many parts of the nervous system.

How do the nerves of the ANS differ from spinal nerves? As we have seen, once they leave the spinal cord, spinal nerves may extend for considerable distances, but eventually they all make direct synaptic connections with the muscles or sense organs they serve. The nerves of the ANS are organized in a different way. In the first place, not all autonomic nerves originate from the spinal cord—some autonomic fibers originate within the brainstem and then exit it as parts of certain cranial nerves. And second, before they reach the organs they serve, all autonomic nerves, whether they originate from the spinal cord or from the brainstem, make synaptic connections with clusters of nerve cells known as *ganglia* (the singular form is ganglion) that are located *outside* the central nervous system.

Judged by anatomy and function, the nerves of the ANS are of two different kinds, *sympathetic* and *parasympathetic.* As shown in Figure 6-26, sympathetic nerves originate from the middle region of the spinal cord and then form synapses with ganglia located in a chain that extends the length of both sides of the vertebral column, not far from the spinal cord. From this chain of ganglia, sympathetic fibers are then sent out to the internal organs. On the other hand, parasympathetic nerves originate both from the brainstem and from the terminal portion of the spinal cord. Also, the ganglia with which parasympathetic fibers form synapses are not organized into chains, but rather are located on or near the surfaces of the internal organs.

Fortunately, the complicated anatomy of its sympathetic and parasympathetic divisions is not the most interesting thing about the ANS. Much more engaging is the way in which all of the complicated circuits interact under the modulating influence of the brain to control the functions of the internal organs and other body structures. As shown in Figure 6-26 sympathetic and parasympathetic nerves often have opposite influences on the functions of the organs

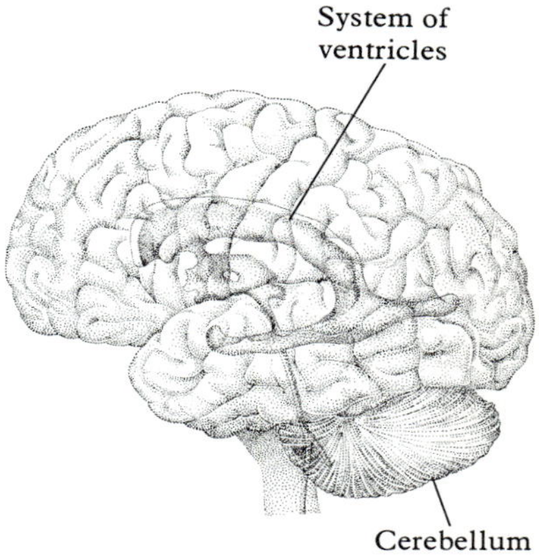

6–24
The brain and spinal cord are covered by protective membranes and bathed by cerebrospinal fluid (CSF). A system of ventricles underlies the cerebral hemispheres. CSF fills the ventricles and circulates in and around the brain and spinal cord.

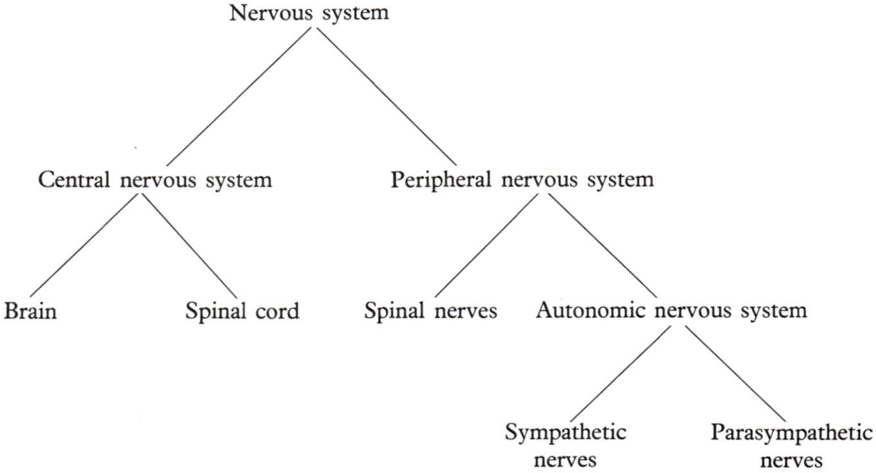

6–25
The anatomical divisions of the nervous system. (Each division or organ is discussed in this chapter, and the anatomy of the brain is further discussed in Chapter 3.)

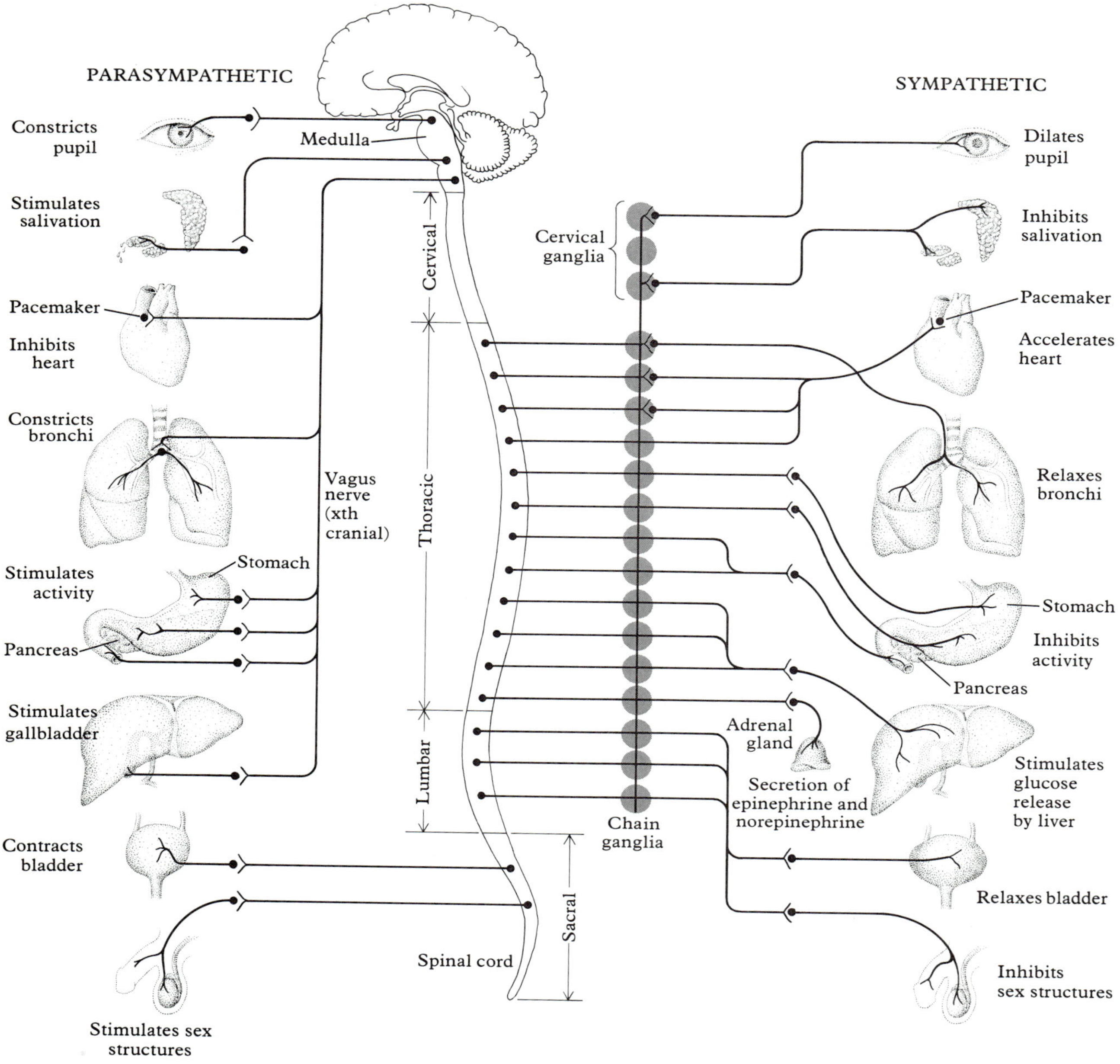

6–26
The major features of the anatomy, and some of the functions of the two divisions of the autonomic nervous system (ANS).

they serve. For example, stimulation by sympathetic nerves results in dilation of the pupils of the eyes, the secretion of scanty thick saliva, an accelerated heart rate, and a widening of the smaller air passages of the lungs. In contrast, parasympathetic stimulation leads to constricted pupils, the production of copius thin saliva, a slowing of the heart rate, and a narrowing of the smaller tubes of the respiratory tree. With a few exceptions, the normal functioning of many body organs can be viewed as a result of the balanced opposition between the activities of the two divisions of the ANS.

In recent years it has become clear that the balanced opposition between sympathetic and parasympathetic nerves can be traced all the way to individual molecules. It has long been known that sympathetic and parasympathetic nerves release different neurotransmitters at synapses (see Figure 6-7). For sympathetic nerves the neurotransmitter is *norepinephrine* (NE), and for parasympathetic nerves it is *acetylcholine* (ACh). ACh also brings about the contraction of skeletal muscles when it is released at the synapses that connect motor nerves and muscle fibers. Only in the past decade have we come to understand how the neurotransmitters of the ANS operate in cells and how upsets in their normal functioning can sometimes be correlated with certain diseases. As an example, let us consider the relatively common and oftentimes frustrating condition known as asthma.

There are several kinds of asthma, and what is said here is true for most, but not all, kinds. Asthma is a disorder of breathing that results when the smaller air passages of the lungs are narrowed and plugged by thick secretions. The conditions are reversible. (The allergic and psychological factors that influence this condition are discussed later.) Anyone who suffers from asthma usually feels short of breath during an attack and the characteristic wheeze associated with the reversible airway obstruction is often clearly audible. If you recall that parasympathetic stimulation of the lung results in narrowing of the smaller airways, you will not be surprised to know that is has recently been discovered that the reversible airway obstruction of asthma does indeed result from an *imbalance* between the activities of the two divisions of the ANS.

What is the nature of this imbalance? To simplify a little, asthma results because the lungs of affected people are in a state of "parasympathetic dominance" and "sympathetic blockade." In other words, the lungs of people with asthma act immune to the effects of the NE released by sympathetic nerves, and at the same time they overreact to the presence of ACh released from parasympathetic nerves. This automatic imbalance makes the airways of the lungs unstable—or, perhaps better, twitchy—in that they are hyperreactive and apt to become constricted in response to relatively minor physical or emotional stimuli.

The autonomic imbalance in asthma is also evident in biochemical changes that occur inside lung cells. The parasympathetic dominance results in a greater concentration of cyclic guanosine monophosphate (cGMP) in the cell. In excess, this substance causes the smooth muscles in the walls of the smaller air passages to contract and also promotes the release of certain chemicals that cause the epithelial lining of the airways to swell, thus further obstructing airflow. In healthy lungs, the activity of cGMP is balanced by that of cyclic adenosine monophosphate (cAMP). Among other widespread affects, this substance results in the relaxation of smooth muscle, and it blocks the release of the chemicals that cause the airway to become further obstructed.

It is known that NE, ACh, and some other neurotransmitters affect cell membranes so that the concentrations of cAMP and cGMP inside the stimulated cells are changed. Because of their effects on cell membranes, neurotransmitters are considered chemical messengers that strongly influence what goes on biochemically inside certain cells. As we shall soon discuss, many of the body's hormones also act as chemical messengers and are known to influence concentrations of cAMP and cGMP within the cell. Neurotransmitters and hormones are thus in some ways functionally similar; this reflects the fact that nerves and hormones are closely allied in their regulatory efforts.

But before we discuss the regulatory effects of hormones, we must return to the brain, this time to the *hypothalamus.* This important region not only provides a direct link between the brain and hormones, but also contains many clusters of neurons that cooperate with the autonomic nervous system and with other brain centers in regulating the body's internal environment.

The Hypothalamus—Some Reflexes and Appetites

The hypothalamus is the small, irregularly shaped upper end of the brainstem. It lies on the undersurface of the brain, very near to where the crossover of certain fibers in the optic nerves takes place (Figure 6-23). The hypothalamus consists of about sixteen rather ill-defined clusters of small neurons and supporting cells that are concerned with regulating the secretion of various hormones and with regulating certain reflexes and appetites. The hypothalamus is richly interconnected with the limbic system. As you will recall from Chapter 3, this complicated neural circuit underlies the cerebral hemispheres and is known to be associated with the production of emotions. Mostly because of the interconnections between the hypothalamus and the limbic system, emotional states are often accompanied by changes in the functioning of those body parts regulated automatically by the ANS, especially its sympathetic division.

We are all familiar with body changes that occur automatically. For example, have you ever suddenly found yourself confronted by a frightening object? Your heart begins to race, your blood pressure rises, the pupils of your eyes dilate, and the hairs on your head and body may bristle defensively. All of these reflex reactions serve the purpose of making you ready for either fight or flight, and they are all brought about by the stimulation of sympathetic nerves. (Frightening stimuli also lead to the reflex release of the hormone epinephrine, or adrenalin, from the adrenal glands. This compound is structurally similar to NE, the sympathetic neurotransmitter.) The same reactions of the body can be produced by direct electrical stimulation of some of the clusters of neurons in the hypothalamus. Thus, the hypothalamus provides a link between emotions arising in the limbic system (or elsewhere) and the body changes that often accompany our subjective emotional states.

Other clusters of neurons in the hypothalamus are concerned with regulating certain appetites, including hunger and thirst. Electrical stimulation or brain damage in certain areas of the hypothalamus can result in a marked increase or decrease in the quantity of food eaten and the quantity of water consumed by experimental animals. It is believed that the normal functioning of the "feeding center" depends on the difference in the concentration of glucose in certain arteries and veins within the hypothalamus. Water consumption, on the other hand, depends on the activity of certain hypothalamic neurons that are sensitive to the concentration of water in body fluids and that produce the sensation of thirst to assure that fluid intake is adequate.

The hypothalamus also plays a part in regulating body temperature. Different clusters of neurons respond to increases or decreases in the temperature of the blood as it circulates through the hypothalamus. The firing of these neuron sets into motion a series of events that either conserves or dissipates body heat, as appropriate. (The regulation of body temperature is discussed further in Chapter 12, where we consider the role of the skin in the economy of body heat.)

In recent years it has become clear that some of the cells in the hypothalamus are specialized for yet another function—the secretion of chemical messengers (hormones) directly into the bloodstream. Within the hypothalamus lies the bridge between the body's two main ways of regulating the internal environment—nerves and hormones. In order to better understand the interworkings or nerves and hormones we now turn our attention to the relations between the hypothalamus and the *pituitary gland.*

The Hypothalamus—Relations to the Pituitary Gland

The pituitary gland is a pea-sized structure located at the end of a short stalk that arises from the hypothalamus (Figure 6-23). The gland is nestled in a bony pocket in the floor of the skull, and in spite of its small size it is of extreme importance because it cooperates with the hypothalamus and with certain other body parts in regulating the concentrations of many of the body's hormones. The pituitary gland has two distinct parts, or lobes—one anterior (front) and one posterior (back). Each portion of the pituitary gland releases hormones, but they are connected to the hypothalamus in different ways. Let us first discuss the connections and functions of the posterior pituitary.

The posterior pituitary is connected to the hypothalamus primarily by means of axons. These axons originate from cell bodies located in certain clusters of neurons in the hypothalamus and terminate around the walls of small blood vessels supplying the posterior pituitary. The cells in the hypothalamus produce two hormones, antidiuretic hormone (ADH) and oxytocin. Both of these hormones are stored in granules that are passed down the axons and stockpiled in the posterior pituitary. From there the axons release the hormone granules directly into the bloodstream whenever the appropriate stimulus is encountered. (The functions of both of these hormones are discussed in following chapters.)

By way of contrast, the anterior pituitary receives very few axons from nerve cells in the hypothalamus (or elsewhere). Rather, the two areas are connected primarily by blood vessels (Figure 6-27). The anterior pituitary is made up of interlacing cords of cells that are specialized for producing one or more of the hormones released by this remarkable little gland. As we shall soon discuss, the hormones released from the anterior pituitary are chemical messengers that regulate the secretion of other hormones from various endocrine glands. In recent years it has been discovered that the hypothalamus also has an active role in helping to regulate the concentrations of the crucial hormones released from the anterior pituitary. The hypothalamus exerts its influence by releasing hormones itself. The hypothalamus produces hormones known as *releasing factors* that are secreted into the blood vessels of the stalk and then rapidly carried to the anterior pituitary, where each releasing factor either promotes or inhibits the secretion of one of the hormones manufactured in that part of the gland. The blood vessels connecting the hypothalamus and the anterior pituitary thus provide the means by which the regulatory effects of the brain are translated into the release of important hormones by the anterior pituitary gland. The hormones of the anterior pituitary are of crucial importance, and we must discuss them further.

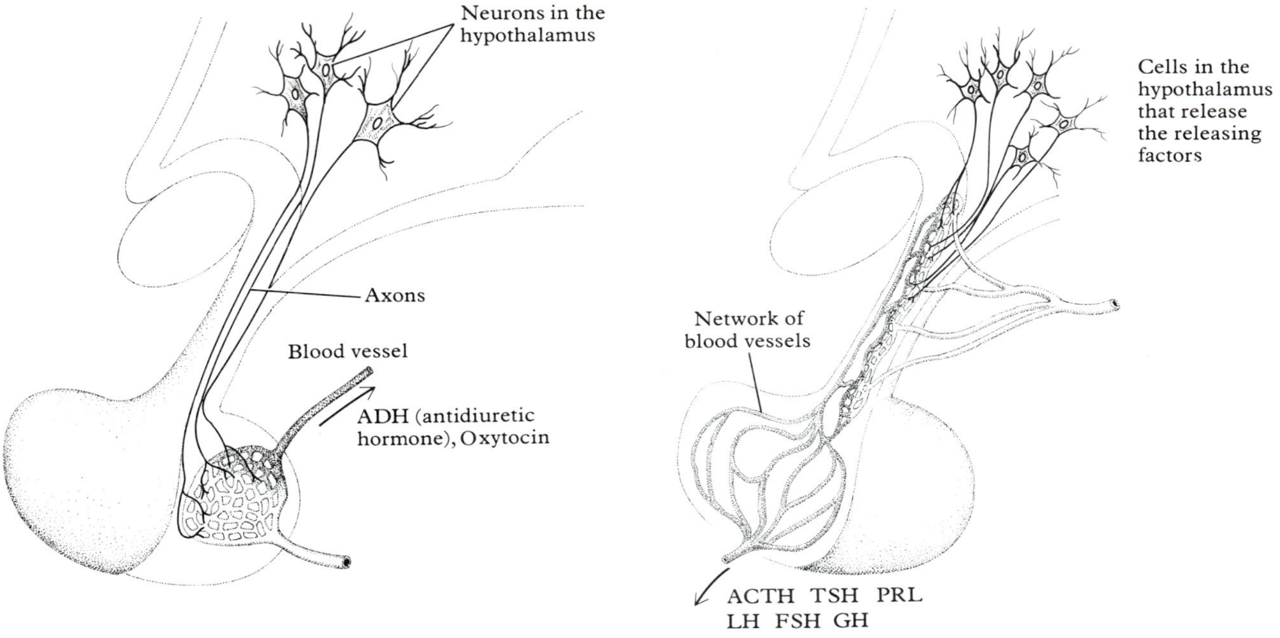

6–27
Left, neural connections between the hypothalamus and the posterior lobe. The only significant nerve fibers connecting the hypothalamus and the pituitary run from two hypothalamic centers to the posterior lobe. They transmit oxytocin and antidiuretic hormone (ADH), two hormones manufactured in the hypothalamus and stored in the posterior lobe. Right, vascular connections between the hypothalamus and the anterior lobe consist of a network of capillaries between the base of the hypothalamus and the anterior pituitary. Small hypothalamic nerve fibers deliver to the capillaries releasing factors that stimulate the secretion of the anterior-lobe hormones. (From "The Hormones of the Hypothalamus," by Rogert Guillemin and Roger Burgus. Copyright © 1972 by Scientific American, Inc. All rights reserved.)

The Hormones of the Anterior Pituitary—How They Work

Like other hormones, the chemical messengers of the anterior pituitary are secreted directly into the bloodstream and are then carried rapidly throughout the body where they come into contact with the cells, tissues, and organs that are influenced by them. The pituitary is but one of several body parts that release hormones directly into the bloodstream. Until quite recently hormones were thought to be secreted *only* by the ductless glands of the endocrine system, the thyroid, parathyroids, adrenals, ovaries, testes, and special groups of cells in the pancreas, as well as the master gland of the endocrine system, the pituitary. We now know that hormones are also released by the lining of the stomach, the lining of the duodenum, and the hypothalamus, and undoubtedly by other body parts. Table 6-2 is a summary of the human body's major hormones, the body parts that release them, and some of their effects.

As shown in Table 6-2, the cells of the anterior pituitary manufacture and release at least six hormones, each of which helps to regulate a different function.

Except for growth hormone (GH), which has a growth-promoting effect on many body parts, each hormone from the anterior pituitary is primarily concerned with stimulating a particular organ, its target organ. Figure 6-28 shows the target organs of the chemical messengers released by the anterior pituitary. Notice that more than one hormone can affect a given target organ. For example, the testes are stimulated by both follicle-stimulating hormone (FSH) and luteinizing hormone (LH). In general, organs that can feel the effects of several hormones respond to each one in a specific way. Thus FSH stimulates the testes to manufacture sperm cells, whereas LS stimulates them to produce the sex hormone testosterone.

With so many kinds of hormones, target organs, and regulatory effects, it is not surprising that our understanding of many aspects of the hormonal control of body functions is incomplete. Yet the past few years have brought some spectacular advances in our knowledge, especially of how hormones interact with the cells of their target organs. Most hormones, no matter where they are manufactured, stimulate their target organs by interacting with cell membranes. (Steroid hormones, whose effects are summed up in Table 6-2, are notable exceptions; they pass through the membrane and enter cells.) Furthermore, it appears that virtually all hormones, in spite of their differing regulatory roles, affect the biochemical machinery of the cell in very similar ways.

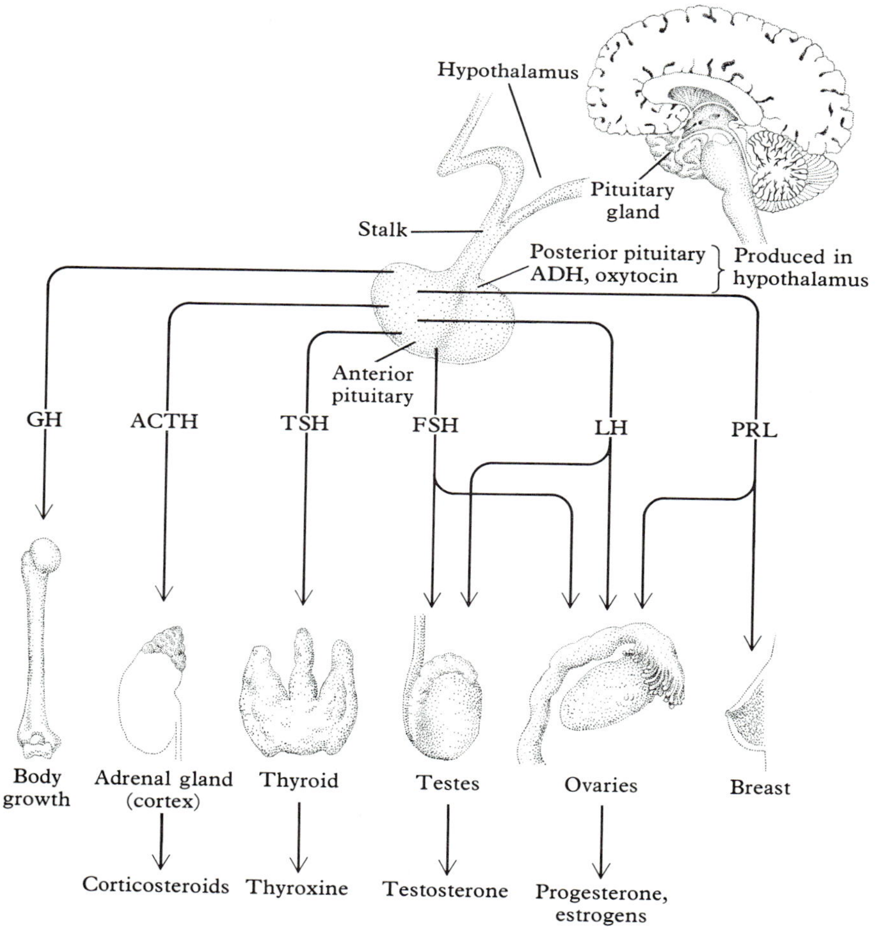

6–28
Target organs. The anterior lobe secretes a number of hormones: growth hormone (GH), which promotes statural growth; adrenocorticotropic hormone (ACTH), which stimulates the cortex of the adrenal gland to secrete glucocorticoids; thyroid-stimulating hormone (TSH), which stimulates secretions by the thyroid gland; and follicle stimulating hormone (FSH), luteinizing hormone (LH) and prolactin (PRL), which in various combinations regulate lactation and the functioning of the gonads. (From "The Hormones of the Hypothalamus," by Roger Guillemin and Roger Burgus. Copyright © 1972 by Scientific American, Inc. All rights reserved.)

TABLE 6-2
The sources and effects of some major hormones.

PARTS OF THE BODY FROM WHICH HORMONES ARE RELEASED	HORMONES	FUNCTION
Hypothalamus	Releasing factors (at least 8 kinds)	Regulate the secretion of hormones produced by the anterior pituitary.
Posterior Pituitary gland (releases hormones produced by the hypothalamus)	Antidiuretic hormone (ADH)	Stimulates the kidneys to reabsorb water and smooth muscle cells to contract.*
	Oxytocin	Stimulates the uterus to contract and the breasts to release milk.*
Anterior Pituitary gland	Growth hormone (GH)	Stimulates the growth of many body parts.*
	Thyroid-stimulating hormone (TSH)	Stimulates the release of hormones from the thyroid gland.
	Adrenocorticotrophic hormone (ACTH)	Stimulates the cortex of the adrenal gland to release various hormones.
	Follicle-stimulating hormone (FSH)	Stimulates the maturation of eggs in the ovary and the production of sperm in the testes.*
	Luteinizing hormone (LH)	Stimulates the ovary to produce a *corpus luteum* and stimulates the release of sex hormones from the ovaries and testes.*
	Prolactin (PRL)	Stimulates the breasts to produce milk.*
Thyroid gland	T-4 and T-3 (thyroxine and triiodothyronine)	Stimulate the metabolic rate as measured by the uptake of oxygen.
	Calcitonin	Helps to regulate calcium in the blood.
Parathyroid gland	Parathyroid hormone (PTH)	Helps to regulate calcium and phosphate in the blood.
Adrenal gland, medulla (central part)	Epinephrine (Adrenalin)	Has widespread effects, all of which prepare the body for "fight or flight."
	Norepenipherine (NE or noradrenalin)	Results in responses similar to those produced by epinephrine, especially the constriction of blood vessels; also acts as a neurotransmitter.

TABLE 6-2 (continued)
The sources and effects of some major hormones.

PARTS OF THE BODY FROM WHICH HORMONES ARE RELEASED	HORMONES	FUNCTION
Adrenal gland, cortex (peripheral part)	Glucocorticoids (several kinds of steroid hormones)	Slow down the rate of protein synthesis and help to maintain normal blood sugar by stimulating the production of glycogen, among other metabolic effects.
	Mineralocorticoids (several kinds)	Influence chemical reactions that regulate the sodium and potassium ions in body fluids.
	Sex hormones (produced in much smaller quantities than in the ovaries and testes)	Stimulate the development and maintenance of secondary sexual characteristics.*
Ovaries	Estrogens (several kinds)	Influence the lining of the uterus and stimulate the development and maintenance of female secondary sexual characteristics.*
	Progesterone	Has effects similar to those of estrogen; also maintains pregnancy.*
Testes	Androgens, especially testosterone	Stimulate the development and maintenance of male secondary sexual characteristics.*
Lining of the stomach	Gastrin	Stimulates specialized cells in the stomach to secrete a watery fluid rich in hydrochloric acid.*
Lining of the duodenum (uppermost part of the small intestine)	Secretin	Stimulates the pancreas to release enzymes involved in digestion.*
	Cholecystokinin	Stimulates the gallbladder to release bile.*
	Enterogastrone	Inhibits the secretion of hydrochloric acid by cells in the lining of the stomach.*
Pancreas	Insulin	Stimulates the uptake of glucose by most body cells; has other widespread effects.
	Glucagon	Stimulates the conversion of glycogen to glucose.
Kidney	Erythropoetin	Stimulates the production of red blood cells by the bone marrow.*

*Functions marked by an asterisk are discussed in other chapters.

Hormones and Second Messengers

It has long been known that hormones are extremely potent substances. This is because most hormones do not get inside cells. Rather, each hormone becomes attached to a specific receptor on the cell membranes of its target organ, and only a few of these receptors need be occupied in order for the cell to respond biochemically to the presence of the hormone on its surface. When a hormone comes in contact with its receptor, the cell membrane responds by releasing a *transducer molecule* that in turn stimulates the release of a second messenger. The second messenger sets into motion the appropriate biochemical responses within the cells.

There are probably only two second messengers, and both are relatively simple molecules. The two known messengers are cAMP and cGMP, which you might recall from our discussion of the biochemical imbalance that causes asthma. (Figure 6-29 illustrates the structures of cAMP and cGMP.) It appears that these two substances, acting in opposition to one another, can influence a large number—perhaps hundreds—of the myriad biochemical reactions that take place inside all living cells.

As shown in Figure 6-30, when the hormones ACTH, LH, TSH (thyroid-stimulating hormone), ADH, and PTH (parathyroid hormone), among others, interact with their specific receptors on the cell membrane they all stimulate the production of cAMP. The presence of cAMP then activates specific enzymes that result in the appropriate biochemical changes in each cell type. (How this occurs in most cell types remains to be explained, but the mechanism by which cAMP stimulates certain liver enzymes is well known. See p. 128.) Where does cGMP fit in? Although the concentrations of cGMP are generally much smaller than those of cAMP, elevated concentrations of cGMP generally result in biochemical effects that are the opposite of those that result from elevated concentrations of cAMP. It has recently been reported that certain hormones—oxytocin and insulin, for example—directly affect the concentration of cGMP within the cell. Thus, although we have much to learn of the details, cAMP and cGMP, by virtue of their opposite effects, may be the yin and yang of the cellular realm. Through the intervention of these two molecules, the

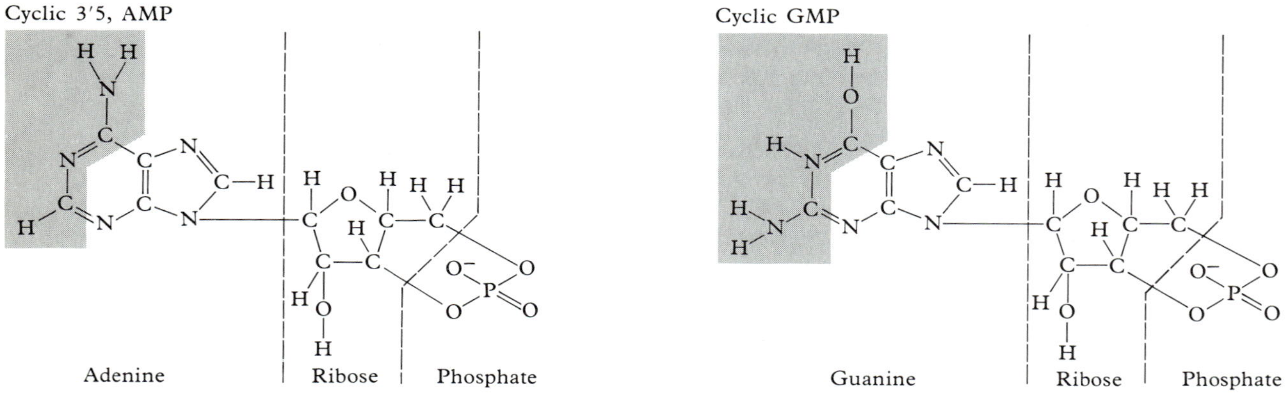

6–29
The structures of cyclic adenosine monophosphate (cAMP) and cyclic guanosine monophosphate (cGMP).

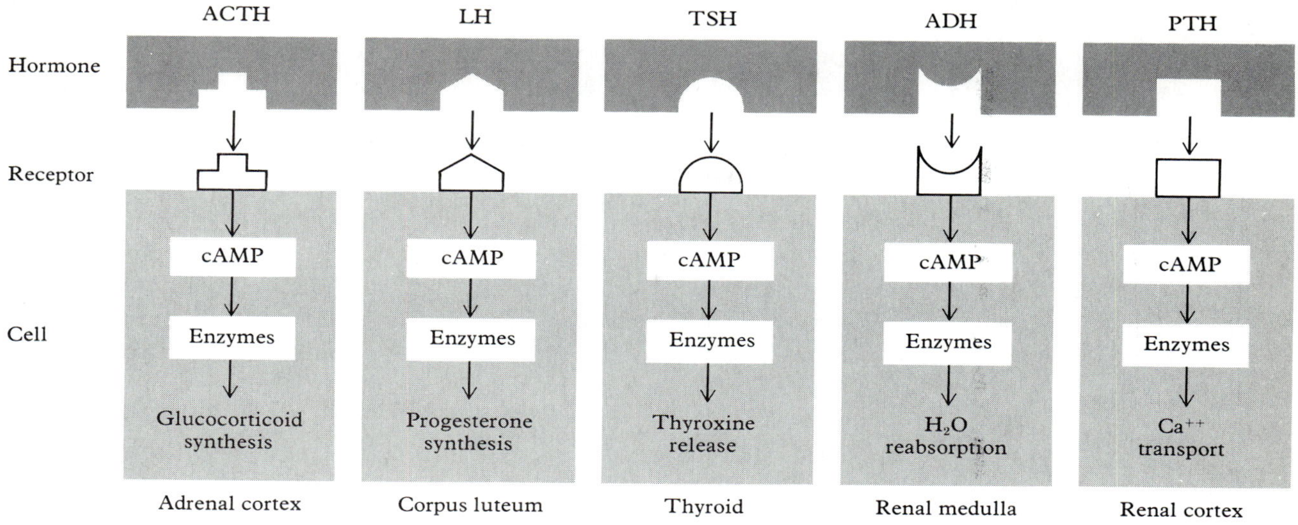

6–30
Many hormones, including those shown here, stimulate the production of cAMP. The hormones become attached to specific receptors on the surfaces of various types of cells. This causes each cell type to produce cAMP, which then stimulates various systems of enzymes.

presence of hormones and neurotransmitters on the cell surface is translated into appropriate biochemical activity within the cell, and there is reason to believe that the full range of the regulatory effects of these two compounds is yet to be discovered.

From this discussion of the cellular events set into motion by the presence of hormones on the cell surface, we now turn our attention to how the constantly changing concentrations of hormones are regulated. Once again, we will focus our attention on the hormones of the anterior pituitary; most of what is said about them applies to a wide variety of hormones.

Hormones of the Anterior Pituitary—Negative Feedback

You will recall that certain neurons in the hypothalamus manufacture releasing factors that are secreted into the blood vessels of the stalk and carried to the anterior pituitary. Although there are at least eight releasing factors, the anterior pituitary releases only six major hormones. What accounts for the difference?

We now know that each of the six hormones of the anterior pituitary has a corresponding releasing factor that *stimulates* its secretion. On the other hand, probably only two hormones, growth hormone (GH) and prolactin (PRL), also have releasing factors that *inhibit* their secretion. How can we explain the presence of hypothalamic inhibitory factors for only two of the six major hormones released by the anterior pituitary gland? The answer has to do with a control mechanism called *negative feedback.*

As shown in Figure 6-28 four of the six target organs influenced by hormones from the anterior pituitary (the thyroid, adrenals, testes, and ovaries) respond to hormonal stimulation by releasing hormones. The rates at which they release them depend on the circulating levels of the hormones secreted by the anterior pituitary and *vice versa.* In other words, there are feedback loops between the hormones released by the target organs and those secreted by the pituitary gland. Because *these systems offset or negate any changes in the levels of circulating hormones,* they are called negative feedback loops.

Negative feedback loops tend to stabilize a system. For example, if the concentration of hormones produced by the adrenal cortex (glucocorticoids) should drop for some reason, the anterior pituitary responds by secreting more ACTH, which stimulates the production and release of glucocorticoids. Similarly, an increase in the quantity of glucocorticoids produced by the adrenal glands inhibits secretion of ACTH, which in turn inhibits production of corticosteroids, and so on. But what about growth hormone and prolactin? The secretion of these two hormones does not result in the release of other hormones, but rather stimulates body growth and milk production, among other effects. Because information about body growth and milk production is not conveyed to the anterior pituitary through a hormonal feedback loop, the hypothalamus has, in effect, taken over the role of regulating the concentrations of these hormones by secreting inhibitory as well as stimulating releasing factors.

Each of the feedback loops that connect the hypothalamus, the anterior pituitary, and the target organs has its own special characteristics. This is also true of the negative feedback loops that regulate the release of hormones from other body parts, as well as the concentrations of certain enzymes and other intracellular substances. We will conclude our discussion of how the body coordinates its parts by briefly considering the feedback loops that control the release of hormones by the thyroid gland.

The thyroid gland is located in the neck, where it is wrapped around the trachea just below the larynx, or voice box. This pink-colored gland is made up of collections of epithelial cells that are specialized for manufacturing the two main thyroid hormones, thyroxine (T-4) and triiodothyronine (T-3). These hormones stimulate the uptake of oxygen by most body cells, and thus result in an acceleration of the metabolic rate, among other effects. (The thyroid also produces the hormone *calcitonin,* which helps to regulate the concentration of calcium in body fluids. See Chapter 11.)

These are the negative feedback loops that control the release of thyroid hormones (Figure 6-31): First, the hypothalamus releases thyrotropin releasing factor (TRF), which causes the anterior pituitary to release TSH into the bloodstream. TSH then stimulates the thyroid gland to produce and release T-4 and T-3. The levels of T-4 and T-3 in turn feed back to, and for the most part regulate, the secretion of TSH, thus completing the negative feedback loop.

The manufacture of T-4 and T-3 depends on the presence of iodine. As shown in Figure 6-32, a chronic shortage of iodine in the diet can result in overstimulation of thyroid tissue by TSH, which causes goiter. This happens because in the presence of reduced iodine less T-4 and T-3 are manufactured. Reduced concentrations of T-4 and T-3 in turn result in elevated TSH production and excessive stimulation of the thyroid gland, which cannot increase its output of T-4 and T-3 as long as insufficient iodine is present in the diet.

The day-to-day maintenance of the secretion of thyroid hormones probably depends mostly on the negative feedback loop between the levels of T-4 and T-3 and the amount of TSH released. What, then is the role of the hypothalamic-anterior pituitary (TRF-TSH) feedback loop? Apparently, this part of the control mechanism is adjusted by the hypothalamus under certain special conditions: namely, exposure to chronically hot or cold temperatures. Exposure to cold is known to increase the production of TRF, and therefore to stimulate the release of TSH, whereas hot temperatures have the opposite effect. Thus the

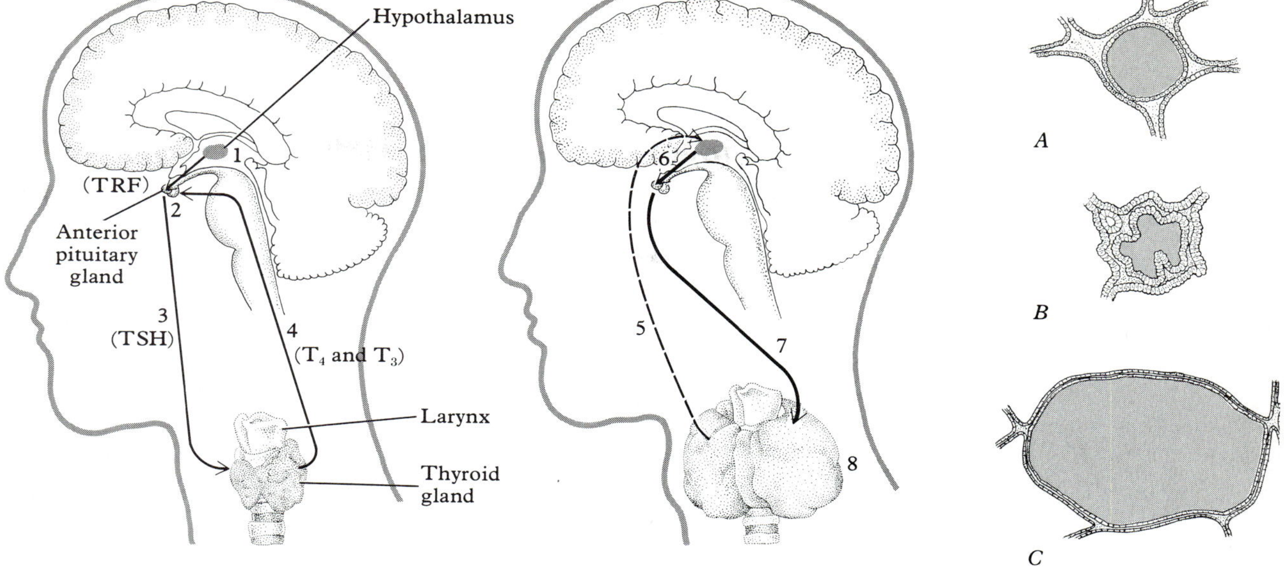

6–31

The negative-feedback system controlling the production of thyroid hormones begins with the neurosecretion from the hypothalamus (1) of thyrotropin-releasing factor (TRF), which goes directly to the pituitary (2) and causes it to release thyroid-stimulating hormone (TSH), into the bloodstream (3). In the thyroid gland (4) TSH acts to bring about the synthesis and secretion into the circulation of the thyroid hormones thyroxine and tri-iodothyronine (T-4 and T-3); the amount of thyroid hormones reaching the anterior pituitary in turn controls the secretion of TSH, thus completing the negative-feedback loop. In the absence of iodine, an essential component of thyroid hormones, insufficient amounts of T-4 and T-3 are produced and excessive amounts of TRF (6) and TSH (7) are therefore secreted, which stimulates the iodine-depleted tissue to grow (8). A normal thyroid follicle, in which hormones are synthesized and stored, consists of an envelope of cells containing colloid, A. *In the absence of iodine, TSH causes the cells to proliferate and become more columnar* B, *and then to produce more colloid, so that the follicles become distended* C, *forming a goiter. (From "Endemic Goiter," by R. Bruce Gillie. Copyright © 1971 by Scientific American, Inc. All rights reserved.)*

releasing factors of the hypothalamus provide a link not only between internal conditions and the regulatory effects of the brain, but also between external conditions and the secretion of some of the hormones that cooperate to control and regulate the internal environment.

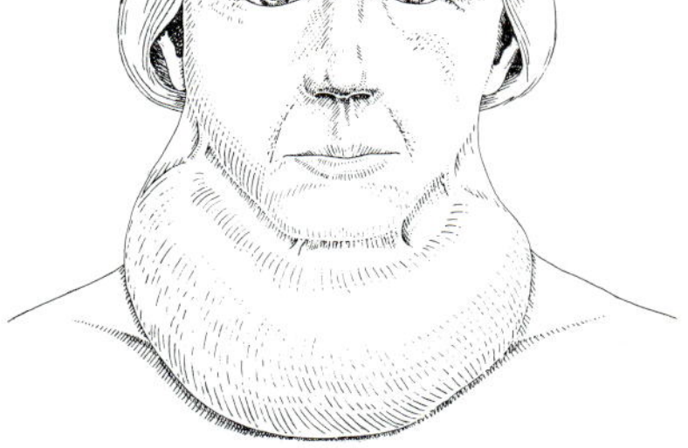

6–32

Large goiters are seen frequently in regions of iodine deficiency. This drawing is based on a photograph made in the Alps near Innsbruck, Austria. Goiters weighing four or five pounds, some of them hanging below the chest, have been reported. (From "Endemic Goiter," by R. Bruce Gillie. Copyright © 1971 by Scientific American, Inc. All rights reserved.)

Summary

Members of the animal kingdom coordinate the workings of their body parts mainly by means of nerves and hormones. The cablelike nerves consist mostly of axons that originate from the cell bodies of neurons in various locations, especially within the brain and spinal cord, and hormones are chemical messengers that are released into the bloodstream by certain glands and other structures.

Neurons transmit information to one another through electro-chemical impulses that are generated because of the selective movement of sodium and potassium ions across the cell membrane. Nerve cells are connected by synapses, most of which release neurotransmitters that either stimulate or inhibit the firing of other neurons.

Highly specialized receptor neurons that respond to specific physical or chemical signals are associated with each of the special senses. Sensations of pain and temperature at the body surfaces travel from spinal nerves to the spinal cord, where they cross to the opposite side and then ascend by a complicated pathway to well-localized areas of the primary sensory cortex of the cerebral hemispheres. Impulses originating in the motor area of the cortex are concerned with the production of voluntary movement.

The brainstem is not only the place of origin of all but one of the twelve pairs of cranial nerves, but is also the location of the reticular formation and a center for the control of various reflexes. The medulla of the brainstem helps to regulate respiration, blood pressure, and heart rate, and it also coordinates certain protective reflexes, such as sneezing and vomiting.

The autonomic nervous system helps to regulate the internal organs and has two divisions, sympathetic and parasympathetic. The two divisions of the ANS have generally opposite effects, and each employs a different neurotransmitter. Autonomic neurotransmitters affect the concentration of the compounds cAMP and cGMP inside cells, as evidenced by the biochemical upset that results in asthma attacks.

The hypothalamus is made of clusters of neurons that regulate certain reflexes and appetites, including hunger and thirst. Connections between the hypothalamus and the sympathetic division of the autonomic nervous system mediate the fight or flight reactions associated with stimuli that are emotion-charged. The hypothalamus bridges the gap between nerves and hormones in that some of the neurons of the hypothalamus either manufacture or release chemical messengers.

The pituitary gland has two lobes, and they are connected to the hypothalamus in different ways. The hypothalamus is connected to the posterior lobe by axons and to the anterior lobe by blood vessels. Certain hypothalamic neurons secrete releasing factors into the blood vessels that supply the anterior lobe and thereby promote or inhibit the release of pituitary hormones.

The anterior pituitary releases six hormones, each of which helps to regulate a function. Most hormones affect their target organs by combining with specific receptors on cell membranes. The presence of the hormone on the cell surface leads to alterations in the concentration of cAMP and cGMP inside the cells of the target organ.

The secretion of most hormones is regulated by negative feedback loops, which tend to offset or negate changes in the concentrations of circulating hormones. The secretion of hormones by the anterior pituitary is regulated by feedback loops between the gland and its target organs and between the gland and the hypothalamus.

Suggested Readings

1. "The Synapse," by Sir John Eccles. *Scientific American,* Jan. 1965, Offprint 1001. Discusses the structure and function of the areas where nerve cells are connected.

2. "Neurotransmitters," by Julius Axelrod. *Scientific American,* June 1974, Offprint 1297. How the molecules released from nerve-fiber endings are chemical messengers that enable nerve cells to communicate with one another.

3. "Smell and Taste," by A. J. Haagen-Smit. *Scientific American,* March 1952, Offprint 404. How these two senses depend upon chemical rather than physical stimuli.

4. "The Neurophysiology of Binocular Vision," by John D. Pettigrew. *Scientific American,* Aug. 1972, Offprint 1255. Discusses how the perception of objects in three dimensions may be related to the activity of single nerve cells in the visual cortex of the brain.

5. "The Reticular Formation," by J. D. French. *Scientific American,* May 1957. Offprint 66. How this area of the brainstem acts as a selective filter for impulses traveling toward the cerebral cortex.

6. "The Human Thermostat," by T. H. Benzinger. *Scientific American,* Jan. 1961, Offprint 129. How certain nerve cells in the hypothalamus are directly involved in the maintenance of human body temperature.

7. "Learning in the Autonomic Nervous System," by Leo V. Di Cara. *Scientific American,* Jan 1970, Offprint 525. Recent evidence suggests that many of the body's autonomic functions may be brought under conscious control by means of learning.

8. "Cyclic AMP," by Ira Pastan. *Scientific American,* Aug. 1972, Offprint 1256. Discusses the role of this second messenger in bringing about the widespread effects of several hormones.

9. "The Hormones of the Hypothalamus," by Roger Guillemin and Roger Burgus. *Scientific American,* Nov. 1972, Offprint 1260. How the release of hormones from the anterior pituitary depends in part on the release of hormones from the hypothalamus.

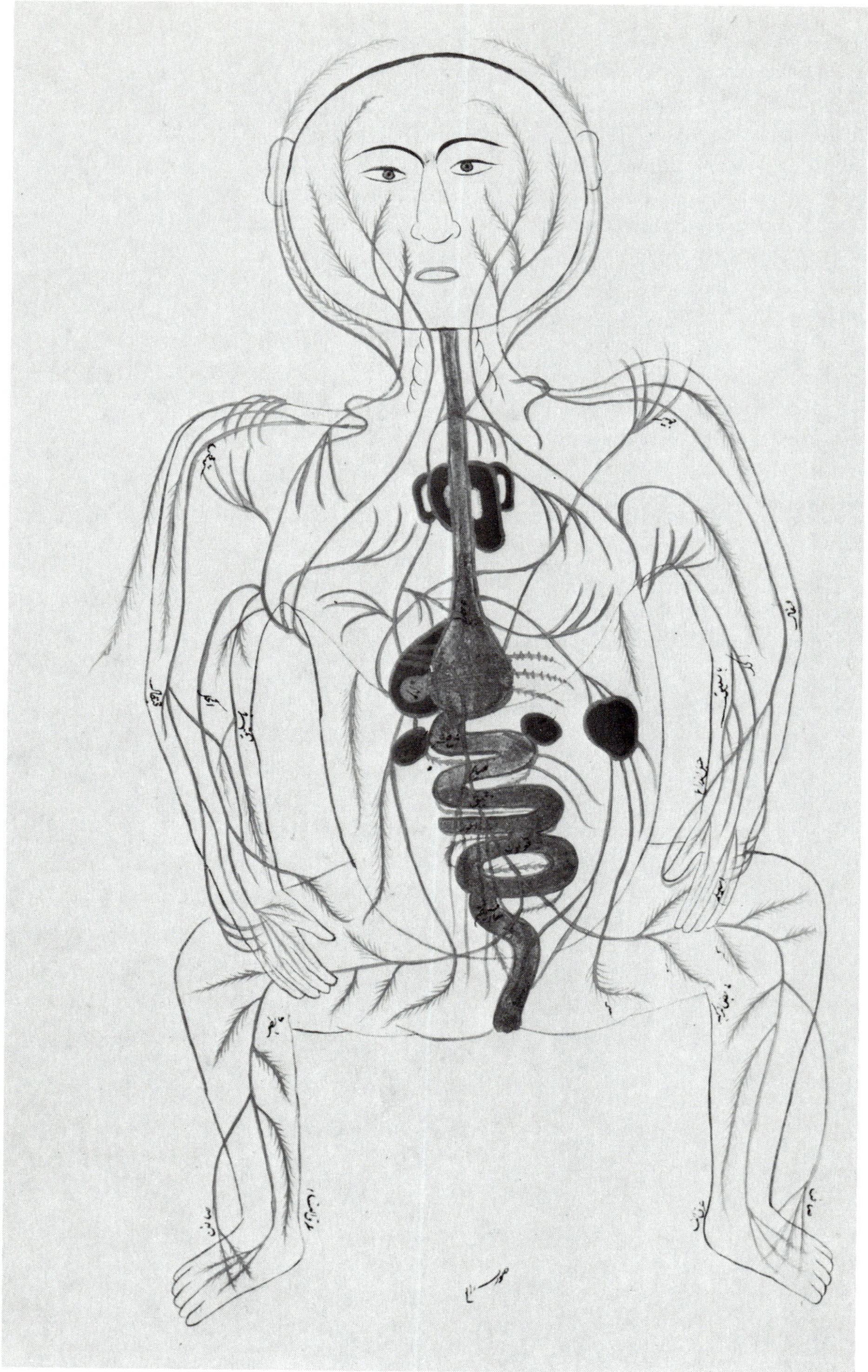

Persian diagram of the circulatory system, from a manuscript dated 1672. (Reproduced by permission of the Director of the India Office Library and Records.)

CHAPTER 7

Blood and Circulation

All cells of the human body require certain nutrients and must eliminate certain waste products. Nutrients and waste products alike are transported to and from cells by means of the blood, which is propelled through a remarkable system of circulatory vessels by the rhythmic pumping action of the heart. Blood also helps to provide protection against infection and it initiates the repair of damaged vessels within the myriad channels of the circulatory system. In this chapter we discuss the structure and function of blood and the circulatory apparatus, as well as some important environmental factors that can strongly influence the circulatory system's proper functioning. Let us begin by turning our attention to the composition of human blood.

Composition of the Blood

Blood is composed of *formed elements* (red cells, white cells, and platelets; see Figure 7-1) and the liquid in which the formed elements are carried, the *blood plasma.* Under usual circumstances, human blood is slightly less than 50 percent formed elements and slightly more than 50 percent plasma. Membrane-surrounded particles in the bloodstream are called formed elements rather than cells because blood platelets are not cells, but rather pinched off bits of large cells, called megakaryocytes, that exist not in blood but in the bone marrow (see below).

When blood is removed from the body and allowed to stand, the formed elements settle to the bottom and the clear, slightly straw-colored plasma remains on top. The plasma contains many kinds of molecules, including nutrients, waste products, hormones, and proteins. The plasma proteins perform three essential functions: first, together they provide the osmotic pressure that is essential for proper delivery of nutrients and removal of wastes. Second, certain of the plasma proteins, the antibody molecules, protect against infection. Finally, certain other proteins, the clotting factors, are essential for blood coagulation.

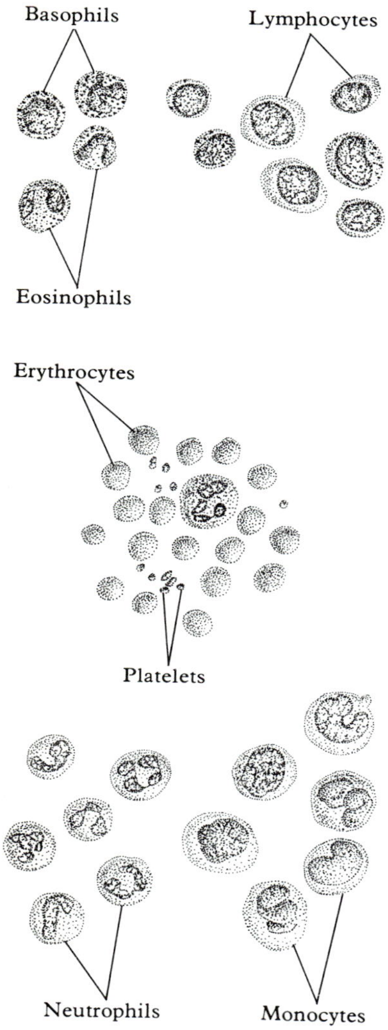

7–1
Normal human blood cells. (After R. O. Greep, Histology, *2d ed. Copyright © 1966 by McGraw-Hill, Inc., New York. Used with permission of McGraw-Hill Book Company.)*

Exchanges Between Plasma and the Body's Cells

All living cells of the body are located close enough to the smallest units of the circulatory system, the *capillaries,* to be able to receive what they need from the plasma that passes out of the capillaries. Even in the epithelial tissues of the body, such as the skin's epidermis, where there are no blood vessels or capillaries, the cells are close enough to the capillaries for exchange to take place. Exchange between cells and plasma depends upon *diffusion,* which, as you may know, is the tendency of dissolved substances to pass from areas of greater concentration to areas of lesser concentration. The cells that form the lining of the capillaries permit the diffusion of the gases oxygen (O_2) and carbon dioxide (CO_2) and of small molecules (amino acids and glucose), but allow only gradual passage of large molecules such as proteins.

When we recall that cells consume oxygen and glucose and give off carbon dioxide (Chapter 5), it is clear that the extracellular fluid, which surrounds the cells, will become depleted of oxygen and glucose and enriched in carbon dioxide. Thus oxygen and glucose, which are present in relatively high concentrations in plasma, will diffuse out of the capillaries and into the tissue fluid and cells. Conversely, carbon dioxide will diffuse away from the cells and tissue fluid and into the capillaries, where it is present in lesser concentrations (Figure 7-2).

In addition to diffusion, two other forces contribute to the exchange between cells and plasma. These forces are *blood pressure* and *osmotic pressure,* and they work in opposition to each other. Blood pressure is generated by the pumping action of the heart. It is highest in the large arteries, but is still present to some extent in all major vessels and even in the capillaries. As the blood enters the capillaries from the smallest arteries its pressure is equivalent to about 35 millimeters of mercury (abbreviated 35 mm Hg, which means that there is sufficient pressure inside the capillary to push a column of mercury up 35 millimeters). By the time the blood has passed through the capillaries to the point where it enters the veins, the blood pressure has dropped to about 15 mm Hg. Thus, the blood pressure tends to drive plasma out of the capillaries and into the tissue fluid and it is greater at the arterial end of the capillary than at the venous end (see Figure 7-3).

The blood pressure is counteracted by osmotic pressure, which owes its action to the fact that there is much more protein inside the capillaries than in the surrounding tissue fluid. Osmosis is the tendency of water to flow from regions of greater water concentration (and lesser solute concentration) to regions of lesser water concentration (and greater solute concentration), and the concentration of water in plasma is less because of the presence of large amounts of protein in the plasma. Thus water (and small molecules dissolved in it) tends to flow from the tissue fluid into the plasma because of osmotic pressure. Because the cells lining the capillaries permit only very gradual passage of the large protein molecules, osmotic pressure (unlike the blood pressure) remains relatively constant at about 25 mm Hg from one end of the capillary to the other.

Inspection of the pressure relationships forged by the opposing forces of blood pressure and osmotic pressure (Figure 7-3) reveals that plasma tends to pass *out* of the capillary at the arterial end of the capillary (where blood pressure exceeds osmotic pressure) and to pass *into* the capillary at the venous end (where osmotic pressure exceeds blood pressure). The exit of plasma into the tissue fluid at the arterial end of the capillary is highly beneficial because the arterial blood, having just come from the lungs, has a great concentration of oxygen, which the cells require. Similarly, the entry of tissue fluid into the capillary at the venous end of the capillary is beneficial because the tissue fluid will contain higher concentrations of cell waste products such as CO_2. The interaction of these two forces thus aids diffusion in providing for exchange between the cells and the plasma.

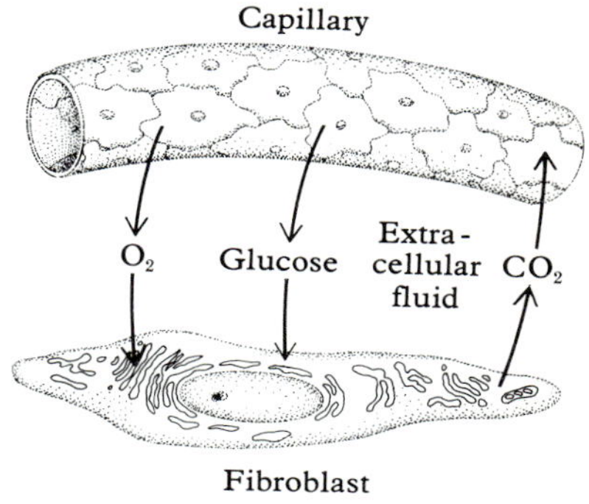

7–2
Cells, such as the fibroblast depicted here, use up O_2 and glucose and release CO_2. As a result, the extracellular fluid around the cells tends to become depleted of O_2 and glucose and enriched in CO_2. The balance is restored by diffusion of O_2 and glucose from the capillaries and the diffusion of CO_2 into the capillaries. The arrows indicate the direction in which the various molecules diffuse.

Red Blood Cells and the Transport of Oxygen

The red blood cells, or *erythrocytes,* are particularly important in oxygen transport because the hemoglobin contained within them greatly increases the quantity of oxygen that can be carried by the blood. If the blood were only plasma, it would carry only about 3 percent of the oxygen necessary to sustain life. There are many more red cells in the blood than white cells or platelets (Table 7-1) and the blood of an average-sized person (about four liters) contains 20×10^{12} (20 trillion) cells. Red cells are manufactured in the bone marrow, where during maturation they lose their nucleus, synthesize large amounts of hemoglobin, and become disc-shaped, with slight concavities on each side (Frontispiece, Chapter 5). After a red cell has spent about four months in the circulatory system it begins to break down and is removed from the blood by

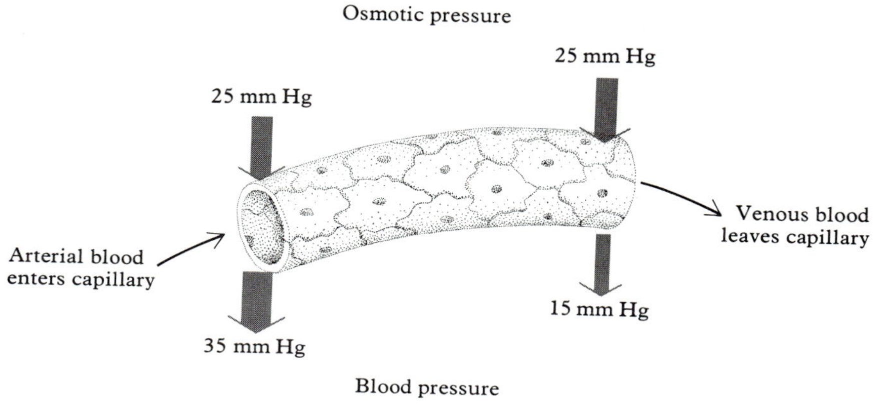

7–3
Pressure relationships at arterial and venous ends of the capillary. Blood pressure exceeds osmotic pressure at the arterial end, which drives plasma rich in oxygen and nutrients out of the capillary. Osmotic pressure exceeds blood pressure at the venous end, which drives extracellular fluids that have picked up waste products back into the capillary.

phagocytic cells present in the spleen and liver. There its main component, hemoglobin, is recycled.

The components of a hemoglobin molecule are iron, heme, and protein. During the recycling process, the protein portions are broken down into amino acids, which are used for the synthesis of new proteins. The heme component, with its attached iron, is converted to the yellowish pigment *bilirubin* (one of the chief pigments of bile), which is excreted by the liver into the small intestine. About 95 percent of the bilirubin is reabsorbed from the intestine.

People who have fewer than normal red cells have *anemia.* They often feel weak and out of energy because they can't get enough oxygen to their cells to support normal cell activities such as muscle contraction. Anemia can result either from greater than usual loss of red cells or from their decreased production. Thus anemia may result from direct blood loss, such as from a duodenal ulcer, or from accelerated breakdown of red cells, which sometimes occurs as a consequence of drug therapy. For example, a drug such as penicillin may become attached to the surface of red cells, and if the body synthesizes antibody molecules that combine with the drug the red cells become coated with the antibodies and may eventually be destroyed. Fortunately, this reaction to drugs is relatively rare. Anemia that results from a severe failure of red cell production in the bone marrow is known as *aplastic anemia.*

White Blood Cells and Immune Defenses

White blood cells, or *leukocytes,* are important in the body's immune defenses. There are five kinds of white cells, as Figure 7-1 illustrates: *lymphocytes,* which synthesize antibodies (see Figure 5-5); *monocytes,* which are phagocytic (they engulf foreign particles); and three kinds of *granulocytes.* One kind, *eosinophils,* help to defend the body against certain parasitic infections; their numbers are

TABLE 7-1
Formed elements found in human blood.

FORMED ELEMENT	NUMBER (PER ML OF BLOOD)
Red cells	5,000,000,000 (5×10^9)
White cells	8,000,000 (8×10^6)
Platelets	300,000,000 (3×10^8)
WHITE-CELL TYPE	**PERCENT OF WHITE-CELL POPULATION**
Granulocytes	
Neutrophils	50–70
Eosinophils	1–4
Basophils	0.1–0.8
Lymphocytes	20–40
Monocytes	2–8

greater in people who have parasitic infections or allergies. The second, *neutrophils,* are phagocytic, and their numbers increase rapidly during the early stages of bacterial infection. The third, *basophils,* are phagocytic, but their specific functions are unknown.

The three kinds of granulocytes have much in common, including nuclei that are multilobed. Because of this, they are sometimes called *polymorphonuclear leukocytes.* All granulocytes also contain large numbers of granules in their cytoplasm, and their names are derived from the fact that the granules of the three different types stain differently: eosinophil granules take acidic dyes (such as eosin), basophil granules take basic dyes, and the neutrophil granules are not stained by either acidic or basic dyes. Neutrophils make up about 90 percent of the polymorphonuclear leukocytes and they are exquisitely sensitive to the products of microorganisms. When microorganisms invade the tissues, the neutrophils burrow through the walls between the cells lining the capillaries and enter the infected tissues, where they engulf and digest the invaders. *Pus,* the whitish material that accumulates in an infected area, is composed mostly of neutrophils. During a bacterial infection the number of neutrophils in the circulation increases because they are rapidly released into the circulation from large bone marrow reserves.

White blood cells are also components of the *lymphatic system* (Figure 7-4), which consists of lymphatic vessels, lymph fluid, lymph nodes and white blood cells. The lymphatic system returns proteins and white blood cells that are forced out of the capillaries back to the circulatory system. In fact the lymphatic system and the circulatory system interact in a number of ways. The lymphatic system, like the circulatory system, contains numerous vessels (the lymphatics) that serve almost all parts of the body, and the interaction of the two systems is worth discussing further.

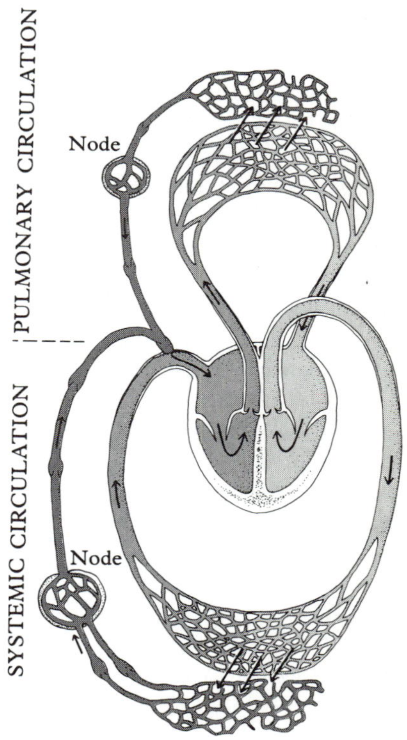

7–4
Two circulatory systems, the blood and the lymphatic, are related in this diagram. Oxygenated blood (light gray) is pumped by the heart through a network of capillaries, bringing oxygen and nutrients to the tissue cells. Venous blood (dark gray) returns to the heart and is oxygenated in the course of the pulmonary (lung) circulation. Fluid and other substances seep out of the blood capillaries into the tissue spaces and are returned to the bloodstream by the lymph capillaries and larger lymphatic vessels. (From "The Lymphatic System," by H. S. Mayerson. Copyright © 1963 by Scientific American, Inc. All rights reserved.)

The Lymphatic System

Lymphatic vessels exist throughout most of the body (Figure 7-5). The tiny lymphatic vessels originate in the tissues as blind-end tubes and pass through *lymph nodes,* which are nodular accumulations of cells (mostly lymphocytes and macrophages). They then unite with other lymphatics to form larger lymphatic channels. All lymphatics from the trunk and lower extremities finally join to form the largest lymphatic vessel in the body, the *thoracic duct,* which eventually empties its contents into a large vein—the left subclavian vein—at the junction where the veins that serve the left arm and the left side of the head and neck meet. The lymphatics that serve the left side of the upper body (head, neck and arm) join the thoracic duct just before it empties into the subclavian vein. On the other side of the body, lymphatics serving the right side of the upper body eventually empty into the right subclavian vein. So overall, the lymphatic vessels are a one-way drainage system that originates in the tissues, then filters through lymph nodes and eventually rejoins the blood circulatory system.

The lymphatics are particularly important in the transport of proteins, white cells, and fatty acids. Many white cells are very active and can even squeeze through the membranes of other cells, thus entering and leaving capillaries and the circulatory system (Figure 7-6). Some of these errant white cells are returned to the circulation through the lymphatics. Proteins, which pass

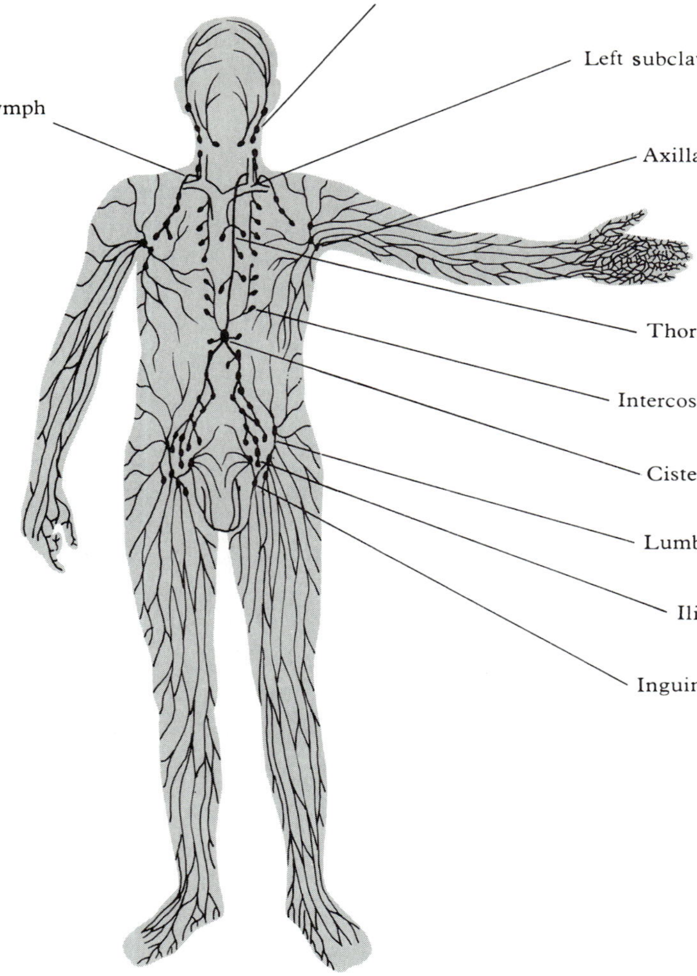

7–5
Lymphatic vessels drain the entire body, penetrating most of the tissues and carrying back to the bloodstream excess fluid from the intercellular spaces. This diagram shows only some of the larger superficial vessels, which run near the surface of the body, and deep vessels, which drain the interior of the body and collect from the superficial vessels. The thoracic duct drains most of the body and empties into the left subclavian vein. The right lymph duct drains the heart, lungs, part of the diaphragm, the right upper part of the body, and the right side of the head and neck, emptying into the right subclavian vein. Lymph nodes interspersed along the vessels trap foreign matter, including bacteria. (From "The Lymphatic System," by H. S. Mayerson. Copyright © 1963 by Scientific American Inc. All rights reserved.)

through capillary walls, though slowly, are also returned to the circulation through the lymphatics (Figure 7-7). If the lymphatics are blocked, as sometimes occurs when people are infected with certain nematode roundworms, *elephantiasis* may result. In this condition the proteins are not removed from the tissues, and affected regions can become enlarged (elephantine in proportions) by the accumulation of tissue fluid rich in protein (Figure 7-8). Elephantiasis in its early stages can be reversed by drugs that kill the roundworms. Lymphatics also absorb large fatty acid molecules from the small intestine. The mechanisms by which proteins, fatty acids and other substances enter the lymphatics is incompletely understood, although it is thought that entry may take place in part through the extensive overlapping junctions between the cells that line the lymphatics (Figure 7-9).

In the event of infection, or some other kind of tissue damage, the lymphatics remove the foreign microorganisms, white cells, and tissue debris that accumulate at the site of injury. During infections, the lymphatic vessels may become inflamed, which causes their channels to become visible as red streaks. The lymphatics serve the very important function of delivering

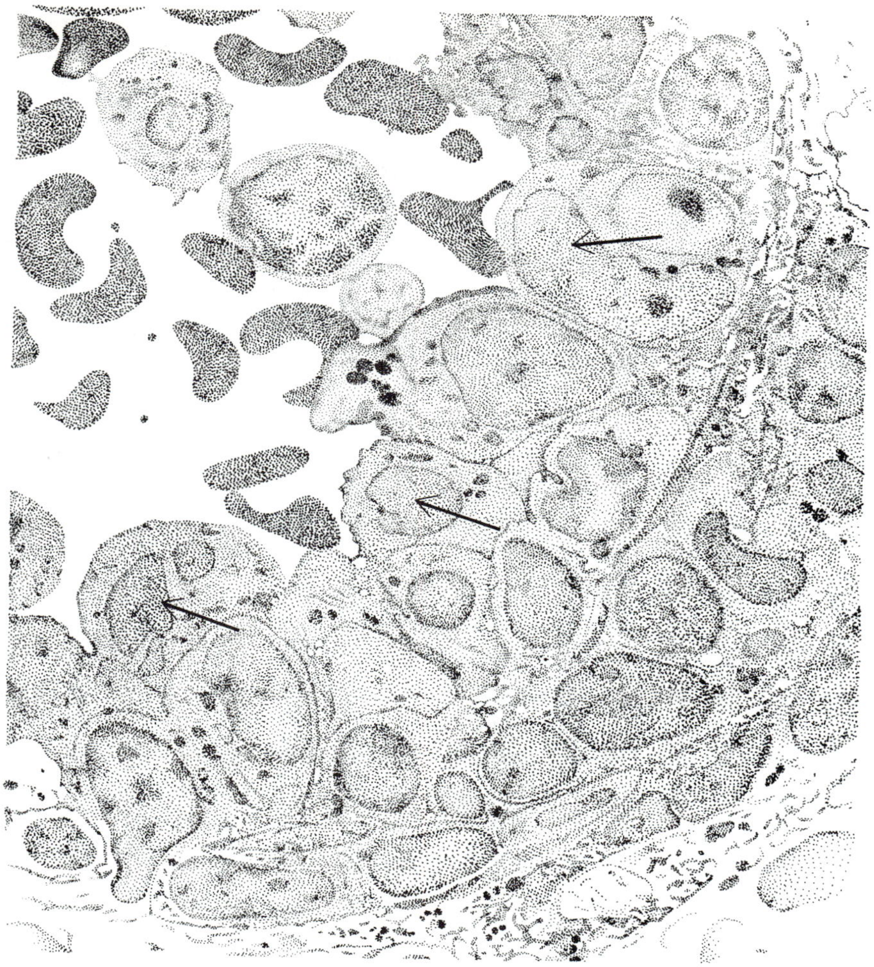

7–6
The passage of lymphocytes through postcapillary venules of a normal laboratory rat (×3240). The nuclei of several lymphocytes (dark gray shapes) can be seen in the cytoplasm of endothelial cells, the larger, lighter nuclei of which are indicated with arrows (the darkest cells in the white lumen at the upper left are red blood cells). Most of the lymphocytes are outside the endothelium and are visible in the crescent extending from the upper right side of the picture to the lower left.

infectious organisms and their products to the lymph nodes, where the immune responses are initiated by contact between the invading microbes and the macrophages and lymphocytes in the lymph nodes (see Chapter 13). As a result of the intense cellular activity associated with coping with an infection, the lymph nodes that drain an infected region frequently become considerably enlarged.

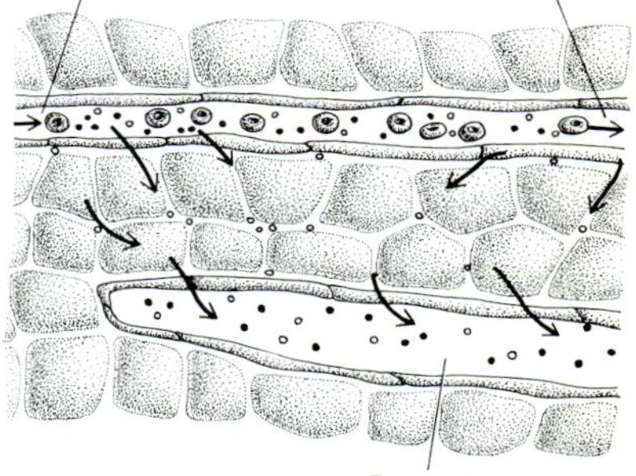

7–7
Large molecules such as proteins (black dots) and large fatty acids (open circles) leave the blood capillaries along with fluid and salts. Some of the fluid and salts are reabsorbed; the excess, along with large molecules that cannot re-enter the blood capillaries, is returned through the lymphatic system. (From "The Lymphatic System," by H. S. Mayerson. Copyright © 1963 by Scientific American, Inc. All rights reserved.)

Platelets and the Clotting Mechanism

Platelets are the smallest formed elements in the blood. Platelets have no nucleus, like red blood cells, but they do contain numerous small granules. Platelets are produced from very large cells, megakaryocytes, which reside in the bone marrow. Megakaryocytes have fingerlike cytoplasmic processes that penetrate into the capillaries in the bone marrow, and the platelets are believed to be formed when the ends of these processes break off inside the capillaries.

When a blood vessel is severed, the damaged ends draw back and contract, thereby reducing the diameter of the vessel. For reasons that are not completely understood, platelets are attracted to the damaged area, where contraction of their microfilaments enables them to change shape so that they adhere to the damaged cells and to each other, forming a *platelet plug* that can seal off the end of a small vessel. During this process the platelets discharge their granules, which contain *serotonin,* a potent constrictor of the smooth muscle that encircles small vessels.

If larger vessels are injured, the platelet plug alone is inadequate to stop the exit of plasma and formed elements. But a more secure and permanent plug is provided by a *blood clot,* which is formed by the conversion of a dissolved circulating protein, *fibrinogen,* into an insoluble meshwork of strands of the protein *fibrin* (Figure 7-10). The steps that lead to the formation of fibrin from fibrinogen are complex and they require the interactions of at least twelve different clotting factors. The absence of even one of these clotting factors can lead to *hemophilia,* in which clotting cannot occur or takes place very slowly. At one time, hemophilia debilitated a number of the royal families in Europe, as we discuss in Chapter 17. Although there is some debate about the precise workings of these clotting factors, it is generally believed that they are part of a chain reaction and that the conversion of fibrinogen to fibrin is merely the final event in that reaction (Figure 7-11).

Although blood clotting is a highly beneficial process, it must be regulated so that the clot does not spread beyond the area where it is initiated and needed to prevent blood loss. Spreading is prevented partly by another set of chemical reactions known as the *fibrinolytic system* (Figure 7-11). In the fibrinolytic system another blood protein, *plasminogen,* is converted to *plasmin* by the action of plasminogen activator, which is found in many cells and in the body fluids. Plasmin in turn breaks down the fibrin meshwork into inactive degradation products. A newly formed clot is in a state of equilibrium, because there is a balance between clot formation and fibrinolysis. On occasion, however, the clotting mechanism may become overactive, so that the normal fibrinolytic system cannot clear away clots as rapidly as they are formed. *Thrombosis* is the term used to describe clotting that occurs inside an intact vessel, and it is a dangerous condition because clots can block the circulation to important regions. Thrombosis usually results because clots are formed more rapidly than the body can dissolve them.

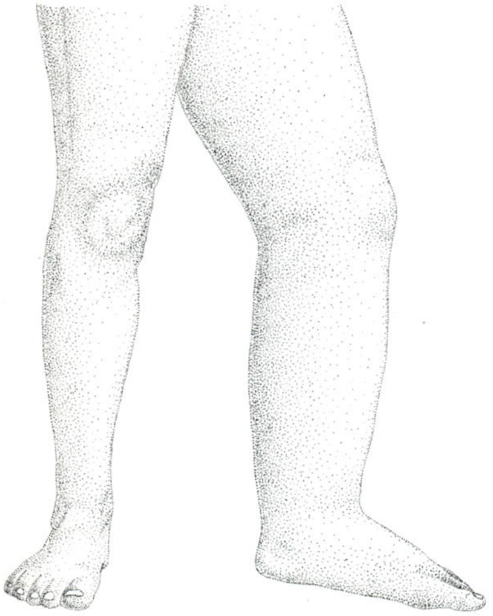

7–8
Elephantiasis resulting from blockage of lymphatic channels can cause gross deformity of a limb and even disability. The drawing is based on a photograph of an 11-year-old girl whose leg began to swell when she was seven, probably because of an insufficiency of lymphatic vessels. Her condition was greatly improved by an operation in which the tissue between skin and muscles was removed. (From "The Lymphatic System," by H. S. Mayerson. Copyright © 1963 by Scientific American, Inc. All rights reserved.)

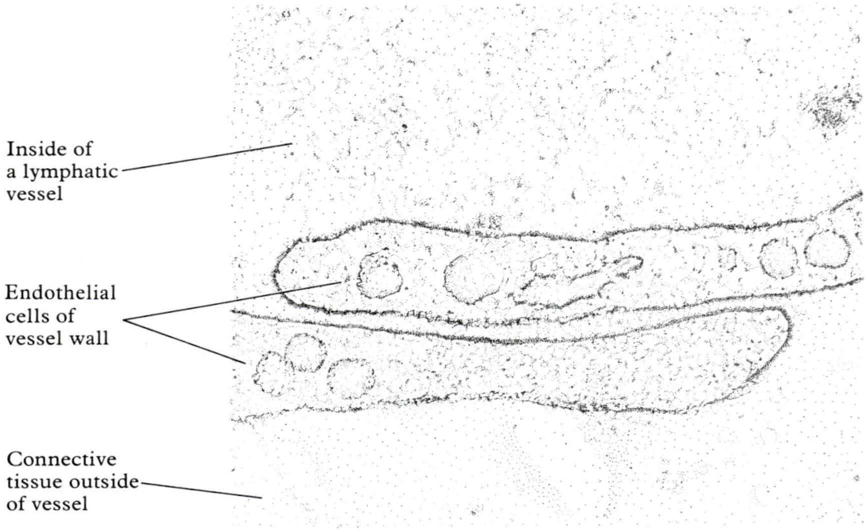

7–9
Loose junction *between the overlapping ends of two endothelial cells of a vessel wall is seen in this drawing from an electron micrograph of a lymphatic in mouse intestinal tissue. The lymph vessel is at the top, connection tissue at the bottom. (After an electron micrograph by Johannes A. G. Rhodin of the New York University School of Medicine.)*

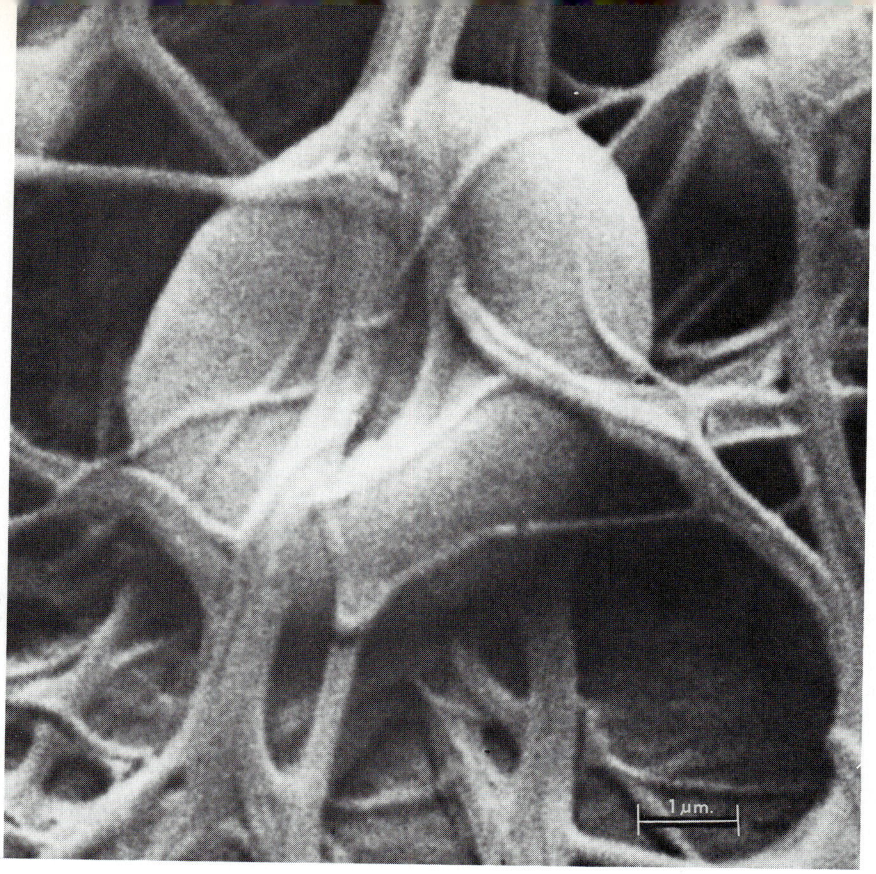

7–10
An erythrocyte enmeshed in strands of fibrin. (Scanning electron micrograph courtesy of Emil Bernstein, The Gillette Company Research Institute.)

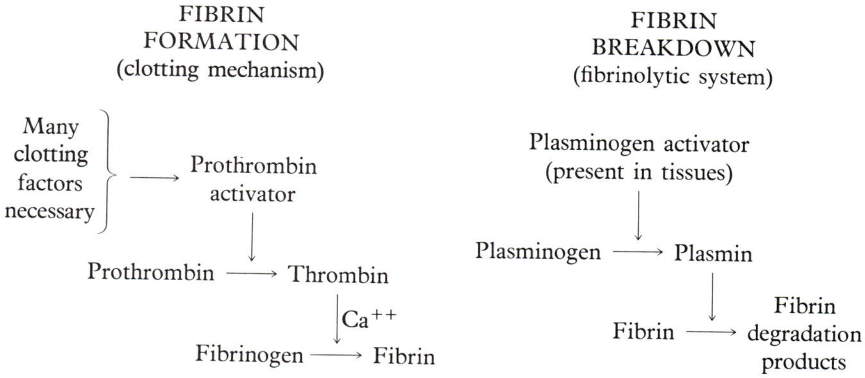

7–11
Formation and breakdown of a blood clot. The last two steps in the clotting mechanism (above left) lead to fibrin formation, and can occur only if many clotting factors have led to the generation of prothrombin activator. Breakdown of the clot is accomplished by the fibrinolytic system (above right), which is brought to action by plasminogen activator. A newly formed clot is in a state of equilibrium, with both processes occurring simultaneously. The fibrinolytic system prevents the spread of the clot throughout the vascular system.

Circulation of the Blood

Blood moves continuously throughout the body, propelled by the powerful contractions of the heart. The term *artery* is applied to vessels that carry blood away from the heart, and the term *vein* to vessels that carry blood toward the heart. Blood that has picked up oxygen in the capillaries of the lungs collects in the pulmonary veins, which empty into the left side of the heart (Figure 7-12). In turn, the heart pumps blood under pressure into the aorta, the largest of the arteries. All major oxygen-carrying arteries branch from the aorta and continue

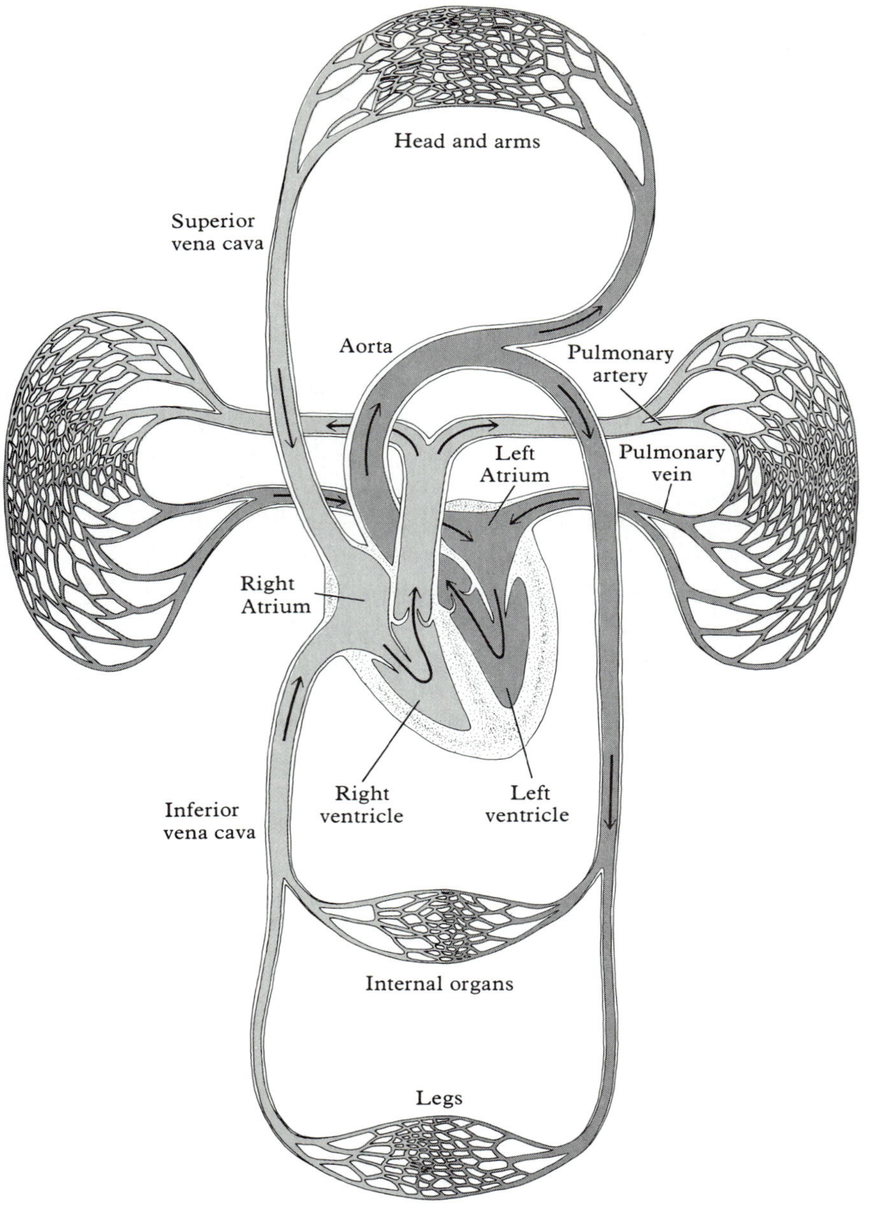

7–12
The heart and its relationship to the circulatory system. The arterial blood is represented in dark gray; venous blood, in a somewhat paler gray. The capillaries of the lungs are represented at the left and right; the capillaries of the rest of the body, at top and bottom. (From "The Heart," by Carl J. Wiggers. Copyright © 1957 by Scientific American, Inc. All rights reserved.)

to branch into smaller and smaller arteries, the smallest of which are called *arterioles.* Like larger arteries, arterioles have smooth muscle fibers in their walls that allow them to expand or contract under the influence of blood pressure or the effects of either the sympathetic or parasympathetic nervous system. Arterioles empty into capillaries, where the exchanges of nutrients and wastes take place both through and between the single layer of cells that line the capillaries (Figure 7-13). After passing through the capillaries, the blood collects first in the smallest veins, or *venules,* and then in larger and larger veins until the largest veins empty into the right side of the heart. The body's two largest veins, both of which empty into the right side of the heart, are the *superior vena cava,* which collects blood from the head, neck and arms, and the *inferior vena cava,* which collects blood from the rest of the body. The right side of the heart then pumps its oxygen-poor blood into the pulmonary artery, whose branches terminate in the capillaries of the lungs, thus completing the circuit.

Notice that although the heart is a single organ, its two sides are separated from one another by a thick wall of muscle. The right side of the heart pumps blood to the capillaries of the lungs and the left side pumps blood to the capillaries of the rest of the body. Because of this separation, the circulatory system is often described as two circulations, the *systemic circulation* (to most of the body) and the *pulmonary circulation* (to the lungs), as Figure 7-4 illustrates. The pulmonary arteries and veins are the only ones in the body in which the veins contain oxygen-rich blood and the arteries do not.

Whereas the pumping action of the heart keeps the blood moving in the arteries, there is no equivalent pump that keeps the blood in the veins moving toward the heart. Instead, flow in the veins depends upon two things: the pumping action of muscles that squeeze the veins that run inside them, and the pumping action of the diaphragm, which causes changes in pressure in abdominal veins as it rises and falls during breathing. Sometimes a person who has stood relatively still for a long time faints because the muscle-pumping is not sufficient to propel enough blood back to the heart and thence to the brain.

Within the veins are *valves,* which are crucial to the flow of blood because they keep it moving in the right direction (Figure 7-14). The valves prevent backflow by closing completely. If they do not close completely the backflow of blood can stretch the veins behind the valves—a condition known as *varicose veins.* Varicose veins can occur in the superficial veins of the leg if the veins receive backflow from the deep veins of the leg. This backflow occurs when the valves of communicating veins do not close, causing engorgement of the superficial veins (Figure 7-15). Varicose veins of the leg occur in both sexes, but are more common in women who have had children. This is because there is more pressure in the abdomen during pregnancy, which tends to stretch the veins below the abdomen, including the communicating veins. Stretched veins tend to close incompletely, so the blood they carry backs up into the superficial veins.

Portal Systems

Although blood leaving the capillaries usually flows into increasingly larger veins to be delivered back to the heart, in two regions of the body there are special portal systems in which the blood leaving capillaries flows into veins terminating in *other capillaries* before the blood is returned to the heart. These special capillary-to-capillary circulatory arrangements are found in the abdomen, where the *hepatic portal circulation* connects the abdominal portion of the

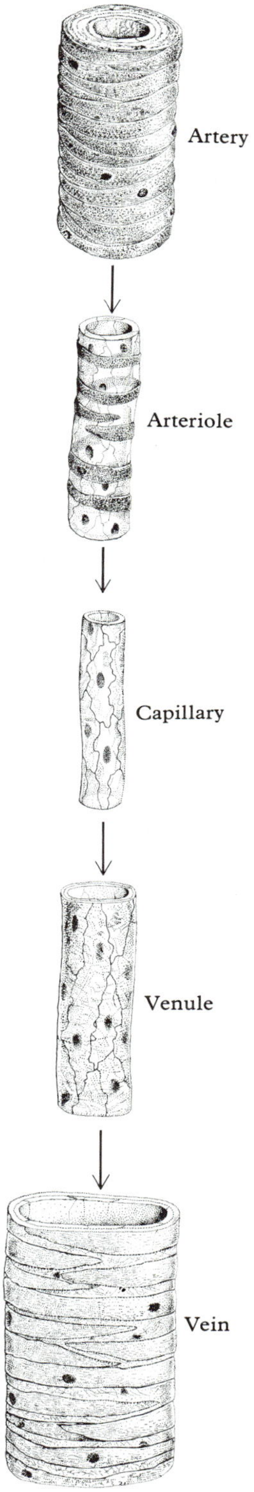

7–13
Walls of blood vessels of various kinds. The wall of an artery consists of a single layer of endothelial cells sheathed in several layers of muscle cells interwoven with fibrous tissues. The wall of an arteriole consists of a single layer of endothelial cells sheathed in a single layer of muscle cells. The wall of a capillary consists only of a single layer of endothelial cells. The wall of a venule consists of endothelial cells sheathed in fibrous tissue. The wall of a vein consists of endothelial cells sheathed in fibrous tissue and a thin layer of muscle cells. Thus a layer of endothelial cells lines the entire circulatory system. (From "The Microcirculation of the Blood," by Benjamin W. Zweifach. Copyright © 1958 by Scientific American, Inc. All rights reserved.)

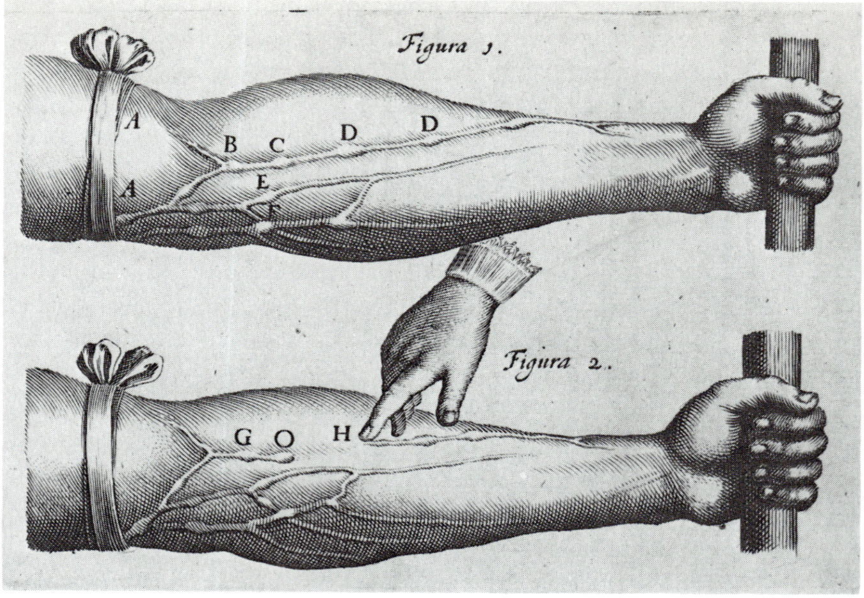
Figura 1.
A
A
B
C
D
D
E
F
Figura 2.
G
O
H

7–14
Plate from a manuscript written by William Harvey in 1628 showing the superficial veins of the forearm. The locations of venous valves are marked by bulges in the veins. The lower figure demonstrates that valves maintain one-way flow in the veins. Blood will not flow back from the section of vein labeled G-O into section O-H if the blood is pressed out of section O-H. (Courtesy the National Library of Medicine, Bethesda, Maryland.)

gastrointestinal tract with the liver (Figure 7-16); and in the brain, where the *hypophysial-portal circulation* (Figure 7-17) connects the hypothalamus to the anterior portion of the pituitary gland. In these portal systems the circulatory system facilitates important functions by providing direct region-to-region connections. In the hepatic portal circulation, the blood from the intestinal capillaries, which contains nutrients absorbed from the gastrointestinal tract, is delivered directly to the liver, where the nutrients are used as the building blocks for the body's proteins, carbohydrates, and fats. Toxic molecules such as drugs, alcohol, and pesticide residues are also delivered to the liver, where they are

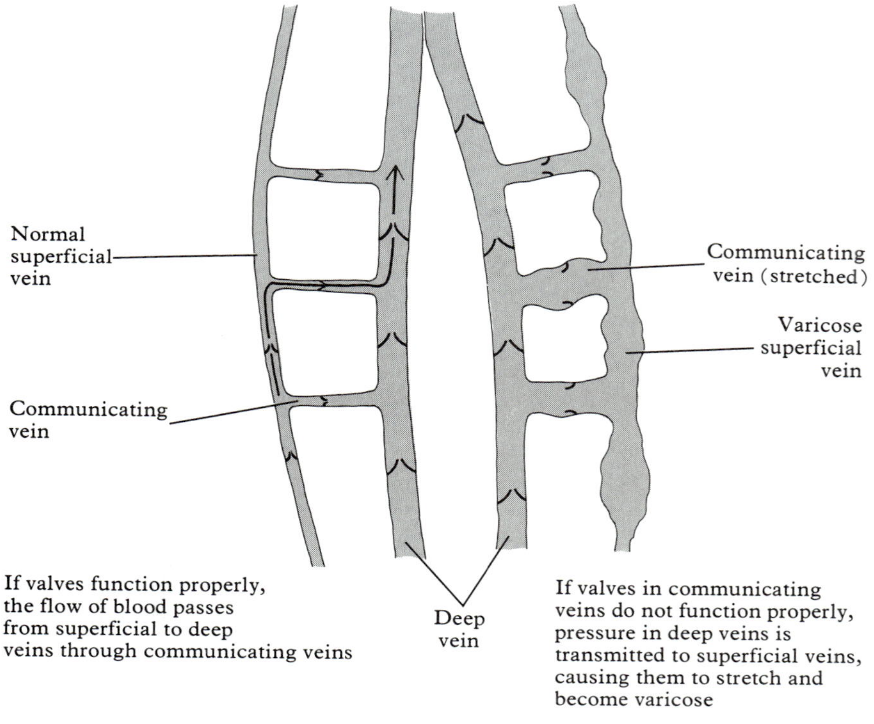
Normal superficial vein
Communicating vein
Communicating vein (stretched)
Varicose superficial vein
If valves function properly, the flow of blood passes from superficial to deep veins through communicating veins
Deep vein
If valves in communicating veins do not function properly, pressure in deep veins is transmitted to superficial veins, causing them to stretch and become varicose

7–15
Relationship between superficial, deep, and communicating veins in the leg. Backflow of blood into superficial veins may cause varicose veins to form. Backflow occurs when communicating veins have been stretched so that their valves will not close.

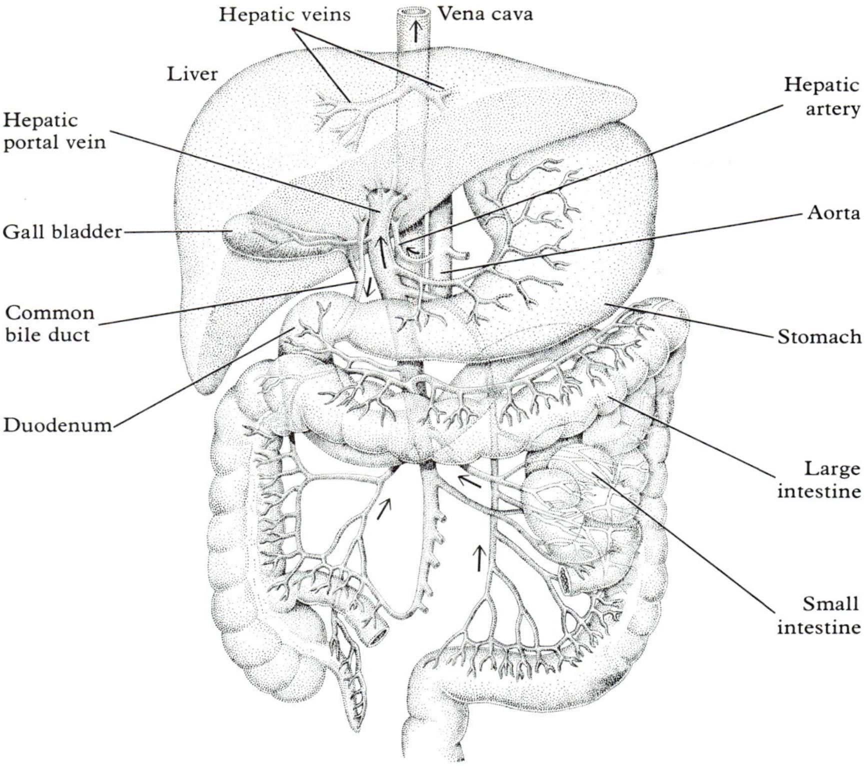

7–16
The liver is a primary site of metabolism of substances entering the body through the gastrointestinal tract. All the capillaries that absorb nutrients and other substances through the wall of the intestines come together and enter the liver through the hepatic portal vein. The liver's supply of oxygenated blood from the heart enters through the hepatic artery. After passing through the capillaries of the liver the blood is collected by the hepatic veins, which feed into the vena cava. (From "How the Liver Metabolizes Foreign Substances," by Attallah Kappas and Alvito P. Alvares. Copyright © 1975 by Scientific American, Inc. All rights reserved.)

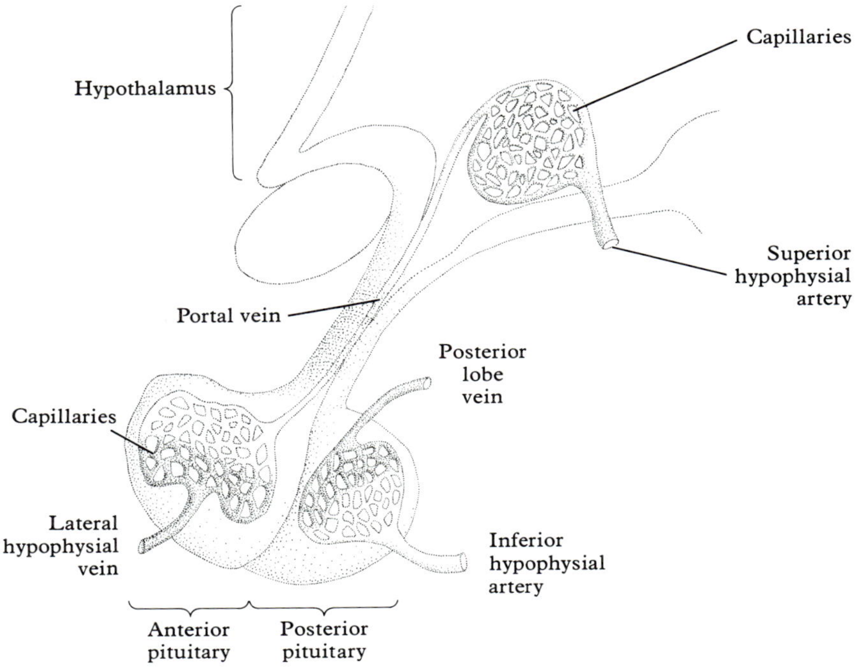

7–17
The hypophysial portal system connecting the hypothalamus with the anterior portion of the pituitary gland. Releasing factors from the hypothalamic neurons are emptied into the capillaries of the hypothalamus and are transported to the pituitary by the hypophysial portal system.

detoxified by liver enzymes. From the liver, the blood collects in the hepatic veins and journeys back to the heart. (The liver, which is very active in the synthesis of many kinds of molecules, requires oxygenated blood as well, and this is supplied by the hepatic artery, which enters the liver directly. Approximately 20 percent of the blood that passes through the liver comes from the hepatic artery, and 80 percent comes from the portal vein.) In the brain, the hypophysial-portal system delivers hormone-releasing factors to the anterior pituitary and is a crucial link in the feedback loops that regulate the concentrations of circulating hormones.

The Heart

The human heart has four chambers, two on its left side and two on its right (Figures 7-12 and 7-18). Each side has an atrium on top, into which the venous blood first enters, and a thick-walled ventricle underneath that pumps the blood into the arteries. The left ventricle, which pumps blood to most of the body, generates a much higher pressure (120 mm Hg) than the right ventricle (15 mm Hg), which pumps blood only to the lungs. The direction of flow through the heart is determined by valves (Figure 7-18). Between the right atrium and right ventricle is the *tricuspid valve,* and between the left atrium and left ventricle is the *mitral valve.* When open each of these atrioventricular (A-V) valves allows blood to flow from the atrium into the ventricle. When closed each prevents the flow of blood back into the atrium. Between the right ventricle and the pulmonary artery is the *pulmonary valve,* and between the left ventricle and the aorta is the *aortic valve.* When these valves are open they allow blood to flow into the pulmonary artery and the aorta, respectively, and when closed they prevent backflow into the ventricles. Thus, both sets of heart valves function to keep the blood moving from the veins to the arteries.

Whether a given valve is open or closed depends upon the relative pressures applied to its two sides, and these pressures change during the *cardiac cycle.* At a heart rate of 75 beats per minute, each cycle (which includes a phase of

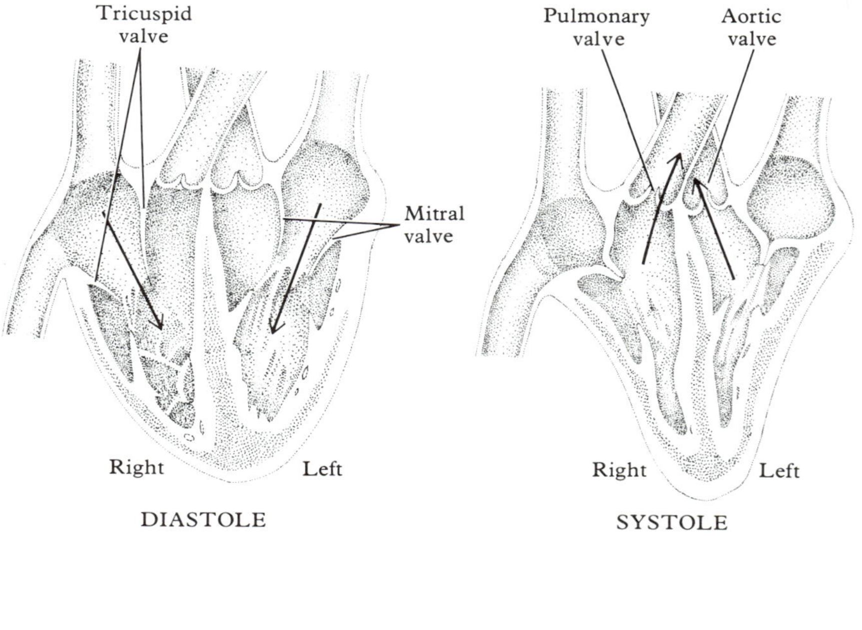

7–18
Systole and diastole is the pumping rhythm of the heart. At left is diastole, in which the ventricles relax and blood flows into them from the atria. The inlet valves (mitral and tricuspid) of the ventricles are open; the outlet valves (pulmonary and aortic) are closed. At right is ventricular systole, in which the ventricles contract, closing the inlet valves and forcing blood through the outlet valves. (From "The Heart," by Carl J. Wiggers. Copyright © 1957 by Scientific American, Inc. All rights reserved.)

contraction and a phase or relaxation for each chamber of the heart) lasts an average of 0.8 second. During the relaxation period, or *diastole,* of the cycle the atria and ventricles are relaxed, the A-V valves are open, and the aortic and pulmonary valves are closed. When the relaxation phase is over, the atria contract in unison and propel additional blood into the ventricles through the open A-V valves. Contraction is known as *systole,* and atrial contraction is thus *atrial* systole. Finally the ventricles contract in unison (*ventricular* systole), which causes the A-V valves to close. When the pressure in the contracting left ventricle exceeds the pressure in the aorta and the pressure in the right ventricle exceeds the pressure in the pulmonary artery, the aortic and pulmonary valves open and blood is ejected into the aorta and pulmonary artery. Thus there are three phases to the cardiac cycle: diastole, atrial systole, and ventricular systole.

The closure of the heart valves is responsible for the familiar "lub-dup" heart sounds heard during each cardiac cycle. The "lub," or first heart sound, is caused by the closing of the A-V valves, and the "dup," or second sound, by the closure of the aortic and pulmonary valves. Heart murmurs, whose symptoms are rumbling, buzzing, or whisperlike sounds, are often associated with blood flow across valves that leak when closed or that cannot open fully (Figure 7-19). If a heart murmur can be heard during ventricular systole (between the first and second heart sounds), when the A-V valves are closed and the pulmonary and aortic valves are open, it means that it may be due either to a leaky A-V valve (*insufficiency*) or to some abnormality of either the aortic or pulmonary valves. If a murmur is heard during ventricular diastole (after the second heart sound), when the A-V valves are open and the pulmonary and aortic valves are closed, the murmur may result from various abnormalities of the A-V valves or from insufficiency of either the aortic or pulmonary valve.

The familiar lub-dup of a beating heart seems to have a soothing effect upon newborn infants, as evidenced by the way mothers usually hold and carry their infants. A study of 255 right-handed mothers showed that 83 percent held their newborn infants on the left side of their chests or on their left shoulder, where the heartbeat is best heard, whereas 17 percent held the baby on the right side. The percentages were nearly the same for left-handed mothers. A study of 466 paintings and sculptures from museums and galleries in which a child was held by an adult revealed that 80 percent depicted the child being held on the left side (Figure 7-20). On the other hand, women leaving a supermarket were found to be equally likely to carry a large package in either their left or their right hand.

In an experiment conducted to find out whether babies respond to human heart sounds, a group of newborn infants were exposed to a recorded heart-beat sound in a hospital nursery. Compared to a control group of infants who were not exposed to the heart-beat sound, the exposed infants were less likely to cry and more likely to gain weight between days one and four after birth. During the fetal period, the infant is constantly exposed to the rhythmic beating of a mother's heart, and this early exposure may explain why heart sounds are comforting and why the beat of many musical rhythms is similar to the beat of the human heart. It seems possible that this early sensory experience may even be related to the universal appeal of music.

The cardiac cycle depends on electrical discharges to initiate each cycle. Each electrical discharge first appears in a region of the heart known as the sinoatrial (S-A) node, which is in the wall of the right atrium near the point of emptying of the superior vena cava (Figure 7-21). From the S-A node, which is the heart's normal *pacemaker,* the electrical excitation spreads through both

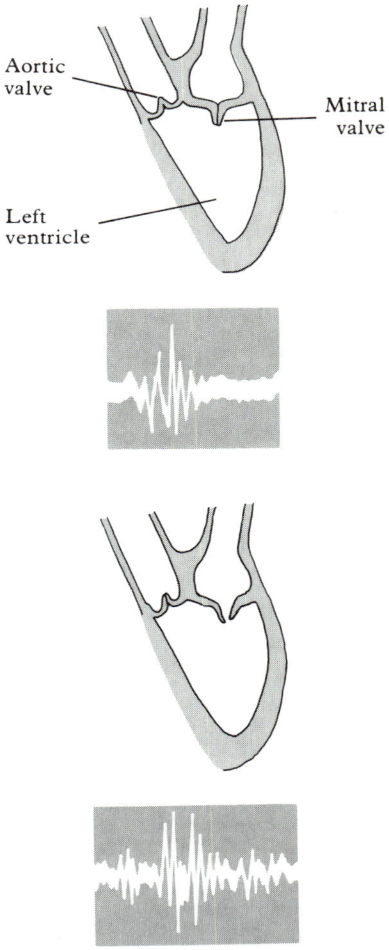

7–19
Heart sounds indicate the normal and abnormal functioning of the heart valves. The two drawings show the left side of the heart in cross section. Blood enters the left ventricle through the mitral valve and leaves through the aortic valve. The drawing at the top shows the normal closing of the mitral valve; the trace below it records the sound made by this closing. The bottom drawing shows the partial closing of a leaky mitral valve (insufficiency); the trace below it records the murmur which results when blood continues to flow through the valve. (From "The Heart," by Carl J. Wiggers. Copyright © 1957 by Scientific American, Inc. All rights reserved.)

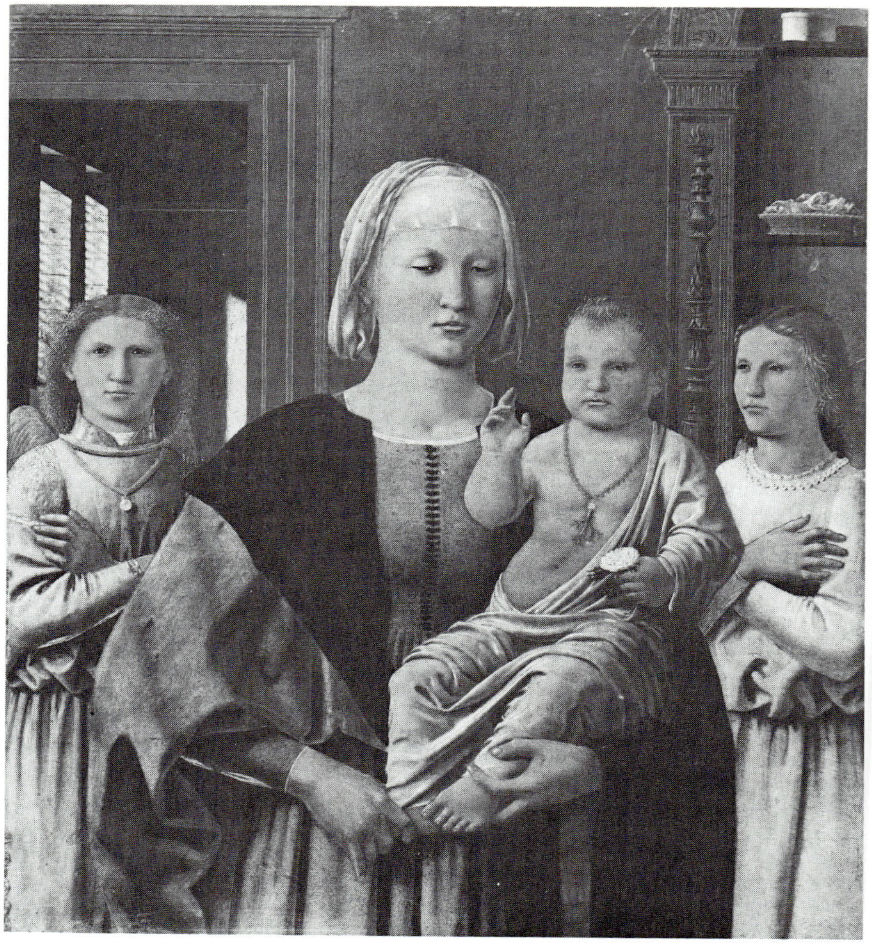

7–20
"The Virgin and Child with Two Angels," by Piero della Francesca, the fifteenth-century Italian painter. Works of art from all parts of the world and from all historical periods that depict an infant held by a woman tend to show the infant being held on the woman's left side. This tendency is particularly noticeable in paintings of the Madonna and the Christ child. (Courtesy of the Galleria Nazionale della Marche, Urbino.)

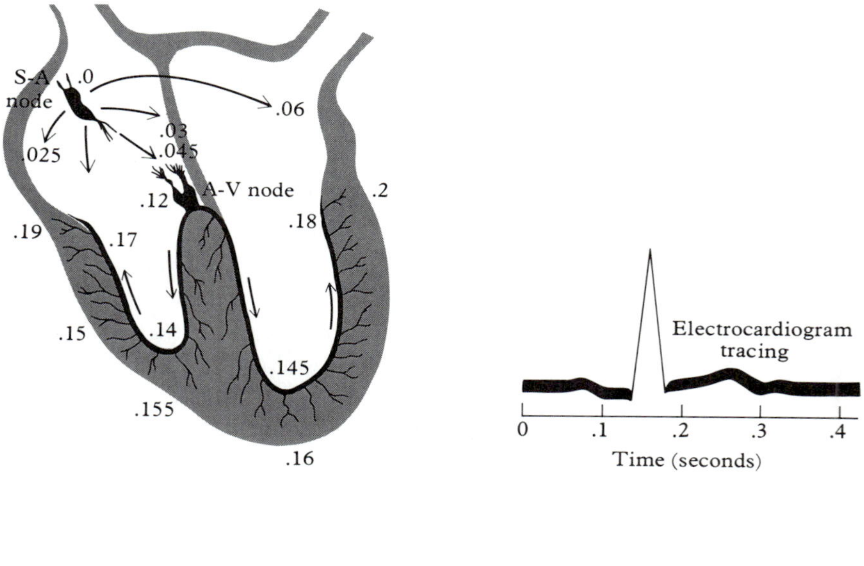

7–21
The heart's natural pacemaker system generates and transmits the impulses that cause the contraction of the heart muscles. The impulse is generated in the sinoatrial (S-A) node and spreads across the atria, causing their contraction and stimulating the atrioventricular (A-V) node. This in turn stimulates the rest of the system, causing the rest of the heart to contract. The numbers indicate the time (in fractions of a second) it takes an impulse to travel from the S-A node to that point. The electrocardiogram curve at the right indicates the change in electrical potential that occurs during the spread of one impulse through the system. (From "The Heart," by Carl J. Wiggers. Copyright © 1957 by Scientific American, Inc. All rights reserved.)

atria, initiating atrial contraction. The excitation then converges on another specialized area that is located in the septum between the two atria and is called the A-V (atrioventricular) node. From the A-V node the excitation spreads to the ventricles along special conducting fibers and fans out to initiate contraction of both ventricles. The pattern of spread of electrical excitation through the heart can be recorded and displayed as an *electrocardiogram* (Figure 7-21).

All cells of the heart are able to generate electrical impulses spontaneously, but in the normal heart the S-A node acts as a pacemaker because its spontaneous rate of discharge is more rapid (about 72 per minute) than the spontaneous rate of discharge in the rest of the heart. If for some reason the function of the S-A node is depressed or its connections to the A-V node are blocked, the A-V node may take over as pacemaker. It is relatively common for healthy people to have occasional "extra" ventricular systoles, which result from spontaneous discharges within the ventricles themselves, and in general this condition is not a serious one. Much more dangerous and almost always associated with heart disease is *ventricular fibrillation,* in which the ventricles contract in an irregular and ineffective way, probably because of a rapid discharge in several regions of the ventricles. Normal contraction can sometimes be reinitiated during ventricular fibrillation by applying an electrical shock across the chest by means of a machine called a defibrillator.

The heart rate of a normal person fluctuates considerably. The fluctuations are mediated in part by the sympathetic and parasympathetic nervous systems, which send fibers to the S-A node. Thus anger, excitement, and exercise, which are associated with stimulation of the sympathetic nervous system, accelerate the heart rate through the sympathetic nervous system and also through an elevation of circulating epinephrine that is released from the adrenal gland. The discharge rate of the S-A node is also influenced by temperature and drugs. A rise in temperature elevates the discharge rate, and this probably contributes to the increased heart rate associated with fever.

In spite of the enormous amount of blood that passes through the heart's chambers, the heart muscle is so thick and active that it requires its own separate blood supply, which is provided by the *coronary arteries* (Figure 7-22). The right

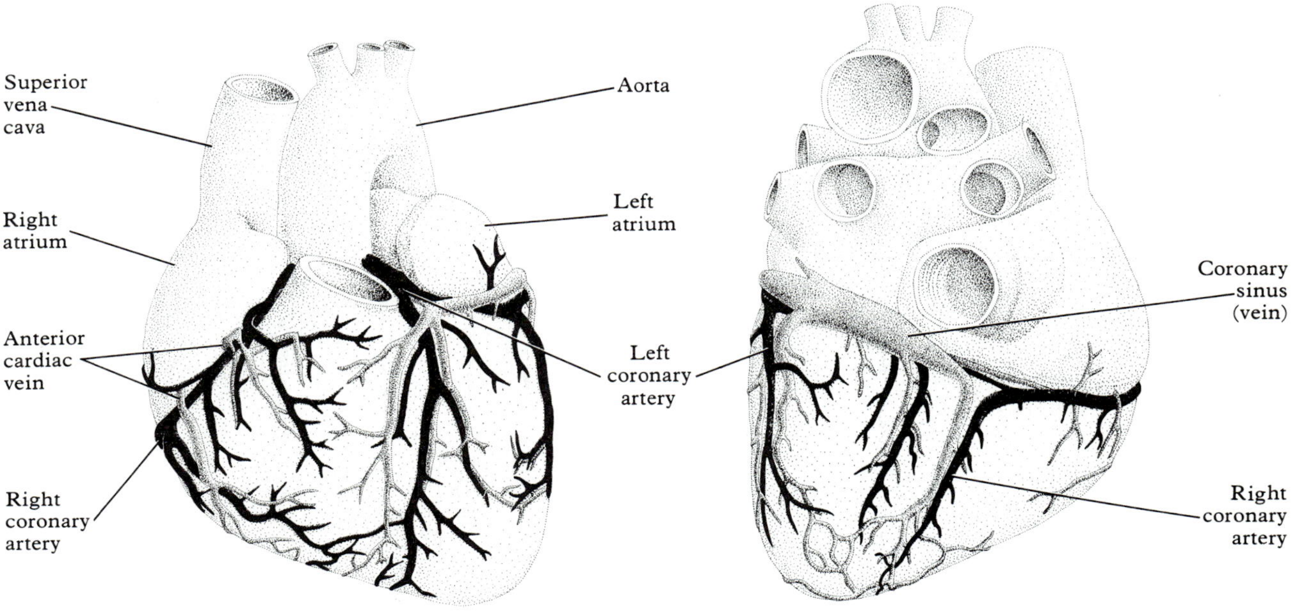

7–22
Coronary circulation. Arteries and veins that carry blood to and from the muscles of the heart are shown from the front (left) and back (right). The arteries are black; the veins, lighter gray. (From "The Heart," by Carl J. Wiggers. Copyright © 1957 by Scientific American, Inc. All rights reserved.)

and left coronary arteries are the first branches of the aorta, opening off the aorta just beyond the aortic valve. These arteries are of crucial importance: if a branch of a coronary artery becomes blocked, the heart muscle served by the branches of the artery will die. Blockage of a coronary artery is called a *heart attack,* and the resulting region of dead heart muscle is called a *myocardial infarction* (*myocardium* is heart muscle; *infarction* is tissue destruction). After a myocardial infarction the spread of electrical impulses in the heart is considerably altered, and these alterations can be detected by means of an electrocardiogram. The most common cause of death following a myocardial infarction is ventricular fibrillation.

Atherosclerosis

The most common finding in an area of the heart damaged by a heart attack is that a blood clot has formed in a coronary artery that was severely affected by atherosclerosis (Figure 7-23). *Atherosclerosis* is the deposition of fats (mostly cholesterol) and the growth of fibrous tissue in the vessel wall. *Arteriosclerosis* (*sclerosis* means hardening) is hardening of the arteries, and it always accompanies advanced atherosclerosis. It has been estimated that one in every 20 people in the United States has coronary heart disease associated with atherosclerosis and that 500,000 deaths per year in the U.S. are attributable to atherosclerosis. Atherosclerosis is an ancient disease; arteries of Egyptian mummies show clear signs of its presence (Figure 7-24).

The mechanism by which atherosclerosis can lead to a heart attack is straightforward: the fat deposits and fibrous thickening both increase resistance

7–23
Photomicrographs of human coronary arteries (× 13). A, a normal artery in cross section; B, a diseased artery, in which the channel is partially occluded by atherosclerosis; C, deposits harden within the arterial wall; D, a life-threatening stage, at which atherosclerosis has advanced and a blood clot fills the constricted channel. (Courtesy American Heart Association.)

A B

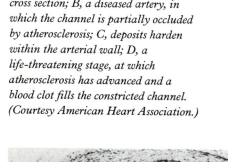

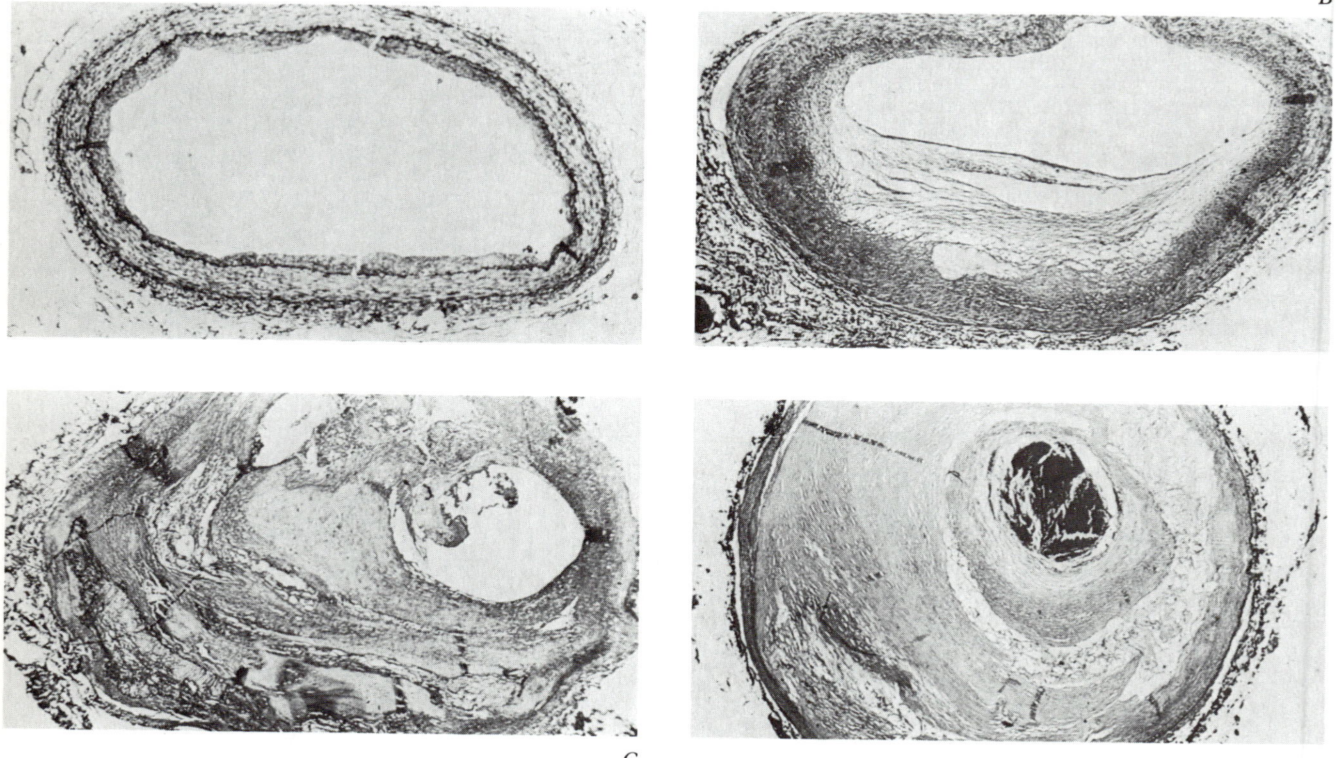

C D

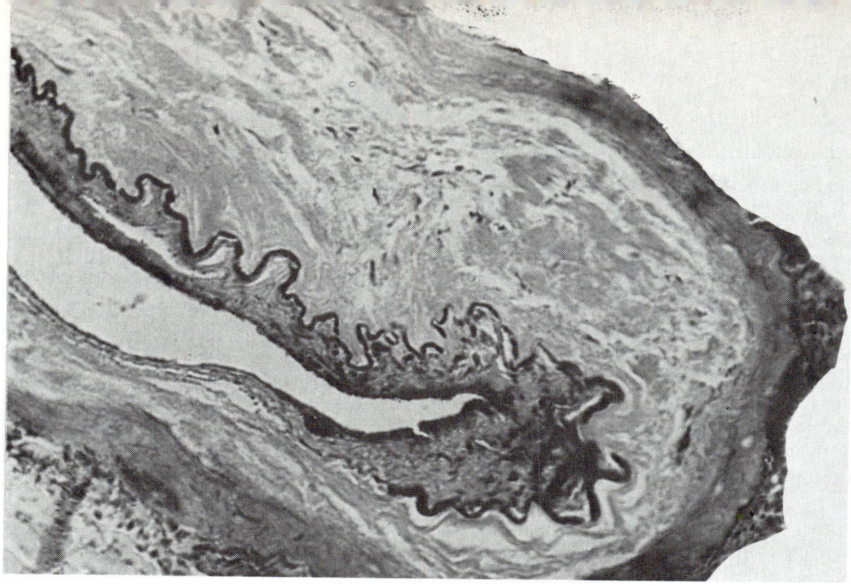

7–24
Cross section of an artery from the leg of the mummy of an elderly female. Excess deposits of fibrous tissue in the wall of the vessel are evident. Atherosclerosis is an ancient condition. (Courtesy Dr. A. T. Sandison.)

to blood flow, and sudden occlusion of a coronary artery may occur if a blood clot forms in the roughened inner surface of the artery. Gradual occlusion may also occur, and is less severe, because new accessory vessels supplying the same area sometimes develop before a full occlusion occurs. People who have gradual but increasing occlusion of a coronary artery often experience intense pain in the area of the heart and in the left arm, known as *angina pectoris.*

In 1954 the United States Public Health Service began a study in Framingham, Massachusetts, to uncover any facts about heart disease that possibly could be of clinical use. This most famous of all heart studies showed that the incidence of new coronary disease in a group of men in early middle age, followed over an eight-year period, was more than four times as great in those who had plasma cholesterol concentrations of 240 milligrams per 100 milliliters of blood or more as compared with those with cholesterol concentrations lower than 220 milligrams per 100 milliliters. In addition, it was found that the risk of heart disease was compounded by other factors such as heredity, obesity, hypertension, and cigarette smoking.

Because high concentrations of cholesterol seem to pose a risk of coronary heart disease, the practical question is: how can concentrations of cholesterol be reduced if they are too high? One way might be to reduce cholesterol intake, but this means eliminating from the diet many foods that are otherwise good for the body, such as eggs. Furthermore, cholesterol is synthesized within our own bodies, and this cholesterol, as well as cholesterol supplied in the diet, can be deposited in the coronary arteries.

Another approach to reduce concentrations of cholesterol that seems promising but is only beginning to receive attention is based on the notion that other dietary nutrients, if present in inadequate amounts, can lead to elevated concentrations of cholesterol, even if cholesterol intake is normal. For example, monkeys deficient in vitamin B_6 (pyridoxine) develop atherosclerosis. When they are fed diets supplemented with B_6 they have lower cholesterol levels and reduced atherosclerosis even though they consume more cholesterol. Other studies, which will require more follow-up work, have suggested that deficiencies of lecithin, vitamin C, and magnesium also facilitate atherosclerosis. The other side of the nutritional coin is that excess sugar consumption is statistically associated with elevated cholesterol levels and heart attacks. These studies seem to be leading us to the sensible conclusion that good general nutrition, even in the face of considerable cholesterol intakes, can help prevent elevated cholesterol levels and atherosclerosis.

Studies of persons under natural conditions show that factors other than cholesterol intake can exert a profound influence on cholesterol levels and heart disease. For example, a study of 400 Masai men in east-central Africa, whose diet consists mostly of meat and milk and is very high in cholesterol, revealed that they have very low concentrations of cholesterol in their blood and no evidence of heart disease. Another study comparing men from northern and southern India revealed no difference in concentrations of cholesterol in their blood, even though the consumption of animal fats and cholesterol was about ten times higher in the men from northern India. In fact, the incidence of heart disease among the northern Indian population is about 15 times less than that among the southern Indian population.

In addition to elevated concentrations of cholesterol in the blood, certain behavioral patterns known as *Type A behavior* are associated with heart disease. Type A people are active, energetic, perfectionistic, achievement-oriented, tense, and unable to relax. A study was recently undertaken to determine whether such behavior, identified *beforehand,* was associated with an increased heart attack rate. About 2,750 men free from heart disease answered a computer-scored questionnaire to see whether they had a Type A personality. They were then followed the next four years to see which of them had heart attacks. Among the men who scored high on Type A behavior, nearly twice as many had heart attacks as those who scored low, and those who scored in the intermediate range had an intermediate number of heart attacks. Although this study strongly implicates behavioral patterns as influencing the rate of heart attacks, it provides no clues as to the mechanism by which behavioral patterns are linked to incidence of heart attack.

Blood Pressure

Blood pressure is described by two numbers (for example, 120/70). The greater number represents the highest pressure reached in major arteries. Because this high point in pressure is reached during ventricular systole, when the left ventricle is forcing blood out into the aorta, it is called the *systolic pressure.* The lower number reflects the lowest pressure maintained in the arteries, and it is called the *diastolic pressure* because it occurs during left ventricular diastole. Blood pressure is usually measured by listening through a stethoscope to the sound in the brachial artery (in the arm) as the pressure is released from an inflated pressure cuff encircling the upper arm. The cuff, which has a gauge that reads the pressure it exerts, is pumped until the pressure is well above the systolic pressure. The pressure in the cuff is then gradually released, and when the pressure falls to just below the systolic pressure, blood begins to flow past the cuff, creating a popping sound. This point is recorded as the systolic pressure. The popping sound, which is probably due to the artery opening and then rapidly collapsing, disappears as the pressure in the cuff falls below the diastolic pressure (and the artery no longer collapses). The point at which the sound ceases is recorded as the diastolic pressure.

The blood pressure in the major arteries is carefully regulated by the body, and the main mechanism of regulation assures the maintenance of adequate pressure (and therefore blood flow) to the brain. There are special stretch-sensitive and pressure-sensitive areas, or *baroreceptors,* in the carotid arteries (in the neck) and in the most headward point of the aorta, the aortic arch. When blood pressure drops, the drop is sensed in these pressure-sensitive receptors, and their rates of discharge decrease. This reduced rate of discharge is transmitted over nerves to a special center in the medulla of the brainstem called the *vasomotor center.* The vasomotor center then initiates increased activity of the sympathetic nervous system (which has the effect of constricting the smooth muscle of the arterioles) and decreased activity of the parasympathetic nervous system (which also facilitates constriction of arterioles, because the parasympathetic nervous system usually relaxes the smooth muscle of the arterioles). As a result of these changes, the blood pressure returns toward normal as the constriction of arterioles reduces the flow of blood from the arterioles to the capillaries. Because the arterioles in the brain are much less affected by constriction than the arterioles elsewhere, the blood supply to the brain is favored by these compensatory mechanisms. The mechanisms that control blood pressure work in the opposite direction as well. An increase of blood pressure increases the rate of discharge of the pressure receptors, and the vasomotor center responds by decreasing sympathetic activity and increasing parasympathetic activity to the arterioles.

The concentration of the blood gases O_2 and CO_2 can also influence the vasomotor center. A decrease in O_2 concentration or an increase in CO_2 concentration is detected by the chemoreceptors, which send fibers to the vasomotor center to initiate arteriolar constriction by the mechanism we have just discussed. The resultant rise in blood pressure helps to send more blood to the brain, a welcome side effect of the main function of chemoreceptor stimulation, which is to increase the rate and depth of breathing.

Finally, the flow of blood to the kidneys can strongly influence blood pressure. If the pressure in one of the renal arteries decreases, the kidney releases an enzyme *renin* that initiates the formation of a molecule called *angiotensin II,* which is a potent constrictor of arterioles anywhere in the body. This constriction of arterioles causes the blood pressure to rise. (Renin also stimulates the release of the hormone aldosterone, which promotes reabsorption of salt and water by the kidneys. This leads to an increased blood volume, which tends to elevate blood pressure by filling up the vascular system.)

Hypertension

Hypertension is a sustained elevation of systemic blood pressure above normal. Of particular importance is the diastolic pressure, which in health usually does not exceed 90 (millimeters of mercury). Although blood pressure values tend to rise with increasing age, resting diastolic values of 90 or more in people under 60 years old are generally considered to be evidence of hypertension. Among the causes of hypertension are a few physical abnormalities, including constriction of a renal artery (which causes elevated blood pressure by the renin mechanism), a tumor of the adrenal medulla (which can cause a rise in blood pressure by secreting large amounts of epinephrine, the main hormone of the sympathetic nervous system), and an overactive adrenal cortex (which can cause a rise in blood pressure by releasing large amounts of corticosteriod hormones). Nevertheless, about 90 percent of the incidences of hypertension are not associated with a demonstrable physical abnormality. These cases are known as *essential hypertension.*

A clue to the nature of essential hypertension may lie in the fact that higher centers in the brain send nervous connections to the vasomotor center. There is a rise in blood pressure associated with anger, pain, and sexual excitement, all of which trigger increased vasomotor stimulation of the sympathetic nervous system. It seems possible, therefore, that emotional situations in everyday life could exert an influence over the vasomotor center as well. Because of the likely connection between emotions and elevated blood pressure, newer methods of treatment of essential hypertension attempt either to reduce emotional tension through relaxation (for example, yoga) or to put people more in touch with the factors inside themselves that alter their blood pressure (for example, biofeedback). Yogic exercises, which can produce complete mental and physical relaxation, have been shown to significantly reduce blood pressure in hypertensive patients. Biofeedback provides continuous visual or auditory displays of normally involuntary functions (such as blood pressure) while the hypertensive person attempts to influence them by mental or emotional means. These methods have met with some success in the treatment of essential hypertension. In one study of hypertensives who were being treated with drugs, the use of a combination of yoga and biofeedback allowed patients to stop using chemical therapy altogether in some instances.

Among the most serious dangers of hypertension is the chronic strain it places on the left ventricle. Because the left ventricle must generate increased pressure, its muscle fibers hypertrophy (increase in size). The oxygen consumption per unit of muscle remains normal, but the total oxygen consumption must be increased because there is more muscle mass, and the coronary arteries must therefore supply more oxygen. This means that any decrease of coronary blood flow is much more serious in hypertensive people, and even a moderate narrowing of the coronary arteries, which would not affect a normal person, may lead to a heart attack in a hypertensive person. Hypertension also increases the possibility that a person will have a *stroke,* which is an impairment of function caused by the occlusion or bursting of a blood vessel in the brain. Hypertension should be treated, because treatment has been shown to reduce the incidence of both strokes and heart attacks.

Summary

Blood is composed of formed elements—red cells, white cells, and platelets—and plasma, in which the formed elements are transported. Red cells transport oxygen, white cells aid in providing immune defenses, and platelets help to stop blood flow after injury to a vessel. If a small vessel is severed, platelets are attracted to the area and they aggregate together to form a plug that can seal off the vessel and prevent the excessive loss of blood. If a larger vessel is severed, the clotting mechanism may also be required to seal off the vessel. Blood clotting is the conversion of the soluble plasma protein fibrinogen into an insoluble meshwork of fibrin, a process that requires the concerted action of many clotting factors. During clot formation the clot is prevented from spreading beyond the immediate area by the action of the fibrinolytic system, which converts fibrin into breakdown products.

Important exchanges occur between plasma and the body's cells. Exchanges of gases and other small molecules take place across capillary walls by diffusion, which is the tendency of dissolved substances to pass from areas of higher solute concentration to areas of lower solute concentration. In addition to diffusion, the balance between blood pressure and osmotic pressure facilitates these exchanges. Osmotic pressure, which tends to draw plasma into the capillary, remains relatively constant from one end of the capillary to the other. Blood pressure, which tends to drive plasma out of the capillaries, is considerably greater at the arterial end of the capillary than at the venous end. The balance between osmotic pressure and blood pressure therefore results in plasma tending to pass out of the capillary at its arterial end and into the capillary at the venous end.

Large molecules, such as proteins and large fatty acids, may leave the circulatory system through the capillaries, but are returned to it through the lymphatic system, which includes a series of vessels throughout the body that begin as blind sacs, pass through lymph nodes, and then unite to form larger vessels that empty into the subclavian veins. Lymphatic vessels also serve the important function of delivering infectious microorganisms to the lymph nodes, where immune responses are initiated.

Blood from the capillaries of the lungs collects in the pulmonary veins, which empty into the left atrium of the heart. From the left atrium the blood enters the left ventricle, which pumps it out into the largest artery in the body, the aorta. From the aorta the blood passes into arteries, then into arterioles, and finally into the capillaries of the body. Blood returning to the heart from these capillaries passes first into venules, then into veins, and finally into the vena cavae, which empty into the right atrium of the heart. From the right atrium the blood enters the right ventricle, which pumps it out into the pulmonary artery to the lungs.

Although blood leaving capillaries usually flows into veins to be delivered back to the heart, in two regions of the body there are special portal systems that transport blood from one set of capillaries to another before the blood is returned to the heart. In the abdomen the hepatic portal circulation connects the abdominal portion of the gastrointestinal tract with the liver, and in the brain the hypophysial portal circulation connects the hypothalamus to the anterior pituitary.

Blood flows toward the heart in the veins because of the presence of the venous valves, which maintain one-way flow, and because of the pumping action of skeletal muscles and the diaphragm.

The direction of blood flow in the heart is also maintained by valves: each ventricle has a valve at its inlet (an atrioventricular, or A-V, valve) and a valve at its outlet (the pulmonary valve for the right ventricle and the aortic valve for the left ventricle). The cardiac cycle encompasses a period of relaxation of the heart muscle (diastole), a period of atrial contraction (atrial systole), and a period of ventricular contraction (ventricular systole). During diastole the ventricles fill with blood, and their inlet valves are open while their outlet valves are closed. During ventricular systole the inlet valves to the ventricles are closed and the outlet valves open as the ventricular pressure rises.

The familiar "lub-dup" sound of the heart results from closure of the heart valves. The first heart sound ("lub") is generated by the closing of the A-V valves and the second sound ("dup") by the closing of the aortic and pulmonary valves. Heart murmurs often occur when valves either do not open fully or do not close properly.

The sound of the beating heart seems to be soothing to newborn infants, and the mother of a newborn infant usually holds her child next to the left side of her body, where the heartbeat is more easily heard than on the right.

The cardiac cycle is initiated by electrical discharges generated in the sinoatrial (S-A) node, which is the natural pacemaker of the heart. From the S-A node the electrical excitation spreads through the heart in a systematic fashion so that the events of the cycle occur in proper sequence. This electrical activity can be recorded and displayed in an electrocardiogram.

The heart muscle receives its blood supply from the coronary arteries. Blockage of a coronary artery is called a heart attack, and death of heart muscle a myocardial infarction. The coronary arteries of a number of people are damaged by atherosclerosis, which is the deposition of fats and fibrous tissue in the vessel wall. People who have high concentrations of cholesterol in their plasma have more of a tendency to have heart disease and atherosclerosis. Cholesterol levels seem to be influenced both by diet and by patterns of behavior.

Blood pressure is described by two numbers (for example, 120/70). The larger number, the systolic pressure, represents the highest pressure reached in a major artery at the time of measurement, and the smaller number, the diastolic pressure, represents the lowest pressure in that artery. Under normal conditions the blood pressure remains relatively stable because of reflexes initiated in pressure-sensitive areas of the aorta and the carotid arteries. Nervous impulses are sent from these pressure-sensitive areas to the vasomotor center of the medulla of the brain, which in turn regulates blood pressure through the activities of the sympathetic and parasympathetic nervous systems.

A diastolic pressure of greater than 90 mm Hg in a person under 60 years of age is usually defined as high blood pressure, or hypertension. Most cases of hypertension are not associated with a demonstrable physical abnormality, and in such cases the elevated pressure is called essential hypertension. A clue to the nature of essential hypertension may lie in the fact that higher centers in the brain send nervous connections to the vasomotor center. Emotional factors may therefore influence blood pressure through higher brain centers. Newer methods of treatment of essential hypertension, which have met with some success, include yoga, in which an attempt is made to reduce emotional tension through relaxation, and biofeedback, in which an attempt is made to put people in touch with the factors inside themselves that alter their blood pressure.

Selected Readings

1. "Yoga and Bio-Feedback in the Management of Hypertension," by C. H. Patel, *The Lancet, ii* (1973), 1053-1055. A study of blood pressure control by means of yogic relaxation and biofeedback techniques.
2. "Prediction of Clinical Coronary Heart Disease by a Test for the Coronary-Prone Behavior Pattern," by C. D. Jenkins, R. H. Rosenman, and S. J. Zyzanski. *The New England Journal of Medicine, 290* (1974), 1271-1275. A study of the relationship between Type A behavior and coronary heart disease.

3. "The Heart," by C. J. Wiggers. *Scientific American,* May 1957. An excellent discussion of the anatomy and physiology of the heart.
4. "The Microcirculation of the Blood," by B. W. Zweifach. *Scientific American,* Jan. 1959. A good discussion of the capillaries and how they provide nourishment for the tissues.
5. "The Role of the Heartbeat in the Relations between Mother and Infant," by Lee Salk. *Scientific American,* May 1973. A thorough discussion of the importance of the human heartbeat.
6. "Atherosclerosis," by D. M. Spain. *Scientific American,* Aug. 1966. A good overview of the process of atherosclerosis.
7. "The Lymphatic System," by J. S. Mayerson. *Scientific American,* June 1963. A comprehensive description of the lymphatic system and its importance in health and disease.

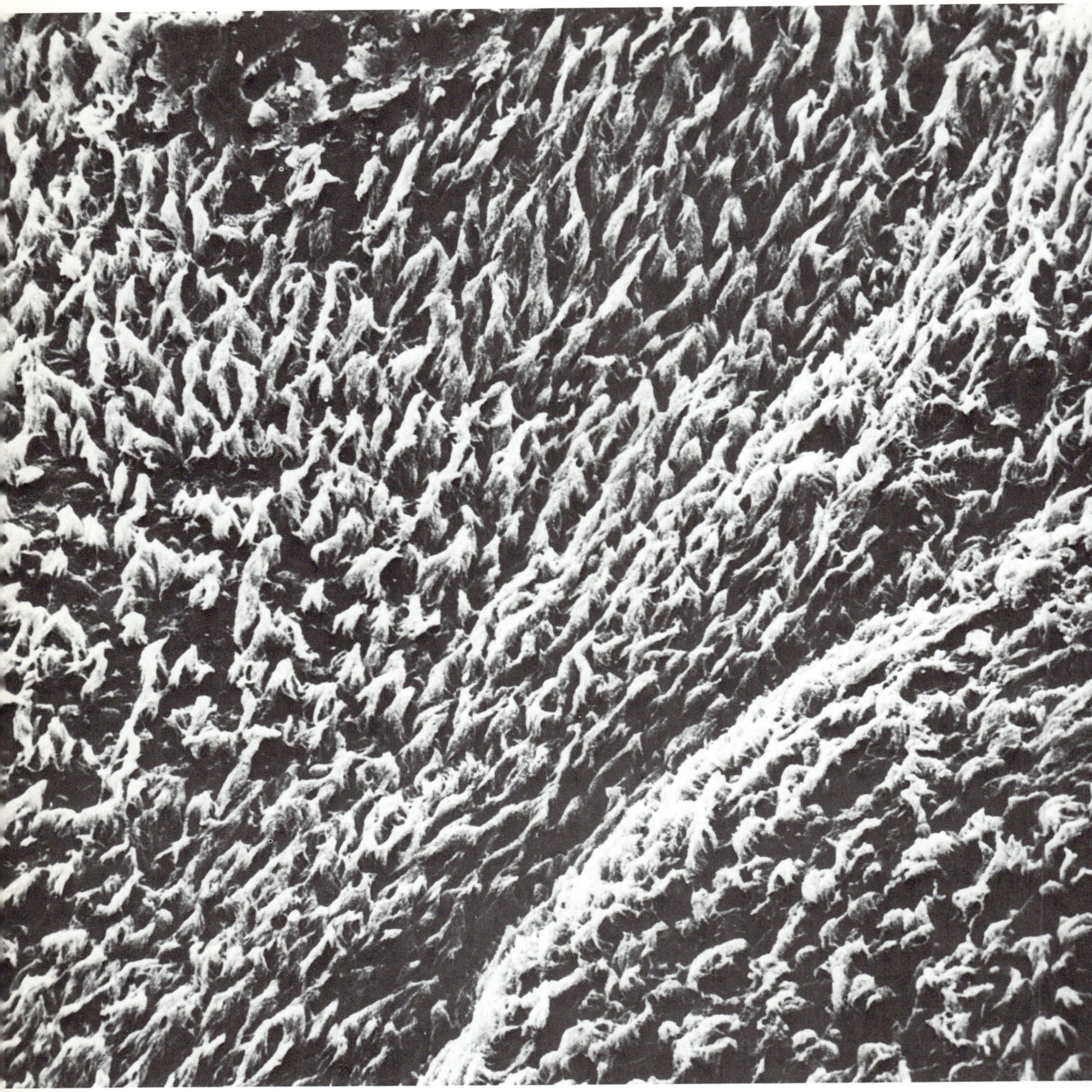

Cilia protrude from cells that line the air passages. They are clearly visible in this scanning electron micrograph because their covering has been removed. (From Lennart Nilsson, Behold Man. *Little, Brown and Company, 1974. Courtesy Albert Bonniers Förlag, Stockholm.)*

CHAPTER 8

Respiration

Breathing is essential for most living things because almost all of them must have a supply of oxygen in order to obtain usable energy. The exchange of gases with the environment that occurs during breathing is known as *external respiration.* The energy-yielding chemical reactions associated with breathing are known as *internal respiration.* In general, internal respiration, which occurs inside cells, consumes both oxygen and intracellular fuel. Most often the fuel is a sugar, glucose, that in the presence of oxygen from the atmosphere is chemically burned within the cell to yield usable energy. Of all the processes that take place inside cells internal respiration is probably the most obviously significant, because every activity of the cell—indeed, every activity associated with life itself—requires an almost constant input of energy.

In order for internal respiration to take place efficiently oxygen must be supplied and carbon dioxide gas must be disposed of; consequently, the processes concerned with *gas exchange* between the external environment and an organism's cells are of the utmost importance. In this chapter we will first discuss how the breathing apparatus of the body works to deliver O_2 and to eliminate CO_2 and how breathing itself is controlled. We will then discuss how gases are transported back and forth between the lungs and the cells. Finally, we will discuss the energy-yielding process of internal respiration within the cell and see how it is connected to gas exchange.

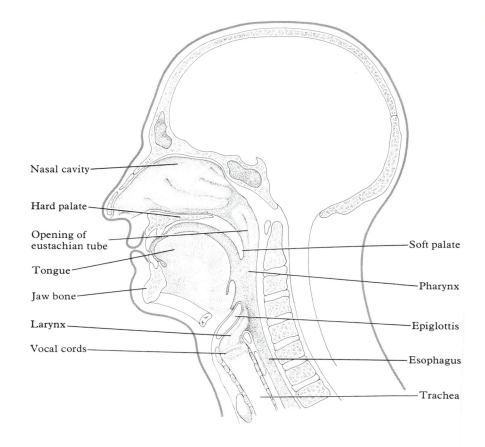

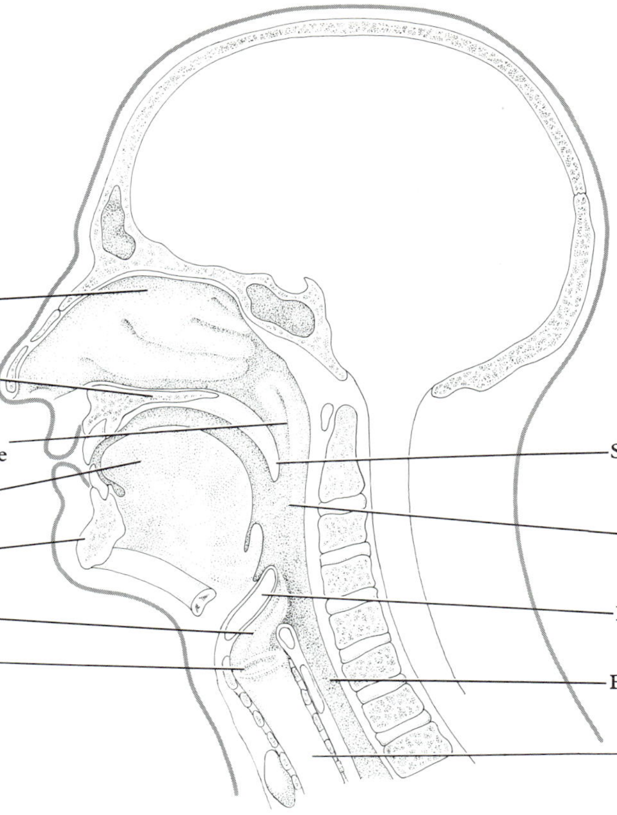

8–1
Air passages in the head and neck. The trachea, which carries air to the lungs, lies in front of the esophagus, which conducts food to the stomach.

The Breathing Apparatus

The air passages of the human body (Figures 8-1 and 8-2) are well designed for their tasks. After entering through the mouth and nose, air first passes through the throat (pharynx) and into the larynx (or voice box), which lies at the entry to the trachea. The larynx contains the vocal cords and it is protected from above by a structure called the epiglottis. The larynx not only allows us to speak what is on our minds, but it also keeps food out of our air passages during swallowing (Figure 8-3). As food enters the back of the mouth, the swallowing reflex is initiated (under the control of the medulla of the brain). During the act of swallowing, the vocal cords close tightly, and the epiglottis automatically leans backward, thus covering the closed vocal cords as the food slides by.

Just below the larynx is the *trachea,* the main windpipe in the neck and chest. The trachea is held open by a number of circular rings of cartilage. Certain cells lining the trachea secrete mucus, which traps microscopic particles inhaled with the air. Other cells lining the trachea have *cilia* (frontispiece), which beat rhythmically in such a way that particles trapped by mucus are forced upward, eventually into the back of the throat where they are swallowed. Some bacteria that cause lung infections (for example, *Mycobacteria,* which cause tuberculosis) are often best searched for by examining the contents of the stomach, particularly early in the morning after a person has been swallowing all night on an empty stomach.

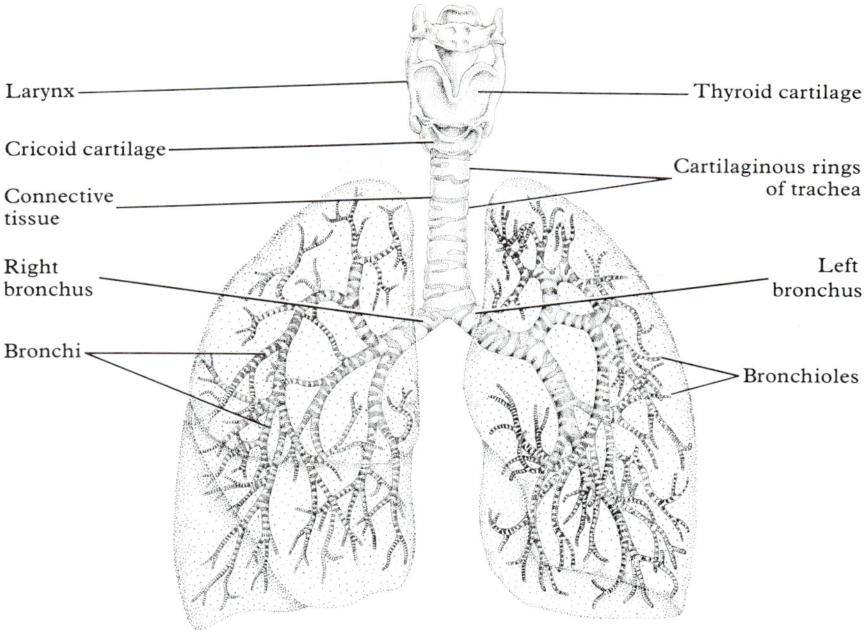

8–2
The air passages from larynx to lungs. The trachea, which leads to the right and left bronchi, is supported by cartilaginous rings. The bronchi branch into bronchioles, which finally terminate in alveoli (too small to be visible in this drawing).

Bronchi and Bronchioles

The trachea branches first into bronchi and then into smaller bronchioles, the tubules that lead into the tiny air sacs known as alveoli (Figure 8-4). The two main branches of the trachea are the right bronchus and the left bronchus. Each of these divides into about 20 bronchial branches and finally into the bronchioles. More than a million bronchioles serve the 300 million alveoli of the human lungs. Mucus glands and cartilage are present in bronchioles as small as 1 millimeter in diameter. Bronchioles smaller than 1 millimeter in diameter are surrounded by smooth muscle that may constrict or dilate, and in this way either discourage or encourage the flow of air. *Bronchitis* is an inflammation of the bronchi and bronchioles, a common ailment whose symptoms are increased production of mucus, coughing, and changes in the cells that line the bronchi.

8–3
Swallowing. At rest the air passages are usually open, and the esophagus is kept closed by the action of a sphincter present just below the larynx (A). During swallowing the glottis closes and seals off the trachea (B). As food passes by the larynx the epiglottis leans backward to protect the entrance to the larynx (C). After completion of the swallow, closure of the esophageal sphincter helps to prevent regurgitation of food into the air passages (D).

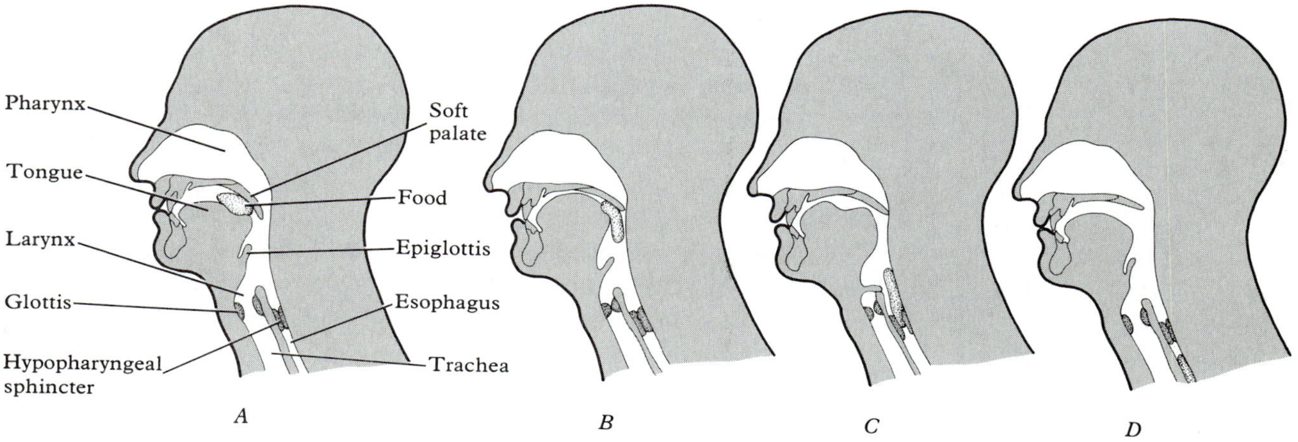

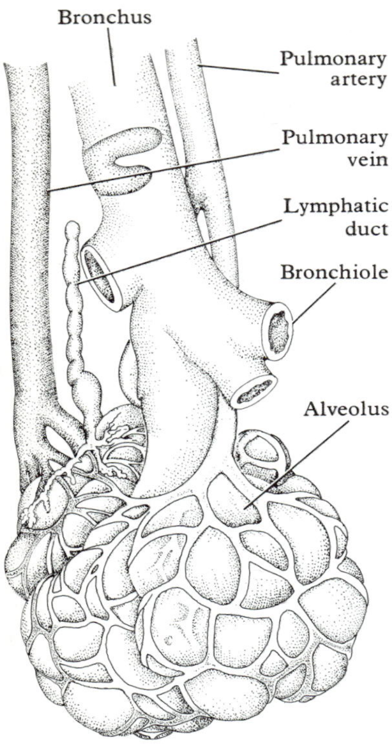

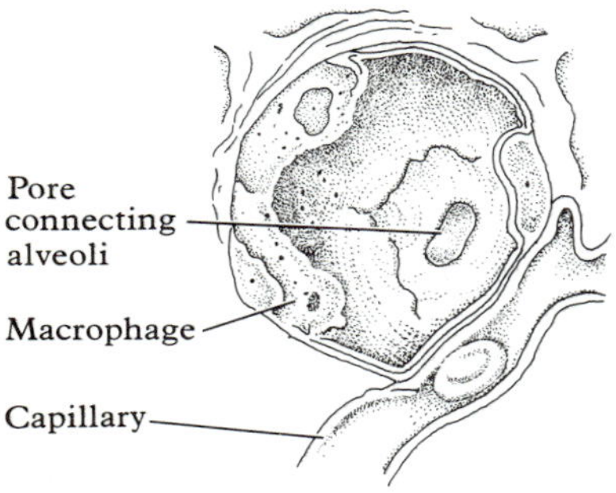

8–4
Clusters of alveoli are at the end of each bronchiole. The alveoli are the tiny air sacs where gas exchange takes place between air and blood. The alveoli are in intimate contact with capillaries and are served by lymphatic vessels as well. A single alveolus seen in cross section (below) reveals a pore that connects it to a neighboring alveolus, a macrophage in its lumen, and a nearby capillary. (From "The Lung of the Newborn Infant," by Mary Ellen Avery et al.

Asthma is a disease in which breathing becomes difficult for two reasons. First, bronchioles are constricted by contraction of the surrounding smooth muscle. Second, there is an overproduction of mucus, which tends to plug the bronchioles. During an attack of asthma it is difficult to pass air through the bronchioles, and expiration is particularly difficult. Inhaling (inspiration) is easier than exhaling (expiration) during an attack of asthma because during inspiration, small bronchioles become somewhat dilated and elongated as the lungs increase in size with the expansion of the chest. The mucus that lines them adheres to their walls and allows the passage of air. During expiration, the bronchioles (already constricted) become shorter and narrower as the lungs decrease in volume, so that mucus fills up and blocks the entire airspace within the bronchiole.

The constriction of bronchioles and the secretion of mucus during asthma are both accentuated by the release of *histamine,* a substance found in *mast cells* that are present in abundance in the tissues around the bronchioles. The mast cells can be triggered to release histamine by either immune or psychological factors (see Chapter 13).

The chemical treatment of asthma is directed toward relieving bronchiolar constriction, either by preventing the release of histamine and other substances from mast cells (sodium cromoglycate and diethylcarbamzine work in this way) or by directly relaxing bronchiolar smooth muscle (epinephrine, isoproterenol and aminophylline have this effect). These treatments, although often successful in relieving symptoms, do not affect the underlying allergic, immunological or psychological disturbances that cause the asthmatic condition in the first place.

Alveoli

The smallest bronchioles open into *alveoli,* the sacs where gas exchange occurs (Figure 8-4). Some of them are dead-end sacs, but many alveoli are connected to each other by pores. Alveoli have a rich supply of blood capillaries and thin walls through which gas exchange takes place. The alveoli are tiny; there are about 150 million of them in each lung.

Figure 8-5 shows the wall that separates an alveolus from a capillary, as viewed through the electron microscope. Note that the oxygen must cross three barriers to get from the alveolus into the bloodstream. These barriers are: (1) the epithelial cell lining the alveolus, (2) the basement membrane, and (3) the endothelial cell lining the inside of the capillary. The actual distance covered is very small—only 0.2 to 0.6 micrometers in all. In fact, before the use of the electron microscope, it was not known that the cytoplasm of the epithelial and endothelial cells had to be traversed, because the cytoplasm of these cells is too thin to be seen through the light microscope.

Emphysema is a degenerative condition in which the alveolar walls break down and are replaced by large air-filled spaces (Figure 8-6). This seriously reduces the total surface area available for gas exchange. Emphysema is a disease of the elderly that is more common in people chronically exposed to air pollution. In an attempt to compensate for the lost alveoli, people who have emphysema use all their breathing muscles and may even elevate their shoulders in order to keep their lungs relatively fully inflated. In severe cases, oxygen and bronchodilator drugs may be needed to sustain life.

Another condition affecting the alveioli is *heart failure,* in which the left ventricle of the heart becomes unable to expel blood with the force necessary to

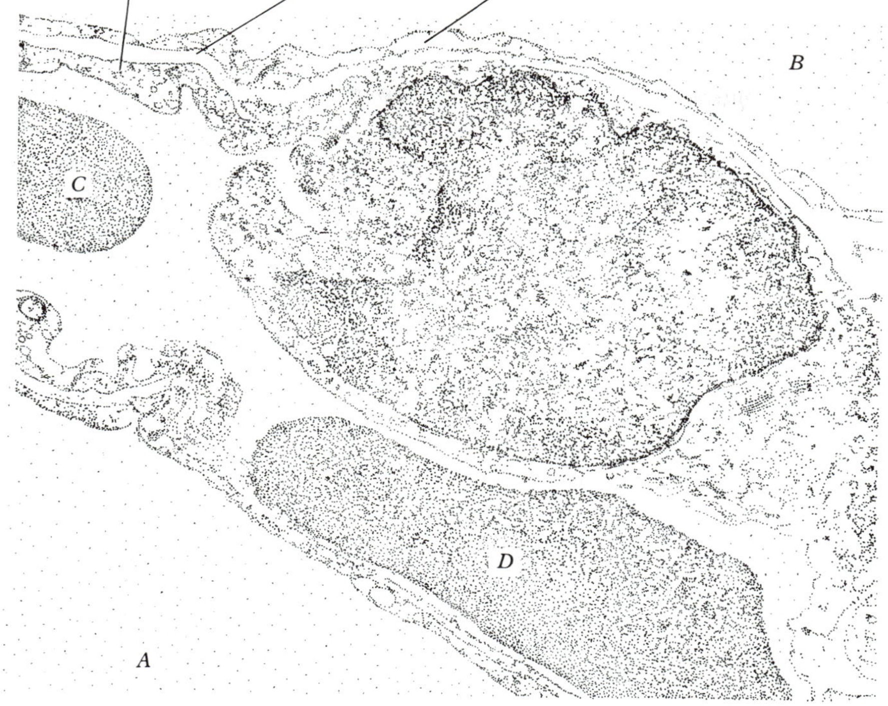

8–5

Separation between alveolus and capillary. In this drawing made from an electron micrograph, A and B are air spaces inside alveoli, and C and D are two erythrocytes inside a capillary. Gases must pass across the wall that separates the alveolus from the capillary. This wall consists of the cytoplasm of a cell lining the alveolus, a basement membrane, and the cytoplasm of a cell lining the capillary. The total distance may be as small as 0.2 micrometer. (Drawing after electron micrograph by Julius H. Comroe, Jr.)

drive it throughout the body. When this happens, the blood backs up in the capillaries of the lungs behind the left atrium. This results in increased pressure inside these capillaries, which in turn leads to the leakage of fluid across the alveolar wall into the alveoli. The presence of fluid in the alveoli because of heart failure is known as *pulmonary edema.* Pulmonary edema is a dangerous condition because the leakage of fluid, if unchecked, will eventually fill the alveoli and prevent gas exchange, thus resulting in suffocation. Left ventricular heart failure can be treated by drugs such as digitalis that help the heart muscle to contract more forcefully and by diuretics which cause the kidneys to eliminate some of the excess fluid from the circulatory system.

Keeping the Air Passages Clean

Each of us takes more than 4 million breaths each year, so practically everyone's lungs encounter dirty air at some time or other. Not only does wind whip up particles of sand, leaves, and dust, but in the more recent stages of human history the air of cities has become especially befouled. For example, it has been estimated that each liter of average city air contains several million particles and that a typical city-dweller may inhale 20 billion such particles in a day.

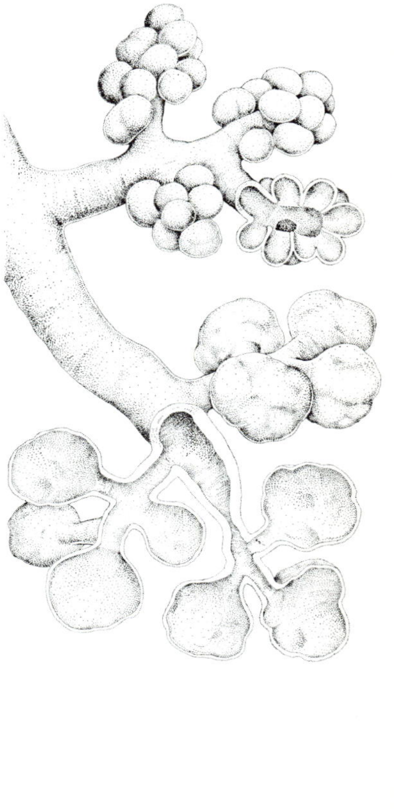

8–6

Emphysema is a degenerative lung disease in which the walls of normal alveoli, top, break down and become large spaces with much less surface area available for gas exchange, bottom. (From "Air Pollution and Public Health," by Walsh McDermott. Copyright © 1961 by Scientific American, Inc. All rights reserved.)

Several defense mechanisms help to keep the air passages clean in the face of such an onslaught. When air first enters the air passages larger particles (above 10 micrometers in diameter) are removed by hairs in the nose and by the mucus of the nasal turbinate bones (Figure 8-7). Particles smaller than 10 micrometers diameter usually settle in the trachea, bronchi, and bronchioles and are expelled by the action of cilia, which beat rhythmically with a wavelike motion. The beating of cilia takes place under a covering of mucus and the particles are moved along in the mucus until they reach the throat, where they are swallowed. Another mechanism of dealing with inhaled particles is triggered by sensory receptors that are scattered throughout the air passages. These receptors, when stimulated, initiate protective reflexes such as coughing and sneezing.

Some inhaled particles, particularly those less than 2 micrometers in diameter, eventually reach the alveoli, which have no ciliated or mucus-producing cells. These particles are consumed by phagocytic cells of the

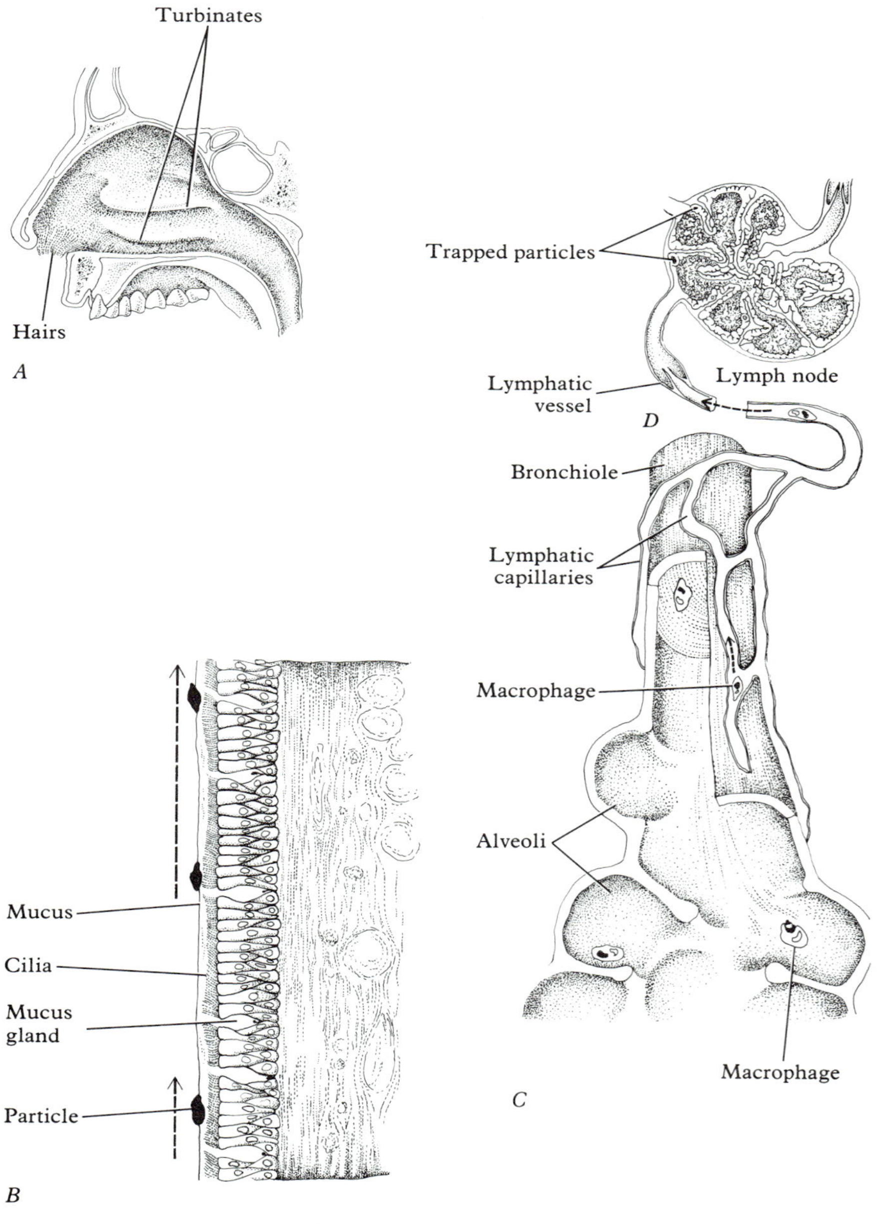

8–7
Defense mechanisms that help to keep the air passages clean. Nasal hairs and the projecting turbinate bones help trap larger particles (A). Smaller inhaled particles settle on the mucus of the trachea, bronchi, and bronchioles (B) and are swept upward by mucus-covered cilia to the throat for expulsion. Particles less than two micrometers in diameter may reach the alveoli (C), where they are picked up by macrophages and carried in lymphatic vessels to the lymph nodes of the lung (D). (From "The Lung," by Julius H. Comroe, Jr. Copyright © 1966 by Scientific American, Inc. All rights reserved.)

immune system (see Chapter 13), which enter the alveoli to engulf them. Some phagocytes exit with the engulfed particles in the mucus of the bronchioles, and others enter the tissue of the lung and carry the particles through the lymphatic vessels to lymph nodes in the lung where immune responses can be initiated (Figure 8-7). Some particles, however, are deposited in the tissues of the lungs, which becomes apparent when the darkened lungs of smokers or coal miners are revealed during surgery or at autopsy. Large quantities of inhaled particles may cause serious damage. Epidemiological studies, in which large populations of people are analyzed, have shown that people chronically exposed to air pollution have a higher incidence of all types of respiratory diseases, including asthma, bronchitis, emphysema, and cancer.

Prolonged exposure to specific inhaled substances may also have deleterious effects. For example, particles of free crystalline silica can cause overgrowth of fibrous tissue which leads to the serious lung disease *silicosis.* The common form of silica (SiO_2) is quartz, which is found in abundance in the earth's surface. Activities such as mining, sandblasting, rock-drilling, and glass manufacture have the potential for exposing a worker to relatively large quantities of free silica. Lung damage begins when silica particles enter the lysosomes of phagocytic cells (Figure 8-8). The silica somehow destroys the membranes of the lysosomes, which leads to the death of the phagocytic cells and to the release of a substance that initiates the growth of fibrous tissue. With prolonged exposure to silica, fibrous tissue may become so widespread that it seriously compromises gas exchange in the lung.

Cigarette smoking is also known to have deleterious effects on the air passages. People who smoke are many times as likely as nonsmokers to die from lung cancer, and the likelihood of developing cancer increases with the number of cigarettes smoked. Fortunately, the data that exist indicate that the probability of developing lung cancer returns toward normal after one has stopped smoking for eight years. Cigarette smoking also increases the probability of dying from coronary artery disease by about 70 percent, and because coronary artery disease is much more common than lung cancer, more than three times as many smoking-associated deaths result from coronary artery disease as from lung cancer.

Smoking causes remarkable changes in the epithelial cells of the bronchi (Figure 8-9). Cilia disappear, and the basal cells of the epithelium, which are usually confined to a single layer, proliferate and form many layers. Under the continued stimulus of heavy smoking these cells may become cancerous; they may break through the basement membrane, invade lung tissue, and spread throughout the body. (Stages *B* and *C* of Figure 8-9 may represent reversible stages of the proliferative process.)

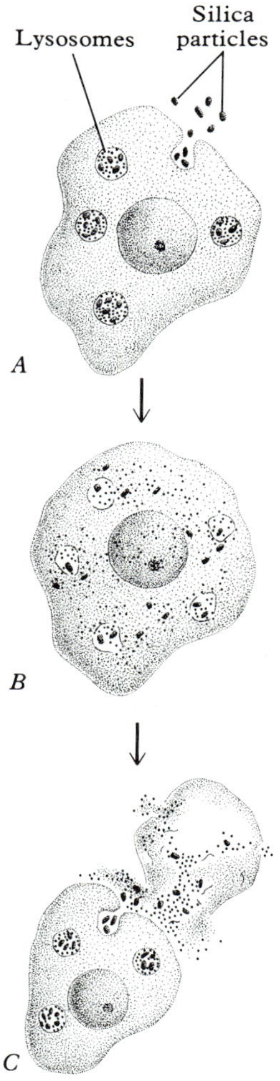

8–8
Silicosis, a lung disease, begins with the phagocytosis of silica (SiO_2) particles by phagocytic cells in the lung (A). Once inside lysosomes, the silica particles damage the lysosomal membrane (B). Enzymes released from ruptured lysosomes kill the phagocytes and release the silica particles, which enter other phagocytes and initiate a new cycle of the process (C). The dying phagocytic cells release a factor that can stimulate the overgrowth of fibrous tissue in the lungs. (From "Lysosomes and Disease," by Anthony Allison. Copyright © 1967 by Scientific American, Inc. All rights reserved.)

Breathing

The importance of breathing has not always been appreciated. An illustration from the textbook of Tobias Cohn, published in 1707 (Figure 8-10), compares the human body to a house, in which the stomach is the kitchen, the eyes are the windows, and the lungs are simply the ventilators. A demonstration by Robert Boyle in 1725 convinced the skeptics that air was necessary to sustain life. Boyle, who had invented a vacuum pump, showed that the animals he studied could not survive without breathing air.

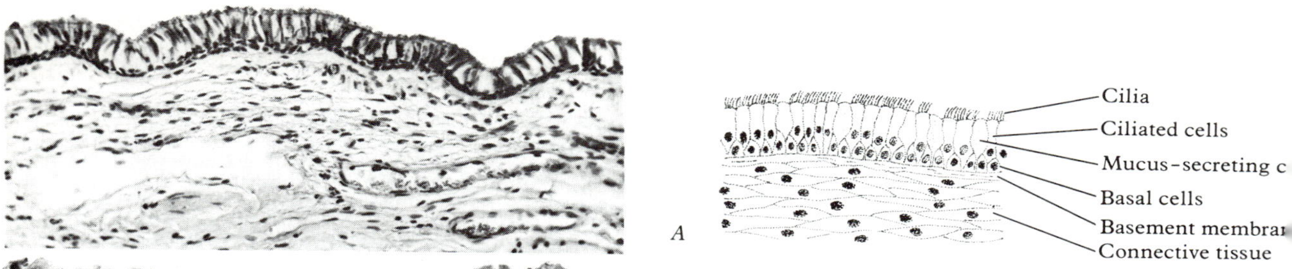
Cilia
Ciliated cells
Mucus-secreting c
Basal cells
Basement membra
Connective tissue

A

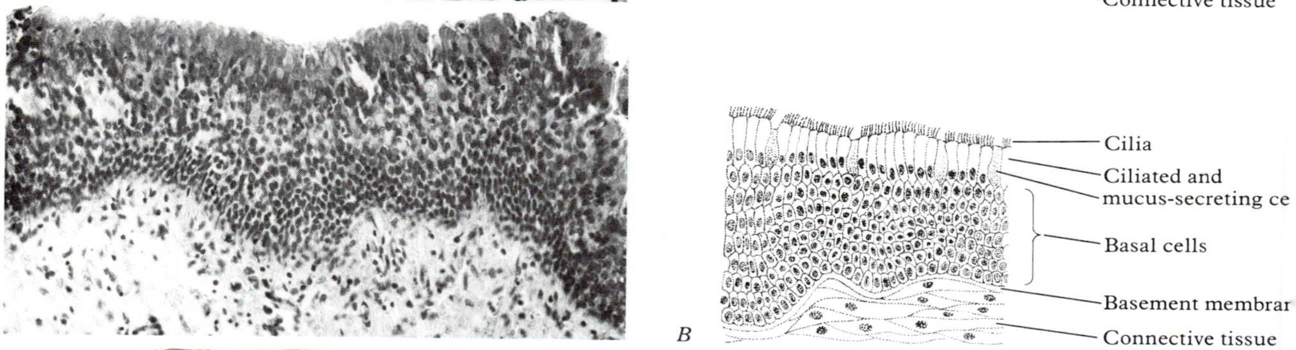
Cilia
Ciliated and
mucus-secreting ce
Basal cells
Basement membra
Connective tissue

B

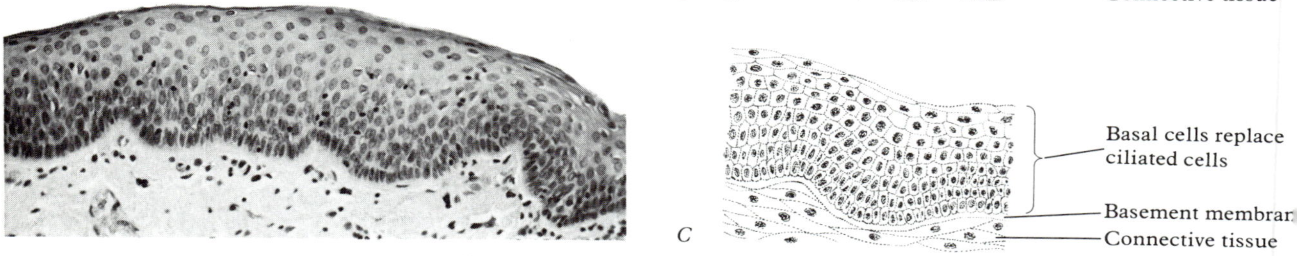
Basal cells replace
ciliated cells
Basement membra
Connective tissue

C

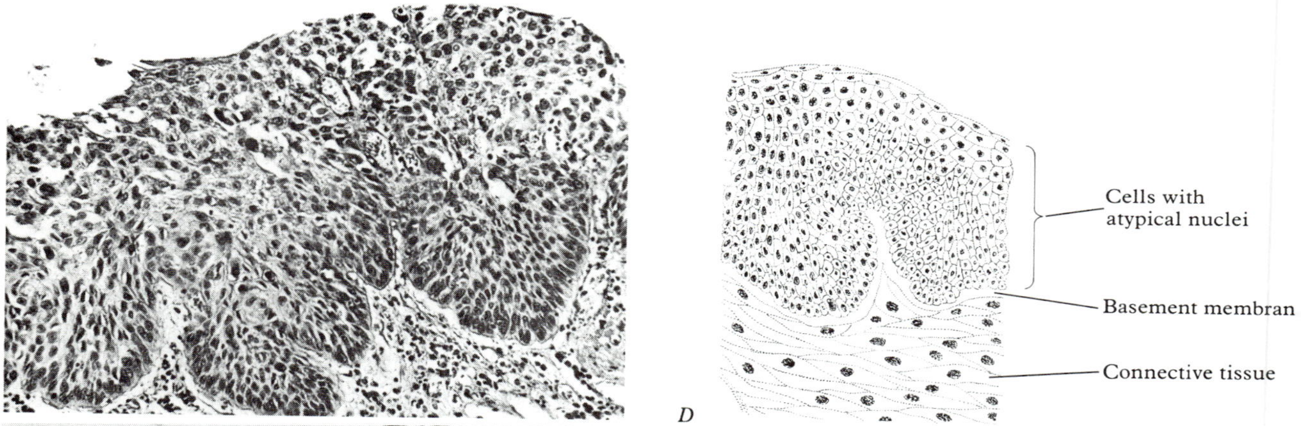
Cells with
atypical nuclei
Basement membran
Connective tissue

D

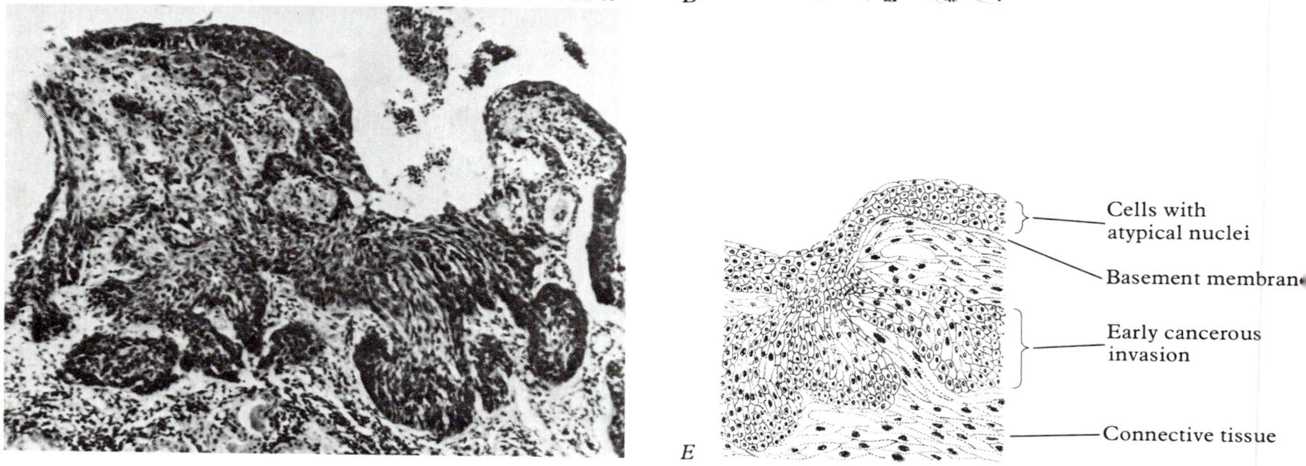
Cells with
atypical nuclei
Basement membran
Early cancerous
invasion
Connective tissue

E

8–10
Illustration from a textbook, written in 1707 by Tobias Cohn, comparing the human body to sections of a house. The lungs were considered to be a ventilation system. It was not generally realized at the time that air was essential for life. (Courtesy National Library of Medicine, Bethesda, Maryland.)

The act of breathing is carried out by muscles that are not located within the lungs. The lungs contain no skeletal muscle and therefore cannot expand by their own efforts. In fact, the lungs are quite elastic, and if the chest cavity is opened to the outside world (such as through a knife wound), the lungs tend to collapse in much the same way that the gills of a fish collapse when it is removed from the water.

8–9
Bronchial epithelium is the original site of almost all lung cancer. One of the first effects of smoke on normal bronchial epithelium (A) is proliferation of the basal cells (B). Ciliated cells disappear from the lining (C), and the proliferating basal cells become disorganized in appearance (D). Finally groups of epithelial cells invade the basement membrane (E). These cells may spread within the lung and sometimes throughout the body as well. (Photomicrographs courtesy of Dr. Oscar Auerbach of the Veterans Administration Hospital, East Orange, N.J. Drawings from "The Effects of Smoking," by E. Cuyler Hammond. Copyright © 1962 by Scientific American, Inc. All rights reserved.)

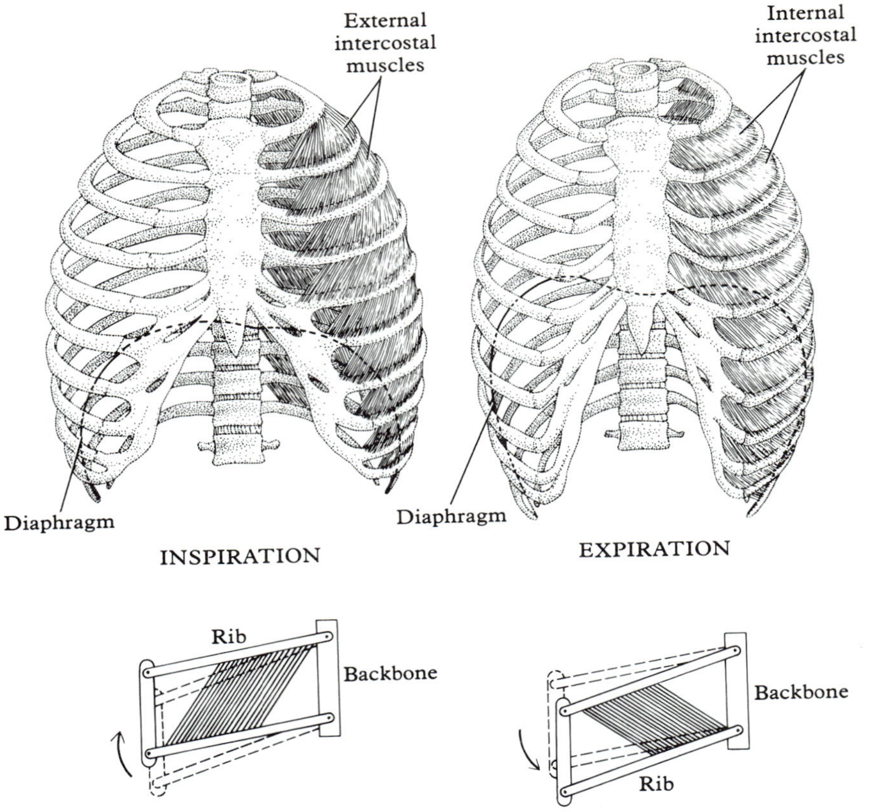

8–11
The thoracic cavity is bounded by the rib cage and the diaphragm. The diaphragm, shown by the dotted lines, separates the thoracic contents (principally the heart and lungs) from the abdominal contents. During inspiration the diaphragm is pulled downward as it contracts and the external intercostal muscles elevate the ribs (left). Both of these actions expand the thoracic cavity. During expiration the diaphragm rises as it relaxes and contraction of the internal intercostal muscles lowers the ribs (right). These actions decrease the volume of the thoracic cavity. (From "The Lung," by Julius H. Comroe, Jr. Copyright © 1966 by Scientific American, Inc. All rights reserved.)

The natural tendency of the lungs to collapse is counteracted by the thoracic cavity, by which they are enclosed (Figure 8-11). The external covering of the lungs (the visceral pleura) fits against the inner lining of the chest cavity (the parietal pleura). The two pleural linings are separated only by a thin layer of liquid, and when the chest cavity expands or contracts, the lungs follow suit. During inhalation, or *inspiration,* the diaphragm, a muscular partition that separates the chest from the abdomen, contracts. When this happens, the central portion of the diaphragm moves downward, and the external muscles of the rib cage contract, which draws the rib cage up and out. As the air in the lungs expands, its pressure drops below atmospheric pressure, and air from outside flows into the lungs.

Exhalation, or *expiration,* is a more passive process that requires less action of the muscles. Only during very active breathing do the muscles of the chest wall actively draw the rib cage down and in, forcing exhalation. Generally, the elastic lungs tend to recoil on their own from the stretch of inhalation. Relaxation of the diaphragm also facilitates exhalation.

In *artificial respiration* the amount of air in the lungs is increased periodically by artificial means that either raise the external air pressure or lower the pressure inside the lungs. An iron lung lowers the pressure of the air which is in direct contact with the chest so that the chest and thoracic cavity expand. As the air in the lungs expands, its pressure drops and outside air (at atmospheric pressure) flows in through the mouth and nose, which are not inside the iron lung. In mouth-to-mouth or mouth-to-nose resuscitation, the external air pressure is increased by the muscular action of the thorax of the person who administers it.

A resting person inhales and exhales about one-half liter of air with each breath (a liter is equal to 1.057 quarts) and breathes about 10 to 14 times per minute. This means that the amount of air entering and leaving the lungs each minute is between 5 and 7 liters. A considerable increase in breathing is possible, however, up to 150 to 200 liters per minute, which is even above the 80 to 120 liters per minute actually breathed during strenuous exercise. We next consider how breathing is regulated so that the amount of air supplied is attuned to the body's needs.

The Control of Breathing

The normal rhythm of breathing depends on special respiratory centers located in the medulla of the brain. From these centers two groups of neurons, inspiratory neurons and expiratory neurons, send out impulses that initiate contraction of the muscles associated with breathing (Figure 8-12). Activity of one group of neurons is always followed by activity of the other.

During inspiration the inspiratory neurons become active. Their activity stimulates the motor neurons that innervate the diaphragm and the external intercostal muscles, thus causing these muscles to contract. At the same time they activate interneurons that inhibit the expiratory neurons. The activity of the inspiratory neurons soon wanes, however, and as a result the inspiratory muscles relax and the inhibition of the expiratory neurons is released. The expiratory neurons then become active. They inhibit the inspiratory neurons by means of interneurons, and may stimulate the nerves that serve the chest muscles, whose contraction facilitates the mostly passive process of expiration. The activity of the expiratory neurons in turn wanes and the inspiratory neurons once again take over.

What regulates the rate and depth of breathing? Because the most important function of breathing is to take in oxygen and to expel carbon dioxide, it is not surprising that the concentrations of O_2 and CO_2 have an important influence (Figure 8-13). When breathing is inadequate, the O_2 concentration of the blood falls and the CO_2 concentration rises. Under either of these conditions, breathing is stimulated. Reduced O_2 stimulates chemical receptors *(chemoreceptors)* present in specialized regions of the largest artery (the aorta) and in the major arteries leading to the brain (the carotid arteries). From these chemoreceptors, impulses are sent through nerves to the respiratory centers in the brain. Increased CO_2 does not affect the chemoreceptors, but has a direct effect on the respiratory centers. Stimulation of the respiratory centers, either directly by CO_2 or indirectly by O_2, results in a more rapid rate of oscillation between inspiratory and expiratory neurons (which increases the rate of breathing) and in an increased rate of discharge of the neurons (which results in more forceful muscle contraction). Through this mechanism the rate and depth of breathing are adjusted in response to blood gas concentrations.

Certain drugs, such as narcotics, exert a direct depressant effect upon the respiratory center in the brain. Someone who has become unconscious because of taking an overdose of a narcotic breathes primarily because of the indirect stimulus of decreased oxygen at the chemoreceptors; increased CO_2 cannot provide a direct stimulus to the incapacitated respiratory center. Thus administering oxygen to someone whose respiratory center has been depressed by an

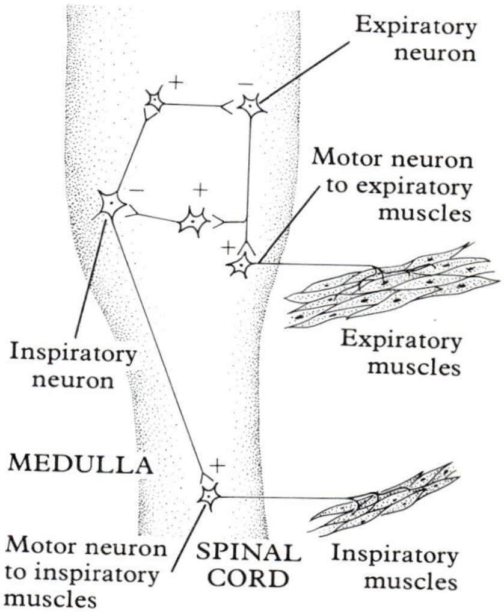

8–12

Neurons of the medulla help to control breathing. In the diagram (+) = stimulation and (−) = inhibition. Inspiratory neurons stimulate motor neurons leading to inspiratory muscles and also stimulate interneurons that inhibit expiratory neurons. Expiratory neurons stimulate interneurons that inhibit inspiratory neurons, and may also stimulate motor neurons to expiratory muscles (during strenuous conditions). The fact that inspiratory and expiratory neurons inhibit each other helps to maintain the alternation between inspiration and expiration.

overdose of narcotics can be lethal, because the artificially supplied O_2 removes the last remaining stimulus for breathing.

The most dramatic increase in rate of breathing occurs during exercise. Yet we know very little of the mechanisms by which exercise increases the rate and depth of breathing. It is reasonable to assume that exercise would alter O_2 and CO_2 concentrations in the blood, yet moderate exercise neither increases the amount of CO_2 in the blood nor decreases its O_2 content. Also, the increase in breathing occurs immediately when exercise begins, before the changes in O_2 and CO_2 could have affected breathing. Two mechanisms have been suggested to explain these facts. First, as it sends impulses to the muscles during exercise, the motor cortex may also send stimulatory impulses directly to the respiratory center. Second, movement of body parts may stimulate receptors inside the joints and muscles (proprioceptors) that initiate impulses that in turn are relayed to the respiratory center. (One reason for thinking that joint and muscle receptors are involved is that passive movement of the limbs may result in a several-fold increase in breathing.) Both of these mechanisms are attractive explanations because their neural circuits could act rapidly enough to explain the almost instantaneous effect of exercise on breathing. But neither has been proven.

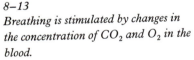

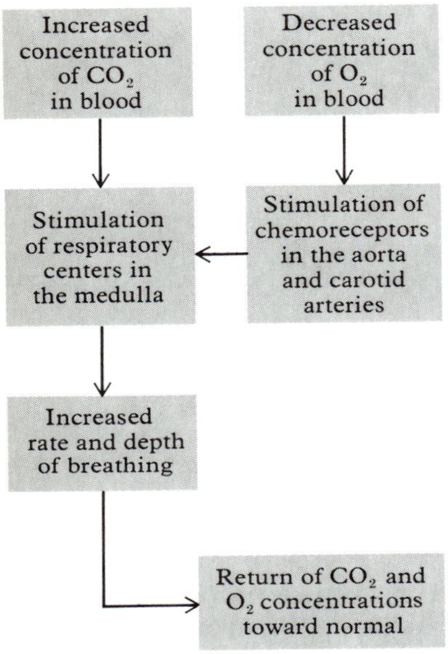

8–13
Breathing is stimulated by changes in the concentration of CO_2 and O_2 in the blood.

Breathing also has a voluntary component that can override the involuntary control of breathing. People who become unusually excited in certain stressful situations are apt to overbreathe, or *hyperventilate.* Hyperventilation leads to muscle spasms and to tingling sensations in the arms, legs and elsewhere, because it results in a lower-than-normal concentration of calcium ions in the blood. This occurs as follows: first, hyperventilation leads to a reduced CO_2 content of the blood because more CO_2 than normal is being expelled by the rapidly breathing lungs. This leads to a decrease in acidity of the blood by causing the following chemical reaction to favor the production of the products on the right hand side of the arrows.

$$\underset{\text{bicarbonate ion}}{H^+ + HCO_3^-} \xrightarrow{\text{hyperventilation}} \underset{\text{carbonic acid}}{H_2CO_3} \xrightarrow{\text{hyperventilation}} H_2O + CO_2 \uparrow$$

As the reaction proceeds, H^+ is used up. Because there is less H^+ present, the blood is less acid. Under these circumstances the proteins in the blood have more of their negative electric charges exposed, because fewer of the negative charges on the proteins are neutralized by the positive charges on hydrogen ions. When this happens, the proteins bind to positively charged Ca^{++} ions in the bloodstream. Hence, the available Ca^{++} in the blood is reduced and the symptoms of low calcium result (see Chapter 11). Hyperventilation can be treated by having the person who is hyperventilating breathe into a paper bag. This simple procedure speeds up the return of the concentration of CO_2 to normal, thus reversing the biochemical problem and relieving the physical symptoms.

The voluntary control of breathing is very important during speech, and especially during singing, when exactly the right amount of air must pass over the vocal cords for a pleasing tone and volume to be achieved. Breathing is also voluntarily controlled during urination and defecation. Here, the glottis and epiglottis are first closed (so that air will not escape), and then the muscles of the chest wall, the diaphragm, and the abdominal muscles all contract at the same time. Contraction of these muscles results in an increase in the intraabdominal

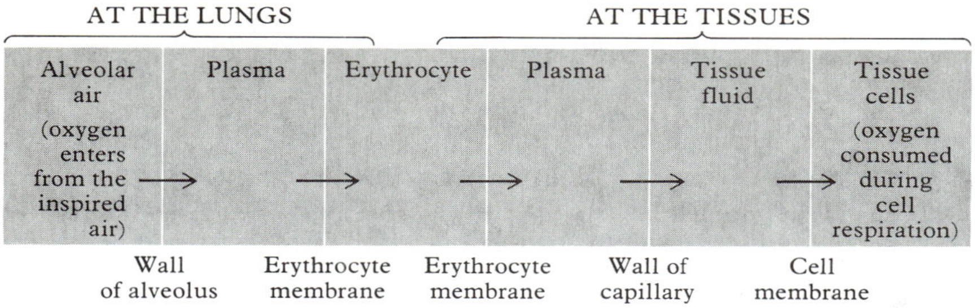

8–14
Oxygen passes from alveoli to tissue cells entirely by passive diffusion. At the lungs, oxygen diffuses from the alveoli to the erythrocytes of the lung capillaries. At the tissues, oxygen diffuses from erythrocytes to tissue cells.

pressure. Coughing is also partially voluntary. When we cough the glottis and epiglottis are at first closed, and contraction of the internal intercostal muscles forcibly compresses the air in the lungs to the point where the glottis and epiglottis are suddenly forced open, expelling a blast of air. In this way a foreign body may be expelled from the trachea or bronchi before it reaches the bronchioles or alveoli.

Transfer of Gases Between Alveoli and Tissue Cells

Oxygen and carbon dioxide are the most important gases exchanged between the alveoli and the tissues. Let us begin with a discussion of oxygen.

Oxygen

Once the oxygen of inspired air has flowed into the alveoli, it passes first into the plasma of the surrounding capillaries, after which most of the O_2 is taken up by red blood cells (Figure 8-14). The red blood cells (erythrocytes) then carry the oxygen to the capillaries of the body, where the oxygen passes back out into the plasma, crosses endothelial cells of the capillary wall into the tissue fluid, and finally enters the cells of the body, where it is utilized. This oxygen transfer across membranes occurs entirely by diffusion, which, as you will recall, is the tendency of dissolved molecules to pass from an area of higher concentration to an area of lower concentration. For example, the pressure of oxygen in the alveoli averages about 107 mm Hg (millimeters of mercury), whereas the pressure of oxygen in the blood entering the capillaries of the lung is about 40 mm Hg. With this large pressure difference, oxygen readily diffuses from alveoli to capillaries, and by the time this blood leaves the alveolar capillaries to be distributed throughout the body once again its oxygen pressure has risen to about 100 mm Hg. Similarly, after the blood has reached the tissues, oxygen readily leaves the capillaries and enters the tissues, where the pressure of O_2 is lower because it is constantly being consumed by cells during chemical respiration.

The hemoglobin molecules contained within red blood cells are especially important for oxygen transport. Hemoglobin increases the oxygen-carrying capacity of the blood more than thirty-fold over that of hemoglobin-free plasma. Hemoglobin has this effect because it enters into a loose, easily reversible association with oxygen. This association has some remarkable properties that become evident on close inspection of the hemoglobin *dissociation curve* (Figure 8-15). Notice that this graph, which relates oxygen pressure to the percent

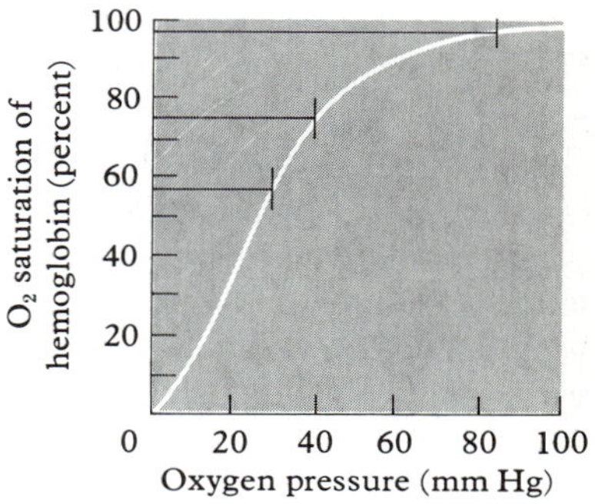

8–15
The hemoglobin dissociation curve. The S-shaped curve expresses the relationship between oxygen pressure and percentage saturation of hemoglobin with oxygen. Hemoglobin unloads about 25 percent of its oxygen when the oxygen pressure drops from 100 mm Hg (blood leaving alveoli) to 40 mm Hg (blood leaving tissue capillaries). If the oxygen pressure drops from 40 to 30 mm Hg hemoglobin unloads almost another 20 percent of its oxygen.

saturation of hemoglobin by oxygen, has an S shape. At the top of the curve (at high oxygen pressures) the curve falls off gradually, but farther down it becomes quite steep. The top of the curve describes what happens in normal circumstances when the oxygen pressure of the blood drops from 100 mm Hg (in arterial blood) to about 40 mm Hg, after it passes through the tissues. Under such circumstances hemoglobin releases about 25 percent of its oxygen. The steep part of the curve applies to *strenuous* exercise, when cells use extra oxygen and the blood oxygen pressure drops, perhaps to 30 mm Hg. This additional drop of 10 mm Hg (from 40 to 30 mm Hg) causes the hemoglobin molecules to release almost another 20 percent of their oxygen. Thus, the S-shape indicates that hemoglobin molecules are capable of delivering a considerable amount of extra oxygen when the demand for it arises.

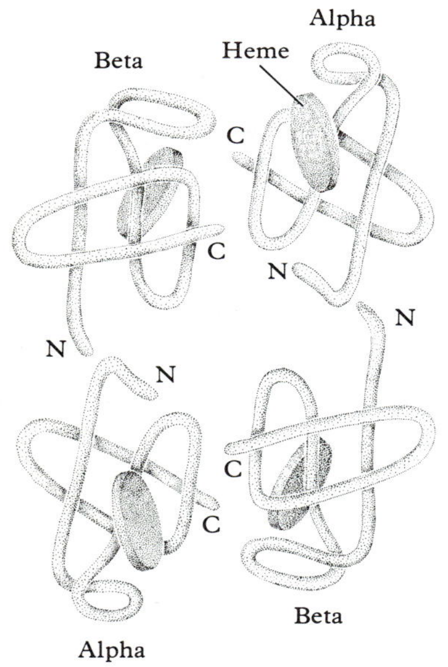

8–16
A hemoglobin molecule (hemoglobin A). The hemoglobin molecule has four subunits that interact with each other to determine the shape of the dissociation curve with oxygen. Each subunit consists of a winding polypeptide chain and a heme group (depicted as a wafer), which carries oxygen. The two alpha polypeptide chains are identical, as are the two beta polypeptide chains. The designations N and C at the ends of each polypeptide chain indicate which is the NH_2 and which is the COOH end of the chain. The four subunits are farther apart from each other in this drawing than they are in the actual molecule.

Hemoglobin has an S-shaped dissociation curve because of special properties resulting from the fact that it is a tetramer, which is to say that the hemoglobin molecule is made up of four subunits (Figure 8-16). Each subunit of the molecule consists of a heme group and a polypeptide (protein) chain, and each heme group can associate with a molecule of oxygen. Thus, a fully saturated hemoglobin molecule carries four oxygen molecules. When a heme group in one of the subunits loses an oxygen, it causes another subunit to change shape, which makes the other subunit either more or less likely to lose its oxygen molecule. For example, if a hemoglobin molecule is combined with four oxygens (for example, at an oxygen pressure of 100 mm Hg), the blood pressure must drop 60 mm Hg before one oxygen is lost. But if the hemoglobin is combined with only three oxygens (at an oxygen pressure of 40 mm Hg), a drop of only 10 to 15 mm Hg is necessary before it gives up an additional oxygen. The tetramer structure of hemoglobin accounts for the delivery of especially large amounts of oxygen to oxygen-depleted tissues.

The four heme groups in hemoglobin are identical, but there are two kinds of polypeptide chains. In adult hemoglobin there are two alpha and two beta chains. In fetal hemoglobin there are two alpha and two gamma chains. (See Chapter 18.) The gamma chains of fetal hemoglobin enable it to become more saturated with oxygen at lower oxygen pressures than adult hemoglobin (Figure 8-17). This is useful because fetal hemoglobin can become fully saturated with oxygen, as if the fetus were breathing, even though the fetal hemoglobin receives its oxygen by diffusion from the blood of the mother at a lower oxygen pressure than alveolar oxygen pressure.

Another important property of hemoglobin is that its affinity for oxygen is decreased when the blood becomes more acid (Figure 8-18). This is indicated by a shift of the S-shaped curve to the right (the Bohr effect). Under acid conditions, more oxygen is released at a given oxygen pressure. Acid conditions usually occur during strenuous exercise, when large amounts of lactic acid are released by the muscles. The release of lactic acid triggers the release of oxygen to meet the needs of the muscles for more oxygen during strenuous exercise.

The properties of the hemoglobin molecule that make it combine readily with oxygen also allow it to combine readily with certain other gases, including carbon monoxide (CO). Carbon monoxide is an extremely toxic gas formed by the incomplete combustion of hydrocarbons. Its toxicity stems from the fact that it reacts with hemoglobin to form *carboxyhemoglobin,* and once this has happened the hemoglobin can no longer carry oxygen. The affinity of CO for hemoglobin is about 210 times as strong as the affinity of O_2 for hemoglobin; as a consequence a sustained concentration of as little as 0.1 percent CO in the air is enough to eventually inactivate about half of the hemoglobin molecules in the body. Death usually occurs when slightly over half of the hemoglobin molecules in the body have become inactivated. The best treatment for carbon monoxide poisoning is breathing fresh air, or even better, air that has been enriched by a high concentration of oxygen.

Let us now turn our attention to carbon dioxide, and in particular to how this gas, once it has been generated as a waste product of metabolism, passes from cells to alveoli and is finally eliminated in expired air.

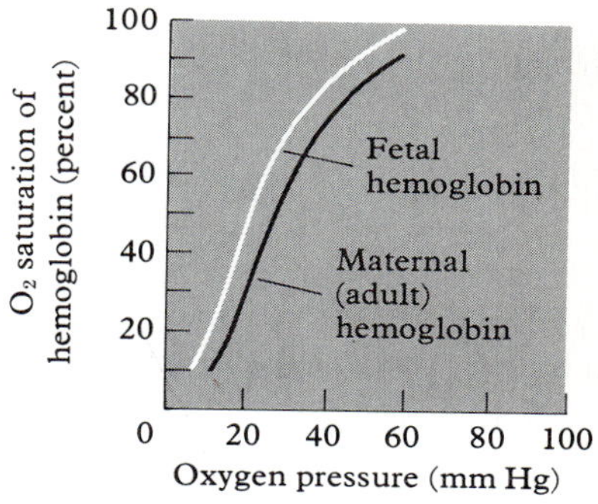

8–17
Hemoglobin dissociation curves for fetal and adult hemoglobin. Fetal hemoglobin is more fully saturated with oxygen at any given oxygen pressure than adult hemoglobin. Thus fetal hemoglobin is almost fully saturated when it leaves the placenta, even though the oxygen pressure at the placenta is not as high as alveolar oxygen pressure.

Carbon Dioxide

Like oxygen, carbon dioxide gas (CO_2) passes between the cells of the tissues and the alveoli entirely by means of diffusion. CO_2 produced by tissue cells reaches red blood cells by diffusion across three barriers: the tissue cell membrane, the capillary wall, and the membrane of the red blood cell (Figure 8-19). Once it has entered the red blood cell, only a small amount (about 8 percent) of the CO_2 remains dissolved as CO_2. About a quarter of it attaches to the heme portion of hemoglobin and is transported the same way that oxygen is. The remaining 67 percent of the CO_2 combines with water (H_2O) to form carbonic acid (H_2CO_3), which spontaneously dissociates into H^+ and HCO_3^- according to the following equation:

$$CO_2 + H_2O \xrightarrow{\text{carbonic anhydrase}} \underset{\text{carbonic acid}}{H_2CO_3} \longrightarrow H^+ + \underset{\text{bicarbonate ion}}{HCO_3^-}$$

A large part of the HCO_3^- that is generated inside the red cells by this reaction then passes back out into the plasma. This reaction is useful because HCO_3^- is much more soluble in the blood than CO_2, and the quantity of CO_2 released from the body's cells is so large that it simply could not all be eliminated as dissolved CO_2. The reaction is driven by the high CO_2 pressure in the tissues, and it takes place in erythrocytes because the enzyme that speeds up the reaction, carbonic anhydrase, is present there in abundance.

When the blood arrives at the capillaries of the lungs, carbon dioxide diffuses from erythrocytes into the alveoli because the pressure of CO_2 is comparatively low in the alveoli (Figure 8-19). When CO_2 leaves the erythrocytes, the chemical reactions we have just described are driven in the reverse direction, and much of the HCO_3^- and $H+$ are reconverted to CO_2 and H_2O. Alveolar CO_2 is then exhaled into the atmosphere.

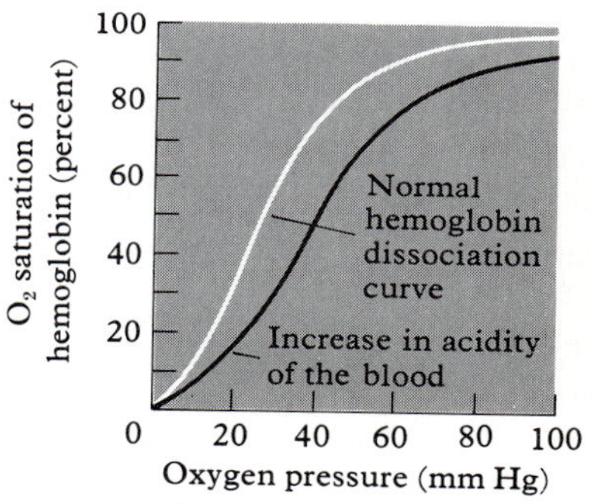

8–18
Effect of increased acidity of the blood on hemoglobin. The white line is a normal hemoglobin dissociation curve. As the dark line indicates, an increase in acidity (H^+ ion concentration) causes the dissociation curve to shift to the right (the Bohr effect). This means that in active tissues, where the H^+ ion concentration is the highest, hemoglobin delivers more oxygen at a given O_2 pressure.

High Altitudes

Being adventurous, human beings tend to travel into places where the atmosphere is very thin, by climbing or by flight. We know that the body can compensate for a minor decrease in O_2 pressure by increasing the rate and depth of breathing, but the major reduction in oxygen concentration that people experience at very high altitudes cannot be compensated for and can have serious effects on normal brain function. The metabolism malfunctions seriously because of lack of oxygen at about 7000 meters (20,000 feet), and consciousness is difficult to maintain at heights above 7900 meters (26,000 feet). G. L. Mallory

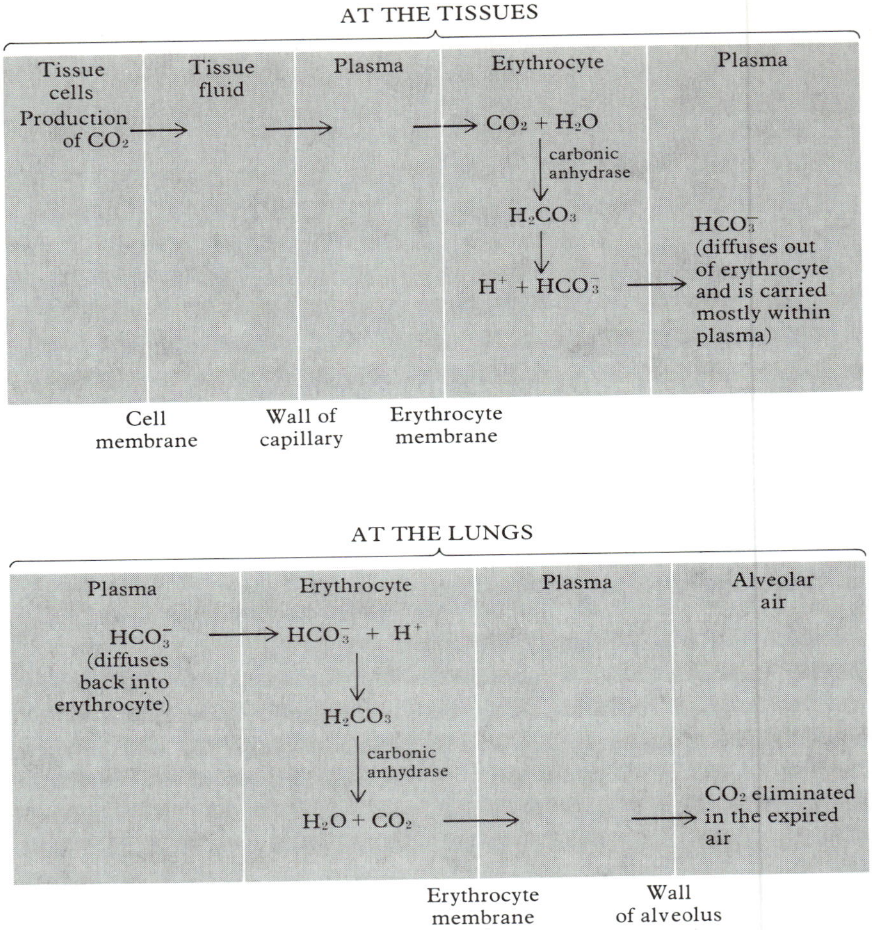

8–19
Carbon dioxide passes from tissue cells to alveoli entirely by passive diffusion. Most CO_2 that diffuses into tissue capillaries passes into erythrocytes and is converted to HCO_3^- (bicarbonate) under the influence of the enzyme carbonic anhydrase. At the lungs HCO_3^- is reconverted to CO_2 as CO_2 enters the alveoli and is blown off from the lungs. The presence of HCO_3^- increases the carrying capacity of the blood for CO_2 because HCO_3^- is more soluble in the blood than CO_2.

and A. C. Irvine, who were attempting to climb Mount Everest without oxygen supplies in 1924, reached 28,000 feet but did not return. At the summit of Everest, 8848 meters (29,028 feet), the pressure of oxygen (about 40 mm Hg) is less than a third its pressure at sea level (about 142 mm Hg in humid air). The oxygen pressure of blood returning *from* the tissues is about 40 mm Hg at sea level, so a profitable gas exchange is obviously not possible at the top of Mount Everest.

At 19,500 meters (64,000 feet) the *total* atmospheric pressure (46 mm Hg) is less than the pressure exerted by water in the moist air of the alveoli (47 mm Hg). Because boiling occurs when the pressure of water exceeds the total atmospheric pressure, water will start boiling out of the blood supplying the lungs at that elevation, a most unpleasant (and fatal) experience.

The highest permanent human settlements, in the Andes, are at almost 5500 meters (18,000 feet). At these high altitudes, the body makes another adjustment to increase the oxygen content of the blood by stepping up the production of erythrocytes. In people who live at such altitudes, the formed elements of the blood (principally erythrocytes) may account for as much as 70 percent of the blood volume (roughly 50 percent is normal for people who live at low altitudes).

People living on Pike's Peak (Colorado), at 4250 meters (14,000 feet), also have a greater erythrocyte content in their blood.

The mechanism by which the number of erythrocytes increases at high altitudes seems to depend upon the hormone *erythropoetin*, which stimulates the production of red blood cells and is normally present in very small quantities in the bloodstream. Erythropoetin is produced by specialized cells in the kidneys, and its production is stimulated by a decrease in oxygen delivery to the tissues. The increase in the number of erythrocytes means that more oxygen can be delivered to the tissues.

Pressure Changes During Scuba Dives

Under certain circumstances, the human body must cope with gases at greater than normal atmospheric pressure. For example, gas pressures increase rapidly during a dive made with scuba gear. The pressure exerted on the human body increases by 1 atmosphere (the usual pressure at sea level, 760 mm Hg) for every 10 meters (33 feet) of depth in sea water so that at 30 meters (100 feet) in sea water a diver is exposed to a pressure of about 4 atmospheres. The pressure of the gases being breathed must equal the external pressure applied to the body; otherwise breathing is very difficult. Therefore all of the gases in the air breathed by a scuba diver at 100 feet are present at four times their usual pressure. Nitrogen (N_2, which composes 80 percent of the atmosphere) usually causes a balmy feeling of well-being (euphoria) at this pressure. At depths of 5 atmospheres or more (40 meters) nitrogen causes symptoms resembling alcohol intoxication that are called *raptures of the deep*, or, more technically, *nitrogen narcosis*. Nitrogen narcosis apparently results from a direct effect on the brain of the large amounts of N_2 dissolved in the blood. Deep dives are less dangerous if helium is substituted for N_2, because under these pressures helium does not exert a similar narcotic effect.

As a scuba diver descends, the pressure of N_2 in the alveoli increases. N_2 then diffuses from the alveoli to the blood, and from the blood to the tissue. The reverse occurs when the diver surfaces: the alveolar N_2 pressure falls, and N_2 diffuses from the tissues into the blood, and from the blood into the lungs according to the pressure gradient. If this return to the surface is too rapid, N_2 in the tissues and blood cannot diffuse out rapidly enough, and N_2 bubbles are formed, causing the symptoms of *the bends* (also known as *decompression sickness* and *caisson disease*). Bubbles in the tissues cause severe pains, particularly around the joints, and bubbles in the blood stream, which occur in more severe cases, may even obstruct arteries to the brain, causing paralysis, respiratory failure, and death. Treatment is a prompt reestablishment of high pressure in a pressure chamber, followed by a *slow* return to normal atmospheric pressure.

Another complication may result if the breath is held during ascent. During ascent from a depth of 33 feet (about 10 meters), the volume of air in each alveolus will double because the air pressure at the surface is only half of what it was at 33 feet. This change in volume may distend and even rupture the alveolar walls. If air escapes from the ruptured alveoli into the veins of the lung and from there to the rest of the circulation, it may be fatal. This condition, called *air embolism*, has occurred during rapid ascent from depths of less than 15 feet. To avoid this tragic event, a scuba diver must ascend slowly, never at a rate exceeding the rise of exhaled air bubbles, and must exhale during ascent.

During a free dive without scuba gear there is no danger, either from ascending without exhaling or from the bends. The reason is that the air in the lungs has been taken in at the surface, under just 1 atmosphere of pressure. Therefore the volume of air in the lungs is correctly adjusted for life at the surface. Similarly, nitrogen is not taken in under greater than normal pressure and it will not build up in the blood and tissues in excess of its normal concentration during ascent from a free dive.

So far in this chapter we have discussed the respiratory gases and how they are transported, but we have not yet discussed why cells take in oxygen and excrete carbon dioxide. We now turn to a discussion of this process, internal respiration, and we shall see that gas exchange is tied to the body's need to supply its cells with energy.

Internal Respiration and Energy

All living cells in the body consume oxygen and give off carbon dioxide (Figure 8-20). The overall process by which this occurs, internal respiration, can be summed up in the following chemical equation:

$$\underset{\text{Glucose}}{C_6H_{12}O_6} + 6O_2 \rightarrow 6CO_2 + 6H_2O + 38\text{ATP} + \text{heat}$$

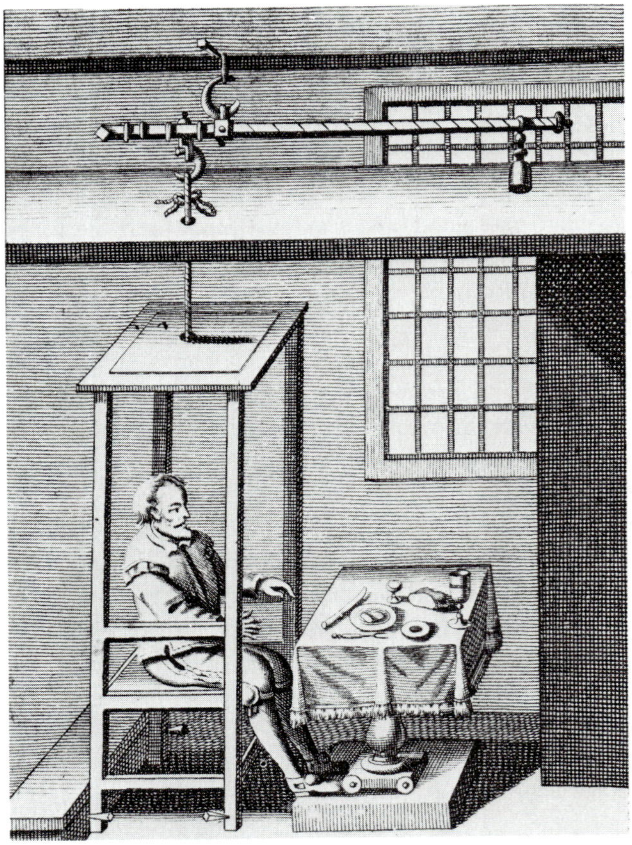

8–20
Santorio Santorio (1561–1636), an Italian physician, conducted numerous experiments on himself. This illustration, from a work published in 1626, shows him seated in a chair suspended from a beam balance. He measured his weight loss under a variety of conditions (asleep, awake, during emotional stress) as he sat in the chair. He attributed the weight loss to "insensible perspiration." We now know that the weight loss is due to the fact that the products of respiration (CO_2 and H_2O) that are given off outweigh the O_2 that is taken in. The weight loss is equivalent in many respects to an automobile losing weight as its fuel is used up during the process of combustion. (Courtesy of the New York Academy of Medicine Library.)

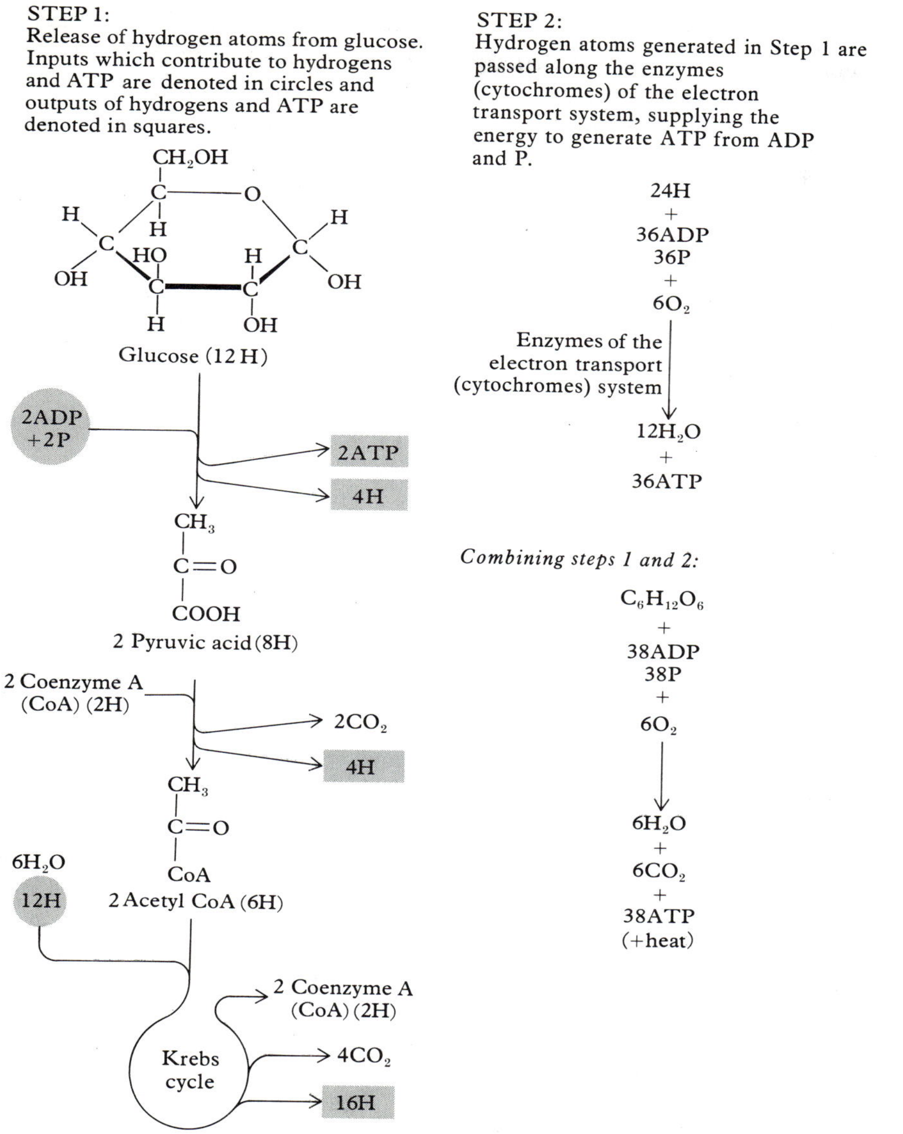

8–21

Cellular respiration: the utilization of glucose to produce ATP. In Step 1 hydrogen atoms are released from glucose as it is broken down. Step 2 is the passage of the released hydrogen atoms through the electron transport system.

The cells obtain their energy from the 6-carbon sugar, glucose, when the chemical bonds that hold the atoms of the molecule together are broken apart. A good portion of this energy is transferred into ATP, a high energy molecule whose energy is directly available to all parts of the cell for its crucial metabolic processes. Some of the energy from this reaction is irreversibly lost as heat. The breakdown of one glucose molecule allows the formation of 38 ATP molecules.

How is ATP, a usable and readily available source of energy, generated? Its production is a two-step process. In the first step, glucose is broken down and two ATP molecules are generated directly when the 6-carbon glucose molecule is converted into two 3-carbon pyruvic acid molecules (Figure 8-21). The remaining 36 ATPs are generated during the second step when the energy resulting from breaking the remaining bonds is captured as *the 12 hydrogen atoms of glucose and 12 hydrogen atoms from water are passed along a series of enzymes known as the cytochromes.*

Oxygen is essential in the process because its presence allows the hydrogen atoms released from glucose to be disposed of (Step 2 of Figure 8-21). The overall result is that hydrogen combines with oxygen to form H_2O. If hydrogen is prevented from combining with oxygen, respiration stops and the result is the death of the cell. This is illustrated by the effects of cyanide poisoning.

Cyanide Poisoning

"Cyanides" are a class of chemicals containing a nitrogen atom that is attached solely to a carbon atom (a CN^- group). Cyanide combines readily with many metals and has widespread industrial applications. For example, it is used to separate gold from its ores. In the human body, cyanide acts as an extremely potent poison because it prevents the hydrogen atoms of glucose from combining with oxygen. How does cyanide do this?

Cyanide blocks the action of enzymes that normally transport hydrogen to oxygen. Cyanide does this by combining readily with the triple-charged (Fe^{+++}) iron atoms of the enzymes of the cytochrome system and thereby rendering these crucial carrier molecules unable to function. The treatment of cyanide poisoning is designed to counteract this blockage of enzymes by allowing cyanide to combine with the iron of hemoglobin rather than the iron of the cytochrome enzymes. This treatment consists of injecting sodium nitrate ($NaNO_2$) into the bloodstream of the person who has been poisoned. There the $NaNO_2$ changes the chemical state of hemoglobin's iron from its usual Fe^{++} (ferrous) state to the Fe^{+++} (ferric) state. After the change in state, some of the cyanide will shift from the Fe^{+++} of the cytochrome enzymes to the Fe^{+++} of the hemoglobin. This procedure can protect against up to 40 times the lethal dose of cyanide. The actual removal of cyanide from the body is then achieved by the intravenous administration of sodium thiosulfate ($NO_2S_2O_3$). It reacts with cyanide to form thiocyanate (SCN^-), which is readily excreted in the urine:

$$\underset{\text{Sodium thiosulfate}}{Na_2S_2O_3} + \underset{\text{cyanide}}{CN^-} \rightleftharpoons \underset{\text{thiocyanate}}{SCN^-} + \underset{\text{sodium sulfite}}{Na_2SO_3}$$

Obtaining Some Energy Without Oxygen

In the absence of sufficient oxygen, there is an alternative process by which cells can get some energy. This alternative process, known as *anaerobic glycolysis,* is not sufficient to sustain the life of a cell for long, but it can provide a stopgap when there is not quite enough oxygen. For example, during strenuous exercise anaerobic glycolysis occurs inside muscle cells. The chemical formula for anaerobic glycolysis is this:

$$\underset{\text{Glucose}}{C_6H_{12}O_6} \rightarrow 2\begin{array}{c} CH_3 \\ | \\ CHOH \\ | \\ COOH \\ \text{lactic acid} \end{array} + 2ATP + \text{heat}$$

Notice that no hydrogen atoms are released when glucose is split into two lactic acid molecules. Because no hydrogen atoms are given off, there is no need for oxygen to combine with them. Of course, this also means that there are no hydrogen atoms available to generate abundant supplies of ATP. As shown in this equation, only two ATP molecules are produced for each glucose, compared to 38 produced during internal respiration. So anaerobic glycolysis is only $\frac{2}{38}$, or less that 6 percent, as efficient as respiration in providing usable energy in the form of ATP.

The Similarity Between Internal Respiration and Combustion

The release of energy through burning is called *combustion.* This process, which refers equally to the burning of a log in the fireplace or gasoline in an internal combustion engine, is in many ways similar to internal respiration. Both combustion and respiration use up oxygen and release energy stored in chemical bonds. In fact, when glucose is the fuel consumed the chemical formula for combustion is identical to that for respiration, except that no ATP is generated.

At first glance, the body's chemical reactions for obtaining energy do not seem very similar to the burning of a log in the fireplace or to the consumption of gasoline in an engine. For example, burning does not seem to produce water. However, water *is* produced by burning fuel; we may not see it because water is a gas at temperatures above 100° C (the temperatures usually associated with combustion). If a gasoline automobile engine is started on a cold morning, liquid can sometimes be seen dripping out of the exhaust pipe. The liquid is water, produced by combustion and cooled enough in the cold exhaust pipe that it changes from a gas into a liquid. Furthermore, the trail made in the sky by a jet airplane is composed of the water of combustion that has frozen into ice crystals at very cold temperatures in the upper atmosphere.

In spite of the fundamental similarity between internal respiration and combustion, the processes differ in four principal ways. First, respiration can occur at various body temperatures, whereas combustion occurs only at high temperatures. Second, energy is liberated gradually during respiration and rapidly during combustion. Third, respiration proceeds by ordered, step-by-step reactions that are catalyzed by enzymes, whereas combustion proceeds by less orderly, random reactions. Fourth, respiration is connected to ATP generation, whereas combustion is not connected to other chemical processes. In comparison to combustion, then, the respiratory process of the living cell is a delicate and precise process, carefully regulated in order to sustain the energy needs of the cell and of the entire organism.

Summary

The air passages conduct gases between the external environment and the alveoli of the lungs. In the alveoli gases are exchanged between blood and air across the thin walls that separate the alveolar air sacs from the capillaries of the lung.

Several defense mechanisms help to keep most particulate matter out of the air passages and the alveoli. Larger particles (more than 10 micrometers in diameter) are filtered out by hairs at the entrance to the nose and the mucus of the nasal bones, and smaller particles are captured by mucus, which is present throughout the air passages. Once particles have been trapped by mucus, cilia sweep the particles toward the throat, where they are usually swallowed. Reflexes such as coughing and sneezing also help to expel particles that trigger off the reflexes by stimulating sensory receptors. Particles that manage to reach the alveoli are usually engulfed by wandering phagocytic cells. In spite of these defense mechanisms, the constant exposure to large amounts of particulate material may have serious consequences, including silicosis in those who inhale silica and lung cancer in those who smoke.

The rhythm of breathing is controlled by impulses sent from respiratory centers in the medulla of the brain. In the respiratory centers inspiratory neurons and expiratory neurons alternate in activity. The lungs themselves have no skeletal muscles and cannot expand by their own efforts.

The rate and depth of breathing are controlled in part by the concentrations of O_2 and CO_2 in the blood. Elevated concentrations of CO_2 directly stimulate the respiratory centers. Reduced O_2 levels are detected by chemoreceptors located in the aorta and the carotid arteries, which send stimulatory impulses to the respiratory centers through nerves. The dramatic increase in breathing that accompanies moderate exercise is not due to changes in concentrations of O_2 and CO_2, but apparently to stimulation of the respiratory centers as a result of messages from the motor cortex and from joint and muscle receptors. Excitement can lead to overbreathing (hyperventilation), which sometimes results in muscle spasms and sensations of tingling because of a reduction of ionized calcium in the blood. The voluntary control of breathing enables us to speak as we do.

In breathing oxygen passes from alveoli to erythrocytes in the lungs, and from erythrocytes to tissue cells in the capillaries of the tissues. This flow of O_2 takes place by the process of diffusion, which is the tendency of dissolved molecules to pass from an area of higher to an area of lower concentration. The oxygen-carrying capacity of the blood is increased more than 30-fold by hemoglobin molecules present within the erythrocytes. Each hemoglobin molecule is made up of four subunits, and each subunit contains a heme group and a polypeptide chain. A hemoglobin molecule can carry four O_2 molecules, one attached to each heme group. The amount of oxygen carried by hemoglobin depends upon the oxygen pressure. In the lungs, where the oxygen pressure is high, hemoglobin becomes nearly 100 percent saturated with oxygen. In the tissues, where the oxygen pressure is low, hemoglobin gives up oxygen and becomes less saturated. The relationship between oxygen pressure and percent saturation of hemoglobin is spelled out by the S-shaped hemoglobin dissociation curve. Because of the nature of the S-shaped curve a small drop of oxygen pressure in the tissues (as during strenuous exercise) results in a proportionately large increase in the delivery of oxygen. The dissociation curve also undergoes a shift to the right if the blood becomes more acid (as occurs during strenuous exercise). This shift to the right has the effect of causing the hemoglobin to deliver more oxygen at a given pressure of O_2 in the tissues. Hemoglobin can also combine with carbon monoxide; its affinity for CO is about 210 times its affinity for O_2. This gas is a poison because it prevents hemoglobin from carrying oxygen.

Carbon dioxide is a product of cell respiration that must be removed from the body. It passes from the cells where it is released into tissue capillaries, and is then transported in the blood to the lungs, where it enters the alveoli. Like oxygen, CO_2 passes from tissue cells to the alveoli entirely by diffusion. When

CO_2 enters red blood cells, most of it is converted to HCO_3^- under the influence of the enzyme carbonic anhydrase. HCO_3^- is much more soluble in blood than CO_2, and its presence therefore increases the capacity of the blood to carry CO_2. At the lungs HCO_3^- is converted back to CO_2 as CO_2 diffuses into the alveoli.

At very high altitudes, atmospheric pressure and oxygen pressure are low. Above 7000 meters (20,000 feet) serious malfunction of the body metabolism may occur. Many people who live at high altitudes have a greater-than-normal number of red blood cells in their blood, an effect that is mediated by the hormone erythropoetin. The increase in the numbers of red blood cells helps to compensate for the low oxygen pressure.

At the other extreme are the high atmospheric pressures encountered by scuba divers. At 5 atmospheres of pressure a diver breathing air may develop a toxicity due to nitrogen under pressure, known as nitrogen narcosis or raptures of the deep. The bends is a hazard that may occur if a diver ascends so rapidly that bubbles of nitrogen form in the tissues.

The needs to take in oxygen and eliminate carbon dioxide stem ultimately from the need for energy. The energy-yielding process of the body, internal respiration, uses glucose as its major fuel and converts the energy stored in the chemical bonds of glucose into 38 molecules of ATP and heat. ATP is a form of energy that the cells can readily use and heat is a by-product of the process. The conversion of the chemical bond energy of glucose into 38 ATP molecules requires that the 12 hydrogen atoms of glucose be passed along a series of enzymes known as the cytochrome system. Cyanide is toxic because it prevents some of these enzymes from combining with hydrogen and hence stops the body's cells from obtaining energy.

When the chemical bonds of glucose are broken, the carbon and hydrogen atoms that are released must be disposed of. After passing through the cytochrome system, the hydrogen atoms combine with oxygen to form H_2O. The carbon atoms combine with oxygen to form CO_2. Oxygen is required for internal respiration because only through its action can the products of the process be disposed of.

Cell respiration and combustion are similar in that both use oxygen and fuel and both release CO_2, H_2O, and energy. But unlike combustion, cell respiration is an orderly, step-by-step process connected to ATP production that can occur at body temperatures.

Selected Readings

1. *Human Physiology: The Mechanisms of Body Function,* by A. J. Vander, J. H. Sherman, and D. S. Luciano. McGraw-Hill, 1970. Contains an excellent section on respiration (Chapter 11, pp. 301–334).

2. *Review of Medical Physiology,* 7th ed., by W. F. Ganong. Lange Medical Publications, 1975. Several chapters are devoted to respiration, and Chapter 37, "Respiratory Adjustments in Health & Disease," is exceptionally well-done.

3. "The Lung," by J. H. Comroe, Jr. *Scientific American,* Feb. 1966. An excellent description of the anatomy and physiology of the air passages.

4. "The Effects of Smoking," by E. C. Hammond. *Scientific American,* July 1962. Presents a clear description of the structural and functional changes caused by smoking.

Mochican maize god from Peru depicted on a water vessel. The vessel is estimated to date from 500-1000 A.D. *The replicas of the god on either side are said to signify fertility. Corn was the staple of the Peruvian diet and its importance is symbolized in this sculpture. (Courtesy of The American Museum of Natural History.)*

CHAPTER 9

Nutrition

Like all other animals, people must have food. Our bodies require it for two reasons. First, our bodies need energy to run our chemical machinery and to maintain proper physiologic control over the internal environment. This energy is found within the chemical bonds of the food we eat. Second, food provides us with the chemical building blocks from which we can make new molecules for growth or to replace those that are worn out.

In this chapter we discuss human nutrition in terms of what constitutes a good diet and why. Other topics that are relevant to this discussion are considered elsewhere. In Chapter 4 we discuss how closely dependent all living things are on one another as sources of food and energy; we also discuss our prospects of supplying a rapidly growing world population with enough food to ensure good health. Our main concerns in this chapter are with what constitutes good nutrition and with how nutrition relates to health.

Most of the food we eat contains proteins, carbohydrates, and fats. During the process of digestion, they are broken down into their component parts:

Carbohydrates ⟶ 6-carbon sugars
Proteins ⟶ amino acids
Fats ⟶ fatty acids and glycerol

These component parts pass through the wall of the intestine and enter the blood stream. They become the building blocks of our own carbohydrates, proteins, and fats or are used in chemical reactions that provide our bodies with energy.

Food for Energy

The energy requirements of most cells are met largely by the breakdown of a 6-carbon sugar, *glucose* (Figure 9-1), and the energy stored in its chemical bonds is released in the process. This breakdown process, *internal respiration,* occurs in all of the cells of our bodies, and, as you will recall from Chapter 8, it yields 38 molecules of the energy-rich compound ATP per molecule of glucose consumed.

Certain kinds of cells normally require more glucose than others. The cells that consume the most glucose are nerve cells, and this is reflected by the fact that the adult human brain, which is less than three percent of the body by weight, consumes two-thirds of the total circulating glucose.

The brain's energy requirements can be met by between 100 and 145 grams of glucose per day, and most diets contain enough carbohydrate to provide the brain and other tissues with this amount of glucose on a daily basis. But even if a person's diet changes so that it never contains enough carbohydrate to provide this much glucose, the concentration of glucose in the blood will not change. This is because natural selection has provided our bodies with a means of manufacturing it from molecules other than carbohydrates. The end result is a nearly constant concentration of glucose in the blood despite wide variations in dietary intake of carbohydrates. The way in which the body goes about regulating the proper concentration of glucose is a fascinating and complex process. For our discussion it is enough to know that the concentration of circulating glucose is closely regulated by homeostatic mechanisms.

Glucose is so important in meeting the energy requirements of the brain and other organs that we might expect our bodies to be able to store large amounts of

GLUCOSE
A monosaccharide present in fruit juices and honey

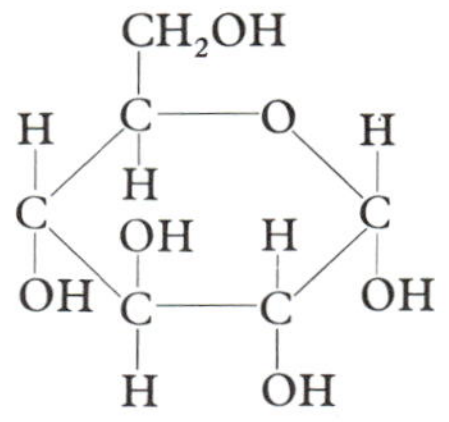

FRUCTOSE
A monosaccharide present in fruit juices and honey

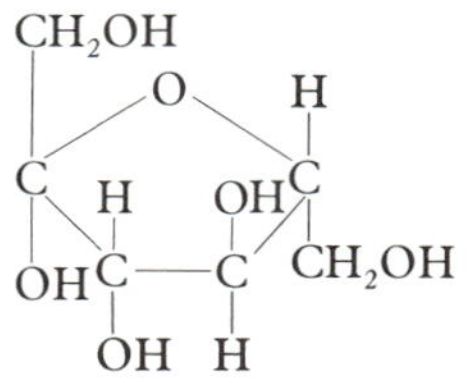

GALACTOSE
A monosaccharide produced by breakdown of lactose

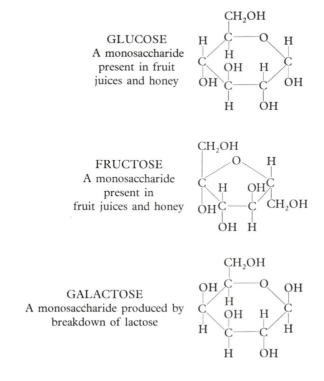

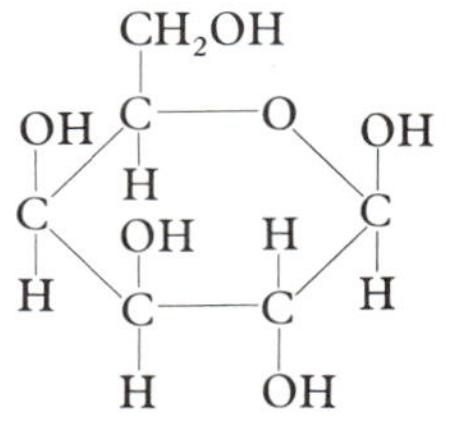

9–1
Carbohydrates are composed of carbon, hydrogen, and oxygen atoms. The principal dietary carbohydrates are polysaccharides, disaccharides, and monosaccharides. In the gastrointestinal tract the larger carbohydrates are broken down into monosaccharides that are rapidly absorbed from the small intestine. The polysaccharide glycogen is synthesized in the body from glucose; its structure is the same as that of starch, except that it is more branched.

(continued)

LACTOSE (milk sugar)
A disaccharide
composed of galactose and
glucose

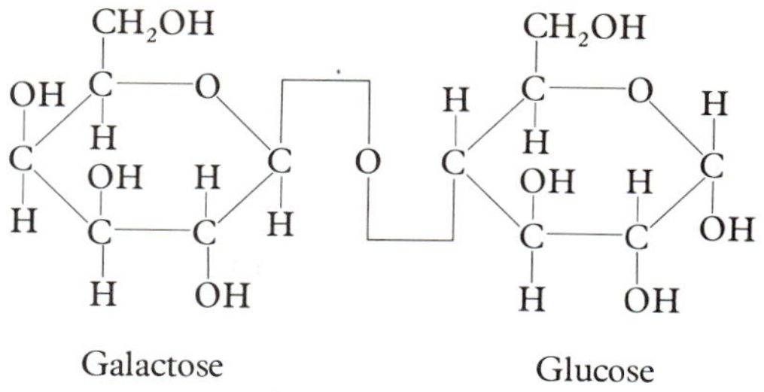

SUCROSE (table sugar)
A disaccharide composed of
glucose and fructose; produced
in sugar cane and beets

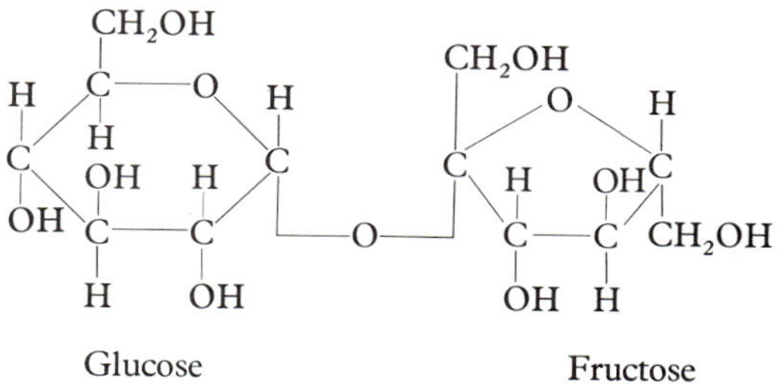

MALTOSE
A disaccharide composed of
two molecules of glucose;
produced by starch digestion
and present in malt

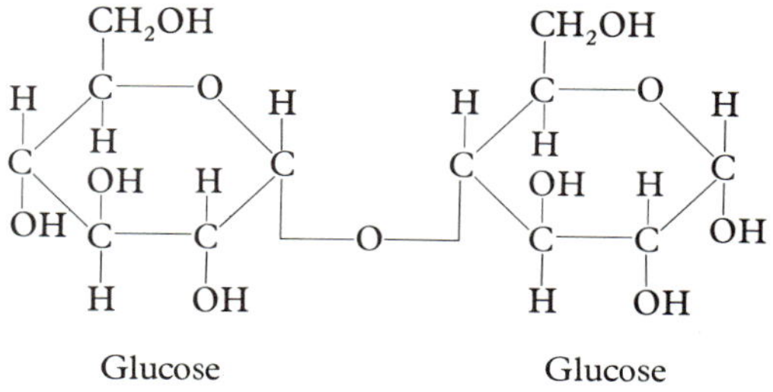

GLYCOGEN
A branched polysaccharide
composed of thousands of
repeating glucose subunits

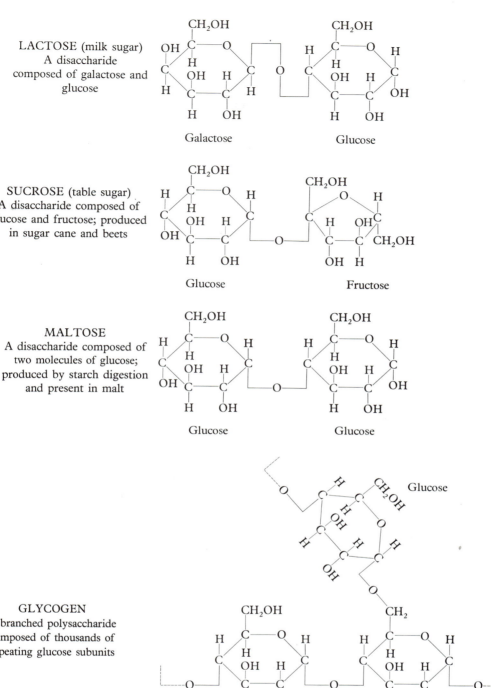

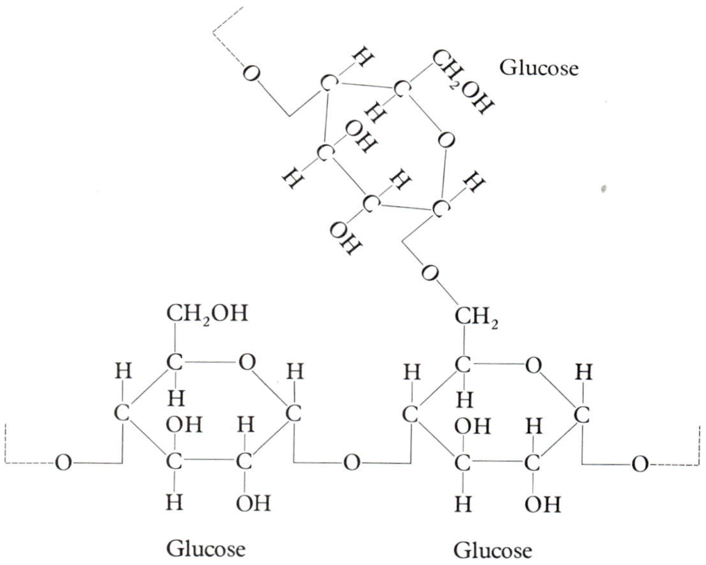

this sugar, and to simply draw on this reservoir to maintain blood glucose. But this is not true. Some glucose is stored, mostly in the liver and muscles, in the form of *glycogen,* a large carbohydrate molecule composed exclusively of glucose subunits (Figure 9-1). But the amount of glucose on reserve as glycogen is surprisingly small—less than 100 grams are stored in the liver, and the muscles and other tissues contain even less than that. If we do not eat for only 24 hours,

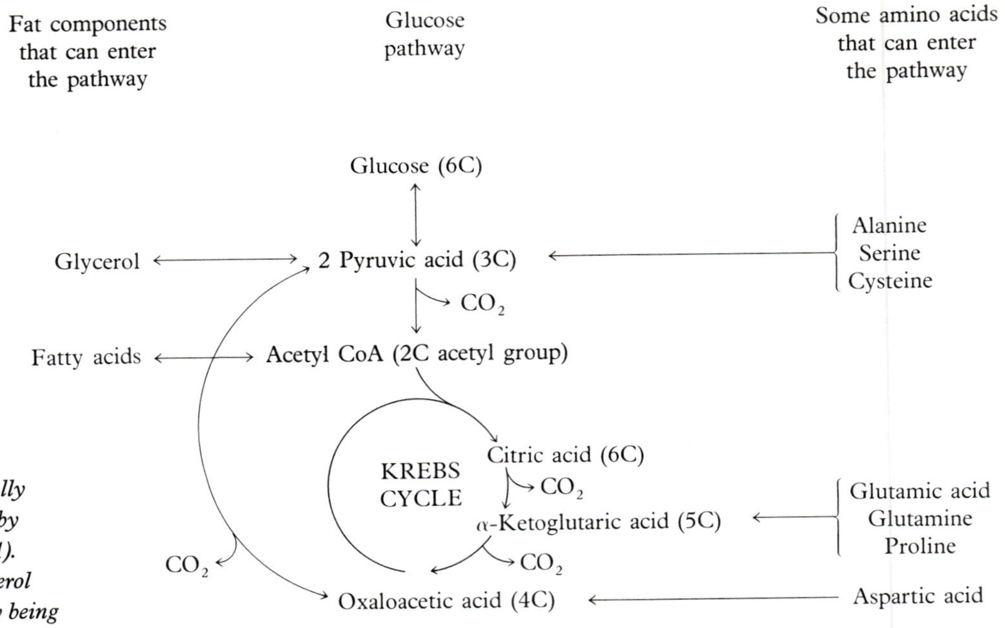

9–2
Energy needs of the body are usually met when glucose is broken down by this pathway (also see Figure 8-21). Amino acids, fatty acids and glycerol can also serve as energy sources by being converted into compounds along the pathway. The amino acids shown are converted into compounds containing the same number of carbon atoms. (Only some carbons of other amino acids can enter the pathway.) Fatty acids enter the pathway as two carbon fragments.

almost all of our available glycogen is used up. Obviously, our bodies cannot rely for long on stored glucose to maintain proper levels of this important sugar in the blood.

In the absence of dietary carbohydrate or stored glycogen, most of the glucose in the circulation is manufactured from substances that can be converted directly into pyruvic acid and then into glucose—namely, certain amino acids found in protein and the glycerol component of fat (Figure 9-2). The conversion of these dietary components into glucose is so effective that Masai warriors, Eskimos, South American gauchos, and other peoples may live for long periods of time on foods that are exclusively of animal origin and that contain almost no carbohydrate. These people are vigorous and healthy, and we learn from their eating habits that carbohydrate is *not* a dietary essential, because they have normal levels of glucose in their blood in spite of the fact that their diets contain almost no carbohydrate.

Some of the components of proteins and fats that enter the glucose pathway are shown in Figure 9-2. Protein can serve as a source of glucose; some of its amino acids, such as alanine, serine and cysteine, are first converted into pyruvic acid and then into glucose. Other amino acids cannot be converted as directly into glucose, but enter the pathway at various points and can reach glucose through the conversion of oxaloacetic acid to pyruvic acid. Fat serves as a direct source of glucose through its glycerol component, and fatty acids can also be used because they are broken down into 2-carbon fragments that enter the pathway as acetyl-Coenzyme A. Fatty acids can also be broken down into β-hydroxybutyric acid (CH_3—CHOH—CH_2—COOH), which can serve as an alternative to glucose as an energy source. In times of starvation, the brain can alter its metabolism so that as little as 30 percent of its energy needs are supplied by glucose; under these circumstances, most of the brain's energy is supplied by β-hydroxybutyric acid.

Fats (Lipids)

Fat and *lipid* are used to refer to a variety of compounds that are, for the most part, insoluble in water. The majority of the fat molecules in the body are *neutral fats* (triglycerides) that consist of three fatty acid molecules attached to glycerol (Figure 9-3). The neutral fats are stored within the cells of the fatty tissues of the body, where they serve as an emergency source of energy and also provide insulation against cold.

Fats are said to be *saturated* (with hydrogen atoms) if their fatty acids contain no double bonds (Figure 9-4). The fatty acids of *unsaturated fats* have at least one double bond, and they associate with fewer hydrogen atoms. Animal fats are high in saturated fatty acids, whereas plant fats and oils contain more unsaturated fatty acids. The incidence of atherosclerosis is associated with a high intake of saturated fatty acids, so some people prefer to cook with plant oils rather than with animal fats such as butter. However, a cause-and-effect relationship has not been established between intake of saturated fatty acids and atherosclerosis.

A *phospholipid* is formed if one of the fatty acid molecules of a neutral fat is replaced by a phosphate group. Phospholipids are a second major group of fats, and they are especially important as structural components of cell membranes. One portion of a phospholipid molecule is usually charged, enabling it to interact with water. The third major group of fats is the *steroids,* which are very different structurally from the neutral fats and phospholipids. Steroids include cholesterol, Vitamin D, the steroid hormones, and the bile acids. Cholesterol is synthesized by most of the cells of the body, and (along with phospholipids) it is an important component of most cell membranes as well as a precursor of other steroids, such as the hormones and the bile acids.

Calories and Growth

In order to compare the amount of energy contained within the chemical bonds of different kinds of food, nutritionists employ units called *calories.* The small calorie (cal) is defined as the amount of heat required to raise the temperature of one gram of water by one degree Celsius. The kilocalorie (kcal) is equal to 1000 small calories, and is sometimes referred to as the Calorie (Cal with a capital C) in studies of nutrition. The caloric content of foods is found by determining how much heat is given off when various kinds of food are burned up in the presence of oxygen. The energy content of fat (9 kcal per gram) is slightly more than twice that of carbohydrate (4 kcal/g) or protein (4 kcal/g). Adults generally require between 1600 and 3000 kilocalories per day, depending on their body weight and the amount of physical activity in which they engage.

A newborn baby, which weighs three kilograms (kg), has a daily requirement of about 48 kcal/kg, nearly twice the requirement of an adult on the basis of weight. The higher requirement of the newborn is caused by the energy requirements of growth. As the growth rate declines, the daily requirement of a youth approaches that of an adult—about 27 kcal/kg.

In spite of the greater kilocalorie requirements of young children, they nonetheless require *fewer total* calories than do adults. This fact is important

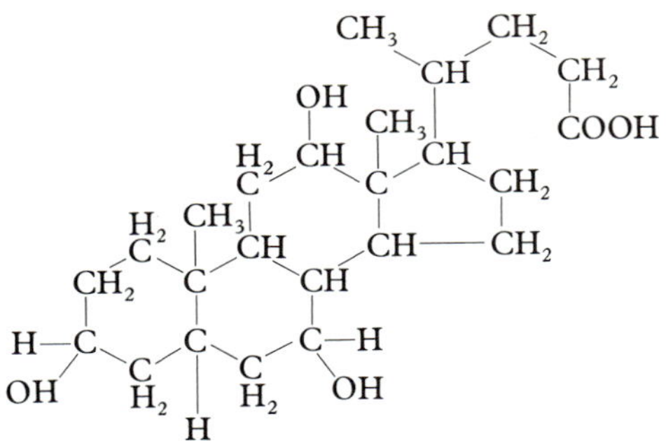

Cholic acid
(a bile acid)

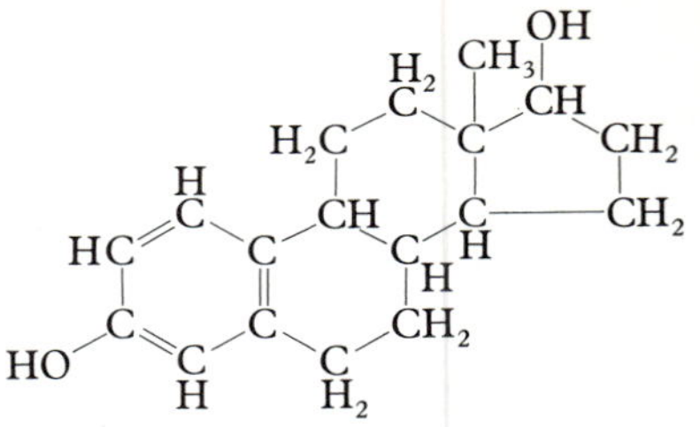

Estradiol
(an estrogen)

STEROIDS

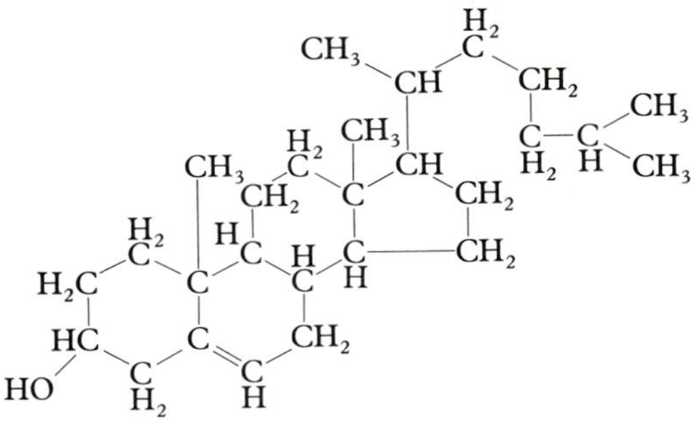

Cholesterol

9–3
Fats (lipids) are composed mainly of carbon and hydrogen atoms, although oxygen, phosphorous, and nitrogen atoms are present as well in small amounts. The three major classes of fats are neutral fats (triglycerides), phospholipids, and steroids. In the gastrointestinal tract neutral fats are broken down into glycerol and three fatty acid molecules. Phospholipids have charged atoms that allow them to interact with water molecules.

NEUTRAL FAT
(triglyceride)
Storage compound

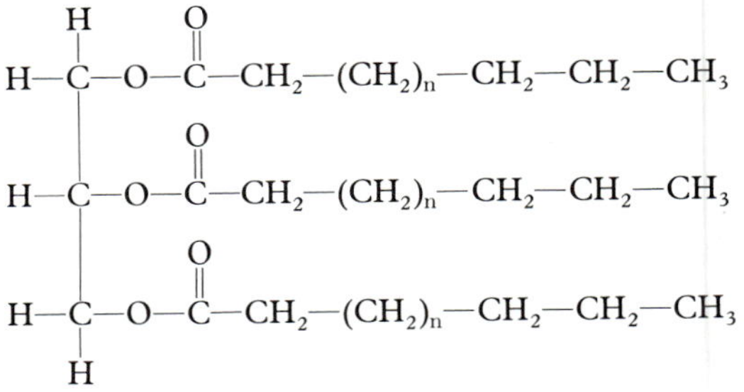

COMPONENTS OF
NEUTRAL FAT

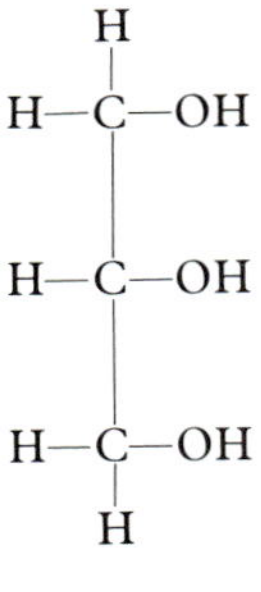

Glycerol

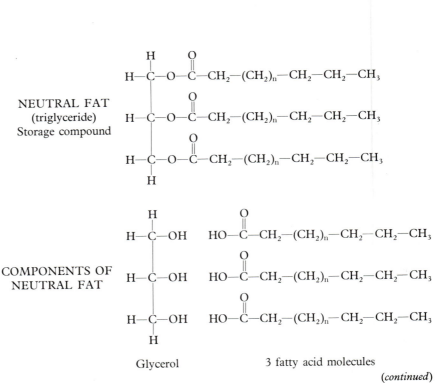

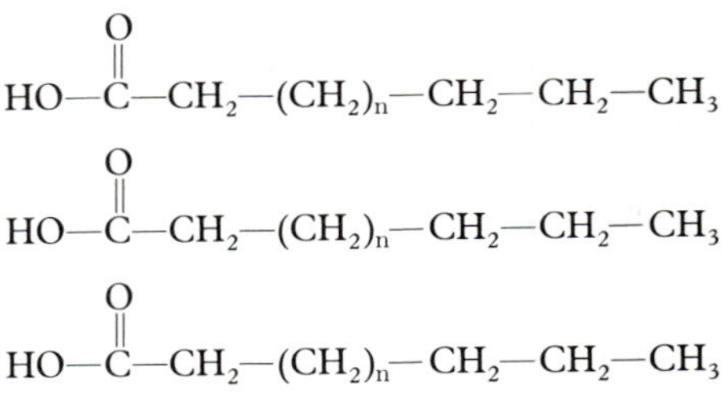

3 fatty acid molecules

(*continued*)

PHOSPHOLIPID
Important structural component of membranes

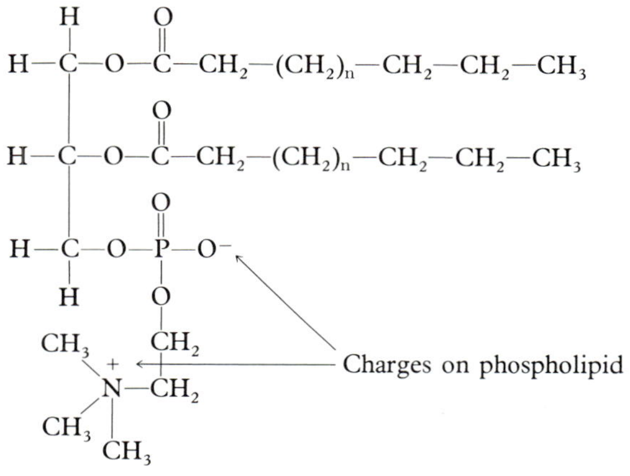

because even in countries in which there is no increase in birth rate, more food is required to feed the population as the children grow up. In areas where food is in short supply, this presents a problem. As infant mortality is reduced in these areas because of increased availability of medical treatment, the total population will itself increase. Starvation will become more widespread unless food availability also increases.

Food for New Body Substance

As we mentioned before, food is required not only as a source of energy, but also as a source of the building blocks from which we can make new molecules to replace the ones that are worn out. Most of our body components are maintained in a steady state—that is, they are constantly being both synthesized and degraded. This is especially obvious in certain protein molecules (Figure 9-5). For example, albumin, which makes up about 45 percent of the protein in our blood plasma, is replaced at the rate of about 3 percent per day. Fibrinogen, a plasma protein important in blood clotting, is replaced at the rate of about 25 percent per day. The cells that form the inner lining of the small intestine undergo complete renewal every two to four days. If the body were a perfect recycling center, the proteins that are broken down would themselves supply the amino acid building blocks for the replacement proteins being synthesized.

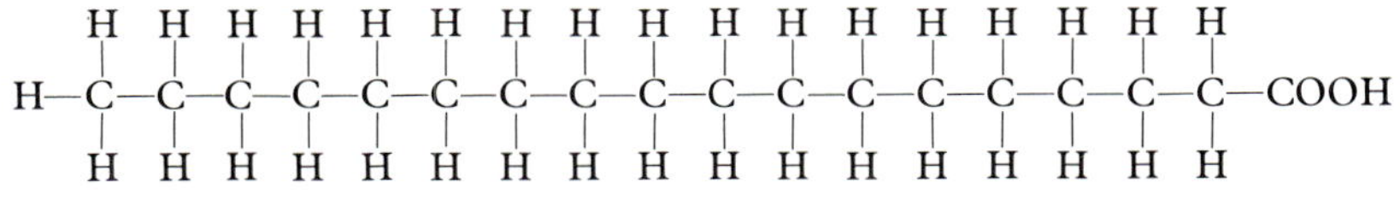

Stearic acid ($C_{17}H_{35}COOH$)

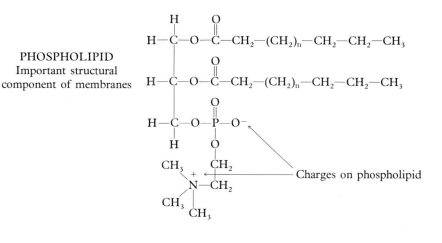

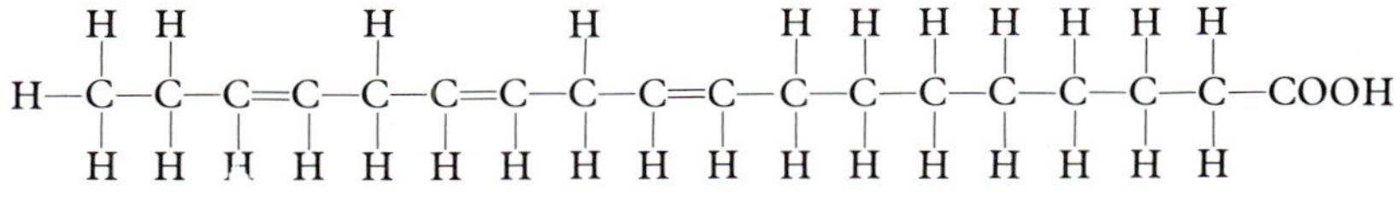

Linolenic acid ($C_{17}H_{29}COOH$)

9–4
Saturated fatty acids (stearic acid is an example) have no double bonds between their carbon atoms. Unsaturated fatty acids (linolenic acid is an example) have one or more double bonds; those with more than one are also called polyunsaturated.

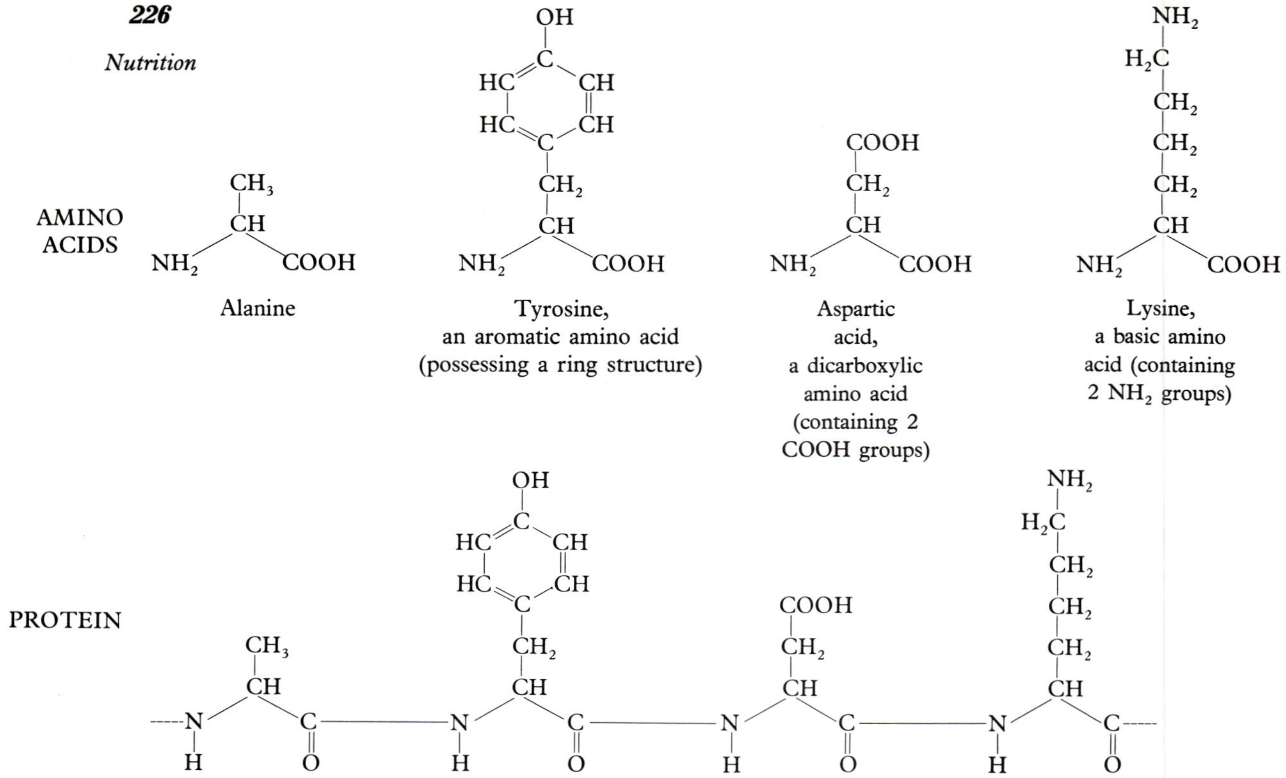

9–5
Proteins are made up of amino acid building blocks. Every amino acid has a carboxyl (COOH) and an amino (NH_2) group separated by a single carbon atom. In proteins, these two groups link each amino acid to its neighbors.

However, animals are not perfect recycling centers, and their bodies constantly lose some of the amino acid building blocks necessary for protein synthesis. Amino acids are lost in the urine, in the feces, and from the skin. When amino acids are broken down in the body, their nitrogen atoms are removed. Most of these nitrogens become part of the urea molecule and are expelled from the body in the urine. Even under conditions of starvation, amino acids continue to be lost as urea. Amino acids in the feces are derived partly from food, partly from digestive secretions, and partly from the cells of the lining of the small intestine that have been sloughed off. Amino acids are lost from the skin as its outer layer, formed mostly of the protein keratin, is continuously worn away.

The Need for Dietary Protein

Of the three major classes of organic compounds, proteins, carbohydrates, and fats, protein is the most essential, because our bodies cannot manufacture certain amino acids present in protein from either carbohydrate or fat, even if there is a supply of nitrogen available. Protein is unique in this regard, because our bodies can make fat from carbohydrate and carbohydrate from either protein or fat (Figure 9-2). This distinction becomes more significant as it becomes increasingly apparent that protein is in short supply in many parts of our planet.

Inadequate nutrition is almost always associated with a deficiency of protein. Protein malnutrition is almost certainly the largest single cause of human disease in the world. In the poorer countries, many children receive too little protein in a diet that consists mainly of carbohydrate of cereal origin. The chronic lack of protein produces a severe wasting disease, which in East Africa is called *kwashiorkor.* Kwashiorkor is characterized by apathy, loss of appetite, accumulation of fluid in the tissues (edema), wasting, and changes in the skin and hair. Protein deficiency is especially devastating to children because amino acid building blocks are required for the synthesis of many structural proteins that produce growth.

Although we understand the seriousness of protein malnutrition, it is not sufficiently well-recognized by many people as a cause of death. This is because many deaths, particularly those of children, are attributed in statistics to a variety of specific infectious diseases that are the immediate cause of death, even though it is clear that these infectious diseases would not have been fatal to a well-nourished person. The connection between protein malnutrition and susceptibility to disease is explored further in Chapter 13.

The Eight Essential Amino Acids

We have become accustomed to the rather egocentric notion of evolution as a progression from the simple to the complex, during which new organisms appear that are capable of a variety of activities unknown to their ancestors. Yet this is not true of amino acid synthesis: most animals, including people, have *lost* the ability to synthesize some amino acids, whereas plants and many bacteria can synthesize them all. Humans cannot synthesize eight of the twenty-two amino acids of which our proteins are composed. These eight amino acids are called *essential* amino acids (see Table 9-1), because we must obtain them from dietary sources. *Good nutrition* requires a balanced intake of protein that will supply all eight of the essential amino acids.

TABLE 9-1
The eight essential and fourteen nonessential amino acids found in human proteins.

ESSENTIAL	NONESSENTIAL
tryptophan	glycine
phenylalanine	alanine
lysine	serine
threonine	cysteine*
valine	aspartic acid
methionine	asparagine
leucine	glutamic acid
isoleucine	glutamine
	arginine
	hydroxylysine**
	histidine
	tyrosine
	proline
	hydroxyproline

*Two of these linked together constitute the amino acid cystine.
**Found so far only in collagen.

The high quality of nourishment provided by food of animal origin has been recognized for thousands of years, and early peoples placed great value on animal-capturing activities (Figures 9-6 and 9-7). We now know that food of animal origin is beneficial to humans partly because animals and animal products (like milk and cheese) possess all eight of the essential amino acids in adequate quantities. Plant proteins, on the other hand, are not so well endowed with amino acids, and are often deficient in one or two of them, as shown in Table 9-2. For example, the amino acids tryptophan and methionine are lacking in most legumes (beans, peas, and lentils), and the amino acids lysine and isoleucine are often lacking in grains, cereals, nuts, and seeds. Notice that the eight essential amino acids are easily obtained without eating any animal product, if the proper combination of plants are eaten. However, they must be eaten together at the same meal.

Some cultural traditions incorporate plant proteins into the diet in ways that allow for the deficiency of essential amino acids in one plant protein to be balanced by the proteins derived from other vegetable sources. For example, beans and rice, the traditional dietary staples of Mexican people, complement one another, not only in taste, but also in essential amino acids.

9–6
A prehistoric cave painting from Spain depicting the hunting of game animals. (Courtesy of The American Museum of Natural History, New York.)

Too Much Animal Protein

Good nutrition does not require the consumption of as much animal protein as is consumed by most people in the United States. Further, plant protein is converted to animal protein very inefficiently. About 5.5 kilograms (12 pounds) of plant protein are required to produce 1 kilogram (2.2 pounds) of chicken protein, 8 kilograms (18 pounds) of plant protein to produce 1 kilogram (2.2 pounds) of pork protein, and 21 kilograms (46 pounds) of plant protein to produce 1 kilogram (2.2 pounds) of beef protein. The least costly products according to this measure are eggs and milk, which require 4.4 and 4.3 kilograms (9.7 and 9.5 pounds) of plant protein, respectively.

The inefficiency of converting plant protein to animal protein can be stated another way: an acre of land devoted to meat production can produce only one-fifth as much protein as an acre of cereals, one-tenth as much as an acre of

TABLE 9-2
*The deficiencies of some plant foods in essential amino acids.**

CATEGORY	FOOD	DOES THIS FOOD SOURCE SUPPLY THE MINIMUM NECESSARY AMOUNT OF THIS AMINO ACID?			
		TRYPTOPHAN	METHIONINE	LYSINE	ISOLEUCINE
legumes	kidney beans	no	no	yes	yes
	lima beans	no	no	yes	yes
	soy beans	yes	no	yes	yes
	peas	no	no	yes	yes
	lentils	no	no	yes	yes
grains and cereals	rice	yes	yes	no	no
	wheat	yes	yes	no	no
	oats	yes	yes	no	no
	rye	no	yes	no	no
nuts and seeds	sesame seeds	yes	yes	no	no
	sunflower seeds	yes	yes	no	yes
	brazil nuts	yes	yes	no	no
	cashews	yes	yes	yes	yes
	peanuts	yes	no	no	no

*These four of the eight essential amino acids are lacking in certain plant foods.

legumes, and one-fifteenth as much as an acre devoted to leafy vegetables. Statistics compiled by the U.S. Department of Agriculture indicate that enormous quantities of protein that could be eaten by people are routinely fed to animals: in 1968 U.S. livestock consumed 20 million metric tons of protein that could have been eaten directly by people. Of this 20 million metric tons, only 2 million metric tons were retrieved as animal protein for human consumption. It has been estimated that the 18 million metric tons that were lost could have made up about 90 percent of the world protein deficit that year—enough to provide 12 grams (.4 ounce) of protein *per day* for every person in the world or about 40 percent of the daily minimum requirement.

Food and Cancer of the Digestive Tract

Epidemiological studies in which large numbers of people are studied in an attempt to relate certain diseases to the ingestion of particular kinds of food have yielded interesting but largely unexplained correlations. For example, with respect to cancers of the digestive tract, which may in some cases correspond to diet, the data include these facts. Oral (mouth) cancer, especially cancer of the lips, is rare in the United States (3 to 4 percent of all cancers), but more common in India (up to 70 percent of all cancers), where the chewing of betel nuts is common practice. Cancer of the esophagus is rare in the United States and

9–7
Hunting wild fowl in the Nile marshes, about 1450 B.C. *(Copyright The British Museum, by permission of the Trustees of the British Museum.)*

Europe, but common in Japan, where the frequency of stomach cancer is also significantly higher than in Europe or the United States. It has been suggested that the ingestion of large quantities of smoked foods (especially fish), or the practice of consuming food and drink while they are still scalding hot, may be responsible for the increased incidence of cancer of the esophagus and stomach in Japan. But although we may be able to establish a connection between eating certain foods and the likelihood of developing cancer of the esophagus or stomach, this is *not* evidence of a cause-and-effect relationship.

The same reservation applies to the connection between beef-eating and cancer of the colon (large intestine) that has received a great deal of publicity lately. The data include the following facts: (1) There is an overall correspondence between animal protein consumption of various countries and mortality rate from cancer of the colon. (2) There is a greater incidence of cancer of the colon in Scotland than in England, and the Scottish each more beef (but less meat) than the English. (3) There is a very high incidence of cancer of the colon in New Zealand, where the beef consumption is much greater than in the rest of the world. (4) The incidence of this disease has declined in Australia as beef consumption also declined. (5) No population that has a low beef consumption has a high incidence of cancer of the colon. These data may seem convincing at first glance, but they should not be alarming, even to people who eat a lot of beef. The number of people who develop cancer of the colon is small, even in New Zealand, where the consumption of beef is the greatest. Once again, these kinds of studies merely establish a connection between beef-eating and cancer of the colon, not a cause-and-effect relationship.

It is quite possible, in fact, that the connection between beef-eating and colon cancer has nothing to do with beef at all, but rather is related to some other aspect of a beef-eater's diet. For example, in parts of the world where the diet is high in beef it is also relatively low in fiber content. The combination of sugar from the sugar bowl and bread from refined flour has practically no fiber content, whereas sugar from fibrous fruits and bread from fibrous whole grains are high in fiber content. Low fiber diets lead to less fecal material and a longer transit time for food in the gastrointestinal tract, and under these conditions cancer-promoting substances (carcinogens) in the diet might become more concentrated. Also, they would be in contact with the intestinal wall for a longer time. It is possible that a diet low in fiber content leads to colon cancer, but we simply do not know what dietary factors actually contribute to colon cancer.

The Vitamins

In addition to protein, fat, and carbohydrate, the body also requires vitamins and minerals. *Vitamins* (Table 9-3) are organic molecules that are necessary in small amounts for normal body function and growth. About 20 essential vitamins have been identified, and we know about them because dietary deficiencies of them have occurred from time to time. Vitamins were originally named by the Polish biochemist Casimir Funk, who thought they were all vital amines. It is now known that only a few of the vitamins are actually amines (possessing an amine or $-NH_2$ group), but the name has stuck, although the final "e" has been removed. The vitamins as a group have no chemical structure in common, but in function most of them serve as *coenzymes,* molecules that combine with certain enzymes to make them functional. In the twentieth century most of the vitamins have been identified chemically and given chemical names, although their earlier letter names (A, B, C, D, etc.) are still in use. Six major diseases are attributable to vitamin deficiencies: scurvy, beri-beri, pellagra, dry-eye disease (keratomalacia), rickets, and pernicious anemia. Let us now discuss these diseases and the vitamins assciated with them.

Vitamin C (Ascorbic Acid) and Scurvy

The great breakthrough in recognition of vitamin deficiency diseases came as a result of the investigations of James Lind, an English physician of the eighteenth century. Lind made one of the simplest but most celebrated of therapeutic suggestions: that sailors should be given a ration of fresh fruit to prevent scurvy. His findings were published in 1757 in his "Essay on the most effectual means of preserving the health of seamen."

TABLE 9-3
Major vitamins.

VITAMIN	SOURCES	DEFICIENCY DISEASE	EFFECTS OF DEFICIENCY
A (retinol)	Butter, cheese, milk, liver, precursors in yellow and green vegetables	Dry eye disease (keratomalacia)	Dry eyes, degeneration of corneas, dry skin, night blindness
B_1 (thiamine)	Whole grains, liver	Beri-beri	Muscle weakness and soreness
B_2 (riboflavin)	Milk, liver		Sore tongue, cracked lips
Niacin (nicotinic acid)	Milk, meat, eggs, yeast	Pellagra	Dermatitis, diarrhea, mental changes
B_6 (pyridoxine)	Yeast, meat, corn, liver, cabbage		Irritability, convulsions
Pantothenic acid	Liver, yeast, green vegetables		Dermatitis, disorders of the gastrointestinal tract, adrenal insufficiency
Biotin	Liver, tomatoes, egg yolk		Dermatitis, disorders of the gastrointestinal tract
Folic acid	Green vegetables		Anemia, disorders of the gastrointestinal tract
B_{12} (cyanocobalamin)	Liver, meat, eggs, milk	Pernicious Anemia	Weakness, sore tongue, numbness, tingling of extremities
C (ascorbic acid)	Fruits, leafy green vegetables, tomatoes	Scurvy	Bleeding of gums and skin, defective wound healing
D (calciferol)	Fish liver, sunlight on the skin	Rickets	Fragile bones, twisted bones in children
E (tocopherols, a group of chemicals)	Milk, eggs, leafy green vegetables		Leg cramps(?), childhood anemia(?)
K (naphthoquinone and its derivatives)	Leafy green vegetables		Bleeding, internal hemorrhaging

Scurvy is a disease caused by a deficiency of Vitamin C in which bleeding is likely to occur anywhere in or on the body—in the skin, giving the appearance of large bruises, in the gums, which appear swollen and bleeding, or in the internal organs, which is often fatal. Wound healing is also greatly impaired because the body cannot synthesize the protein *collagen,* which is the major protein found in healing tissue. Therefore, even small abrasions remain as open sores.

Before Lind's time, thousands of sailors had died from scurvy. In 1497 on Vasco de Gama's voyage to Portugal around the Cape of Good Hope to India, 100 of his 160 men died from this disease. In 1535, when the French explorer Jaques Cartier attempted to colonize Newfoundland, 100 of his 110 men developed scurvy. Fortunately for them, the natives of Newfoundland knew what to do and gave the men a tea made of spruce needles (which, as it happens, are rich in Vitamin C).

Lind was a very careful investigator. He took 12 men who had scurvy and whose cases were as similar as any he could find. He then divided the men into six pairs, and gave each pair a different treatment food along with a diet that was known to be useless, by itself, in the treatment of scurvy. The two who received two oranges and a lemon every day made rapid recoveries, whereas pairs who received ordinary seawater, dilute sulfuric acid, vinegar, cider, or a special medicinal paste didn't fare nearly as well. On the basis of this investigation, Lind recommended fresh fruit to prevent scurvy. His findings were adopted by Captain Cook, who provided a ration of lemon juice for his men, and Cook's sailors remained free of scurvy throughout their voyage of over three years in the South Pacific in the years 1768–1771.

However, Lind's recommendation was apparently too far ahead of its time to be adopted by the British Admiralty, who weren't impressed by his studies. "Some persons cannot be brought to believe that a disease so fatal and so dreadful can be cured or prevented by such easy means," Lind complained. "They would have more faith in an elaborate composition dignified with the title of 'an antiscorbutic golden elixir' or the like." Lemon juice was eventually supplied to a naval squadron in the year of Lind's death, 1794, and about ten years later lemon juice became a standard daily issue of the British Navy. The English at that time called lemons "limes," which is where the American term "limeys" for English sailors came from. We now know that Vitamin C was the important ingredient of lemon juice that prevented scurvy.

In his book entitled *Vitamin C and the Common Cold,* Linus Pauling questioned whether the dose of Vitamin C needed to prevent scurvy should be used as a guideline for setting standards for daily consumption. He pointed out that although 60 milligrams of Vitamin C per day, the recommended daily allowance, is well above the 10 milligrams required to prevent scurvy, it still might be below the dosage needed to promote good health; in particular, he suggests that higher doses of Vitamin C might protect against respiratory infections. To investigate Pauling's ideas, "double-blind" studies have been carried out by several groups of researchers. Subjects received either Vitamin C pills or placebo pills (which contained no Vitamin C or any other pharmacologically active substances). The subjects were blind in the sense that they didn't know which pills they were receiving. This was possible because the Vitamin C and the placebo preparations were made indistinguishable from one another in appearance and taste. The researchers gathering the information were also blind because they also didn't know who was taking what. Thus the double-blind nature of the study served to eliminate the biases of researchers or subjects who might want to influence the outcome of the study.

Two studies clearly showed that a dosage of 1000 milligrams of Vitamin C per day reduced the *number of days of disability* resulting from respiratory illness. In a Canadian study in which there were more than 400 persons in each group, those who took 1000 milligrams per day had a 30 percent reduction in days of disability, although there was no significant reduction in the total number of respiratory illnesses they contracted. In a U.S. study carried out among children at a Navajo boarding school, children received either 0, 1000, or 2000 milligrams per day. The two groups receiving Vitamin C showed roughly similar 30 percent reductions in number of days sick with respiratory illnesses, but again there was not significant difference in the number of respiratory illnesses. Although they had colds and other respiratory ailments, their ailments were less severe.

Two other studies did not show a protective effect of Vitamin C. In a second double-blind study of over 2000 volunteers carried out by the same Canadian researchers, there were no significant differences between any of the groups taking Vitamin C and those taking placebo. And a study of children in four Dublin boarding schools, in which 200 and 500 milligrams of Vitamin C were given daily to different groups for 35 weeks, revealed no consistent effect of Vitamin C on either frequency or duration of colds.

Thus some studies suggest that large doses of Vitamin C reduce the severity but not the number of respiratory illnesses, whereas others reveal no effect of Vitamin C at all. It is not clear why the studies gave different results. It does seem clear, however, that Vitamin C does not reduce the number of respiratory infections, and that the Vitamin C controversy will continue to be with us for some time.

Vitamin B_1 (Thiamine) and Beri-beri

From the earliest times in the Far East there has been a disease characterized by muscle tenderness and weakness of the legs, followed by loss of power in the muscles of the hands and arms. This disease came to be known as *beri-beri.* Our understanding of the nature of this disease began in the 1890s when a Dutch physician, Christiaan Eijkman, showed that pigeons developed a disease that closely resembled beri-beri when they were fed a diet of polished rice. The disease was prevented if the birds were given a small portion of the outer pericarp (or germ) of the rice grain, which usually is removed in the milling process. In 1911 Casimir Funk (as you may recall, he coined the name "vitamines") concentrated the ingredient that prevented beri-beri from the material discarded from a rice-polishing mill. It was first called Vitamin B and subsequently Vitamin B_1 to distinguish it from other vitamins also present in the concentrate. It was found to contain an amine group and a sulfur atom and was named thiamine. A person suffering from beri-beri can be relieved of its symptoms within an hour of taking as little as 5 milligrams of pure thiamine.

Niacin and Pellagra

The word *pellagra* is derived from the Italian *pelle* (skin) and *agra* (sour), and it is caused by a deficiency of niacin. A prominent characteristic of the disease is a severe inflammation of the skin, or *dermatitis.* The inflammation is especially severe in the regions of the body exposed to sunlight; in areas of the world where niacin is in short supply, pellagra tends to recur every spring as a result of increasing exposure to the sun. Other symptoms associated with pellagra include diarrhea, mental disorders, and sore tongue.

Outbreaks of pellagra have occurred primarily in the areas of the world where corn has been widely cultivated and used as the dietary staple because corn is deficient in both niacin and the amino acid tryptophan, which the body can convert to niacin. That pellagra is a deficiency disease was proven by an American physician, Joseph Goldberger, who showed that pellagra could be prevented by the addition of milk, meat, or eggs to the diet. The pellagra-preventing action of these foods is due primarily to their high content of the amino acid tryptophan.

Although corn is a poor source of niacin, rice is an even poorer one. Yet pellagra is not often found among people who eat rice as a dietary staple. The explanation is that rice contains much more of the niacin precursor tryptophan than does corn.

Vitamin A (Retinol) and Dry-Eye Disease (Keratomalacia)

Vitamin A helps to maintain epithelial tissues such as skin. If the body does not get enough of it the skin becomes dry and susceptible to invasion by microorganisms. Furthermore, the epithelial surfaces cease to shed their dead cells, and they become dry and thickened. This is especially damaging when it affects the cornea of the eye, which becomes opaque so that light cannot pass through. This severe condition is called *dry eye disease,* and if untreated it leads to total degeneration of the cornea, to infection, and eventually to loss of the eye.

Vitamin A also forms part of the visual pigment *rhodopsin,* found in the retina of the eye. When light hits the retina rhodopsin is split into opsin and Vitamin A. The rhodopsin must then be regenerated from the Vitamin A, and this regeneration process is inadequate when Vitamin A is in short supply. The result is night blindness, a condition in which it is difficult to perceive dim light. When a person has a shortage of Vitamin A that is not severe enough to cause nightblindness, testing of responses to dim light sometimes reveals it. Vitamin A is also necessary for normal bone formation.

Vitamin A is present only in certain foods of animal origin, such as butter, cheese, and milk. It is also found in large quantities in fish-liver oils and in enormous quantities in polar-bear liver. Meat contains only traces of Vitamin A.

The body does not require the presence of Vitamin A itself, because Vitamin A can be synthesized by the body from the *carotenes,* which are widespread in fruits and vegetables, particularly yellow and yellow-orange ones such as carrots, peaches, sweet potatoes, and apricots. Carotenes are absorbed through the intestinal wall about 50 percent as efficiently as Vitamin A. Vitamin A deficiency occurs in areas where fruits and vegetables are scarce and there are few productive dairy cattle. The recommended daily allowance is 5000 international units per day for adults. An international unit is a measure of biological activity rather than of weight.

Very high doses of this vitamin can create serious health problems. Cases of *hypervitaminosis-A* have been reported in people receiving around 100 times the recommended daily allowance, usually in the form of fish-liver oils or vitamin pills. Hypervitaminosis-A is characterized by gastrointestinal upset, dermatitis, loss of hair, and pain in the bones.

Vitamin D (Calciferol) and Rickets

Vitamin D is one of the most fascinating of the vitamins, partly because it is not essential to the diet if the skin is exposed to sufficient sunlight (Figure 9-8). In fact, Vitamin D should be considered a hormone because, like other hormones, it is synthesized in an organ (the skin, under the influence of ultraviolet irradiation from the sun) and it has effects elsewhere in the body, mainly on bone. Although Vitamin D is a hormone, its hormonal name calciferol is only used rarely, so we will continue to refer to it as Vitamin D.

Vitamin D came to be considered one of the vitamins primarily because fish and fish oils (such as cod-liver oil) were found to prevent rickets when there was insufficient sunlight. The beneficial effects of cod-liver oil had been known to people in France, Germany, Holland, and England since the early 1800s, and in

1917 the careful studies of A. W. Hess proved that cod-liver oil prevented rickets. It was natural enough to assume that if Vitamin D cured rickets, then rickets was a vitamin-deficiency disease.

When there is both insufficient sunlight and insufficient Vitamin D in the diet, the body cannot absorb enough dietary calcium, and as a consequence calcium is removed from bone in order to maintain the necessary concentration of calcium in the blood (this topic is discussed further in Chapter 11). Severe losses of bone calcium lead to *rickets,* a condition characterized by weakened bones that become twisted and bent. The word rickets comes from the Anglo-Saxon "wrikken," which means to twist.

Rickets became especially prevalent—in fact, it became an epidemic—in the industrial cities of Northern Europe in the seventeenth and eighteenth centuries. There were three reasons for this: (1) the further north the city was located, the less sunlight it received, particularly in the winter; (2) narrow city streets and alleyways cut down further on sunlight penetration; and (3) coal smoke produced in abundance during the industrial revolution prevented penetration of what little sunlight remained. With the sunlight almost gone, only those lucky enough to get fish or fish oils would escape rickets. The fish on Friday custom of Roman Catholics must have saved many a child from rickets. It has even been suggested that the custom of June weddings came about in part because the first offspring would probably be born the following spring, when there would be plenty of sunlight available for the first few months of the baby's life.

Vitamin D probably influenced the development of differences in skin pigmentation of different races. We know that, in general, more lightly pigmented people live in regions far away from the equator, where there is less sunlight, whereas more heavily pigmented people live (or their ancestors lived) in regions near the equator, where there is relatively more direct exposure to sunlight. We also know that a more lightly pigmented skin will absorb more ultraviolet light. It seems likely that as people moved away from the equator, those whose skin had less pigment would have had a selective advantage. Indeed, rickets was most devastating in Europe to people who had more heavily pigmented skins.

For people who live in the tropics, a heavily pigmented skin is an advantage; it protects them from the burning effects of sunlight, and it might prevent synthesis of *too much* Vitamin D. It has been calculated that an exposed, lightly pigmented European would absorb enough sunlight to synthesize eight times the toxic amount of Vitamin D in a day (if tanning didn't occur). A heavily pigmented person would absorb 95 percent less. (This topic is discussed further in Chapter 19.) The toxic amount is around 100,000 units per day; and the recommended daily allowance is 400 units per day.

Vitamin B_{12} (Cyanocobalamin) and Pernicious Anemia

Prior to 1926, pernicious anemia was an invariably fatal disease in which the early symptoms were weakness, sore tongue, and numbness and tingling of the arms and legs. Anemia was severe; the number of circulating red blood cells usually dropped to about 20 percent of normal. In 1926 Dr. George R. Minot first recognized that pernicious anemia was a disease caused by deficiency of a substance that is now called Vitamin B_{12}, or cyanocobalamin. His treatment consisted of supplying patients with at least one pound per day of raw or lightly cooked liver. Sometimes this treatment relieved the symptoms.

9–8
Egyptian king and queen Aknahton and Nefertiti holding their children up to the rays of the sun. Limestone relief from about 1300 B.C. Ultraviolet rays from the sun penetrate the skin and convert 7-dehydrocholesterol into Vitamin D. (Courtesy of the Egyptian Section of the State Museums, Berlin. Picture Archives of the Prussian Cultural Trust.)

Most people who have pernicious anemia cannot *absorb* the B_{12} from the diet, because the digestive juices in their stomachs do not provide a substance (intrinsic factor) that is necessary for absorption. Even when massive amounts of the vitamin are supplied (such as in a pound of raw liver), many people who have pernicious anemia still cannot get the necessary one-millionth of a gram of B_{12} across the two millimeters of the gut wall to supply their bodies' needs. Treatment consists of circumventing the faulty absorption process by giving injections of the vitamin.

Because all animal products contain B_{12}, only strict vegetarians develop pernicious anemia as a result of inadequate dietary intake.

Food Fads and Vitamin Doses

Throughout history human beings have sought an elixir for life; in modern times, vitamins are thought by many to be that elixir. Extravagant claims are made concerning the benefits of taking vitamins in large doses, and when little is known about a vitamin, these assertions are hard to refute. Vitamin E is an example. We know relatively little about Vitamin E for several reasons: assessment of intake is difficult because it exists in many foods, there are eight known chemical forms of Vitamin E that differ in their biological activities, and there is no specific disease caused by a deficiency of it. Yet Vitamin E has been touted as an aphrodisiac and has been heralded as a cure for a variety of conditions, including heart disease. In spite of all the hoopla, however, it has proven helpful mainly in the treatment of leg cramps and muscle spasms in adults and in the treatment of anemia and malnutrition in infants.

A note of caution has been sounded by the finding that rats fed a stock diet supplemented with Vitamin E for six months had greater concentrations of cholesterol deposits in the aorta than did control rats. Also, a recent report revealed that a controlled study of the effects of high doses of Vitamin E upon humans had to be discontinued when two volunteers developed severe weakness after taking 800 units daily for three weeks. In sum, an analysis of the available information does not justify any extravagant claim, but rather suggests that we should refrain from forming an opinion about Vitamin E while more information is gathered.

Even when we do know quite a bit about a vitamin, it is easy to be misled about its capabilities. A good example is Vitamin A, which has been popularized as a possible anticancer agent. The title of a recent article in a popular magazine suggested that Vitamin A was winning another fight against cancer. This article in turn quoted an article in a scientific journal that asserted that Vitamin A inhibited the induction of cancer. However, a look at the information in the scientific article gives a different picture. First, Vitamin A stimulated cellular immunity, which plays a role in defenses against cancer (see Chapter 13), but it did this in mice given doses that—when adjusted for body weight—would be toxic to people. Vitamin A was not *fed* in the diet, but was applied to the skin after being mixed with a cancer-promoting agent. Therefore the results of the study have nothing to do with diet. Finally, when carotenes were used in place of Vitamin A, the results were reversed: a greater number of tumors was produced in the mice. The message in this is that we must examine data carefully and thoughtfully before accepting it. And these data are available to anyone who takes the trouble to look for them.

Minerals and Trace Elements

A variety of minerals are essential for normal human growth and maintenance. Among these are sodium, needed to maintain blood pressure; potassium, needed for nerve conduction and muscle contraction; calcium, essential for bone; and iron, the component of hemoglobin to which oxygen attaches.

Trace elements are those that are present in the body in very small amounts: less than 1 part in 20,000. These include zinc, copper, iodine, cobalt, manganese, selenium, chromium, molybdenum and fluorine. That they are needed by the body in small quantities makes them no less essential.

Among the trace elements iodine is of special interest because it is an essential component of *thyroxin* and other products of the thyroid gland (the thyroid gland is discussed in Chapter 6). In the absence of iodine, the thyroid gland enlarges considerably (reflecting a biochemical attempt to compensate for the deficiency) and a *goiter* is formed (see Figures 9-9 and 6-32). The iodine content of food varies a great deal, and it depends upon the iodine content of the soil in which the foods were grown. Iodine is found in high concentration in seawater and in soils that have been recently uplifted from the sea. As was mentioned in Chapter 6, iodine deficiency and goiter are especially common in high mountain areas such as the Andes and the Himalayas. But goiter also occurs in Africa and it used to occur frequently in the Great Lakes area of the U.S. before the addition of potassium iodide to table salt. Sea fish eaten once a week will also supply enough iodine to prevent iodine deficiency.

Fluoride is a trace element whose effects, both beneficial and toxic, are well known. Epidemiological studies have shown that dental caries (tooth decay) are more common where drinking water contains less than 1 part per million (ppm) of fluoride. Therefore, fluoride has been added to the water supplies of many areas to bring the concentration up to 1 ppm. This procedure has drastically reduced the decay of children's teeth. However, if the fluoride content of water reaches 3–5 ppm, the teeth develop an unsightly (but not unhealthy) brown mottling. At 10 ppm the water causes bones to become dense, and calcium may be deposited in ligaments. The back is especially likely to be affected, and the excess of fluoride can even result in fusion of spinal vertebrae, making it very difficult to walk.

9–9
Cleopatra's goiter. Iodine is a component of thyroid hormones. If there is insufficient iodine in the diet, reduced concentrations of thyroid lead to compensatory enlargement of the thyroid because of increased output of TRF and TSH (see p. 164). (Courtesy of the Bettmann Archive.)

Food and Good Health

Nutritionists are students of foods as they contribute to "good health." But health is a characteristic of the whole organism, and is a very difficult, if not impossible, quality to measure in any given experiment. So nutritionists usually use a *rate of body growth* as a measurement from which they derive their standards. This use of body growth as an indicator of proper nutrition makes sense historically, because the field of nutrition developed from the study of the worst deficiency states, which are usually accompanied by severe stunting of growth.

Yet there is ample evidence that body growth does not necessarily go hand in hand with good health. For example, European food promotes body growth, but a study conducted in South Africa showed that people who were too poor to obtain European food had only 40 percent as many dental caries as more prosperous people. Another study showed that people living on the island of Tristan da Cunha had essentially no dental caries in 1932, before the introduction of European food, but that by 1955 European food was abundant and caries were as common as in Western Europe.

Laboratory experiments are even more revealing. In one study, groups of young, growing mice were fed special diets for eight days before being inoculated with infectious bacteria. With the exception of protein, the diets were complete, containing all known essential nutrients. In one group, the diet was supplemented with protein only, and in another group it was supplemented with both protein and amino acids. The mice receiving protein and amino acids gained no weight in the eight days (perhaps because they did not like the diet), but all survived the infection. The mice receiving protein gained over 30 percent in body weight during the eight days, but all succumbed to infection. It is clear from these results that weight gain does not reflect good health as measured by resistance to infection.

Individual Variation in Requirements

It is very difficult to develop standards such as the "minimum daily requirements" that can truly apply to everyone. For one thing, these standards are usually based on the amount of a nutrient necessary to prevent obvious growth deficiency, and the prevention of deficiency is not the same as the promotion of health. But, even more important, any single standard value will not reflect the importance of individual variations.

Individual variations in nutritional requirements have been acknowledged for years by workers in nutrition. Recognizing this, a table of "recommended daily allowances" has been formulated that provides for larger amounts of some nutrients than do the minimum daily requirements. But undoubtedly even the recommended daily allowances are not enough for some people. Although data on healthy people are difficult to obtain, the meager data we do have indicate that for calcium, and for several amino acids, five-fold variations in requirement are relatively common.

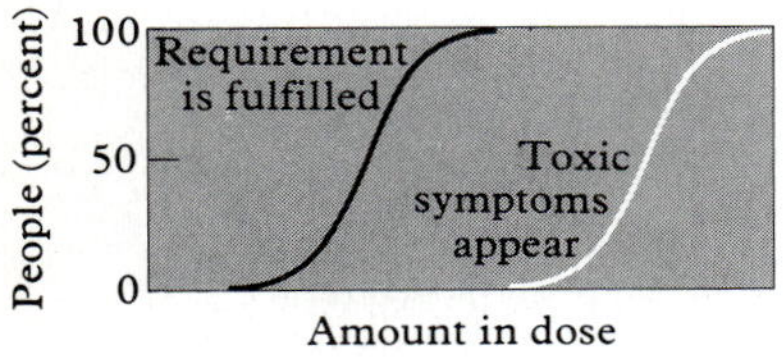

9–10
Relationship between the dose of a nutrient that fulfills a minimum requirement and the dose that is toxic. Curves such as these apply to nutrients such as Vitamin A, Vitamin D, and fluoride.

Those who have intestinal disease have increased nutritional requirements because they do not absorb nutrients well. For example, even ten times the recommended daily allowance of Vitamin A a day, given orally, may not restore normal night vision in people who have intestinal disease.

There are several factors to which variations among normal people can be attributed: differences due to different microorganisms in the intestine, genetic differences, and habituation.

Some strict vegetarians require no Vitamin B_{12} at all, probably because some helpful intestinal bacteria are synthesizing it. There is also evidence that intestinal bacteria supply a portion of the essential amino acids for some people. However, intestinal organisms are not always helpful, and they may be the creators of shortages as well as the providers of benefits. Certain forms of anemia from which Africans suffer have been cured by penicillin, presumably because penicillin eliminates bacteria that are *using up* Vitamin B_{12}. But antibiotic ingestion can also destroy beneficial intestinal microorganisms.

An extreme example of individual variation determined by a genetic difference is the condition known as *Vitamin D-resistant rickets.* People who have this condition require between 125 and 1500 times the recommended daily allowance of Vitamin D to prevent rickets, the Vitamin D deficiency disease. Yet for most people these doses of Vitamin D would be toxic.

Serious problems can apparently result from the tendency of the body to develop a habituation (or addiction) to high doses of nutrients. For example, it has been reported that some people who stopped taking high doses of Vitamin C developed scurvy (Vitamin C deficiency) when they returned to a more "normal" daily intake. Apparently their metabolism had not only adjusted to a higher dose, but had come to *require* the higher dose.

Figure 9-10 illustrates the relationship between the dose of a nutrient and the percentage of people for whom that dose fulfills the daily requirement. But many nutrients (Vitamin A, Vitamin D, fluoride) can be toxic. Fortunately, the dose required by biological necessity is usually much less than the dose that is toxic.

Food and Culture

The relationship between people and their food is often much more profound than scientific terms indicate. Consider, for example, the following description about the relationship of the Hopi Indians to corn:

> For many corn-growing Indians, the cornfield was traditionally an altar and the act of cultivation was an act of worship. If we were to describe the agriculture of the Hopi Indians as it was in traditional times, we would have to describe the entire culture: the ceremonials, the ball games, the dancing, the maskmaking, the scrambling up to the aeries of the eagles, the silent retreats to the kivas, the long journeys to the Pacific Ocean in search of turtle shell, the joyous cries of the playing children (yes, even these) and the smiles of the parents. All these found their meaning and their function in the growing of corn, which was ultimately an act of worship. Corn itself meant life. . . . From "Food and Human Existence," by D. Lee. *Nutrition News; 25* (1962) 9–10.

It is remarkable that healthy, flourishing cultures could depend so strongly upon the cultivation of a single grain. Yet corn (maize) was under cultivation 5000 years ago in central Mexico and strongly contributed to the rise of the great Mesoamerican civilizations. Ancient sculptures from Mexico remain as mute testimonials to the important role of corn. Today, corn still predominates as a chief source of nutrition in Mexico and Central America, and it is the largest single crop in the United States.

There are two major drawbacks to the use of corn as the main dietary staple: corn contains very little of the amino acid lysine and of the vitamin niacin. Also, the absorption from the intestine of what little niacin there is depends upon the ratio of leucine to isoleucine in the diet, and in corn this ratio is nearly three times greater than that considered optimal for such absorption. Therefore, corn's nutritional value as the major food in the diet is at best marginal, and any human population depending on it might be expected to suffer from at least some degree of malnutrition, whether protein deficiency (kwashiorkor) from the lysine shortage or pellagra from the niacin deficiency. In fact, in a number of places where corn has been widely cultivated—Spain, Italy, the Balkans, Egypt, and in isolated areas of the Americas—pellagra has been widespread.

A special cooking technique, in which corn is cooked with alkali, solved the problem long ago. In Central America, tortillas are made by heating dried corn in a 5 percent solution of limestone in water; after cooling, the liquid is poured off and the corn is washed, drained, and made into the dough from which the tortillas are prepared. Different corn-dependent societies obtain the alkali from different sources: some use lye, some wood ashes, and some ashes from the cremated shells of freshwater mussels. Alkali cooking greatly increases the digestibility of the corn protein *glutelin*, which contains two-thirds of the small amount of lysine in corn. People are normally unable to digest glutelin to obtain the essential lysine. In a recent study alkali-cooked corn and uncooked corn were subjected to digestion by pepsin, an enzyme from the stomach that breaks down protein into amino acids. The results of the study showed the following beneficial effects of alkali cooking: (1) the amount of lysine that was released increased 2.8 times, (2) the amount of isoleucine increased 2.1 times while leucine increased only 1.1 times (thereby greatly increasing the absorption of niacin), and (3) the availability of niacin and its precursor tryptophan also increased. Thus we see that alkali cooking is a remarkable cultural adaptation that greatly increased the nutritional value of corn and thereby assisted the rise of civilizations that depended upon it as the dietary staple. When corn was exported from Mexico, the cultural cooking tradition was not exported with it. Had it been, the resultant numerous outbreaks of pellagra could have been avoided.

Obesity

When the availability of food is seasonal and undependable, the body finds it useful to be able to store up energy in the form of fat for use during the lean times. Fat is particularly well suited as a vehicle for energy storage, because, as we have discussed, its energy content (9 kcal/g) is more than twice that of either protein or carbohydrate. However, where there is an almost unlimited supply of food, and the conditions of life do not require vigorous activity, there is the ever-present possibility that the advantage of energy storage may become transformed into the burden of obesity.

Obesity (defined as body fat in excess of 20 percent of normal body weight) usually goes hand in hand with being overweight (defined as excess weight for a given height) but this is not necessarily so. It is possible to be obese without being overweight, as for example is an inactive person who has poorly developed muscles. Nevertheless, the two are usually found together, and the risks of being obese first came to light some years ago when insurance companies began to study the hazards of being overweight. Data from many sources now clearly establish that people who are 20 percent or more overweight have at least a 20 percent greater chance of dying prematurely than do people of normal weight. The excess deaths of overweight men and women are especially associated with diabetes, heart disease, brain hemorrhage (stroke), and diseases of the digestive tract. Atherosclerosis, which predisposes one to both heart disease and stroke, is the deposition of fats (including cholesterol) and fibrous tissue in blood vessel walls. The relationship between atherosclerosis and heart disease is discussed in Chapter 7.

Studies of insured people in the United States have revealed that the incidence of obesity increases with age, and that fully one-third of both men and women in the 50–59 year age group are at least 20 percent overweight, so the problem seems worthy of some attention. We will begin by taking a look at some of the factors that might contribute to obesity, particularly food intake and exercise, because these are subject to environmental control.

Some obese people are compulsive eaters, but the notion that overeating applies to all obese people is not supported by the data. Studies of obese people have shown that their intake of food is similar to that of normal people. Overeating may occur, however, during the *developing phase* of obesity; once a person is already obese the condition will remain even if energy input and energy output are equal.

The control of obesity requires three changes to take place: increasing motivation to lose weight, decreasing energy (calorie) intake, and increasing energy output. The importance of motivation is underscored by the success of self-help groups such as TOPS (Take Off Pounds Sensibly), in which people provide emotional support for each other and compete with each other to see who can lose the most weight.

Dieting is the traditional method of combatting obesity, and a variety of special diets have been suggested. One of these, the low-carbohydrate diet, has some interesting features. First, it allows people to eat until they are full, because they can have unlimited amounts of foods such as meat, fish, eggs, cheese, butter, margarine, oils, cream, and leafy vegetables—all of which are low in carbohydrate. Second, it does not require detailed knowledge of calorie values, because the main carbohydrate-rich foods are well known: sugar, candy, cakes, cookies, soft drinks, ice cream, bread, and pasta. Third, when the diet is effective it seems to work because people don't make up in protein and fat calories what they are leaving out in carbohydrate calories. Finally, many very healthy people spend most of their lives on a diet that is naturally low in carbohydrates, which indicates that it must have beneficial effects. However, a difficulty with all specialized diets, including this one, is that they may not be safe for all people, because they demand sudden drastic changes in the body's established metabolic patterns. There is an advantage to sticking with one's usual diet, if it is nutritious, but eating less of it.

Exercise combats obesity by increasing energy output, and it also helps to maintain physical fitness. About half of an average person's energy is used for muscular activity; very active people may use three-fourths. Therefore muscular exercise can play an important role in weight control programs. Information about the relative energy values of various kinds of exercise is set out clearly in a number of books. (*Aerobics,* by K. H. Cooper is one example: the author assigns a number value to each kind of exercise (see Table 9-4).) Information of this kind can help to suggest ways to achieve increased energy output and physical fitness even when exercise possibilities are limited.

TABLE 9-4
Different forms of exercise are equivalent in maintaining physical fitness. Each of these five exercises is assigned a value of 6 points. To maintain physical fitness, 30 points per week are suggested.

TYPE OF EXERCISE	QUANTITY
Walking	3 miles
Bicycling	6 miles
Running	$1\frac{1}{2}$ miles
Swimming	800 yards
Golf	36 holes

SOURCE: From *Aerobics,* by K. H. Cooper. M. Evans, 1968.

Summary

Food is required both as a source of energy and as a source of building blocks from which new molecules are manufactured. New molecules are required for growth and to replace those that are used up during the normal turnover processes.

The energy needs of the body are met mainly by the breakdown of the 6-carbon sugar glucose. If dietary glucose is in short supply, glucose may be obtained from stored glycogen or from components of proteins and fats. During starvation the body may also use β-hydroxybutyric acid, a breakdown product of fatty acid molecules, as a substitute for glucose.

The terms *fat* and *lipid* refer to a variety of molecules that are, for the most part, insoluble in water. These include the neutral fats, the phospholipids, and the steroids. Neutral fats are said to be saturated (with hydrogen atoms) if their fatty acids contain no double bonds, and unsaturated if their fatty acids contain at least one double bond. Animal fats are high in saturated fatty acids, whereas plant fats and oils contain more unsaturated fatty acids. There is a correlation between atherosclerosis and a high intake of saturated fatty acids. The phospholipids are especially important as structural components of cell membranes. The steroids include cholesterol and compounds derived from cholesterol such as vitamin D, the steroid hormones, and the bile acids.

The kilocalorie (kcal) is a unit of energy commonly used to compare the amount of energy contained within the chemical bonds of different kinds of foods. The energy content of fat (9 kcal/g) is greater than that of carbohydrate (4 kcal/g) or protein (4 kcal/g). The daily requirement of a newborn infant is nearly twice the requirement of an adult per kilogram of body weight. The higher requirement of the newborn is caused by the energy requirements of growth.

Certain amino acids present in proteins cannot be synthesized within the body. These essential amino acids must be supplied in the diet. Dietary protein from animal sources generally contains all eight of these essential amino acids in adequate amounts, whereas protein from individual plants is usually lacking in one or two of them. All eight essential amino acids may be obtained by eating protein from the proper combination of plants.

Protein malnutrition is almost certainly the largest single cause of human disease in the world. The total supply of protein available for human consumption is reduced when large amounts of edible plant protein are fed to animals in order to obtain animal protein.

Vitamins are organic molecules that are necessary in small amounts for normal body function and growth. Six major human diseases are attributable to the lack of specific vitamins: scurvy (Vitamin C), beri-beri (Vitamin B_1), pellagra (niacin), dry-eye disease (Vitamin A), rickets (Vitamin D), and pernicious anemia (Vitamin B_{12}).

Minerals and trace elements are also dietary essentials. Among the principal minerals are sodium, potassium, calcium, and iron. Trace elements include zinc, copper, iodine, cobalt, manganese, selenium, chromium, molybdenum, and fluorine.

Good health is difficult to measure. Good body growth, which is easy to measure, does not always go hand in hand with other measures of good health, such as good teeth and the ability to resist infection. It is also difficult to develop standards of daily nutritional requirements that will apply to everyone, primarily because there is considerable variation in requirements from person to person.

Food and culture are usually interwoven. Corn was a staple of the diet for thousands of years in the western hemisphere, and it was accorded a special symbolic significance. The nutritional value of corn was increased by the development of special cooking techniques. When corn was exported without the special cooking techniques, outbreaks of pellagra occurred among people who made corn the staple of their diet.

Obesity, defined as excess body fat, increases one's chances of developing diseases such as diabetes, heart disease, and stroke. The control of obesity requires attainment of three goals: increasing motivation to lose weight, decreasing energy (calorie) intake, and increasing energy output through exercise.

Selected Readings

1. *Diet for a Small Planet,* by F. M. Lappé. Ballantine Books, 1971. Explains how feeding edible plant protein to animals reduces overall protein available for human consumption. Points out which plant foods complement each other in amino acid composition.
2. *Man Adapting,* by R. Dubos. Yale University Press, 1965. Excellent discussion of the difficulties of measuring good nutrition. Many insights concerning how people adapt to diverse diets.
3. *Nutrition Against Disease,* R. J. Williams. Pitman Publishing, 1971. Discusses a variety of human diseases and how nutrients may affect them.
4. *Aerobics,* by K. H. Cooper. M. Evans, 1968. The value of exercise is discussed and ways of keeping fit are presented.
5. *Vitamin C, the Common Cold, and the Flu,* by Linus Pauling. W. H. Freeman, 1976. The first edition of this book, *Vitamin C and the Common Cold,* stimulated many investigations about the effects of Vitamin C on people.
6. "Vitamin C and the Common Cold: A Double-blind Trial," by T. W. Anderson, D. B. W. Reid, and G. H. Beaton. *Canadian Medical Association Journal, 107* (1972), 503–508. A double-blind study of the effects of Vitamin C.
7. "Vitamin C Prophylaxis in a Boarding School," by J. L. Coulehan, K. S. Reisinger, K. D. Rogers and D. W. Bradley. *New England Journal of Medicine, 290* (1974), 6–10. A double-blind study of the effects of Vitamin C.
8. "Ascorbic Acid for the Common Cold." *Journal of the American Medical Association,* 231 (1975), 1038. Reviews the results of several studies of the effects of Vitamin C.
9. "Vitamin E in Clinical Medicine," by M. H. Briggs. *The Lancet,* Feb. 9, 1974, p. 220. Points out problems of using high doses of Vitamin E.
10. "Traditional Maize Processing Techniques in New World," by S. H. Katz, M. L. Hediger, and L. A. Valleroy. *Science 184,* (1974), 765–773. Discusses the cooking techniques developed in the western hemisphere that increased the nutritional value of corn.
11. "The Physiology of Starvation," by V. N. Young and N. S. Scrimshaw. *Scientific American,* Oct. 1971, Offprint 1232. Points out the capacity of the body to utilize stored molecules for energy during starvation.
12. "Rickets," by W. F. Loomis. *Scientific American,* Dec. 1970. Discusses the relationship between sunlight, dietary Vitamin D, and rickets.

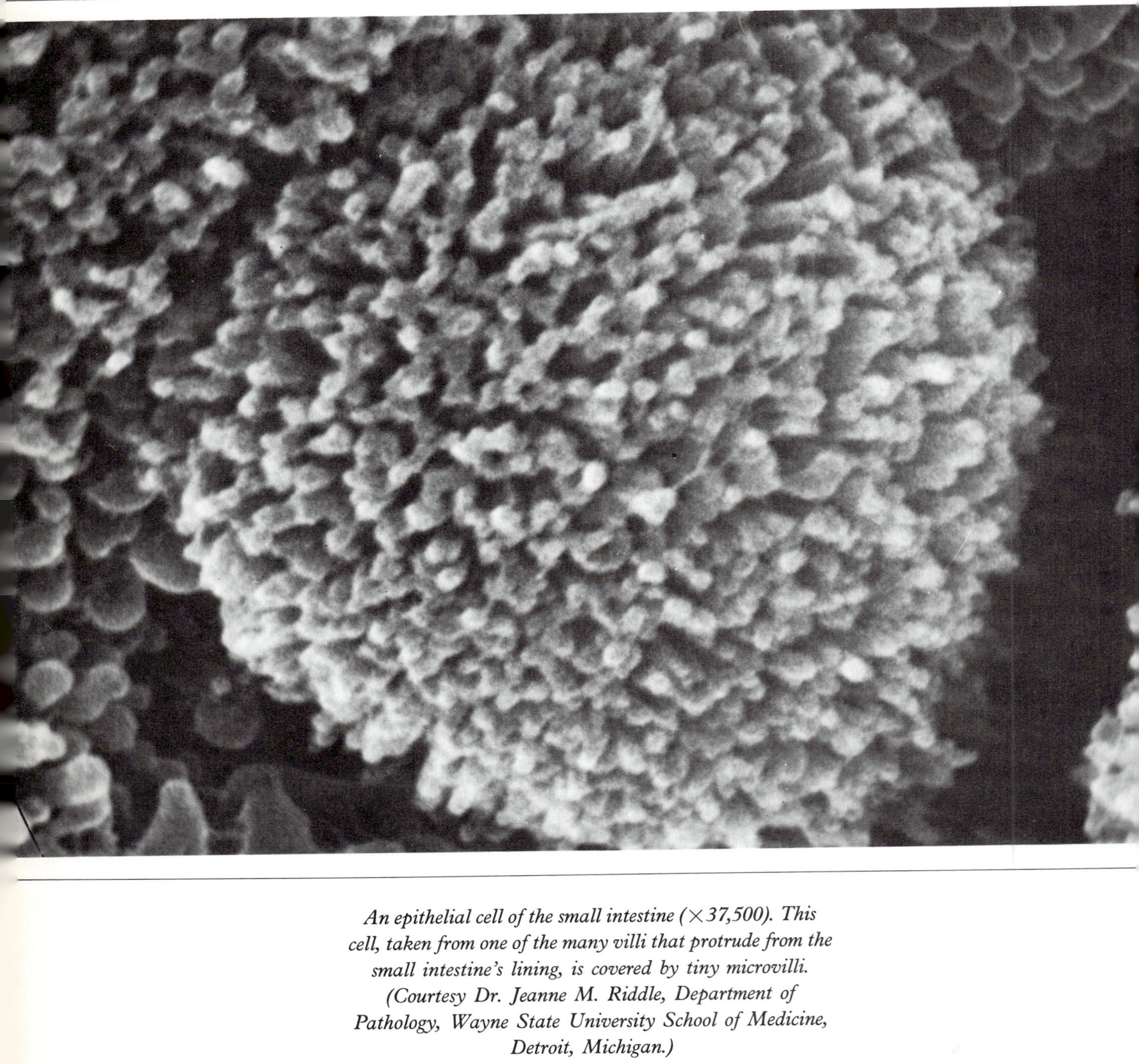

An epithelial cell of the small intestine (×37,500). This cell, taken from one of the many villi that protrude from the small intestine's lining, is covered by tiny microvilli. (Courtesy Dr. Jeanne M. Riddle, Department of Pathology, Wayne State University School of Medicine, Detroit, Michigan.)

CHAPTER 10

Intake and Outgo: The Gastrointestinal Tract and the Kidneys

Day to day, the overall chemical composition of the human body remains relatively constant. Yet a great variety of solids and liquids enter and leave the body. Food and drink enter through the mouth, and oxygen enters through the lungs.

The unusable residue of what was eaten exits through the anus, water and carbon dioxide gas leave through the lungs, and water and various salts leave the body both as sweat through the skin and as urine produced by the kidneys. The kidneys are also responsible for excreting into the urine the waste products of nitrogen metabolism. Because the body's overall composition does not change much from day to day, we can conclude that the daily intake of all substances usually equals the outgo.

But this does not mean that substances that enter the body go out unchanged. In fact, change is the rule. Consider what happens when glucose from a glass of fruit juice is absorbed into the body from the gastrointestinal tract. The equation for respiration sums up the chemical reactions that occur:

$$6O_2 + C_6H_{12}O_6 \rightarrow 6H_2O + 6CO_2 + \text{energy}$$

When we look carefully at this equation, we see that the number of C (carbon), H (hydrogen), and O (oxygen) *atoms* entering the body as oxygen and glucose exactly equals those leaving as water and carbon dioxide. Thus respiration, like most other physiological processes, does not change the atomic composition of the body. When the body is in a state of health, the overall balance of different kinds of atoms that enter and leave it is maintained—even though a particular substance may undergo extensive chemical changes between the time it enters the body as raw material and the time it exits as some kind of waste product.

In this chapter our main concern is with the intake and outgo of solid and liquid materials from the human body. (The intake and outgo of gases through the lungs is discussed in Chapter 8.) The gastrointestinal (G-I) tract is particularly important to intake because it is responsible for breaking down food and drink into the small molecules that can be absorbed into the body's tissues. As you will see, digestive activities are closely tuned to the chemical composition of what is eaten. The kidneys are especially important to outgo because they exert most of the body's control over the excretion of ions, water, and nitrogen. We begin our discussion of the human body's input and outgo by considering the major route of input, the gastrointestinal tract.

The Gastrointestinal Tract

The gastrointestinal tract is a long tube, continuous at both ends with the skin, that consists of the mouth, pharynx, esophagus, stomach, small intestine, large intestine, rectum, and anal canal (Figure 10-1). It is responsible for all digestion and absorption of food, and it plays an important role in excretion and protection as well. In this section, we will first discuss the several functions of the gastrointestinal tract and then discuss the mouth, which is its most accessible portion. Finally, we will discuss the regulation of digestion and will attempt to explain three difficulties related to digestion: ulcers, gallstones, and milk-intolerance.

The Passage of Food

Once food has been swallowed it must pass down the gastrointestinal tract unobstructed so that food can be digested and the products of digestion can be absorbed. The passage of food from esophagus to the anal canal is accomplished both by waves of contraction, or *peristalsis,* and by segmental contractions. Peristalsis, which results from the coordinated contraction of smooth muscle present in the gut wall, is illustrated in Figure 10-2. Peristalsis is the only means of propulsion of food in the esophagus, and its effectiveness is shown by the fact that we can swallow and pass food to the stomach while standing on our heads. Segmental contractions are small ring-like contractions which divide the intestine and its contents into sausage-like segments. These contractions are especially prominent in the small intestine where they mix the intestinal contents and aid the relatively weak peristaltic waves to propel the food.

Digestion

Digestion is the breaking down of large molecules present in food and drink into small molecules. These smaller molecules then pass through the intestinal wall into the bloodstream, and they eventually meet the body's needs for energy and

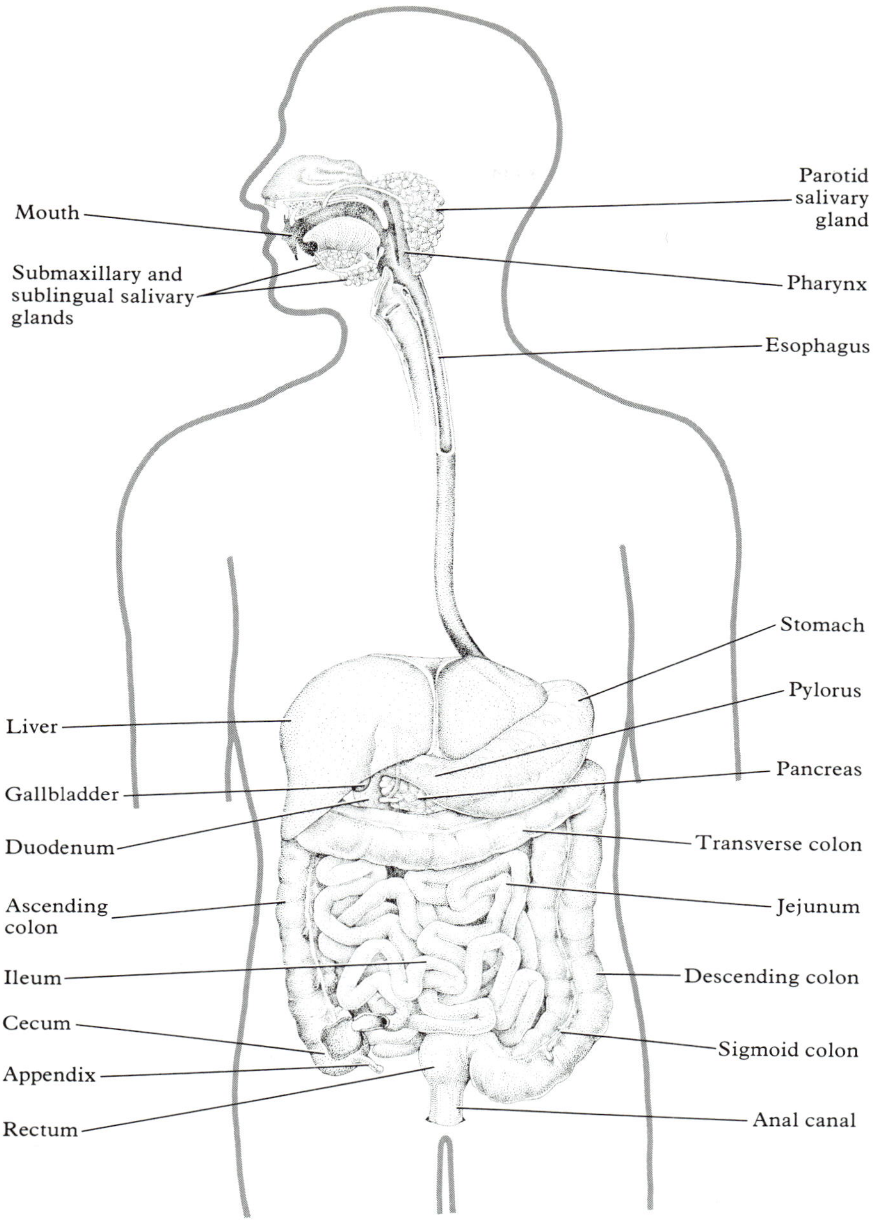

10–1
The gastrointestinal tract. Its main components are the mouth, pharynx, esophagus, stomach, small intestine (duodenum, jejunum and, ileum), large intestine (ascending colon, transverse colon, descending colon, and sigmoid colon), rectum, and anal canal.

new body substances. This breakdown is carried out by the *digestive enzymes* that are released into the upper portions of the gastrointestinal tract: the mouth, stomach, and small intestine. Digestion is almost completed by the time food has passed halfway through the small intestine.

Digestive enzymes are manufactured by enzyme-secreting cells, some of which are found scattered throughout the lining of the gastrointestinal tract (Figures 10-3 and 10-4), and some of which are concentrated in special organs: namely, the liver, the pancreas, and the salivary glands. The major digestive enzymes and their activities are listed in Table 10-1. The action of at least two enzymes is required to degrade large molecules into components that can be absorbed. For example, starch is broken down by salivary and pancreatic

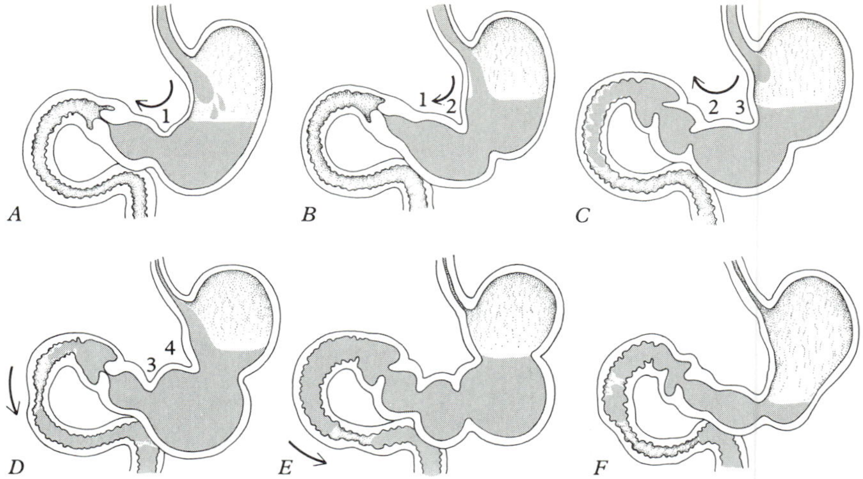

10–2
Waves of muscular contraction known as peristalsis facilitate the passage of material in the gastrointestinal tract. Here peristaltic waves are shown in the stomach and duodenum. A, a weak peristaltic contraction (1) moves across the stomach. B, a second stronger wave (2) is generated as this first wave reaches the muscular outlet of the stomach (the pyloric sphincter). C, the sphincter opens when this second contraction approaches, and some of the stomach contents empty into the duodenum. D, a wave of contraction occurs in the duodenum as new waves (3 and 4) are generated in the stomach. E, peristaltic waves continue to propel the contents of both the stomach and the duodenum until (F) the stomach is almost empty about four hours later.

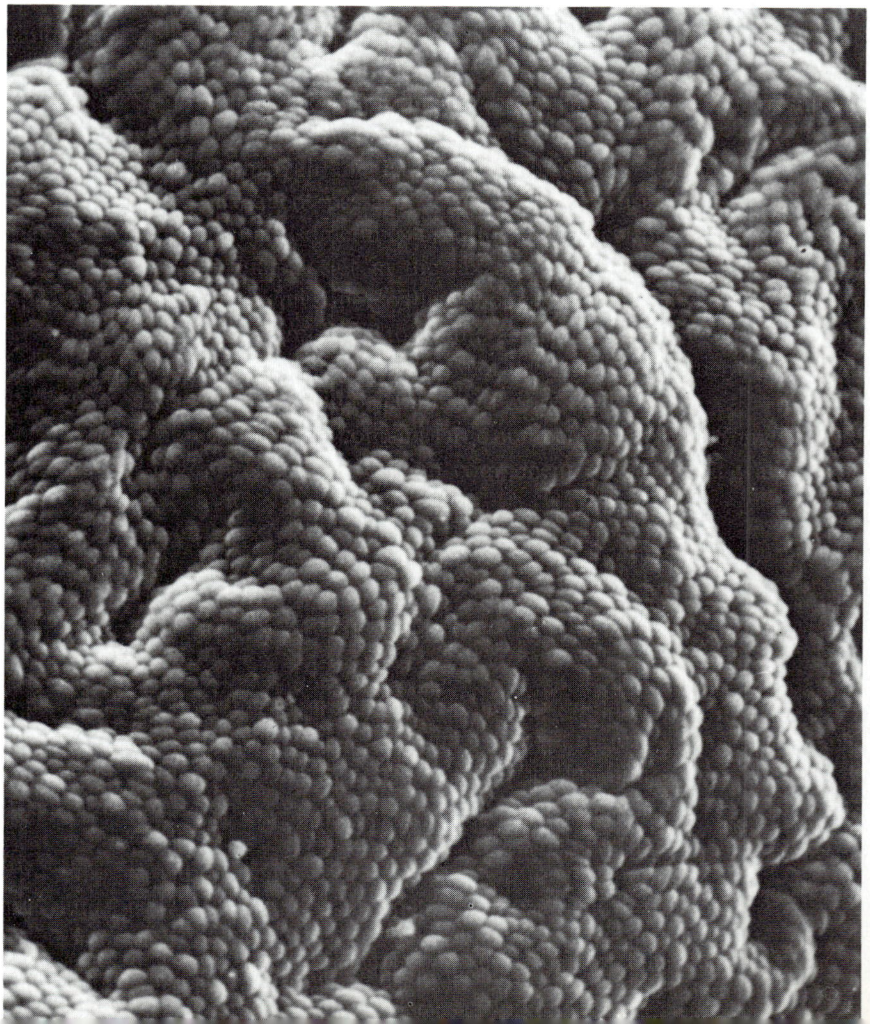

10–3
Surface of the stomach (× 700) as seen through an electron microscope. The folds of the inner lining are clearly visible, as are the individual epithelial cells (small bumps) that secrete mucus. The holes that penetrate the surface are gastric pits, which lead to the cells secreting pepsinogen and hydrochloric acid. (Courtesy Dr. Jeanne M. Riddle, Department of Pathology, Wayne State University School of Medicine, Detroit, Michigan.)

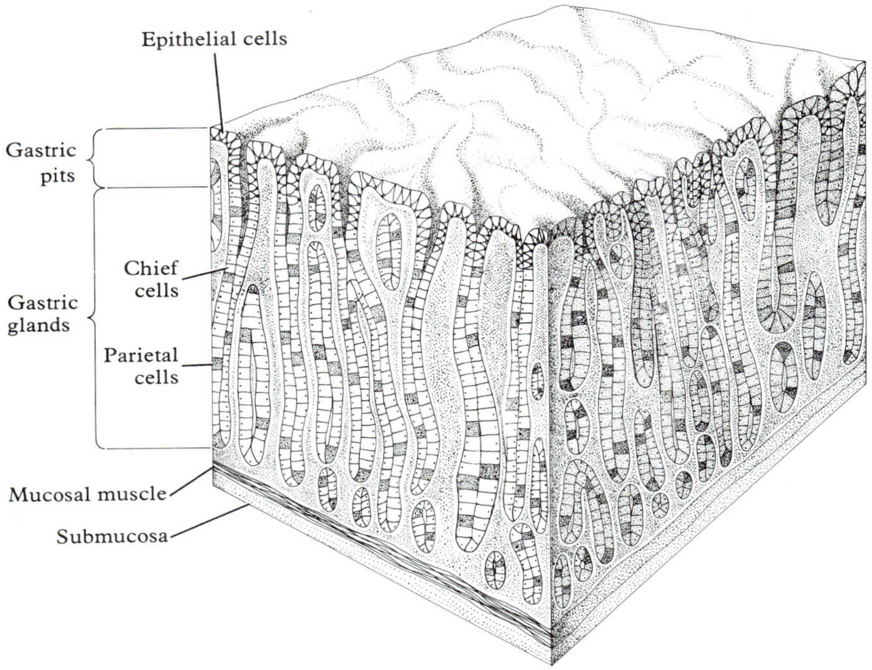

10–4
Cross section of stomach wall reveals the deep tubules that lead from the gastric pits. The epithelial cells that line the gastric pits secrete mucus. Scattered throughout the deeper portions of the tubules are chief cells (lighter color), which secrete pepsinogen, and parietal cells (darker color), which secrete hydrochloric acid. (From "Why the Stomach Does Not Digest Itself," by Horace W. Davenport Copyright © 1971 by Scientific American, Inc. All rights reserved.)

amylase into the disaccharide maltose, and the maltose is further broken down by maltase from the small intestine into its absorbable glucose components. Each digestive enzyme can break only certain specific chemical bonds. Thus the protein-digesting enzyme pepsin attacks many proteins, but only at certain points (Figure 10-5).

Because the cells that secrete these digestive enzymes are themselves composed of digestible large molecules, the body has evolved a mechanism to prevent digestive enzymes from digesting the cells that produce them.

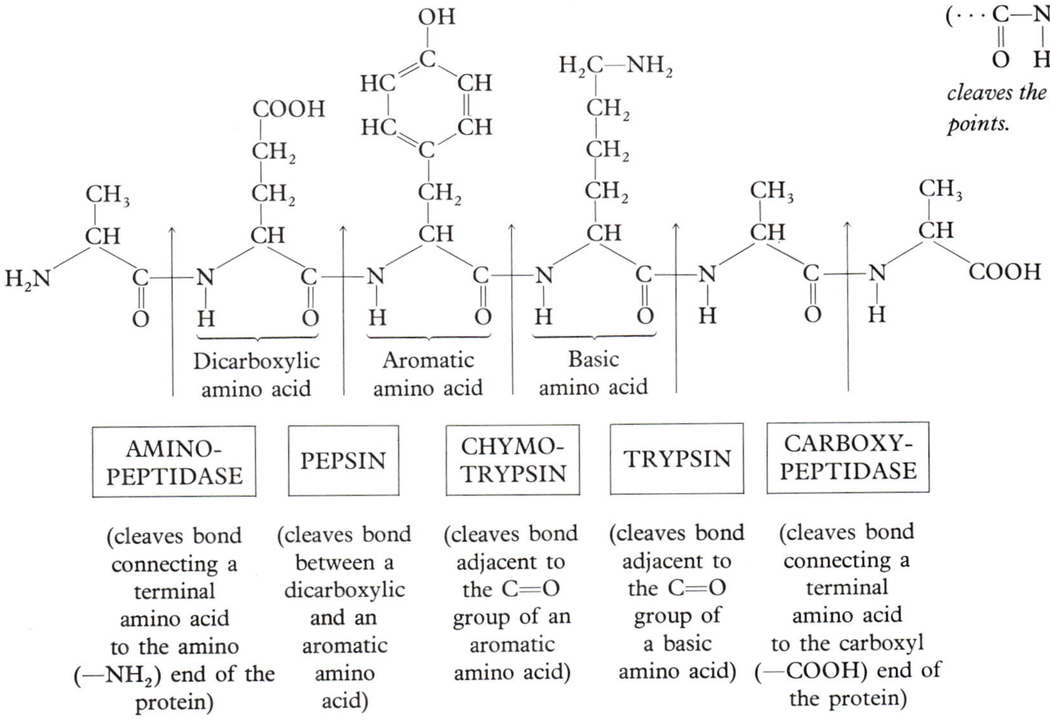

10–5
A protein molecule, showing points at which various enzymes cleave its peptide $(\cdots\underset{\|\atop \mathrm{O}}{\mathrm{C}}-\underset{|\atop \mathrm{H}}{\mathrm{N}}\cdots)$ *bonds. Each enzyme cleaves the protein at only certain points.*

TABLE 10-1

Some major digestive enzymes. These enzymes break down the large molecules found in food: proteins, carbohydrates, fats, and nucleic acids. The first six enzymes listed below, which are secreted into the uppermost portion of the gastrointestinal tract, produce products that require further digestion before they are absorbed. Enzymes secreted into the small intestine beyond the pancreas complete the digestive process.

SECRETED BY	THIS ENZYME	ACTS ON	TO RELEASE THESE PRODUCTS
Salivary glands	Amylase	Starch (carbohydrate)	Maltose (a disaccharide)
Stomach	Pepsin	Proteins	Peptides (small proteins)
Pancreas	Trypsin	Proteins	Peptides
	Chymotrypsin	Proteins	Peptides
	Deoxyribonuclease	DNA (nucleic acid)	Nucleotides
	Ribonuclease	RNA	Nucleotides
	Lipase	Triglycerides (lipids)	Fatty acids and Glycerol
Cells lining the small intestine	Maltase	Maltose	Glucose
	Lactase	Lactose (a disaccharide)	Glucose, galactose
	Sucrase	Sucrose (a disaccharide)	Glucose, fructose
	Aminopeptidase, Carboxypeptidase, Dipeptidase, Tripeptidase	Peptides	Amino acids
	Nucleases	Nucleotides	Pentoses and Nucleic Acid Bases

Interestingly enough, digestion of the body's own cells is prevented because the digestive enzymes are secreted in an inactive form. They become activated only after they have reached the gastrointestinal tract. Thus the protein-digesting enzyme *pepsin* is secreted in the inactive form *pepsinogen,* which is activated only after it encounters the acidic environment of the stomach. Similarly, *trypsin* is released from the pancreas as inactive *trypsinogen,* which is converted to active form only when it meets another enzyme (enterokinase) in the small intestine.

Hydrochloric acid (HCl), released by the stomach's parietal cells, might also be expected to digest the stomach's lining: it is one of the most corrosive of acids. Much of the protection from digestion by acid apparently is provided by sheets of fat molecules in the membranes of the cells lining the stomach. This fatty layer can be broken down by detergents such as the bile acids, which are secreted into the duodenum just past the stomach. Stomach ulcers, which are areas in

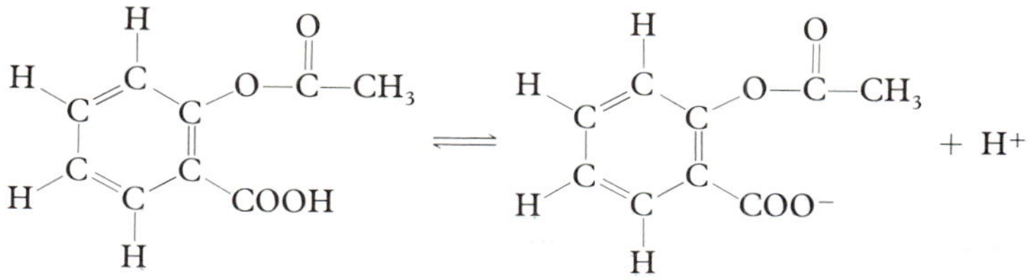

Aspirin in acid solution is uncharged and fat soluble

Aspirin in neutral solution becomes ionized (charged) and is insoluble in fats

10–6

Aspirin, or acetylsalicylic acid. In the acid environment of the stomach it is uncharged and fat soluble (left). In this state it can pass through the fatty layers of cell membranes and into the cells that line the stomach. In the neutral environment inside the stomach's cells aspirin is charged and insoluble in fat (right). Therefore it cannot readily leave the inside of the cells and it accumulates there, where it can cause damage.

which the lining of the stomach has been damaged, usually occur in the portion of the stomach nearest the duodenum, and it seems likely that regurgitation of bile acids may contribute to the formation of these ulcers. Many people who develop stomach ulcers are found to have bile acids in the stomach.

The fatty membranes of the stomach's cells can be penetrated by certain substances such as *aspirin,* or acetylsalicylic acid (Figure 10-6). Aspirin ingestion frequently causes some bleeding from the lining of the stomach, about .5 to 2 milliliters in most people who take two 5-grain aspirin tablets. However, aspirin can cause considerable bleeding in the gastrointestinal tract. The majority of people who are treated in hospital emergency rooms for massive bleeding in the gastrointestinal tract have taken aspirin within the past 24 hours. How does aspirin enter the stomach's cells and cause damage? In the acid environment of the stomach, aspirin is un-ionized (uncharged) and fat soluble (Figure 10-6). Therefore, it passes through the fatty barrier of the stomach's cells. Once it has entered the cells, however, the environment is more neutral, and it becomes ionized (charged) so that it is no longer fat soluble and cannot pass back out of the cells. It accumulates inside the cells, where its presence is injurious (Figure 10-7). Alcohol is a fat solvent: when taken with aspirin it increases the rate of penetration of aspirin into the stomach's cells. Taken together the two can be very damaging to the stomach's lining.

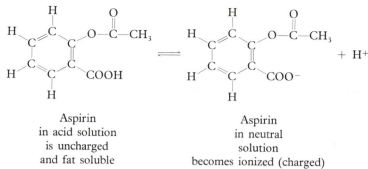

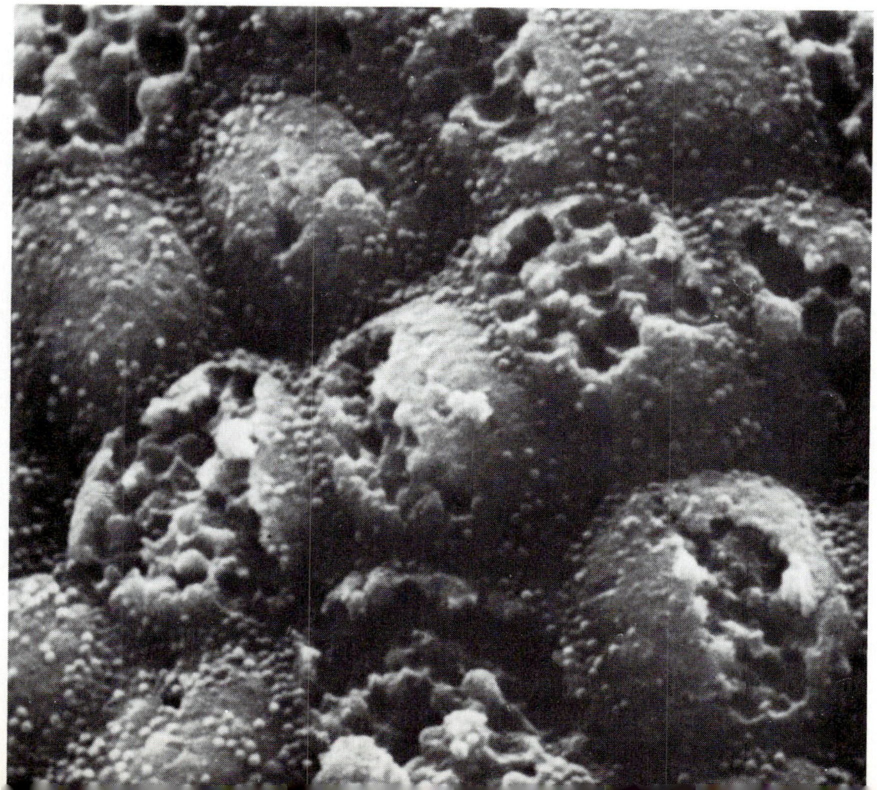

10–7

Several epithelial cells of the stomach's surface (×12,000). Destruction of parts of the outer membrane is apparent. (Courtesy Dr. Jeanne M. Riddle, Department of Pathology, Wayne State University School of Medicine, Detroit, Michigan.)

Absorption from the Small Intestine

Most of the lower part of the small intestine (the jejunum and ileum, which are illustrated in Figure 10-1) aids in the body's absorption of digested materials. This area provides an enormous surface area for absorption, because its inner lining consists of a series of folds covered with tiny fingerlike projections called *villi* (Figure 10-8) that in turn are covered with smaller microvilli (Frontispiece). This extensive absorptive area is crucial in providing nourishment for an animal with as large a volume as that of a human being. Some small, wormlike aquatic animals have such small bodies that they don't require a gastrointestinal tract. For example, animals of the phylum *Pogonophora,* as well as many flatworms, have no gastrointestinal tract. These animals have a small bulk to feed, and their exterior absorptive surfaces allow them to get enough nourishment directly from their seawater environment without one.

The sugars, amino acids, fatty acids, and glycerol that are absorbed into the bloodstream from the small intestine are carried in solution directly to the liver through the portal vein (see Chapter 7). The liver in turn serves as a metabolic factory that uses these nutrients as raw materials for the manufacture of the body's own carbohydrates, proteins, and fats. After the small intestine has absorbed most of these nutritional molecules, the large intestine absorbs most of the water and salts that remain.

Absorption and Excretion from the Large Intestine

The large intestine, or colon, has primary responsibility for the absorption of water and ions. About 500 milliliters of intestinal contents enter the colon each day, of which about 350 milliliters are absorbed into the body. The ability of the large intestine to absorb water is based upon the capacity of its cells to transport sodium from the intestinal contents into the blood. The osmotic absorption of water accompanies the active transport of sodium salts. The cells lining the large intestine are highly specialized in their functions, however. If nutrients such as glucose or amino acids enter the large intestine they will not be absorbed because the cells cannot transport them.

The amount of water reabsorbed by the large intestine depends mainly upon how rapidly the intestinal contents pass through. If the passage through the large intestine is very rapid the result is *diarrhea,* the frequent passage of relatively fluid feces. Certain foods, such as prunes, stimulate the peristaltic activity of the large intestine and thereby encourage rapid passage. Peristalsis is also stimulated by products of some disease-producing bacteria. Diarrhea can result in dehydration and the loss of needed ions from the body.

If the passage through the large intestine is slow, the result may be *constipation,* the infrequent passage of feces that are usually relatively dry. Laxatives (cathartics) are sometimes used to counteract constipation. Magnesium sulfate, the principal salt in the laxative milk of magnesia, is hardly absorbed from the large intestine, so the osmotic absorption of water is partially prevented. Other agents employed to counteract constipation include castor oil, which acts by irritating the smooth muscle of the colon, and mineral oil, which coats the feces so that they can be eliminated more easily.

The large intestine always contains large numbers of bacteria. Some of these bacteria can digest cellulose, the main component of plant cell walls. (The

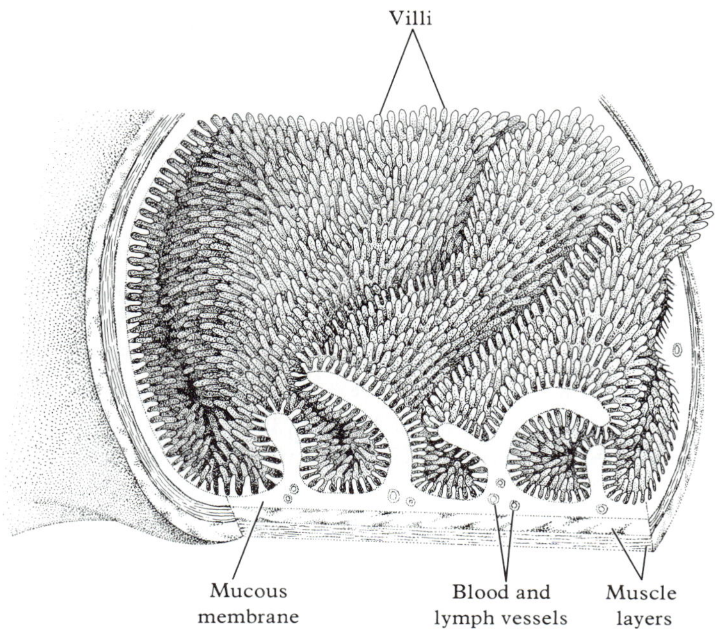

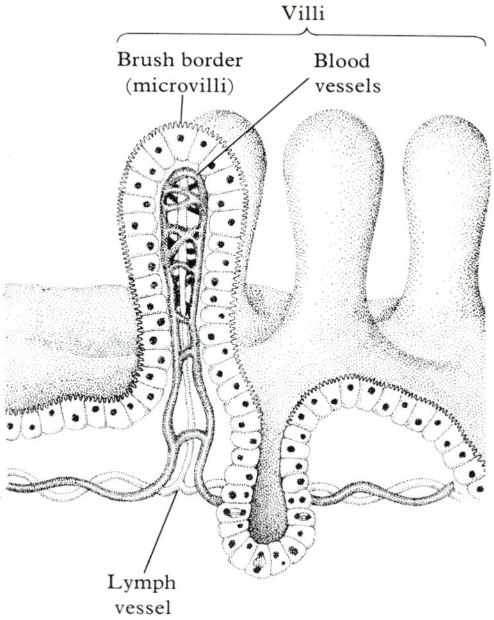

10–8
Surface of the small intestine is thrown up into folds covered with small villi (left). The outer surface of each villus (right) is made up of epithelial cells that have a "brush border" composed of microvilli. The villi and microvilli greatly increase the surface area of the small intestine available for absorption of nutrients. (From "Lactose and Lactase," by Norman Kretchmer. Copyright © 1972 by Scientific American, Inc. All rights reserved.)

human body does not synthesize an enzyme capable of digesting cellulose.) The small amount of glucose released from cellulose by bacterial action is inconsequential in human nutrition, but it can be used by the bacteria of the small intestine as an energy source. Intestinal bacteria also synthesize vitamins, some of which can be absorbed, and this source of vitamins probably accounts for the fact that some people do not need to obtain Vitamin B_{12} in their food.

When the large intestine has finished absorbing most of the salts and water, only the compacted feces remain within the gastrointestinal tract. By weight, the feces contain mostly materials that have not been absorbed by the body, such as roughage from foods (including indigestible plant cellulose) and bacteria (about one third of the feces by weight). However, a few products of the body's own tissues are also present in the feces, including epithelial cells (which have been sloughed off by the inner lining of the gastrointestinal tract), digestive enzymes, bile pigments (which give the feces their characteristic brownish color), bile acids, and cholesterol. The epithelial cells and digestive enzymes present in the feces account for part of the body's daily outgo of protein molecules.

An understanding of the excretory processes related to the gastrointestinal tract can aid in the diagnosis of human diseases. For example, the yellowish pigment *bilirubin* is a product of the breakdown of hemoglobin, which is changed into an excretable form by the liver and is then passed into the small intestine through the bile duct. When the liver is not functioning normally, bilirubin is not cleared from the bloodstream as it should be. The feces therefore become very light in color, and bilirubin builds up in the blood, causing the body to take on the yellowish color known as *jaundice*. (In dark-skinned people, jaundice can be seen in the whites of the eyes.) The presence of jaundice may instead indicate that a major bile duct is blocked, or that the liver is damaged so that its cells cannot convert the bilirubin into an excretable form. The latter often occurs during virus infections of the liver, or *hepatitis*.

The Protective Role of the Gastrointestinal Tract

In addition to being the site of digestion, absorption, and excretion, the gastrointestinal tract also provides protection for the neighboring internal tissues. These tissues must be protected both from the attack of digestive enzymes and from invasion by the swarms of microorganisms that inhabit the gastrointestinal tract.

The need for these protective functions serves as a reminder that the gastrointestinal tract is, strictly speaking, *outside* of the body. The epithelium of the intestines, like the epithelium of the skin, marks the boundary between inside and outside. The swarms of microorganisms that occupy the surface of the intestine's epithelium, like the skin's microorganisms, must be prevented from entering the body's tissues. Thus the body is like an elongated irregular doughnut, and the hole in the center is the gastrointestinal tract. Any breaks in the intestine's epithelium, such as those that may occur as a consequence of ulcers or appendicitis, are very serious events.

The gastrointestinal tract is a little difficult to understand, perhaps because we are unable to observe most of it in action. The mouth, however, is readily available for inspection, and it has much in common with the rest of the tract, as well as some notable differences. Because the mouth is so accessible and familiar, we will now discuss its physiology further.

The Mouth

The pink color of the mouth is a characteristic of the entire lining of the gastrointestinal tract, and it is caused mainly by the reddish hemoglobin of the blood underneath the transparent epithelium. The mouths of all people are pinkish, irrespective of their skin pigmentation, because melanocytes, the cells responsible for skin pigmentation, are found only on the exposed surfaces of the body. Similarly, the wetness of the mouth is characteristic of the entire tract. This wetness in the mouth enables food to be moved without sticking; farther down the tract the liquid environment allows the small molecules produced by digestion to move into contact with the extensive absorptive surfaces of the small intestine. Much of the mouth's wetness is supplied by three pairs of salivary glands, which produce about 1.5 liters of saliva each day. With the use of a mirror and a flashlight, you can easily see the openings of the ducts of these glands (Figure 10-9).

The mouth participates in the breakdown of food most obviously through the use of the teeth, which grind large chunks of food into swallowable pieces. The teeth shred the cell walls of plants, which is fortunate because the cell walls of plants are composed mostly of indigestible cellulose. After their cell walls are broken the contents of the plant cells can be digested and absorbed. The mouth's saliva lubricates the food for further passage and contains digestive enzymes, of which the most important is *amylase,* which breaks down certain kinds of carbohydrates (see Table 10-1).

Because the mouth is at the very beginning of the gastrointestinal tract, it is not surprising that it has almost no role in absorption of food. The surface of the mouth is smooth and it lacks the protruding villi and microvilli that characterize the small intestine. But the surface of the tongue does have some villi, which provide an expanded surface area for contact with the taste buds, and the receptor cells of the taste buds possess microvilli.

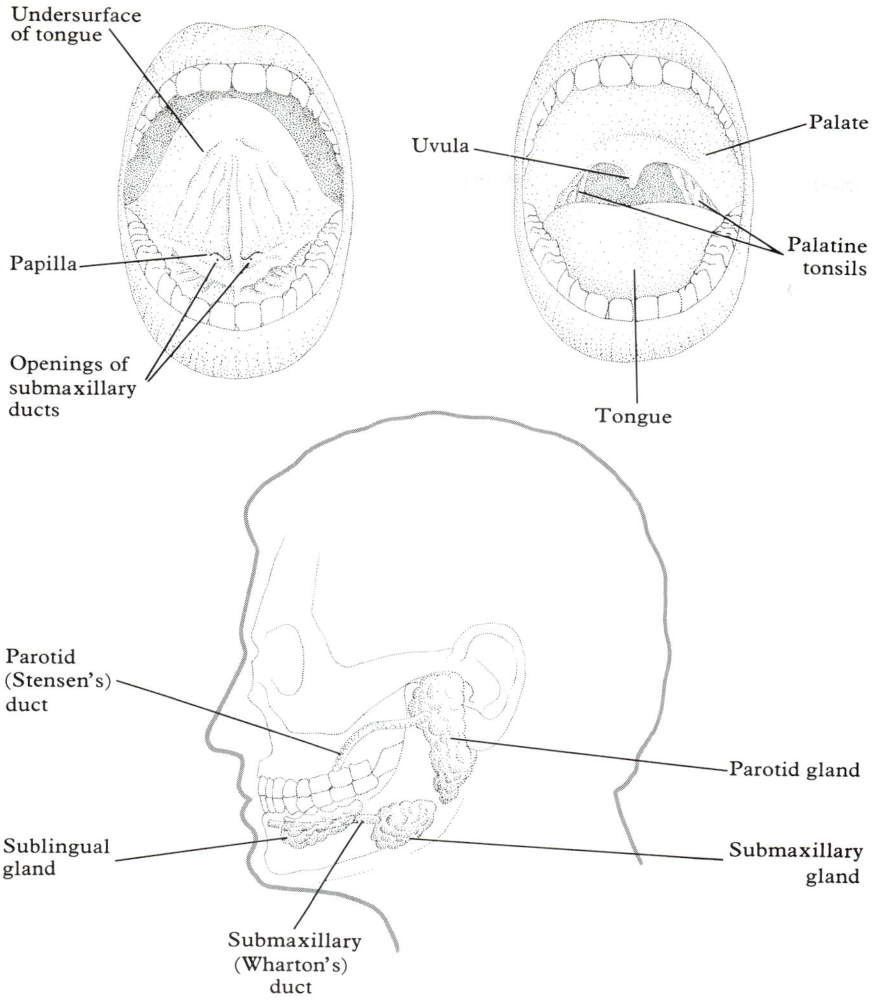

10–9
The tonsils and the three pairs of salivary glands. The largest salivary glands (the parotid glands) open into the mouth through ducts on the cheeks across from the upper molars. The smaller salivary glands open into the mouth beneath the tongue.

The ways in which the mouth moves food along the gastrointestinal tract, namely through tongue movements and swallowing, are quite different from peristalsis. However, both swallowing and peristalsis provide a means of keeping the partially digested food relatively well compartmentalized. During swallowing, the tongue rises in the back of the mouth to meet the soft palate, an action that limits the entry of food into the esophagus and prevents food from getting into the respiratory tract. Once food has entered the stomach it usually remains there for four to six hours, which allows plenty of time for protein digestion to take place; food then passes into the small intestine whenever the pyloris (the tight muscular sleeve at the entrance to the small intestine) relaxes enough to allow peristalsis to push the food through.

The mouth also has a protective role, because it is supplied with two tonsils that guard the entrance to the throat (Figure 10-9). These tonsils are packed with lymphocytes, which counteract unwanted invaders by synthesizing antibody molecules. While usually beneficial, the tonsils sometimes over-respond to invasion by microbes. This over-response is the cause of *tonsillitis,* in which the tonsils themselves become severely inflamed and infected. Throughout the rest of the gastrointestinal tract, there are tonsil-like areas packed with lymphocytes. (One of the best known is the appendix, situated at the junction between the small and large intestines. Like the tonsils, the appendix can become severely inflamed and infected. In fact, appendicitis, or inflammation of the appendix, is very much like tonsillitis.) The saliva secreted into the mouth provides additional protection against invading organisms because it contains antibody molecules as well as the enzyme *lysozyme,* which digests bacterial cell walls. In all, the mouth has quite an impressive immunological arsenal to unleash on a would-be bacterial invader!

Let us now consider how the process of digestion is regulated once food and drink have been swallowed.

Regulation of Digestion

Digestion must be regulated by the body. There are several reasons for this: first, people don't eat all of the time, and it would be wasteful to have the digestive processes continuously active when food is available only periodically. Second, meals vary considerably in composition, for example in their carbohydrate, protein, and fat content, and it is economical for the body to release specific enzymes only when there is something for them to digest. And finally, special features associated with the digestion of protein and fat require a somewhat different handling of each of these classes of molecules.

The Digestion of Protein

Because the human body depends on dietary protein as the source of essential amino acids, its digestion is the function of one of the main digestive organs, the stomach. The two major products secreted by the stomach, hydrochloric acid (HCl) and pepsin, both participate in protein digestion. Pepsin, which is secreted in its inactive form pepsinogen, is the major enzyme the stomach secretes in large quantities. HCl initiates the conversion of pepsinogen to pepsin and provides an optimal pH for the activity of pepsin.

The regulation of acid and pepsin secretion by the stomach proceeds at least in part as we might expect: the presence of protein in the stomach (but not carbohydrate or fat) stimulates the secretion of HCl and pepsin (Table 10-2).

TABLE 10-2

Factors affecting the secretion of hydrochloric acid and pepsin by the stomach.

THIS FACTOR	ACTS ON THE	WHICH LEADS TO	SO THAT THE SECRETION OF ACID AND PEPSIN
Emotional states (worry, conflict) Thought, sight, smell of food	Brain	Stimulation of the vagus nerve (which innervates the stomach)	Increases
Stretching of the stomach Substances in the stomach (protein, caffeine, alcohol)	Stomach lining	Release of gastrin into the blood	Increases
Fats in the duodenum	Duodenum	Release of enterogastrone into the blood	Decreases

The mechanism by which protein stimulates acid and pepsin secretion is indirect. A hormone, *gastrin*, which is synthesized by cells in the wall of the stomach, is released into the bloodstream when protein and partially digested protein are present in the stomach. Gastrin then travels in the blood back to other cells of the stomach, where it initiates the secretion of HCl and pepsin. Gastrin release is also stimulated by the presence of alcohol and caffeine in the stomach, as well as by the stretching of the stomach that occurs after a large meal.

Acid and pepsin secretion can also be initiated by stimulation of the *vagus nerve*, one branch of which directly connects the brain to the stomach. Through this nerve pathway, the sight, smell, taste, or thought of food can result in increased acid and pepsin secretion. Emotional states such as worry and conflict are known to increase secretion by means of this nerve. Thus there are two ways in which acid and pepsin secretion can be stimulated: through the hormone gastrin and through the vagus nerve.

The presence of fat in the first part of the small intestine slows the secretion of acid and pepsin. Fat not only slows secretion, but it also slows the rate at which the stomach empties itself. The slowdown is advantageous in that fat digestion takes longer than the digestion of either protein or carbohydrate (see the following section on the digestion of fat). This slowdown is strongly influenced by the hormone *enterogastrone*, which is released from the wall of the small intestine when fat is present. Enterogastrone passes through the bloodstream to the acid- and pepsin-secreting cells of the stomach, where it inhibits their secretion.

Duodenal Ulcers

The duodenum is the first 20 centimeters of the small intestine, and is located immediately beyond the stomach in the digestive tract. Probably because of its close proximity to the stomach it is the site of a malady that affects five percent of the people in the United States—the *ulcer*. Ulcers are shaggy craters with rough edges where the normal lining of the intestine has been eroded away. Figure 10-10 shows the eroded surface of a cell in a duodenal ulcer crater. Most ulcers occur very close to the outlet of the stomach, probably because the lining of the

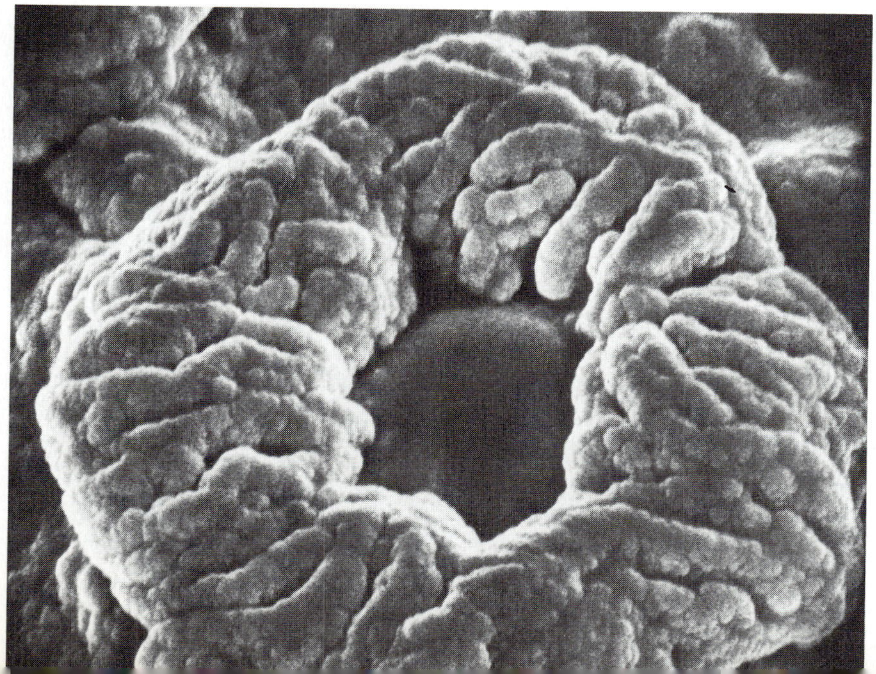

10–10
Cell in a section of ulcerated duodenum (×3080). Notice that the central area of the cell surface is completely denuded. The surrounding area is covered by well-preserved microvilli. (Courtesy Roche Laboratories, Division of Hoffman-La Roche, Inc., Nutley, N.J.)

duodenum, unlike the lining of the stomach, is not designed to tolerate the acid that sometimes enters it from the stomach. The causes of duodenal ulcers are complex, because they are influenced by hard-to-define emotional factors on the one hand and the presence of hydrochloric acid in the duodenum on the other. We cannot say definitely that elevated acid secretion by the stomach is a *cause* of duodenal ulcers, for two reasons. First, many people who have ulcers do not have elevated secretion of acid; second, even when acid secretion is elevated, other factors may have caused this elevation. It is clear, however, that acid usually aggravates a duodenal ulcer once it has become established. People who have duodenal ulcers commonly experience pain beginning four hours after meals or just before breakfast, when the stomach contains the least food to neutralize its acids. Pain occurs because the acid irritates the exposed nerve endings in the ulcerous area.

With these points in mind, we can understand some of the rationale behind the prevailing dietary approach to the treatment of duodenal ulcers. This approach consists of giving many small servings of antacids and milk (provided the person's digestive system in not intolerant to milk) and excluding alcohol, coffee, and tea from the diet. Milk and cream have a high concentration of fat (and cholesterol) and this reduces acid secretion by stimulating the secretion of enterogastrone. Frequent small servings keep something in the stomach constantly to neutralize the acid, and the small size of the servings prevents the stretching of the stomach, which would stimulate the gastrin mechanism. Alcohol and caffeine are omitted from the diet because they increase the secretion of acid through gastrin, even though neither of these substances is a protein. Such a diet, while frequently helpful in relieving symptoms, may not eliminate the ulcer. This is because the special diet does not deal with the various factors, such as stress, that led to the formation of the ulcer in the first place.

The Digestion of Fat

Fat cannot be digested unless it is first pretreated in the gastrointestinal tract because fats aren't soluble in water, and the enzymes that break down fat aren't soluble in fat. What is needed is a detergent that can make the fats soluble in water.

The liver synthesizes and secretes the necessary detergent in the form of the *bile acids.* These molecules (Figure 10-11), which are synthesized from cholesterol, serve as a bridge; one side of it interacts with the hydrogen ions of the water, and the other side interacts with fat. Because of this detergent action, the fat is brought into close contact with the water, so that the fat-digesting enzymes can get at the fats.

Most of the bile acids are not eliminated from the intestine once they have been transported there from the liver. About 95 percent of them are reabsorbed through the intestinal wall and are eventually secreted by the liver once again, a neat bit of nature's recycling.

The *gall bladder* is a pouch that lies on the path between the liver and the small intestine and that serves as a storehouse for the *bile,* which is rich in bile acids. When fats enter the small intestine, the gall bladder contracts in response to the presence of enterogastrone and releases bile acids into the duodenum. As discussed before, enterogastrone reduces stomach acid secretion and it also slows the emptying of the stomach. This slowdown allows time for the bile acids to act as detergents.

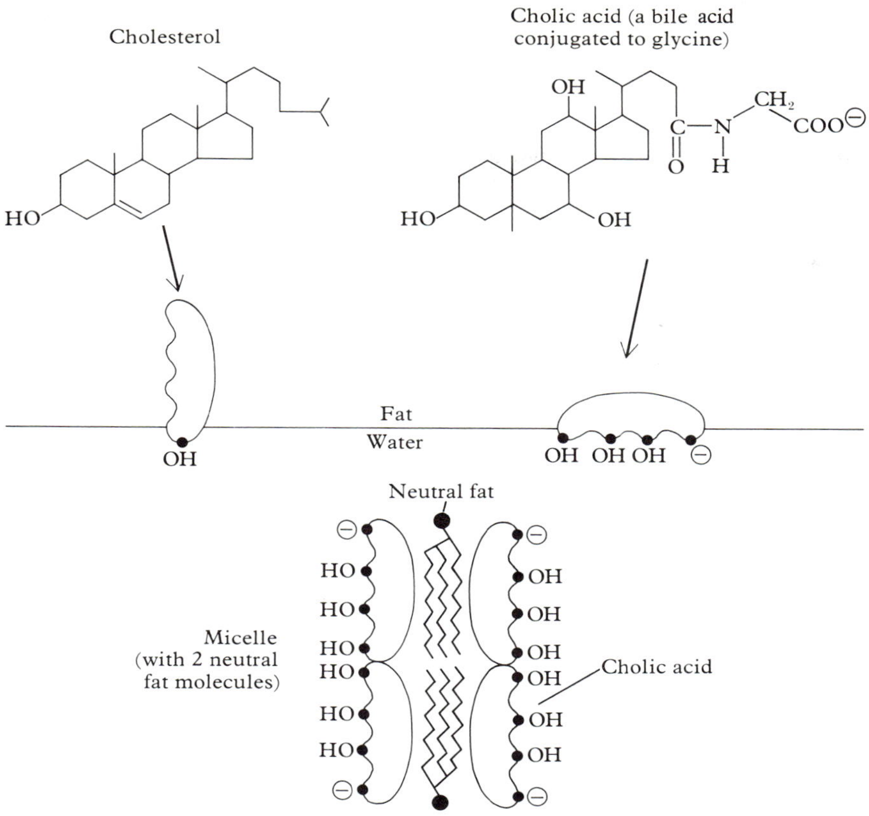

10–11
Bile acids aid fat digestion by serving as a bridge between fat and water. They interact with water through their negatively charged side chain and through their hydroxyl (OH) groups. (In contrast, cholesterol has only one OH group.) The bile acids aggregate in groups of four to ten molecules called micelles. Fats enter the micelles because the fat-soluble portions of the bile acid molecules face inward.

The gall bladder also stores cholesterol, which enters the small intestine along with the bile acids. Cholesterol, like fat, is mainly insoluble in water, as it has only one hydroxyl (-OH) group that can enter into association with the hydrogen ions of water (Figure 10-11). Cholesterol is an important component of cell membranes, and is a precursor of many important molecules, including the bile acids, the sex hormones, and the hormones produced in the adrenal cortex.

Gallstones

In general, cholesterol is not stored in the gall bladder, because the bile acids there make the cholesterol partially soluble in water. Sometimes, however, cholesterol crystals do form, and they may develop into much larger *gallstones* that sometimes fill the gall bladder. In the United States, cholesterol is the major constituent of gallstones in about 90 percent of the cases. Gallstones may occasionally pass into the common bile duct (which connects the gall bladder to the small intestine), thus resulting in intense pain. The pain of gallstones characteristically appears after a fatty meal, because the fat in the small intestine causes the gall bladder to contract on its stony contents.

It has been estimated that 15 million people in the United States have gallstones and that treatments for surgical removal of the gall bladder cost about one billion dollars per year. Although we are uncertain about the causes of gallstones, one of the clues we have is that people who develop gallstones have been found to have only about half of the normal amount of total bile acids.

Perhaps this shortage of bile acids allows crystals of cholesterol to form, thus initiating gallstone formation.

Milk-Drinking and Lactose Intolerance

For those of us who grew up believing that milk is nature's best food for infants and for adults alike, it is a surprise to learn that most people around the world think that milk is unfit for adult consumption. At one time Western nutritionists sent powdered milk to various parts of the world where the people were undernourished in hopes of helping to improve their diets. They were surprised to find that Balinese people used the powdered milk as a laxative, Guatemalans used it to whitewash their houses, Navaho Native Americans threw it away, and the Kanuri people in Nigeria refused to use it because it contained "evil spirits." The fact is that cow's milk makes *most* members of the human race above the age of five feel sick.

The reason is that most adults cannot digest *lactose,* the main sugar present in all animal milk. When lactose remains undigested, it passes on into the large intestine, where it interferes with the absorption of water, thus resulting in diarrhea. In the large intestine it is fermented by bacteria, which causes large amounts of intestinal gas to form. Lactose is a disaccharide (di means two; saccharide, sugar) formed by the chemical bonding of a molecule of the sugar glucose to a molecule of the sugar galactose (Figure 9-1). To be absorbed from the intestine, lactose must first be broken down into glucose and galactose by the enzyme *lactase.* This enzyme is present in almost 100 percent of human infants, but it drops to low concentrations in many people by the time they reach five years of age. This low concentration of enzyme is insufficient to digest large amounts of lactose. However, there is much less lactose in milk products that have been fermented by microorganisms, such as cheese and yogurt. People who have low concentrations of lactase can tolerate these products (Figure 10-12).

10–12
Woman offers nono, *a yogurtlike milk drink, for sale in a Nigerian town. She is a member of a dairying ethnic group from northern Nigeria, the Fulani, whose adult members drink fresh milk. The nono she offers is partially fermented, so its lactose content is reduced and it can be tolerated by townspeople who cannot digest milk. (Courtesy Norman Kretchmer. Published in* Scientific American, *October 1972.)*

Only 10 percent or less of adult Thais, Taiwanese, Andean Indians, Eskimos, and other peoples have high concentrations of the enzyme lactase. By way of contrast, high concentrations are found in about 90 percent of adults of Northern European extraction. Among people who live in the United States, high concentrations of lactase are present in 80–90 percent of adults of predominantly European extraction, but only in about 30 percent of adults of African extraction.

Adults who have high concentrations of lactase and who can drink milk without problems have in general about 10 times as much lactase in their intestines as people who have low concentrations and who consequently cannot tolerate milk. Studies of familes have shown that lactose intolerance is inherited as a recessive trait. The dominant form of the gene determines high lactase production, so only a person who has two recessive genes for low lactase production will produce low concentrations of the enzyme. (Dominant and recessive genes are discussed in Part III.)

It has been suggested that before people began to consume the milk of domestic animals, most adults had little lactase in their digestive tracts. Presumably, this is because the production of lactase beyond the cessation of breastfeeding would have offered no evolutionary advantage. However, when some groups of people began to consume animal milk the few persons who had the dominant gene for high lactase production would be able to utilize milk, and thus might have a selective advantage over those who could not. Indeed, the populations in which many people have high lactase production are found in areas where cow's milk has been available for many generations.

The Kidneys, Regulators of the Composition of Body Fluids

In the previous section we saw that the regulation of digestion by the gastrointestinal tract requires that it make appropriate responses to the various foods that are eaten. But the G-I tract is mostly concerned with utilizing what it is given at the moment. Much of the job of eliminating excesses and retaining what is necessary for normal function is carried out by the kidneys.

Like other large land animals, human beings face a potential problem of not being able to get enough water to maintain the constancy of their internal environment. This is particularly a problem of those people who live in hot, dry desert areas. Thus it is not surprising that human beings have evolved methods for conserving water by regulating water output and that we have also evolved a mechanism that alerts us to obtain water when we need it—thirst. A look at the daily balance of water in the body reveals that water enters the body only through the G-I tract, but that it exits through four routes: skin, lungs, G-I tract, and kidneys (Table 10-3).

As the input column of this table indicates, not only do we take in water in food and drink, but we also *manufacture* water by breaking down fats and carbohydrates.This manufactured water, called metabolic water, provides enough water for some animals that they can survive without drinking. Consider, for example, the kangaroo rat. If it eats the right food, it can get the amount of water it needs from food and metabolic water alone. Although it lives in the desert, it is completely nocturnal, coming out onto the surface of the desert

TABLE 10-3

Daily balance of water in the body.

INPUT (MILLILITERS PER DAY)		OUTPUT (MILLILITERS PER DAY)	
G-I Tract		Skin (evaporation)	500
Drink	1200	Lungs	500
Food	500	G-I Tract	100
Metabolic	400	Kidneys (urine)	1000
	2100		2100

only during the cool of night. It has no sweat glands and its kidney is better able to conserve water than ours. Now look at the four output routes for water: of these only the output of the kidneys can be regulated to any great degree. Thus the workings of the kidney are crucial in human water balance.

In addition to water, the other chemical inputs of the body are oxygen, which enters through the lungs, and carbohydrates, fats, amino acids, nucleic acids, and ions, which enter through the G-I tract. The fates of these various molecules once they enter the body are shown in Table 10-4. Notice that the kidneys are principally responsible for the elimination of many ions and for the elimination of the nitrogen atoms derived from amino acids and nucleic acid bases. The kidneys are also important for the *regulation* of the excretion of three kinds of molecules: water, urea, and ions. Let us now examine in some detail how the kidneys accomplish their regulatory functions.

How the Kidneys Perform Their Task

The mammalian kidney is the product of a long evolutionary history. The first air-breathing vertebrates probably evolved from fish that lived in fresh water.

TABLE 10-4

The fate of molecules that enter the body.

THESE SUBSTANCES ENTERING THE BODY	CONTAIN THESE ATOMS	WHICH ARE ELIMINATED AS	FROM THESE PARTS OF THE BODY
Oxygen Fatty acids Sugars, glycerol	O C, H C, H, O	CO_2 and H_2O	Lungs Lungs, skin, gastrointestinal tract, kidneys
Amino acids, some nucleic acid bases*	N (also C, H, O)	$(H_2N)_2C{=}O$ (urea)	Kidneys
Ions	Na^+, Cl^-	Na^+, Cl^-	Kidneys, skin
	K^+, Mg^{++}, Ca^{++}, H^+, HCO_3^-, $PO_4^{\equiv}$	K^+, Mg^{++}, Ca^{++}, H^+ HCO_3^-, $PO_4^{\equiv}$	Kidneys

*Certain nucleic acid bases are eliminated as uric acid, but uric acid accounts for only one-thirtieth of the body's nitrogen excretion.

Fresh water fish must conserve sodium and excrete water to survive. Life in fresh water requires a mechanism to conserve sodium because the concentration of sodium is higher inside the fishes' bodies than in the water, and thus it naturally tends to be lost into the environment. At the same time, fresh water fish must constantly excrete water, because the concentration of water is higher in the environment than in the fish's body, and this results in a strong natural tendency for water to enter the body, not to leave it.

The fresh water fish solved these dual problems by evolving a filtering device, the *glomerulus,* in which a coil of capillaries is butted up against the blind end of a hollow tube that opens to the outside of the body. The arterial pressure in the capillaries causes water, sodium, and other small molecules to cross the capillary walls and enter the tube, where selective absorption of sodium and other molecules takes place before the remainder of the filtrate, which is mostly water, is drained to the exterior.

In the mammalian kidney, the formation of urine begins when glomeruli filter the blood plasma from a tuft of capillaries into the tubules of the kidney. Each day almost fifteen times the volume of blood plasma (about 75 liters) is transferred by filtration from the vascular system to the kidney tubules. But the requirements of a land-dwelling mammal are very different from those of a freshwater fish, and water *conservation,* rather than water elimination, is the kidney's main preoccupation. Therefore, most of the large volume of fluid transferred to the kidney tubules is reabsorbed in mammals, whereas it would be excreted by freshwater fish.

The kidney can simultaneously conserve water, excrete urea, and regulate the composition of many of the body's ions. It manages these tasks by employing three mechanisms to transport substances across biological membranes:

1. *Diffusion.* As you will recall from earlier chapters, the tendency of dissolved molecules to pass from areas of high concentration to areas of lower concentration is called diffusion. It is a passive process, which requires no energy from the body. (The tendency of *water* to pass from an area of higher concentration to an area of lower concentration by diffusion is known as osmosis.)
2. *Filtration.* The use of physical pressure (such as blood pressure) to push small molecules and water across membranes is called filtration.
3. *Active transport.* An active, energy-requiring process by which dissolved molecules can be transported from areas of lower to areas of higher concentration is called active transport.

Conservation of Water by the Kidney

The conservation of water by the kidney is a complex process, because it depends both on the anatomy of the kidney and on a mechanism by which ions become concentrated in the inner portion, or medulla, of the kidney. The anatomy of the urinary tract is shown in Figure 10-13. In the kidney the blood plasma is filtered from capillaries to tubules in the glomeruli, which are located in the outer area (cortex) of the kidney. A glomerulus and its tubule together compose the basic functional unit of the kidney, the *nephron.* Beyond the glomeruli the tubules descend deep into the medulla where they make a hairpin turn, doubling back to ascend into the cortex once again (Figures 10-14 and

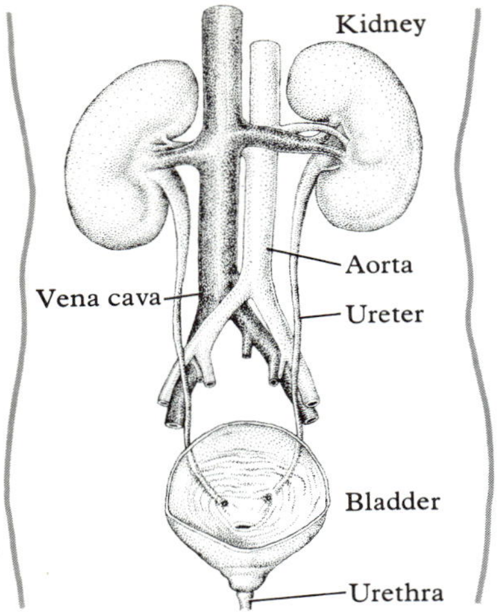

10–13
Human urinary tract. The kidneys are located behind other abdominal organs. Descending from the kidneys are the ureters, which empty urine into the bladder. Urine is stored in the bladder until it is released through the urethra. Blood vessels that connect the kidneys with the circulatory system are also shown.

10-15). At the cortex they then make *another* hairpin turn, descending once again into the medulla, where they empty into the collecting ducts. From the collecting ducts the urine empties into the renal pelvis and travels in the ureters to the bladder, where the urine is stored until it is released through the urethra.

When there is a special need for water conservation, more water than usual is removed from the collecting ducts. Because water cannot undergo active transport, the additional water removed from the collecting ducts must diffuse back into the kidney tissue through osmosis. This can occur only if the concentration of water is less in the kidney tissue of the medulla than in the collecting ducts. Accordingly, the medulla of the kidney has evolved a mechanism for reducing its water content. It accomplishes this task by accumulating ions, particularly sodium (Na^+) ions (which represent 90 percent of all the extracellular positive ions) and chloride (Cl^-) ions. The mechanism by which the kidney's medullary tissue concentrates ions is the subject of controversy at the present time. It seems likely that ion concentration depends primarily upon the active transport of chloride ions and a special mechanism known as the *countercurrent* mechanism.

The countercurrent mechanism can cause a high concentration of ions to collect in the medulla and thus cause most of the water filtered by the kidney to be reabsorbed in the following way. As a tubule begins to ascend from its first dive into the medulla, chloride ions are pumped by active transport out of the tubule into the surrounding kidney tissue, and sodium ions follow (Figure 10-16). Thus the tissue of the medulla becomes more concentrated in ions than the tissue of the cortex. Some of the sodium and chloride ions will diffuse into the descending tubule, making the fluid in the descending tubule more

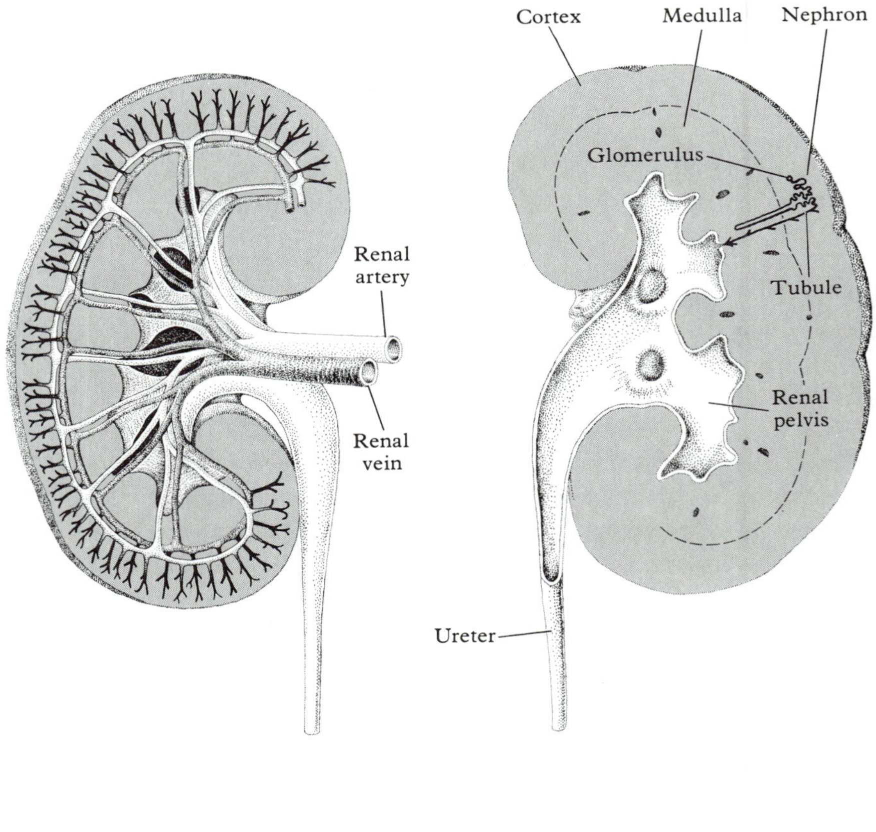

10–14
Cross sections of the kidney. The circulation of the kidney is depicted on the left. Both arteries and veins send branches into the cortex (outer area) of the kidney where the glomeruli are located. The location of one nephron is shown on the right. The tubules, which originate at the glomeruli, descend twice into the medulla (inner area) of the kidney and discharge urine into the renal pelvis. Each kidney contains about one million nephrons. (From "The Kidney," by Homer W. Smith. Copyright © 1952 by Scientific American, Inc. All rights reserved.)

concentrated than before. This fluid in the descending tubule passes the hairpin turn in the medulla, and as it does so more chloride is once again pumped by active transport out of the ascending tubule. Only there is an important difference now, namely that the tissue of the medulla can become even more concentrated in ions than before, because the active transport mechanism starts with a higher concentration of chloride in the ascending tubule than was there before. As a result of the cycling of ions a greater concentration of ions in the medulla is produced than would be produced by active transport alone. The term *countercurrent* refers to the passage of ions from the ascending tubule, where the current of flow is toward the cortex, to the descending tubule, where the flow is in the opposite or counter direction. Irrespective of what mechanism is employed, it is clear that the high ionic concentration in the kidney's medulla allows water to diffuse from the collecting ducts to the medulla; from the medulla the water reenters the plasma. In this process the ionic concentration of the urine becomes greater than that of the plasma.

Maximum water conservation by the kidney is not needed when water intake is adequate, however, and control over the amount of water to be conserved is in large part the responsibility of two hormones, *antidiuretic hormone (ADH),* and *aldosterone* (Figure 10-17). ADH is synthesized by cells in the hypothalamus of the brain and released by the posterior pituitary. It

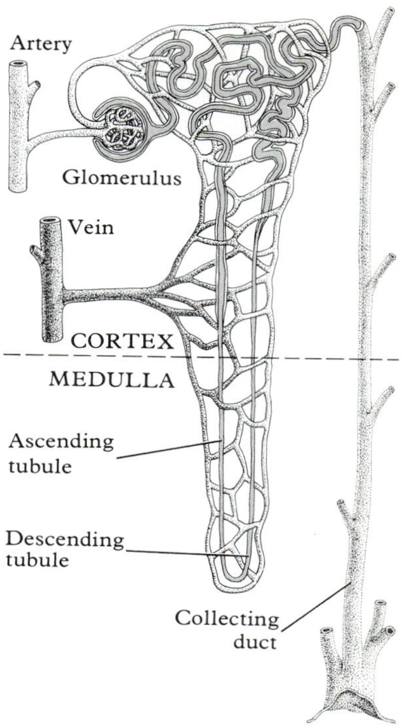

10–15
Nephron consists of the glomerulus, the ball of capillaries at upper left, and the tubule, which twists from the glomerulus to the collecting duct, right. Blood plasma is filtered out of the glomerulus into the tubule. In the tubule most of the water and other useful substances are reabsorbed, leaving behind urea and other waste products. (After "The Kidney," by Homer W. Smith. Copyright © 1952 by Scientific American, Inc. All rights reserved.)

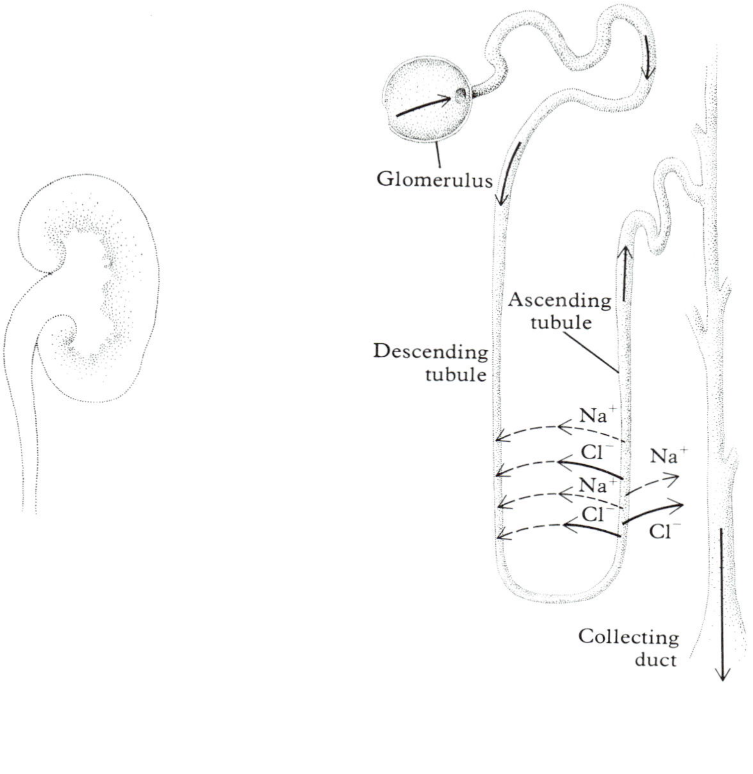

10–16
Ion concentration by the kidney. The shading of the kidney on the left indicates the progressive increase in ion concentration from cortex to the depths of the medulla. The greater the shading, the higher the ion concentration. The concentration depends in large part upon the active transport of chloride. Chloride (Cl^-) is actively transported out of the ascending tubule and into the tissue of the medulla (⟵ indicates active transport). Sodium (Na^+) passively follows the chloride (⟵ indicates passive movement).

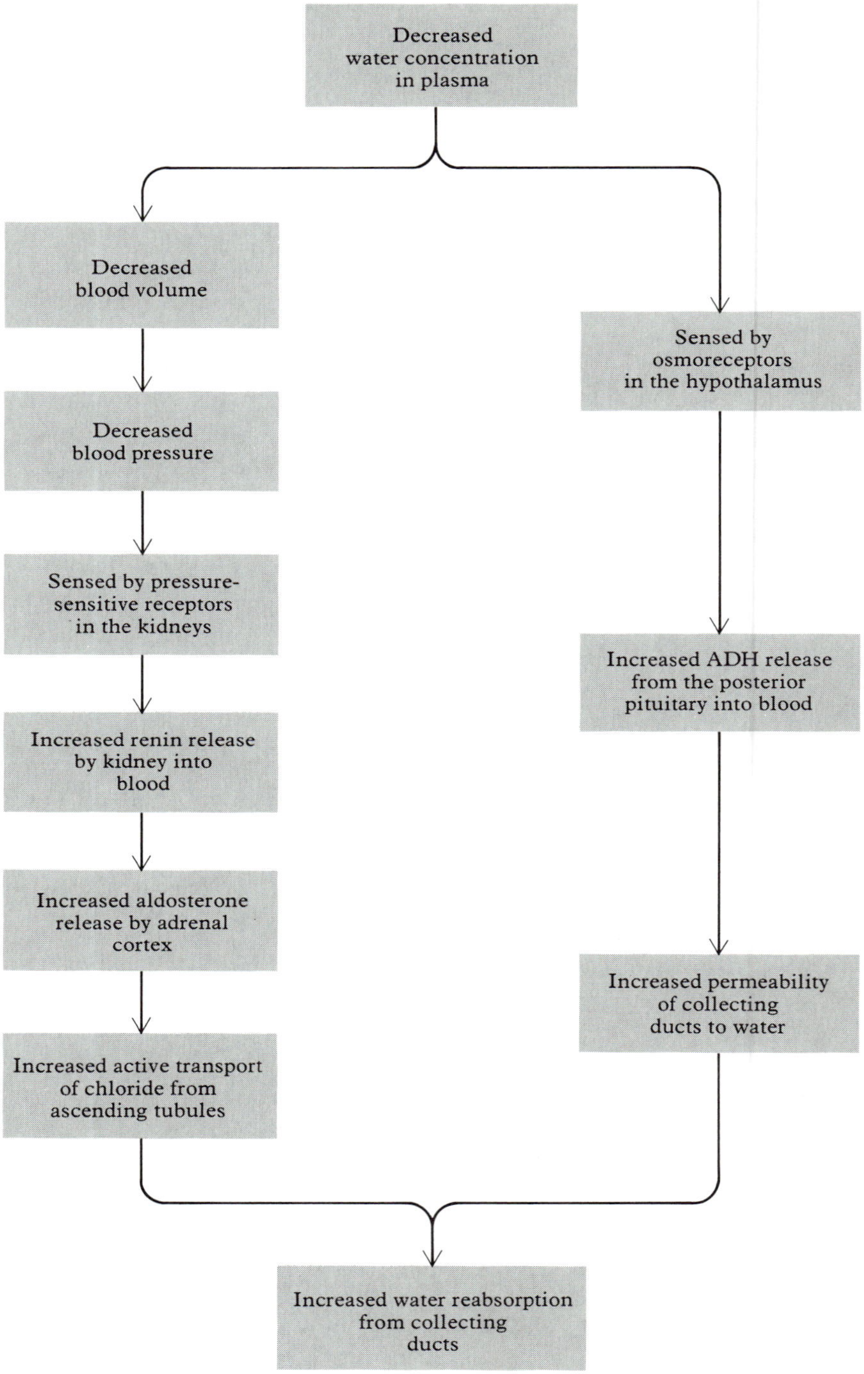

10–17
The kidneys and water conservation.

increases the permeability of the collecting ducts to water, and thus allows water to be reabsorbed very efficiently. Aldosterone is synthesized and released by the adrenal glands, located one each on top of the two kidneys. This hormone promotes active transport of chloride and thus allows the medulla of the kidney to develop its maximum ion concentration.

The main stimulus for the release of ADH is dehydration. When the water content of the plasma drops even slightly, the concentration of ions correspondingly increases. Even a one to two percent increase in ion concentration of the plasma in the main arteries serving the brain causes an increase in the release of ADH from the posterior pituitary. There are receptors in the brain that are sensitive to small changes in ion concentration. These receptors, appropriately called *osmoreceptors* because they detect small concentration differences that would result in osmosis, regulate the output of ADH. If there is damage to the pituitary, interfering with the production or output of ADH, the condition known as *diabetes insipidus* can result.* This condition is characterized by the outpouring of large amounts of dilute urine, because the relative absence of ADH makes the collecting ducts relatively impermeable to water so that water cannot be reabsorbed in normal quantities. The most effective thereapy for diabetes insipidus is the administration of ADH.

The hormone aldosterone helps the body to conserve both ions and water. When the concentration of ions is low or the blood pressure drops (which may occur when the concentration of water in the blood is low) the kidney releases an enzyme *renin*, which through an indirect mechanism stimulates increased release of aldosterone from the adrenal gland. Aldosterone is released directly into the bloodstream and when it reaches the cells of the kidney tubule it promotes the active transport of chloride. Thus ion conservation will always occur in the presence of increased concentrations of aldosterone. Water conservation, on the other hand, depends upon the presence of aldosterone (though it need not be present in large quantities) *and* ADH.

The kidney regulates the concentrations of ions other than sodium and chloride (Table 10-4). Regulation of potassium (K^+) and hydrogen (H^+) ions, is accomplished by either increasing or decreasing the secretion of these ions from the tubular cells into the tubular fluid, where they are exchanged for sodium (Figure 10-18).

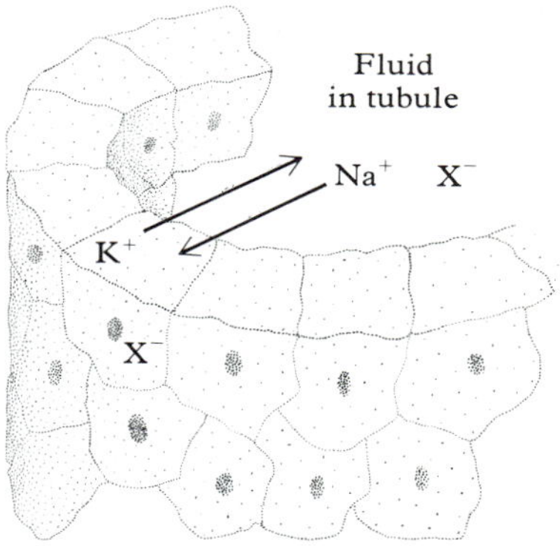

10–18
A mechanism by which the kidney regulates the concentration of ions in the body. Potassium (K^+) is shown being secreted by a tubular cell in exchange for sodium (Na^+) from the tubular field. Negative ions are represented as X^-.

Urea

Urea is the chemical compound (NH_2—CO—NH_2) principally responsible for the excretion of nitrogen atoms. The excretion of urea is crucial to life; and without kidney function, people die within two to three weeks as a result of urea reaching toxic concentrations in their bodies. Fortunately, the body easily tolerates a two- to three-fold increase in urea, so very precise regulation is not required. Urea diffuses rapidly through most biological membranes, including those in the kidney, so when urine flow is profuse, urea excretion is also profuse

*Diabetes insipidus and diabetes mellitus are unrelated diseases characterized by a common symptom, excess urine formation. The ancient Greek physician Aretaeus described the diabetic condition and gave it its name, *diabetes,* which means in Greek "to run through a siphon."

because there is less time for the filtered urea to flow back into the kidney tissue. Conversely, when urine flow is low, there is less urea excretion because there is more time for urea to be resorbed. In general, impaired urea excretion results only when kidney function is severely impaired, and an elevation in the concentration of urea in the blood (measured as the blood urea nitrogen, or BUN) is an indication of whether the kidneys may have suffered some serious injury.

Uric Acid

The urine also contains uric acid, a nitrogenous excretory product derived from the turnover of two nucleic acid bases: adenine and guanine. Uric acid only accounts for one-thirtieth as much nitrogen excretion as urea, but its excretion is important. When concentrations of uric acid in the blood become too great, the serious disease known at *gout* may result. In gout salts of uric acid (urates) are deposited in joints, kidneys, and other tissues, where they cause great pain. The most commonly afflicted joint is the one that allows the great toe to bend on the foot. The elevated concentrations of uric acid that are characteristic of gout are usually the result of inherited metabolic defects that either bring on excess formation of uric acid or cause a failure of excretion. Gout can be treated with drugs that reduce the concentrations of uric acid in the blood by increasing its urinary excretion.

Summary

The overall chemical composition of the body remains relatively constant, and yet a great variety of solid and liquid materials enter and leave the body every day. The gastrointestinal tract is particularly important to intake because it is responsible for breaking down food and drink into the small molecules that are absorbed into the body's tissues. The kidneys are especially important to outgo because they exert most of the body's control over the excretion of ions, water, and nitrogen.

The gastrointestinal tract is a long tube, continuous at both ends with the skin, consisting of the mouth, pharynx, esophagus, stomach, small intestine, large intestine, rectum, and anal canal. Once food has been swallowed it is passed along the tract mostly by waves of smooth muscle contraction, or peristalsis, aided in the small intestine by segmental contractions. Digestion is facilitated by the digestive enzymes. Each digestive enzyme is capable of breaking only specific chemical bonds. The cells that produce the digestive enzymes are not themselves digested because they release enzymes in inactive forms. The enzymes are activated only after they reach the lumen of the intestine.

Absorption of digested material takes place mostly in the lower portion of the small intestine. The surface area available for absorption is large because the inner lining of the small intestine is covered by fingerlike villi that are in turn covered by microvilli.

The large intestine has primary responsibility for absorption of water and ions, and it cannot absorb the products of digestion such as sugars and amino acids. Some of the bacteria present in the large intestine can digest cellulose, which is the main structural component of plant cell walls. Glucose released from digested cellulose is used for energy by the bacteria. These bacteria also synthesize and release vitamins that can be absorbed into the body. The feces consist mostly of bacteria and roughage, but a few products of the body's tissues are present as well, including epithelial cells, digestive enzymes, bile pigments, bile acids, and cholesterol.

Strictly speaking, the gastrointestinal tract is outside of the body, and any break in the lining of the intestine is potentially life threatening. Many lymphoid cells are present in the tissue surrounding the gastrointestinal tract, and large aggregations of lymphoid cells are found in the tonsils and the appendix. These lymphoid cells are part of the body's immune system and they help to counteract unwanted organisms that might otherwise invade through the wall of the intestine into the tissues.

Digestive processes are regulated to accommodate the amount and kind of food that is eaten. For example, the presence of protein in the stomach causes the release of the hormone gastrin from cells present in the stomach wall. Gastrin travels in the blood back to the stomach, where it stimulates the secretion of hydrochloric acid and pepsin, which digest protein.

The digestion of fat is a two-step process. First, the bile acids bring the fats into contact with the digestive enzymes; second, the digestive enzymes break down the fats. The presence of fat in the duodenum causes the release of the hormone enterogastrone, which stimulates the gall bladder to release bile acids into the duodenum. Enterogastrone also allows more time for fat digestion by slowing down the emptying of the stomach.

The gall bladder, which contains cholesterol as well as bile acids, can become filled with stones known as gallstones. Sometimes these stones block the duct between the gall bladder and the duodenum, resulting in considerable pain. Cholesterol is frequently the major chemical constituent of gallstones. People who develop gallstones have on the average only half of the normal amount of bile acids; this decrease in bile acids may allow cholesterol crystals to form, thus initiating the formation of gallstones.

Human infants can digest lactose, the principal sugar in milk, but many adults cannot. The inability to digest lactose, known as lactose intolerance, occurs when only small quantities of the enzyme lactase are released into the small intestine. Lactase breaks lactose into its glucose and galactose components. When lactose remains undigested it can cause diarrhea and the production of excess intestinal gas.

The kidneys are mainly responsible for regulating the amount of water, urea, and ions present in the body. The formation of urine begins in the glomeruli of the kidney cortex, where the blood plasma is filtered into kidney tubules. The tubular fluid follows a circuitous route from cortex to medulla to cortex again before it finally empties into the collecting ducts of the medulla. From the collecting ducts the urine empties into the renal pelvis and travels in the ureters to the bladder, where it is stored until its release through the urethra.

When the body has a special need for water conservation, more water than usual is absorbed from the collecting duct fluid into the tissue of the medulla. This causes the ionic concentration of the urine to become higher than that of plasma. Fluid can be absorbed into the medullary tissue because the concentration of water is low in the medullary tissue, and water will diffuse from an area of high concentration (in the fluid of the collecting duct) to an area of low concentration (in the medulla). In the medulla the concentration of water may be kept low by replacing water with ions. The active transport of chloride (with sodium passively following) from the tubular fluid into the medulla accounts in large part for the high ionic concentration of the medulla.

The maximal absorption of water from the tubular fluid depends upon the actions of two hormones: aldosterone and antidiuretic hormone (ADH). Aldosterone promotes the active transport of chloride from the tubular fluid into the medulla, thereby increasing the ionic concentration of the medulla and decreasing its water content. ADH increases the permeability of the collecting ducts to water, thus allowing water to be absorbed efficiently. The principal stimulus for ADH release from the posterior pituitary is dehydration, which is detected by osmoreceptors located in the brain. Only small amounts of aldosterone are necessary for water conservation.

Aldosterone plays a part in conserving both ions and water. In response to a low ionic concentration in the blood, the kidney releases renin, an enzyme that indirectly stimulates the release of aldosterone from the adrenal glands.

Urea, the compound that contains most of the body's nitrogen wastes, is eliminated in the urine. Urea is constantly being formed in the body as nitrogen atoms are released from amino acids and certain nucleic acid bases during metabolism. An elevation in the concentration of urea in the blood is an indication of impaired kidney function.

A smaller fraction of nitrogen wastes is eliminated as uric acid. Gout, a disease in which elevated concentrations of uric acid in the blood lead to painful deposits of uric acid salts, can be treated by drugs which reduce blood uric acid levels by increasing urinary uric acid excretion.

Selected Readings

1. "Why the Stomach Does Not Digest Itself," by W. H. Davenport. *Scientific American,* Jan. 1972. Discusses how the stomach usually avoids being damaged by the hydrochloric acid it releases.
2. "Lactose and Lactase," by N. Kretchmer. *Scientific American,* Oct. 1972. Discusses why many adults cannot digest milk.
3. "Recessive Inheritance of Adult-type Lactose Malabsorption," by T. Sahi, M. Isokoski, J. Jussila, K. Launiala, and K. Pyorala. *The Lancet, ii* (1973) 823–825. Evidence from family studies is presented which suggests that the inability to digest lactose is genetically determined.
4. "Aspirin Use in Patients with Major Upper Gastrointestinal Bleeding and Peptic-ulcer Disease," by M. Levy. *The New England Journal of Medicine, 290* (1974), 1158-1162. An investigation of the relationship between aspirin usage and serious gastrointestinal problems.
5. "The Kidney," by H. W. Smith. *Scientific American,* Jan. 1953, Offprint 37. Reviews the anatomy and physiology of the mammalian kidney and its evolutionary history.

6. "Renal Tubular Chloride Transport and the Mode of Action of Some Diuretics," by M. Burg and L. Stoner. *Annual Review of Physiology, 38* (1976), 37–45. Presents recent evidence that supports the view that the active transport of chloride, not sodium, is primarily responsible for the concentration of ions in the kidney's medullary tissue.

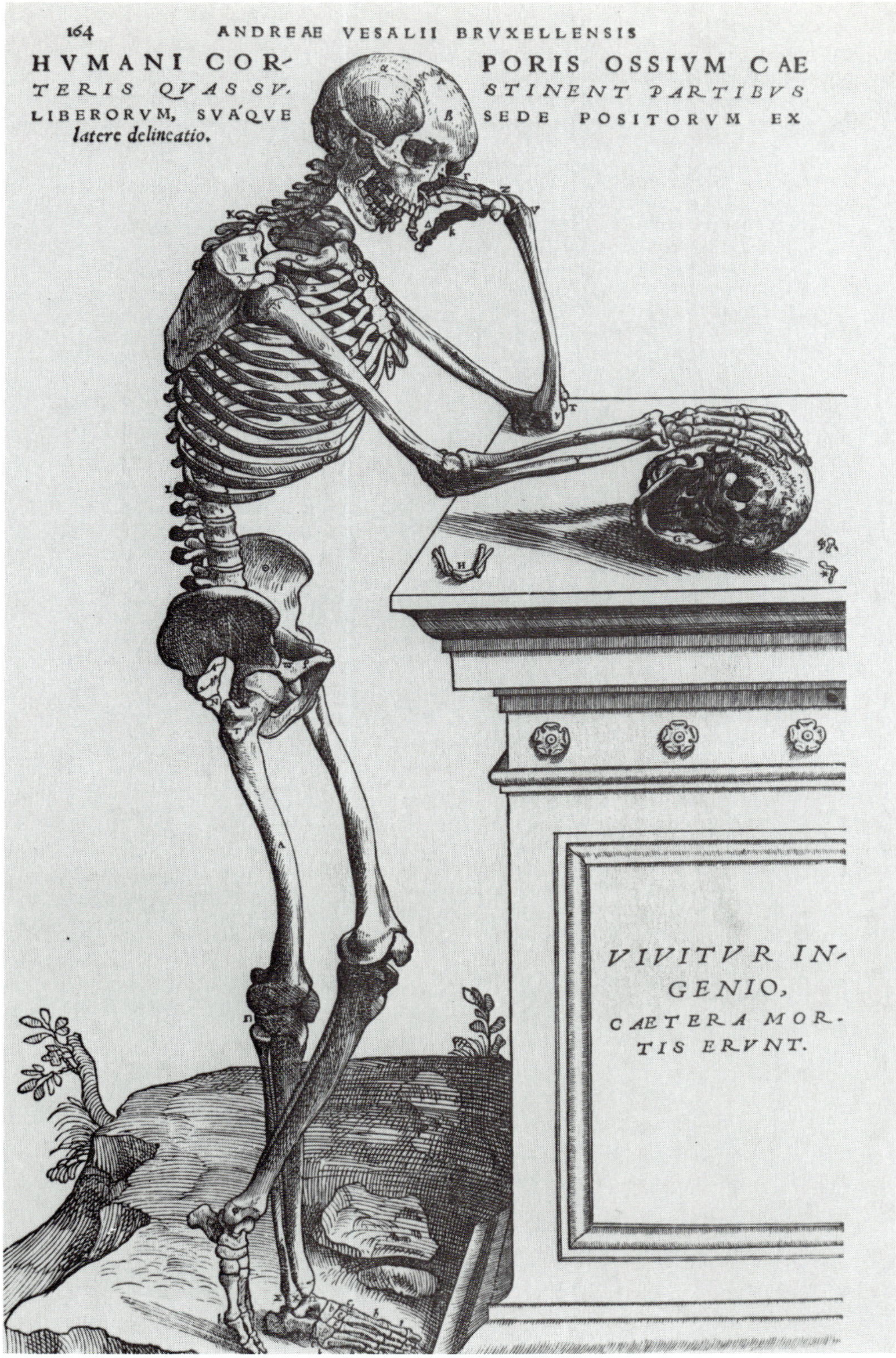

Drawing of the skeleton by Andreas Vesalius, from De humani corporis fabrica, *published in 1543. The inscription on the tomb translates "The spirit lives on, all else is death's portion." (Courtesy the National Library of Medicine, Bethesda, Maryland.)*

CHAPTER

11

Bones, Calcium, and Muscle

In some ways, it is obvious that the muscles and skeletons of our bodies make up a system. Muscles are organs whose function is to shorten themselves, and, because the ends of many muscles are attached to two rigid bones, their contraction can result in the movement of a complicated system of bony levers connected to each other by various joints. The musculoskeletal system provides us not only with a means of locomotion, but with protection for internal organs and support against gravity. Other equally important functions of the musculoskeletal system are not so obvious. Normal muscular contraction depends in part on the maintenance of proper concentrations of calcium in the blood, and the skeleton in turn is a major component of the homeostatic mechanism that keeps the concentration of calcium in the blood within the narrow limits compatible with life. So muscle and bone are closely connected, not only physically by means of ropelike tendons, but also chemically by metabolic interaction.

In this chapter we discuss the structure and function of the human skeleton and its important role in the mechanism of calcium homeostasis. We also discuss the structures that hold the skeleton together—joints—and the organs that move skeletal levers—muscles.

The Human Skeleton

The first accurate and detailed description of the human skeleton to be published was written by Andreas Vesalius in 1543 (frontispiece). Most human skeletons consist of about 206 bones. The number 206 is tentative for two reasons. First, some people may have extra bones, such as an extra pair of ribs near the neck, called cervical ribs (see Figure 11-1). Second, because the bones of the human coccyx, or "tailbone," are fused and indistinct, there is disagreement about their number.

The bones of the body have many shapes and sizes, depending on the mechanical function in which they participate. Some bones, like those of the wrist and ankle, tend to be roughly cuboidal. Others are irregular in shape (such as those of the vertebral column and pelvis) or thin and flat (such as ribs and the bones of the cranial vault). The large bones of the arm and leg, or *long bones,* are the parts of the skeleton with which we are most familiar; by examining one of them we can see most clearly how the architecture of the skeleton is related to gravitational forces encountered by land-dwelling animals.

Bone Architecture

The *femur,* or thigh bone, is the longest and heaviest of human bones. If we split a femur down the middle and examine the cut surface we see that the bony texture is not uniform (Figures 11-2 and 11-3). The middle portion of the long bone, the *shaft* (the part you can see in cross section at the local supermarket as the top of a round steak), is a hollow tube. The bone on the outside of this tube is very compact; hence it is called *compact* bone. In contrast, the bony tissue at each end of the split long bone is much more porous and is referred to as *spongy bone.* (You can see an example of spongy bone at the supermarket in a sliced knuckle

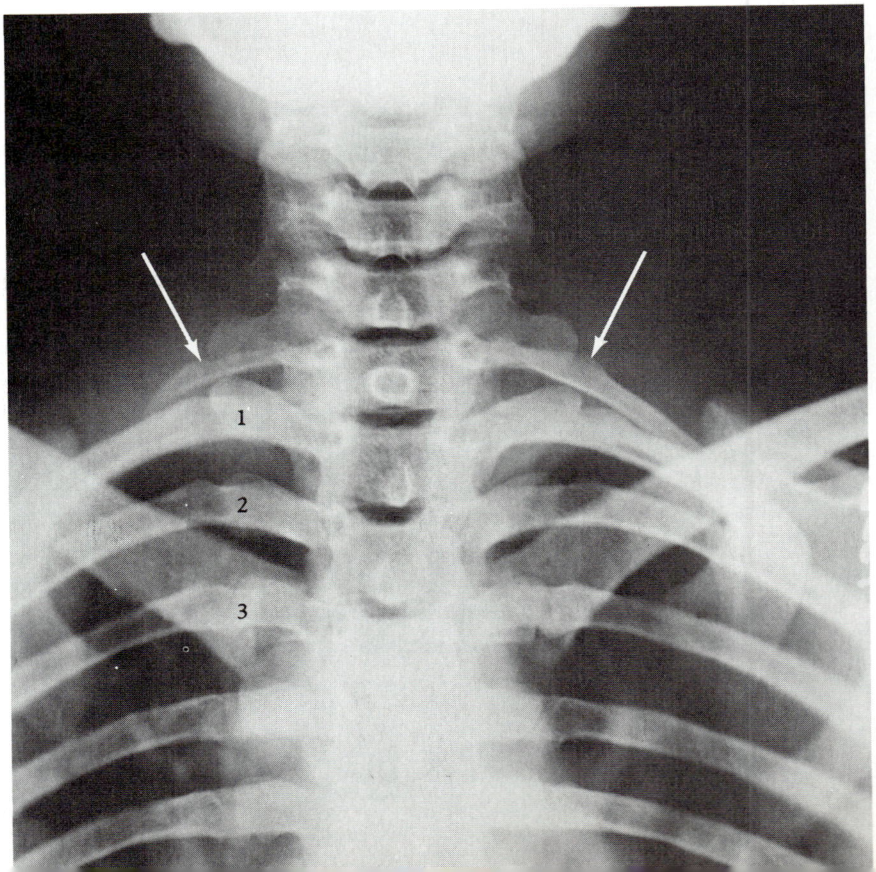

11–1
X ray showing cervical ribs. Cervical ribs are indicated by white arrows. Numbers indicate usual first, second, and third ribs. (Courtesy Lucy F. Squire.)

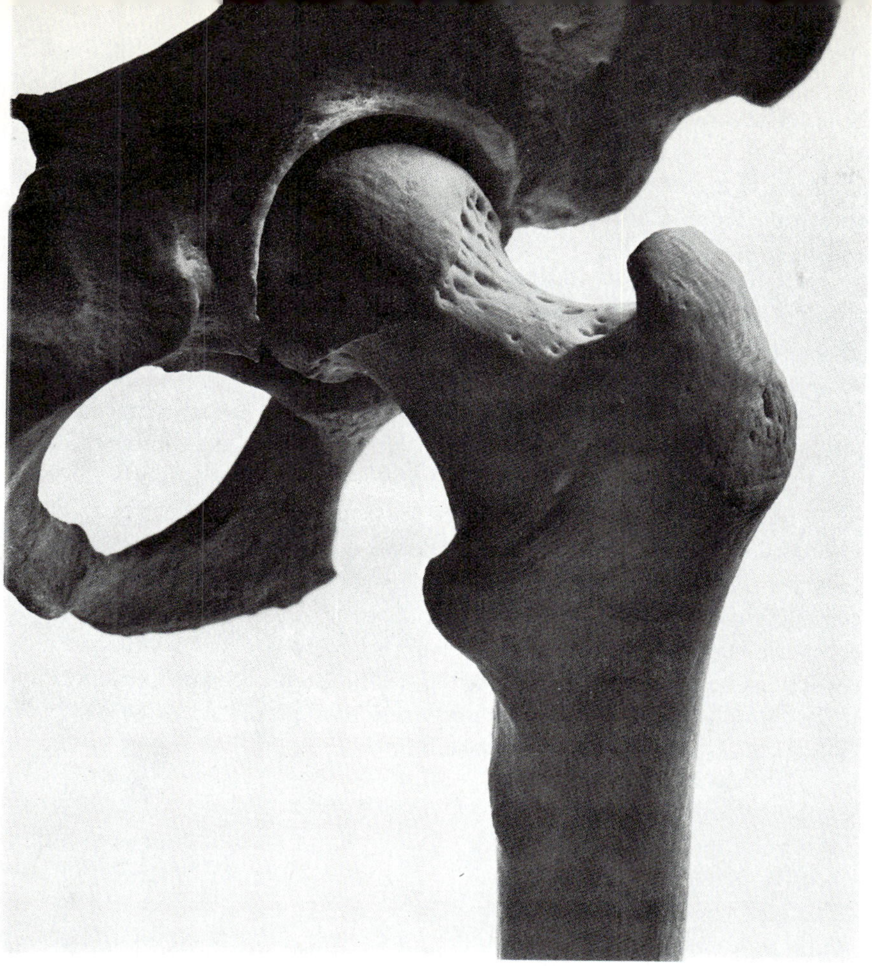

11–2
The hip joint and upper femur. The surface of the neck of the femur is normally pitted. (Courtesy Andreas Feininger.)

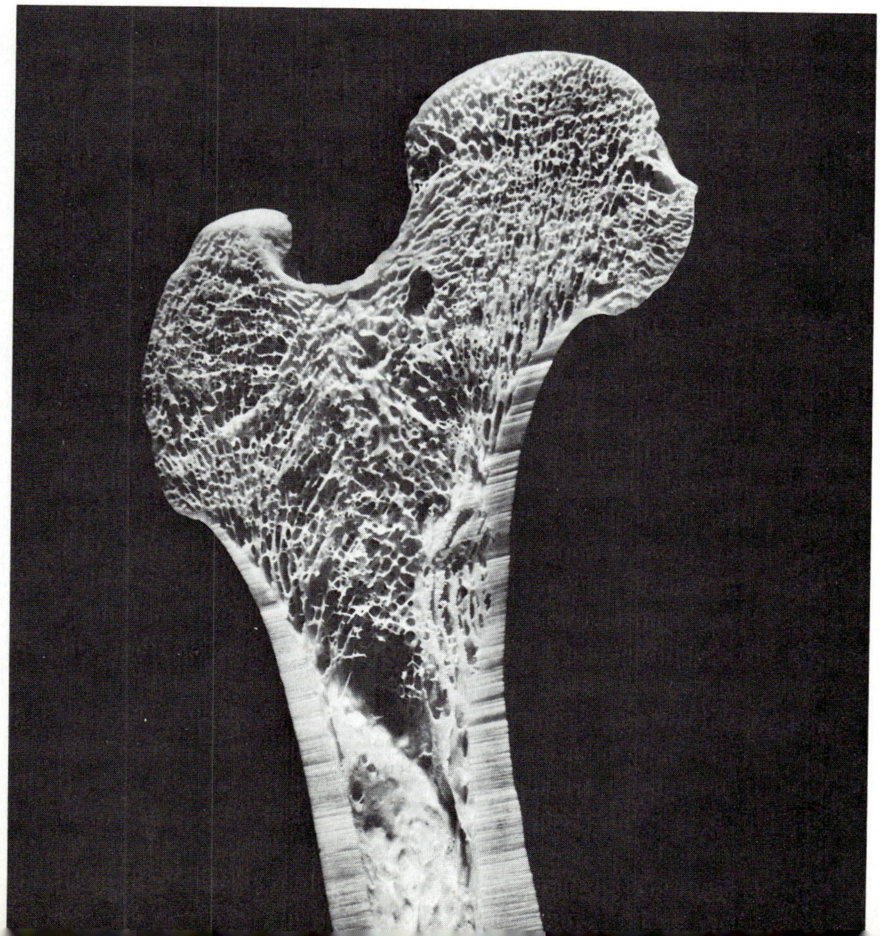

11–3
The upper femur, split down the middle. (Photograph by Gjon Mili.)

bone.) The compact bone of the shaft of the femur is stronger than the spongy bone of the ends; this is a reflection of the fact that the shaft of the femur is subjected to the strongest stresses that are produced by gravity. The shaft is the part of the femur that is broken most frequently.

Bone is an excellent structural material because it has strength to resist forces of stretching and compression almost equally well. But bone derives additional strength from its architecture. The spongy portion at the ends of the femur consists of a fine latticework of bony walls, or *trabeculae* (Figure 11-3). Analysis of the orientation of these short bars and plates of bone reveals that their spatial arrangement is far from random. The trabeculae are arranged in one of two patterns: either they follow what structural engineers call the "lines of force" acting on the entire bone because of the weight it bears, or else they tend to be arranged at right angles to other trabeculae along the lines of force, an arrangement that supports the other trabeculae and adds to the weight-bearing capacity of the bone. The second arrangement is the most efficient one for providing support against the force, or shearing stress, that otherwise would break the bone. As we might expect, the engineers say that if a structure must resist a bending force that comes from all directions, the construction best suited to the task is a strong hollow tube reinforced at the center—exactly the construction of the human thigh bone.

11–4
Atrophy of the leg depicted on limestone stele of the Egyptian Priest Ruma, XVIII Dynasty, about 1365 B.C. *Ruma carries a walking stick. There has probably been paralysis of the leg since early childhood, resulting in failure of bone growth and in atrophy of the leg muscles. This may be an example of poliomyelitis, in which there is destruction of the nerve cells that are needed to work the muscles of the leg. (Courtesy The Ny Carlsberg Glyptotek, Copenhagen.)*

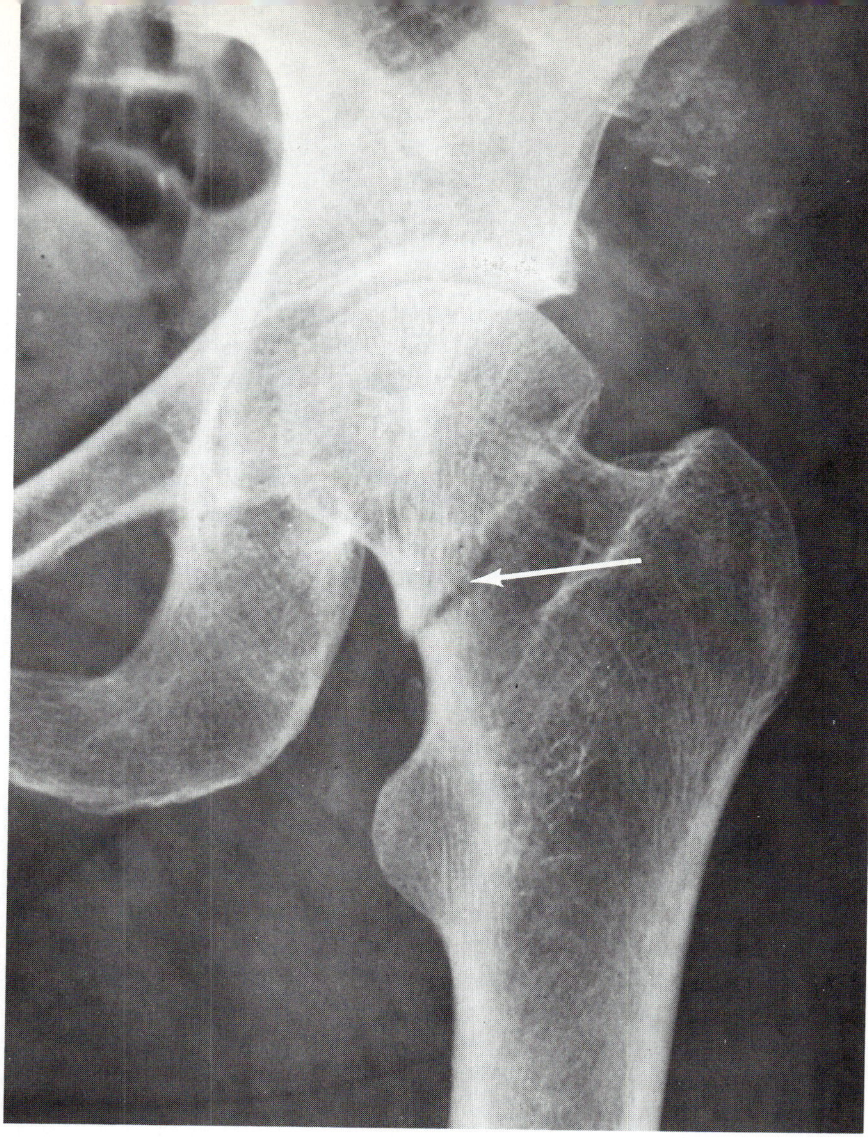

11–5
Fracture of the neck of the femur. Notice the displacement of the bone at the fracture site. (Courtesy Dr. Lucy F. Squire.)

Trabecular patterns in various bones and their relation to physical activity have been studied by X-ray examination over periods of up to ten years. It has been found that when white-collar workers become physically active through manual labor and exercise, the trabecular patterns of their skeletons change to reflect new forces they encounter. Likewise, when laborers and athletes assume sedentary jobs, their skeletons change to reflect the absence of strains or forces to which they previously had been accustomed. Bone responds to a change in physical demand (as exerted by the different use of a particular muscle) by changing its internal architecture and by increasing or decreasing its bulk at the point where greater or lesser force is applied.

Many organs or parts of organs become enlarged in response to increased use. The enlargement of muscle and bone is due to an increase in the size of individual cells (and to the amount of structural material deposited in bone) and is an example of *hypertrophy.* The ability to undergo hypertrophy is genetically determined, but where it occurs depends on the particular stresses the body encounters. Conversely, if the skeleton is not subjected to many physical forces through its tendons, or if particular muscles are not used over a period of time, then their individual cells decrease in size and finally the size of the entire bone or muscle also decreases (Figure 11-4). This wasting away because of disuse is called *atrophy,* and bone and muscle are not the only body tissues to respond to disuse by shrinking in size. The capacity to undergo atrophy also is genetically determined, and where it occurs depends on the forces and stresses the body encounters.

Musculoskeletal atrophy caused by inactivity can be quite extreme. If a long bone is broken, or *fractured* (Figure 11-5), an external cast of supportive material is applied to the injured limb in order to allow the broken ends to grow together again and reestablish normal strength (Figure 11-6). After a cast has been in place for several months and is then removed, usually the muscle mass of the casted limb is obviously smaller than that of the normal extremity that has been active during the period of casting. The muscle and bony tissues rapidly regain their normal bulk with increased use. It has been found that the bones of astronauts also tend to become less dense during long stays in space because the normal effect of gravity on the skeleton and muscles is reduced or absent during low gravity space flight. Regular exercise helps to overcome this condition, but musculoskeletal wasting is still a major physiological problem facing would-be space travelers.

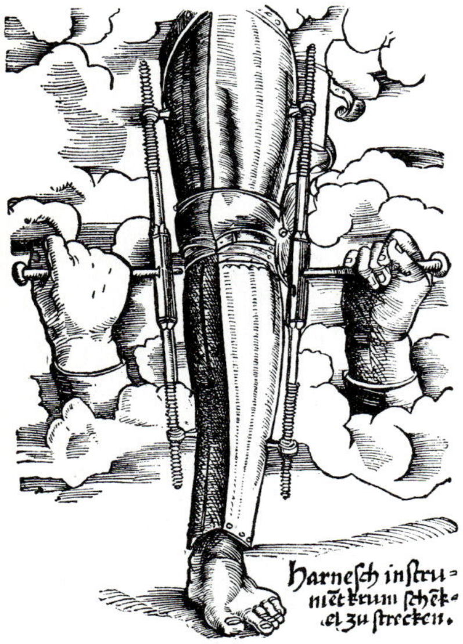

11–6
Sixteenth-century German leg splint. Wars created an increased demand for bonesetting devices. (Courtesy The Bettmann Archive.)

Even if it is normally strong and nearly perfectly engineered, any bone will break if enough physical force is applied to it. A land animal that stands and moves by the use of its legs (either two or four) is limited in its capacity for growth by the laws of physics. The ability of a long bone to bear weight depends on its cross-sectional *area,* but the amount of weight it actually has to hold up depends on the *volume* of the entire animal. For example, if an animal were magnified so that it were four times taller, its limbs would then be capable of supporting 16 times more weight, but it would actually be 64 times heavier. As animals become large, body weight increases much more rapidly than the ability of bone to bear weight, and this simple fact places a definite upper limit to the size of any land animal.

The tallest human giants have been about nine feet tall. The record is held by Robert Wadlow, of Alton, Illinois, who attained a height of 8 feet, 11.1 inches (Figure 11-7). But the abnormal stature of giants like Mr. Wadlow is due to the action of an abnormally high concentration of growth hormone in the body fluids during the active period of skeletal growth. Because growth hormone is secreted by the anterior (front) part of the pituitary gland (see Chapter 6), these people are called *pituitary giants.* The tallest non-pathologic giants (who attain their size having normal concentrations of growth hormone) reach a height of about seven feet nine inches. Proper skeletal growth depends not only on the presence of the proper amount of growth hormone but on nutrition as well. There must be adequate protein, minerals, and Vitamin D in the diet for normal growth of the skeleton to take place.

Bone Growth

Most of the bones of the human skeleton are preceded during development by cartilage of the same shape. This reflects the fact that skeletons of bone evolved from skeletons made entirely of cartilage. During growth the cartilage erodes, and *ossification,* the formation of bony tissue, takes place. The principal chemical differences between cartilage and bone are summed up in Table 11-1. As the table indicates, bone contains almost 200 times as much calcium and about 40 times as much magnesium as does cartilage.

The process of ossification begins in each bone at one or more distinct points, or *ossification centers.* Most long bones have one primary and two

11–7
Robert Wadlow, a pituitary giant, standing with his father. In 1939, when this photograph was taken, Robert was 2.67 meters (8 feet 10¾ inches) tall and wore size 37 shoes. Pituitary giantism results from an oversupply of growth hormone during the active period of skeletal growth. (Courtesy Wide World Photos.)

secondary ossification centers. The primary center is in the shaft, and the secondary centers are toward each end of the bone. The secondary centers are responsible for all long bone elongation and they are arranged as one disc-shaped *epiphysis* at each end (Figure 11-8). Most of the epiphyses do not appear until childhood or adolescence, but growth from primary centers in the region of the shaft begins as early as the second fetal month. Almost all of the newborn's skeleton is formed from primary centers, whereas growth after birth, especially

TABLE 11-1
Composition of bone and cartilage (per kilogram of fresh tissue).

CONSTITUENT	RELATIVE AMOUNT BY WEIGHT	
	IN CARTILAGE	IN BONE
Protein	170	200
Water	740	250
Potassium	12	40
Sodium	170	230
Calcium	48	9000
Magnesium	10	390
Chloride	40	30
Polysaccharide	80	30

Source: Adapted from *A Companion to Medical Studies,* Volume 1. F. A. Davis, 1968.

during the adolescent growth spurt, proceeds from secondary centers under the influence of growth hormone. The epiphyseal discs are arranged in such a way that one flat surface faces the shaft and the other side of the same disc faces the end of the long bone. The disc produces cartilage but no bone on the side away from the shaft. The side of the disc facing the shaft behaves differently. Here the disc does produce bone, using the cartilage already present as its model. So bone is added only on one side of the plate and as new bone is deposited, the plates are pushed farther and farther away from the center of the shaft. Thus the bone grows longer.

The chemical process of replacing cartilage by bone is a complicated one, and normal children have different concentrations of enzymes in their blood for bone-building than do normal adults. Growth is essentially completed by the age of twenty, when the epiphyses stop functioning as growth centers and become permanently inactive; only a telltale line that is revealed on X-ray films demonstrates their former presence (Figure 11-9).

11–8
X ray of the knees of an adolescent showing epiphyseal lines, where bone growth is occurring. (Courtesy Dr. Lucy F. Squire.)

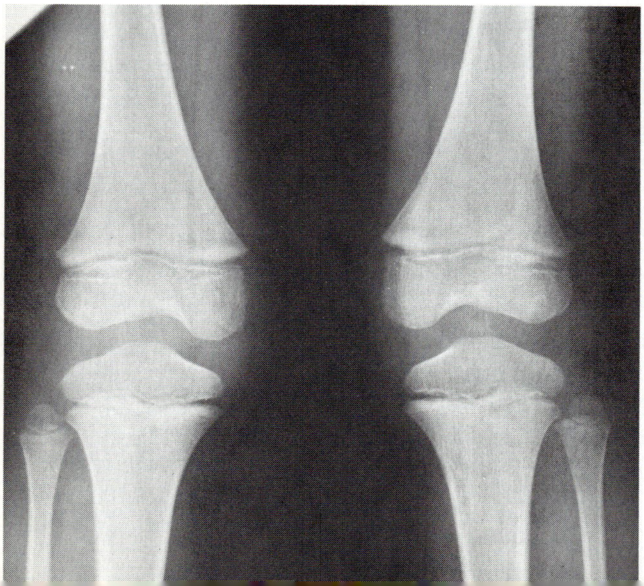

11–9
X ray of the knees of an adult. Growth of the long bones has stopped and only small lines indicate where the epiphyses were. (Courtesy Dr. Lucy F. Squire.)

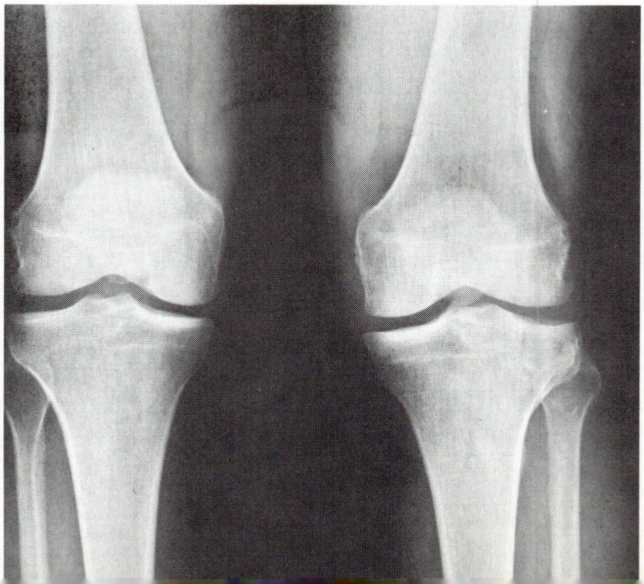

11–10
The development of acromegaly. The features of this man are normal in the left-hand photograph, which was taken when the man was 24 years old. The second photograph was taken when he was 29, at the time of onset; the third, at 37 years; and the fourth, at 42 years, when acromegalic changes were very pronounced. (From The Pituitary Body and Its Disorders, *by Harvey Cushing. J. B. Lippincott, 1912.)*

If for some reason (usually the presence of a tumor in the anterior pituitary gland) growth hormone is secreted in large quantities *after* the epiphyses have become inactive, the result is a condition called *acromegaly.* People who have this disease experience changes in body configuration and body chemistry (Figures 11-10 and 11-11). The hands and feet become broad and generally enlarged. The lower jaw (mandible) may become so massive and protruding that it interferes with chewing. Usually the bones and joints degenerate as well.

Unlike other hormones secreted by the anterior pituitary, growth hormone does not have a specific target organ but rather produces growth of all organs and

11–11
Limestone portrait head of the Egyptian king Akhenaten (Amenhotep IV), from XVIII dynasty, about 1365 B.C. *Medical historians have discussed and argued about Akhenaten's features, and it is difficult to disregard an acromegalic element in his appearance. Pointing to acromegaly is the exceptional length of the lower jaw, and the slight enlargement of the nose and upper jaw. Against acromegaly, Akhenaten's feet and hands were normally proportioned. We are left with an interesting diagnostic dilemma. (Courtesy Ägyptisches Museum, Berlin. Bildarchiv Preussischer Kulturbesitz.)*

tissues (see Chapter 6). It exerts its major influence on skeletal growth before epiphyses become inactive, but it also affects metabolism throughout life.

Once the skeleton has achieved maximum size it continues to be an active organ system. Bone tissue is continually destroyed and resynthesized throughout life. The process of bone deposition and reabsorption (which normally go hand in hand) is called bone *remodeling.* Our skeletons are constantly being remodeled. In general, reabsorption of old, well-established bone is followed by replacement with new bone. The rate of reabsorption and deposition varies widely from bone to bone and in different parts of the same bone. The spongy bone of the femur may be completely replaced by new bone in a few months, whereas the compact bone of the shaft may not turn over completely once in an entire lifetime. To understand how remodeling takes place we need to know some facts about bone as a tissue.

Freshly dried compact bone consists of about 50 percent organic matter by weight and 50 percent inorganic salts. It is said to be composed of organic and mineral *phases.* If a rib is submerged in a fairly strong solution of acid most of the minerals are removed from the tissue, but the organic phase, or *organic matrix,* remains. The bone still looks like a rib but it is flexible enough to be tied in a knot! The flexibility contributed by the organic matrix is a crucial property of bone that enables it to withstand stresses without fracturing. Table 11-2 compares the flexibility of bone with that of cast iron, steel, and mahogany. The cells that not only produce the organic matrix but also deposit minerals in the newly formed matrix are called *osteoblasts.* All bony surfaces are covered by a thin layer of these cells. The major bone-eroding cells are large, multinucleated *osteoclasts,* which for the most part are found in areas where bone is being reabsorbed. Through the controlled interaction of osteoblasts and osteoclasts, new matrix is synthesized and minerals are deposited and reabsorbed.

The Mineral Phase of Bone

The mineral phase of bone is responsible for its amazing hardness and for its capacity to resist compression (see Table 11-2). The shaft of the femur of a young adult is so strong that it can withstand a vertical load of nearly a ton, which is about twice the load associated with the most demanding athletic activity.

TABLE 11-2

Capacity of bone to withstand compression and bending.

MATERIAL	CAPACITY TO WITHSTAND COMPRESSION (KG/MM^2)	FLEXIBILITY (MODULUS OF ELASTICITY)*
Human compact bone	14 to 21	1.6
Cast iron	42 to 100	20.0
Steel	45	30.0
Mahogany	4	1.3

*A *low* value for modulus of elasticity indicates *greater* flexibility and less likelihood to fracture.

Source: Adapted from *A Companion to Medical Studies,* Volume 1. F. A. Davis, 1968.

The major mineral component of bone is hydroxyapatite [$Ca_4(CaOH)(PO_4)_3$], which is composed primarily of calcium and phosphate that have precipitated out of the body fluids in a complex arrangement. Crystals of hydroxyapatite are formed from calcium and phosphate ions in the body when (1) a certain concentration is reached, and (2) there is a nucleus around which crystals can be formed.

Hydroxyapatite can be formed not only by calcium but also by related metals such as lead, radium, or strontium combined with phosphate ions. A radioactive form of the element strontium called strontium 90 (^{90}Sr) is one of the products of nuclear fission—which occurs in the explosion of atomic bombs in war or in peace. ^{90}Sr may be absorbed directly into the blood by breathing contaminated air. It also may make its way into the diet in plants that have extracted and concentrated ^{90}Sr from the air or in animals that have fed on them. Once the radioactive element enters the body it is quickly deposited in bone, where it is stored for variable periods of time. It can cause cells in the surrounding tissue, especially in the surrounding bone marrow, to be damaged by the radiation it emits. Strontium is deposited in bone because its atoms are similar to those of calcium and magnesium, the minerals normally found in large amounts in bone. Because of similarities in their atomic structures, all three of these elements tend to enter into similar chemical reactions. Cells of the bone marrow of people subjected to fallout of ^{90}Sr are very prone to cancer because the skeletons of which they are a part have stored this radioactive element.

Accumulation of large amounts of lead in the molecule of hydroxyapatite also produces serious illness that may be fatal. Automobile exhaust fumes contain considerable amounts of lead because lead is added to gasoline. People living in highly industrialized areas may have a greater-than-normal chance of developing lead poisoning because they are chronically exposed to high concentrations of lead in the environment. The present concentrations of lead in various industrial areas are considered "safe," but the physiological effects of lifelong exposure to significant (and increasing) amounts of pollution-generated environmental lead remain to be determined.

Bone Matrix Formation

The orderly deposition of minerals in bony tissue (mineralization) depends on the presence of normal bone matrix. This is demonstrated by some abnormalities that result from deficiencies of hormones and vitamins that participate in the production of healthy matrix.

Vitamin C (ascorbic acid) is necessary for normal production of *collagen,* which composes much of the body's connective tissue (about 95 percent of the organic matrix of bone is collagen). As you may recall from Chapter 9, a chronic lack of Vitamin C in human diets results in a condition called scurvy. Other body systems are also affected by a lack of Vitamin C, but the main result in the skeleton is the production of a bone matrix that is not mineralizable, because it is not a suitable nucleus for the deposition and growth of hydroxyapatite crystals.

Sex hormones also exert an effect not only on the composition but also on the growth and development of organic bone matrix. Female estrogens and male androgens—both of which are present in the bodies of men and women—promote growth and maturation of matrix, whereas other steroid hormones secreted by the adrenal gland oppose it. Normally these hormones balance matrix formation and destruction. However, the concentrations of androgen and estrogen tend to drop as people age; there are also an overall decrease in the rate of new bone formation and a consequent reduction in the total quantity of bone deposited (Figure 11-12). Older people are more likely to fracture their bones for two reasons. First, there is less mineral present in their bones; and second, their bones are less elastic because of a reduction in the quality of the organic matrix that results from aging.

Vitamin A is essential for normal matrix formation, and bone remodeling ceases in its absence. Unlike most other vitamins (except Vitamin D), Vitamin A is toxic if consumed in excess. Chronic excess of Vitamin A can result not only in brittle bones that do not grow properly but also in enlargement of the liver and spleen, a lowering of the blood count, and the production of coarse, sparse hair, among other effects. People can acquire toxic amounts of Vitamin A in several ways. Most commonly, they can be fed too many vitamins by well-meaning parents. Or Vitamin A toxicity can occur from eating the livers of polar bears and various seals. Eskimos (and even their dogs) avoid eating the livers of these prey animals because they are unpleasant to taste. This is clearly adaptive behavior in that these organs usually contain very high (toxic) concentrations of Vitamin A. The precursors of Vitamin A are substances called carotenes, which are manufactured only by plants. Animals obtain carotenes in their diets and transform them into Vitamin A, which is stored in the liver. In the Eskimos' case, the abundant carotenes from free-floating microscopic plants called diatoms are transformed into Vitamin A. As the food chain leads from diatoms to shrimp to seals and finally to polar bears and to people (see Chapter 4), the Vitamin A reaches toxic concentrations.

In healthy bodies the hormones and vitamins we have discussed (as well as Vitamin D, whose role we will discuss later) interact to produce normal bone matrix that is quickly mineralized as long as proper concentrations of calcium

11–12
Section of normal compact bone (left) and a section showing osteoporosis in an elderly person (right). The reduction in total amount of cortical bone is striking. These sections are taken from the same area of the femur. The outer surface of the bone is at the top. (Courtesy Marshall R. Urist.)

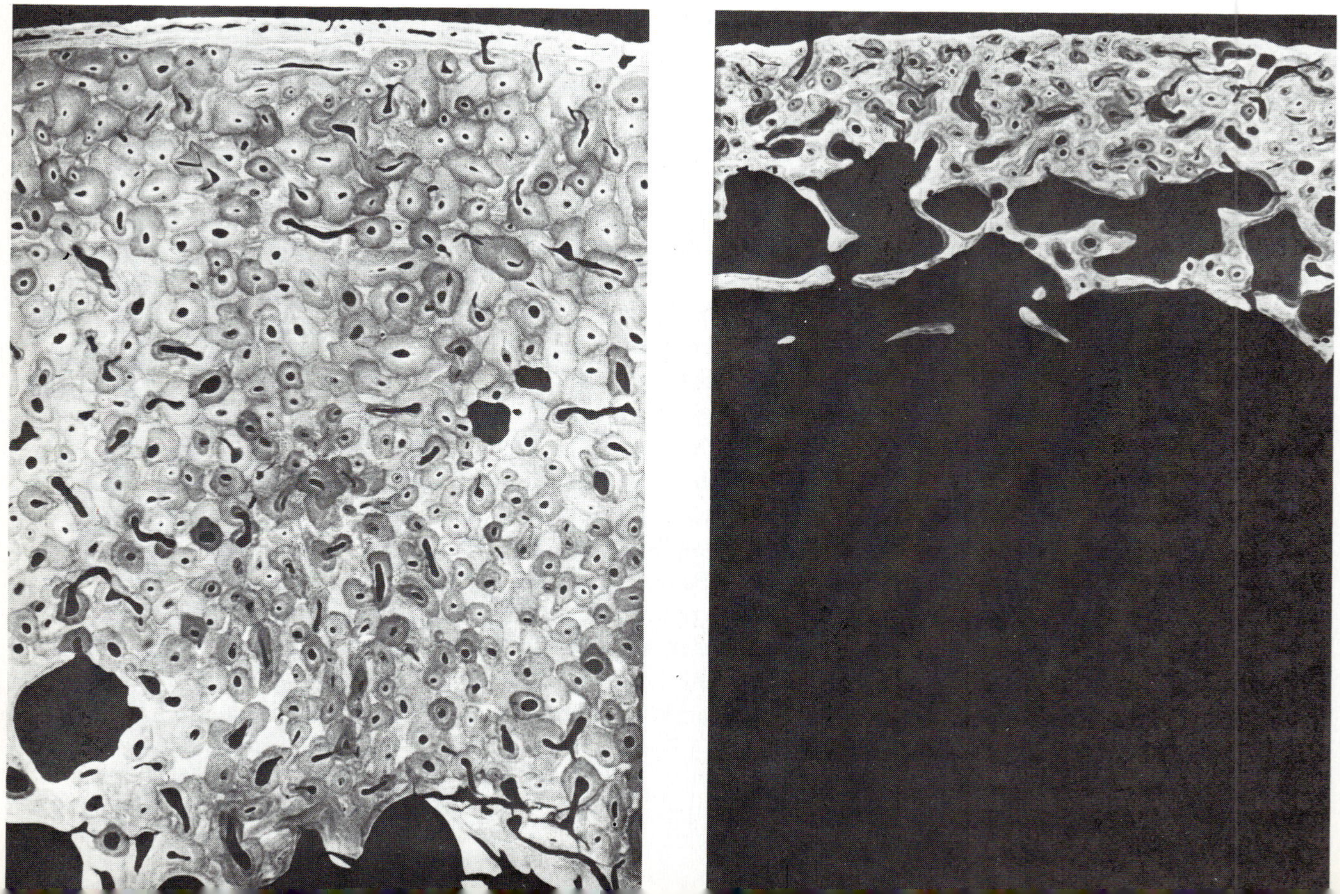

and phosphorus are maintained in the blood. Structural support depends upon proper calcification of the skeleton, but if our bodies are required to make a choice between maintaining structural support or maintaining a constant concentration of calcium ions in the blood, they favor the latter and the structural role of skeletal tissue may be compromised.

Keeping the Concentration of Calcium Ions Constant—Calcium Homeostasis

The close control of the concentration of circulating calcium ions reflects the fundamental importance of calcium throughout the body. Proper calcium concentration is essential not only for mineralization of healthy bone matrix but also for such processes as the coagulation of blood, the maintenance of normal permeability and excitability of cell membranes, and the proper functioning of various enzyme systems. The contraction of muscles and the integrity of various structural proteins are also dependent on proper calcium concentration.

The mechanism of calcium homeostasis is this: The skeleton serves as an enormous reservoir of calcium—99 percent of the body's total calcium content of about 1200 grams is on store there. The body is designed to efficiently absorb and store dietary calcium in the skeleton and to mobilize calcium from it as necessary to maintain normal concentrations of it in the blood (about 10 mg/100 ml). Calcium mobilization from the skeleton is brought about by *parathyroid hormone* (PTH), which induces bone reabsorption.

The *parathyroids* are four tiny (about pea size) endocrine glands located two each at the top and bottom of the backside of the thyroid gland (in the neck). There is a simple negative feedback loop (see Chapter 6) between calcium in the blood and PTH, so that low concentrations stimulate and high concentrations inhibit production of the hormone.

The parathyroid mechanism would respond to a sudden drop in calcium in the following ways in order to adjust the concentration back to normal (Table 11-3). The low concentration of calcium stimulates the parathyroid glands,

TABLE 11-3
Calcium homeostasis.

STIMULUS	HORMONE RESPONSE	EFFECT OF HORMONE	RESULT
↓ Ca^{++} in blood	↑ PTH released from parathyroids	↑ Ca^{++} (more reabsorbed from bone storehouses) ↑ Ca^{++} (more intestinal absorption) ↑ Ca^{++} (less excreted by kidney into urine)	Concentration of calcium in blood increased to normal
↑ Ca^{++} in blood	↓ PTH released from parathyroids	↓ Ca^{++} (less reabsorbed from bone) ↓ Ca^{++} (less absorbed from intestine) ↓ Ca^{++} (more excreted by kidney into urine)	Concentration of calcium in blood reduced to normal

which respond by producing more PTH. This has three effects. First, the increased PTH causes increased bone reabsorption by stimulating osteoclasts, so that the calcium concentration in the blood begins to rise. Second, should there be any calcium in the intestine, it is more efficiently absorbed in the presence of PTH. And third, PTH acts on the kidney so that less calcium is lost in the urine.

Diseases of the parathyroid glands are rare but serious. *Hyperparathyroidism* is accompanied by the overproduction of PTH. This is usually caused by the presence of a hormone-producing tumor in one of the parathyroid glands. The gland stops responding to the negative feedback signal of calcium concentration, so homeostasis is lost. Under the influence of excess PTH the concentration of calcium in the blood is chronically elevated (Table 11-4). Because the calcium is in high concentrations, it tends to be deposited in abnormal places—kidney stones occur commonly in people who have hyperparathyroidism. As we would expect, they are usually composed of calcium salts. Meanwhile, the skeleton loses its calcium reserves because of excess PTH production. X-ray studies may reveal irregular areas where calcium has been reabsorbed and the bones may fracture easily. Treatment consists of removing the tumor that is producing the excess PTH.

Although overproduction of PTH is rare, the underproduction of the hormone, which is *hypoparathyroidism,* is rarer still. Lack of human PTH most often occurs because of inadvertent removal of the parathyroid glands during surgery on the thyroid gland, but a few spontaneous instances of it have been reported. People who have hypoparathyroidism have very low concentrations of calcium in their blood. They consequently tend to experience a general increase in excitability of their nervous and muscular tissue. They often experience muscle cramps, spasms of the muscles in the hands and feet, and they may undergo personality changes. PTH is available for replacement, but is expensive and must be administered intravenously, so treatment usually consists of large daily doses of calcium and Vitamin D by mouth.

Vitamin D in Calcium Homeostasis

Vitamin D affects calcium homeostasis by helping to provide calcium in two ways. First, the efficient absorption of dietary calcium from the intestine is largely dependent on the presence of Vitamin D. Second, Vitamin D is required for the proper reabsorption of bone under the influence of PTH. The physiology of Vitamin D is discussed further in Chapters 9 and 19.

TABLE 11-4
Lack of calcium homeostasis in hyperparathyroidism and hypoparathyroidism.

	HORMONE ALTERATION	EFFECT OF HORMONE	RESULT
Hyperparathyroidism	More PTH released from parathyroids	Ca^{++} concentration increased by all three mechanisms shown in Table 11-3	Elevated blood Ca^{++}, with deposition of calcium crystals in abnormal places such as kidney and cartilage
Hypoparathyroidism	Less PTH released from parathyroids	Ca^{++} concentration decreased by all three mechanisms shown in Table 11-3	Reduced blood Ca^{++}, resulting in muscle cramps, spasms, and personality changes

There are several forms of Vitamin D that are closely related chemically. We can obtain small amounts of the preformed vitamins in the diet—fish livers and fortified milk products contain them. However most of our D vitamins are supplied in the diet as D protovitamins, which are widely distributed in both plant and animal tissue. When they are subjected to ultraviolet light they become active D vitamins and function fully. This activation normally occurs in our skins when we are exposed to the ultraviolet component of sunlight; hence the name "sunshine" vitamins. *Rickets,* the disease resulting from chronic lack of D vitamins in the diet, is known to be more prevalent among children who are not exposed to sunlight than among those who are. In fact, rickets may have been the first significant human disease caused by air pollution. The first descriptions of the appearances of the disease in England date from about 1650, a time when the introduction of soft coal as fuel for fires of the Industrial Revolution resulted in a haze of coal smoke that filtered out the healthful ultraviolet rays of sunlight. As the Industrial Revolution spread throughout Europe, so did its smoke, and so did rickets. This disease results in relative demineralization of the skeleton, and the bones may bow, even painfully, under the weight of the affected person.

Vitamin D (like Vitamin A) can be toxic if present in very high concentration. Very large quantities of Vitamin D mobilize calcium from the entire skeleton and the result is much the same as when there is high calcium due to PTH excess. Body tissues that do not normally do so become mineralized, and the skeleton may take on a rather moth-eaten appearance reminiscent of that caused by hyperparathyroidism.

Vitamin D ensures the efficient absorption of calcium, but even when there is an adequate amount of Vitamin D in the diet the amount of calcium absorbed from food varies markedly from person to person. In part, this has to do with the composition of individual meals because certain lipids and some other normal dietary components decrease the amount of calcium absorbed. If the dietary intake of calcium is low, then it tends to be more efficiently absorbed. Also, normal children absorb proportionately more calcium than do normal adults.

Vitamin D also acts to provide calcium by allowing it to be reabsorbed from bone. PTH, even in large quantities, cannot bring about the reabsorption of bone in the absence of Vitamin D. So the absorptive and reabsorptive aspects of calcium homeostasis are closely connected with the sunshine vitamins.

Calcitonin

In concluding our discussion of calcium homeostasis we must briefly mention the role of the hormone *calcitonin.* This hormone is released by the thyroid gland, whose physiology is further discussed in Chapter 6. Compared to most other hormones, calcitonin is produced in extremely small quantities. It is released from the thyroid gland in response to the presence of an abnormally high concentration of calcium. Calcitonin decreases the concentration of calcium in body fluids by increasing the rate at which calcium is deposited in bone. Calcitonin thus has the opposite effect of PTH on bone, and it brings about its effects much more quickly—a few minutes for calcitonin as compared to a few hours for PTH.

The role of calcitonin in normal physiology is not clear, but the hormone probably functions to offset the effects of sudden increases in the concentration of calcium that may occur following the ingestion of a meal that is rich in calcium. Calcitonin may thus be relatively more important to infants and

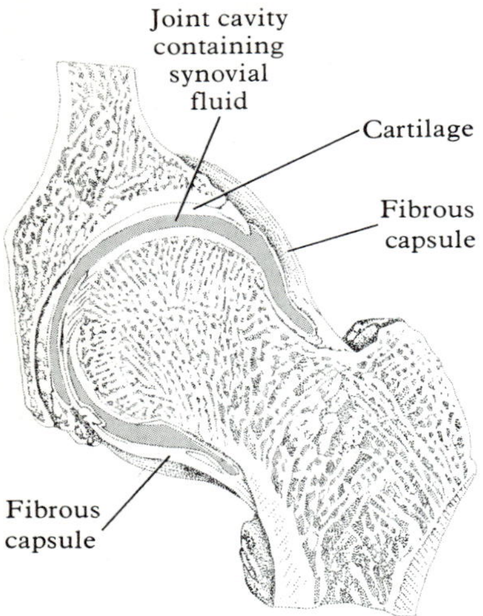

11–13
Cross-section of the hip joint, showing the location of the articular cartilage.

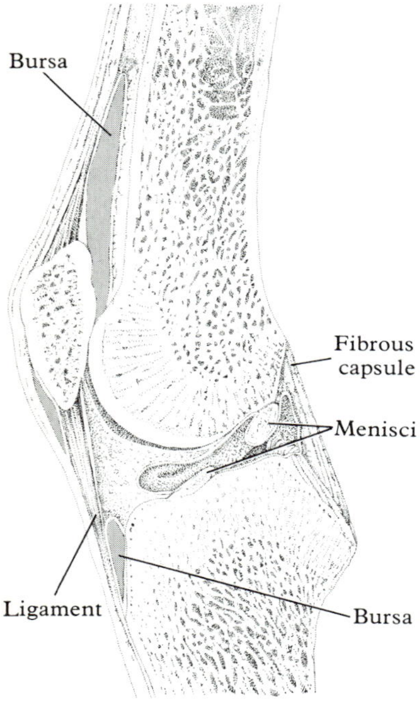

11–14
Cross-section through the knee joint, showing the two menisci.

children than to adults. The smaller bodies (and smaller skeletons) of infants and children make them more apt to experience the adverse effects of an elevated concentration of calcium following a calcium-rich meal.

From a consideration of the role of calcitonin in calcium homeostasis we now turn our attention to how the individual bones of the human skeleton are fastened together.

Joints

The bones of the skeleton are connected by structures called *joints.* Some joints, like those connecting the bones of the skull, are immovable, and they help to provide protection for underlying organs by reinforcing the points of contact between adjacent bones. But most joints allow some movement to occur between the bones they connect, and how much and what kinds of movement are possible depend on the structure of the joints themselves.

In joints that move freely, there is a *joint cavity* between the ends of the bones that are connected and *articular cartilages* (which normally are not calcified) cover each bony end. (Figure 11-13). A *joint capsule,* composed of strong connective tissue, encloses the joint like a collar. The innermost layer of the capsule lines the joint cavity and produces a thick lubricating liquid called *synovial fluid.* Both the articular cartilages and the synovial fluid help to reduce the friction that would otherwise be produced by the grinding of bone on bone as movement occurs within the joint.

Joints are structurally reinforced by *ligaments,* which are bands of dense connective tissue whose function is to withstand the mechanical stresses to which individual joints are subjected. Some large joints, like that of the knee, not only are surrounded by a tough capsule of connective tissue and by bands of ligaments, but also contain built-up cushions of cartilage within the joint cavity itself. Each cushion, or *meniscus,* lies within an outpocketing of the synovial membrane and projects into the joint cavity, where it helps to reduce wear on the articular cartilages. (Figure 11-14). The knee joint contains two such cushions, and when it is subjected to excessive use (as by some athletes), one or both of the menisci may degenerate and even extend out of the joint cavity. There they may become trapped between the articular cartilages, a painful condition known as locked knee in which the knee is capable of little or no normal function.

Motion around any freely movable joint is bound to produce friction, not only in the joint itself, but in the surrounding tissues as well. At various places in the body where friction results from bones or muscles rubbing against one another, there are saclike structures containing a viscous fluid. Each of these sacs is called a *bursa;* there are thirteen around the knee joint alone. For reasons that are largely unknown, deposits of calcium sometimes accumulate in the connective tissues of the bursal sacs. This causes an inflammatory response, known as *bursitis,* the symptoms of which are pain, tenderness, and limited motion in the area of the joint whose surrounding bursal sacs are inflamed. Chronic pain in the shoulder, in particular, is often due to this condition.

The term *arthritis* refers to many disorders of the joints in which pain, swelling, and stiffness are caused by some abnormality of the joint itself. There are *many* kinds of arthritis, but all kinds include inflammation or degeneration of the bone or cartilage of the joint itself (Figure 11-15). The term *rheumatism* is

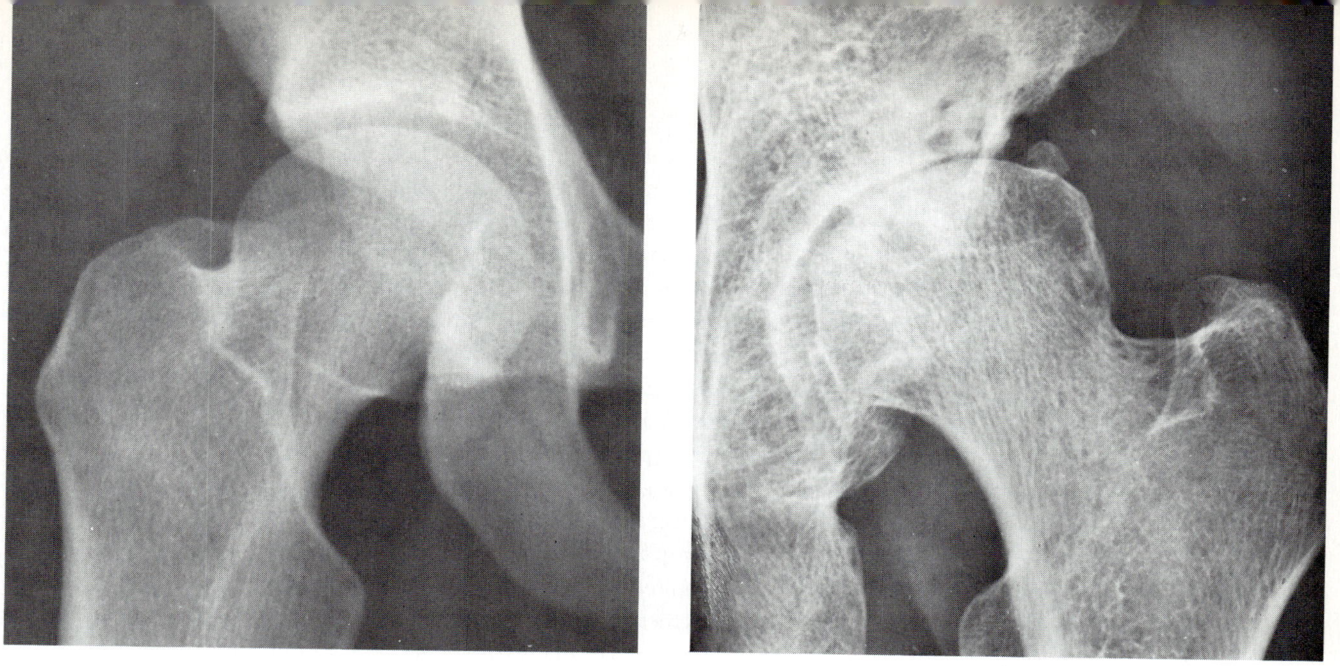

11–15
X rays of a normal hip joint (left) and a hip joint affected by arthritis (right). The space within the arthritic joint is decreased. The white densities on both sides of the arthritic joint indicate the growth of new bone in response to the arthritic process. (Courtesy Dr. Lucy F. Squire.)

popularly used to describe pain and stiffness of any joint from any cause. Although this term does not specify a cause of the symptoms, the fact that we all know what it means indicates that painful joints are a common human ailment.

Joints, Levers, and Muscles

All forms of voluntary movement—whether locomotion (moving the entire body from place to place) or the movement of specific parts of the body relative to one another while the rest of the body remains stationary—are accomplished primarily by bending and straightening the appendages or the trunk. This bending and straightening can best be described by considering the skeleton as a system of levers.

A lever is a device for transmitting force. In the context of body movement, it is a rigid bone that can move around a fixed point (or fulcrum)—the joint—when the mechanical force of muscular contraction is applied to it. For example, the force required to hold a heavy object, such as a bowling ball, in the palm of the hand is transmitted from the biceps muscle to the forearm by means of a lever that can move around the elbow joint. Almost all forms of voluntary movement can be attributed to the use of skeletal levers. Muscles are attached to bone by cablelike structures, or *tendons,* that transmit the mechanical energy for moving skeletal levers. Muscles that play a part in producing voluntary movements are known as *skeletal muscles,* and they vary widely in size and shape, depending on the functions they serve. For example, a single muscle (the gastrocnemius) forms the bulk of the calf of the leg and helps to propel the body for walking, running, and leaping as it contracts across the knee and ankle joints. In contrast, the tiny stapedius muscle of the middle ear (the body's smallest muscle) is only two to three millimeters long and weighs about one-tenth of a gram. Contraction of the stapedius produces a dampening effect on the movement of the tiny bones of the middle ear. This contraction occurs when a loud noise (like that from an air hammer) reaches the eardrum, and the resulting decreased movement of the three tiny bones of the middle ear protects the sensitive inner ear from injury by loud noises.

The coordinated movement of skeletal levers is not brought about by single, isolated muscles contracting across a joint, but rather by the interaction of groups of muscles working together. Virtually every muscle acting on a joint is matched by another muscle that has an opposite action. The muscles that act directly to bring about a desired movement are called *prime movers,* and those that have opposite action are called *antagonists.* For example, the prime mover for flexion (bending) of the elbow joint is the biceps muscle, and the antagonist that causes extension (straightening) is the triceps muscle, which attaches to the elbow on the side opposite from where the biceps attaches. When prime movers contract, their antagonists help to restrain them, and thus aid in the production of smooth, coordinated movement.

In more complicated kinds of activities, additional groups of muscles may assist the prime movers, especially if contraction of the prime movers can result in movement across several joints at once. For example, in making a fist, the prime movers are the flexors of the fingers and thumb. But these muscles also produce flexion of the wrist, which is an undesirable effect if one simply wants to make a fist. To counteract the undesirable effects of the prime movers, other muscles located in the hand itself contract, and their contraction prevents flexion of the wrist. The muscles that prevent the undesirable effects of contraction of the prime movers are called *synergists.* As the prime movers and synergists contract in the process of making a fist, other muscles of the arm and shoulder also come into play. These muscles are called *fixation muscles,* and they function to hold the rest of the arm steady while the prime movers and synergists form the hand into a fist. The incredible dexterity of the hands and the varied movements of which they are capable are the result of the interaction of many groups of muscles from the shoulder to the fingertips, all working together to produce the desired movement. As we have discussed in Chapter 2, the muscular control that people have over their hands (and in particular, their thumbs), has been extremely important in human evolution.

But not all skeletal muscles help to move bony levers. For example, the muscles of facial expression, which allow us to smile and otherwise directly communicate our emotions to others, are not attached to bones, but rather attach directly to the skin of the face. Other skeletal muscles may form circular bands called *sphincters* (nonskeletal muscles often form sphincters, too,), which open and close natural orifices like the anus or the opening of the urinary bladder. But all skeletal muscles, whether or not they are attached to bony levers, produce movement by contracting. We now turn our attention to the process of muscular contraction and to its biological basis in muscle proteins.

Muscle Contraction

Muscles are made up of long, thin, multinucleated cells that are specialized for the production of specific protein molecules. Within the muscle cell, these protein molecules are arranged in a very orderly fashion so that they can *move* relative to one another and produce reversible tension in the entire muscle as they do so. Muscle contraction thus has its molecular basis in the orderly movement of protein molecules within individual muscle cells.

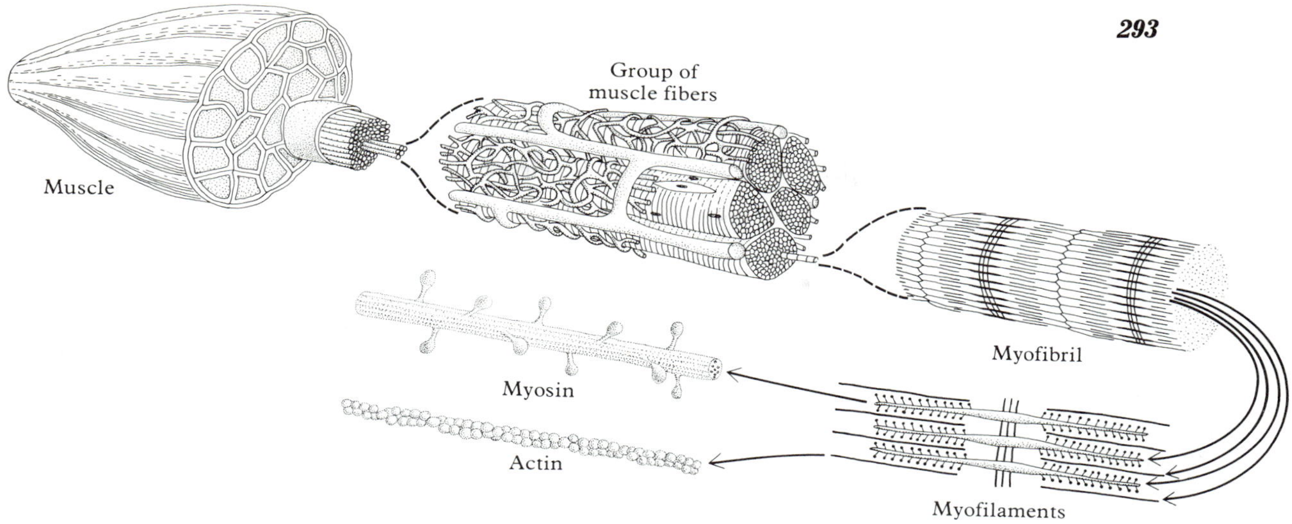

11–16
The organization of skeletal muscle.

If we cut a muscle in two across and examine the cut surface, we see that the organ consists of bundles of *muscle fibers* bound together by connective tissue. Each one of these muscle fibers is actually a single muscle cell that may be up to 30 centimeters long in some muscles and whose many nuclei are situated just under the cell membrane near the periphery of the fiber. Within the muscle fibers are bundles of still smaller threadlike structures called *myofibrils,* and these in turn are made up of still smaller threads called *myofilaments.* This arrangement of smaller filaments within larger ones is shown in Figure 11-16. Notice that the myofilaments are of two kinds, thick and thin, and that the thin filaments are about half the diameter of the thick ones.

When seen through the light microscope, skeletal muscle fibers appear to have crossbands, or striations, running across the width of the fibers. These striations result from the orderly arrangement of thick and thin myofilaments within the muscle cell.

The thick and thin filaments are actually protein molecules, and the electron microscope reveals that they are arranged into functional units or *sarcomeres,* as shown in Figures 11-17 and 11-18. The thin filaments, composed of the protein *actin,* are arranged so that they alternate with the thick filaments, which are made up of the protein *myosin.* The thin filaments are attached at one end to a structure called the *Z* line, and an individual sarcomere consists of the thick and thin filaments between two *Z* lines. The darker bands in striated muscle (as seen through the light microscope) are areas where the thick and thin filaments overlap, and the lighter bands are made up of thin filaments alone (the *Z* line is too small to be seen through the light microscope).

When a muscle contracts, the *length of the sarcomere changes* as the thin filaments *slide* towards each other (see Figure 11-17). The movement of the thin filaments towards the center of the sarcomere during contraction results from the successive making and breaking of cross-connections between the thick filaments of myosin and the thin filaments of actin in such a way that the thin filaments are *pulled* towards the center of the sarcomere. Because a muscle is composed of many sarcomeres connected in tandem, the entire muscle shortens in length as the thin filaments are pulled towards the center of each sarcomere. When the cross-connections between the thick and thin filaments are broken, the thin filaments slide back to their original uncontracted position once again.

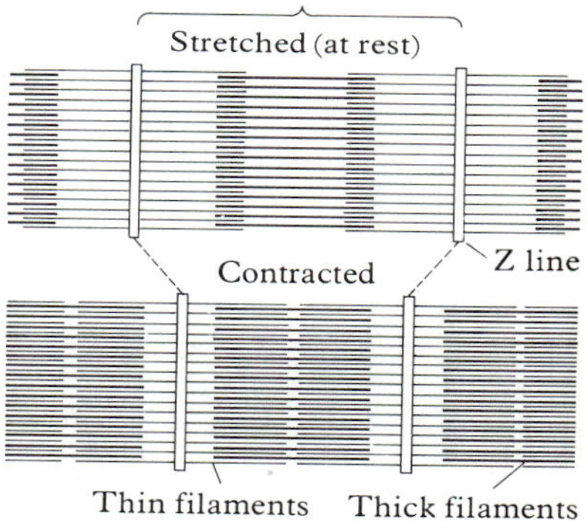

11–17
Diagram showing one sarcomere (the unit of muscle between two Z lines) and parts of the two adjacent ones, at about the same scale as the electron micrograph in Figure 11–18. The top diagram shows the muscle fiber in a relaxed state; in the bottom diagram the muscle is contracted. (From "The Cooperative Action of Muscle Proteins," by John M. Murray and Annemarie Weber. Copyright © 1974 by Scientific American, Inc. All rights reserved.)

The energy required to move the thin filaments towards the center of the sarcomere during contraction is supplied by the energy-rich molecule, ATP (see Chapter 8). In order for ATP to be utilized in the making and breaking of cross-connections between the thick and thin filaments, calcium ions must be available to activate the filaments before contraction can occur. Calcium ions are stored in a system of tiny tubules that surrounds the sarcomeres. They are released from these tubules in response to the nerve impulses that initiate muscular contraction. (How nerve impulses initiate muscular contraction is discussed in Chapter 6.) Shortly after calcium ions are released into the area of the sarcomere, they are pumped back into the tubules again. The muscle then relaxes because actin and myosin cannot form cross-connections in the absence of calcium ions. This reinforces the importance of maintaining a proper concentration of calcium ions in the body fluids. Bone and muscle are thus connected not only by tendons, but by metabolism because, as we have discussed earlier in this chapter, the skeleton is a major component of the homeostatic mechanism for keeping the concentration of calcium ions in the body fluids relatively constant.

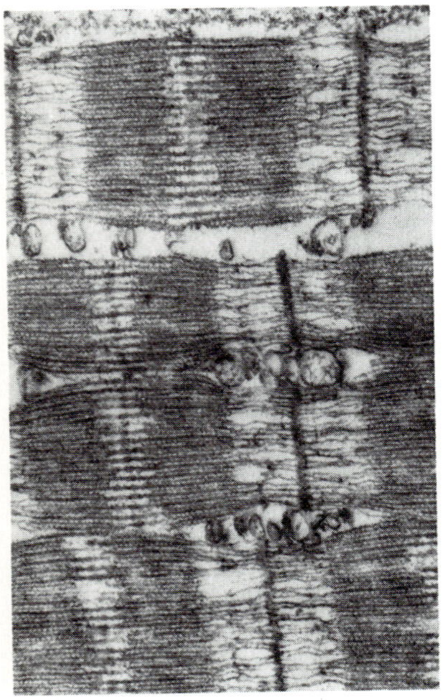

11–18
Electron micrograph of muscle fiber from skeletal muscle (×25,000). The specimen is a very thin longitudinal section. The fiber is an assembly of thin filaments (the finest horizontal elements) that extend in two directions from disklike protein structures called Z lines (dark vertical lines). Between the thin filaments, centered between Z lines, are the thick filaments. (Courtesy Dr. H. E. Huxley of the Medical Research Council Laboratory of Molecular Biology in Cambridge, England.)

Nonskeletal Muscle

In addition to the skeletal muscle we have just described, there are two other kinds of muscle tissue in the human body, whose structure and physiology differ from that of skeletal muscle in several important ways. The first of these, *cardiac muscle,* is found only in the heart. Cardiac muscle is striated and somewhat similar to skeletal muscle in appearance, but it differs in its response to electrical stimulation and in the way in which its cells are connected.

The remaining type of muscle tissue is called *smooth* or *nonstriated muscle,* because it lacks the characteristic striations of skeletal muscle that can be observed under the light microscope.

Smooth muscle cells differ markedly from those of skeletal muscle. In general, they have only one nucleus and are spindle shaped. Their cytoplasm contains many thin filaments that closely resemble the actin of striated muscle and many thick filaments of myosin. Most of the actin filaments of smooth muscle cells have a rather random orientation, and this is probably related to the fact that smooth muscle is quite different from striated muscle in the way in which it contracts.

Compared with striated muscle, smooth muscle produces a slow contraction that is maintained longer. (However, all skeletal muscles do not contract at the same rate.) Smooth muscle is found primarily in the walls of hollow internal structures like the stomach, intestines, blood vessels, uterus, and bladder, where sustained contractions are important in the regulation of the internal environment. Unlike those of skeletal muscle, small clusters of smooth muscle cells may undergo rhythmic contractions even if they are not stimulated to do so by nerve impulses. But nerve impulses can enhance or depress both the rate and force of those contractions. How the interaction of the smooth muscle cells and the nerve impulses reaching them helps to regulate the internal environment is discussed in Chapter 6.

Summary

The human skeleton is not only a precisely engineered scaffold on which we hang our other organs and systems, but a dynamic participant in the maintenance of the internal environment. Bone architecture reflects the physical stresses the body encounters, and bone growth takes place at ossification centers.

Bone tissue consists of organic and mineral phases and it is continually being destroyed and resynthesized throughout life. In order to function properly, the organic phase of bone depends on the presence of adequate amounts of Vitamins C, A, and D, as well as on various hormones.

Calcium homeostasis depends on the huge reservoir of calcium in the skeleton, on Vitamin D, on calcitonin, and on parathyroid hormone, which affects not only the skeleton but the intestine and the kidneys too.

Individual bones are connected by joints, which allow for different kinds of movements. The contraction of skeletal muscles provides the force to move bony levers, and coordinated body movements usually depend on the interaction of many muscle groups.

Muscle contraction is accomplished by the orderly intracellular movement of the protein molecules actin and myosin, which compose the filaments, of which muscle is made. Muscle contraction depends on the presence of calcium ions and on the making and breaking of cross-connections between actin and myosin molecules.

Suggested Readings

1. "Bone," by Franklin C. McLean. *Scientific American,* Feb. 1955, Offprint 1064. A brief review of the structure and function of bony tissue.
2. "Electrical Effects in Bone," by C. Andrew Bassett. *Scientific American,* Oct. 1965, Offprint 1021. Discusses how the mechanical deformation of bones can generate a small electric current.
3. "The Parathyroid Hormone," by Howard Rasmussen. *Scientific American,* April 1961, Offprint 86. How PTH acts on bones, kidneys and the intestinal tract to regulate the concentration of calcium in body fluids.
4. "Calcitonin," by Howard Rasmussen and Maurice M. Pechet. *Scientific American,* Oct. 1970, Offprint 1200. How this recently discovered hormone secreted by the thyroid gland helps to regulate the concentration of calcium by inhibiting the breakdown of bone.
5. "The Mechanism of Muscular Contraction," by H. E. Huxley. *Scientific American,* Dec. 1965, Offprint 1026. Discusses how muscle contraction depends on the movement of thick and thin intracellular filaments made of protein.
6. "The Sources of Muscular Energy," by Rodolfo Margaria. *Scientific American,* March 1972, Offprint 1244. How the energy required to bring about the contraction of muscles depends immediately on ATP and ultimately on what one eats.
7. "The Protein Switch of Muscle Contraction," by Carolyn Cohen. *Scientific American,* Nov. 1975, Offprint 1329. Discusses how calcium ions "turn on" the contraction of muscle by triggering changes in the shapes of certain protein molecules.

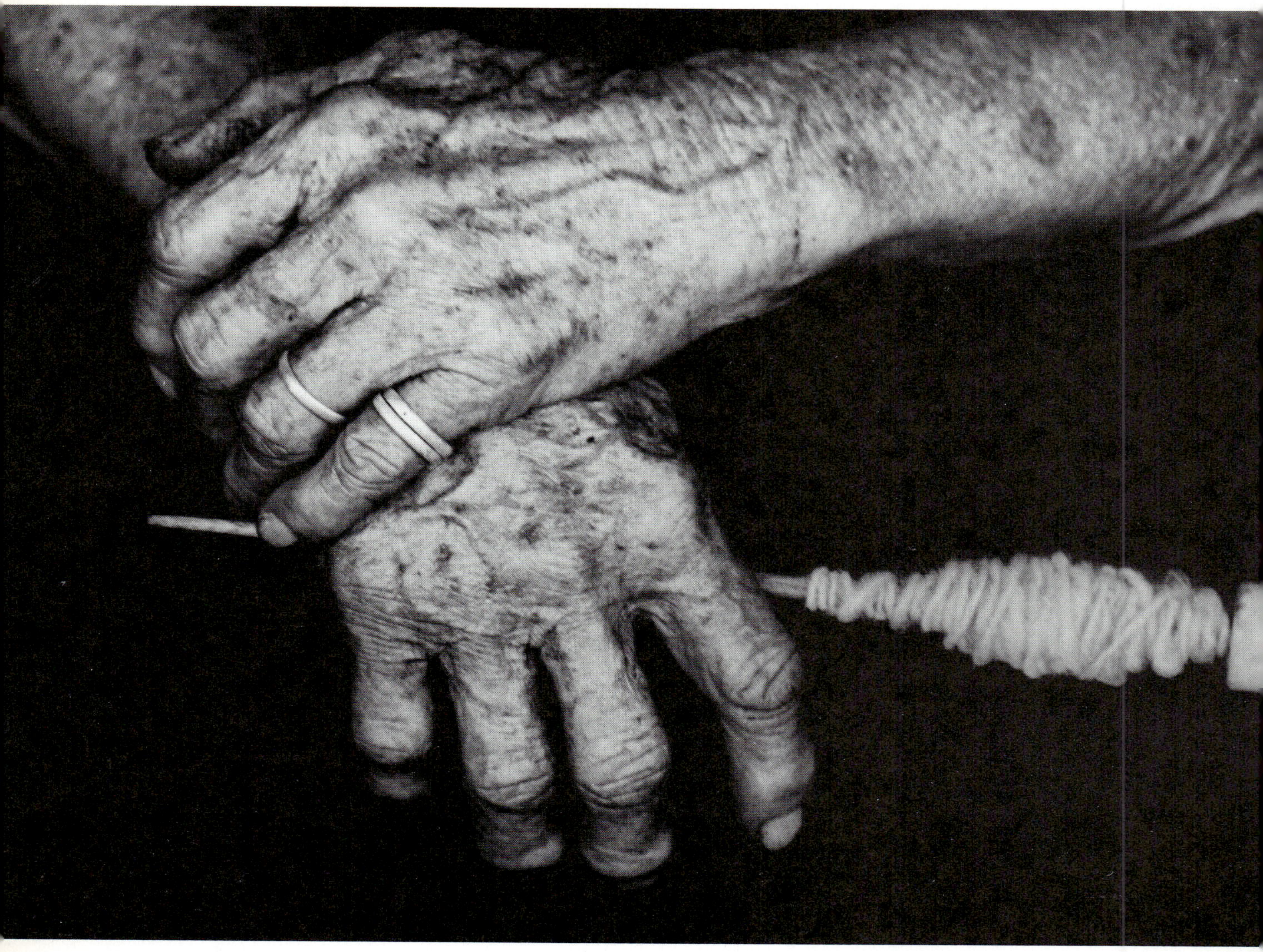

The hands of an elderly working woman from Ecuador's Vilcabamba region. Their blotchiness reflects variations in activity of aging melanocytes in the epidermis, and the wrinkling is the result of changes in elastin and collagen fibers of the dermis brought about by aging. (Photograph by John Launois. Courtesy Black Star Publishing Company.)

CHAPTER 12

The Skin

The skin is the largest human organ, and it is also the most thinly spread. Skin is 7 percent of the body by weight and its thickness ranges from 0.6 centimeters over the back to only 0.05 centimeters over the eardrum and eyelid. Skin covers a greater surface area than does any other organ of the body. In fact, because it is not found all in one grand lump, like the liver or spleen, its status as an organ has tended to be overlooked until recently. But an organ it is, possessing both epithelial and connective tissue elements that together carry out highly specialized functions.

In many respects, the skin is second only to the brain in determining our human characteristics. It gives us our body odors, our rather odd hair distribution, and the highly individualized surface markings and facial expressions that enable us to identify each other. It also provides an expression of emotion through responses such as blushing and paling.

As the outermost layer of the human body, the skin occupies the buffer zone between the environment on the outside and the trillions of living cells on the inside. Because of its location it is often the first part of the body to receive news of changing environmental conditions.

The skin is exquisitely sensitive to touch, the most widespread of the senses, and it has other functions as well. It helps to regulate body temperature, it provides nutrition (in milk-giving), and it protects the body (as a barrier against microorganisms, a storehouse of fat for emergencies, and a provider of a mechanism for coping with harmful effects of the sun's rays). But in order to sort out and understand these many functions, we must first examine the anatomy of the skin in some detail.

Like all organs, the skin is composed of both epithelial and connective tissues. On the outside, covering the body, is the epithelium, or *epidermis,* and inside just underneath the epidermis is the connective tissue that supplies the epithelium with nourishment, the *dermis* (Figure 12-1). The epidermis and dermis are separated from each other by a basement membrane.

12–1

Zones of the skin. The underlying dermis, the thickest part of the skin, is supported by a fat-rich subcutaneous stratum. Intermingled with the cells of the dermis are fine blood vessels, tactile and other nerves, the smooth muscles that raise the hair when contracted, and a variety of specialized glands. Above the dermis are the twin layers of the epidermis: a lower layer of living cells capped by a horny layer of dead cells filled with the fibrous protein keratin. Melanocyles, the pigmented cells that produce the granules responsible for varying skin colors, lie at the base of the epidermis. (After "The Skin," by William Montagna. Copyright © 1965 by Scientific American, Inc. All rights reserved.)

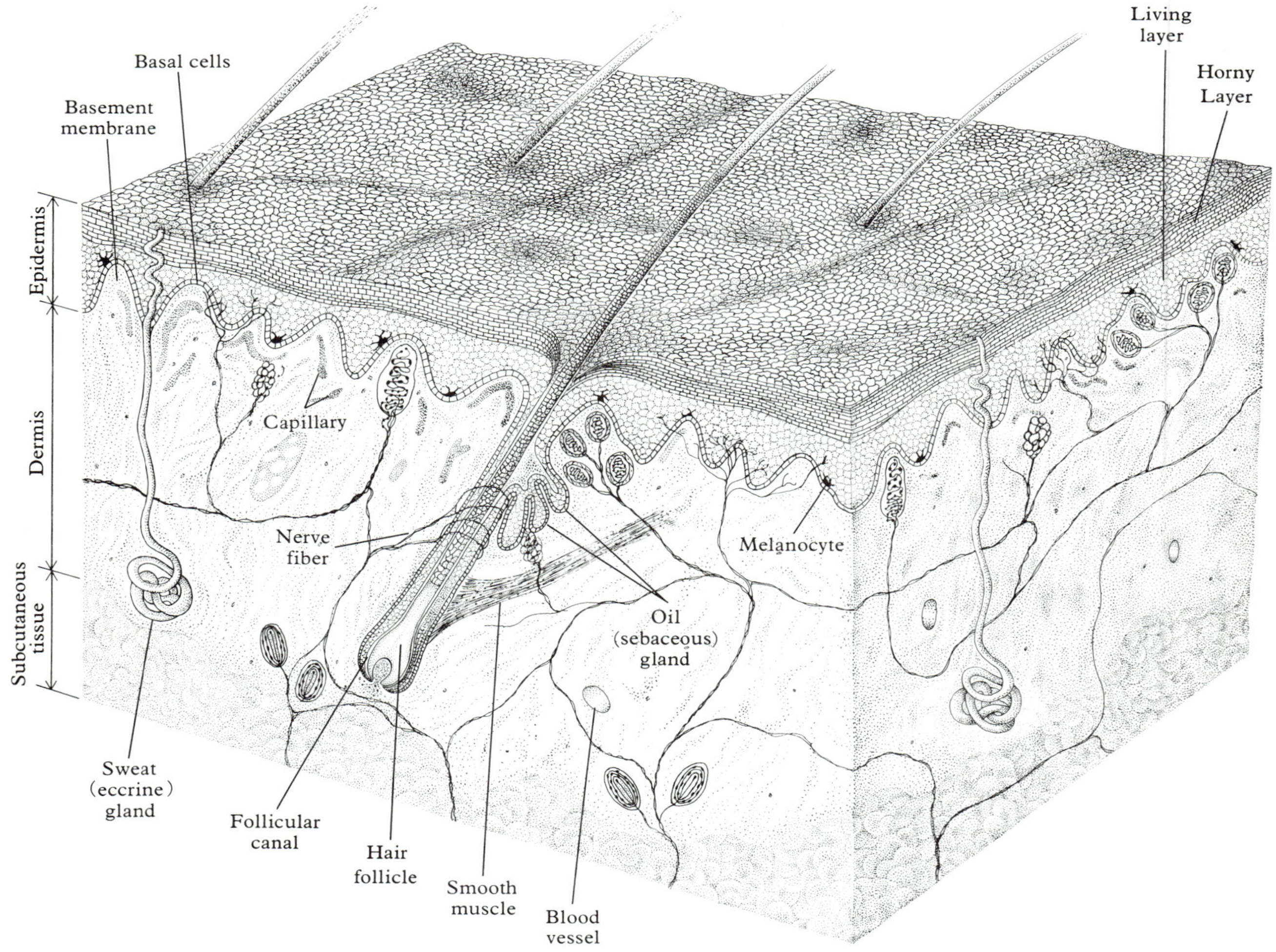

Epidermis

The epidermis is composed of both living and dead cells, but the outermost layer of epidermis, which is in contact with the air, consists entirely of dead cells. The dead cells of the skin's surface are continuously being rubbed and flaked off into the environment (Figures 12-2 and 12-9). You can see this process by vigorously rubbing your arm in a sunny room; your cells will be dispersed into the air like fine particles of dust. The skin of the scalp sometimes flakes in larger particles known as dandruff. Replacements for these dead cells are provided by the innermost layer of epithelial cells, the *basal cells,* which are just above the basement membrane. These cells divide regularly by mitosis (Figure 12-1). Thus in the epidermis there is cell production in the deepest layer, cell death in the middle layer, and flaking off at the surface. It takes about four weeks for a cell produced in the basal layer to be discarded, although this time is reduced if the skin is stimulated to repair itself by an injury.

It is essential for the epidermis to maintain a steady state in which the rate of new cell production equals the rate of cell death. If production lags behind, the epidermis degenerates. If the production greatly exceeds the death, the result is cancer. However, there are two special conditions in which the production of new cells normally exceeds the death of old cells for a *limited* and *brief* period of time: during wound healing, when the normal number of cells must be regained by the injured area, and during stimulation of the epidermis by constant rubbing, which leads to callous formation. In the second condition, the epidermis becomes thicker by an increase in the total number of cells rather than by a thickening of the cells themselves.

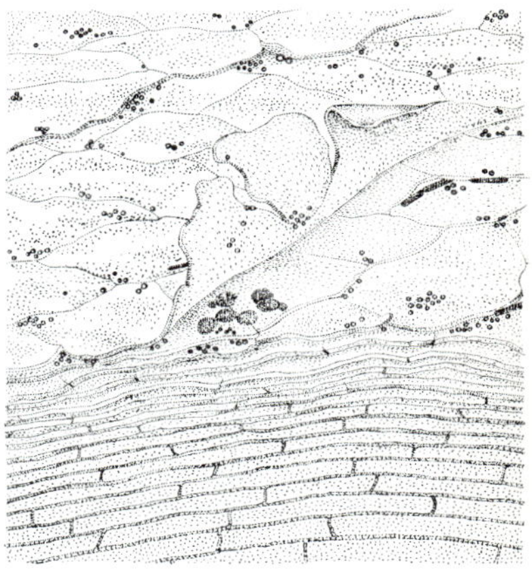

12–2
Outermost layer of epidermis, the horny layer, consists of many flat scales. Each scale is the remnant of an epidermal cell and is composed mostly of the protein keratin. Lodged on, under and around the scales are bacterial and fungal cells, shown as spheres and rods. (From "Life on the Human Skin," by Mary J. Marples. Copyright © 1968 by Scientific American, Inc. All rights reserved.)

A major biochemical activity of epidermal cells is synthesis of the protein *keratin.* By the time the cells die, they consist mostly of keratin. These keratin molecules adhere closely to one another and form a tough pliable membrane over the surface of the body. Keratin imparts to the skin its ability to keep out water and microorganisms and to keep in the moisture of the living tissue underneath. Indeed, the skin is quite a remarkable barrier, not only against the penetration of water and microorganisms, but against the penetration of many chemical agents as well. However, there are some substances that can penetrate readily, such as various pesticides, nickel, and the molecules that are responsible for poison oak and poison ivy allergies.

The epidermis forms a kind of indented mat, penetrating deeply down into the dermis to completely surround each hair follicle, and it also lines the glands that open directly onto the skin's surface. Hence the epidermis of the skin is actually an unbroken covering layer, even though it appears to be punctured here and there by glands and hair follicles.

The epidermis that lines the hair follicles and glands is a particularly valuable resource, because it can serve as a reservoir to regenerate the surface epithelium after an injury. After a severe burn, for example, the only remaining epidermal cells may be those that survived because they were far below the surface (around a hair follicle). Thus regenerating epidermis in a severely burned area may appear first in patches, each patch representing an outgrowth from a single hair follicle.

The epidermis also gives animals a remarkable array of visible surface structures, including horns, hooves, and claws (Figure 12-3), as well as hair and nails. Let us now discuss hair and nails as examples of structures derived from epidermis.

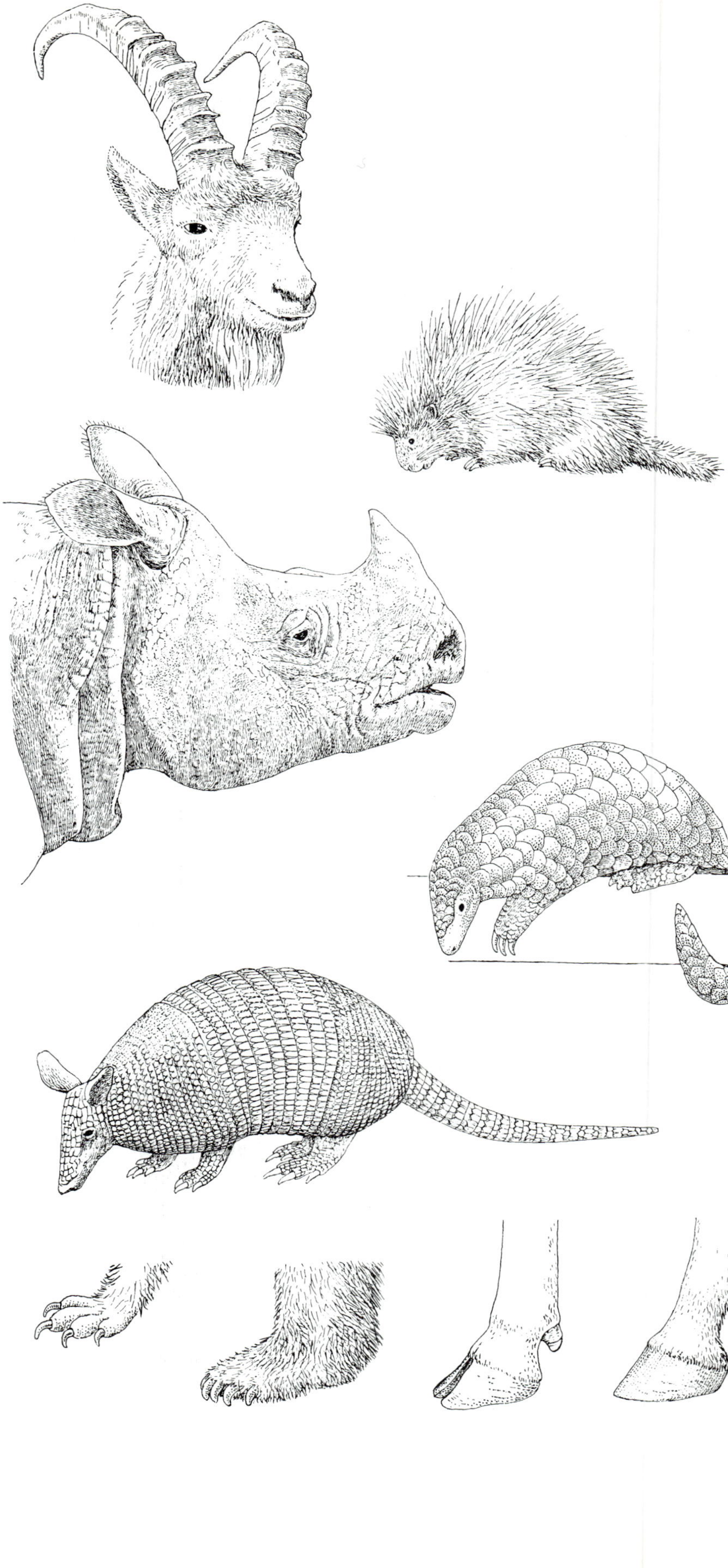

12–3
The versatile epidermis gives rise to a wide variety of superficial structures; among the mammals these include, from top to bottom, the bold horns of the ibex, the quills of the porcupine, the hairy horn of the rhinoceros, the scales of the pangolin and the armadillo, and various claws and hooves. Like all of these, human hair and nails are special skin structures. (From "The Skin" by William Montagna. Copyright © 1965 by Scientific American, Inc. All rights reserved.)

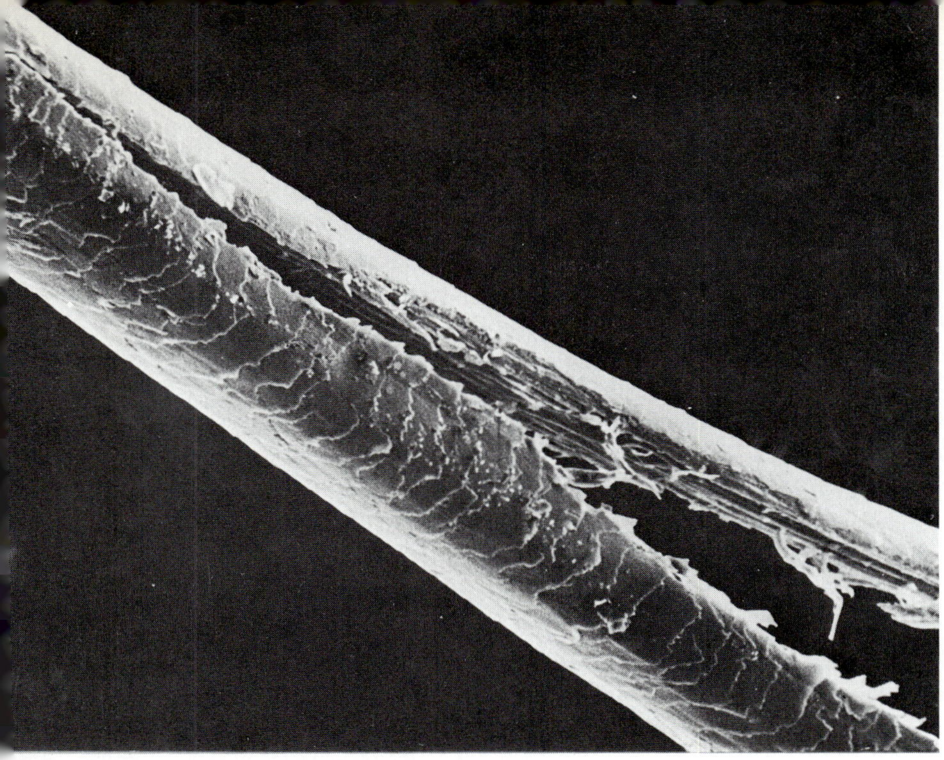

12–4
The shaft of a human hair (×1200) as seen through a scanning electron microscope. This hair has a split end. (From Lennart Nilsson, Behold Man. *Little, Brown and Company, 1974. Courtesy Albert Bonniers Förlag, Stockholm.)*

Hair

Each hair grows from a group of rapidly dividing epidermal cells found at the bottom of the tube of epidermis surrounding the hair itself. Hair is like the rest of the skin's surface in that it is composed of compacted dead cells that are made almost entirely of keratin (Figure 12-4). The way in which the keratin molecules are superimposed on each other determines whether hair is curly or straight. Heat applied to the hair in an attempt to make it curl or uncurl simply forces the keratin molecules into different alignments with one another.

Each hair tends to grow to a specific length; after spending some time at that length, it is shed, and another hair starts in the same follicle. These repeating cycles of growing, resting, shedding, and growing again account for the fact that hair eventually returns to the same length after it has been shaved or singed. The lifespan of individual hairs varies according to where they are on the body: hair of eyebrows, arms, and legs lives for about four months before it is shed, whereas scalp hair has a lifespan of four years. The longest hairs are those on the human scalp; they commonly reach a length of 60 to 75 centimeters (24 to 30 inches) before being shed. Scalp hair can be much longer, but this is rare. Usually there is a limit of about 75 centimeters, beyond which even it will not grow.

The days of our being hairy creatures, like our fellow primates, seem to have gone forever. But our lack of hairiness is *not* due to a lack of hairs: we have as many hairs on our bodies as do far hairier primates. But the hairs over much of our bodies have become small, relatively colorless, and unobtrusive.

It appears that human beings once had a hair pattern more like that of most other primates. For example, most primates have hairy foreheads, whereas ours is naked. But the embryonic human is well endowed with forehead hair, and up until the fifth month the forehead of the fetus is covered with hairs that are as luxuriant as those found on the rest of the scalp. Before birth these hairs gradually disappear, except in very rare instances.

Around each hair follicle, surrounding it in the dermis, is a net of nerve fibers, that makes each hair a sensitive receiver of sensory information. This should become apparent if you stroke your hair without touching the surface of the epidermis itself. Some hair follicles are more richly endowed with nerves than others; in humans those of the face and the anogenital area are especially sensitive. So besides making us look different, the absence of luxurious hair over much of our bodies probably means that we cannot experience the full joys of grooming and preening known by our primate relatives.

Nails

The nails provide protection for our crucially important fingers and toes, and fingernails also serve as tools for certain activities, such as picking and scraping. Nails are formed by a process almost identical to that by which hair and epidermis are formed. A deep layer of dividing epidermal cells around the nail root gives rise to cells that die after they have synthesized keratin (Figure 12-5). But for nails, the dead cells become compacted to form the translucent protective plates that are attached to the ends of our fingers and toes.

The layers of dividing cells that give rise to the nail are not of uniform thickness; hence the nail has tiny stripes running from base to tip that represent the irregularities in thickness. These nail cells cannot be regenerated from surrounding epithelial cells, so if they are damaged—as frequently happens during an injury to the nail—the new nail usually has visible imperfections.

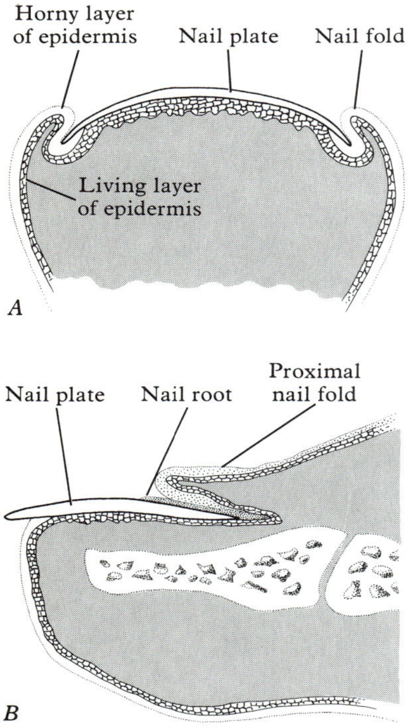

12–5
Diagram of the nail as seen in cross-Section from the front (A) and from the side (B). The nail itself, which consists of the nail plate and nail root, grows out from the epidermis surrounding the area of the nail root. The nail is quite adherent to the underlying epidermal cells, which apparently move toward the tip of the finger along with the nail itself but do not become keratinized.

Because nails are translucent, they are a window through which we can view the color of the blood. If you press on the tip of your nail, you will notice that the blood is squeezed out of the tissue under the nail. The *lunula* or half-moon, a crescent-shaped whitish area that in some people can be seen only in the thumbs, lies in the area of the nail root and is caused in part by the opacity of the nail root.

Skin Pigmentation

The cells that produce the main pigment of the skin, *melanin*, are scattered among the cells of the epidermis. These cells, the *melanocytes* (Figure 12-6), are responsible for the differences in pigmentation that characterize each human race. The melanocytes do not originate within the epidermis, but rather migrate into it from ectoderm in the region of the spinal cord during fetal life. A pigmented mole is an area of skin that has more than its share of melanocytes. Birth marks are areas in which the skin is either lighter than normal (hypopigmentation) or darker than normal (hyperpigmentation). In general, the normal number of melanocytes is present in such areas, and the differences in pigmentation reflect differences in the amount of melanin they produce. Similarly, freckles result from a localized increase in melanocyte synthesis rather than an increase in their numbers. As people get older the activity of their melanocytes varies considerably from place to place and the skin takes on a patchy quality that reflects these variations (frontispiece).

The differences in amount of pigmentation exhibited by different races are also due to differences in amount of melanin synthesized rather than differences in melanocyte number. All human races have approximately the same number of melanocytes in the skin, but the amount of melanin produced by each melanocyte is much greater in people who have heavily pigmented skin than in people whose skin is lightly pigmented. Many different molecules are called collectively "melanin pigments." All are composed of proteins linked to

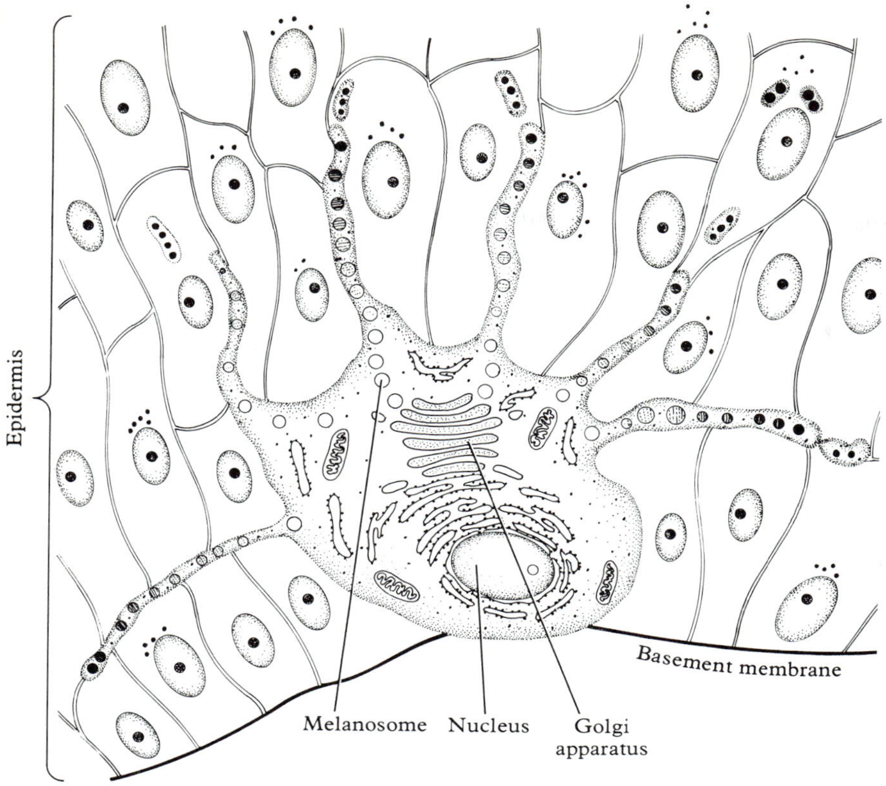

12–6
Melanocyte and basal cells of the epidermis. The precursors of melanin are synthesized by the ribosomes of the melanocyte's rough ER and released into its sacs. The synthesis of melanin inside the sacs proceeds as the sacs become melanosomes and move out to the tips of the melanocyte's cell processes. The tips of the cell processes enter neighboring cells and are pinched off. Most melanosomes are deposited in the region of the cytoplasm between the nuclei of the basal cells and the surface of the skin.

derivatives of the amino acid tyrosine. Differences in the composition and light-absorbing qualities of these molecules probably also account in part for some racial and individual differences in skin color. Other molecules that influence the color of the skin are in the dermis: the *carotenes,* which impart a yellowish color to the skin, and *hemoglobin,* which imparts a reddish color.

Although the numbers of melanocytes do not differ significantly from race to race, different numbers are found in different regions of the body. For example, there are approximately twice as many melanocytes per square millimeter on the forehead as on the arm, leg, or back. The forehead of a light pink European has twice as many melanocytes per square millimeter as the arm of a black African.

Pigmentation and the Sun

As we have mentioned before, lightly pigmented skins may have come into existence as people took up residence in temperate areas, where there was less sunshine than in the tropics, where the human species probably evolved. With less pigment, more synthesis of sunlight-dependent Vitamin D in the skin would be possible. If this idea is correct, the loss of skin pigmentation was an evolutionary change designed to allow the body to receive the benefical effects of the sun's rays. But the sun's rays also exert damaging effects; they can damage the dividing cells of the basal layer of the epidermis in such a way that cancer is the result. Indeed, skin cancer is the most common human cancer. It occurs mainly in people who work outside in the sun on the parts of their skin that are exposed constantly to the sun's rays. In order to counteract this, and possibly other damaging effects of sunlight, the tanning reaction occurs. The skins of people of all races become more deeply pigmented as a result of prolonged exposure to sunlight. In the tanning reaction the sun's rays activate the melanocytes to produce and distribute more melanin, which exerts its protective effect by absorbing the sun's rays before they penetrate the nuclei of the dividing basal cells.

The way that the melanocyte acts to spread its pigment is remarkable (Figure 12-6). After melanin is synthesized in sacs called *melanosomes,* the melanosomes move out into the fingerlike processes of the cell. The tips of the cell processes then penetrate epithelial cells and are pinched off inside them. Most melanosomes end up in the area of the epithelial cell that lies between the nucleus and the skin's surface—perfect placement for protection of the genetic material from the sun's rays.

Cancers of the basal cells of the skin are usually slow growing and curable by surgery because they tend not to break off and spread throughout the body. But cancers of the melanocytes, or *malignant melanomas,* act very differently. These cancers, which often originate from pigmented moles, are highly malignant and they almost always spread. The tendency of melanomas to spread may be a reversion by the melanocytes to their earlier behavior, manifested in fetal life, in which they spread themselves throughout the body from the region of the spine.

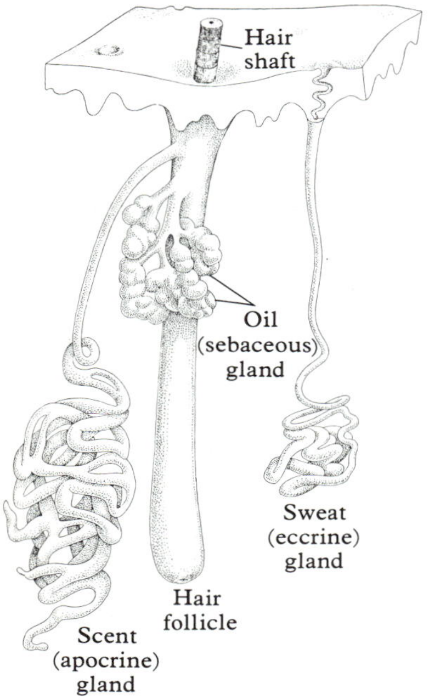

12–7
Glands of the skin include oil and scent glands, some of which open into hair follicles, and sweat glands, which open onto the skin's surface. (After "The Skin," by William Montagna. Copyright © 1965 by Scientific American, Inc. All rights reserved.)

Glands of the Skin

There are four major kinds of glands that discharge their products on the skin's surface: the sweat glands, the scent glands, the oil glands, and the mammary glands.

Sweat Glands The sweat, or *eccrine,* glands open onto the skin separately from the hair follicles (Figure 12-7) and release a clear fluid that is mostly water but that also contains a significant amount of body salts (especially sodium chloride, or table salt). Sweat glands are distributed over most of the body, but their function differs with their location. Those on the palms of the hands and soles of the feet serve as a lubricant (see Figure 12-8): primates have many sweat glands on all skin surfaces that are subject to friction. Fingers that are slightly moist are better able to sense the environment, and are less likely to slip while grasping. On the arms, legs, and torso, sweating occurs in response to heating of the body, and the sweat serves mainly to cool us as it evaporates. Sweating as a way of cooling the body is much better developed in people than in the other primates (which sweat hardly at all in hot environments), and people are more richly endowed with sweat glands than the other primates as well.

Scent Glands The scent, or *apocrine,* glands are usually associated with hair follicles and are found in large numbers only in a few areas of the body—the ear canals, the nipples, the axillae (underarm areas), the navel, and the anogenital area. The product of these glands is a rather thick and sticky fluid very different from sweat. Scent-gland fluid is odorless at the moment of release from the gland, but after release it is rapidly broken down by the skin's bacteria into the molecules that give these areas their characteristic human odors.

The body odors of most animals have sexual and social significance, and it seems likely that this is true for people as well, at least in societies where the natural body odors are appreciated. Indeed, the scent glands are closely

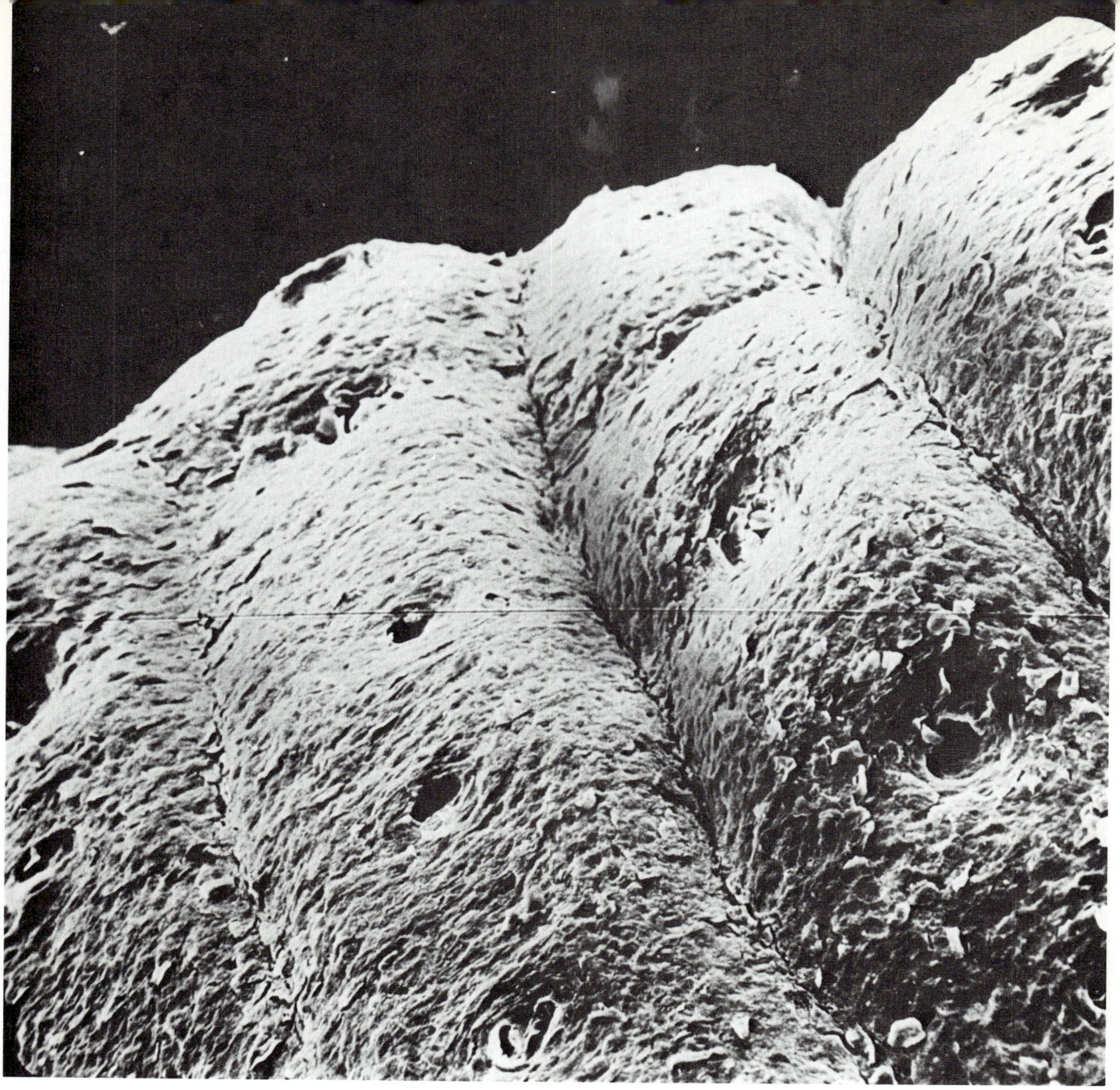

12–8
Human fingertip ($\times$ 150), as seen through the electron microscope. The scales of the surface of the skin and the openings of the sweat pores are clearly present. (From Lennart Nilsson, Behold Man. *Little, Brown and Company, 1974. Courtesy Albert Bonniers Förlag, Stockholm.)*

associated with sexual development, enlarging and becoming more active in both sexes about the time of puberty, and they secrete in response to sexual stimulation.

The bacteria that create body odors by breaking down the products of scent and oil glands belong to a category known as *gram positives.* The designation *gram positive* refers to a staining reaction in which the bacteria, after being stained by a violet dye, stay stained after they are placed in an iodine solution. Gram-negative bacteria, which do not produce odor, lose the dye when placed in the iodine solution. Studies have shown that the surface of the skin in the underarm is normally dominated by the odor-producing gram positives. Deodorants exert their odor-reducing effect by upsetting the ecology of the area so that the gram-negative bacteria become dominant. When deodorant usage is discontinued, the gram-positive bacteria once again take over. The major problem associated with deodorant usage has nothing to do with bacterial imbalance, but rather with the allergic rashes that some people develop to the deodorants themselves (see Chaper 13).

Oil Glands The oil, or *sebaceous,* glands are distributed over most of the body and some are associated with hair follicles. They are found in particularly high concentrations in areas of the body likely to be moist some (or all) of the time: around the nose and mouth, inside the cheeks, on the inside edge of the eyelids, around the anus, and in the genital areas (on the labia minora and clitoris of women and on the glans penis and inner surface of the foreskin in men). The oil glands on the flared portion of the nose are so large that their ducts are easily seen with the naked eye.

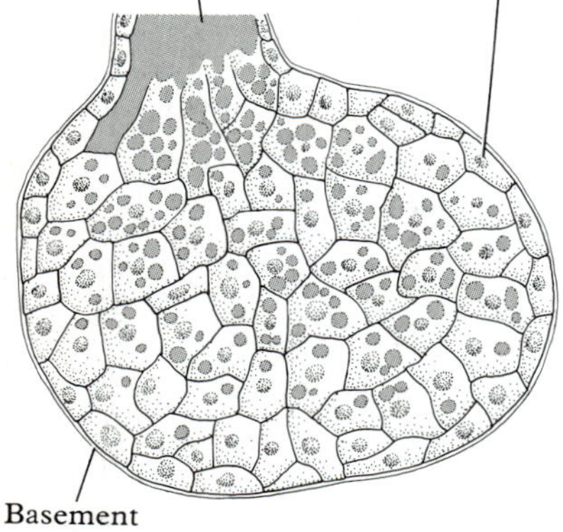

12–9
Production of sebum, a mixture of fatty acids, triglycerides, waxes and cholesterol, occurs as the sebaceous gland cells become choked with accumulated fat globules (gray circular areas), die, and disintegrate. As the sebum accumulates in the gland's duct, the pressure from skin movements bring the semiliquid material to the surface of the skin. (After "The Skin," by William Montagna. Copyright ©1965 by Scientific American, Inc. All rights reserved.)

The product of the oil glands, *sebum,* contains large amounts of fat, fatty acids, waxes, and cholesterol, all of which contribute to its oiliness. Because the oil glands (like the other glands of the skin) are derived from the skin's epidermis, it is perhaps not surprising that the production of sebum is very much like the production of keratin. The cells in the gland are derived from the basal layer by mitosis, and they go about their task (of synthesizing sebum); finally they become bloated with their product and die, to be extruded upon the skin's surface (Figure 12-9).

There is no agreement among investigators about the functions of the oil glands, but it seems that they are important in the prevention of skin infections caused by fungi. Certainly the saturated fatty acids found in sebum are strongly inhibitory to many fungi. Furthermore, the action of sebum in inhibiting fungi is interfered with by moisture, and fungal infections are particularly likely to occur wherever there is a wet environment, as for example the growth of athlete's-foot fungus in the moist environment of the sweaty shoe. If containing fungal growth is indeed an important function of sebum, then possibly large numbers of oil glands have come to be associated with wet areas of the body (for example, inside the cheeks and eyelids) as a means of coping with the possibly deleterious effects of excess moisture.

Oil glands arise during the fifth fetal month, and by the time of birth the skin is covered by a white cheesy coating (known as the *vernix caseosa*), which consists of sloughed off epithelial cells and the sebum that has accumulated during the last four months of pregnancy. The vernix caseosa probably provides needed protection for the newborn infant, whose skin is being colonized by microorganisms for the first time.

There are some problems associated with the presence of oil glands, however, including blackheads and acne. *Blackheads* are oil glands that have become clogged with sebum and dirt. *Acne* is an inflammation of the oil glands that typically occurs around the time of puberty, when the glands greatly increase their activities under the influence of the androgenic hormones. Acne is more common and generally more severe among males than females, probably because males have much higher concentrations of the androgenic hormones. The only drugs available at present to decrease oil gland activity are the estrogens, which are usually referred to as "female hormones" because they are present in higher concentrations in women than in men.

Mammary Glands The mammary glands, or breasts, are present in both sexes, but in the male they remain rudimentary and functionless throughout life.

The mammary glands of the female are highly responsive to hormonal changes, and because of this responsiveness they undergo conspicuous changes at puberty, during pregnancy, and during lactation (Figure 12-10).

The glands and ducts of the breast develop from the epidermis of the embryo at a very early stage. During the second month of fetal life two thickened ridges of epidermis, the milk lines, develop and extend between the arm bud and the leg bud on each side of the body. In humans, the milk lines soon disappear except for the portions in the area where the two breasts will develop. In other mammals, who have more than two breasts, other areas along the milk line also develop. Occasionally, humans develop extra, or *supernumerary,* breasts; 90 percent of them are found along this line. During the third fetal month the two remaining portions of the milk line sink into the tissue below and begin their development into the glands and ducts of the breast.

The breasts of the female provide nourishment for the infant, but breast-feeding is more than just the provision of milk. It is also a mutual human relationship during which the infant receives warmth, contact, and support.

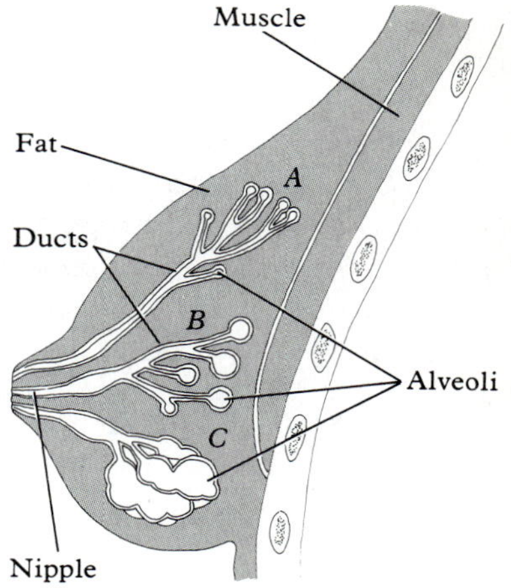

12–10
Adult female breast showing the mammary glands as they appear (A) in a nonpregnant woman, (B) in mid-pregnancy, and (C) during lactation.

Some of the benefits of this relationship have been revealed by a number of studies. In one investigation of 173 children followed from birth to 10 years of age, children who had not been breast-fed developed four times as many respiratory infections, 20 times as many attacks of diarrhea, and 21 times as many cases of asthma as breast-fed children. Similarly, premature infants who are breast-fed have fewer infections and a lower death rate than those who are not.

Although it is possible that breast-feeding is advantageous mainly because of the warmth and contact provided by the relationship, a major advantage of breast-feeding is that human milk is an outstanding food for the human infant. This fact should come as no surprise, because the milk of each species is well adapted for the young of that species. For example, whale milk is loaded with high-calorie fat, especially suitable for animals that must generate heat to withstand cold waters, and rabbit milk is particularly rich in protein, possibly related to the very rapid growth of rabbit babies.

A brief comparison of human milk with cow's milk will illustrate the point that human milk has many advantages (see Table 12-1). First, cow's milk contains cow's proteins, which are foreign to the human infant and which sometimes cause skin or intestinal allergies. Second, cow's milk has about three times the mineral content of human milk, which is too much for the human newborn's kidney to manage, unless the milk is diluted. Studies of the individual minerals have shown that the high phosphorous content of the cow's milk causes a reduced absorption of the calcium from the intestine, which can lead to the tetany (see Chapters 11 and 8) of low blood calcium. Indeed, the convulsive muscle spasms of tetany occur almost exclusively in infants who are not breast-fed and who do not receive enough calcium to keep the calcium in their blood at normal concentrations. Third, both cow's milk and formulas derived from cow's milk contain more protein than human milk, and that protein (mostly casein) absorbs about four times as much acid from the stomach as does human protein. This leaves the stomach with a less acidic environment, and therefore with less capacity to kill potentially harmful bacteria before they can pass on into the intestines. This loss of stomach acidity could be responsible in part for the fact that bottle-fed infants have more gastrointestinal infections than breast-fed infants. Finally, enzymes that break down fats (lipases) are found in

TABLE 12-1
Composition (per 100 grams) of human milk and cow's milk.

MILK ELEMENTS	MATURE HUMAN MILK*	MATURE COW'S MILK*
Protein	1.2 g	3.3 g
Fat	3.7 g	4.3 g
Lactose	7.0 g	4.8 g
Sodium	15 mg	58 mg
Potassium	57 mg	138 mg
Magnesium	35 mg	125 mg
Phosphorus	15 mg	96 mg
Vitamin A	280 IU**	180 IU**
Vitamin C	5 mg	1.5 mg
Vitamin D	5.0 IU**	2.5 IU**
Food energy	65 kcal	65 kcal
Water	87.5 g	86.0 g
pH	7.3	6.8

*Mature milk refers to milk secreted after the first few days.

**An international unit is a unit of biological activity, not a unit of weight. The international unit is used when several different molecular forms of a vitamin are active.

both human and cow's milk, but they are largely destroyed by the heating or boiling associated with preparation of the cow's milk; therefore, the fat in cow's milk is not digested as well or absorbed as well from the intestine. And the vitamins present in the fat, A and D, are not absorbed as well either.

All in all, the facts about milk itself would suggest following nature's way, if possible. But often there are factors that make breast-feeding unhandy or impossible, in which case it should be remembered that formulas are constantly being improved and that *most* children who are not breast-fed are healthy and vigorous. The association between breast-feeding and lack of disease is a statistical one, which says very little about the special features of an individual mother–child relationship. For some, *not* breast-feeding is the best way, and it leads to more happiness between mother, child, and other loved ones.

One other product of the mammary glands deserves special mention: *colostrum,* a lemon-colored fluid that is delivered to the newborn during the first two days of life. Colostrum contains large quantities of antibody molecules (see Chapter 13) capable of inactivating disease-producing microorganisms. When the infant is born, there are no bacteria in its intestine, and these antibody molecules help to determine which bacteria will be able to colonize. Because the antibody molecules present in the colostrum have been synthesized by the mother, they protect against organisms in the mother's environment, which also is the environment where the child will be. Therefore, colostrum from *one's own mother* is more likely to be protective than colostrum from other sources. Colostral antibodies are also absorbed by many mammals into the bloodstream from the intestine, but this does not happen in people.

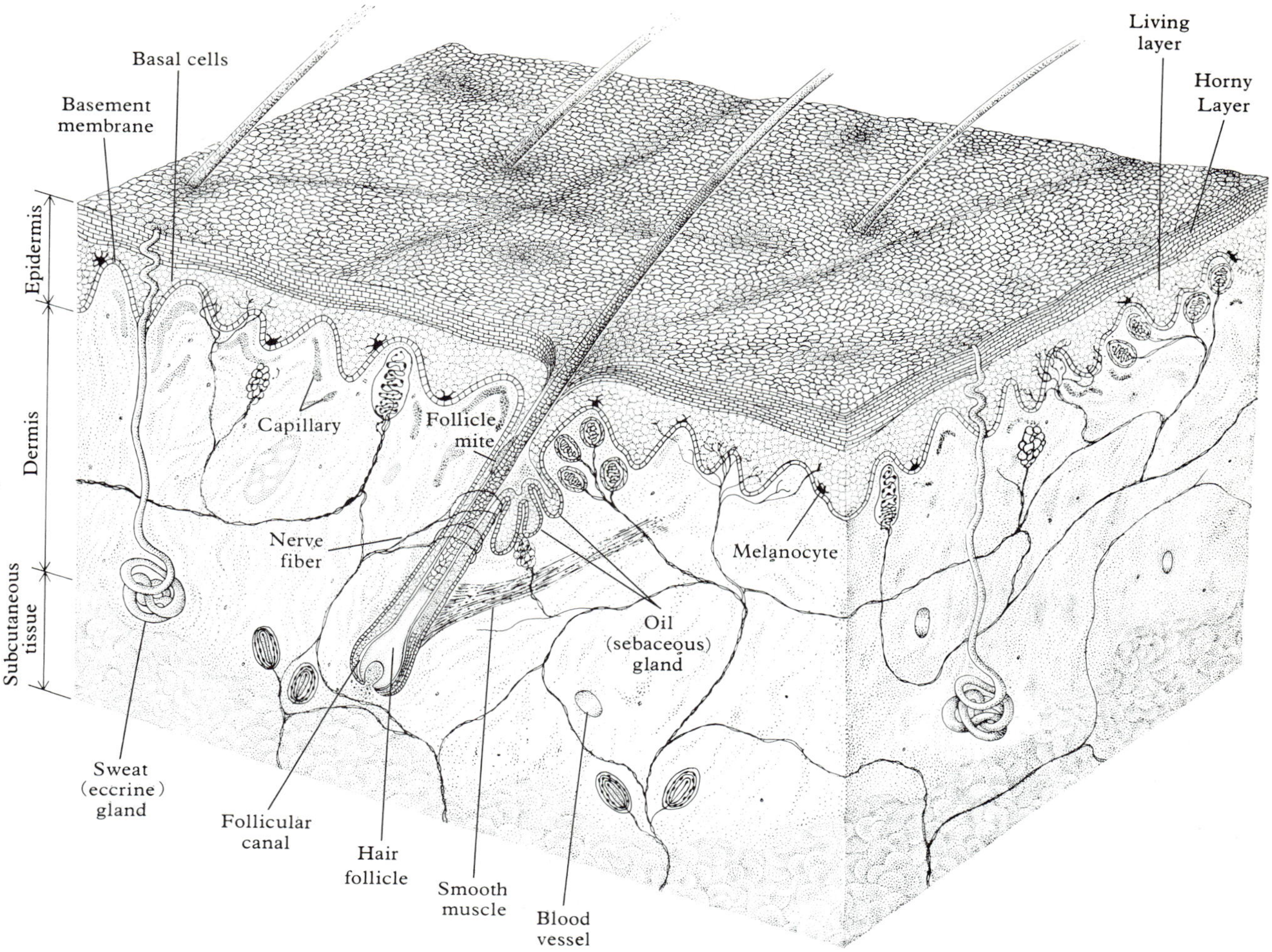

12–11

Habitat of normal microorganisms (×50). Demodex folliculorum, *the follicle mite, is the only animal that resides in healthy, undamaged skin. Scarcely visible to the unaided eye, it is 400 times larger than an average bacterium.* Demodex *inhabits the skin of most adults. (After "Life on the Human Skin," by Mary J. Marples. Copyright © 1968 by Scientific American, Inc. All rights reserved.)*

Organisms of the Skin

The skin's surface provides a home for many different microorganisms with many different lifestyles (Figures 12-11 and 12-12). This is possible because the skin provides a great variety of environments, ranging in temperature from warm to cool and in humidity from moist to dry. There also is considerable variation in the numbers of microorganisms from place to place on the skin's surface: studies by Peter Williamson and collaborators have shown that the most heavily populated area is the underarm, where there are an average of 2.41 *million* bacteria per square centimeter of epidermis. In contrast, on the back and on portions of the forearm, there are only a few hundred bacteria per square centimeter.

Among the normal inhabitants of the skin's surface are many types of bacteria, some fungi, and a single type of animal: a mite (relative of the spiders) named *Demodex folliculorum* (Figure 12-11). These mites live only in a few areas of the facial skin, including the outer flared areas of the nose and among the eyelashes, where some of them are inhabitants of hair follicles. Mites are quite small and cannot be seen with the naked eye.

Some of the bacteria and fungi of the skin's surface have the potential to become the agents of disease under special circumstances. They are discussed in Chapter 13.

Dermis

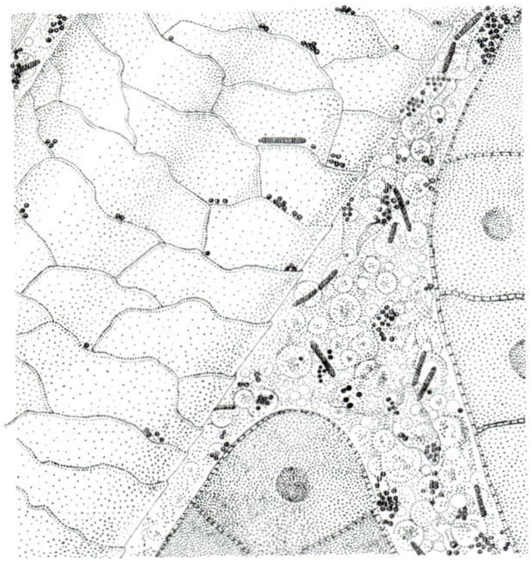

12–12
Hair follicle houses many bacteria, depicted here as spheres and rods (×1000). (From "Life on the Human Skin," by Mary J. Marples. Copyright © 1968 by Scientific American, Inc. All rights reserved.)

Having considered the structures and functions of various elements of the epidermis, we now turn our attention toward functions of the skin in which the *dermis* plays a crucial role. The dermis aids the skin in its functions as a sense organ and a food reserve, and in its participation in temperature regulation. As an introduction to the subject of temperature regulation, which is achieved partly through the control of blood flow, we will begin by looking at the skin's circulation.

Circulation in the Skin

The blood vessels of the skin are only in the dermis. Neither blood vessels nor capillaries are in the epidermis. Therefore, the skin's epidermal cells and glands receive their nourishment entirely by diffusion from the capillaries in the dermis below. For this reason, superficial wounds and scrapes of the epidermis ooze a clear rather than a bloody fluid. Similarly, *blisters,* which are irritations *within* the epidermis, accumulate the clear fluid that moves upward from below to bathe the epidermal cells.

Lymphatic vessels (Figure 7-5), which are well represented in the dermis, do not enter the epidermis either. These vessels normally return proteins and white blood cells that have entered the tissues to the circulation, and during an infection they also carry bacteria and foreign material out of the infected area to the lymph nodes, where the antibodies are made. The lymphatic vessels may become visible if an infection begins to spread: red streaks may extend up the arm from the point of a serious local infection. These streaks are lymphatic vessels that appear red because there is more blood in the vessel walls when the vessels are helping to cope with an infection.

The body has many more vessels than are needed to supply blood for nourishment of the skin's cells. The excess vessels are used mainly to help regulate the body temperature. We shall now consider the mechanism by which the body regulates its temperature and the role of the skin in this regulation.

Temperature Regulation

Among animals, birds and mammals are *warm blooded,* which means that they maintain a nearly constant body temperature most of the time. With only a few exceptions other animals are considered to be *cold blooded* because their body temperature fluctuates widely according to the temperature of the environment. An animal's reactions and responses are more rapid at higher temperatures, but a warm-blooded animal has the advantage of being able to operate at peak responsiveness even in a cold environment. This difference should be apparent to anyone who has observed the sluggishness of a lizard (a reptile) on a cold morning.

The warm-blooded animals have evolved special mechanisms in order to maintain a relatively constant body temperature. In humans, changes in outside temperature are sensed by receptors in the skin and by a temperature-sensitive region of the hypothalamus. The hypothalamic receptors are especially important because it is the core temperature of the body, not the surface temperature, that is being maintained within relatively narrow limits. When the core temperature rises even slightly, the hypothalamus sends impulses along the nerves to the smooth muscle fibers surrounding the smallest arteries, or arterioles, of the skin. As a result, the arterioles dilate, and the blood flow to the skin increases enormously. These nervous impulses also activate the sweat glands, which pour sweat out onto the skin. As the sweat evaporates, the skin's surface is cooled, and the cooling easily extends the short distance down into the dermis to cool the blood. The cool blood is then returned to the rest of the body, and it eventually cools the temperature-sensitive region of the brain to normal, shutting off the cooling mechanism.

The brain watches vigilantly for any change in body temperature and is equally ready to respond to a drop in temperature (Figure 12-13). When the environment is cool and the temperature of the blood drops slightly, the drop is again sensed within the hypothalamus. The hypothalamus this time activates different nerves that constrict the smooth muscle surrounding the arterioles in the skin. Thus the blood flow is greatly reduced to cool the skin, allowing the core of the body to remain warmer. We have all experienced this response on a cold day–not only do our fingers and toes cool down a bit at first (because the environment is cold), but the body's response is to let them cool even further by shutting off their blood supply. This is the explanation of *frostbite*–a combination of a cold environment and the body's response to it–in which the body maintains the central core temperature necessary for survival, but sacrifices the skin blood flow that is necessary to maintain the temperature of the fingers and toes.

If the reduced blood flow to the skin does not restore the cool blood back to normal, the body will employ other mechanisms as well–such as elevating the hairs–in an attempt to provide more insulation for the skin. This mechanism of hair elevation is relatively ineffective, because human beings do not have quite enough hair to be able to elevate it into a thick bushy insulating coat. What we get are *goose-pimples,* each of which represents a rumple of the skin's surface as the hair is drawn into an upright position by the smooth muscle that inserts into it (Figure 12-1). If this mechanism too is ineffective, as it usually is, shivering will begin, releasing heat as a by-product of supplying ATP to the muscles. Finally, if conditions are exceptionally cold, only a strong *voluntary* effort to keep the body constantly moving will supply enough heat to maintain life. The hypothalamus is also activated by cold to stimulate the thyroid gland, which releases thyroid hormone that increases the metabolic rate.

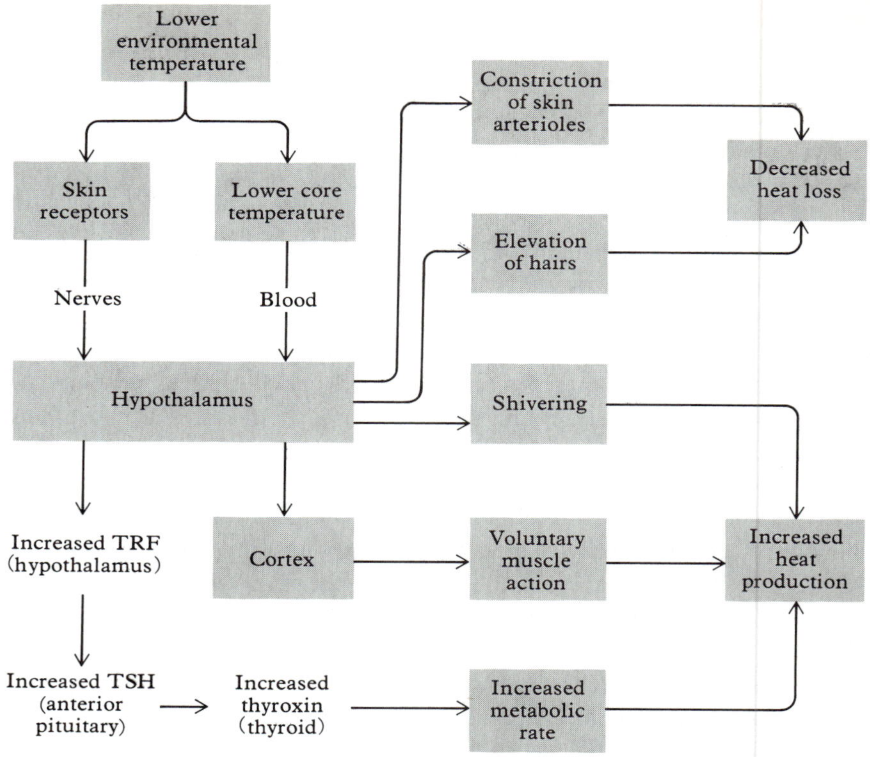

12–13
A drop in environmental temperature activates mechanisms to decrease heat loss and increase heat production. The hypothalamus triggers these mechanisms.

Skin as a Storage Organ

The dermis normally contains a store of fat that can be utilized readily for energy during times when food is in short supply. Fat is especially advantageous as a source of energy, as it contains nine kilocalories of energy per gram, compared to four kilocalories per gram of either protein or carbohydrate. In cold climates, this dermal fat layer also serves as insulation. That "spare tire" of fat around the human waist is indeed like an automobile spare tire in that it is ready and available for an emergency.

The skin also is used for the short-range storage of the sugar glucose. This storage function is well documented by experiments in which the sugar concentration of the blood is suddenly increased by rapid intravenous injection of glucose. Following this procedure, about 15 percent of the glucose enters the dermis. Whereas the glucose returns to preinjection concentrations in the blood within one hour, the dermal glucose does not return to its original concentration for two to three hours. Thus the dermis serves as a glucose reservoir, helping mainly to prevent the blood glucose from dropping too far by slowly leaking glucose back into the blood. As a storage organ, the skin is second only to the liver in helping to maintain concentrations of glucose in the blood.

In *diabetes mellitus* not only is the concentration of glucose in the dermis unusually high, but after a meal the concentration, which is further elevated, returns much more slowly to that of the blood. Thus there is a pool of glucose just under the epithelium, capable of providing a rich growth medium for bacteria or fungi. Access to this rich medium can occur only if there is a break in the epidermis, but these breaks happen fairly often, and diabetics commonly have recurrent skin infections. Thus a condition such as diabetes can convert the benefits of dermal glucose storage into a danger of infection simply by altering the amount of glucose present in the skin.

Skin as a Sense Organ

As a sense organ, the skin has no equal. The sense of touch is not only more widely dispersed over the body than the other senses (of smell, sight, hearing, taste), but it is undoubtedly the most necessary of the senses. A person can manage without one or two of the other senses, but there is no known instance of anyone surviving without the sense of touch. The dramatic story of Helen Keller, who became deaf and blind in infancy, but who eventually was able to communicate with the world fully and completely, emphasizes the point. It was only after she had learned the finger alphabet, communicated through the skin, that Helen escaped from social isolation and entered into the world of symbolic communication.

To the biologist the importance of the sense of touch is underscored by its exceptionally early development in the fetus. Touch develops during the second fetal month, when the fetus is less than three centimeters (about one inch) long, at which time it will withdraw from a gentle stimulus applied to the face.

Touching, which takes the humanly significant forms of handling, fondling, and caressing, is especially important for the development and even the survival of the human infant. Early in the 1900s, the death rate for children living in orphanage institutions was nearly 100 percent, and it is possible that many of these deaths can be attributed to a lack of loving handling. The situation improved dramatically when the practice was begun of picking up, handling, and fondling the infants regularly. In a large hospital, the death rate for infants under one year of age fell from 30 percent to less than 10 percent when loving handling was regularly provided for the children.

The anatomical basis of the sense of touch is to be found in the rich networks of nerves interweaving throughout the dermis of the skin. These nerves send naked sensory nerve endings up into the epidermis, adding to the skin's exquisite sensitivity (Figure 12-1).

Perhaps the term "sense of touch" does not do justice to the variety of experiences that touching can provide. Although many measurable factors enter into the sense of touch—such as the temperature, movement, pressure, direction, and duration of a touch stimulus—the experience of a tender, gentle caress seems somehow more than a combination of these factors.

Summary

The skin, our largest organ, is composed of an outer layer, the epidermis, and an underlying layer of connective tissue, the dermis. The epidermis and dermis are separated from each other by a basement membrane.

The outermost layer of epidermis, which is in contact with the air, consists of dead cells that are constantly being rubbed off into the environment. Replacements for these dead cells are provided by mitosis of the basal cells that lie just above the basement membrane. A major activity of the epidermal cells is the synthesis of the protein keratin, and by the time the epidermal cells die they are mostly a mass of keratin. Keratin molecules are responsible in large part for the ability of the skin to keep out water and microorganisms.

Hair and nails are products of the epidermis. Each hair grows out of a group of rapidly dividing epidermal cells that is located at the bottom of the hair follicle. The hair follicles are surrounded by networks of nerve fibers that make the hairs sensitive receivers of sensory information. The nails serve as tools for such activities as picking and scraping, and are derived from epidermal cells located around the nail root. Both hair and nails are composed primarily of keratin molecules.

The skin's principal pigment, melanin, is derived from melanocytes located in the epidermis. Melanocytes send out cell processes that penetrate surrounding epidermal cells and inject them with melanin-containing sacs called melanosomes. In each epidermal cell melanosomes come to lie in a special position in the epidermal cells—between the nucleus and the surface of the skin—where the melanin can protect the nucleus from the potentially damaging effects of sunlight. Melanocytes are stimulated by sunlight to increase their activity, and the skins of people of all races become more deeply pigmented as a result of prolonged exposure to sunlight.

The epidermis also gives rise to four different kinds of glands: sweat (eccrine) glands, scent (apocrine) glands, oil (sebaceous) glands, and mammary glands. The sweat glands release a clear watery fluid that lubricates surfaces such as the palms of the hands and the soles of the feet and also cools the body when it evaporates. The scent glands release a rather thick and sticky fluid that is odorless when released but that is rapidly broken down by the skin's bacteria into odorous molecules. The body odors are probably of both social and sexual significance. The oil glands—which are found in high concentrations in areas of the body likely to be moist some (or all) of the time—release an oily product, sebum, that probably helps to inhibit the growth of fungi.

The mammary glands, which are present in both sexes, remain rudimentary in the male. In the female they release colostrum and milk after delivery. Colostrum, which contains antibodies, helps to determine which bacteria will colonize the newborn's gastrointestinal tract. A comparison of human and cow's milk reveals that there are considerable species differences in composition and that human milk is superior for the human infant. Breast-feeding is a mutual human relationship that provides the infant with warmth, contact, and support.

The skin's surface is covered with microorganisms, most of which are bacteria and fungi. The one animal present on the skin is the mite, *Demodex folliculorum,* which inhabits hair follicles.

The skin's dermis serves several functions. It is entirely responsible for the nourishment of the epidermis, because it contains all the blood and lymphatic vessels of the skin. It serves as a sense organ and a storage organ and plays an important role in temperature regulation.

Temperature changes in the body are sensed by temperature-sensitive receptors in the skin and in the hypothalamus of the brain. The hypothalamic receptors are especially important because it is the core temperature of the body that is maintained within relatively narrow limits. Cooling is accomplished by a combination of increasing the blood flow to the skin and sweating. As the sweat evaporates, the skin's surface is cooled and the blood near the skin's surface is cooled as well. The cool blood is then returned to the rest of the body. In a cold environment, several mechanisms may play a part in warming the body. These include: reduction of blood flow to the skin, elevation of hairs, shivering, voluntary muscle activity, and increased output of thyroid hormone.

The dermis stores the body's fat and provides short-term storage of glucose. Fat can be utilized as an energy source when food is in short supply, and it also insulates the body against a cold environment. Glucose, which enters the dermis after a meal, is released into the blood stream gradually over two to three hours.

The sense of touch is also a function of the dermis, which is supplied with a rich network of sensory nerve endings. Touching is necessary for the normal development of the human infant, and takes humanly significant forms such as handling, fondling, and caressing. Whereas a person can live without one or two of the other senses, there is no known instance of anyone surviving without the sense of touch.

Selected Readings

1. *Touching. The Human Significance of the Skin,* by A. Montague. Perennial Library, Harper & Row, 1972. A well-written book devoted to the sense of touch.
2. *Man,* 2nd ed., by R. J. Harrison and W. Montagna. Appleton-Century-Crofts, 1973. A fine book with an excellent chapter devoted to the human skin (pp. 167-197).
3. "Life on the Human Skin," by M. J. Marples. *Scientific American,* Jan. 1969, offprint 1132. Discusses the organisms of the skin and their ecology.
4. "To Breast-Feed or Not to Breast-Feed," by H. Vorherr. *Postgraduate Medicine,* June (1972), 127-134. A thorough comparison of human milk and cow's milk.

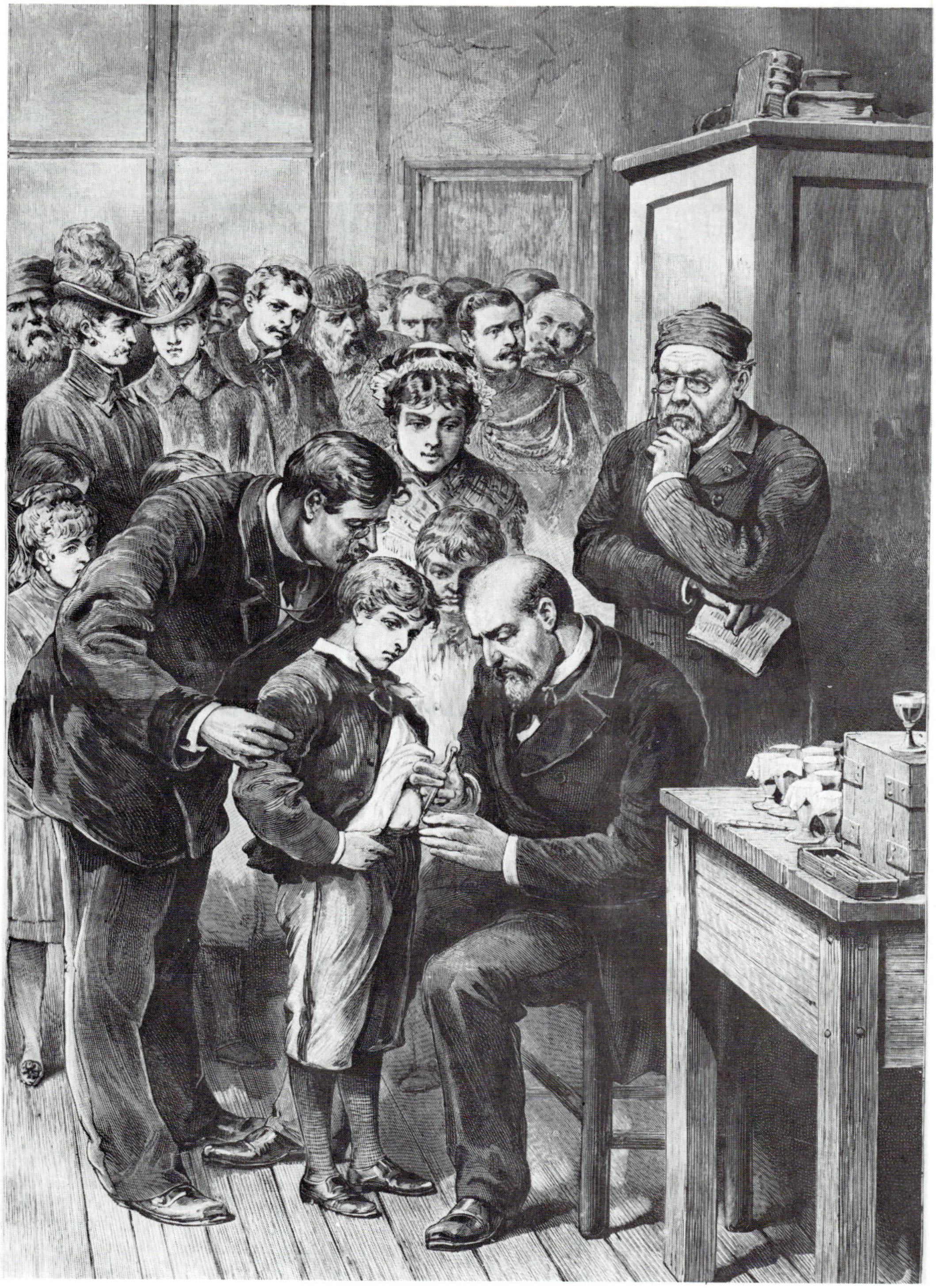

Louis Pasteur (standing at right) observes as young Joseph Meister is inoculated with an extract of rabies virus prepared by Pasteur. Meister was the first person to be deliberately immunized against rabies; before this time (1885) the bite of a rabid animal was usually fatal. (Courtesy Roger-Viollet, Paris.)

CHAPTER

13

Immunity

In general, immunity is the capacity to resist infectious diseases. It is a complex subject, both because many factors help to determine disease resistance and because many kinds of organisms "cause" disease (viruses, bacteria, protozoans, fungi, etc.). It is also a subject about which advances in knowledge have led to the prevention and control of many formerly devastating human diseases.

In spite of existing accomplishments, however, a number of infectious diseases have failed to yield. Among these are the two infectious diseases that at present cause more worldwide sickness and death than any others: *schistosomiasis,* caused by a flatworm, and *malaria,* caused by a protozoan (these diseases were also discussed in Chapter 4). It is estimated that over 200 million people suffer from schistosomiasis; perhaps 150 million suffer each year from attacks of malaria. In Africa, malaria kills about a million children under the age of seven every year. Among widespread bacterial diseases, there are still some for which no immunizations now exist. Syphilis and gonorrhea, which have become more prevalent recently in the United States and in Europe, are examples. So a great deal remains to be done.

For noninfectious diseases as well, the potential human benefits of immunological knowledge have by no means been realized. For example, we have learned that immunity can play an important role in the prevention and control of cancer, but we have not yet learned how to go about taking advantage of the knowledge. We have also found that the immune system can cause "autoimmune" diseases by attacking the body's own parts. Here again, we have not succeeded in prevention, but there is the hope that work being done now will lead to this in the near future.

Because immunity affects so many different diseases, we will take the liberty in this chapter of broadening the scope of discussion to include not only infectious diseases, but others as well, such as cancer and autoimmunity. We will begin, however, by looking at two factors that greatly influence resistance to *all* diseases: genetic background and good general health.

Genetic Factors and Disease Resistance

Experiments on laboratory animals have shown that genetic factors are important in disease resistance: animals can be bred easily for either resistance or nonresistance to countless infectious diseases. However, laboratory experiments on humans are not always necessary because nature has already performed many experiments in her own way.

Malaria

Malaria provides a good example of one of nature's experiments. We can observe two natural environments, one in which malaria is ever present, and another in which there is no malaria. In the malaria-infested region (for example, parts of Africa), the only people who survived to have offspring were those who had *resisted* the disease, and thus the genes of these resistant people were the ones that were passed on. Most people now living in the region are descended from parents, grandparents, and great-grandparents, *none of whom* died at an early age from malaria, and therefore they are likely to carry the resistant genes of their resistant forebears. One of the genes best known for its resistance to malaria is the gene for sickle-cell hemoglobin (Hemoglobin S; see p. 464). In the human body, malarial parasites spend much of their life inside red blood cells. When the red blood cells harboring the malarial parasites contain hemoglobin S, the cells break down more readily, releasing the parasites before they are capable of infecting other cells. Thus the parasite is inhibited, and malaria is much less severe. There is a high incidence of hemoglobin S genes in people from malaria-infested regions, which indicates that immunity can be a selective agent in evolution.

The second group of people in nature's malaria experiment (the control group) has lived for generations in a malaria-free region, developing resistance to many local diseases, but not to malaria. If some of these people venture into a malaria-infested region, or if malaria is introduced into *their* region, the effect can be devastating.

Smallpox and History

One of the great calamities of medical history occurred when a virus disease, smallpox, was first introduced into Mexico in 1519, brought in by one of the Spaniards arriving by ship to meet Cortez's forces near Vera Cruz. The Mexican people had not experienced smallpox before, and there had been no chance for the selection of resistant genes. The smallpox epidemic that ensued is estimated to have killed as many as 3.5 million people. Some even died from starvation because the epidemic struck everyone in a village at once, leaving no one healthy enough to prepare corn. The Spanish had another advantage as well: many of the conquistadors had probably suffered smallpox back in Spain, and were therefore completely immune to it. Most likely smallpox guaranteed Cortez's victories in Mexico.

General Health and Disease Resistance

Diseases can only be warded off by a healthy, normally operating body. To be healthy, one must have proper nutrition, especially adequate protein. The connection between protein malnutrition and susceptibility to disease is a logical one for several reasons. First, we know that all substances manufactured by the body depend for their synthesis on enzymes, which are proteins. So a deficiency of protein leads, for example, to a decreased ability to synthesize the adrenal hormones that are needed to cope with stresses of all kinds, including the stress of an infectious disease. Second, toxic substances that enter the body are degraded by protein enzymes synthesized in the liver. It has been shown in animal experiments that starvation may cause a hundred- or even thousandfold increase in susceptibility to the lethal effect of a bacterial toxin. And finally, the body's defenses against microorganisms depend heavily upon the production of antibody molecules, which are also proteins. People who receive inadequate protein are ill-prepared to meet the challenges posed by a disease-producing microorganism.

Hormonal Imbalance and Disease

The role of slight imbalances in reducing resistance has been clearly shown in experiments in which animals were subjected to adrenalectomy (surgical removal of the adrenal glands). The adrenal glands produce most of the body's steroid hormones, of which cortisone is the most widely known, and these hormones are needed in proper quantities for the synthesis of antibody molecules. Not only is a deficiency of these hormones immunosuppressive (suppressing the synthesis of antibody molecules), but an overabundance is immunosuppressive as well.

In a typical experiment mice are injected with bacteria that have the potential to cause serious disease. The resistance of the mice is measured by the average number of bacteria they can ward off. In one set of experiments, normal mice resisted more than 32,000 bacteria, whereas adrenalectomized mice contracted the disease when exposed to only 1000 bacteria. However, when the adrenalectomized mice were given daily injections of cortisone, they were restored to normal resistance *only if given exactly the right amount of cortisone.* Too much or too little cortisone would not restore normal resistance. The results of the experiment illustrate the common-sense principle that a well-balanced, well-tuned body is essential to disease resistance.

This experiment on the adrenals does more than illustrate a principle, however. It emphasizes the importance of the adrenal glands and gives us an understanding of why people who have either low adrenal function (Addison's disease) or high adrenal function (Cushing's disease) have an increased susceptibility to infections. And in addition, the experiment gives us an insight into one of the reasons that people tend to become sick when they are under stress: in stress situations we have an elevated output of adrenal hormones.

The Capacity of Microorganisms to "Cause" Disease

The best evidence that general good health helps to keep the body resistant to disease comes, not from experiments, but from the fact that many organisms that "cause" disease are present on the healthy as well as on the sick. For example, among bacteria, *Staphylococcus aureus* (Figure 13-1) is an opportunist that takes advantage of hospitalized patients who are already in a weakened condition. This bacterium, a normal inhabitant of the skin, the nostril, and other mucous membranes, very rapidly develops resistance to antibiotics, and so is very difficult to eradicate once it has become associated with an infection. It readily spreads from person to person on flaked-off epithelial cells. It has been estimated that between one and 10 percent of hospitalized patients develop *Staphylococcus aureus* infections.

Other well-known and potentially harmful bacteria are occasional inhabitants of the epithelial surfaces of the throat, where they usually do no harm. Such bacteria include streptococci that can cause throat infections (strep throat); pneumocci, which can cause lung infections (pneumonia); and meningococci, which can cause infections of the membranes surrounding the brain (meningitis). The majority of people who harbor these bacteria do not develop disease or even feel sick; the development of disease requires conditions *in addition to* the presence of the organism. This is why the "ecology" of the body is so important and why it is often incorrect to say that a given organism is the cause of an infectious disease. Factors other than the presence of the organism determine who develops disease and who does not.

The fungi associated with athlete's foot are also frequently present between the toes without causing any problems. However, a change in the ecology of the area—whether a change in footgear, a move to a warmer climate, or the development of a disease such as diabetes—may upset the balance and lead to the spread of the fungus.

13–1
The action of penicillin on Staphylococcus aureus *bacteria. The normal spherical bacteria with smooth outer cell walls appear in the photo at left (×18,135). Bacteria treated with penicillin appear in the right-hand photo (×23,585). They have irregular, lumpy cell walls because penicillin has interfered with cell wall synthesis.* Staph. aureus *is responsible for many human problems, including boils and wound infections. Many strains of* Staph. aureus *have developed resistance to penicillin and other antibiotics, and infections caused by these resistant strains are very difficult to eradicate. (Courtesy David Greenwood,* Science, *Vol. 163, pp. 1076–1077. Copyright © 1969 by the American Association for the Advancement of Science.)*

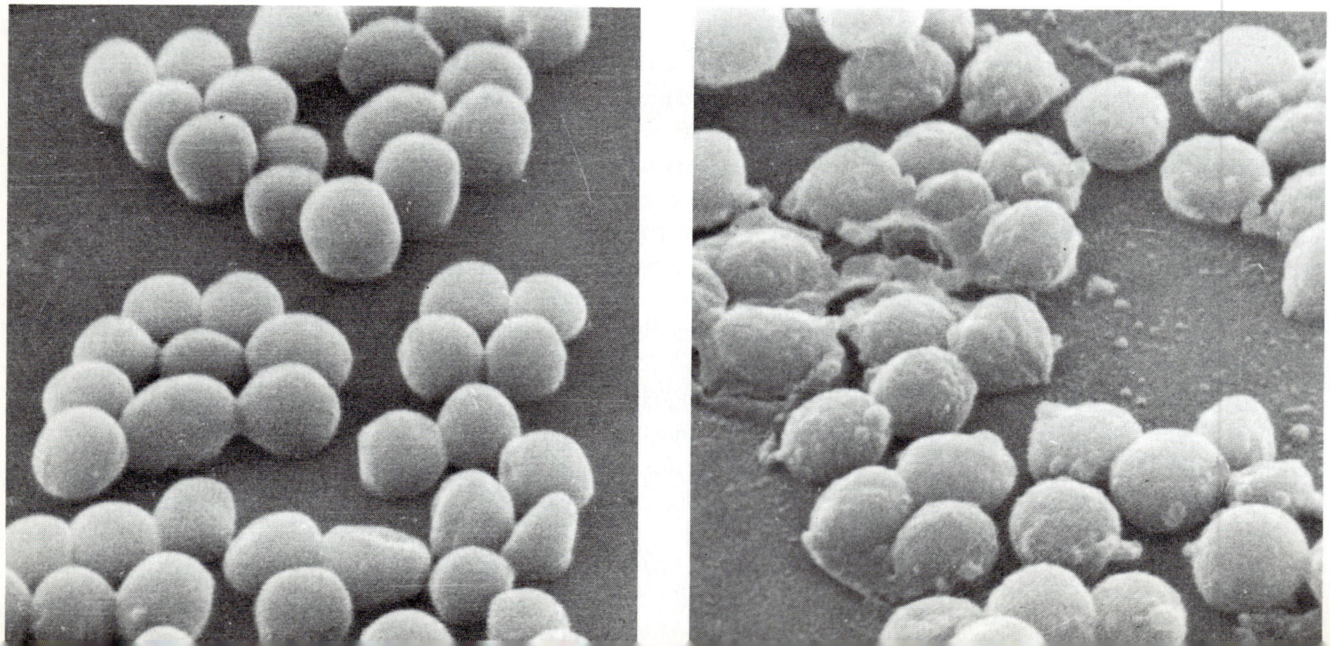

Fundamental Defenses Shared by All Animals

All animals, vertebrates and invertebrates alike, have in common two mechanisms of defense: their external covering layers and their phagocytic cells.

The Covering Layers

In people the covering layers consist of the skin, the lining of the intestine, and the lining of the respiratory passages. All are relatively effective in preventing the entry of most potentially harmful organisms. The outermost layer of skin, for example, is composed of dead cells, which provide an inhospitable environment for the growth of microorganisms. Also, the oil (sebaceous) glands of the skin secrete fatty acids that are strongly inhibitory to many fungi and some bacteria.

Even if the outermost layer of skin is broken, the deeper layers usually inhibit the growth of microorganisms. However, this capacity to inhibit is decreased somewhat during certain diseases; for example, in *diabetes mellitus* an excess of the sugar glucose in the dermis creates a danger of infection (see Chapter 12).

In other animals the nature of the covering layers may be quite different. For example, the sea anemone, a coelenterate that inhabits marine tidepools, continually produces mucus that flows over it from top to bottom, sweeping away any bacteria that might otherwise create problems. This mucus is slightly acidic (pH 5.9), which in itself is inhibitory to almost half of the bacteria isolated from tidepool seawater.

The Phagocytes

As you may recall from Chapter 5, the phagocytes of all animals are capable of recognizing foreign objects, engulfing them, and digesting them with powerful enzymes. This is the fate of most microorganisms that penetrate the covering layers: they end up dead inside the phagocytes of the host.

There is renewed research interest in the phagocytes of the vertebrates, some of which are found in blood (neutrophils; basophils; and monocytes, one of which is shown in Figure 13-2) and others in tissues (macrophages). Disorders of macrophages may account for the inability of some animals to respond fully to some foreign substances, because, as you may remember, macrophages digest foreign substances. Thus a macrophage disorder might account for the differences between two strains of mice ("high" responders and "low" responders) that produce different amounts of antibody to certain antigens. Furthermore, macrophages seem to be the principal cells that destroy tumors, and tumors may gain a foothold by producing a substance that inhibits macrophages. Attempts to find out the nature of the inhibitory substance to see whether is can be counteracted hold some promise in cancer research.

Special Capabilities of Vertebrates

The vertebrates are the only animals known to produce *antibody molecules,* and they are also the only animals that become immunized against a specific

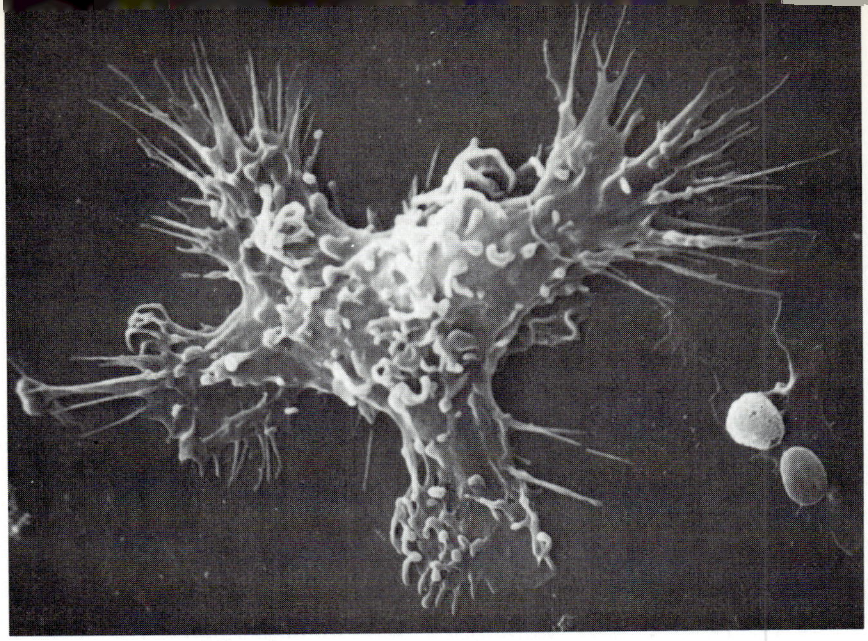

13–2
Scanning electron micrograph of a human monocyte that has been cultured for 24 hours. Notice the microvilli and cytoplasmic processes reaching out from the cell. This is a phagocytic cell. (Courtesy Ralph M. Albrecht, University of Wisconsin.)

infectious disease through having contacted the "causative" organism. These two processes, immunization and antibody formation, go hand in hand because increased numbers of antibody molecules are almost always associated with immunization. The antibody molecules are unique among the body's products in that they combine with certain foreign substances called *antigens.* When antibody molecules combine with the antigens on the surfaces of a microorganism, the result is often inactivation or death of the microorganism.

Although the synthesis of antibody molecules is exclusively a vertebrate phenomenon, the process is nonetheless tied to the not-so-exclusive phagocytes. A phagocyte macrophage is usually the first cell to get hold of a foreign antigen (Figure 13-3). The macrophage attaches the antigen to its surface, where the antigen stimulates a *lymphocyte* to initiate the synthesis of antibody molecules that combine with the antigen.

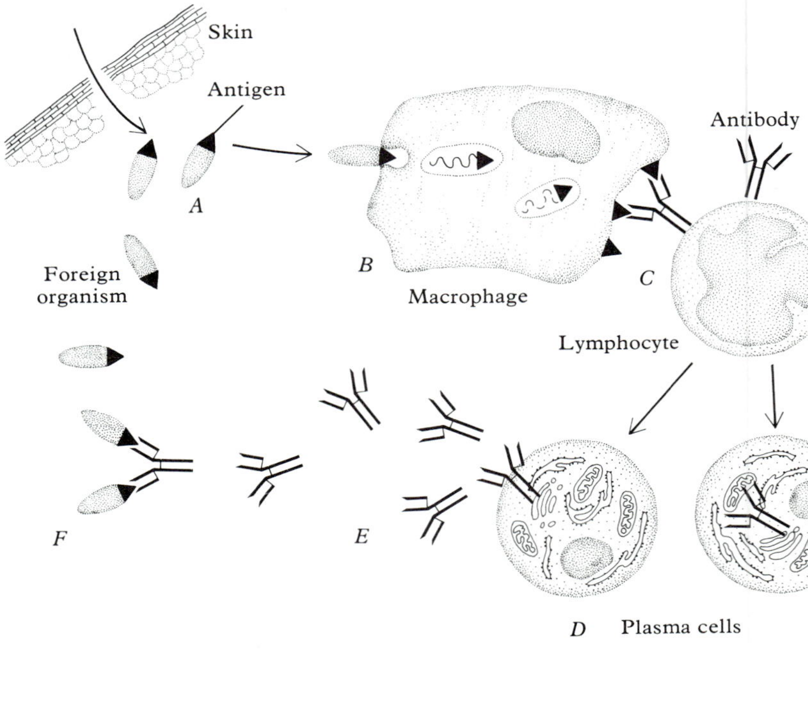

13–3
Immunization: events leading to the synthesis of antibody molecules
A, foreign organism (▶) penetrates the tissue and is taken into a macrophage by phagocytosis. B, the macrophage processes an antigen (▶) associated with the organism and attaches the antigen to its surface. C, a lymphocyte, which has antibodies ( ) on its surface that can combine with antigen, is stimulated when its antibodies contact the antigen. D, the lymphocyte divides and its offspring are transformed into plasma cells that synthesize antibody molecules. E, the plasma cells synthesize and release antibody molecules. F, these antibody molecules combine with the antigen and lead to inactivation of the organism.

In the minds of some researchers, immunization and antibody formation have become synonymous with immunology, and indeed remarkable strides have been made in immunization. But this view is narrow in that it does not acknowledge the importance of the factors we have already discussed, such as genetic background and general health, in determining the outcome of infectious disease processes. It is important to keep in mind that immunity as defined here refers to defense against disease rather than simply to immunization.

History of Immunization

Neither genes that confer resistance nor good general health can alone produce immunity. Until recently, the only people who were *fully* immune to any disease were those who had already had it. The ancient Greek scholars, Thucydides and Hippocrates, had observed that many people who survived the plague developed a lifelong resistance to it. But this natural way of developing immunity to infectious diseases was hazardous: the plague usually killed well over half of those who contracted it. Because of the devastating effects of contracting these diseases under natural conditions, ancient peoples developed methods of deliberately producing mild disease on purpose in order to protect against the more severe form. As long as 2000 years ago the Chinese had discovered that a mild form of smallpox would develop in people who inhaled powdered smallpox crusts from infected people. This mild form of disease, given deliberately, would confer lifetime resistance to smallpox. Another method of immunization, inoculation of the powder containing smallpox virus, was practiced for centuries in the Mideast and in Africa where it was called "buying" the disease.

Europeans and Americans did not learn about inoculation against smallpox until the eighteenth century, when it was introduced by an Englishwoman, Lady Mary Wortley Montagu (Figure 13-4). In the year 1717, Lady Montagu wrote to a friend about immunization as she had observed it in Turkey:

> The small-pox, so fatal, and so general amongst us, is here entirely harmless....People send to one another to know if any of their family has a mind to have the small-pox: they make parties for this purpose, and when they are met...the old woman comes with a nutshell of the matter of the best sort of small-pox, and asks what vein you please to have opened. She immediately rips open that you offer her, with a large needle (which gives you no more pain than a common scratch) and puts into the vein, as much matter as can lie upon the head of her needle....There is no example of anyone that had died in it: and you may believe I am well satisfied of the safety of this experiment, since I intend to try it on my dear little son. I am patriot enough to take pains to bring this useful invention into fashion in England. (From *Smallpox,* by C. W. Dixon. J. & A. Churchill Ltd., 1962.)

13–4
Lady Mary Wortley Montagu, who introduced inoculation for smallpox to England in the early eighteenth century after she had observed it being used successfully in Turkey. (Courtesy the National Library of Medicine, Bethesda, Maryland.)

Lady Montagu was a woman of action as well as a woman of words: she had her son inoculated in 1718, and in 1721 her daughter became the first person to be immunized while in England. We now understand that this immunization procedure worked because the living smallpox virus had been altered by drying, so that it would immunize and yet would not cause serious disease. This procedure of altering an infective agent so that it will confer immunity but not

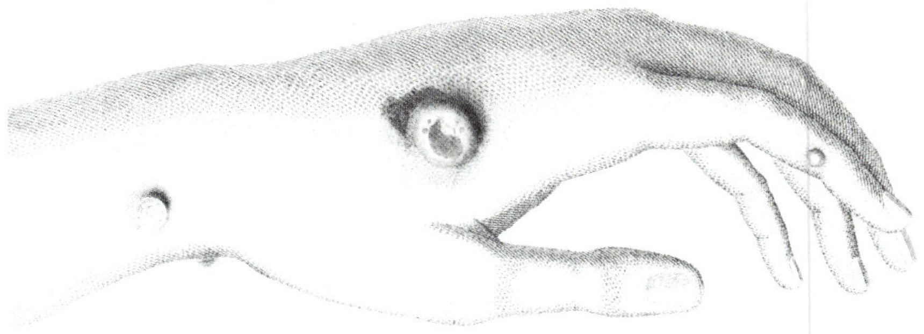

13–5
The hand of Sarah Nelmes, revealing a cowpox lesion. In 1796 Edward Jenner transferred cowpock material from her hand, to the arm of a young man, James Phipps. The cowpock material immunized Phipps so that he was resistant to smallpox. Inoculation with a virus probably derived from cowpox is still the preferred method of immunization against smallpox. (Courtesy the National Library of Medicine, Bethesda, Maryland.)

disease is called *attenuation.* Attenuation is the basis of most immunization procedures carried out today.

There were problems, however, associated with the inoculation of dried smallpox crusts. For one thing, there was not a standardized preparation procedure, and it was by no means clear how to get the "best sort" of smallpox to use. But the main difficulty was that the smallpox caused by the inoculation, although not fatal like smallpox caught during an epidemic, nonetheless could be a severe and disfiguring disease, *and a contagious one,* constituting a hazard to the community's health. For these reasons, inoculation quickly became controversial and was banned in parts of the United States. Nevertheless, when there was a smallpox epidemic, many people sought inoculation, accepting a greater risk for their neighbors in order to secure a lesser risk for themselves.

The Englishman Edward Jenner was responsible for improving the technique of immunization against smallpox. Jenner had heard the rumor that people who developed cowpox, a *mild* pox disease transmitted from cattle, were later immune to smallpox. He investigated and found the rumor to be true: those who tended the herds and milked the cows contracted the mild cowpox but not smallpox. So on May 14, 1796, he initiated an important and hazardous experiment, transferring material from a cowpock on the hand of a young woman named Sarah Nelmes to the arm of a young man named James Phipps (Figure 13-5). Six weeks later he risked the life of young Phipps by inoculating him with material from a fresh smallpock. Phipps did *not* develop smallpox, and Jenner published his findings in 1798. After an initial period in which his technique of inoculation was resisted and ridiculed (Figure 13-6), his procedure was finally accepted. The technique produced effective immunity against smallpox and had two great advantages as well: it (cowpox) produced a far milder disease and it was hardly contagious at all. Inoculation with a virus derived from cowpox virus is still the preferred method of immunization against smallpox. We now know that Jenner's procedure worked for the same reason that the Chinese and Middle-Eastern procedures worked: the cowpox virus caused immunity but no severe disease in people.

It is important to realize that during Jenner's time viruses had not yet been discovered, and no one even suspected that bacteria could cause infections. Perhaps for this reason immunization against smallpox was thought to be a special circumstance; in any case it took 80 years before anyone realized that immunization could be used to protect against other diseases as well.

In the 1880s, Louis Pasteur made the next important discovery. Pasteur had found that in chickens cholera was caused by bacteria, and he had been able to transmit the disease by taking bacteria from infected chickens, cultivating them in culture dishes, and then inoculating them into fresh chickens, which promptly developed the disease. At one point, by mistake, the cultures were allowed to age for an exceptionally long period of time in the culture dishes. In an attempt to recover something from the unusually old cultures, chickens were inoculated, but to no avail: they remained healthy. Finally, when fresh cultures of cholera bacteria were again obtained, they were found to kill normal chickens but to have no effect on the birds previously inoculated with the aged cultures. Pasteur had the critical insight: he realized that the aged culture had protected against the fresh culture. Within a few years, Pasteur extended these observations to other animal diseases as well, and he eventually coined the term *vaccination.* (The term is derived from the Latin word *vacca,* meaning cow, which honors Jenner's earlier work with cowpox.)

One of the most exciting chapters in the history of immunity is the story of Pasteur's first use of immunization as a treatment of human rabies. Rabies, a fatal disease, is caused by a virus that attacks the brain and spinal cord, producing a deranged mental condition. The virus is transmitted to people by bites of rabid animals—animals who have virus in their saliva and whose virus-filled brains bestow upon them unusual behavior. Pasteur had found that he could transmit rabies to a normal rabbit by injecting it with an extract of spinal cord from an infected rabbit. He also learned that if he dried out the spinal cord extract for several days, it caused less disease, and that if he let the extract dry even longer, it would not cause disease at all. He correctly reasoned that a normal animal would not get rabies if he injected it first with inactive extract, next with a weak extract, building up finally to a fully active extract. In rabbits, it worked. The first application of this procedure to a human being began on July 6, 1885, when Pasteur started this same treatment on a boy named Joseph Meister (Frontispiece) who had been seriously bitten by a rabid dog. Pasteur was at first unwilling to give Meister the treatment, but he was finally persuaded by the boy's parents, who knew of Pasteur's promising work with animals and who also

13–6
The inoculation of cowpox to protect against smallpox, which has eradicated epidemic smallpox wherever it has been used, was fiercely opposed and ridiculed. Note the people sprouting parts of cows in this satirical cartoon entitled "The Wonderful Effect of the New Inoculation," published in London in 1802. (Courtesy the National Library of Medicine, Bethesda, Maryland.)

knew that their son would die if untreated. Joseph received twelve injections of the virus-containing spinal cord extract, beginning with the inactive preparation and building up to the active extract. He survived, and within the next year the same treatment was given to hundreds of others who had been bitten by rabid animals. Pasteur's new treatment was a resounding success: among the people he first treated for rabies not a single death occurred.*

In spite of this dramatic breakthrough, the Pasteur treatment for rabies (like the earlier treatment of smallpox by inoculation of smallpox crusts) turned out to be a mixed blessing when applied to large numbers of people. The problem stemmed from the fact that spinal cord material was also inoculated along with the rabies virus, and a small proportion of the people given the spinal cord extract became immune to *spinal cord substances* as well as to rabies with the result that they developed an often fatal *autoimmune disease* (auto means self) in which they systematically began to destroy their *own* brain and spinal cord tissues. Since Pasteur's time, the risk of autoimmune disease from rabies vaccination has been greatly reduced by removal of most of the brain material from the virus preparation. But the risk has been hard to eradicate completely, because until recently the rabies virus has refused to grow outside of the living tissues of the nervous system, making it impossible to completely separate virus from nervous system material. This is the reason why rabies immunization is not given to everybody, but is reserved instead for people who have been bitten. Recently, it has been possible to grow rabies virus in duck embryo cells, which has resulted in a procedure that presents fewer risks. Perhaps soon the last slight risk of autoimmune disease will be eliminated.

How Immunity Works

When an infective agent enters the body, it has on its surface and releases into the environment its own molecules, or *antigens,* which are recognized by the body as being foreign. These antigens trigger events that lead to full immunity. First, a macrophage picks up the antigen (Figure 13-3). Some antigen is then transferred to the surface of the macrophage. Here the antigen can make contact with antibody molecules present on the surfaces of *lymphocytes.* Only one lymphocyte in 10,000 may be able to respond to a given antigen, because each lymphocyte carries only *one* kind of antibody, and there are perhaps 10,000 kinds of antibody molecules.† Nevertheless, when the antigen comes into contact with a lymphocyte carrying an antibody molecule that can combine with the antigen, it stimulates the lymphocyte to divide. Many of its offspring become plasma cells

*As fate would have it, Joseph Meister eventually became the gatekeeper at the Pasteur Institute in Paris, where Louis Pasteur's remains are housed in a crypt. In 1940, when Paris was occupied by Nazi Germany, the German military command requested the keys to Pasteur's crypt. Rather than give up the keys, Meister committed suicide.

†Two theories have been proposed to explain the existence of so many kinds of antibody molecules. The "germ line" theory says that all antibodies are coded for by thousands of genes present in the fertilized egg. The "somatic mutation" theory says that only a few genes in the fertilized egg give rise to thousands in the adult by a process of somatic mutation (which occurs during the lifetime of the individual). There is considerable excitement and controversy about these theories, and the issue is not settled.

that rapidly synthesize antibody molecules (Figure 13-7). Because this lymphocyte and its plasma cell offspring can synthesize just the one kind of antibody, within a week or two there will be billions of newly synthesized antibody molecules, each one capable of grabbing the antigen of the infective agent. The net molecular result of immunization will be a great flood of antibodies into the bloodstream. Immunization also results in a great increase in the number of memory cells ready to respond to this antigen (Figure 13-7). The memory cells are responsible for the long-lasting immunity that frequently results from contact with a given infectious agent.

Primary and Secondary Immune Responses

The synthesis and release of antibody molecules that follows an antigen's first entry into the body is known as the *primary response* (Figure 13-8). For a period of time (the latent period) after the first entry of antigen, no detectible antibody

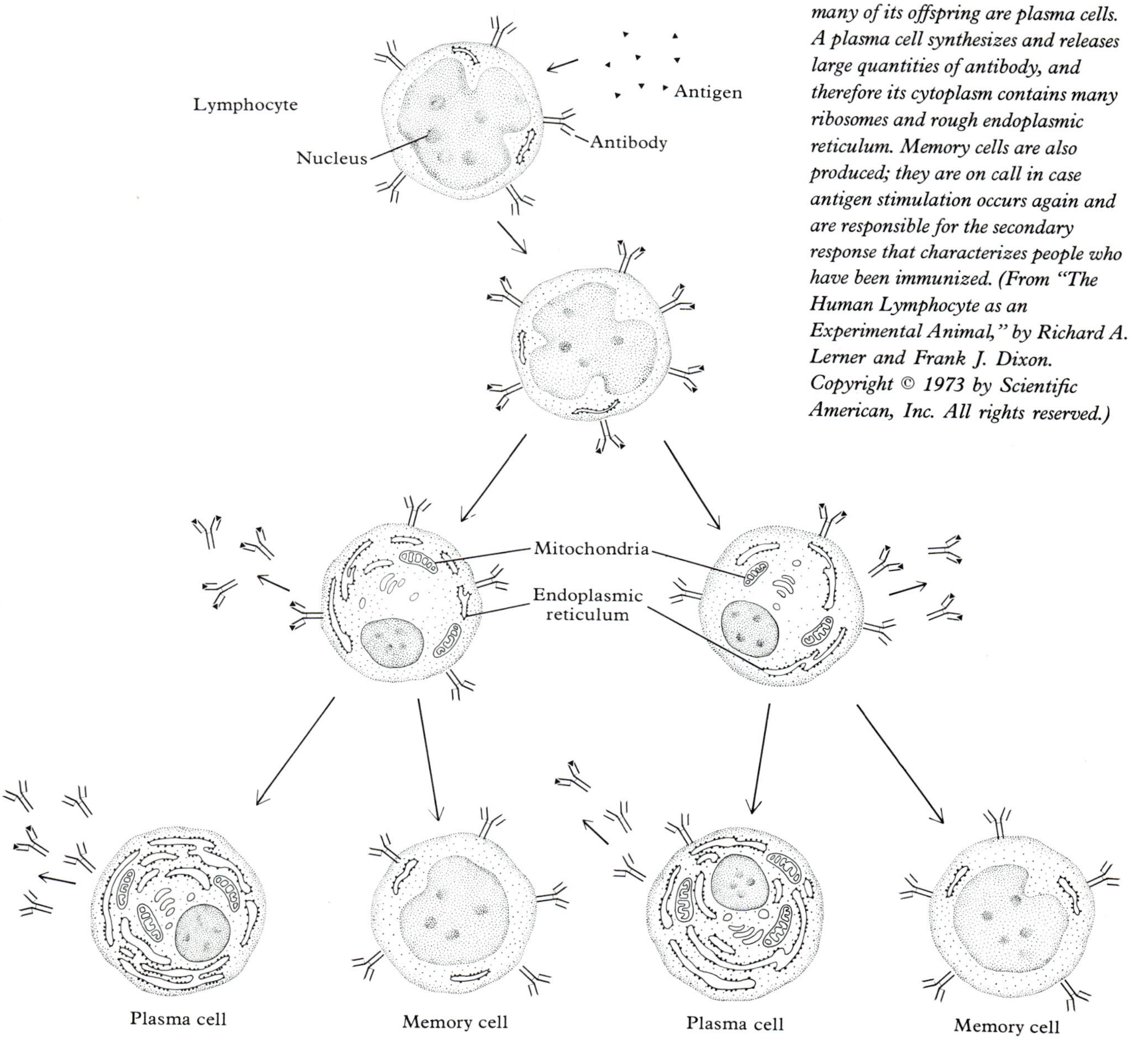

13–7
When a lymphocyte (top) is stimulated as a result of contact between antigen and its surface antibody it divides, and many of its offspring are plasma cells. A plasma cell synthesizes and releases large quantities of antibody, and therefore its cytoplasm contains many ribosomes and rough endoplasmic reticulum. Memory cells are also produced; they are on call in case antigen stimulation occurs again and are responsible for the secondary response that characterizes people who have been immunized. (From "The Human Lymphocyte as an Experimental Animal," by Richard A. Lerner and Frank J. Dixon. Copyright © 1973 by Scientific American, Inc. All rights reserved.)

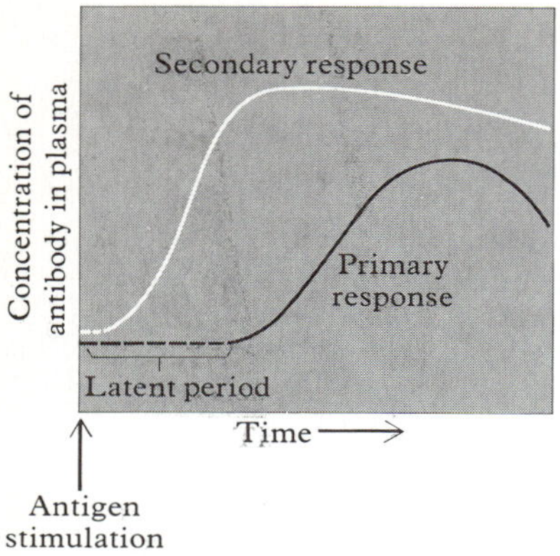

13–8
The primary and secondary immune responses. The primary response occurs after an antigen enters the body for the first time; the secondary response after the antigen enters for the second time. The secondary response is characterized by a shorter latent period, a more rapid rise in the concentration of antibody, a higher peak concentration of antibody, and a more gradual fall-off from peak levels. Memory cells (Figure 13–7) are responsible for the characteristics of the secondary response.

molecules to that antigen are present in blood plasma. During this time the antigen is being picked up by macrophages, recognized as foreign, and passed on to the appropriate lymphocyte. Following the latent period, the number of antibody molecules present in the plasma rises, peaks, and falls off. Upon second exposure to an antigen the *secondary response* occurs. This response is characterized by a shorter or nonexistent latent period, a more rapid rise in the number of antibody molecules present, a higher peak number of antibody molecules, and a more gradual decline. The differences between the primary and secondary responses result mainly from the participation of memory cells in the secondary response, and they account in large part for the fact that the first exposure to many infectious agents results in disease whereas second exposure does not.

One of the ways that antibodies provide protection is illustrated by the disease *malaria.* As you may recall from Chapter 4, malaria is caused by a protozoan that spends parts of its life cycle inside human red blood cells. During its life inside the human body, a malaria parasite enters a red blood cell and reproduces itself there. After a period of maturation, the offspring burst open the red cell and in so doing produce the characteristic malarial fever. The newly released parasites are free in the blood plasma for only a short time before entering other red cells. Antibodies grab the parasites during the short time they spend outside the red cell, and act by covering the parasites (by attaching to their surface antigens) so that the parasites are effectively blindfolded and cannot enter another red cell. This blindfolding is effective because the entry of a parasite into another red cell is totally dependent upon the ability of the parasite's surface receptors to attach it to the red cell. Thus, the parasite is doomed if its surface is covered by antibodies. Unfortunately, people develop this natural immunity to malaria over many years—hence the interest in the development of an antimalarial vaccine that could produce immunity more rapidly.

Antibody Structure and Function

Antibody molecules often do much more than just cover surface antigens: they frequently initiate actions that either damage the infective agent or prevent its

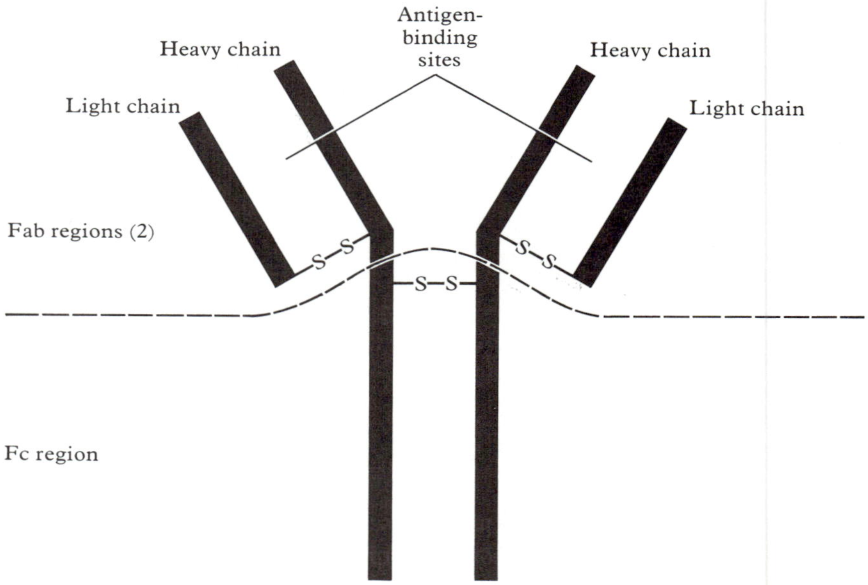

13–9
The structure common to all antibody molecules consists of four polypeptide (protein) chains, joined together by disulfide (—S—S—) bonds. Two of the chains are heavy chains, and two are light chains. Cleaving the molecule along the dotted line divides it into three regions: two Fab regions, both of which combine with antigen, and one Fc region, which determines other properties of the molecules.

Class (Ig = immunoglobulin)	Common structural form	Characteristic	Protective function
IgG	 (monomer)	Crosses the placenta Increases phagocytosis Initiates complement reaction	Protects the fetus and newborn infant Engulfs and destroys foreign microorganisms Leads to lysis (destruction) of foreign organisms
IgM	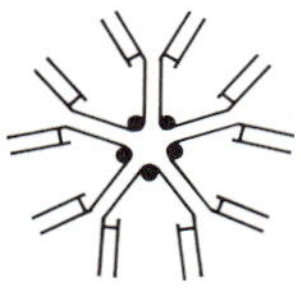 (pentamer)	Increases phagocytosis Initiates complement reaction	Engulfs and destroys foreign microorganisms Leads to lysis (destruction) of foreign organisms
IgE	 (monomer)	Causes histamine release Collaborates with eosinophils	Inhibits penetration of parasites (?) Lysis of parasites (?)
IgA	 (dimer with secretory piece)	Present in secretions (tears, saliva, intestine, milk)	Protection "outside" of the body's tissues
IgD			Unknown

13–10
The five classes of antibody molecules and some of their properties.

entry into the body. In order to understand how antibodies initiate these actions, we must take a look at the antibody molecules themselves.

Figure 13-9 shows the structure common to antibody molecules: two Fab regions (F stands for fragment, ab for antigen-binding) that combine with antigen, and an Fc region (c for crystallizable) that undergoes a change in shape after the Fab regions have combined with antigen. This change in shape of the Fc region sometimes triggers actions against the infective agent. As Figure 13-10 indicates, there are five classes of antibody molecules, each possessing a different Fc region and different characteristics. Thus the immune system may synthesize up to five different classes of antibody that can combine with any antigen. These five give the body more ways in which to fight an infective agent.

The five classes of antibody are usually designated as IgG, IgM, IgA, IgE, and IgD (Ig stands for immunoglobulin). Three actions can be initiated by changes in the shape of the Fc regions: (1) the complement reaction, which culminates in the punching of a hole in the cell membrane of the invader (Figures 13-11 and 13-12); (2) a stepped-up rate of phagocytosis of the antigen, which leads to digestion of the invader, and (3) the release of histamine from mast cells, which triggers an outpouring of mucus and helps protect against invasive organisms (Figure 13-13). Either IgM or IgG can initiate the complement reaction and increase phagocytosis, but only IgE can initiate histamine release. Only IgG antibodies pass through the placenta from mother to offspring, so they are responsible for protecting the infant from any microorganisms it might contact (Figure 13-17). IgA antibodies are present in the body's secretions (saliva, tears, intestinal fluid) and have a special extra molecule, or secretory piece, that makes them resistant to being digested by

13–11

Lysis of a cell by complement. Lysis is initiated when an antibody molecule combines with antigen on a cell surface (A), which causes a shape change in the antibody molecule exposing complement-binding sites (B). The complement-binding sites initiate activation of the nine components of complement in the order indicated. Components C_8 and C_9 are responsible for lysis of the cell because they enzymatically digest a hole in the cell membrane.

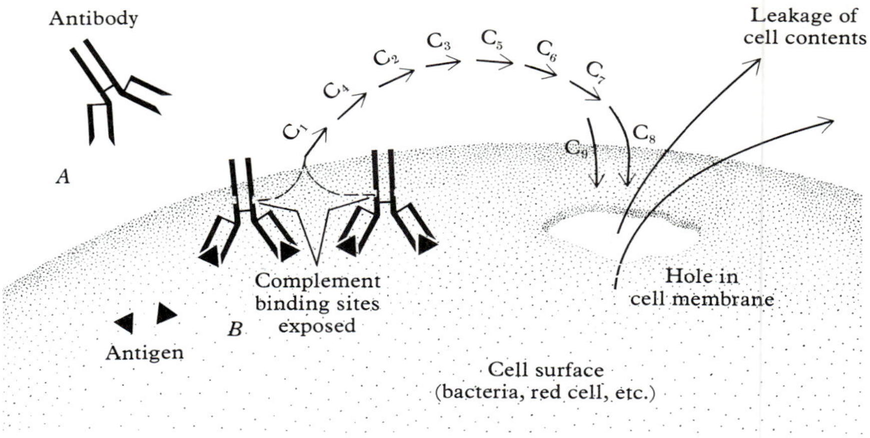

enzymes that would break down the other antibodies (Figure 13-14). IgA inactivates organisms in the body's secretions before they can penetrate into the tissues. Although the function of IgD is still unknown, it is clear that each class of antibody plays a rather specific role in the body's defenses.

In spite of the many beneficial effects of antibodies, they are sometimes responsible for problems. For example, IgE antibodies protect against invaders, but they also play a major role in causing allergies to many substances such as foods, pollens (Figure 13-15), animal fur, feathers, house dust (Figure 13-16), and insect stings.

13–12

Holes in the membrane of a human red blood cell treated with antibody in the presence of complement. The complement reaction, like the blood clotting sequence, is a cascading series of chemical reactions involving many plasma proteins. The last two proteins to be activated are enzymes that dissolve lipid components of the cell membrane. (Courtesy Robert R. Dourmashkin, Clinical Research Centre, Harrow.)

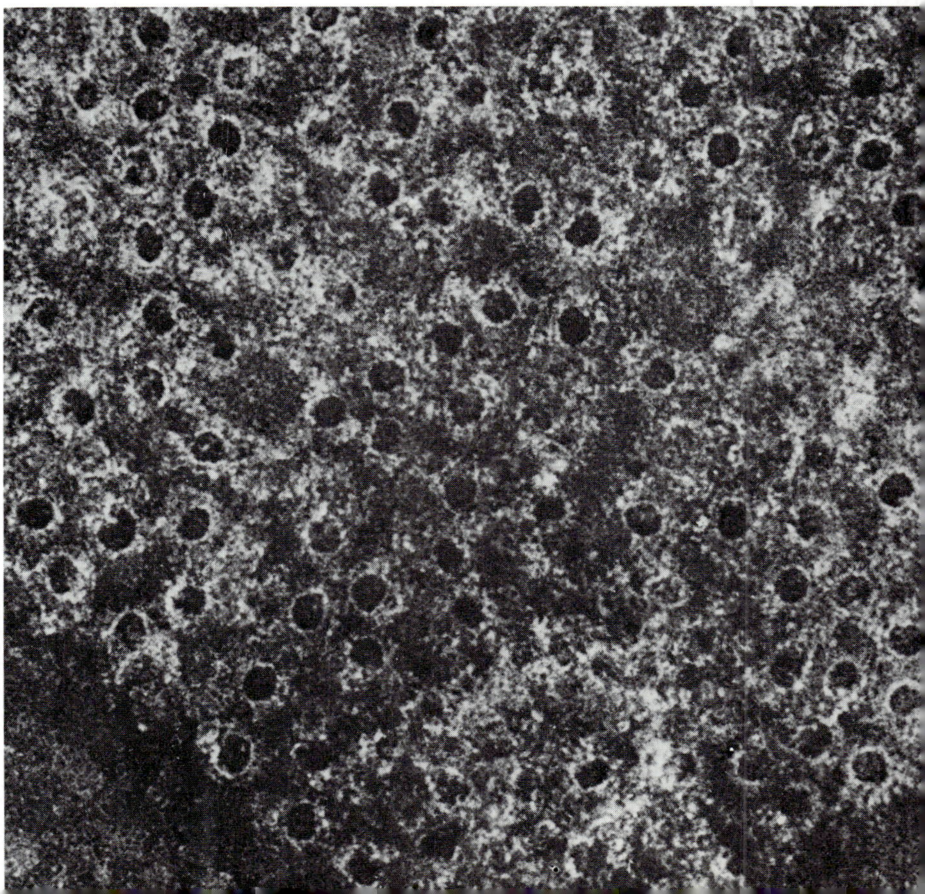

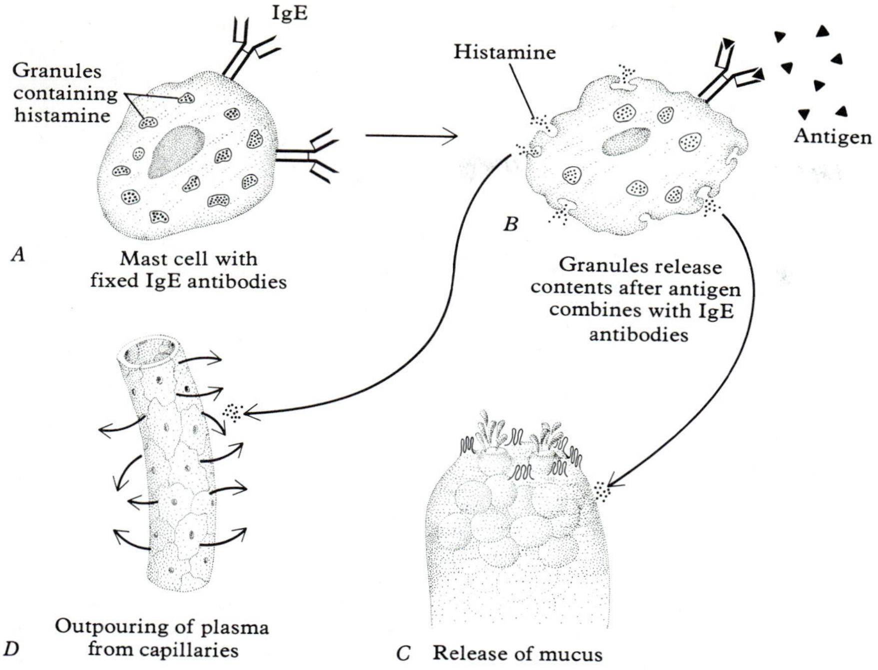

13–13

IgE antibody is fixed onto mast cells, A. When antigen combines with this fixed IgE the mast cell granules release their contents, B. The contents, which include histamine, stimulate mucus-secreting cells to produce a copious flow of mucus, C, and cause capillary epithelial cells to pull apart, resulting in an outpouring of plasma from capillaries, D.

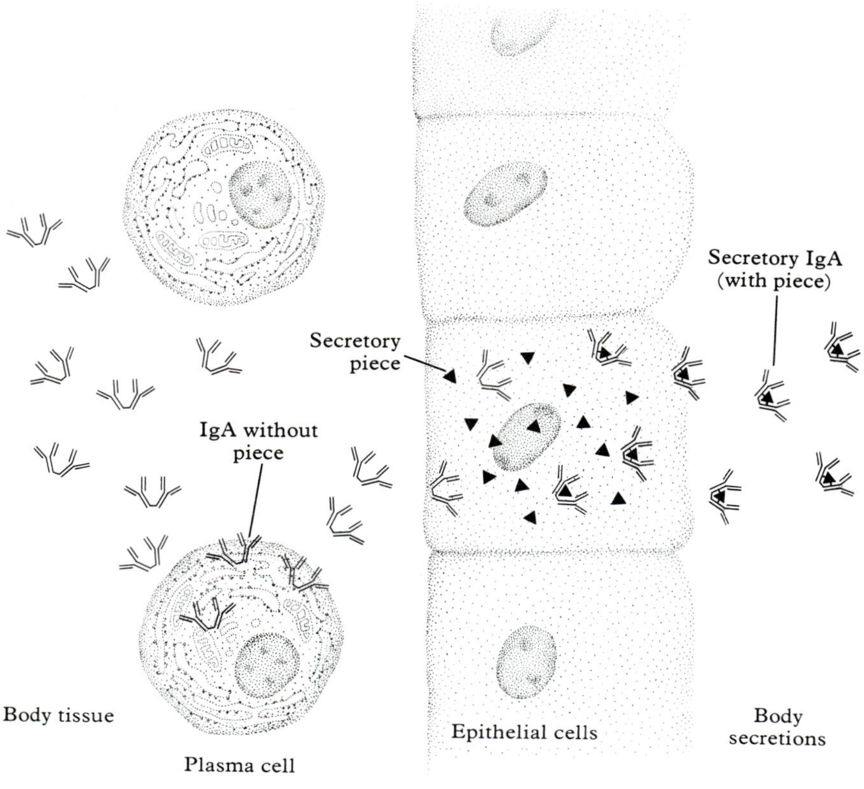

13–14

IgA antibody is found in secretions, even in intestinal secretions. It is not destroyed by protein-digesting enzymes because it is somehow protected by the "secretory piece," which it picks up as it is transported through epithelial cells.

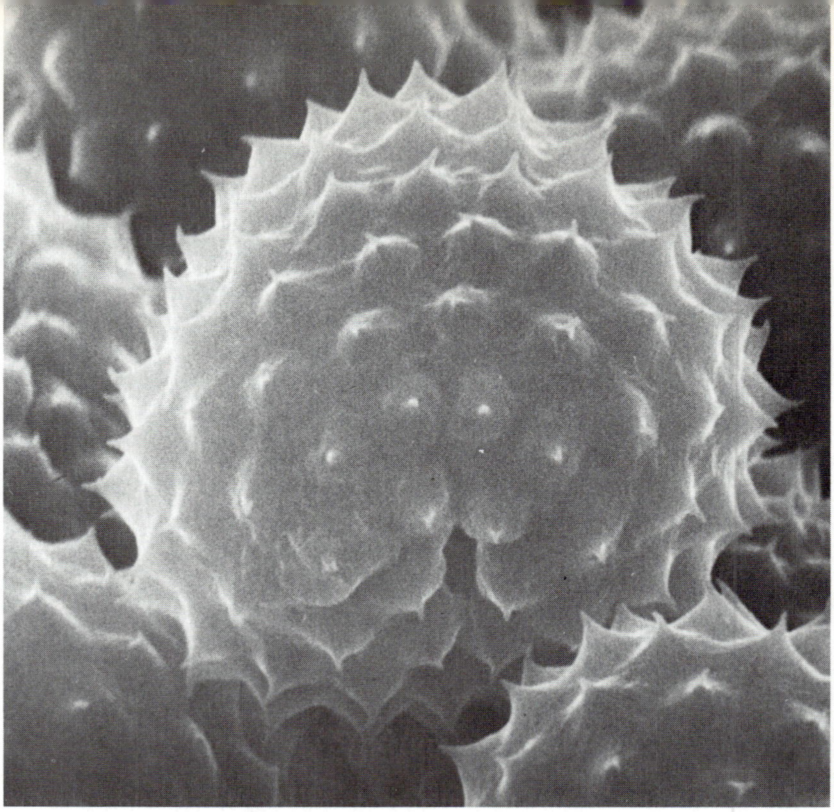

13–15
Ragweed pollen (×3340), as seen through the scanning electron microscope. These pollen grains can cause allergies by eliciting an IgE antibody response. (Courtesy Pat Lincoln, University of California at Santa Cruz.)

Schistosomiasis, Allergies, and IgE Antibodies

Large amounts of IgE antibodies are formed in response to organisms that actively penetrate through the outer epithelial surfaces of the body, whether through the skin, the gut wall, or the lining of the respiratory passages. Among the organisms that evoke the most active IgE responses are several kinds of worms, including the flatworms responsible for schistosomiasis (which penetrate the skin; see Figure 4-12).

The Fc region of an IgE antibody molecule gives it a special affinity for *mast cells,* and because of this affinity most of the IgE is attached to mast cells, which are found in abundance in the dermis of the skin and in the tissue surrounding the gut and respiratory passages. When IgE antibodies combine with antigen, the shape change in the Fc site triggers the mast cell to discharge its granules containing *histamine* as well as other substances. The histamine in turn triggers

13–16
The agent in house dust responsible for many allergies is this house mite, Dermatophagoides farinae *(×436), as seen through the scanning electron microscope. This almost invisible mite is present in house dust from all major regions of the United States. Many people allergic to house dust respond to an antigenic component of this ubiquitous creature by synthesizing IgE antibody molecules. Contact with large numbers of mites then triggers the IgE response in which histamine is released. The symptoms include itchy eyes and runny nose. (Courtesy G. W. Wharton, Ohio State University,* Science, *Vol. 167, pp. 1382-1383. Copyright © 1970 by the American Association for the Advancement of Science.)*

the mucus-producing cells of the gut and respiratory epithelium to secrete large amounts of mucus, which helps to sweep away the invaders. The histamine also causes the pouring out of plasma into the affected area of tissue, bringing all the other classes of antibody to the site. In addition, the IgE antibodies somehow work together with blood eosinophils to injure the invader.

The response to *ragweed pollen* provides an example of the problems that IgE antibodies can cause. The fierce looking appearance of this pollen (as seen through the electron microscope in Figure 13-15) has nothing to do with its troublemaking capacities; in fact the pollen itself is quite harmless and incapable of invading the body. Yet during the pollen season (when the pollen antigens are permeating the environment), allergic people, who have synthesized large amounts of IgE, inhale pollen and suffer from the typical hay fever response: itchy eyes and nose and copious mucus production. This response occurs when antigen molecules shed from the pollen's surface pass through the respiratory epithelium and reach the mast cells below. Because this response is mediated in large part by histamine, the *antihistamine* drugs are often effective in controlling the symptoms. But sometimes the antihistamines are not effective, or are only partially effective, because products of the mast cell other than histamine may also play a major role in the process. In spite of the fact that IgE antibodies should be triggered into action mainly by their corresponding antigens, it is becoming increasingly clear that emotional factors can also trigger these IgE-mediated reactions. The importance of emotional factors was shown in an experiment conducted with asthmatic children in a convalescent hospital. Stories were taped, and the tapes were played to the children. During the story, a parent's voice took over as speaker. At these times, it was found, the child of the speaker sometimes developed an asthmatic wheeze.

The "hives" of food allergies, in which there is an outpouring of clear plasma fluid into localized areas of the dermis, are another typical histamine response mediated by IgE antibodies. Hives are formed when the skin's IgE antibodies combine with food antigens that have been absorbed from the intestine and transported to the skin in the bloodstream.

Allergies to antigens introduced by *insect stings* can be very serious: the sudden injection of antigen molecules into the bloodstream by an insect may cause histamine to be released simultaneously in many parts of the body. This causes so much plasma to suddenly pour out of the bloodstream that the blood pressure cannot be sustained and life is endangered. An injection of the drug epinephrine can restore the blood pressure and may be lifesaving.

Because these IgE antibodies initiate discomforting and even life-threatening effects, a method called desensitization has been developed to rid the body of these allergic reactions. During desensitization the allergic person receives a series of injections of allergy-provoking antigen. This encourages the body to produce large amounts of other classes of antibody, particularly IgG. If enough IgG antibody is synthesized, the next time the antigen enters the body it will be stopped by the IgG antibodies (which don't cause any symptoms when they combine with antigen) before it ever meets the (harmful) IgE antibodies, and there will be no allergic response.

Protection of the Newborn Infant: IgG and IgA Antibodies Synthesized by the Mother

The newborn infant has had no contact with the world's microorganisms and thus no chance to build up its own immunities through experience, so one might expect that it would be particularly susceptible to many diseases. In fact, the newborn is only somewhat more susceptible than older children because it is protected by being born with large amounts of IgG antibody that were received from the mother across the placenta (Figure 13-17). In addition, during the first two days of life the infant who is breast-fed will receive IgA antibodies in the lemon-colored colostrum that precedes the milk. These class A antibodies have a special quality: they cannot be broken down and digested in the intestine like other protein molecules. For this reason they help to determine which organisms will be able to take up residence in the infant's digestive tract.

IgG antibodies from the mother are vital to the protection of the infant. For example, in areas of the world where there is a very great exposure to malaria, newborn infants receive immunity across the placenta that is equivalent to that of an adult who has developed active immunity by having been exposed to the malarial parasite. The incidence of malaria rises as mothers' antibodies gradually disappear during the first six months of life, and the disease reaches a peak incidence among children who are five years old. After five, one begins to develop one's own immunity to malaria, and the incidence of disease begins to decline toward the adult level.

Cellular Immunity and Cancer

The immune system of the body is really two systems: the one we have just discussed, called the *humoral immune system,* which consists of the circulating antibodies found in the blood, and the other, made up of antibody-like molecules (which may be unrelated chemically to the circulating antibodies) that stay fixed to the lymphocytes by which they are produced—the *cell-bound antibodies.* This second system is called the *cellular immune system.*

Cellular immunity has remained relatively mysterious, mainly because it has been difficult to isolate and characterize the cell-bound antibodies that produce it. Nevertheless, we know that cellular immunity is important in many ways: for example, it provides protection against many virus and fungus diseases. Very rarely, people are born without a cellular immune system, and because of this they suffer from repeated infections even though they are often able to synthesize adequate quantities of circulating antibodies.

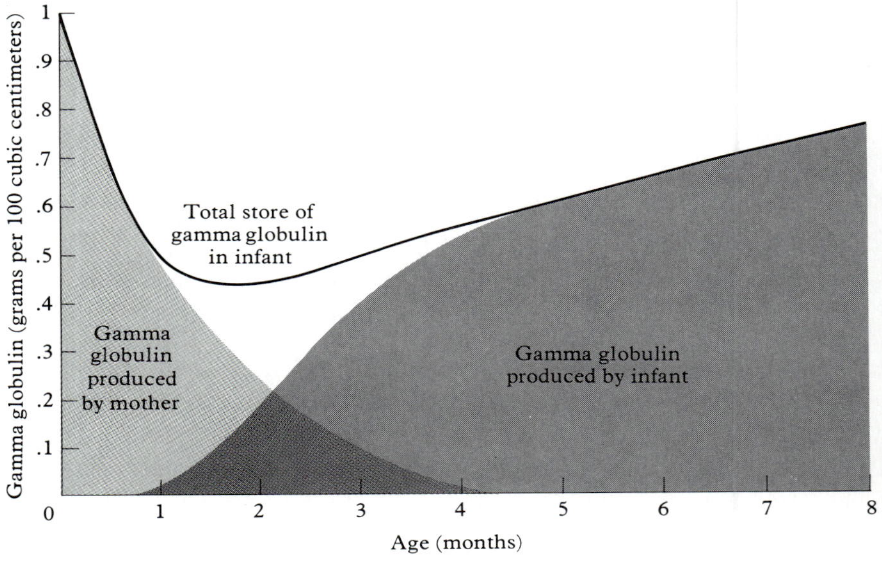

13–17
Infant's antibody supply (measured as gamma globulin) first comes from the mother's blood. This gift is gradually depleted in the first four months of life. The normal infant starts to manufacture its own antibodies before it is one month old and gradually builds up its store. The black line shows the actual concentration of gamma globulin usually found in an infant's serum during its first eight months. (After "Agammaglobulinemia," by David Gitlin and Charles A. Janeway. Copyright © 1957 by Scientific American, Inc. All rights reserved.)

A most exciting recent finding is that cellular immunity is also responsible for the body's defenses against *cancer.* Cancer cells are different from normal cells, and they almost always carry antigens that the body recognizes as foreign. It seems possible that abnormal, potentially cancerous, cells regularly appear within the body as a result of mutations that occur frequently among the body's rapidly dividing cells. If cellular immunity is responsible for the destruction of cancerous cells as they arise, then procedures that stimulate cellular immunity should help to prevent cancer. Apparently they do in some instances, and one example is immunization with BCG, a strain of *Mycobacterium tuberculosis.* BCG was originally developed as a vaccine to promote immunity to tuberculosis, but it was found that BCG also gives a powerful boost to the entire cellular immune system. Recently, investigators have searched through the records of thousands of children who were vaccinated with BCG to see whether these children had a lower incidence of cancer. In particular, they studied the incidence of *leukemia,* a cancer of the white blood cells, which is one of the common cancers of children. One study conducted in Canada revealed that twice as many unvaccinated children developed leukemia as vaccinated children. A study conducted in the United States showed that seven times as many unvaccinated children developed leukemia.

BCG seems to have had some effect in *preventing* cancer, and it has been used to *treat* cancer as well. In treatments, people who have cancer are injected with BCG, either in their arms and legs or directly in their tumor nodules. The results so far have been remarkable in the treatment of malignant melanoma, although occasionally BCG, being a live vaccine, has itself caused an infection in a severely debilitated patient. The prospects for the future of immunotherapy seem bright: it is likely that agents even more effective than BCG will be found. We still have much to learn about BCG, such as the most effective doses and routes of injection to use.

Special Topics

The Thymus Gland and the Development of the Immune System

In human beings, the thymus gland lies behind the breast bone, just below the notch between the collarbones (you can feel this notch) (Figure 13-18). Unlike most glands and organs, the thymus is at its maximum size in teenagers, and after that it declines, becoming a shrunken remnant of its former self by the time a person is 60 or 70. The thymus exerts its influence over the development of the immune system in two ways: by distributing lymphocytes to the rest of the body and by nourishing these lymphocytes with hormones it produces. The lymphocytes it distributes carry out the functions associated with cellular immunity.

During embryological development the thymus is the first place in the body where lymphocytes are found. If the thymus of an animal is surgically removed soon after it has formed in the embryo development of the immune system will not proceed and the animal will later die because of repeated infections. Furthermore, in some people the thymus does not develop; unfortunately most of them die of infections before they are a year old. Thus a properly functioning thymus is essential for the development of a properly functioning immune system.

In the evolutionary history of the vertebrates, the thymus appears as the first organized portion of the immune system. Invertebrates do not have an

13–18
Human thymus (shown here in a child) is a flat, pinkish-gray, two-lobed organ that lies high in the chest, in front of the aorta and behind the breastbone and partly behind the lungs. The thymus is large in relation to the rest of the body in fetal life and early childhood; then it grows less quickly, and by the age of puberty it has stopped growing and then begins to atrophy. This course of events suggested that the thymus completes most of its work early in life, but until recently its exact function was unknown. (From "The Thymus Gland," by Sir Macfarlane Burnet. Copyright © 1962 by Scientific American, Inc. All rights reserved.

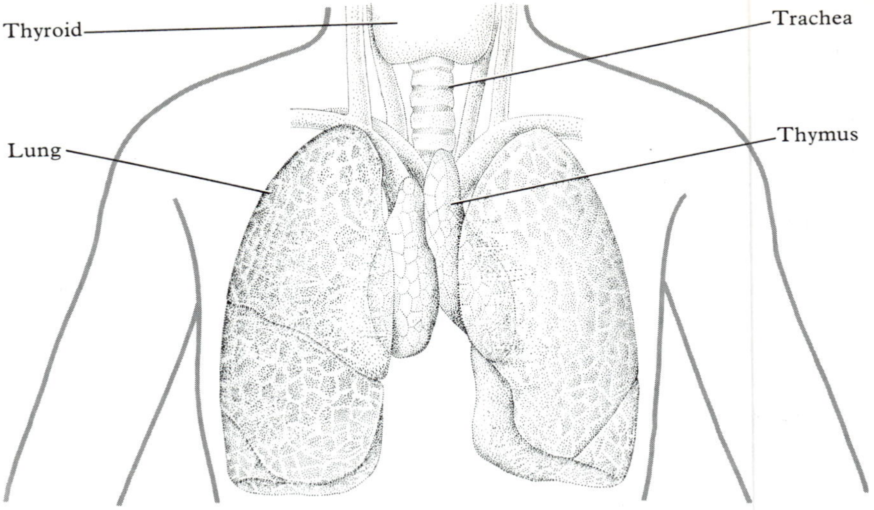

immune system like ours, and cannot be immunized against bacteria and viruses the way we can. Even the sea squirts, our invertebrate relatives that like us are chordates, do not have a thymus. The thymus first appeared in the lamprey, one of the two most primitive vertebrates (the hagfish is the other). All of the more highly evolved vertebrates, such as the bony fishes, amphibians, reptiles, birds, and mammals, have a thymus. So the thymus is the first organ of the immune system, not only in an individual's development, but also in the evolutionary history of our species.

The ABO Blood Group System and Blood Transfusions

In the human ABO blood group system, individual blood type is determined by the distribution of two genes, I^A and I^B, each of which codes for an enzyme that attaches a distinct molecule to the surface of red blood cells. A third gene, I^O, determines a molecule that is present on the red cells of *all* people, regardless of their ABO type. When the O molecule is the only one present (no A or B), then people have type O blood. People who have genes I^AI^A or I^AI^O are type A, people who have I^BI^B or I^BI^O are type B, and people who have I^AI^B are type AB (Table 13-1). People of type O (no A or B) make antibodies to A and B, and similarly, A-type people make anti-B and B type people make anti-A. Because O is found on all red cells, it is not recognized as foreign by anyone, and there are not any anti-O antibodies.

TABLE 13-1
The ABO blood group system.

PEOPLE WHO HAVE THESE TWO GENES (ONE FROM EACH PARENT)	HAVE THIS SUBSTANCE ON THEIR RED BLOOD CELLS	BUT LACK THIS SUBSTANCE, WHICH IS THEREFORE FOREIGN TO THEM	AND SO SYNTHESIZE THIS ANTIBODY
I^AI^O or I^AI^A	A	B	Anti-B
I^BI^O or I^BI^B	B	A	Anti-A
I^OI^O	O	A and B	Anti-A and anti-B
I^AI^B	A and B	Neither	Neither

Antibodies to the A and B blood group substances are produced very early in life, because the A and B substances are widespread in bacteria and in plants, and are therefore always present in foods. Because of these antibodies, if one attempted to transfuse type A blood into an O person, the O person's anti-A antibodies would initiate the rapid destruction of the type A blood cells, which would be punched full of holes through the complement reaction and eaten by phagocytes. This is why incompatible transfusions do not work. Type O blood is the only blood that can be successfully transfused into all recipients (because anti-O antibody does not exist, except in the very few people in the world who by some genetic "mistake" have no gene I^O and *do* make anti-O antibody). People of type O are thus called universal donors. As you can probably imagine, before the ABO blood types were discovered and blood typing developed, most transfusions were disastrous.

Today blood typing for ABO (and Rh—see the following section) is routine. If a drop of anti-A antibodies is mixed with a drop of A or AB red cells it will cause the cells to clump together, or *agglutinate.* Similarly, a drop of anti-B antibodies will agglutinate cells of type B or AB. These agglutination reactions provide a simple and reliable tool for determining blood type. As a check on this method, a person's own plasma antibodies can be tested to see which kind of cells they agglutinate.

Although many type O mothers have type A or B offspring (the I^A or I^B gene comes from the father), mother's anti-A or anti-B antibodies almost never do any damage to the offspring's cells. The reason is that the mother's anti-A and anti-B antibodies are exclusively IgM antibodies, and antibodies of type M do not cross the placenta into the fetal bloodstream. But occasionally maternal antibodies *do* cross into the fetus and then they can create life-threatening problems. The best known example of this is an incompatibility in the Rh blood group system.

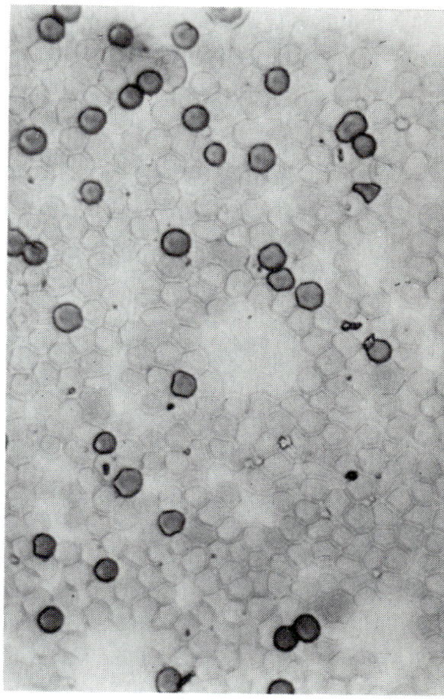

13–19
Fetal cells in the mother's circulation give rise to the Rh problem. In this photomicrograph the red blood cells of the fetus are darker than the mother's red blood cells because of a technique that has washed the hemoglobin out of the adult cells while leaving hemoglobin in the fetal cells that are more resistant to the technique. Rh-positive cells first enter the mother's circulation at delivery. (Photomicrograph by Flossie Cohen of the Child Research Center of Michigan.)

The Rh Blood Group System and Rh Incompatibility Between Mother and Fetus

Almost 90 percent of the problems of immunological incompatibility between mother and fetus result from the presence on the fetus's red cells of a single antigen, the *Rh antigen* (Rh stands for rhesus, the species of monkey where the antigen was first found). In Rh incompatibility, this antigen is present on the fetal but not on the mother's cells. The fetus has this antigen on its red cells because it inherited the gene to manufacture it from its father (Table 13-2). Presence of the antigen is called Rh positive (Rh^+) and absence Rh negative (Rh^-). The Rh antigen is different from the ABO antigens in that it is *not* widespread in nature, and anti-Rh antibodies are rarely produced before pregnancy.

If the Rh antigen is present on the fetus's red cells, it enters the mother's circulation at the time of delivery because some fetal cells pass into the mother's circulation as the placenta (afterbirth) tears away from the uterus shortly after the child is born (Figure 13-19). The entry of the Rh^+ cells stimulates an Rh^- mother to produce anti-Rh antibody of class G. Then if the mother becomes pregnant *again,* with *another* Rh^+ fetus, her IgG antibodies can pass across the placenta and initiate the destruction of the fetal cells (Figure 13-20). Remember that the problem of incompatibility hardly ever affects the *first* Rh^+ offspring, because the mother does not start synthesizing anti-Rh antibody until after the delivery of this offspring when she is first exposed to the Rh^+ cells.

The first synthesis (primary response) of anti-Rh antibody by the Rh-negative mother is initiated only when the Rh antigen enters her body on the surface of fetal red blood cells. The secondary response is different, however, in that it can be initiated by certain products of cell breakdown that contain Rh-antigen. Because these breakdown products leak across the placenta from fetus to mother, an Rh-negative mother carrying a *second* Rh^+ fetus may begin to synthesize large amounts of antibody as early as the fourth month of pregnancy.

When the red cells of an Rh^+ fetus are attacked by anti-Rh antibodies, they are destroyed and removed from the circulation. The fetus compensates for this loss by increasing red cell production. This compensation mechanism results in the appearance of a number of immature red cells, or *erythroblasts,* in the circulation. Hence the problem of Rh incompatibility is known as *erythroblastosis fetalis* (erythroblasts of the fetus). In addition to the adverse effects of the decrease in the normal number of red cells present (anemia), there is an even greater danger to the infant as a side effect of the increased red cell destruction. When the red cells are destroyed, the *heme* portion of hemoglobin is converted to *bilirubin,* the yellow pigment of jaundice. This bilirubin is toxic in high concentrations and can cause severe damage to the newborn's brain. Before birth bilirubin is removed from the fetal circulation by passage into the mother's circulation across the placenta, but after birth its removal depends upon the ability of the newborn's liver to secrete bilirubin into the intestine. A crisis occurs just after birth in many Rh^+ babies born to Rh^- mothers because the liver's capacity to excrete bilirubin is relatively poorly developed at birth, and concentrations of bilirubin may become dangerously high. The treatment of such a crisis is an exchange transfusion, in which the baby's blood (containing antibodies, affected cells, and bilirubin) is removed and replaced with normal blood. This must be performed before the concentration of bilirubin has been high for any length of time because damage to an infant's brain produced by high bilirubin levels is irreversible.

TABLE 13-2

Inheritance of Rh antigen. The Rh problem only occurs if the offspring is Rh positive and the mother is Rh negative. An Rh-negative mother lacks the Rh antigen and can be designated as genotype dd.* *If the father is Rh positive, signifying that he has the Rh antigen, he may have two Rh-positive genes (*DD*), in which case all of their offspring will be Rh positive. Alternatively, an Rh-positive father may be genotype* Dd, *in which case half of the offspring will be expected to be Rh positive.*

GENOTYPE OF RH-NEGATIVE MOTHER		GENOTYPE OF FATHER	GENOTYPE OF OFFSPRING
dd	×	*DD*	All *Dd* (Rh positive)
dd	×	*Dd* (Rh positive)	½ *Dd* ½ *dd*
dd	×	*dd*	All *dd*

*Phenotypes and genotypes are discussed more fully in Part III.

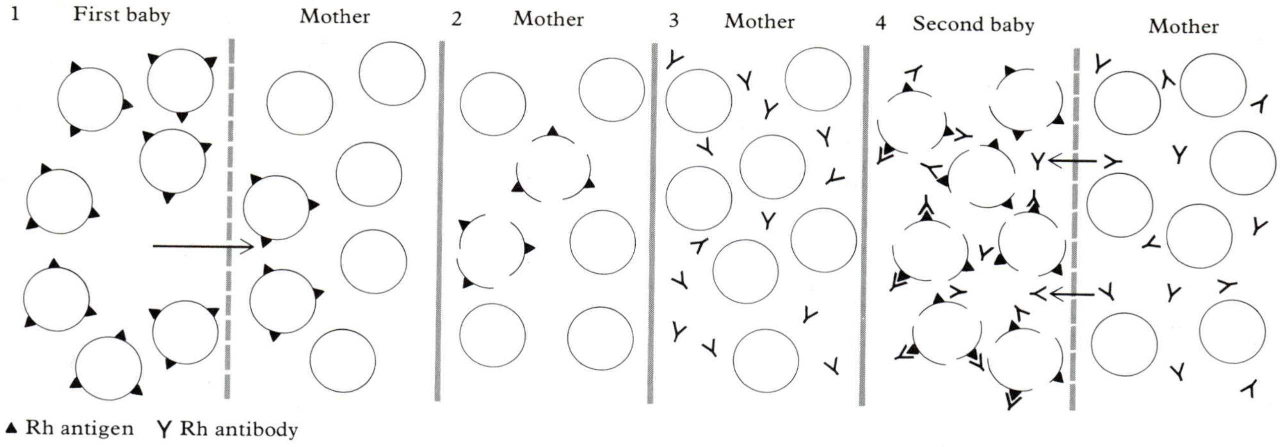

13–20
Steps in the development of the Rh problem begin (1) when an Rh-negative mother has an Rh-positive baby and some of the baby's red blood cells get into her circulation. Although the baby's cells soon disappear naturally (2), the mother may manufacture antibody to the Rh antigen (3). The first baby is not affected, because it has been born by the time the antibody appears. If the mother has a subsequent Rh-positive baby, however, the antibody may attack the baby's red blood cells (4), thereby giving rise to a possibly fatal anemia. (After "The Prevention of 'Rhesus' Babies," by C. A. Clarke. Copyright © 1963 by Scientific American, Inc. All rights reserved.)

It is now possible to prevent the occurrence of erythroblastosis fetalis. This is done by preventing the Rh^- mother from ever making antibodies against the Rh-antigen. At the time of delivery of the first Rh^+ child, a small quantity of concentrated anti-Rh antibodies (called Rhogam) is injected into the mother. These antibodies combine with any Rh^+ fetal cells that may have entered the mother's circulation and destroy them before they can stimulate the mother to synthesize her own antibodies against the Rh^+ cells. This is a simple procedure that is 99.9 percent effective.

Autoimmunity

We have already mentioned that the injection of a small amount of brain material (along with rabies virus) will sometimes cause the body to make anti-brain antibodies that can have destructive effects on the brain and spinal cord. Situations like this, in which the immune system damages the body's own self-components, are called *autoimmune.*

Autoimmunity sometimes results from the breakdown of the body's usually exquisite ability to tell the difference between foreign molecules and its own. An example is *rheumatic fever,* a disease that sometimes develops a few weeks after a streptococcal bacteria infection (such as strep throat). The streptococci stimulate the production of antibodies that react with heart muscle fibers, and the resulting immunological attack on the heart may cause impaired heart function and even death. The problem has been traced to the fact that the cell membranes of these streptococci contain molecules that are very similar to the molecules of the heart muscle fibers; the antibodies which the body synthesizes to attack the former also attack the latter (Figure 13-21). It has been estimated that during outbreaks of strep infections, about 3 percent of those infected will develop rheumatic fever. This is why strep throats are usually treated with antibiotics.

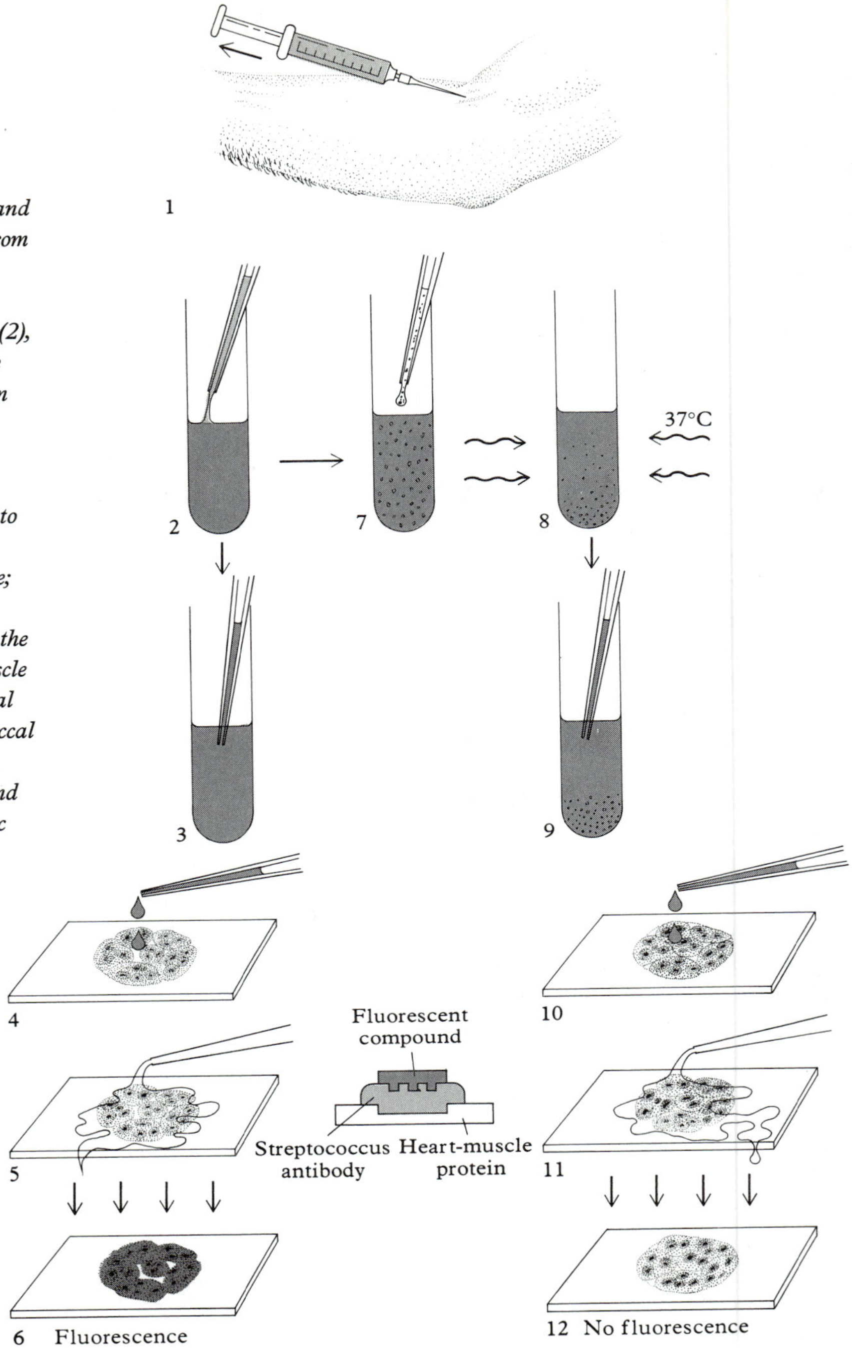

13–21
Immunological relation between the streptococcus and heart muscle is demonstrated by testing antibody from the blood of a rheumatic fever patient (1). The antibody-containing globulin is isolated from the patient's blood, coupled to a flourescent compound (2), and tested on heart muscle (3, 4). It is bound to the muscle, and in spite of washing (5) the heart section glows (6) under ultraviolet light. Is the reaction caused specifically by antibody to streptococcal membrane? An indication that it may be comes from the following procedure: Membrane is added to the globulin solution (7) and the suspension is incubated (8). Antibody is bound by the membrane; when the supernatant is tested on a heart section (9, 10, 11), the muscle fails to fluoresce (12). Thus the antibodies that would have reacted with heart muscle to cause fluorescence have reacted with streptococcal membrane instead. This result shows that streptococcal antigens and heart muscle components are similar. (From "Rheumatic Fever," by Earl H. Freimer and Maclyn McCarty. Copyright © 1965 by Scientific American, Inc. All rights reserved.)

Sometimes the body is damaged by an immune process even though there has not been a breakdown in the capacity to tell self molecules from foreign molecules. For example, during an infection, as the body begins to synthesize antibodies, some of these antibodies may combine *in the bloodstream* with antigens, forming antigen-antibody complexes. Because the kidney's *glomeruli* are constantly filtering the blood, these antigen-antibody complexes may be deposited on the glomeruli. If these antigen-antibody complexes contain IgM or IgG antibodies, as they often do, the complexes may initiate the complement reaction, causing holes to be punched in the glomerular membranes. The resulting condition, *glomerulonephritis,* commonly occurs during malarial infections or (like rheumatic fever) after streptococcal infections. Sometimes such kidney damage occurs only for a brief period during an infection, because the antigen is soon eliminated from the blood as the infection subsides and healing occurs spontaneously. But occasionally kidney damage continues, particularly if the antigen continues to be present, as in long-term malaria infections. When the process continues, drugs such as the adrenal steroids (which suppress immune reactions when given in large doses) are often useful in treatment.

Tissue Transplantation

The immune system will powerfully reject human tissues transplanted from one person to another, unless those tissues are from an identical twin. In fact, our bodies reject human tissues as powerfully as the tissues of other animals, even though one might consider human tissue to be only slightly foreign as compared with the tissues of other animals. The explanation may well be that this rejection capacity evolved in response to the body's need to reject its *own* cells that are slightly aberrant (e.g., precancerous cells).

The transplantation barrier, then, is a very strong one, and with the exception of kidney transplantation, the results of tissue transplantation are not impressive. One of the main problems is that it has not been possible to suppress immune responses *selectively.* Thus, in order to suppress the immune response of rejection, it has been necessary to use drugs that suppress immune responses in general, thereby increasing the risk of infection. Using these immunosuppressive drugs is a delicate process, because if *too little* of the drug is given the immune system may reject the transplant, and if *too much* drug is given infection may occur. Nonetheless, experience with kidney transplants shows that if the kidney can be prevented from being rejected for the first few months, eventually the body may come to tolerate it. And there is hope that new ways may be found to make the body even more cooperative.

An even brighter hope, however, is that the diseases that now require transplants may soon be preventable or at least controllable. Prevention is always more desirable than treatment, not only personally to the prospective patient, but in costs as well, and there is reason to think that prevention will eventually win the day. For example, kidney transplants are often performed to overcome tissue damage resulting from diseases such as chronic glomerulonephritis, and it should be possible to prevent much glomerulonephritis by simply distributing health care more widely so that the infections that trigger the disease can be treated. Also *early* treatment of glomerulonephritis can itself help prevent the need for transplantation.

There is historical precedent for thinking that expensive and long-term treatments may be replaced by inexpensive and effective prevention. Such precedents have been set by diseases such as typhoid fever, polio, and tuberculosis. For example, in the 1940s and 1950s many hospitals were built at great expense to house, treat, and take care of people who had tuberculosis and required lengthy hospitalization. However, with the discovery in the 1950s of new antibiotics that were effective against tuberculosis, low-priced medicines took the place of expensive, long-term care and led eventually to the closing of many of the long-term-care hospitals. Perhaps someday transplantation therapy will similarly be replaced by simple, effective, preventive measures.

Summary

Immunity, which is the capacity to resist disease, depends upon genetic factors, hormonal balance, and adequate nutrition. Protein malnutrition is probably the largest single cause of human disease because people who suffer from protein malnutrition are more susceptible to many infectious diseases.

Vertebrates and invertebrates alike are protected against microorganisms by their external covering layers and their phagocytic cells. However, vertebrates have an immune system that is unique in its capacity to synthesize antibody molecules. The thymus gland is the first organized portion of the immune system to appear during the evolution of vertebrates. The thymus is also the first organ of the immune system to appear during human development.

Antibody synthesis is stimulated by antigens, which are molecules not usually found within the body. Microorganisms entering the tissues are picked up by phagocytes, which put antigens on their surfaces. The antigens then stimulate lymphocytes to transform into plasma cells and synthesize large quantities of antibody molecules. The newly synthesized antibody molecules combine with the antigens and initiate the inactivation or the destruction of the microorganisms. Destructive mechanisms initiated by antibody include the complement reaction, which culminates in a hole being punched in the microorganism, and increased phagocytosis, which results in digestion of the microorganism.

There are three special regions on the typical antibody molecule: two Fab regions that combine with antigen and one Fc region that determines the general properties of the molecule. Each of the five classes of antibody has an Fc region different from the others. The five classes of antibody are IgG, IgM, IgA, IgE, and IgD. IgM and IgG can initiate the complement reaction and increase the rate of phagocytosis of antigen. IgG is the only class of antibody able to cross the placenta from mother to fetus. IgE mediates allergic reactions and defends against worm infections. IgA is found in the body's external secretions, where it is protected from digestion by the presence of a secretory piece. The function of IgD is as yet unknown.

The immune system can be divided into two parts: the circulating antibodies, and the cellular immune system, which is responsible for cellular immunity. Cellular immunity refers to responses of lymphocytes that synthesize antibody-like molecules but do not release them from the cell surface. Both parts of the immune system provide protection against diseases caused by microorganisms. Cellular immunity bears major responsibility for the body's defenses against cancer.

A blood group antigen is a substance found on the red blood cells of some people but not on the cells of others. In the ABO system, A and B substances are antigens for people who lack them. In the Rh system, the Rh antigen is a substance present on the red cells of some people (Rh^+) but not on the cells of others (Rh^-). The presence or absence of these various antigens is genetically determined.

If a fetus is Rh^+ and a mother Rh^-, fetal cells carrying the Rh antigen can enter the mother's circulation at delivery, causing her to synthesize anti-Rh antibodies. If she later has a second Rh^+ fetus, her IgG antibodies may cross the placenta into the fetus, causing a condition known as erythroblastosis fetalis, in which there is an accelerated breakdown of fetal red blood cells and a buildup of bilirubin, a breakdown product of hemoglobin. It is possible to prevent a mother from synthesizing anti-Rh antibodies by giving her an injection of anti-Rh antibodies after delivery of the first Rh^+ child.

Autoimmunity refers to conditions in which the immune system responds to the body's own self-components. Rheumatic fever is an autoimmune process because antibodies are synthesized which combine with heart muscle fibers. Rheumatic fever occurs about 3 percent of the time following severe streptococcal infections. Streptococcal membrane antigens and heart muscle components are very similar, and it is thought that the antibodies formed in response to the streptococcal membrane antigens also attack the heart muscle fibers.

The immune system serves as a barrier which prevents the transplantation of tissue between people unless they are identical twins. Immunosuppressive drugs help to suppress the rejection of grafted tissue, but they also increase susceptibility to infections. There is reason to believe that many of the conditions that are now treated by transplantation will soon be preventable or at least controllable.

Selected Readings

1. *The Body is the Hero,* by R. J. Glasser. Random House, 1976. Very well-written description of the events that have led to our understanding of how the body defends itself against disease. Science and history presented with excitement and drama.
2. *Essential Immunology,* 2d ed., by I. M. Roitt. Blackwell Scientific Publications, 1974. An introduction to the study of immunology.
3. "Macrophage Content of Tumors in Relation to Metastatic Spread and Host Immune Reaction," by S. A. Eccles and P. Alexander. *Nature,* 250 (1974), 667-669, 1974. An experimental analysis of the relationship between macrophages and cancer.
4. "Antiinflammatory Effects of Murine Malignant Cells," by R. M. Fauve, B. Hevin, H. Jacob, J. A. Gaillard, and F. Jacob. *Proceedings of the National Academy of Sciences,* 71 (1974), 4052-4056. Some cancer cells may produce a substance that inhibits macrophages.
5. "BCG and Cancer," by R. C. Bast, Jr., B. Zbar, and T. Borsos. *The New England Journal of Medicine,* 290 (1974), 1458-1469. An excellent discussion of the use of BCG in the immunotherapy of cancer.
6. "BCG Vaccination and Leukemia Mortality," by S. R. Rosenthal, R. G. Crispen, M. G. Thorne et al. *The Journal of the American Medical Association,* 222 (1972), 1543-1544. BCG vaccination may prevent the development of leukemia.
7. "Antibody-Dependent Cell-Mediated Damage to Schistosomula in Vitro," by A. E. Butterworth, R. F. Sturrock, V. Houba, and P. H. Rees. *Nature,* 252 (1974), 503-505. Discusses the role of antibodies in immunity to schistosomiasis.
8. "A Role for the Eosinophil in Acquired Resistance to *Schistosoma mansoni* Infection as Determined by Antieosinophil Serum," by A. A. F. Mahmond, K. S. Warren and P. A. Peters. *The Journal of Experimental Medicine,* 142 (1975) 805. Discusses the role of the eosinophil in immunity to schistosomiasis.
9. "The Prevention of 'Rhesus' Babies," by C. A. Clarke. *Scientific American,* Nov. 1968, Offprint 1126. Describes the history of the discovery that anti-Rh antibody can be used to prevent the development of immunity to Rh-positive cells.
10. "Rheumatic Fever," by E. H. Freimer and M. McCarty. *Scientific American,* Dec. 1965, Offprint 1028. Discusses the relationship between streptococcal infection and damage to the heart.

"Noah's Ark," By Kerstin Apelman Öberg, 1969.
(Courtesy Phyllis and Eberhard Kronhausen.)

CHAPTER 14

The Reproductive Systems and Contraception

All living things eventually die, but the species to which they belong can survive for much longer than a single lifetime because of reproduction. Reproduction adds new organisms to the populations that make up each species and thus helps to compensate for the inevitable death of individual members. Virtually all of the creatures that are familiar to most of us, especially animals, reproduce by means of sex. Individuals are of two kinds, female and male, and each contributes about one half of the genetic information according to which a new creature of the same species as the parents will be elaborated from a fertilized egg. (See Chapter 15.) On the other hand, many less familiar, relatively simple, organisms have been reproducing themselves without sex for billions of years. Most often such creatures simply split into two duplicate copies; only one parent is required. Sex and reproduction are thus two different things. What accounts for the widespread association of sex and reproduction among living things, especially animals?

Sex is a mechanism that has been strongly favored by natural selection for one simple reason: creatures that reproduce sexually, and thus mix their genetic material, produce offspring that are more variable. Because they are more variable they are better able to adapt to a wide range of environmental conditions than are the offspring of creatures that reproduce without sex by merely splitting into two copies of themselves. But although almost all animals reproduce sexually, it is only in the human species that sexual activities have sometimes become so varied and so clearly isolated from their reproductive context.

It is often said that human beings, compared with other species of animals, are extremely interested in sex, if not downright obsessed by the subject. There is surely some truth in this because, unlike most other animals, people frequently engage in sexual behavior with no intention whatsoever of reproducing; in fact, they sometimes go to great lengths to make sure that reproduction does not occur.

In this chapter our main concern is with the physiological basis of human sexuality as it relates to reproduction. Here we shall discuss some of the differences between people of opposite sexes and then consider the workings of both the female and male reproductive systems. We shall also discuss means of contraception that are capable of preventing reproduction, within the limits of chance. But first we should consider how the reproductive patterns of the human species differ from those of other mammals, especially our closest primate relatives.

Human Reproductive Patterns

Among many species of mammals mating occurs only once or twice a year during definite breeding seasons, at which time the female is sexually receptive (that is, willing to participate in mating behavior) and is said to be in heat. Nonhuman female mammals are capable of becoming pregnant only during these periods of heat. (In general, male mammals are sexually active only when the females are receptive, but, unlike females, male mammals of many species can be aroused sexually at any time.) Female dogs, as is well known in many neighborhoods, come into heat for about seven to nine days each spring and autumn. So do female cats, for whom the period of heat of nine or ten days may occur more often. Other female mammals—rats, sows, and cows, for example—have less definite breeding seasons because nonpregnant females go into heat more frequently, from every few days to once a month or so. Most species of our nearest animal relatives, the primates, do not have specific breeding seasons, but nonpregnant females are nonetheless sexually receptive for only a few days each month. Only during this time are they capable of conceiving.

What about human females? As we shall discuss later, their reproductive events, like those of other female primates, occur in cycles. For our species a complete cycle takes place about once each month, and a woman can become pregnant only during a brief time within her monthly cycle. But unlike that of other female primates women's sexual receptivity is *not* restricted to any fixed and hormone-dependent times, but rather depends mostly on circumstances, judgments, and inclinations, not hormones. The relatively more pronounced sexual interest that characterizes females of our species probably arose as an adaptation to ensure the maintenance of the very strong bonds between individual men and women that must be present if human pairs are to stay together long enough to raise families.

The maintenance of strong interpersonal bonds between mates probably also accounts for the intriguing fact that among female animals we know of only women experience orgasm. It is said that mating face to face, which is done only by humans, some chimps, and some gorillas, has contributed to the formation of close interpersonal bonds between mates and has therefore strongly influenced the evolution of human sexual patterns. There is probably something to be said for this, but the insistence of some about the importance of mating face to face probably reflects not biologic relevance, but rather personal opinion.

However it evolved, there is no doubt that the human reproductive pattern has been an enormously successful one. Nearly *four billion* people inhabit our planet, which is reason enough for all of us to be familiar with how human reproduction takes place and with how it can be voluntarily prevented. We now turn our attention to the biological basis of some obvious physical differences between people of different sexes.

Femaleness and Maleness—Sex Hormones

Like all other mammals, human beings are either females or males, and the differences between the two usually are readily apparent (Figure 14-1). What accounts for the obvious differences between the bodies of women and men?

First of all, as we will discuss further in Chapter 17, the two sexes have different sex chromosomes: women are XX and men are XY. Because an individual's sex chromosomes are determined at the time of fertilization, the tendency toward biological femaleness or maleness is also determined at the moment of the union of egg and sperm. The most crucial biological difference between the sexes is that women produce eggs and men produce sperm. Although biological sex is determined at the time of fertilization, human embryos remain sexually undifferentiated until about the seventh week of intrauterine existence, at which time it becomes clear whether the embryos' previously uncommitted gonads will eventually develop into egg-producing *ovaries* or into sperm-producing *testes.* (By the fourth fetal month the external sex organs have developed and the sex of the fetus usually becomes apparent. See Chapter 15.)

We have much to learn about how sex chromosomes bring about their effects and about the details of the early stages of human sexual differentiation, but we do know that sex hormones are of crucial importance in both of these processes. The sex hormones that are most abundant in women are *estrogens* and *progesterone,* and those that are produced in higher concentrations in men are called *androgens,* but members of both sexes normally produce some quantities of both "female" and "male" hormones (Figure 14-2).

The presence of one or another of these sex hormones in sufficient concentration at a particular stage of development probably affects the immature brain of the human fetus in a way that results in further development along unmistakably female or male lines. Sex hormones may even bring about differences in female and male brains, but so far there is no convincing evidence for this.

The important influences of sex hormones on sex are not exerted only before birth. They continue to influence sexuality throughout life, but become of extreme importance once again, as they are during early fetal existence, at the time of *puberty*—those oftentimes painful years during which girlhood and boyhood end and children become biological adults capable of reproducing.

Beginning at puberty sex hormones are secreted in quantities sufficient to strongly influence such characteristics as the texture and distribution of hair on

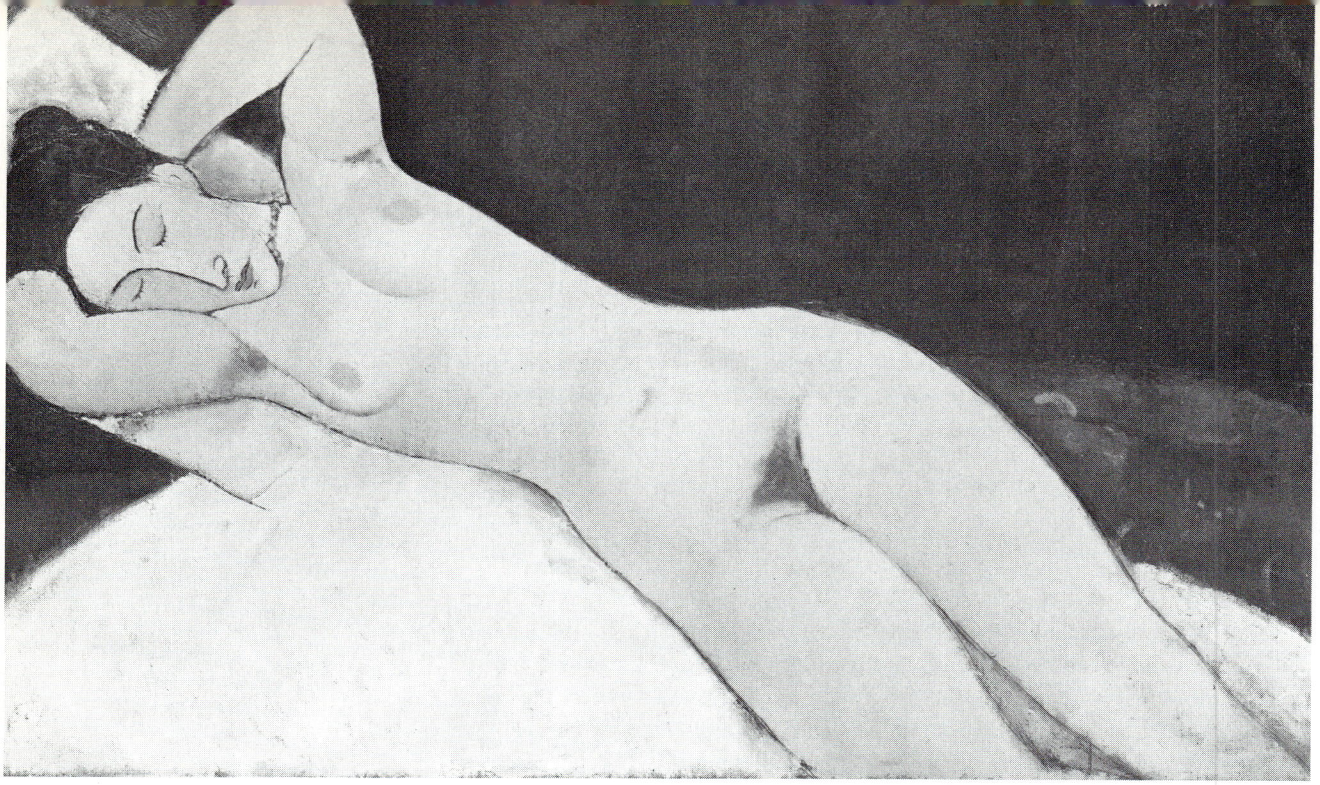

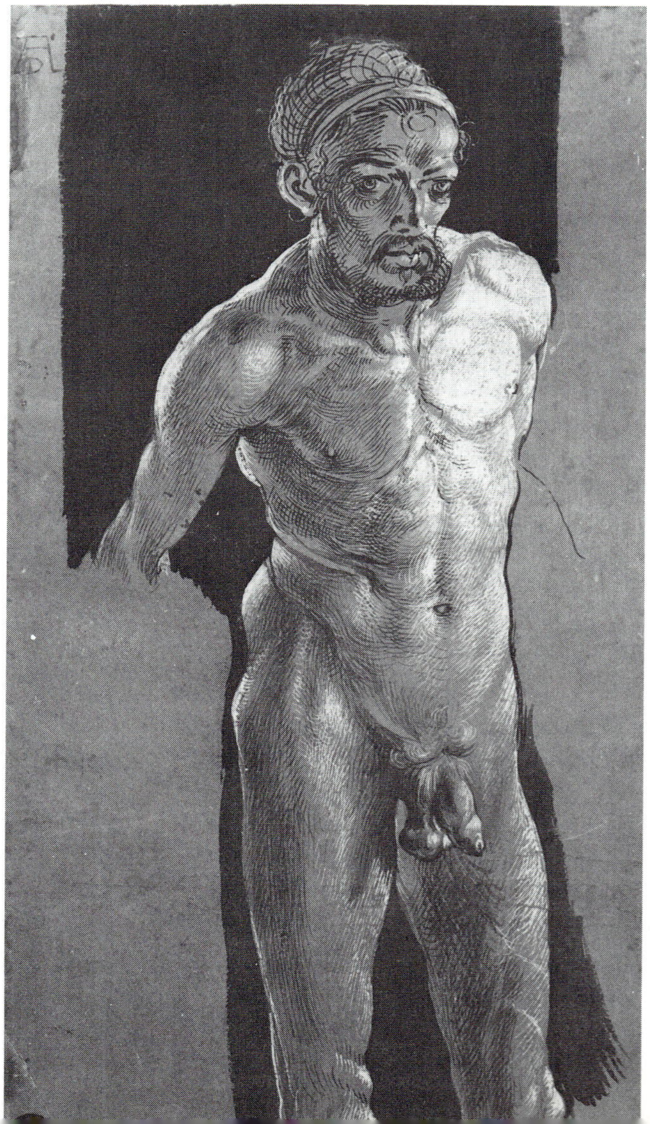

14–1
Top, "Reclining Nude," by Amedeo Modigliani, 1917. Bottom, "Nude" (self-portrait), by Albrecht Dürer, 1502. (Top, courtesy of The Solomon R. Guggenheim Museum, New York. Bottom, courtesy of Schlossmuseum, Kunstsammlungen Zu Weimar-DDR.)

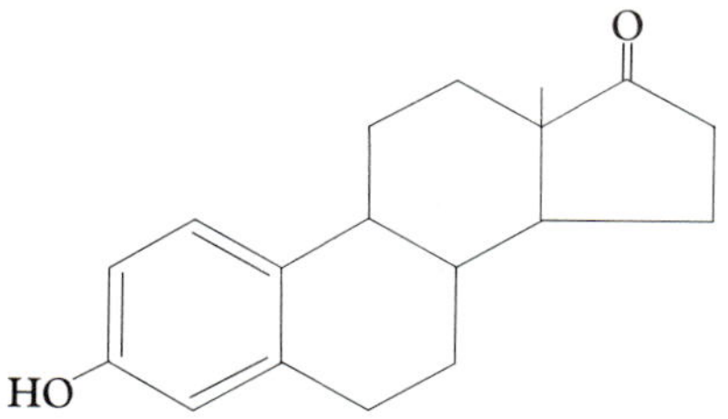

Estrone

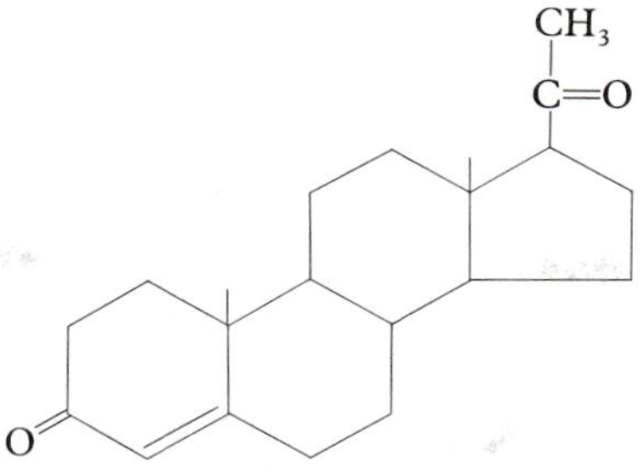

Progesterone

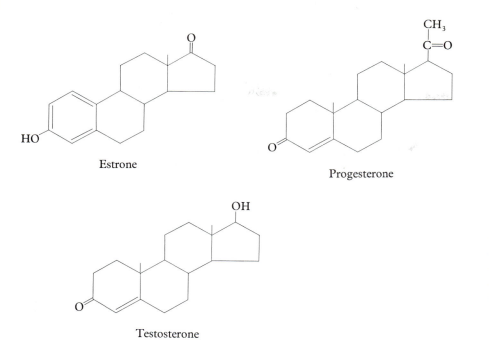

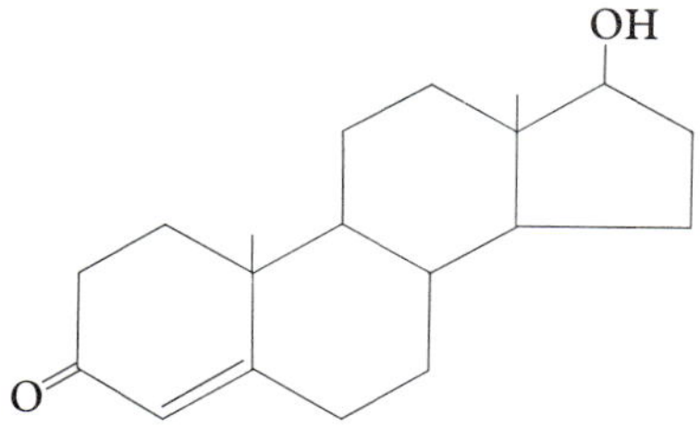

Testosterone

14–2
Some of the main "female" and "male" sex hormones. As you can see, there are relatively few differences between these molecules. Estrone is an estrogen. Testosterone is the major androgen.

the scalp, face, and body; the resonance and quality of the voice; body proportions and contours; the development of breast tissue; fat distribution and muscle mass; and many other body features that consistently are different for the two sexes and that are known as secondary sexual characteristics (Figure 14-3). (The primary sexual characteristic is whether a person has ovaries, and therefore produces eggs, or testes, and therefore produces sperm.) Except for the observation that excess androgens in women often result in a greater sexual appetite, there is no apparent relationship between concentrations of hormone and the desire for sexual activities.

From these general considerations of the physical differences between people of opposite sexes, we now turn our attention to a consideration of the anatomy and physiology of human reproduction. We will now discuss the structure and functions of the human female and male reproduction systems. You should remember that both probably develop under the influence of one or another of the sex hormones, whose widespread effects will become more evident in the pages that follow.

The Anatomy of the Female Reproductive Tract

Figure 14-4 is a diagram of the external female genital organs, which are known collectively as the *vulva.* The most visible part of the female genitalia is a small hillock of fatty tissue, known as the *mons pubis* (mount of Venus) which is covered by skin bearing coarse pubic hairs. The opening into the *vagina* is flanked by two pairs of fleshy folds, one larger than the other, known as the *labia majora* (major lips) and *labia minora* (lesser lips). Above the vaginal opening is the opening of the *urethra,* through which the urinary bladder empties its

14–3
Secondary sexual characteristics of male and female animals are brought out by sex hormones acting in concert with hereditary factors. Estrogen produces the plumage of the female pheasant. Testosterone causes the development of comb and spurs in the white leghorn rooster, the antlers of the white-tailed deer, and probably the mane of the lion. (From "Hormones," by Sir Solly Zuckerman. Copyright © 1957 by Scientific American, Inc. All rights reserved.)

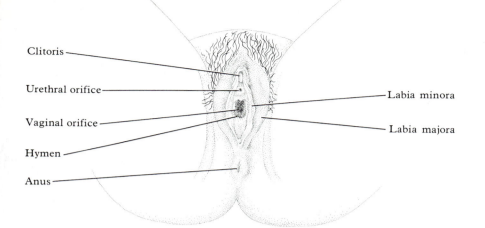

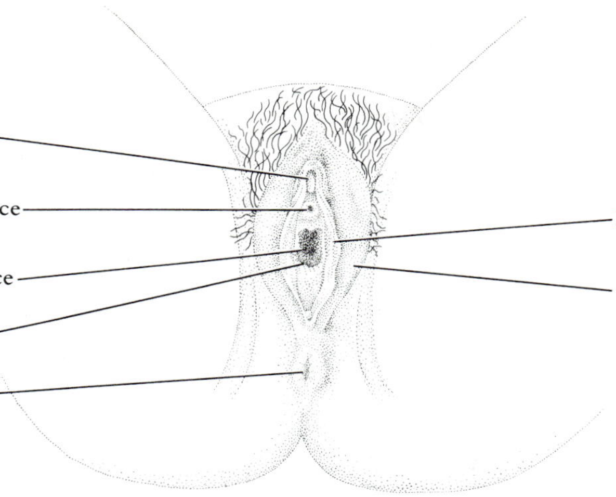

14–4
The external female genitalia.

contents, and above the urethral opening is the *clitoris,* a structure whose sole purpose is to heighten sexual arousal through its extremely rich supply of nerves. The *hymen* is a thin pink-colored membrane that partly occludes the vaginal opening in most women who have not had sexual intercourse, but because this delicate tissue can become torn accidentally during nonsexual activities its presence is not a reliable measure of virginity.

The *vagina* is a muscular tube about 7 to 10 centimeters (3 to 4 inches) long whose walls are usually collapsed on one another but can be stretched to accommodate the head of an infant at birth. As shown in Figure 14-5, the uppermost portion of the vagina is occupied by the *cervix,* which is the lower portion of the womb, or *uterus.* (In a Pap smear, named after George Papanicolaou (he devised the technique), which is a routine screening test to detect precancer or cancer of the cervix, some cells are scraped from the surface of the cervix and then examined for evidence of abnormal cells.)

The uterus is a thick-walled muscular organ that is about the size and shape of a small pear during the reproductive years. The cavity of the uterus is a narrow chamber lined by a tissue known as the *endometrium,* which undergoes a series of cyclic changes during the menstrual cycle, as we shall soon discuss. As shown in Figure 14-5, near its top the uterus is joined by the two *uterine tubes,* also called the *Fallopian tubes* after the sixteenth century anatomist Gabriello Fallopio, who first described them. (He thought they were chimneys for the escape of "sooty humors" from the womb!) Each tube is about 10 centimeters (about four inches) long, slightly swollen and funnel-like at the end, and fringed by fingerlike projections. This arrangement makes it likely that the single egg released each month from one or the other ovary will find its way into the tube and to its outer part, where fertilization by sperm takes place.

The ovaries are located one each near the sides of the pelvis and are suspended by the *broad ligaments,* which reminded early anatomists of the wings of a bat. Like the testes of the male, the ovaries serve the dual purpose of manufacturing sex cells and sex hormones. But there the similarities end: the production of sex cells by women and men are very different processes. Whereas from puberty on men continually manufacture sperm cells in astronomical numbers, each woman usually produces only a single egg each month, or a total of about 400 sex cells during her reproductive years, which extend from puberty to menopause (see the following section). In order to maximize the chances that

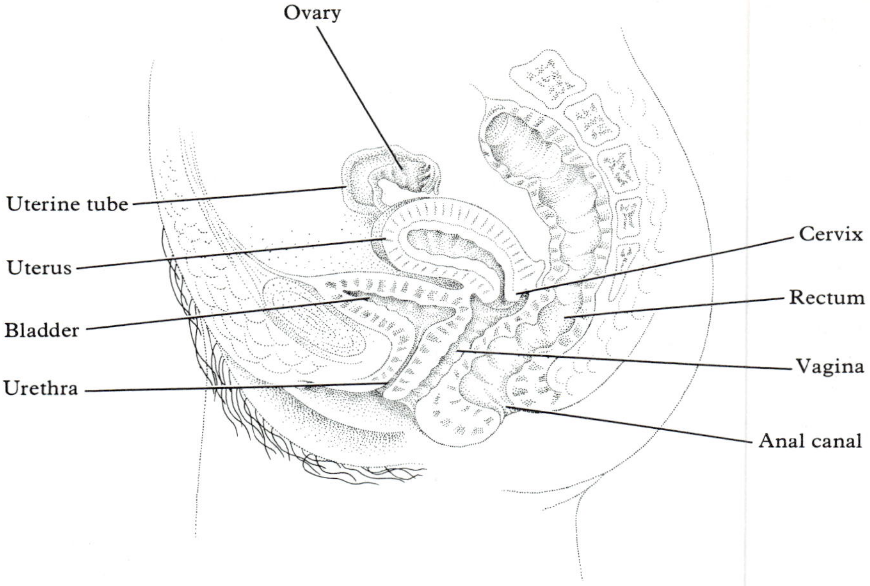

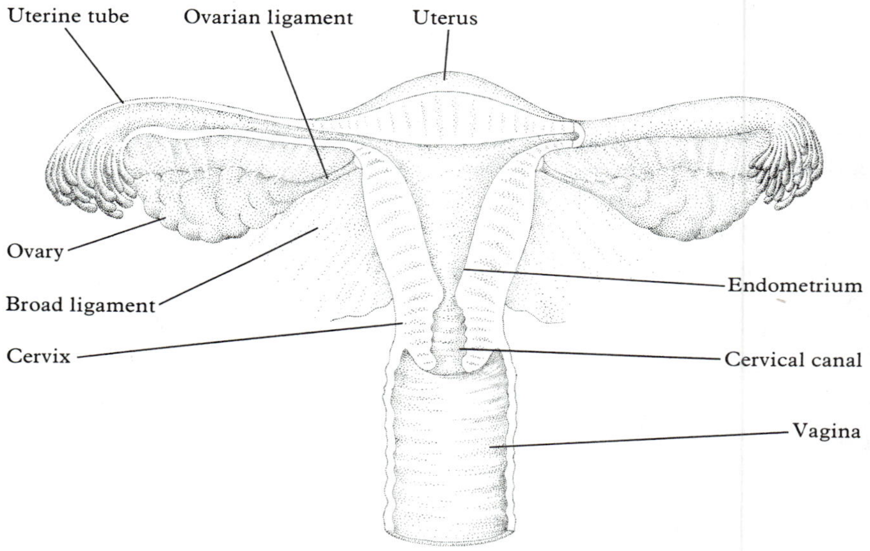

14–5
Top, Side view of the female pelvis. Bottom, Front view of the internal female reproductive organs.

an egg that has been fertilized in the outer uterine tube will not go to waste, our species has evolved a wonderfully complicated mechanism that cyclically affects the endometrium in a way that makes it most favorable to the further development of the fertilized egg at exactly that time of month when a fertilized egg is most likely to be present. This mechanism is known as the *menstrual cycle.*

The Menstrual Cycle

The menstrual cycle is best considered as two inter-related cycles, one of which takes place in the uterus and depends on the other, which takes place in the ovary.

The Uterine Cycle

Menstruation refers to the roughly three to five days a month during which the endometrium of the previous cycle is sloughed off from the uterine cavity and then passes through the opening of the cervix (the *cervical os*) into the vagina and thence out of the body as menstrual blood. The first day of flow is considered the first day of a woman's menstrual cycle (or *period*), which usually takes about 28 to 30 days to complete.

When the flow of menstrual blood has stopped, the endometrium that remains is paper thin, but it becomes progressively thicker over the next two weeks as uterine glands and blood vessels within the endometrium proliferate. As shown in Figure 14-6, this *proliferative* phase of the endometrium depends on the presence of the sex hormone estrogen, whose production by the ovary we shall discuss later. On about day 14 of the menstrual cycle *ovulation* occurs, which means that a single egg is released from one or the other of the ovaries. Following ovulation the sex hormone progesterone becomes increasingly abundant; under the influence of estrogen and progesterone the endometrium becomes thicker, the endometrial glands secrete nourishing fluids, and the blood vessels supplying the endometrium enlarge and become more coiled. Following ovulation the endometrium is said to be in its *secretory* phase. Thus, as shown in Figure 14-6, at the time when a fertilized egg is most likely to reach the cavity of the uterus (about one week following ovulation), the endometrium is best prepared to support its further development. If fertilization does not occur, the egg disintegrates within a few days and before long the circulating concentrations of both estrogen and progesterone fall off sharply. When this happens, the top layers of the endometrium degenerate and are sloughed off as another menstrual flow begins. Thus the cyclic changes that take place each month in the lining of a woman's uterus are designed to maximize a fertilized egg's chances for further development, and these periodic changes depend in large part on the concentrations of circulating sex hormones. Where do these hormones come from, and how are they regulated?

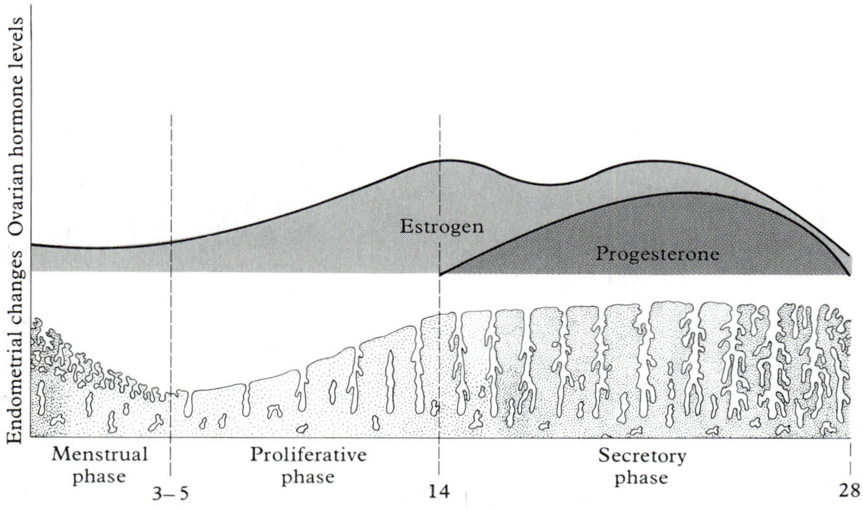

14–6
How the phases of the endometrium depend on the concentrations of estrogen and progesterone. Numbers at the bottom are days of the menstrual cycle.

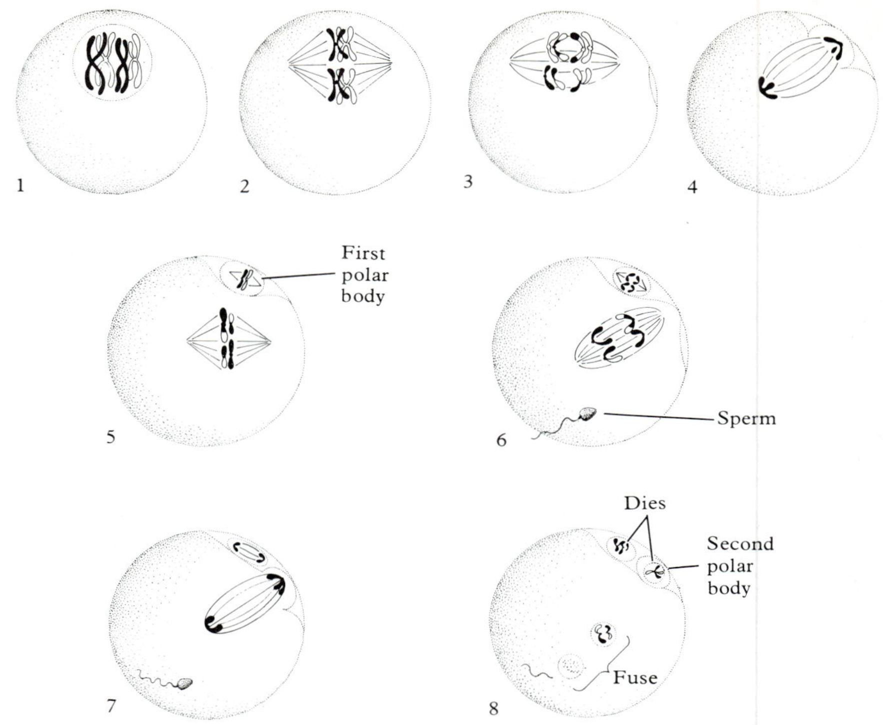

14–7
1–5, a primary oocyte undergoes its first meiotic division shortly before ovulation. This results in the formation of the first polar body (5), which usually undergoes a single division and then dies. 6–8, fertilization (6) causes the remaining oocyte nucleus to divide again, with the formation of a second polar body, which also dies. The nuclei of the oocyte and sperm then fuse (8), thus restoring the original number of chromosomes. (From "Mammalian Eggs in the Laboratory," by R. G. Edwards. Copyright © 1966 by Scientific American, Inc. All rights reserved.)

The Ovarian Cycle—How the Female Sex Hormones and the Production of Eggs Are Regulated

The ovaries of a female infant contain more eggs at the moment of birth than at any other time during her later life, at least half a million. The eggs in a newborn's ovaries are called *primary oocytes* and they originate from primitive germ cells early during the embryo's existence. As we will discuss in Chapter 16, mature sex cells are formed by *meiosis,* cell division in which the members of each pair of chromosomes become separated from one another, so that both mature eggs and sperm have only one-half as many chromosomes as other body cells. The early stages of meiosis are already completed by the time the primary oocytes are formed, but the actual halving of the chromosome number has not yet occurred. That takes place in two steps. The first occurs just before ovulation, at which time a primary oocyte is stimulated to further development by hormones released by the brain. The second division, which results in the halving of the chromosome number, takes place only after fertilization, just before the remaining oocyte nucleus and the nucleus of the sperm fuse, thereby restoring the original number of chromosomes (Figure 14-7).

For reasons that are poorly understood, the total number of primary oocytes steadily declines after birth, rather rapidly during the first ten years of life, then somewhat more slowly until only a few remain at the onset of menopause. Most of these cells simply disintegrate; very few of them are ever ovulated and thus capable of being fertilized. For the first ten or twelve years of a woman's life, those oocytes that do not disintegrate within the ovaries remain there, waiting. Then, beginning at puberty, and at first quite erratically, a precise and closely

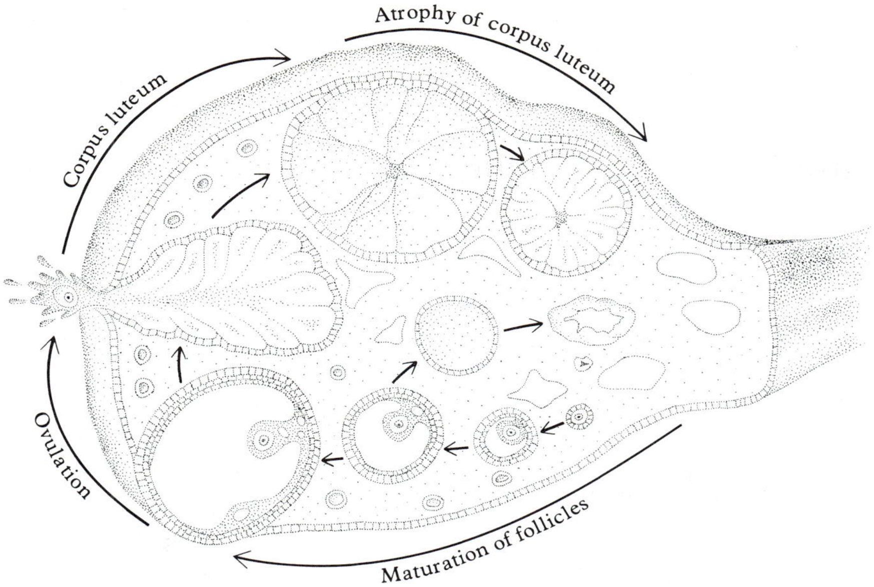

14–8
The maturation of an oocyte within an ovarian follicle. The development of the corpus luteum is discussed later in this chapter.

regulated series of events is set into motion by the brain. That series culminates in the maturation of a single oocyte in one or the other ovary each month for the next three or four decades (Figure 14-8).

The onset of menstruation, and therefore of the reproductive years, is known as the *menarche.* After menarche it often takes several years before a young woman's menstrual cycle begins to occur regularly. As shown in Figure 14-9, in the United States, Great Britain, and Europe the age at which menarche

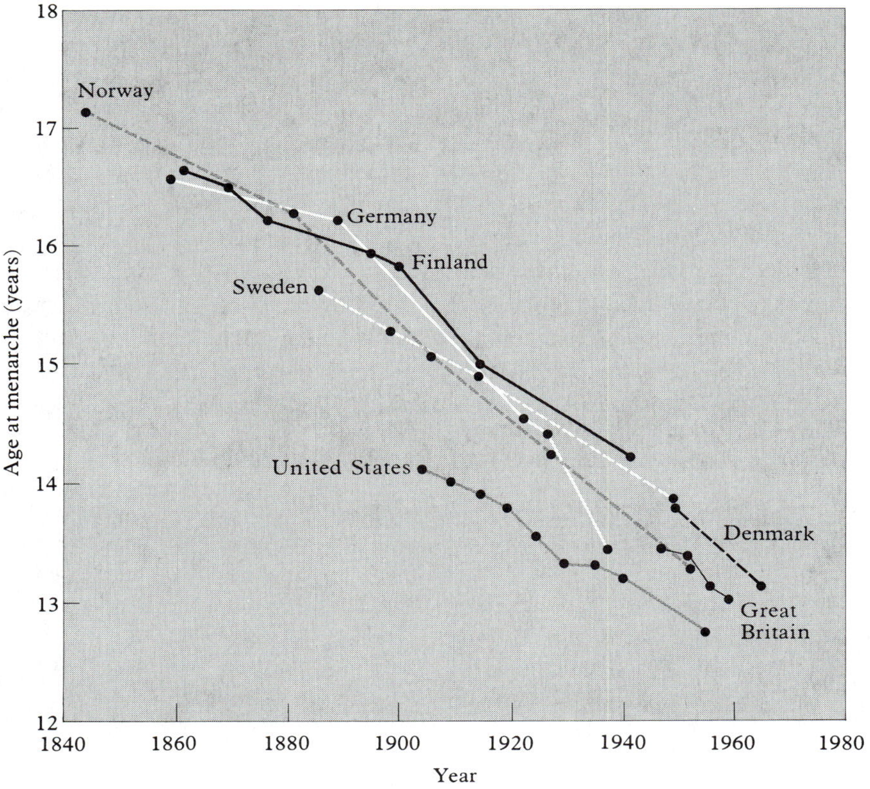

14–9
The age at menarche, or first menstrual period, has steadily declined in the U.S., Great Britain, and Europe. (From "Earlier Maturation in Man," by J. M. Tanner. Copyright © 1967 by Scientific American, Inc. All rights reserved.)

occurs has been steadily decreasing since at least 1840. It is estimated that most young women in these countries begin to menstruate between 2.5 and 3.3 years earlier than young women did a century ago. This trend has probably had some effect on the size of the human population. It is thought that the main reason for the decline in age at menarche is better nutrition.

How does the brain initiate and maintain these monthly cycles? The hypothalamus releases hormones that cause other hormones to be released from the anterior pituitary. These pituitary hormones in turn stimulate the ovary to do two things in cycles: first, to initiate the maturation of a single oocyte a month, and second, to secrete the sex hormones estrogen and progesterone.

Figure 14-10 shows how the concentrations of some of these hormones change during the course of the menstrual cycle and how they are related to

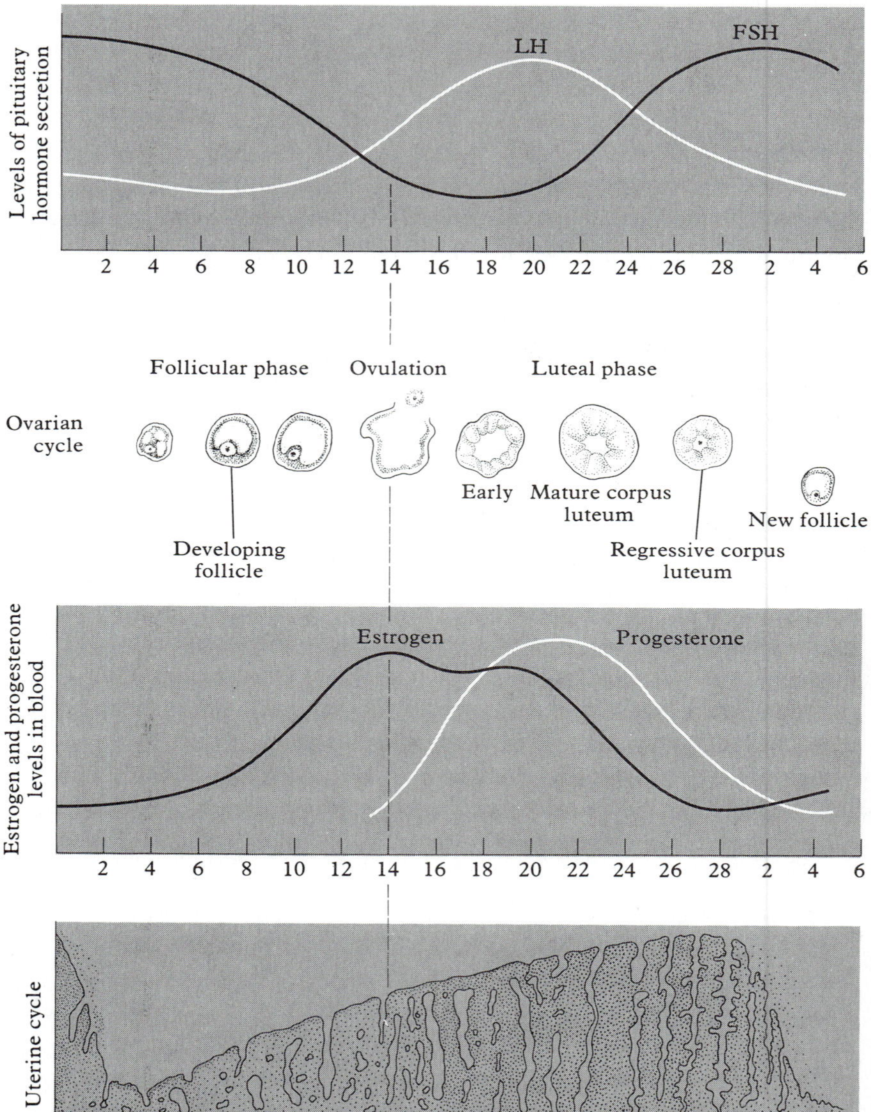

14–10
How hormone concentrations are related to the maturation of oocytes and to the endometrial changes that occur during the menstrual cycle.

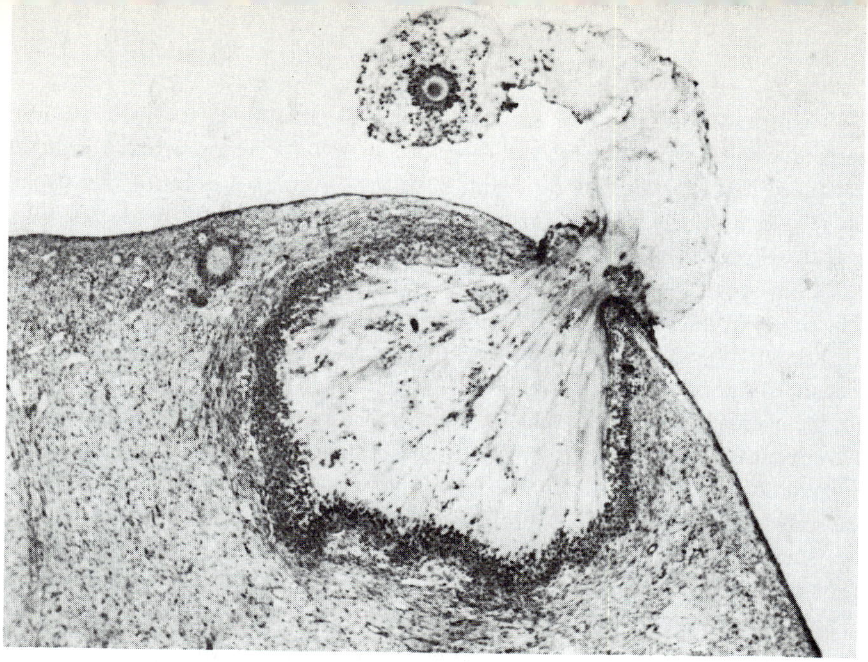

14–11
The moment of ovulation, as it occurs in the ovary of a rabbit. (Photo courtesy of R. J. Blandau.)

oocyte maturation within the ovary. When stimulated by the hypothalamus (or by a lowered concentration of estrogen) the anterior pituitary releases follicle-stimulating hormone (FSH; see Chapter 6), which initiates the maturation of a single oocyte and at the same time causes the ovary to manufacture and release estrogen, whose concentration in the bloodstream therefore begins to rise. As the concentration of estrogen rises it causes less FSH to be released, and when the circulating estrogen reaches a certain concentration it triggers the release of another pituitary hormone, luteinizing hormone (LH). The presence of LH, in combination with the decreasing concentration of FSH, results in ovulation. The oocyte literally bursts forth from the surface of the ovary because of hormone-induced effects on the walls of the tiny bubblelike *follicle* in which it matured (Figure 14-11). (What happens to the egg after ovulation is discussed in the following chapter.)

Following ovulation the empty follicle, which remains within the ovary, soon develops into a cellular *corpus luteum.* (The term *corpus luteum* means "yellow body," and LH is called luteinizing hormone because it supports the formation of the *corpus luteum.*) Under the influence of LH the corpus luteum begins to produce progesterone, as well as more estrogen (Figure 14-10). Just as a high concentration of estrogen results in a lowering of the concentration of FSH, a high concentration of progesterone results in a decrease in LH. When the concentration of progesterone becomes sufficiently high it depresses the concentration of LH so much that LH is no longer present in sufficient quantities to support the *corpus luteum,* which therefore collapses, or, perhaps better, shrivels and ceases to function. When this happens, the concentration of both estrogen and progesterone fall off rapidly, and this causes the endometrium to begin to be shed in another menstrual flow. The relatively low concentration of estrogen at the time of menstruation causes the concentration of FSH to begin to rise again, and the whole cycle starts once more.

Because these interactions are very complicated, it is not surprising that they not infrequently go awry. Disorders of the menstrual cycle are common, and most women experience some sort of menstrual upset at some time during their lives. One common disorder, which is experienced more frequently by younger women, is the failure to ovulate, known as *anovulation.* If ovulation does not occur (and many factors can account for this) a corpus luteum is not formed, and no progesterone, and therefore no secretory endometrium, will be produced that month. This can result in various irregularities in the menstrual flow, especially the production of an abnormally short cycle. But generally menstrual regularity is reestablished within the next few months.

Perhaps the most common of all menstrual disorders are those irregularities that result from emotions. Psychological stress, including anxiety, depression, the loss of a loved one, or even the agonies of final examinations, can result in menstrual irregularities, especially missed periods. This is not surprising: after all, the events of each menstrual cycle are in large part controlled by hormones manufactured and released directly by the brain. Although we have much to learn of how brain chemistry relates to our feelings and mental attitudes, the two are somehow related.

Menstrual periods take place fairly regularly (except during pregnancy) until menopause, which usually occurs when a woman is about 49 years old. (The "normal range" is between 35 and 55 years of age.) At menopause menstruation ceases and the remaining oocytes deteriorate. At this point a woman's reproductive years are over. Menopause can last for a long as a decade, and during this time, as during menarche, the menstrual cycles may be very irregular before they stop altogether. Among other symptoms, women sometimes experience "hot flashes" or "hot flushes" during menopause that result from decreasing concentrations of estrogen, which is produced only in very small quantities when menopause is completed. Curiously, menopause is probably unique to human females. Nothing comparable to it occurs during the lifetime of any other primates. The biological relevance of menopause is an open question, but in spite of the relative absence of estrogen and progesterone, some women enjoy their sex lives more after menopause then they did before, perhaps because menopause relieves them of the fear of conception and the awesome and consuming responsibilities it can bring.

With this discussion of the anatomy and physiology of the reproductive tract of women as background, we now turn our attention to some methods of contraception for which women, because of their anatomy and physiology, accept the primary responsibility.

Contraceptive Methods Employed by Women

The bodies of women are the favored targets of the various contraceptive devices that now exist. In large part this is because of the cyclic nature of reproductive events in women. It is easier to interfere with a cyclical event that occurs only once a month than it is to alter the production of the huge numbers of sperm that men produce continually.

At the present time there are five main methods of contraception used by women: first, the rhythm method; second, controlling the menstrual cycle by the use of hormones in the form of *the Pill;* third, use of a diaphragm and sperm-killing substance to block the passage of sperm through the opening of the cervix; fourth, various intrauterine devices, or *IUDs*; and fifth, by sterilization. Let us now discuss briefly each of these methods as well as some of their risks.

The rhythm method is a simple one: sexual intercourse is avoided on those days during a woman's cycle when it is possible for her to become pregnant. The

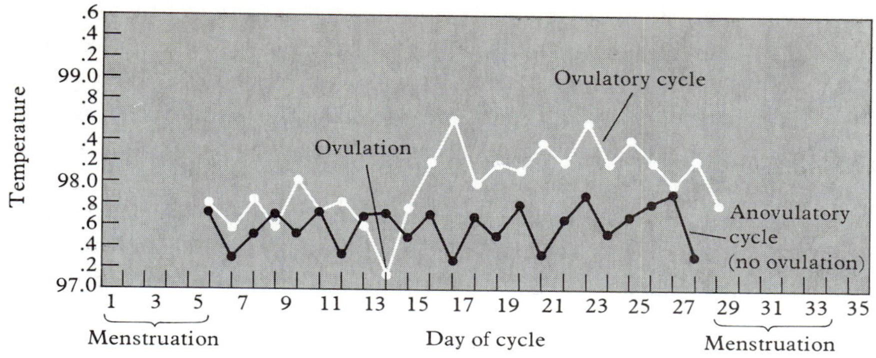

14–12
How basal body temperature changes during the menstrual cycle.

problem is in trying to figure out when those days occur. In order to calculate a woman's "theoretically" fertile period, the day on which she ovulates must be known. This is relatively easy to determine for women who have very regular cycles, especially when basal body temperature charts are kept. As shown in Figure 14-12, at the time of ovulation there is a slight but measurable decrease in body temperature: this is followed the day after by a small but sustained rise in temperature that persists until the end of the cycle. Women who have regular 28-day periods usually ovulate on about day 14, but a few women with cycles of this length ovulate as early as day 12 or as late as day 16. Therefore, allowing for survival times of about one day for an unfertilized egg and about two days for sperm deposited in the female genital tract, the "theoretical" fertile period would last about 8 days (days 10 through 17 of a 28-day cycle), during which sexual intercourse should be avoided. Unfortunately, many women have rather irregular periods, and the times at which they ovulate may vary considerably—from as early as day 8 or earlier to as late as day 19. Reportedly, in rare instances, women have even ovulated during their menstrual periods. Nonetheless, as shown in Table 14-1, which compares the effectiveness of various contraceptive methods, the "temperature-rhythm method" can be 65 percent effective. The one risk is that of becoming pregnant if one miscalculates.

A much more effective method of contraception is the Pill, which is also more risky. Most oral contraceptives in current use consist of a combination of hormones that are closely related to estrogen or progesterone. One pill per day is usually taken from about day 5 through day 25 of the menstrual cycle, and the

TABLE 14-1
Approximate failure rate (pregnancies per 100 woman years).

	THEORETICAL FAILURE RATE	ACTUAL FAILURE RATE
Abstinence	0	?
Tubal Ligation	0.04	0.04
Vasectomy	Less than 0.15	0.15
Oral Contraceptives (combined)	Less than 1.0	2–5
Condom + Spermicidal Agent	1.0	5
Low-Dose Oral Progestin (mini-pill)	1–4	5–10
IUD	1–5	6
Condom	3	15–20
Diaphragm	3	20–25
Spermicidal Foam	3	30
Rhythm (temperature)	15	35
Chance (sexually active)	80	80

resulting concentrations of estrogenlike or progesteronelike hormones (or both) in the blood are sufficient to block the release of the pituitary hormones FSH and LH so that ovulation does not occur. A few days after the last monthly pill is taken menstruation usually begins because the endometrium has had its hormonal support withdrawn. Bleeding continues for about five days or so, after which the next cycle of pills is begun.

Although oral contraceptives are extremely effective they may produce certain serious, even fatal, side effects. Some of the more highly publicized side effects include the formation of blood clots in the veins of the legs and pelvis, mental depression, liver tumors, and increased rates of stroke and heart attack, and there are others. Many countries have recently passed laws that require that a woman be adequately informed of the risks before she decides whether she wants to take the Pill.

As shown in Figure 14-13, the diaphragm method consists of placing a carefully fitted rubber dome over the cervix before engaging in sex, thus preventing the sperm from reaching the uterine tube, where fertilization occurs. The effectiveness of the device is increased by application of a sperm-killing substance to the surface of the rubber dome before it is inserted, and although this method carries with it few risks, it is rather inconvenient. As Table 14-1 indicates, this method is more effective than rhythm but less effective than the Pill.

IUDs are variously shaped objects made of plastic or metal that are inserted directly into the cavity of the uterus, where they come into contact with the endometrium (Figure 14-14). The exact mechanism by which they prevent conception is unknown, but IUDs presumably work by interfering with the implantation of the fertilized and developing egg, perhaps because their contact with the endometrium results in a low-grade infection. Recently developed copper-containing IUDs are probably more effective than plastic ones, as are those that contain small amounts of progesterone.

The obvious advantage of IUDs is that once they are in place they remain there, and the women who can tolerate them can more or less forget about them, at least until the time when they must be replaced. Unfortunately, as many as 25

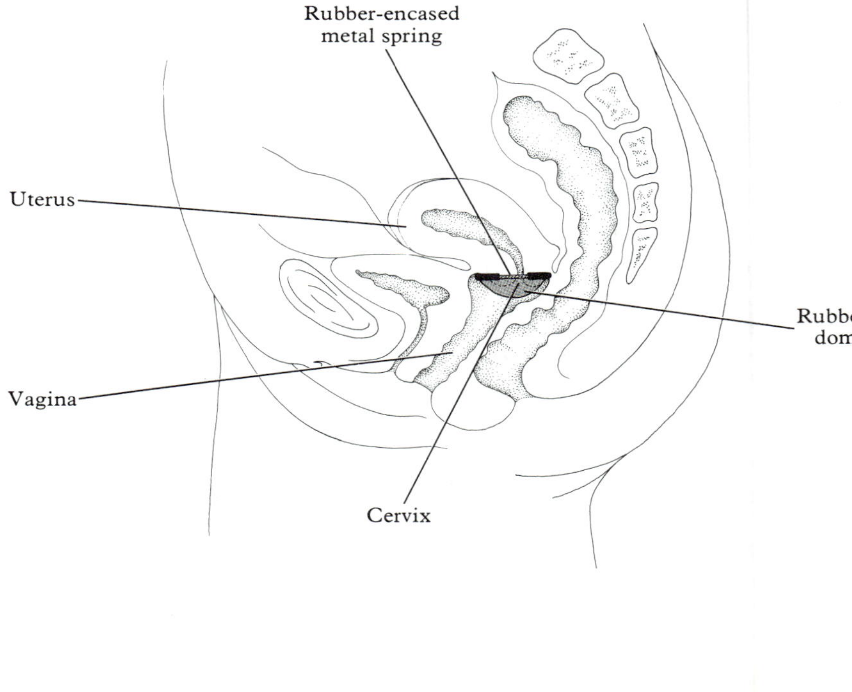

14–13
A properly positioned diaphragm.

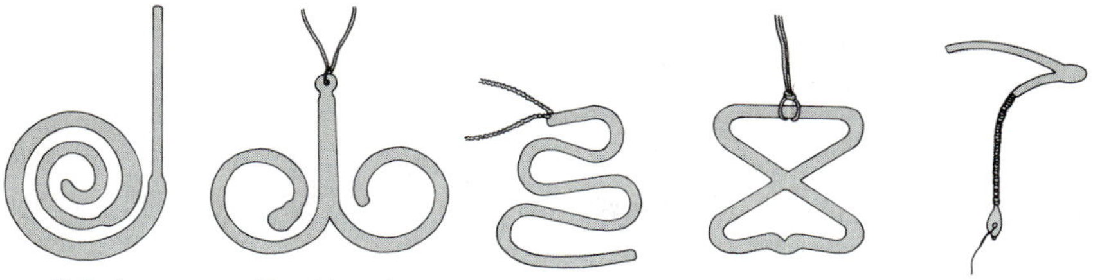

14–14
Some of the kinds of IUDs currently in use in various parts of the world.

percent of normal women cannot tolerate an IUD because of severe cramping, excess menstrual bleeding, infections of the uterus or the uterine tubes, or other reasons.

The most final method of contraception used by women is sterilization by means of tubal ligation, that is, by having their tubes "tied." This method usually consists of cutting out the middle portion of each tube and then sewing the severed ends closed (Figure 14-15). This procedure is irreversible and after it has been performed the possibility of pregnancy no longer exists. The procedure can be done either through surgery or with a tubelike instrument (laparoscope) that is inserted through the wall of the abdomen. Either way, general anesthesia is required and the usual risks of surgery exist.

This concludes our discussion of the anatomy, physiology, and contraceptive methods applicable to the female reproductive tract. We now turn our attention to a consideration of the anatomy and physiology of the reproductive systems of men.

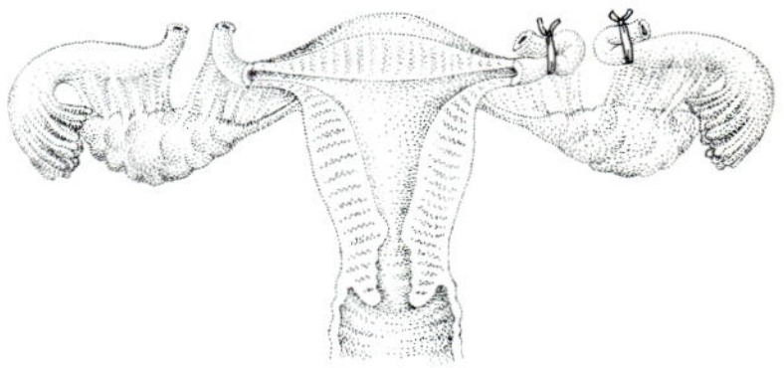

14–15
The best method of tying the tubes (tubal ligation).

The Structure and Function of the Male Reproductive System

The external and internal reproductive organs of the human male are perhaps less complicated than those of the female but are equally interesting and equally vital to reproduction. They are shown in Figure 14-16. The most noticeable parts of the apparatus are the *penis* (about which we will have more to say later) and the rather loose sac of wrinkled skin, the *scrotum,* inside of which the two *testes* (or *testicles*) are suspended in separate compartments at the ends of the *spermatic cords.* (The word "testes" is derived from the Latin word for "witness," because in classical times men often solemnly placed their hands on their genitals when swearing an oath.)

The testes of male embryos develop within the abdominal cavity and then descend into the scrotum at a later time, usually just before but occasionally following birth. The testes of almost all male mammals are suspended in a scrotal sac, whose function is to keep them cooler than the rest of the body. Sperm formation can proceed only at temperatures that are lower than those maintained within the abdomen. However, there are some exceptions to this rule. Whales and elephants, for example, have gigantic testes that always remain inside their abdomens, whereas the testes of bats and of certain kinds of rodents pass into the scrotum only during the breeding season, during which time only they manufacture sperm. But among human beings, testes that fail to descend into the scrotum by puberty are thereafter never able to produce sperm, though

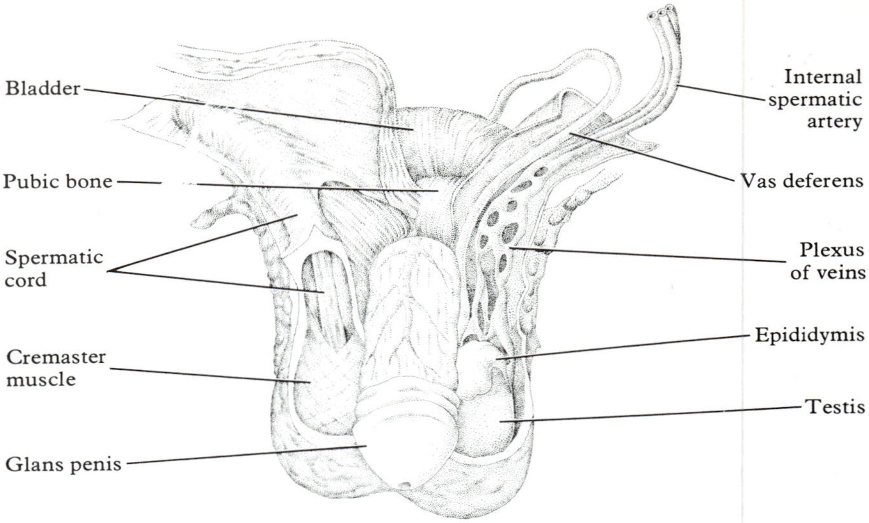

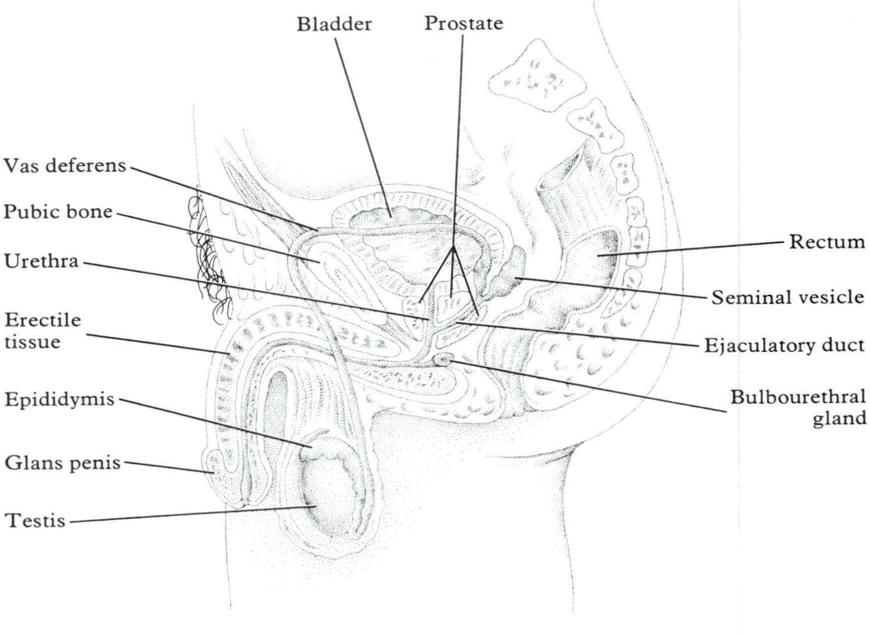

14–16
The male reproductive system. Top, front view (with the front wall of the scrotum partially removed). Bottom, side view of the male pelvis.

they are usually capable of secreting male sex hormones. Also, undescended (or *cryptorchid*) testes are somewhat more likely to develop cancer than are those that descend normally. For these reasons, it is usually recommended that undescended testes be relocated into the scrotal sac by surgery performed before puberty, when sperm cells first begin to be manufactured under the influence of hormones released by the brain.

Because of the extraabdominal wanderings of the testes, the tubal system connecting the various parts of the male reproductive system is more complicated than that of women. In the following section we will discuss the internal structure of the testes and how they produce both sex cells and hormones. But for now let us say that after they are manufactured, mature sperm cells make their ways to a C-shaped storage chamber known as the *epididymis* (Figure 14-16). We shall now trace out the route followed by the 400 million sperm cells that are released each time ejaculation occurs and introduce the structures and secretions that this seminal multitude encounters *en route* to the outside.

The epididymis of each side leads into the *vas deferens,* a highly muscular channel that is one of the components of the spermatic cord. Before it disappears into the abdomen the vas deferens can be felt as a rigid, almost wirelike structure in the upper part of the scrotum. (Other elements of the spermatic cord include arteries, a very rich supply of veins, lymphatic vessels, nerves, and muscles that can protectively draw the testes up toward the body when the temperature around the testes is too cold and during other stressful times.)

Once it has entered the abdominal cavity, each *vas deferens* arches around the urinary bladder and on its lower backside each is joined by a short duct that originates from one of the *seminal vesicles.* The secretions from these sacculated structures provide a supportive medium in which sperm become more motile. As shown in Figure 14-16, each *vas deferens* joins the duct of one of the *seminal vesicles* to form one of two *ejaculatory ducts,* both of which then extend their entire lengths within the substance of the *prostate gland.* The prostate gland is a structure approximately the size and shape of a chestnut that is located at the base of the urinary bladder and that surrounds part of the *urethra,* the tube through which both urine and the ejaculate are discharged to the exterior. (You will recall that in the female reproductive tract the urethra has no sexual function; its only role is as a urinary passage.)

An average man ejaculates about 3.5 ml of the sticky white fluid known as *semen,* but the normal range is from 0.5 to 6 ml or more. Of this, only about one-tenth of the total volume is sperm cells. The rest is made up of secretions from the seminal vesicles and from the prostate gland. Secretions from the latter account for much of the total volume and for the characteristic odor of semen. The secretions of the prostate gland and the seminal vesicles help to protect sperm against the acid environment of the vagina and they also nourish the sperm cells. Male sexual arousal culminates in the rhythmic contraction of muscles within the walls of the vas deferens, within the seminal vesicles, within the ejaculatory ducts, within the prostate gland, and within the pelvis, in that order, with the result that semen is forcefully expelled and the man experiences orgasm. The few drops of cloudy lubricating fluid that may appear on the tip of the penis during sexual arousal are produced by *Cowper's glands,* two small structures located at the base of the penis.

The organ that serves to deliver semen to the female reproductive tract (and that also conducts urine as well as semen to the outside through the urethra contained within it) is the penis, whose internal structure is shown in Figure 14-17. The organ is composed of columns of spongy, expandable tissue, and it becomes hard and erect when a man is sexually aroused. Erection occurs because during sexual excitement blood from arteries pours into the spongy tissue of the penis faster and under higher pressure than it is drawn off by veins. As a result the tissues of the penis become swollen and distended, as does the entire organ, which remains so until sexual excitement has passed, at which time it becomes soft again. (There are no voluntary muscles in the penis; erection is entirely due to pressure changes brought about by the flow of blood.)

From this consideration of how sperm cells take leave of the male reproductive tract we now turn our attention to how sperm cells are produced.

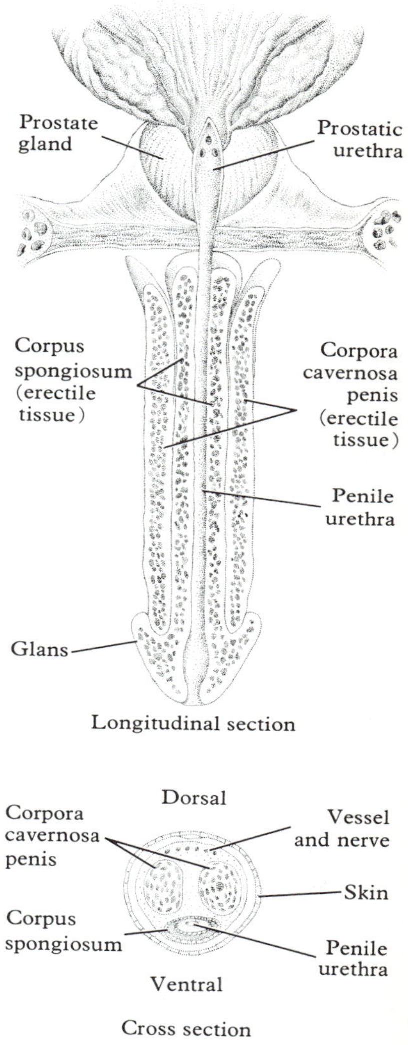

14–17
The internal structure of the penis.

Sperm Cells, and How Their Production Is Related to Sex Hormones

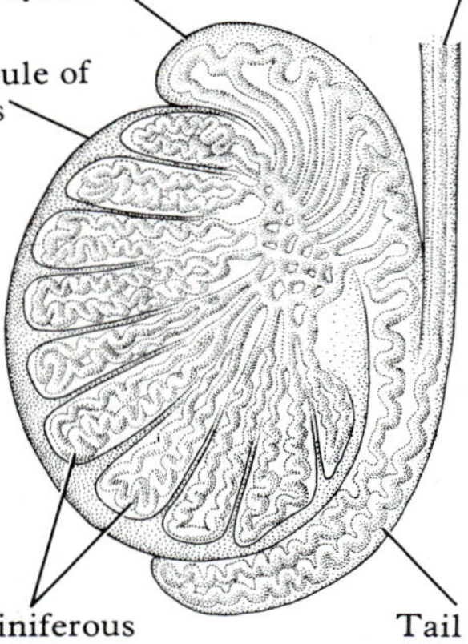

14–18
The structure of the testis and epididymis.

Beginning at puberty the hypothalamus of a young male stimulates the anterior pituitary to manufacture and release the same two hormones that initiate the reproductive process in pubescent females, FSH and LH. But in human males there is no appreciable cyclic variation in the concentrations of these two substances. FSH stimulates the previously quiescent germ cells within the testes to begin to manufacture mature sperm, but in human males FSH does not trigger the release of sex hormones. The release of male sex hormones (or androgens, of which testosterone is the main component) is, however, triggered by the presence of LH. Almost all of the androgens produced by men come from special cells within the testes known as *interstitial* cells, and LH in men is thus usually called interstitial-cell-stimulating hormone, or ICSH. Like the production of sperm, the production of testosterone is fairly constant throughout a man's life, and its circulating concentrations show no cyclic variations, although there is a slight decline in testosterone concentrations among older men.

Sperm production takes place within highly coiled and threadlike structures known as *seminiferous* (sperm-bearing) *tubules.* These delicate structures are tightly packed into several conical lobules within the testes, and when stretched out their combined length is almost one hundred meters, or several hundred feet (Figure 14-18). Before puberty the *seminiferous tubules* contain only two kinds of cells, germ cells and nurse cells, which nourish the developing sperm. (The interstitial cells, which produce testosterone when stimulated by ICSH, are located between the tubules.) Then, under the influence of increased concentrations of FSH at puberty, the germ cells begin to divide and to give rise to primary spermatocytes, which are comparable to the primary oocytes of the female. Primary spermatocytes divide into two *secondary* spermatocytes, each of which divides again to form four *spermatids.* Each spermatid contains half as many chromosomes as the original germ cell. The spermatids become attached to nurse cells and before long they become transformed into mature sperm cells (Figure 14-19).

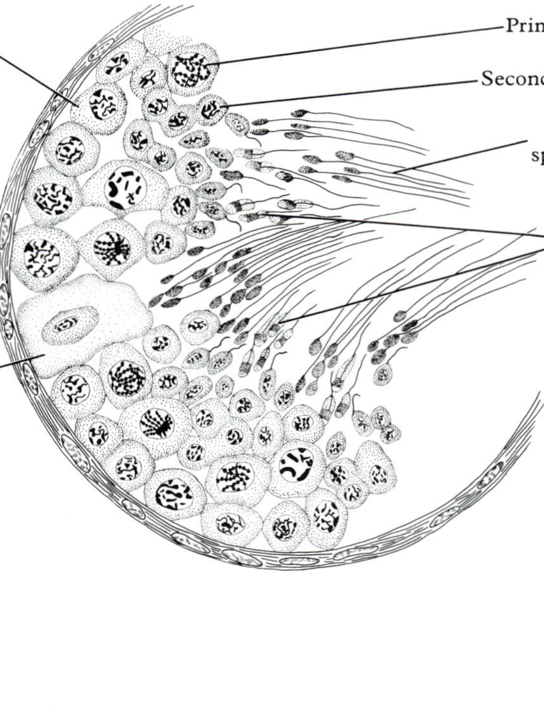

14–19
A cross-section of a seminiferous tubule in which sperm cells are being produced.

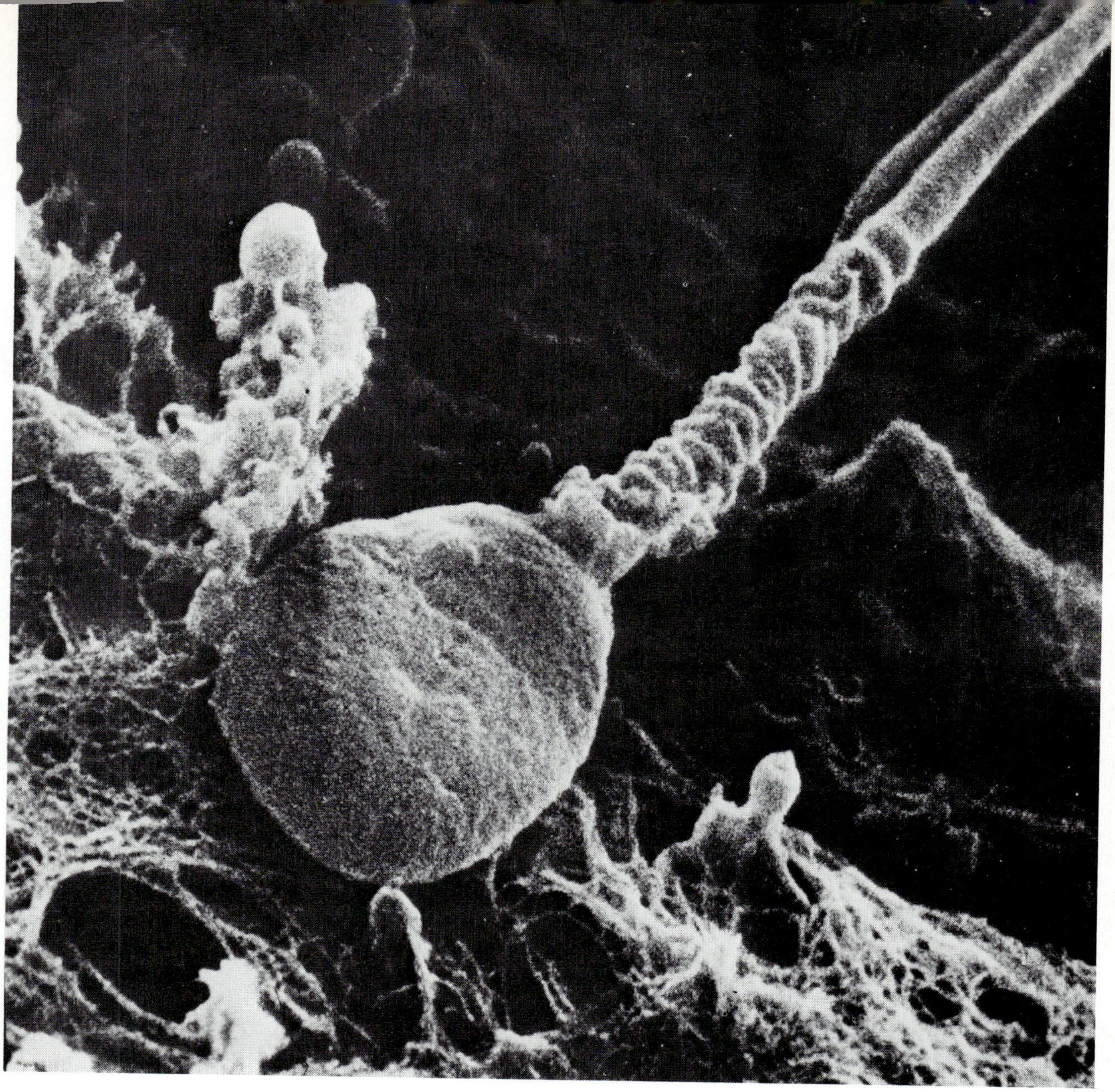

14–20
Human Sperm. (×10,000) as seen through the electron microscope. (From Lennart Nilsson, Behold Man. *Little, Brown and Company, 1974. Courtesy Albert Bonniers Förlag, Stockholm.)*

Mature human sperm resemble tadpoles and are produced at the astonishing rate of about 50,000 per minute. As shown in Figure 14-20, human sperm consist of three basic parts: a pear-shaped head, a middle piece that provides the energy for movement, and a whip-like tail that propels the sperm forward along a spiral path. The heads of the sperm of many species of mammals are distinctive, as is shown in Figure 14-21.

The *seminiferous tubules* ultimately converge upon and empty their contents into the C-shaped *epididymis,* which we have mentioned earlier. Within the epididymis sperm cells become increasingly motile and apparently restless as they travel from one end of this highly convoluted tube to the other, a distance of about 6 meters (about 20 feet). The trip may take three weeks to complete. Those sperm that eventually make it to the outside first pass into the *vas deferens,* then into the ejaculatory ducts, where just before ejaculation they become mixed with seminal and with prostatic secretions. If ejaculation does not take place

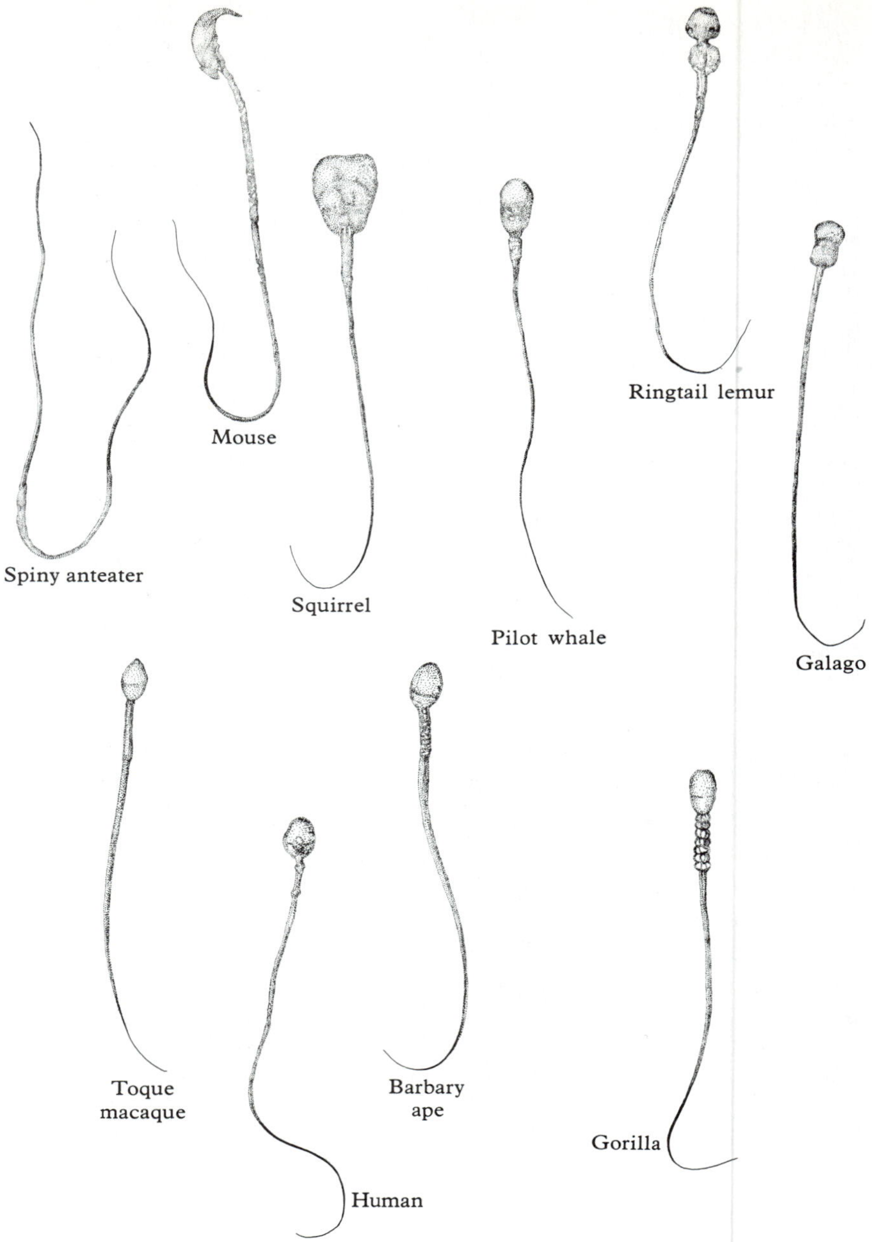

14–21
Some sperm cells from different kinds of mammals.

regularly, many sperm cells degenerate, die, and are then reabsorbed. (The first ejaculate after a period of abstinence usually contains many deformed and degenerating sperm.) We now turn our attention to some of the contraceptive methods employed by men.

Contraceptive Methods Employed by Men

Because sperm are produced continually and in such prodigious numbers, it is much more difficult to interfere with the production of sperm than it is to influence the production of a single monthly egg. In large part, this accounts for the relative paucity of contraceptive methods available to men. At the present time there are only two contraceptive methods used by men, and both of them are mechanical rather than hormonal. These two methods are *condoms* and *vasectomy.*

Condoms, which were first brought into widespread use at the court of Charles II in the 1700's, may be made of rubber or other naturally occurring products as well as of synthetic materials. They are worn over the erect penis during intercourse. As shown in Table 14-1, when used properly condoms can be quite effective. They also have the praiseworthy side effect of helping to prevent the spread of most kinds of venereal disease. Nonetheless, some people object to condoms on the grounds that they reduce the man's pleasure, are messy, or are unaesthetic. The only risk that accompanies use of condoms is occasional mild skin irritations.

Vasectomy is the male equivalent of tubal ligation in women. As presently performed, both processes result in irreversible sterilization by permanently interrupting the tubal systems through which sex cells travel. In performing a vasectomy the surgeon first injects a local anesthetic into both sides of the upper scrotum and then makes two small incisions through which each *vas deferens* is drawn out. About an inch of each *vas deferens* is then cut out, the ends are sealed, and the two small wounds are sewn shut (Figure 14-22). The procedure is relatively painless, and it results in no change in a man's sex drive or in his secondary sexual characteristics. Following vasectomy a period of several months is usually required before a man is effectively sterile. During this time, those sperm downstream from where the *vas deferens* was interrupted either find their ways to the exterior or die. Thereafter sperm are produced in reduced numbers and because they no longer have an avenue of exit, they all eventually die and are reabsorbed.

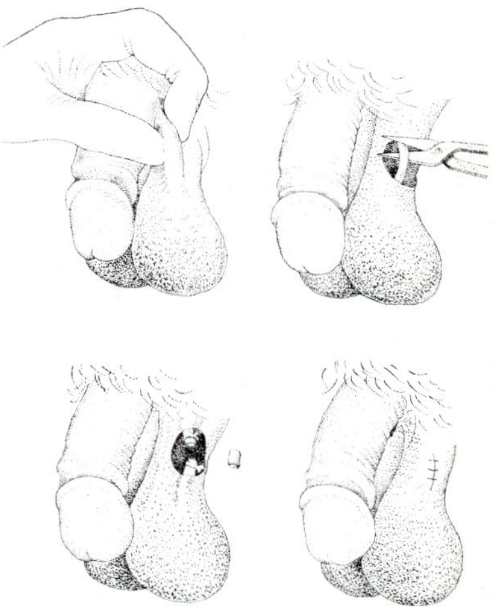

14–22
A vasectomy.

The risks of vasectomy include the rare possibilities of infection or extensive blood clot formation, but overall the procedure is much less risky than tubal ligation in women. Recently several valvelike devices that can be inserted inside the *vas deferens* have been tested, sometimes with promising results. Reversible vasectomy may thus become a reality before too long.

What about male contraception by hormones? The Pill works as well for men as it does for women. This is because the hormones in the Pill stop FSH production about equally well in both sexes, which in men results in the cessation of sperm production. The trouble is that the Pill also causes men to develop larger breasts, higher pitched voices, relatively fatter buttocks, and other characteristics that cause them to feel less like men and to be less attractive to most women. Many hormonal preparations have been tested as possible male contraceptives, and the development of a pill for males may someday become a reality. But so far most hormones tested have not been acceptable. Some of those that at first seemed promising because they eliminated sperm production without producing feminization proved to be unacceptable because they are incompatible with the consumption of alcoholic drinks, a serious drawback to their use in many societies.

More recently, nonhormonal substances such as the sugar 5-thio-D-glucose, have been shown to result in the cessation of sperm production without producing any apparently serious side-effects. Also, nonchemical methods of male contraception, such as temporarily interfering with the production of sperm by means of ultrasound, may soon be feasible.

Whatever means of birth control scientists invent, it seems likely that the human species will continue to reproduce successfully, perhaps too successfully, for many years to come. Nonetheless, the responsibility for contraception, or at least for understanding how the overproliferation of so consuming a species as our own could be prevented, rests squarely on each of us.

Summary

The heightened sexual interest that characterizes our species probably evolved to reinforce the very strong bonds that must exist between men and women if they are to stay together long enough to raise families. The maintenance of strong interpersonal bonds probably also accounts for the human female's ability to become sexually aroused independent of hormone concentrations and for the fact that women experience orgasm.

Sex hormones are of extreme importance during early fetal life, at which time they influence further development in either a male or female direction, and again at puberty, at which time they influence the development of secondary sexual characteristics. Both sexes normally produce some quantities of both "female" and "male" sex hormones, and there is no apparent relationship between hormone concentrations and the desire for sexual activities.

During the menstrual cycle the lining of the uterus changes. It becomes ready to foster the further development of a fertilized egg at exactly that time of month when one is most likely to be present. This periodic change in the uterine lining depends on cyclic alterations in the concentrations of estrogen and progesterone, which in turn are regulated by hormones released by the brain. Hormones from the brain also bring about the maturation and release of one egg cell per month from puberty until menopause. The frequency with which disorders of the menstrual cycle occur reflects the complicated hormonal interactions on which it depends.

The major means of contraception available to women are the rhythm method, the Pill, diaphragm and spermicidal jelly, IUDs, and sterilization through tubal ligation. Some of these methods carry with them much more serious risks than others.

The tubal system of the male reproductive system is more complicated than that of women because the testes develop in the abdomen and descend into the scrotum at a later time. The same hormones initiate puberty in both males and females, but in males their concentrations show no appreciable cyclic variation.

Contraceptive methods available to men are not numerous, mostly because it is difficult to interfere with the huge numbers of sperm cells continually produced by the testes. They include condoms and sterilization by vasectomy, a procedure that may soon become reversible.

Suggested Readings

1. "The Physiology of Human Reproduction," by Sheldon J. Segal. *Scientific American*, Sept. 1974. An excellent summary of the human reproductive process and of some of the contraceptive methods that can be applied to it.

2. "Earlier Maturation in Man," by J. M. Tanner. *Scientific American,* Jan. 1968, Offprint 1091. Discusses the physiologic bases of the progressive decrease in age at menarche of females of certain human populations.
3. "Sex Differences in the Brain," by Seymour Levine. *Scientific American,* April 1966, Offprint 498. Discusses the evidence that patterns of behavior among certain male mammals are induced by the action of testosterone on the brain of the newborn animal.

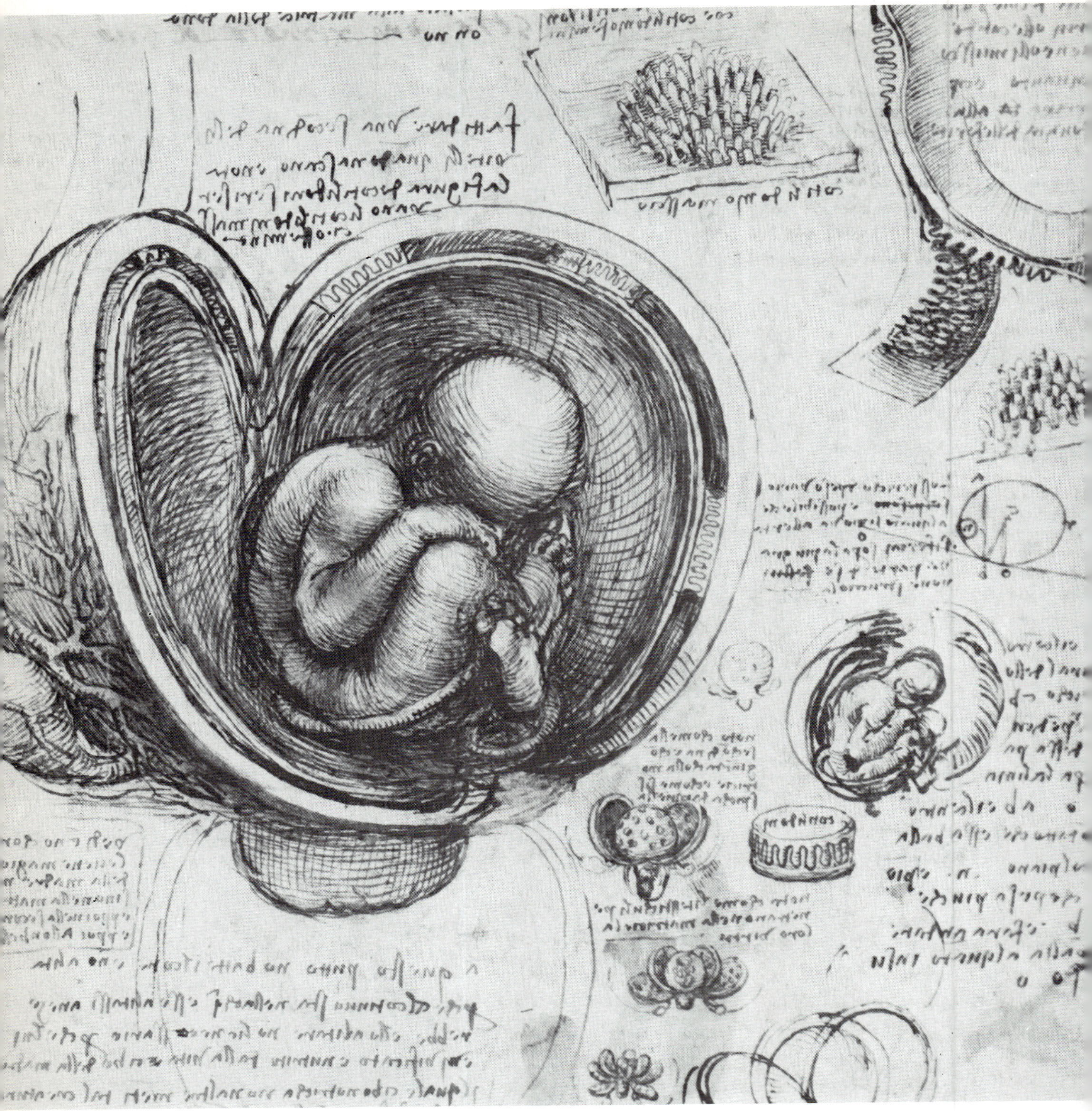

Drawing of a fetus in the womb, accompanied by smaller sketches depicting in detail aspects of uterine anatomy, by Leonardo da Vinci. Leonardo was one of the first people who portrayed human anatomy accurately. (Courtesy Royal Library, Windsor Castle. Copyright reserved.)

CHAPTER 15

The Human Lifetime

Each of the trillions of individual cells that make up an adult human body is ultimately descended from a single cell, the fertilized egg. The union of sperm and egg, known as fertilization, sets into motion a series of intricate and remarkable transformations whose overall result, if all goes well, is experienced as a long and fulfilling life. In this chapter we discuss the major transformations, both before and after birth, that are associated with the unfurling of a human lifetime. We begin by considering life before birth, starting with the initial event that brings together the maternal and paternal halves of the unique genetic program according to which further development proceeds.

Fertilization

In order to get some idea of how improbable the existence of a particular human being—you or me, for example—really is, consider the following facts. Fertilization is the result of the fusion of a single egg and a single sperm. Yet at the time of ejaculation, an average man releases at least 100 million sperm cells, of which about one in ten, or about ten million (10,000,000 or 10×10^6), are capable of fertilizing the egg. On the other hand, the single egg (ovum) that women release each month is one of about one hundred that could have matured and been released. Thus, the overall chances that the two particular sex cells from which each of us originated would join is roughly one in a billion! And since chance plays a role in whether parents meet in the first place, the odds against the existence of any person are higher still.

Is penetration by a sperm cell essential for the further development of a newly released egg? *Yes* is the clear answer for human beings, but for some species of animals an egg can be activated to further development in the absence of a sperm, a process known as *parthenogenesis.* In fact, development of unfertilized eggs is a normal part of the life cycle of some invertebrates, especially insects. For example, the males, or drones, of a bee colony, which have only half as many chromosomes as females, develop parthenogenically. Among the vertebrates parthenogenesis hardly ever occurs, although a strain of turkeys that can develop from unfertilized eggs has been bred. These parthenogenetic turkeys have the normal number of chromosomes because the egg, which begins to develop on its own, does not undergo the second division of meiosis, in which the number of chromosomes would normally be halved. On rare occasions, both human ova and human sperm-producing cells may begin to develop independently, but development never proceeds normally, and the result may be a *teratoma,* a tumor of testis or ovary that may contain a variety of recognizable adult tissues, such as bone, cartilage, glands, and hair.

Most of the sperm that are deposited in the vagina do not reach the Fallopian tube (oviduct) where the newly released ovum is (Figure 15-1). The weeding-out process for sperm begins shortly after they are deposited in the vagina; probably only about a million make it through the cervix into the uterus. Of these, only a few thousand reach the junction between the uterus and the oviduct, and only a few hundred arrive in the region of the ovum. During their trip from vagina to outer oviduct, the sperm cover a distance of about 15 centimeters (6 inches) in approximately 30 minutes, which means that they travel at a rate of about 5 millimeters per minute. Sperm can swim only 3 millimeters per minute at their fastest, so it is clear that they are aided in their movement by muscular contractions of the walls of the uterus and oviduct. These contractions are probably initiated by substances present in the semen.

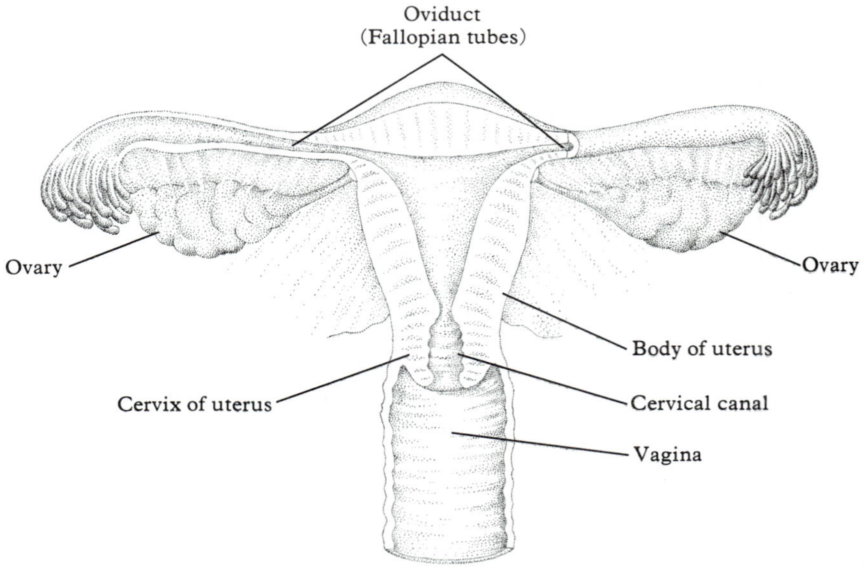

15–1
Portion of the female reproductive tract that must be traversed by sperm.

Although there are few studies of human fertilization, we do know that the process is much like that that occurs in other mammals, and it probably proceeds in the following way: first, a sperm attaches to the outer cell membrane of the ovum, and this is shortly followed by sperm "penetration," which is actually a fusion of the cell membrane of the sperm with that of the egg. Following this fusion, the nucleus of the sperm cell enters the egg and joins with its nucleus, leaving behind the midpiece and tail of the sperm. (Among the products left behind are the mitochondria of the midpiece, which either do not enter at all or soon degenerate if they do enter. This means that our mitochondria are apparently derived solely from our female parent.)

In spite of the small number of sperm that reach the distant regions of the Fallopian tubes, the ovum nevertheless is eventually surrounded by many sperm, and a mechanism must exist to prevent penetration of additional sperm after fertilization. If two sperm penetrated, it could result in a triploid individual, who had 69 chromosomes (three sets of 23) instead of the usual 46 in each cell. This may happen, as shown by the many (20 percent or so) triploid embryos found among miscarriages (spontaneous human abortions) (Figure 15-2). The very few triploid individuals who are born are relatively normal physically but have severe mental defects. (As discussed in Chapter 17, double fertilization may also result in rare individuals known as *genetic mosaics.*)

The mechanism that prevents further sperm penetration seems to derive from the membrane-covered granules, or *cortical granules,* that lie just underneath the outer cell membrane of the ovum. Just after sperm penetration these granules, which contain digestive enzymes, fuse with the outer cell membrane of the egg, releasing their enzymes into the jellylike region just outside the ovum (zona pellucida). It is thought that these enzymes destroy the sites on the surface of the ovum where the sperm attach.

Only if more than one egg is released can more than one sperm succeed in giving rise to a normal human being. When two (or more) eggs are released, and each is fertilized, the result is nonidentical (dizygotic) twins (or some number of multiple births). Nonidentical twins are as different from each other as they are from their other brothers and sisters, because they are the result of fusion of separate eggs and sperm. The other kind of twins (identical or monozygotic) result when a single, already fertilized, egg divides to produce two individuals. These twins are often almost indistinguishable from one another physically because they have the same genetic makeup. Rarely, the separation of potentially identical twins is incomplete, resulting in conjoined, or Siamese, twins (Figure 15-3). Nonidentical twins are about three times as common as identical twins.

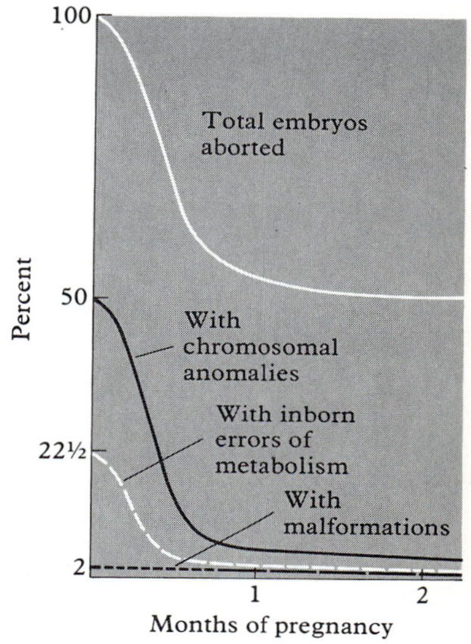

15–2
Most spontaneous abortions occur within the first few weeks of pregnancy, and about half of the embryos aborted have chromosomal abnormalities. Approximately 20% of the embryos that have chromosomal anomalies are triploid.

Development Before Birth

In the normal process of development, the fertilized egg must first make its way down the oviduct into the uterus (Figure 15-4). This voyage to the uterus requires about five days, ending about seven days after fertilization when the developing embryo burrows into the tissue of the lining of the uterus (Figure 15-5). Let us look at some of the changes that take place during this first week, when the fertilized egg is living relatively unattached within the female reproductive tract.

Shortly after fertilization, the ovum begins to divide, first into two cells (this division is completed on day 2), then into four cells, and so on, until day 7 at

15–3
Chang and Eng, conjoined twins from Siam, who lived in the nineteenth century and were the original "Siamese twins" from which the term was coined. Conjoined twins result when the separation of potentially identical twins is incomplete. As illustrated in this Currier and Ives print, the two led active lives; between the two of them they fathered 22 normal children. (Courtesy of the National Library of Medicine, Bethesda, Md.)

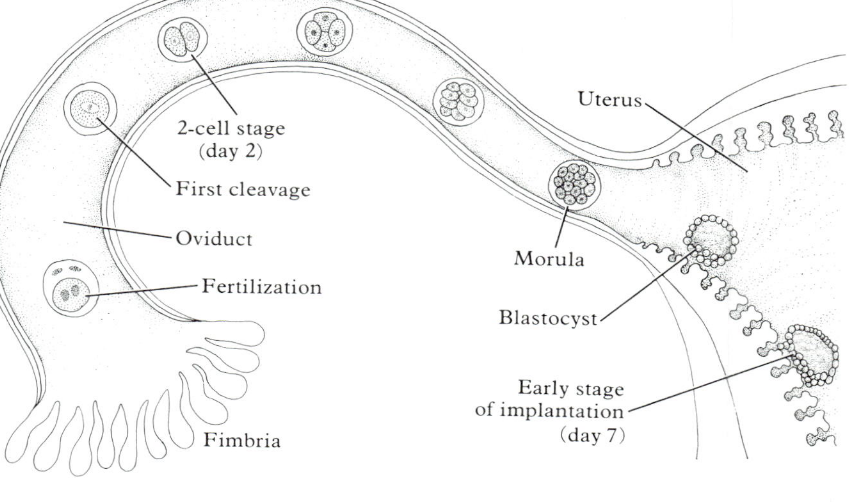

15–4
Development of the human embryo in the reproductive tract.

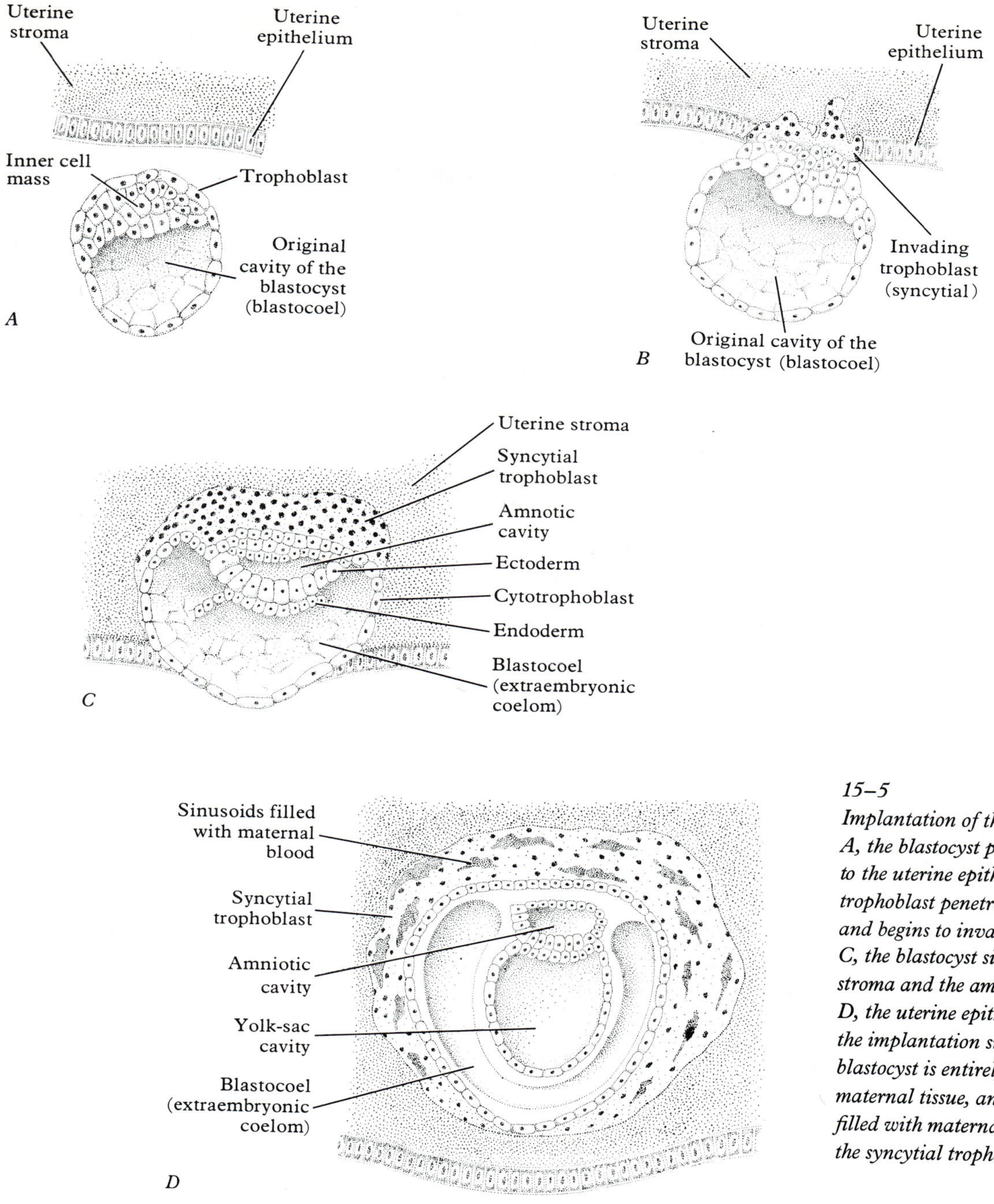

15–5
Implantation of the human embryo. A, the blastocyst prior to attachment to the uterine epithelium. B, the trophoblast penetrates the epithelium and begins to invade the uterine stroma. C, the blastocyst sinks further into the stroma and the amniotic cavity appears. D, the uterine epithelium grows over the implantation site, so that the blastocyst is entirely enclosed in maternal tissue, and the sinusoids, filled with maternal blood, appear in the syncytial trophoblast.

implantation there are over 100 cells present. Yet the total size of the ovum on day 7 is no greater than it was at fertilization, even though it consists of about 100 cells. This is possible because the original ovum was so large that it had enough cytoplasm to be divided among 100 normal-sized cells. However, by day 7 the nutrients stored inside the cytoplasm of the egg are nearly exhausted, and the tiny ball of cells will die if it doesn't implant and obtain maternal nutrients.

Up to day 5 (about the 32-cell stage) the ovum consists of a solid ball of cells known as a *morula* (Figure 15-5). But on day 5 the cells begin to move around, forming a cavity, or *blastocoel,* in their midst. The developing organism is now called a *blastocyst.* The cells of the blastocyst become organized into an outer layer or *trophoblast,* which contains the great majority of the cells and an *inner cell mass.* It is from the few cells of the inner cell mass that the embryo is formed.

During these first seven days the ovum must change from a cell whose exterior receives a single sperm into an organism capable of mounting an invasion into the lining of the uterus. This invasion is accomplished by the trophoblast cells, which become specialized by day 7 so that they release digestive enzymes and can literally digest their way through the lining of the uterus. In addition to digesting, the trophoblast cells also invade by fusing with each other and with the epithelial cells lining the uterus, forming a large mass of cytoplasm that has many nuclei (Figure 15-5). This occurs in the region of trophoblast that first contacts the uterine lining and the region of cell fusion is called the *syncytial trophoblast.* After their entry into the tissue of the uterus, the trophoblast cells continue to release digestive enzymes so that breakdown products of maternal tissues can be used as the blastocyst's first food.

The energies of the blastocyst, which are usually directed toward invading the uterus and using its tissue for nourishment, are sometimes expended on structures other than the uterus. In about 0.3 percent of human pregnancies, blastocysts implant in nonuterine sites. Such *ectopic* pregnancies most commonly occur in the Fallopian tubes, and those that do are called tubal pregnancies. But ectopic pregnancies can occur anywhere in the abdominal cavity. In a tubal pregnancy, growth of the embryo stretches the tube, resulting in severe pain. The embryo must be surgically removed because it would eventually rupture the tube. Ectopic pregnancies that occur in the abdominal cavity are not subject to the same growth constraints, and occasionally such pregnancies proceed normally, except that the baby must be delivered surgically through the abdomen (Caesarean section). The fact that the blastocyst can implant outside the uterus is proof of its extraordinary invasiveness.

In a sense, it is remarkable that mothers tolerate their fetuses at all, even in the uterus, because the fetus possesses many foreign antigens from the father and as such is a graft of foreign tissue. As we have discussed in Chapter 13, grafts of foreign tissue are routinely rejected by the body, except when the graft is a fetus. At least two factors account for the fact that the fetus is tolerated. First, the trophoblast cells seem to have a way of disguising their foreignness. Evidence for this is that a type of tumor derived from trophoblast cells, called *choriocarcinoma,* can spread throughout the mother and can even spread if transplanted to other species of animals without being rejected. Second, in those rare instances in which the mother does develop an immune response capable of rejecting the fetus (cellular immunity), this response seems to be blocked by the other branch of the immune system (humoral immunity). Apparently the humoral antibodies, which are harmless to fetal cells, combine with fetal antigens and cover them so that the lymphocytes of cellular immunity cannot find any antigens with which to react.

Once the blastocyst has implanted, it has solved its immediate nutritional problem, because it can feed on digested maternal materials. But another problem becomes of paramount importance: the prevention of menstruation. If the lining of the uterus were to be sloughed off, it would be the end of the blastocyst and its newly forming individual. The trophoblast cells once again come to the rescue: they prevent menstruation by synthesizing and releasing the hormone *chorionic gonadotropin.* Chorionic gonadotropin maintains the corpus luteum (which is the transformed follicle from which the ovum was discharged), and supports the production of the hormones (estrogen and progesterone) released by the corpus luteum and by the ovaries. Chorionic gonadotropin is needed on day 24 of the menstrual cycle (day 10 after fertilization) because this is the day when the corpus luteum begins to deteriorate, and this deterioration leads in turn to the sloughing off of the uterine lining beginning on the 28th day of the menstrual cycle. By the tenth day after fertilization, the trophoblast cells have become a functional endocrine organ in that they secrete the hormone chorionic gonadotropin.

The tiny human organism accomplishes a great deal in its first ten days: it successfully invades the uterus, it releases digestive enzymes and absorbs nutrients, and it synthesizes a hormone to ensure its continued survival. And all these crucial activities are carried out by trophoblast cells, which do not become part of the embryo itself. They are discarded at birth, but early in the life of the fetus they are essential for establishing connections with maternal support systems.

Just after implantation (day 7-8), changes begin to occur in the inner cell mass, the portion of the blastocyst that will become the embryo. First, on about day 8 a slitlike *amniotic cavity* appears among these cells, and about day 12 a second cavity, the *yolk sac*, appears (Figure 15-5). Between these two cavities lie two of the three layers of cells of the future embryo: a layer of *ectoderm* next to the amniotic cavity and a layer of *endoderm* next to the yolk sac cavity. The ectoderm will give rise to the body's outermost coverings—skin, hair, and nails—as well as to the brain and the rest of the nervous system. The endoderm will give rise to the linings of nearly all the internal organs, including the digestive tract.

The amniotic cavity, which is located next to the ectoderm, remains in contact with the outside of the embryo as it develops. This cavity expands and becomes the fluid-filled cushion or "bag of waters" that surrounds the embryo and protects it by acting as a shock absorber (Figure 15-6). The yolk sac cavity and its contents, next to the endoderm, never grow much, but in the sac are precursors of the red blood cells, the future germ cells, and precursors of the cells of the lymphatic system (Figure 15-7). As development proceeds, the cavity of the yolk sac becomes continuous with the cavity of the gastrointestinal tract.

At about day 16 the third and final layer of body tissue, the *mesoderm*, makes its appearance in the region between the ectoderm and the endoderm (Figure 15-6). Mesoderm gives rise to muscle and connective tissues.

As development proceeds, the amniotic cavity swells and comes to cover the growing embryonic body, crowding out the original cavity of the blastocyst (Figure 15-6). The embryo becomes an aquatic creature, swimming in the amniotic fluid and later swallowing it and urinating into it as well. (The fetal cells present in amniotic fluid can be used for the prenatal diagnosis of genetic diseases, as we will discuss in Chapter 20.)

In order to support its increasing growth rate, the implanted organism must increase its capacities to take in nutrients and to eliminate wastes. To accommodate these needs, the syncytial trophoblast develops projections, or villi, that greatly extend the surface area of the blastocyst over which the transport of nutrients and wastes may take place (Figure 15-6e and f). As the organism increases in size, the villi become branched, which further increases the surface area. Eventually the villi on one side become organized into a true fetal organ, the *placenta* (Figure 15-8), which is connected to the fetus through the umbilical vessels. Once it has developed, the placenta provides for almost all

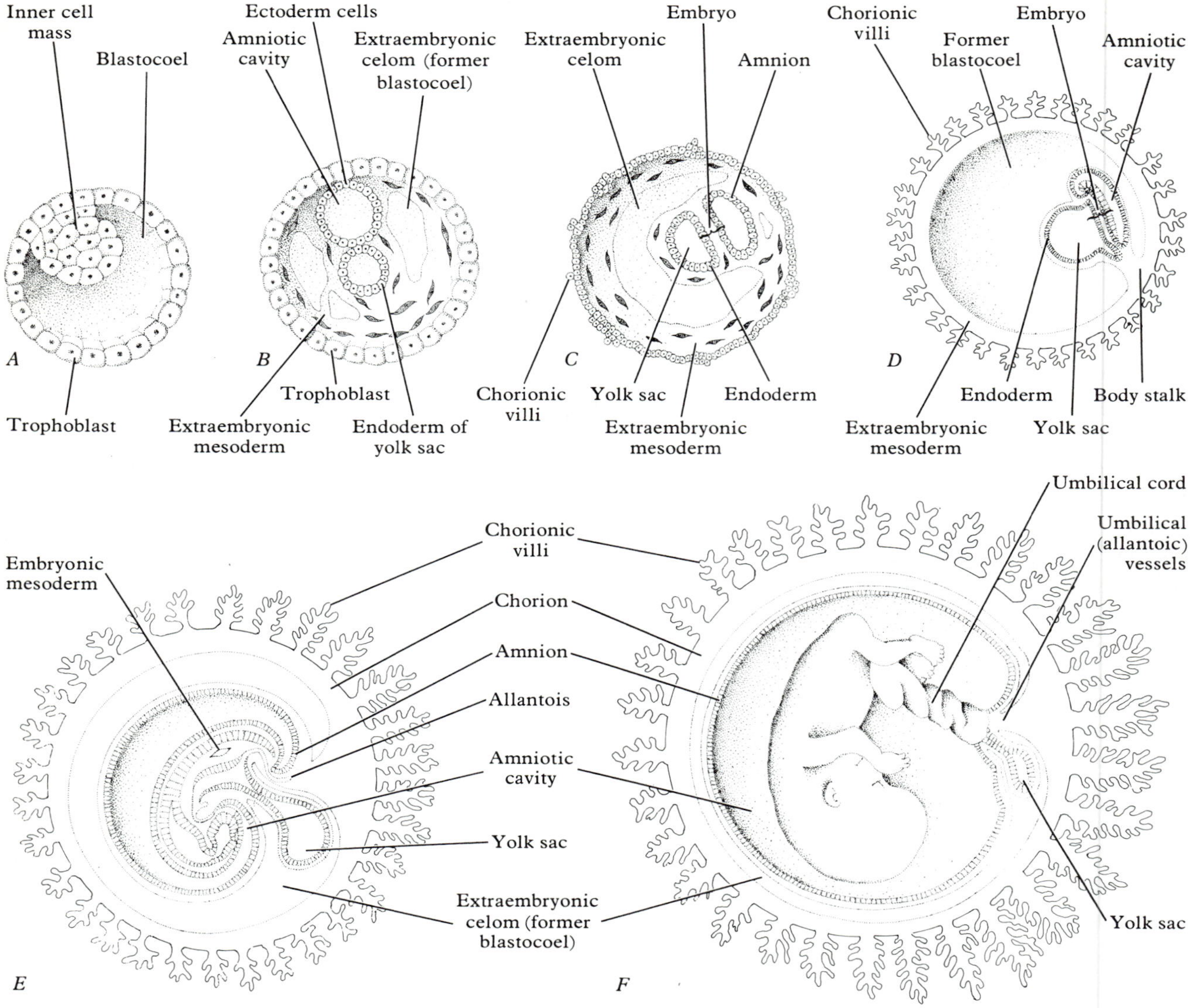

15–6

Early human development. A, the original cell formed by the union of sperm and egg has given rise to an inner cell mass, from which the embryo will emerge, and an outer layer of trophoblast cells. B, the cells of the inner cell mass become two layers, ectoderm and endoderm, each of which encloses a cavity. Ectoderm encloses the amniotic cavity and endoderm encloses the yolk sac. Mesodermal cells (extraembryonic mesoderm) line the inner surface of the trophoblast and the outer surface of the inner cell mass. C, the adjacent portions of ectoderm and endoderm define the area that will become the embryo. Meanwhile the trophoblast cells multiply and begin to put out projections, the chorionic villi. D, the extraembryonic mesoderm forms the body stalk as both embryo and chorionic villi continue to grow. E and F, the chorionic villi begin to expand on one side, where they will become the fetal portion of the placenta. The embryonic mesoderm forms and comes to lie between the ectoderm and endoderm. The amnion becomes the largest embryonic cavity as the extraembryonic coelom and yolk sac decrease in size.

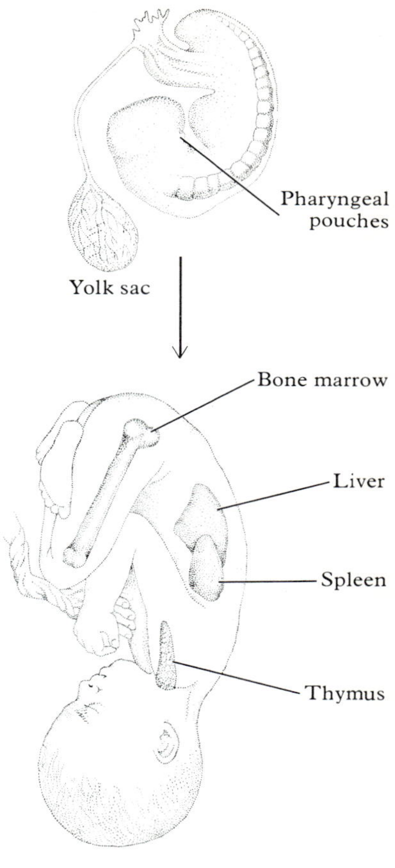

15–7
Derivatives of the yolk sac include precursors of red blood cells and lymphocytes. The precursors of the red cells originate in the yolk sac and migrate to the embryonic liver, spleen, and bone marrow. The precursors of the lymphocytes migrate through the spleen and liver to the bone marrow and thymus. (From "The Development of the Immune System," by Max D. Cooper and Alexander R. Lawton III, Copyright © 1974 by Scientific American, Inc. All rights reserved.)

the needs of the fetus: it delivers nutrients, removes wastes, and carries out gas exchanges. The placenta weighs about 500 grams (slightly more than one pound) at birth and is delivered after the fetus. It is therefore commonly called the *afterbirth.*

The embryo, for its part, develops a functional heart and blood vessel system by 21 days after fertilization, thus providing the necessary link between the embryo and the developing placenta. The importance of establishing good circulatory connections is underscored by the fact that the heart is the first fully functional organ of the embryo itself. When the embryo is between three and five weeks of age the heart is quite prominent; it bulges forward in the region where the chest will form (Figure 15-9). At this stage of development the embryos of many vertebrates look very similar (Figure 15-10), and this similarity is one of many lines of evidence supporting the theory that the various kinds of vertebrates are related to each other by evolutionary descent.

The circulatory system of the fetus is entirely separate from the mother's circulatory system in that there is no mixing of the blood cells. This is important because frequently the red blood cells of the fetus and the mother are incompatible. In order to promote exchange between the two circulatory systems, the capillaries of the maternal circulation pass very close to the capillaries

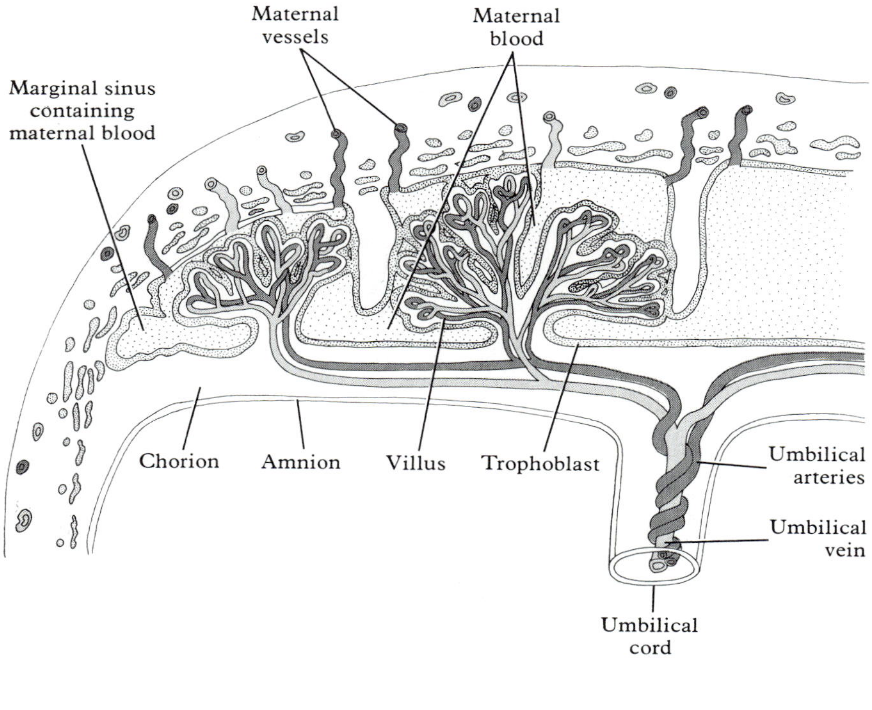

15–8
The human placenta and its circulation.

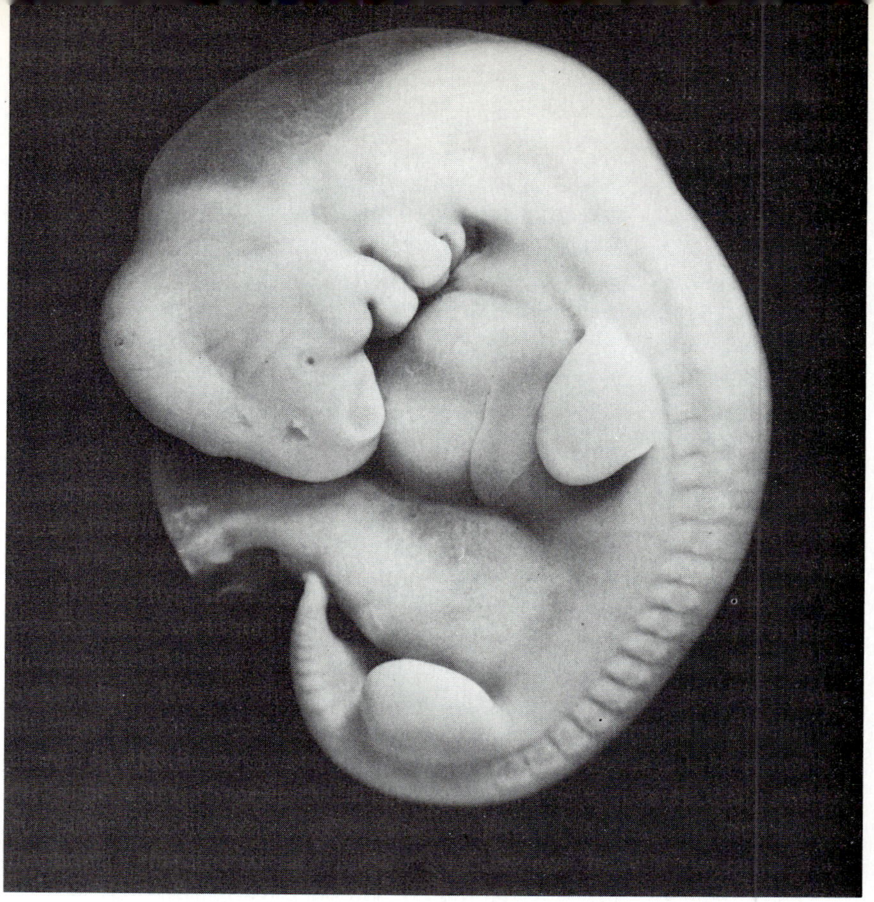

15–9
Human embryo 32 days after fertilization. Note protrusion of the area containing a heart. At this stage the human embryo has gill slits, a tail, and limbs that are just beginning to form. (Courtesy of Carnegie Institution of Washington, Department of Embryology, Davis Division.)

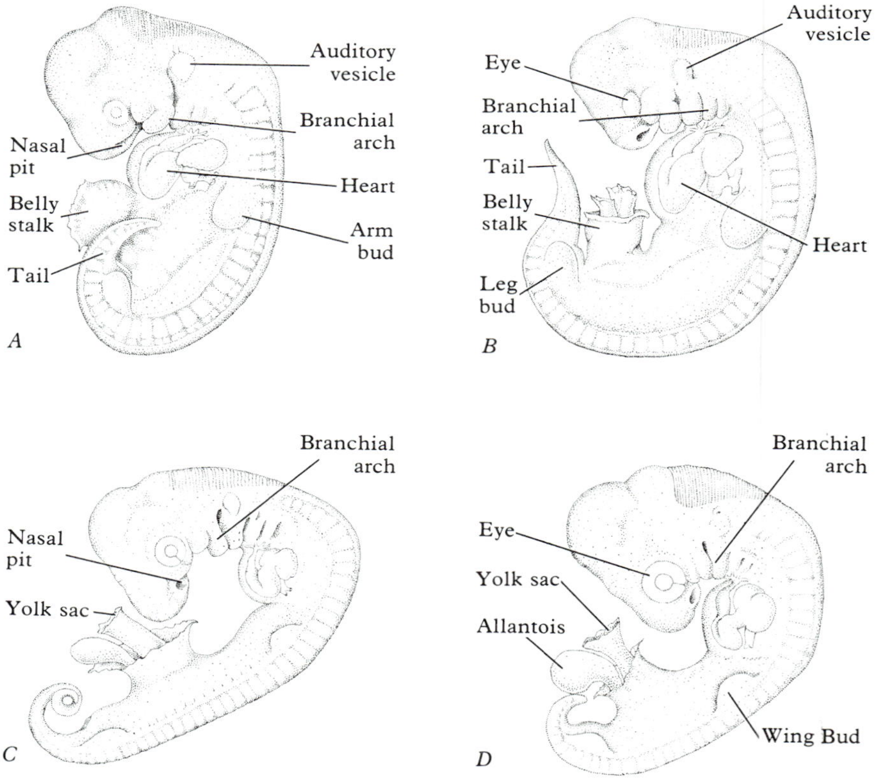

15–10
Embryos of (A) human, (B) pig, (C) reptile, and (D) bird at corresponding stages of development. The resemblance of the embryos to one another reflects the fundamental similarity of their developmental processes. (From Evolution, *by William Patten, Dartmouth College Press, Hanover, N.H., 1922. Courtesy Dartmouth College Press.)*

of the fetal circulation within the placenta. Small molecules (oxygen, glucose, amino acids, carbon dioxide) pass easily between the two circulations by diffusion, and larger molecules are actively transported back and forth. Examples of transported large molecules include bilirubin, a potentially toxic product of hemoglobin breakdown that is transported from the fetal to the maternal circulations; and IgG antibody molecules, which are transported from the maternal to the fetal circulation.

The fetal circulation is different from the adult circulation in two major ways (Figure 15-11). First, fetal blood is circulated to the placenta (entering the placenta in the umbilical arteries and returning to the fetus in the umbilical vein); and second, because they are nonfunctional before birth, the fetal lungs are bypassed by most of the fetal blood. In the adult, blood entering the right side of the heart is low in oxygen and *all of it* goes to the lungs before being delivered to the left side of the heart. In the fetus, blood entering the right side of the heart is relatively high in oxygen (because much of it has just come from the placenta through the umbilical vein) and *only 10 percent of it* goes to the lungs. The remaining 90 percent of the blood bypasses the lungs through two openings: the *foramen ovale* (foramen means hole; this is an oval hole), which connects the right and left atria; and the *ductus arteriosus* (duct of the arteries), a short vessel that joins the pulmonary artery and the aorta. Just after birth, the right to left flow through the foramen ovale and the ductus arteriosus ends as the blood pressure falls dramatically in the pulmonary artery and the right atrium. (This results because the lungs provide much less resistance to blood flow when they expand with air.) The fall in pressure allows a flap of tissue in the left atrium to close over the foramen ovale, and within 24 hours after birth the ductus arteriosus closes as well.

15–11
Differences between fetal and adult circulatory systems. In the fetus (left) most blood that enters the right side of the heart bypasses the lungs through the foramen ovale (a hole between the right and left atria) and the ductus arteriosus (a short vessel joining the pulmonary artery and the aorta). The two umbilical arteries transport blood to the placenta and the one umbilical vein returns blood from the placenta to the fetus. In the adult (right) the foramen ovale and ductus arteriosus have closed, and all blood that enters the right side of the heart passes through the lungs.

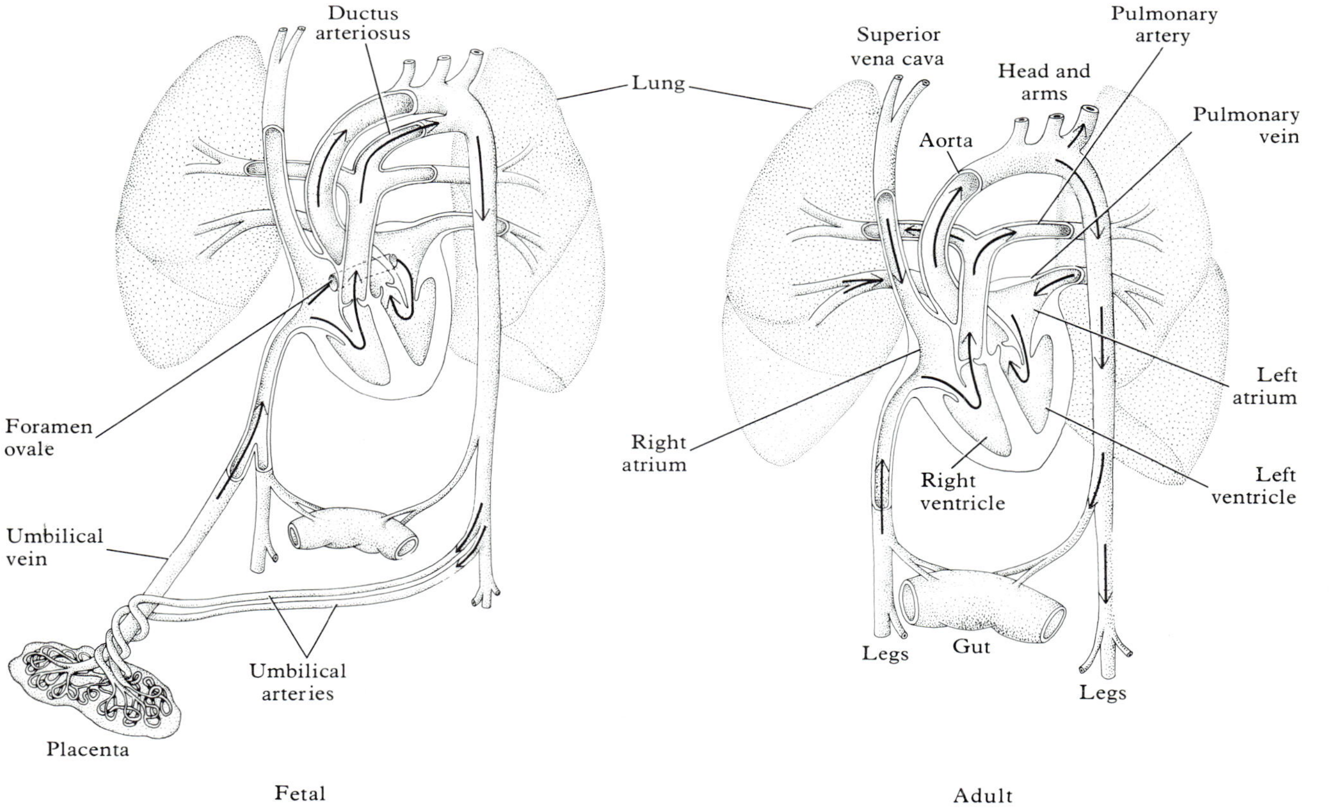

Maternal Changes During Pregnancy

In response to the requirements of the fetus, changes occur in the mother's physiology during pregnancy. Pregnant women have about a 30 percent increase in blood pumped through the heart, a 30 percent increase in volume of blood, a 40 percent increase in air taken in at the lungs, and a 60 percent increase in the amount of filtration by the kidneys. The increase in cardiac output amounts to about 1.5 liters (about a quart and a half) per minute, and of this, about 500 milliliters go to the uterus, 500 milliliters to the skin, and 400 milliliters to the kidneys. The increased flow to kidneys and skin is caused by the need to rid the body of fetal wastes: most soluble wastes from the fetus are eliminated by the maternal kidney, and the waste heat resulting from fetal metabolism is eliminated by the maternal skin.

The weight gain of the mother during pregnancy averages perhaps 12 kilograms (26.4 pounds). Of this total 5.5 kilograms are the uterus and its contents (1,000 grams for the uterus, 500 grams of amniotic fluid, 500 grams for the placenta, 3,500 grams for the fetus), 2.5 kilograms are the extra blood and extracellular fluid, and 4 kilograms are body fat reserves. An increase in smooth muscle mass accounts for the enlargement of the uterus itself, and this is accomplished mainly by an increase in size of existing muscle fibers (hypertrophy). During the first half of pregnancy the muscle fibers of the uterus undergo a twofold increase in width and a tenfold increase in length. During the second half of pregnancy, muscle cell hypertrophy slows down, the total muscle mass of the uterus increases very little, and the uterus stretches so that the thickness of its wall falls from 9 millimeters or so at mid-pregnancy to 6 millimeters at delivery. The uterine muscle retains the strength it needs to expel the fetus at birth.

Labor and Delivery

The average length of pregnancy from fertilization to birth is 266 days, or 280 days from the first day of the last menstrual period. But there is considerable variation, and over 20 percent of births occur more than two weeks on either side of the 280-day average (Table 15-1).

TABLE 15-1
Day of delivery, measured from the first day of the last menstrual period. (From a study of 537 white women in the United States who bore live children.)

DAY OF DELIVERY	PERCENT
Before 266th day	12.7
Second week before (days 266–272)	12.3
First week before (days 273–279)	22.1
On 280th day	3.7
First week after (days 281–287)	24.2
Second week after (days 288–294)	15.6
After 294th day	9.4

We do not understand what initiates *labor,* the process that expels the fetus from its mother's uterus and out onto its own. The onset of labor is heralded by powerful, increasingly regular contractions of the uterus that occur about every ten to fifteen minutes. In fact, uterine contractions occur off and on during the last two to three months of pregnancy, but they are irregular and weak in comparison with those of labor. The presence of these early contractions suggests that during the last few months the uterine muscle becomes increasingly likely to contract in response to several stimuli. It is also during the last few months of pregnancy that the uterus is more likely to contract in response to the hormone *oxytocin,* produced by the maternal posterior pituitary. However, although rising concentrations of oxytocin are crucial for normal delivery, the *onset* of labor cannot be attributed to elevated oxytocin because the concentration of oxytocin does not rise until labor has already begun. Nevertheless, it is likely that changes originating in the mother set into motion labor and delivery, because some experimental animals begin labor when expected even if the fetus has been surgically removed some weeks before.

Labor and birth are usually divided into three stages: the first, during which the cervix and lower portion of the uterus dilate (Figure 15-12); the second, during which the fetus passes through the birth canal and is born and the third, during which the placenta is expelled from the uterus. The first stage of labor is more variable in duration than the others: it lasts 18 to 24 hours for women who are delivering their first baby and six to 12 hours for women who have delivered before. As the contractions of the uterine muscle become increasingly strong, the cervix dilates and the sac containing the amniotic fluid usually bursts (Figure 15-12). In response to dilation of the cervix, nerve impulses are sent from the cervix and surrounding pelvic region to the hypothalamus, which in turn stimulates the release of oxytocin from the posterior pituitary. Oxytocin makes the uterus contract more forcefully, causing more dilation of the cervix and again activating the nerves in the region. This kind of an escalating response is an example of positive feedback, and unlike negative feedback, which helps maintain homeostasis, positive feedback leads to an explosive event—in this instance, expulsion of the fetus.

The baby is usually born head first, but birth can occur buttocks first, which is called *breech delivery,* or feet first, *footling breech delivery.* The head-first delivery is preferred for two reasons. First, the head has the largest diameter of any part of the infant, and if the head has been born, delivery of the rest of the body usually proceeds easily. Second, if the head is not first, then during the delivery of the head the umbilical cord will be constricted between the head and the mother's pelvis; if the passage of the head takes a long time, it can seriously reduce the oxygen supply to the fetus. The lower portion of the female pelvis, the birth canal (Figure 15-13), is widest from side to side at its inlet (which the fetal head passes first) but widest front to back at its outlet. Therefore, the fetal head, which is widest front to back, rotates during delivery, most commonly from a side-facing position at the inlet to nearly a back-facing position at the outlet (Figure 15-14). Fortunately, the bones of the fetal head are not firmly attached to each other, which makes delivery easier because it allows the head to be molded somewhat to suit the shape of the birth canal. The bones may even overlap each other to some extent during delivery without resulting in damage to the infant. Usually a few minutes after the baby is born the placenta separates from the uterine wall and is expelled.

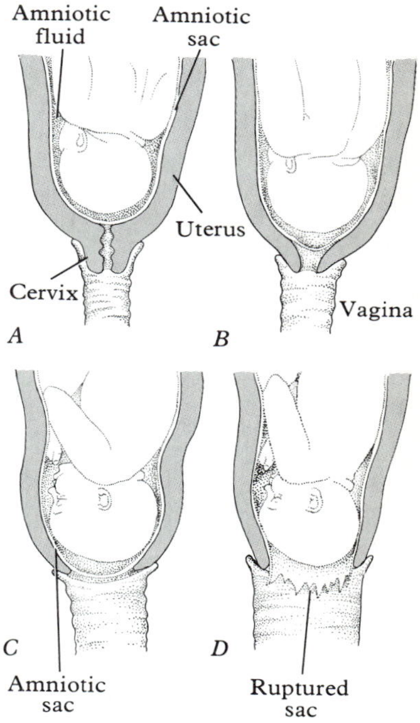

15–12
First stage of labor. The cervix dilates as a result of uterine contractions and the amniotic sac may rupture.

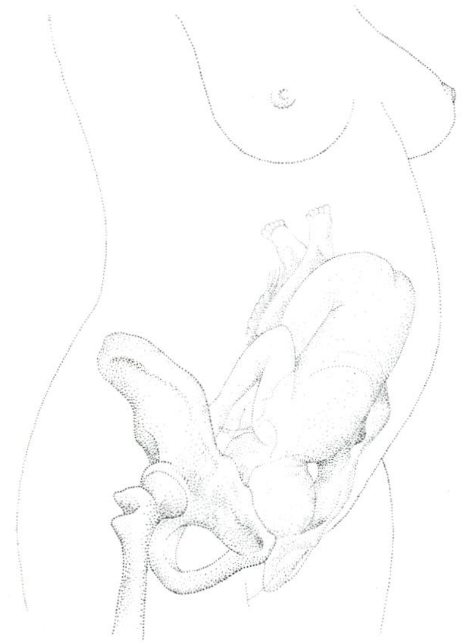

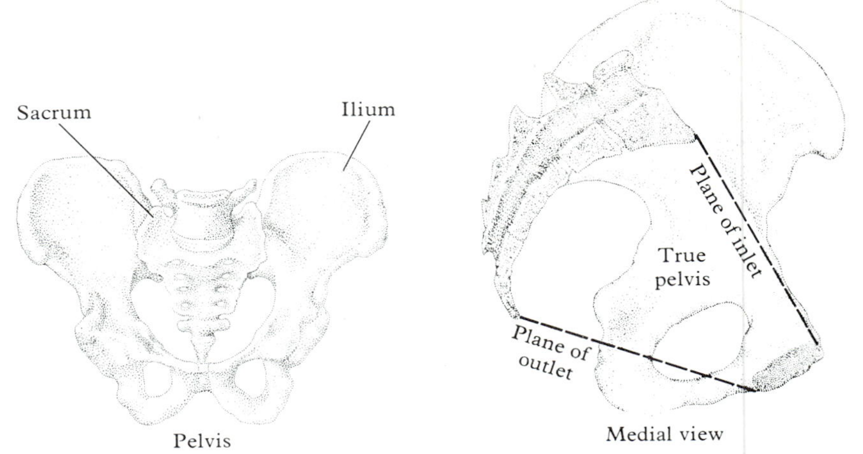

15–13
Human pelvis showing the birth canal. The canal is widest from side to side at the inlet, and widest from front to back at the outlet. Therefore the fetal head must rotate as it passes through the canal.

Infant and Mother after Birth

At the time of birth the infant leaves the constant temperature of its aquatic existence, in which everything has been provided, to enter an environment where it must cope much more for itself. It must breathe for itself, absorb food from its own gastrointestinal tract, rid its own body of wastes, and maintain its own body temperature. The environment itself initiates some of the changes: for example, the drop in skin temperature at birth leads to increased production of thyroid hormone and therefore to increased heat production.

The switchover to the use of the newborn's own systems is not always smooth, however, and during the first week after birth the concentration of bilirubin (the yellowish breakdown product of hemoglobin) normally builds up somewhat because at that time the liver is just beginning to carry out its new function of bilirubin secretion. If there is added demand for bilirubin secretion at this time, as may occur as a result of Rh incompatibility (p. 337), bilirubin can build up to toxic concentrations that may require removal by exchange transfusion (in which blood is transfused both into and out of the infant).

The newborn, which for months has been fed intravenously through the umbilical vein, also enters the new stage of feeding by mouth, either by breast-feeding or milk formula. The breasts (see Figure 12-10) are almost ready to provide nourishment: increasing concentrations of estrogen and progesterone during pregnancy have prepared the secretory cells, alveoli, and duct systems of the breasts for milk-giving. If the infant breast-feeds, during the first two days after birth it receives a special substance, colostrum, from its mother's breasts, and milk is released starting on about the third day. Colostrum and milk are discussed at length in Chapter 12.

Although the breasts are prepared for milk-giving by estrogen and progesterone during pregnancy, two other hormones are needed before the baby can receive the milk: prolactin is required for secretion of the milk into the alveoli, and oxytocin is required for milk letdown, that is, delivery of the milk from the alveoli into the ducts where it is available to the infant. These two

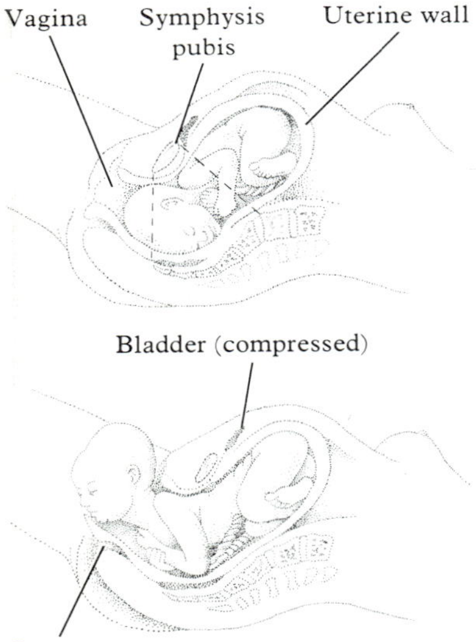

15–14
Rotation of the fetus as it passes through the birth canal. The fetal head faces sideward at the inlet and more towards the mother's spine at the outlet.

hormones are crucial for successful milk-giving, and we should discuss further how the secretion of each is controlled.

Prolactin secretion by the mother's anterior pituitary is initiated at birth by the precipitous fall of progesterone concentration (Figure 15-15). Throughout pregnancy prolactin secretion has been inhibited by the hypothalamus's inhibitory factors, and high concentrations of progesterone during pregnancy have stimulated the secretion of the inhibitory factor. With the fall in progesterone concentration after birth the inhibitory factor is no longer released in large quantities, and prolactin pours forth as inhibition decreases.

Oxytocin secretion by the posterior pituitary, on the other hand, is initiated by the sucking action of the newborn (Figure 15-16). Sucking causes nervous impulses generated in the breast to travel to the hypothalamus, and the hypothalamus responds by sending a message to the posterior pituitary, ordering an increase in oxytocin secretion. Because the event of sucking initiates this neurohumoral reflex, the milk will not flow if sucking does not occur, and milk will also stop being released if a nursing baby stops nursing. Psychological factors also can influence oxytocin concentration and milk flow, and this is perhaps not surprising considering the hypothalamus is closely associated with higher cortical centers.

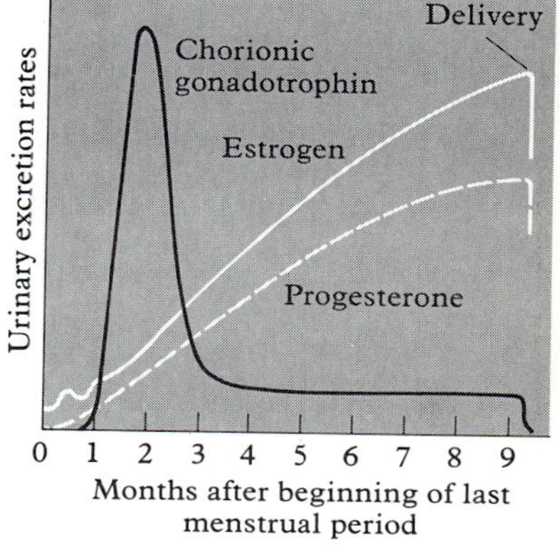

15–15
Urinary excretion rates of estrogen, progesterone, and chorionic gonadotropin during pregnancy. These excretion rates reflect the concentrations of these hormones in the blood.

Growth after Birth

Growth of the infant is quite rapid during its first year after birth, but the rate of growth slows down and stabilizes before adolescence (Figure 15-17). This relatively stable pattern of growth is broken by the adolescent growth spurt, which reaches its peak at about age 12 in girls and two years later in boys. Growth ceases at the end of the growth spurt. Figure 15-17 illustrates the growth curves for North American and Western European youths. Before the adolescent growth spurt the average boy is slightly taller than the average girl, but the average girl becomes taller between the ages of 11 and 13, during her growth spurt. The average boy ends up taller in the long run because he gains more height during his growth spurt, and because his growth spurt occurs later, allowing more total time for growth. A discussion of the "average" boy or girl can be misleading, however, because it tends to mask the fact that there is wide variation between individuals (Figure 15-18). Thus the growth spurt for girls may begin anywhere between 9.5 and 14.5 years of age. For boys it may begin anywhere between 10.5 and 16 years of age.

After birth, growth rates vary for different parts of the body. For example, the brain has grown very rapidly during fetal life, so that at birth it has reached about 24 percent of its adult weight, whereas the newborn body is only 6 percent of its adult weight. The brain continues its rapid pace of growth, so that by the time the child is four years old it has reached 90 percent of its adult weight while the child's body has reached only 25 percent of its eventual adult weight. On the other hand, reproductive organs such as uterus and testes remain less than 10 percent of their final weight until the onset of puberty. The lymphatic system (lymph nodes, thymus, tonsils, etc.) reaches its maximum weight at about age 12, when it weighs nearly *twice* its adult weight. Presumably the lymphatic system

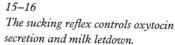

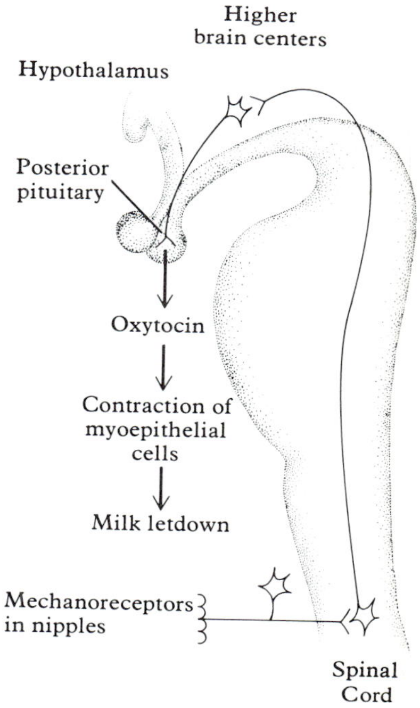

15–16
The sucking reflex controls oxytocin secretion and milk letdown.

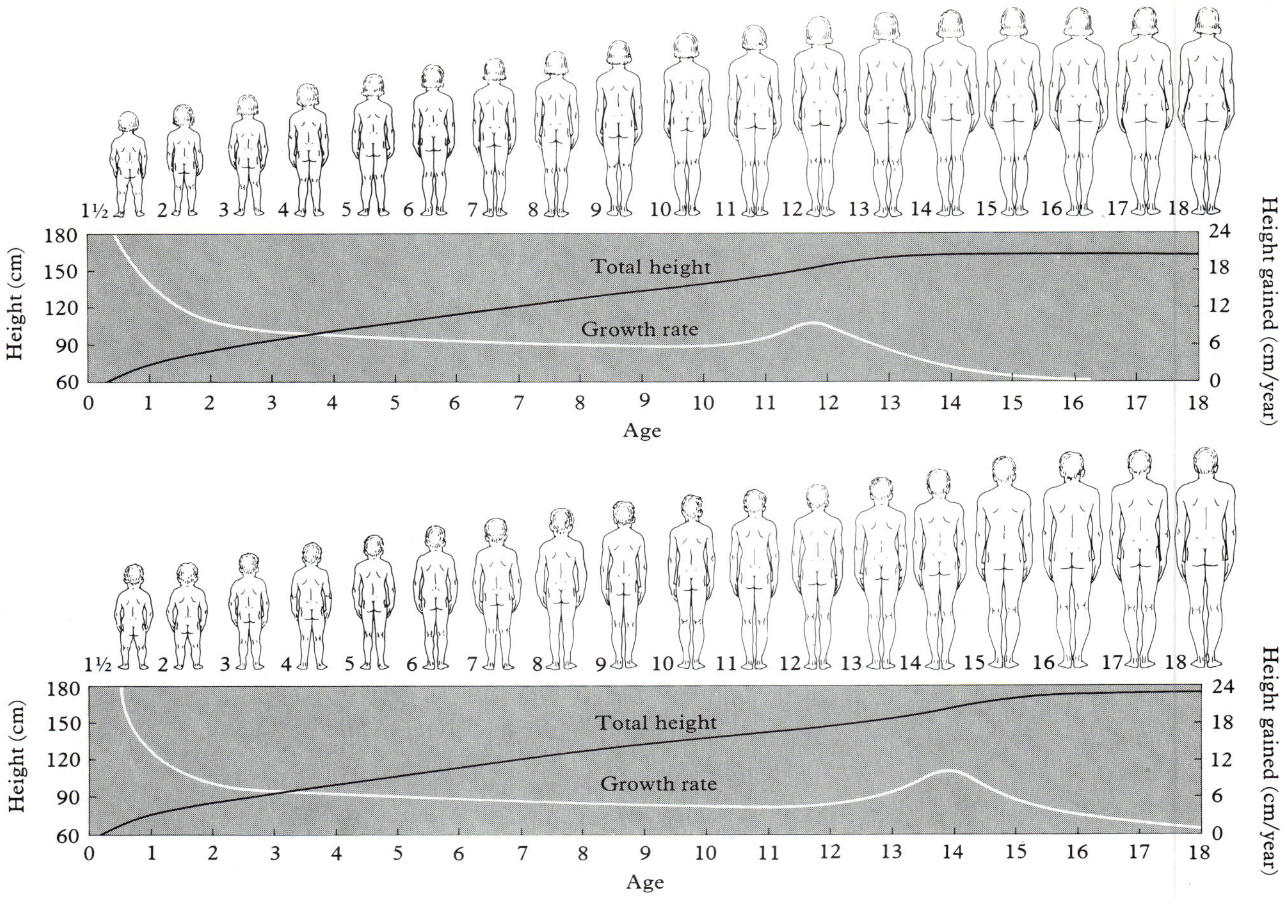

15–17
Girl growing up is shown at regular intervals from infancy to maturity (top). The figure shows the change in the form of her body, as well as her increase in height. The height curve is an average of girls in North America and Western Europe. Superimposed on it is a curve for growth rate that displays the increments of height gained from one age to the next. The sharp peak is the adolescent growth spurt. Boy growing up is shown at the same age intervals as the girl. Again the height curve and growth rate curve are below the figures that show the development of his body. His adolescent growth spurt comes about two years later than the girl's. The human figures in both of these illustrations are based on photographs in the longitudinal-growth studies of Nancy Bayley and Leona Bayer at the University of California at Berkeley. (From "Growing Up," by J. M. Tanner. Copyright © 1973 by Scientific American, Inc. All rights reserved.)

reaches its peak early because this is the time at which immunity to many microorganisms is acquired.

The growth spurt also affects external appearance: the face becomes more angular because of growth of the bones of the forehead, chin, and nose. Somatic differences between males and females become more pronounced as well because the width of the males' shoulders increases proportionately more and the width of the females' hips increases proportionately more. Soft tissue changes include growth of the breasts in the female and testes in the male, as well as the acquisition of pubic and axillary hair by both sexes.

Growth depends upon the concerted action of several hormones, including insulin, growth hormone, thyroid hormone, and (during adolescence) the androgens. Growth hormone acts to cause bone growth through an indirect mechanism: it causes another hormone (as yet uncharacterized) to be produced by the liver, and the latter hormone then stimulates specialized cartilage cells called *osteoblasts* to lay down new bone. Lack of growth hormone leads to short stature (*pituitary dwarfism*) but the condition can be prevented by the injection of human growth hormone, which can be extracted from the pituitaries of persons who have died. Pituitary dwarfism is probably caused by a defect in the hypothalamus of the brain that makes it unable to elaborate enough growth hormone-releasing factor. In fact, the principal control over growth seems to come from the hypothalamus, and because the hypothalamus receives input from the cortex, children who are seriously emotionally disturbed may also not grow, a condition known as *deprivation dwarfism.* Another form of dwarfism, *achondroplasia* (Figure 15-19), is an inherited condition caused by a single dominant gene. In achondroplasia the amounts of growth hormone are normal, but a defect in the cartilage cells, particularly in the bones of the upper arms and thighs, makes the long bones relatively unresponsive to the action of growth hormone. As a result, a person who has achondroplasia has short arms and legs but a trunk of normal size. (See Figure 19-13.)

Not only does the hypothalamus seem to play an important role in growth in general, but it also appears to be critical in initiation of puberty and the adolescent growth spurt. Indeed, tumors of the hypothalamus may cause sexual maturity to occur exceptionally early; this phenomenon is called *precocious* puberty. Other evidence of the control exerted by the hypothalamus is that if the pituitary of an adult rat is surgically removed and the pituitary from a newborn rat grafted into its place, the newborn pituitary quickly begins to release the

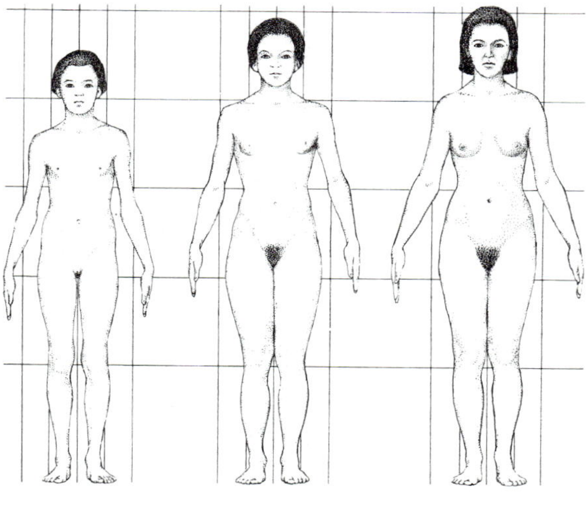

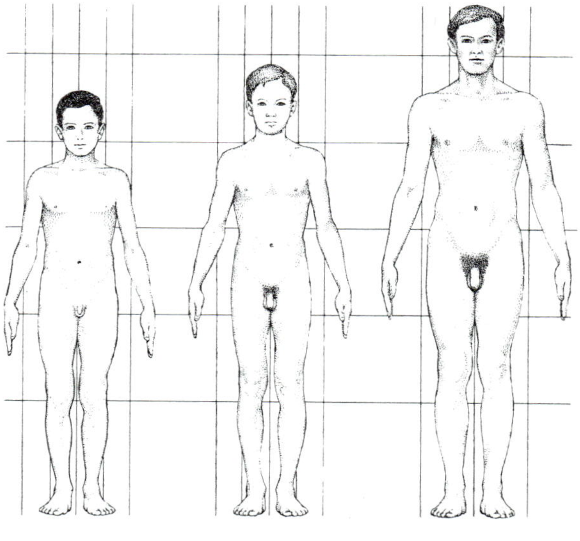

15–18
Three girls, all 12.75 years old, differed dramatically in development according to whether the particular girl had not yet reached puberty (left), was part of the way through it (middle) or had finished her development (right). This range of variation is completely normal. This drawing and the one beside it are based on photographs taken at the Institute of Child Health of University of London. Three boys, all 14.75 years old, showed a similar variation in the range of their development. Some boys of this age have entirely finished their growth and sexual maturation while others have yet to begin. (From "Growing Up," by J. M. Tanner. Copyright © 1973 by Scientific American Inc. All rights reserved.)

15–19
"Las Meninas" (the maids of honor), by Diego Velasquez (1656) shows among other things two conditions of human growth. At left center is the Infanta Margarita Maria, the daughter of Philip IV of Spain and his second wife, Marianna of Austria. The Infanta is a normal child of five. (She is dressed as a miniature adult.) The figure at the far right is a dwarf. She suffers from achondroplasia, in which the growth of the trunk is normal but the limbs are stunted and the face is characteristically altered. A third condition of human growth is depicted in the painting (of which only a part is shown here). A man who appears to be a child (whose foot can be seen here poised on the back of the dog) suffers from the most common form of dwarfism: growth hormone deficiency. In growth hormone deficiency growth is severely limited but the proportions of the body are normal. The Infanta later married Leopold I, the Holy Roman Emperor. Her mother and father appear in the picture as reflections in the mirror to the left of the man in the illuminated doorway. Velasquez, who painted the scene as if he were making a portrait of the king and queen while the Infanta looked on, put himself in the picture at the far left. The painting has a room to itself at El Prado in Madrid. (Courtesy Museo del Prado.)

hormones appropriate to an adult, including the gonadotropic hormones FSH and LH associated with sexual maturity.

Although we have a lot to learn about the details, it is clear that a key event in the initiation of puberty is an increased output by the hypothalamus of the two releasing factors that stimulate the pituitary to release follicle-stimulating hormone (FSH) and lutenizing hormone (LH). FSH and LH, in turn, affect the gonads of both sexes so that they put out increasing quantities of androgens and estrogens. The adolescent growth spurt that accompanies puberty seems to be mainly due to the influence of rising concentrations of androgen. Androgens are produced by the adrenal cortex in females and by the testis and adrenal cortex in males. Androgens stimulate growth considerably more than estrogens, and the higher concentrations of androgen in males during adolescence helps to account for the fact that males grow more during their growth spurt than do females.

Aging

After adolescence there is a period of about 15 years during which most systems of the body operate at about peak performance. However, at about age 30, there begins a gradual and progressive loss of function known as *aging*. Aging affects biological processes throughout the body: extracellular materials such as elastin molecules in the skin become progressively more crosslinked and therefore lose their elastic properties (Figure 15-20), cells show an increase in chromosomal abnormalities, and organs and organ systems function less efficiently (Figure

15–20
Two Rembrandt self-portraits, one from 1633 or 1634, when he was 27 or 28 years of age (from the Uffizi Gallery in Florence), and the other probably from 1658, when he was 52 years of age (from the Kunsthistorisches Museum in Vienna). The effect of aging on the elasticity of the skin is apparent. (Courtesy European Art Color Slides, Peter Adelberg, N.Y. 10023.)

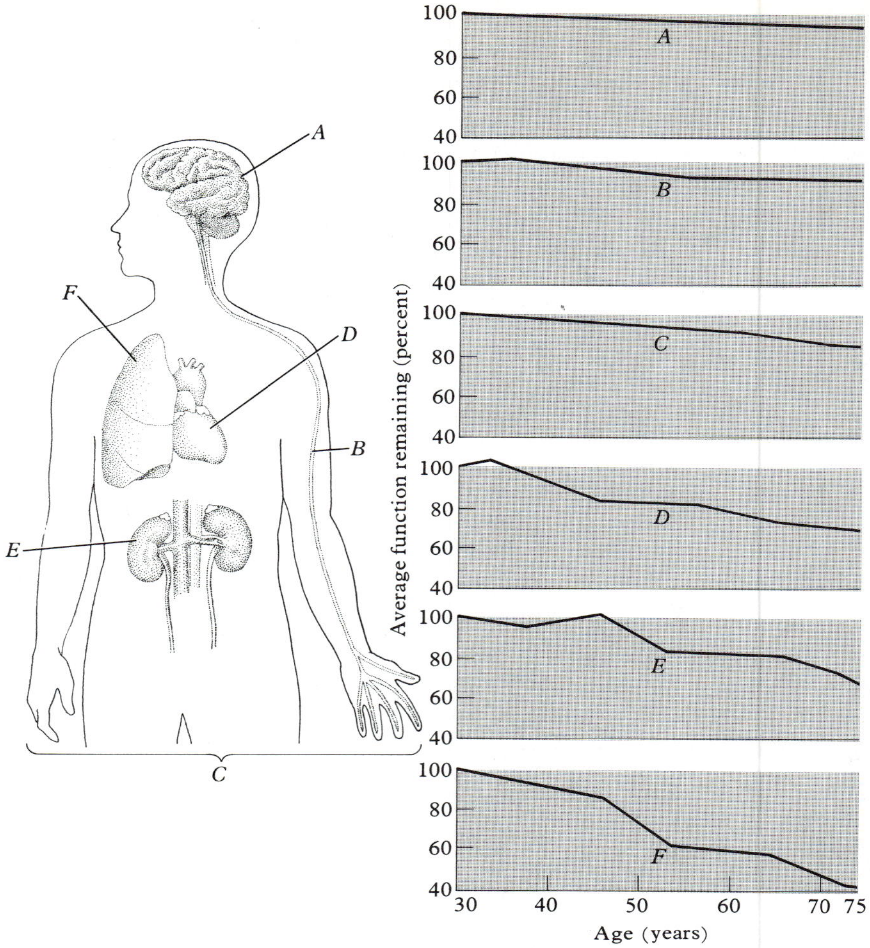

15–21

Loss of function with increasing age. The level of function at 30 years of age is assigned a value of 100 percent. By age 75 brain weight (A) has diminished to 92 percent of its value at age 30 and nerve-conduction velocity (B) has diminished to about 90 percent. Basal metabolic rate (C) has diminished to 84 percent, cardiac output at rest (D) to 70 percent, filtration rate of the kidneys (E) to 69 percent and maximum breathing capacity (F) to 43 percent. (From "Getting Old," by Alexander Leaf. Copyright © by Scientific American Inc. All rights reserved.)

15-21). Homeostasis is also affected: for example, the capacity to maintain a stable body temperature in hot or cold climates is reduced; and the concentration of glucose in the blood rises higher and returns more slowly toward normal after a meal that contains a great deal of sugar.

The problem of trying to find an explanation of biological aging is that everything seems to start to slowly deteriorate at about the same time, making it difficult to determine cause and effect. However, aging does seem to be a programmed stage of life in the sense that it is a developmental process inextricably linked to the earlier developmental processes. For example, when the life expectancy is lengthened by withholding calories early in life, the animals experience delayed aging *and* delayed growth and maturation. It is also pertinent that most developmental events occur in the same fraction of the lifetime in mammals that have very different life expectancies (Table 15-2). Thus all mice and dogs and most human beings attain peak functional capacity during the second quarter of their lifetimes, and all show definite signs of aging in the third quarter. If aging is part of the "program" of life, then certain genes expressed later in life probably bring on and facilitate the aging process.

Although we do not know if this actually occurs, we do at least know that genes *can* be activated late in life. The genes for hairy ear rims (p. 440) are not expressed in some males before age 50.

As a consequence of aging, the occurrence of a variety of diseases and pathological processes becomes more likely. Figure 15-22 shows that the mortality rate for many major diseases rises as one gets older. This is not surprising when you consider that most, if not all, of the systems of the body are weakened by aging.

Since the turn of the century, enormous progress has been made in the fight against infectious diseases. This is clearly shown in Figure 15-23 by the fact that the three commonest causes of death in 1900 in the United States were primarily infectious in origin, whereas no infectious diseases were among the top four causes of death in 1970. Along with the reduction in incidence of infectious diseases there was an increase in *life expectancy* from 47 years in 1900 to 71 years

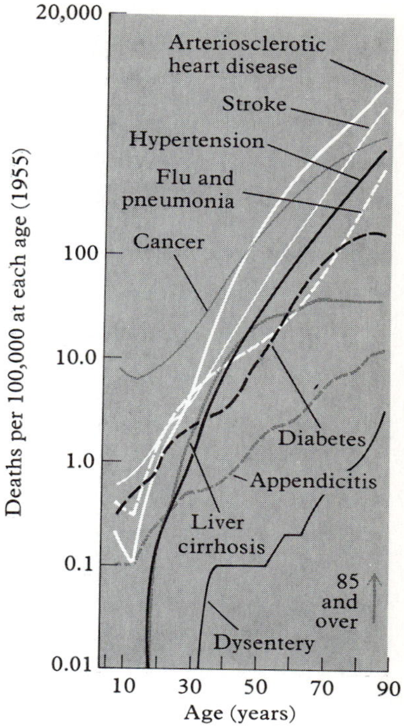

15–22
The mortality rate for many diseases and conditions rises with increased age.

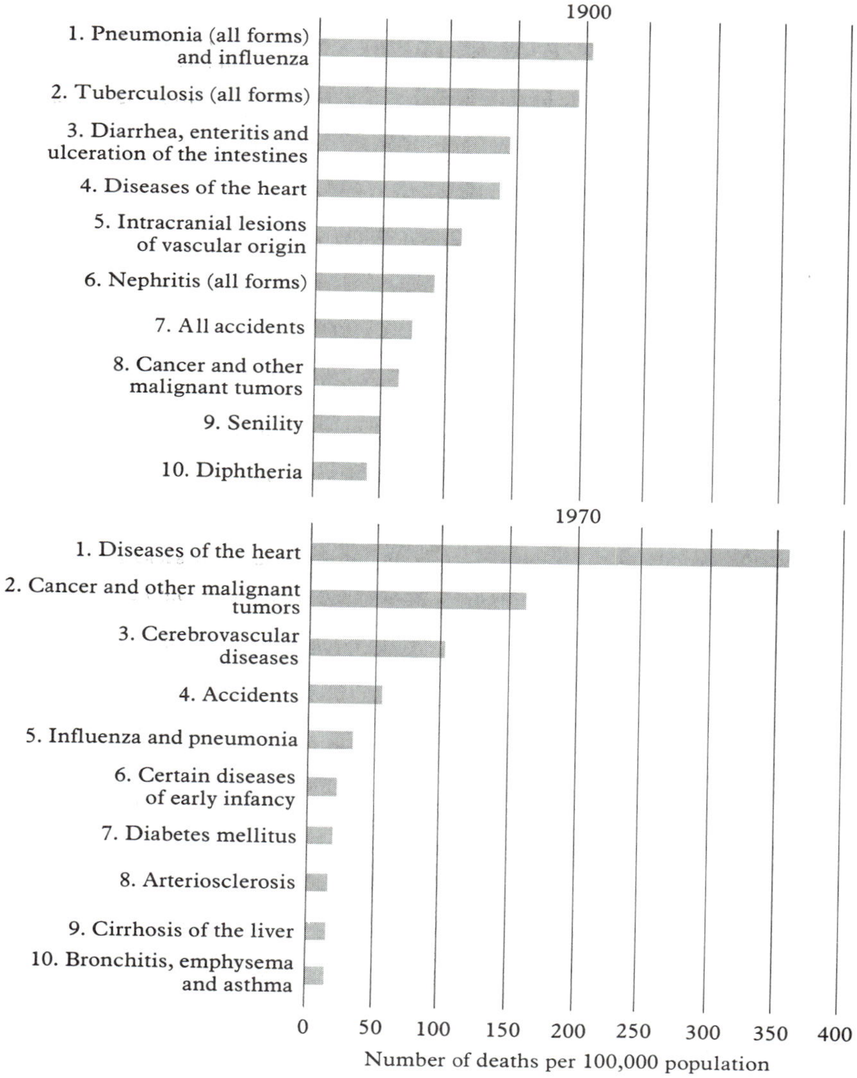

15–23
The major causes of death in the United States in 1970 were very different from what they had been in 1900, and the percentage of deaths from infectious diseases was greatly reduced by 1970. Along with these changes the life expectancy at birth increased from 47 years in 1900 to 71 years in 1970. (After "The Ills of Man," by John H. Dingle. Copyright © 1973 by Scientific American, Inc. All rights reserved.)

TABLE 15-2

Events of development and aging that tend to occur in the same fraction of the lifetime in long- and short-lived mammals.

DURING THIS PERIOD OF LIFE	WHICH OCCURS IN THIS QUARTER OF THE LIFETIME	THESE EVENTS OF DEVELOPMENT AND AGING OCCUR IN MANY MAMMALS	WHEN MICE ARE (YEARS OLD)	WHEN DOGS ARE (YEARS OLD)	WHEN HUMAN BEINGS ARE (YEARS OLD)
Postnatal development	First	Puberty Collagen maturation Chemical maturity: approach to adult body composition	0–0.75	0–3	0–20
Maturity	Second	Achievement of adult body size Maximum immune response Maximum thyroid activity Maximum muscle strength Minimum reaction time Maximum male sex drive	0.75–1.5	3–6	20–40
Middle Age	Third	End of female reproductive period Increase of noninfectious diseases: arterial degeneration, autoimmunity, abnormal growths Declining resistance to temperature stress Onset of osteoporosis Reduced male sex drive Disturbed sleep patterns	1.5–2.25	6–9	40–60
Advanced Age	Fourth	Exponentially increasing mortality	2.25–3.0	9–12	60–80

Source: Adapted from "Regulation of Physiological Changes," by Caleb Finch. *Quarterly Review of Biology,* March 1976.

in 1970. More women than men benefited from this increase in life expectancy—in large part because of the elimination of infections associated with childbirth. Also, as explained further in Chapter 17, more men than women die at every stage of life. The average female in the United States could expect to live eight years longer than the average male in 1970, as opposed to only two years longer in 1900.

If biomedical research and medical practice continue to eradicate major categories of disease (without losing the gains already made), the life expectancy will probably increase even further. However, the elimination of various diseases results in an increase of life expectancy but *not* of lifespan. *Lifespan* refers to the maximum end point of life; it would increase not at all beyond the ages of 90 to 95, where it is now. An increase in life expectancy means that more and more people will actually live out their maximum lifespan of 90 to 95 years.

This important difference is illustrated by the curves in Figure 15-24, which are survival curves for different populations of people at different times in history. Note that with medical advances the curve gets higher, meaning that life expectancy is increasing, but the end point (lifespan) remains about the same. The most we could expect from further knowledge and better practice is a squared-off survival curve (solid line at top of Figure 15-24) meaning that almost everyone would live to be 90 to 95.

Or is 90 to 95 the most we could expect? It has been said that people who live in certain regions of the world live longer than those who live in others: the Vilcabambans in Ecuador, the Hunzas in Pakistan, and Russians from the Caucasus region of the Soviet Union are thought to be blessed with great longevity. Unfortunately, there is no way to verify the assertions of longevity in any of these regions. The Hunzas do not have a written language, the birth records of the Vilcabambans are either nonexistent or unreliable, and there are no valid documents from the isolated Caucasus. Among reasons for caution is the fact that the very old in all these regions are highly respected: the older they become, the more respect and authority they receive. This might stimulate a bit of exaggeration.

Lifespan can be extended in the laboratory through alteration of the diet. If in early life animals are fed a diet deficient in calories (but otherwise balanced) they show delayed growth and maturation, and their lifespans are extended by about 40 percent. However, if a calorie-restricted diet is given after maturation, it does not extend the lifespan. In any case, living to a ripe old age is desirable only if it is compatible with a satisfying life. For this, the Vilcabambans (Figure 15-25) and the Hunzas (Figure 15-26) provide fine models.

15–24
Survival curves for various human populations at various times. If causes of death continue to be eliminated, so that we all live out our maximum lifespan, the 'idealized' solid curve at the top of the figure apply. From Comfort, The Biology of Senescence. *Holt, Rinehart, and Winston, 1956.*

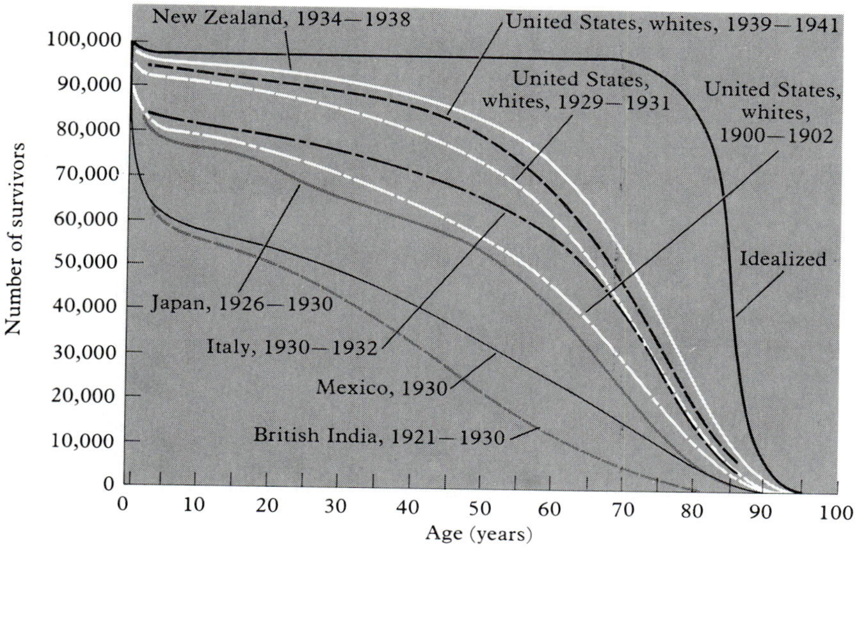

15–25
In the Andes of Ecuador an 85-year-old woman of the village of Vilcabamba works in a cornfield, contributing to the economic welfare of her community. (Photo by John Launois, courtesy Black Star Publishing Company.)

Summary

Fertilization, the union of sperm and egg, sets into motion the events that result in the development of a human being. Of millions of sperm deposited in the vagina during sexual intercourse, only a few hundred reach the region of the egg. The egg must prevent the entry of more than one sperm, and it does this by releasing enzymes contained in its cortical granules: these enzymes destroy sites where sperm attach to the outer membrane of the egg.

During the first week after fertilization, the ovum remains free in the female reproductive tract, passing from the oviduct (Fallopian tube) to the uterus. On about day 2 the ovum begins to divide so that by the end of the week the new organism consists of over 100 cells. This mass is called a *blastocyst* because a fluid-containing cavity is located among its cells. Most of the blastocyst's cells are located around its outside; these are *trophoblast* cells. A few cells composing the inner cell mass eventually give rise to the embryo.

To facilitate entry of the blastocyst into the uterus at implantation, the trophoblast cells release digestive enzymes that enable the blastocyst to digest its way into the uterus. In addition, the trophoblast cells invade by fusing with uterine epithelial cells.

Once implantation has occurred, the digestive enzymes released by the trophoblast break down maternal uterine tissue, and the blastocyst uses the breakdown products for nourishment. By the tenth day after fertilization (day 3 after implantation) the trophoblast cells also synthesize and release the hormone chorionic gonadotropin, which prevents menstruation and therefore ensures the continued existence of the new organism.

About three weeks after fertilization the trophoblast cells develop branching villi that increase the surface area available for fetal–maternal exchange.

The trophoblast cells on one side of the organism eventually develop into the placenta, which provides for almost all the needs of the fetus: it delivers nutrients, removes wastes, and carries out gas exchange. The umbilical vessels of the embryo serve as a transport system between embryo and placenta.

Soon after implantation the inner cell mass develops two cavities—the amniotic cavity and the yolk sac—and two cell layers—ectoderm and endoderm. Important cells and structures are derived from these two cavities and two cell

15–26
In a mountain village on the Pakistan frontier old men of the principality of Hunza winnow threshed wheat. The man on the left says he is 75 years old and the man on the right professes to be 80. Both hard work and a low-calorie diet contribute to the fitness of the Hunzas. (Photo by John Launois, courtesy Black Star Publishing Company.)

layers. The ectoderm gives rise to the body's outermost coverings as well as to the brain and the rest of the nervous system. Endoderm gives rise to the linings of nearly all the internal organs, including the digestive tract. The yolk sac, although small, is the first place where red blood cells, white cells, and germ cells are found. The amniotic cavity expands and becomes the fluid-filled "bag of waters" that surrounds and cushions the embryo. The third cell layer, mesoderm, which gives rise to muscle and connective tissues, appears a few days after the ectoderm and endoderm have formed.

The fetal circulation is different from the adult circulation in that only 10 percent of the blood entering the right side of the fetal heart goes to the fetal lungs. The rest of the blood bypasses the lungs through the foramen ovale, an opening between the right and left atria, and the ductus arteriosus, a duct between the pulmonary artery and the aorta. Both of these passages close at, or shortly after, birth.

Changes in the pregnant mother result from the presence of the fetus: her cardiac output, blood volume, air breathed, and blood filtered by the kidneys all increase. By the end of pregnancy, the elevated concentrations of progesterone and estrogen have prepared her breasts for milk-giving, which will occur after birth under the influence of two additional hormones, prolactin and oxytocin.

Labor is the process that expels the fetus from its mother's uterus and out onto its own. It begins when uterine contractions become strong and regularly spaced. A positive feedback loop facilitates expulsion of the fetus: dilation of the cervix leads to increased release of oxytocin, which in turn leads to more dilation of the cervix, and so on in an escalating sequence. After delivery of the baby, the placenta, or afterbirth, is usually expelled within a few minutes.

Growth after birth requires the concerted action of many hormones, including growth hormone. The adolescent growth spurt and puberty are brought about by an increased output by the hypothalamus of the releasing factors that stimulate the anterior pituitary to release FSH and LH. FSH and LH, in turn, stimulate the gonads to release androgens and estrogens. The adolescent growth spurt that accompanies puberty is mainly due to the added influence of rising concentrations of androgens.

After adolescence there is a period of about 15 years during which most systems of the body operate at about peak performance. About age 30, however, there begins a gradual and progressive loss of function known as *aging.* Aging may be a programmed stage of life in the sense that it is inextricably linked to the earlier developmental processes.

Advances in biomedical research, health, and medical practice extended the life expectancy of people living in the United States from about 47 years in 1900 to about 71 years in 1970, and life expectancy could be extended even further if more diseases that are major causes of death are eradicated. However, lifespan, the maximum age to which one might reasonably expect to live, has not increased and will remain at about 90-95 years of age. As far as is known, in laboratory animals the lifespan can only be increased if calorie-restricted diets are given very early in life so that growth and maturation are delayed.

Selected Readings

1. *Development,* by N. J. Berrill and G. Karp. McGraw-Hill, 1976. A fine discussion of the details of embryology in general and of the human embryo in particular.
2. "Growing Up," by J. M. Tanner, *Scientific American,* Sept. 1973. An excellent discussion of the factors which influence the growth of the human child.
3. 'Getting Old," by A. Leaf. *Scientific American,* Sept. 1973. A presentation of the phenomena of aging.
4. "Why Grow Old?", by L. Hayflick. *The Stanford Magazine,* 3, (1975), 36-43. Discusses theories of aging and the assertions of great longevity.
5. "The Regulation of Physiological Changes During Mammalian Aging," by C. E. Finch. *The Quarterly Review of Biology,* 51 (1976), pp. 49-83. Presents the evidence that suggests that aging is a programmed developmental stage in the human lifetime.

PART III

Human Genetics

The members of the Augustinian monastery in old Brno, Czechoslovakia in the early 1860s. Gregor Mendel is third from the right. (Photo courtesy of Dr. V. Orel of the Moravian Museum, Brno.)

CHAPTER 16

Traits and Chromosomes

In the mid-eighteenth century the city of Paris was the scene of a most unusual mating that aroused widespread public interest. The affair concerned a male rabbit who unexplainably showed great sexual interest in a certain barnyard hen. The hen, for her part in the matter, readily tolerated the rabbit's advances but would have nothing to do with roosters. These two unusual animals, both of whom belonged to a disconcerted clergyman, were observed to "mate" frequently, but the naturalists of the day doubted whether the union of the two was as complete as that of a rooster with a hen or a rabbit with another rabbit. So when the hen obligingly laid six normal-looking eggs, there was great excitement. What would hatch out?

Some people expected long-eared furry chickens to result; others, rabbits with beaks and feathers. But to the great disappointment of most neither rabbit nor chicken nor anything in between emerged. The well-watched eggs merely sat and decomposed.

The eggs failed to develop because rabbits and chickens are different species of animals. As we have discussed in Chapter 1, *species* may be defined as populations of organisms that retain their individuality in nature because they are reproductively isolated from other species around them. In general, reproductive isolation among animals has two important and interrelated aspects: behavior and genetics.

The Parisian observers of the ill-fated "mating" of the rabbit and the hen were as familiar with the behavioral aspects of reproductive isolation as we are. They were well aware that animals of different species generally show no interest whatsoever in mating with one another. But what the Parisians did not realize was that even if the behavioral aspect of reproductive isolation occasionally goes awry, the individuality of a species is still protected because species are genetically distinct from one another. What exactly does this mean?

You already know that animals produce sex cells of two different types—eggs from the female and sperm from the male (Chapter 14). The genetic uniqueness of sexually-reproducing species (and this includes virtually all animals) may be thought of as having its basis in different blueprints, or programs, for the elaboration of different species from fertilized eggs. As we shall see, each egg and each sperm usually contain within their nuclei half of the information necessary to set into motion the complicated process of elaborating a particular kind of animal according to the program of the species to which the parents belong. But in order for proper development to occur, both sex cells that merge to form the fertilized egg, or *zygote,* must contain the same basic program.

When the Parisian rabbit and hen mated it is not likely that their union resulted in a fertilized egg. This is because reproductive isolation operates even in sex cells, and rabbit sperm would be unlikely to penetrate and fertilize chicken eggs. In fact, even if we forced a rabbit sperm to fertilize a chicken egg by accurate mechanical injection of the sperm, the mating still would not produce feathered rabbits or long-eared chickens. A chicken egg fertilized in this way has received two conflicting programs, one for constructing a chicken and one for constructing a rabbit. And the programs for each are different enough that the artificially fertilized egg burns up its supply of intracellular fuel and then dies in the confusion of attempting to initiate the development of a composite creature from conflicting plans.

Actually, rare instances of successful interspecies mating do occur, not only among animals in experimental circumstances, but in nature too. (Also, interspecies crosses are much more common among plants than among animals.) Generally, such crosses occur only between species that are closely related by evolutionary descent, and that therefore presumably have similar genetic

16–1
The mule, left, is a familiar hybrid that is produced by the mating of a female horse with a male donkey. The hinny, right, is more horselike in appearance and results from the mating of a female donkey with a male horse. (From "The Mule," by Theodore H. Savory. Copyright © 1970 by Scientific American, Inc. All rights reserved.)

programs. Perhaps the most familiar animal issuing from an interspecies cross is the mule, the offspring of the mating of a female horse with a male donkey, but viable crosses also occur among closely related species of fish, birds, and porpoises, among others (Figure 16-1). But the animals resulting from these interspecific crosses are often incapable of reproduction themselves and they are clearly exceptions to the rule of reproductive isolation.

Differences in genetic programs between species are in large part responsible for the fact that animal species are usually morphologically distinct: that is, one can usually tell different species apart merely by looking at them. But then it is possible in many instances to distinguish animals from one another *within* a species, too (intraspecific variation). This is especially true of land-dwelling vertebrates, and is nowhere more obvious than in the human species, which is by far the most variable species known. Human beings have various skin colors ranging from almost pure white to jet black, have head and body hair ranging from perfectly straight to tightly kinked, and have unique fingerprints and faces—except for identical twins, who, as we shall see, have identical genetic programs. Yet within the human species, as is true of all others, the characteristics of individuals are not randomly distributed throughout the entire population. There are clear-cut geographic and racial differences as well as differences between related family lines.

The science of genetics is concerned with the study of heritable differences, both how they originate and how they relate to an individual's genetic program. Genetics also concerns itself with the biological basis of the transmission of traits in lineages and with the distribution of heritable characteristics within the populations that make up a given species. In chapters to come we will investigate the biochemical basis of heritable traits in individual people and see how at least some of these traits may have come to have the distributions we observe within the human population today. But in this chapter our main concern is with identifying and explaining the simple patterns of inheritance shown by some rather clear-cut human characteristics that obviously run in families. For the most part, human patterns of inheritance can best be described by some basically simple—yet decidedly unobvious—principles first worked out in the 1860s by an Augustinian monk named Gregor Mendel (frontispiece).

What Mendel Did

Mendel discovered the basic patterns of inheritance by performing carefully planned experiments on the common garden pea, and his success was partly due to his wise choice of experimental subject. Pea plants are good subjects for simple genetic experiments for several reasons. First of all, individual pea plants have clear-cut differences in some easily recognizable alternate characteristics. For example, the ripe seeds may be either smooth or wrinkled, and either yellow or an intense green. Mendel chose to experiment with seven clearly alternate traits in his search for patterns in the way such traits are passed on from parents to offspring (Figure 16-2).

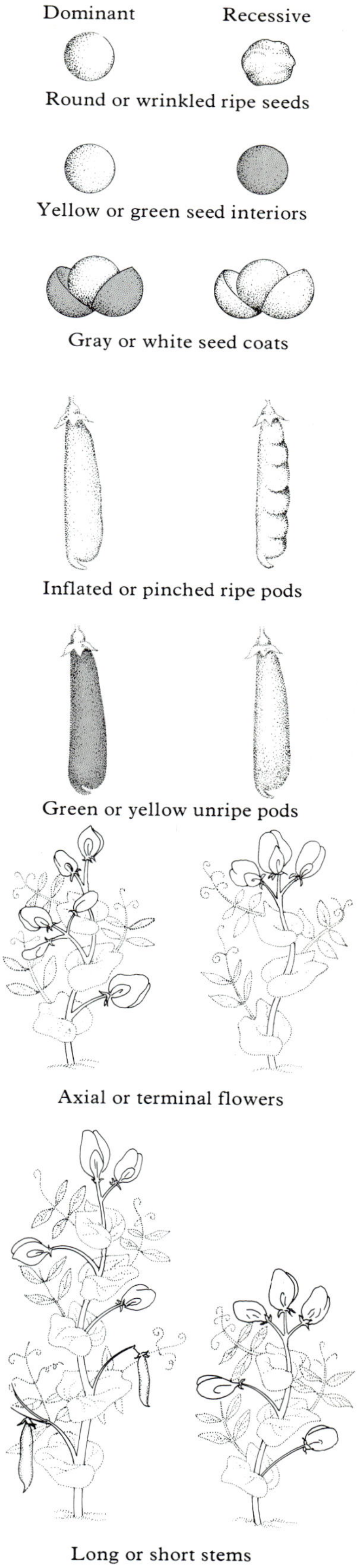

16–2
Mendel's early experiments were conducted upon seven pairs of alternate characteristics of garden pea plants.

Another reason pea plants make good experimental subjects is that they are self-fertilizing. That is, pea blossoms are so constructed that the male sex cells, which are contained in pollen grains, and the female sex cells, or eggs, are located in the same blossoms. (In other species, male and female sex cells may be produced separately, either by separate male and female flowers on the same plant or by flowers on separate male and female plants.) Self-fertilizing plants tend to breed "true," which means that their offspring usually resemble the parents exactly, at least in the alternate traits Mendel observed and recorded. True-breeding plants are thus good subjects for crosses of individual plants that differ in one or more alternate traits.

Mendel crossed plants that bred true for alternate traits and carefully recorded the distribution of these traits in their offspring. What he found is best illustrated by a recounting of some of his experiments.

Mendel began with two varieties of true-breeding pea plants, whose self-fertilized (self-pollinated) offspring had ripe seeds that were either round or wrinkled. He pinched off the pollen-producing parts (anthers) of each blossom on plants that produced only wrinkled seeds, and then fertilized the blossoms with pollen from plants that bred true for round seeds. (He also fertilized some "round blossoms" with "wrinkled pollen" and produced essentially the same results discussed in the following sentences.) Mendel then tied little paper bags over the blossoms to prevent any wind-borne or insect-borne pollen from contacting the artificially fertilized plants. When he opened the pods of his experimental plants he found that all of the seeds were round. The alternate trait, "wrinkled," seemed to have disappeared in the first generation of progeny produced from the cross, the F_1 generation. Mendel then planted the round seeds produced in the cross and allowed the resulting plants to fertilize themselves, as they usually do. When he examined the seeds produced by the second generation (the F_2 generation), he sometimes found round and wrinkled seeds lying together in the same pod (Figure 16-3). To be more exact, he found that about 25 percent of the total number of seeds were wrinkled. The trait "wrinkled," which had disappeared in the F_1 generation, had once again turned up in the F_2 generation about 25 percent of the time.

As shown in Table 16-1 Mendel found the same pattern for all seven of the traits he studied. For example, when true-breeding plants that had yellow seeds were crossed with those whose seeds were green, only yellow-seeded offspring were produced. Accordingly, Mendel called the member of the alternate pair of characteristics that showed up in all of the offspring of the F_1 generation, and in about 75 percent of the offspring in the F_2 generation, a *dominant trait.* And he

TABLE 16-1

Mendel's results from crosses involving some alternate characteristics of the common garden pea.

PARENT CHARACTERISTICS	F_1	F_2	F_2 RATIO
1. Round × wrinkled seeds	All round	5,474 round : 1,850 wrinkled	2.96 : 1
2. Yellow × green seeds	All yellow	6,022 yellow : 2,001 green	3.01 : 1
3. Gray × white seedcoats	All gray	705 gray : 224 white	3.15 : 1
4. Inflated × pinched pods	All inflated	882 inflated : 299 pinched	2.95 : 1
5. Green × yellow pods	All green	428 green : 152 yellow	2.82 : 1
6. Axial × terminal flowers	All axial	651 axial : 207 terminal	3.14 : 1
7. Long × short stems	All long	787 long : 277 short	2.84 : 1

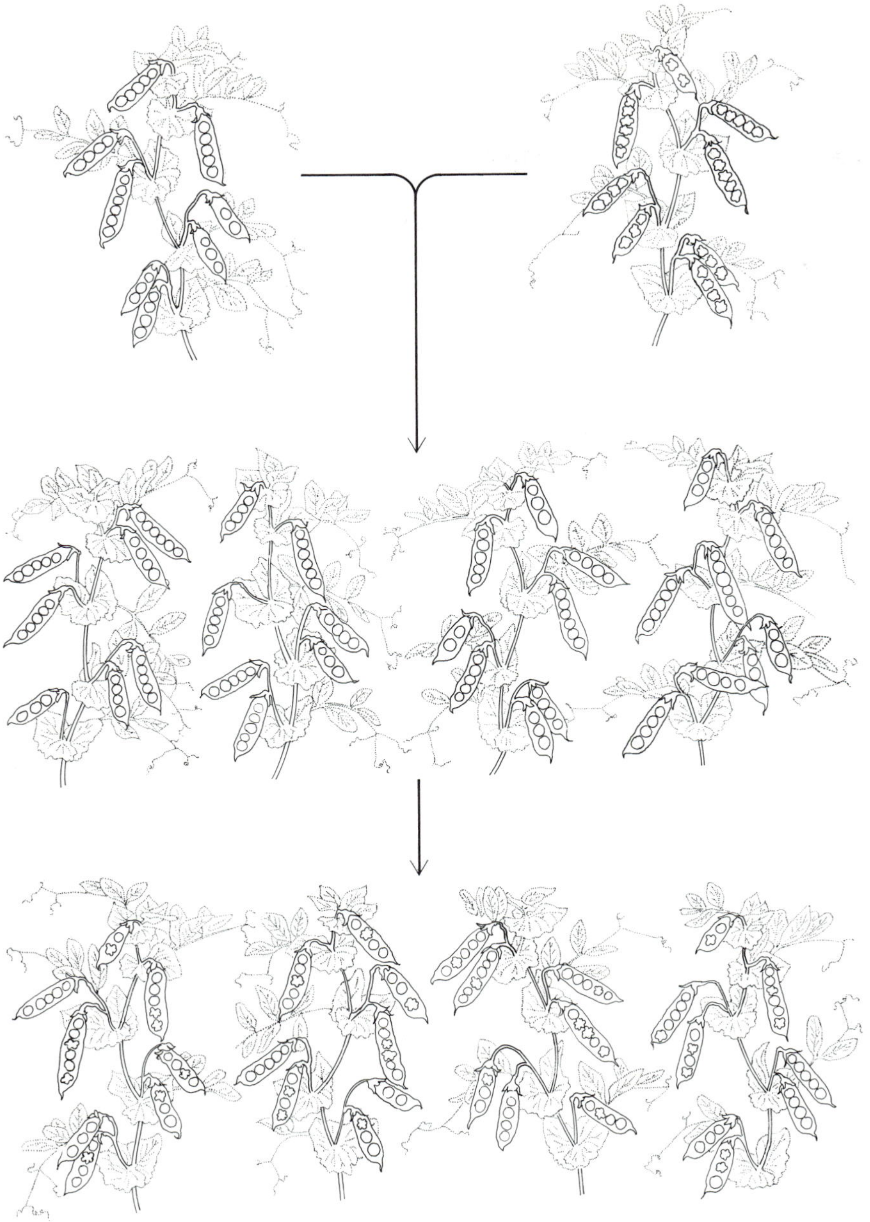

16–3
Mendel found that the wrinkled trait disappeared in the F_1 generation but turned up again in the F_2. (From "The Gene", by Norman H. Horowitz.

named the trait that disappeared in the F_1, only to reappear in about 25 percent of the F_2, a *recessive trait.*

In order to explain the patterns he observed, Mendel proposed that inherited traits are transmitted from parents to offspring by means of independently inherited "factors" that are now known as *genes.* Furthermore, he found that he could predict the results of his experiments if he assumed that true-breeding lines of plants contributed either a dominant or recessive factor to their offspring in the F_1 generation, and that members of the F_1 were therefore *hybrids.* That is, Mendel postulated that each member of the F_1 contained both dominant and recessive factors. This enterprising monk then invented a shorthand notation by which he could follow his hypothetical dominant and recessive factors through various lineages.

Mendel labeled the factor responsible for the dominant trait (round seeds) *A*, and he designated the factor responsible for the recessive trait (wrinkled seeds) *a*. (Geneticists still use capital letters to represent the genes responsible for dominant traits and lower-case letters to represent those responsible for recessive ones. Which letter is chosen to represent a given pair of alternate traits is arbitrary.) When both parents contribute an *A* to their offspring, the offspring are *AA*, and they produce only round seeds. In *aa* plants, which received an *a* from each parent, only wrinkled seeds are produced. Thus, true-breeding lines of plants are either *AA* (round), or *aa* (wrinkled). What happens if two such lines are crossed, as they were by Mendel to produce the F_1 generation? Clearly, all the offspring receive an *A* from one parent and an *a* from the other, so that all members of the F_1 must be *Aa* with respect to the alternate traits "round" or "wrinkled." What do *Aa* plants look like? Because *A* is dominant to *a*, all individuals in the F_1 will have round seeds. The trait "wrinkled" will seemingly have disappeared from the F_1, just as Mendel observed. But, in fact, the factor responsible for the recessive trait has not disappeared; its effects are simply masked by the presence of the factor *A* and, in later crosses, the effect of *a* can become obvious once again.

Consider what happens when the hybrid plants of the F_1 are allowed to self-fertilize and to produce offspring. All of the parents are *Aa*, and can contribute either *A* or *a* to their offspring, and in fact do so in equal proportions. About half the offspring get *A* and half get *a* from *each* parent. This means that three different kinds of offspring can result: *AA* or *aa* plants if both parents happen to contribute the same factor, and *Aa* plants if each parent happens to contribute a different factor.

An easy way to predict what will happen in a given cross is to construct a table that allows us to keep track of all possible combinations of factors. Across the top of the table are listed the factors that one parent can contribute; those from the other parent are listed down the left-hand side. In our example, both parents are hybrids, so they both can contribute either *A* or *a*, and we represent this as follows:

	A	*a*
A		
a		

Then by simply drawing in the boxes and combining the factors we generate the following table. (Capital letters are always written first.)

	A	*a*
A	*AA*	*Aa*
a	*Aa*	*aa*

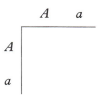

Thus, the combinations we should expect in the offspring are *AA*, *Aa*, and *aa*. Notice that *Aa* appears in the table twice. This means that about two out of every four offspring will be *Aa*. Or, more generally, about 50 percent of the offspring will be *Aa*. Similarly, about 25 percent of the offspring will be *AA* and the remaining 25 percent, *aa*. Looking at it another way, 75 percent of the offspring are either Aa or AA and therefore have round seeds, and 25 percent are aa and have wrinkled seeds. These ratios are exactly those observed by Mendel in the F_2 generation.

Using the same kind of reasoning Mendel predicted what would happen if he crossed plants differing in *two* alternate characters. He crossed plants that had round yellow peas (both dominant) with plants bearing wrinkled green peas (both recessive) and, as predicted, all members of the F_1 were round and yellow. He then allowed the F_1 hybrids to self-fertilize. If the factors he postulated did indeed exist and behave as independent units, then he expected to find four kinds of peas in the F_2: round yellow, wrinkled yellow, round green, and wrinkled green. Moreover, he predicted he would find them in the ratio 9:3:3:1. He performed the crosses and found the actual ratios to be exactly as he predicted, allowing for small deviations introduced by chance (Figure 16-4).

Mendel had discovered the most fundamental patterns of inheritance, and they have stood the test of time to the present day. We shall soon see how they apply to human lineages. He published his results in 1866, but for the most part his manuscript was ignored. By and large this was because Mendel's "factors" could not be seen; they were rather mysterious units for which no physical basis was known. But all that had changed when Mendel's work was rediscovered in 1900, when it was fully appreciated for the first time. What made the difference was that in the interim biologists had discovered what they presumed to be the physical basis of Mendel's mysterious factors. They had discovered chromosomes.

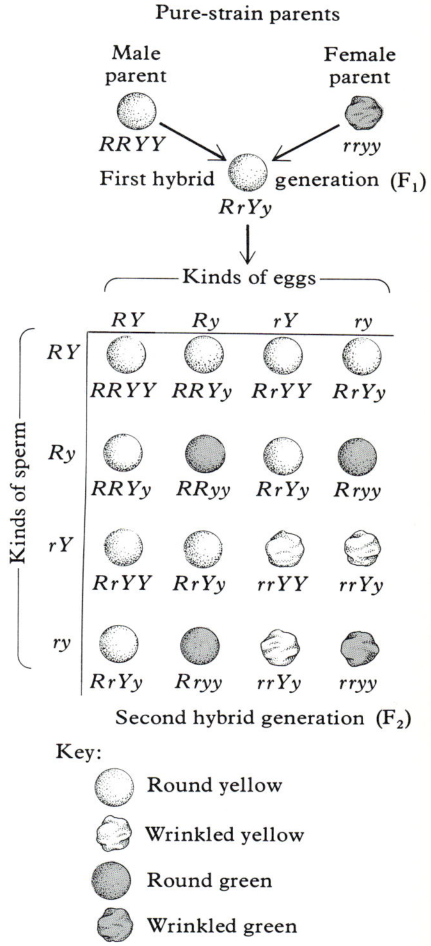

16–4
The results of a cross of pea plants that differ in two pairs of alternate characteristics (round and yellow versus wrinkled and green).

Chromosomes and Mendel's Patterns

When Mendel published his results in 1866 it was well known that cells are the basic building blocks of all living things (see Chapter 5). But at that time these fundamental units were poorly known and largely undescribed, because the manufacturing of microscopes and the preparation of specimens for microscopic study had not yet become highly developed arts. Nonetheless, it was known that most plant and animal cells have a distinct nucleus inside them, and that within the dividing nucleus are rodlike structures called chromosomes. In general, chromosomes are clearly visible only in cells that are in the process of dividing. As we discuss later in this chapter, in resting, nondividing cells, the chromosomes are still inside the nucleus, but they are much thinner and are highly entangled with one another so that individual chromosomes cannot be distinguished. (We should mention here that the relatively simple *prokaryotic* cells of bacteria and blue-green algae lack a distinct nucleus and have a single, unpaired chromosome that is never visible through the light microscope and that is much less complex than the chromosomes of eukaryotic cells that we are discussing here. We will have more to say about the structure of prokaryotic and eukaryotic chromosomes in Chapter 18.)

By the time Mendel's work was rediscovered in 1900 enough was known about the remarkable behavior of chromosomes in dividing cells to suggest that the observed patterns of inheritance could be explained in cellular terms by assuming that Mendel's independent factors were located on chromosomes. The gist of the evidence, as first described about the turn of the century, is this: within the nucleus chromosomes exist in pairs, except in a single, revealing instance. The exception is the sex cells, whose nuclei contain only one member of each pair, or half the number of chromosomes in other body cells. How does this tie in with the inheritance of alternate traits described by Mendel?

Assume that the dominant and recessive forms of a particular factor are located one each on the two chromosomes of a particular pair. Thus, an *Aa* individual has an *A* on one chromosome of a given pair and an *a* on the other. When the individual produces sex cells, pairs of chromosomes separate so that each egg or sperm contains either an *A* chromosome or an *a* chromosome. Then, when self-fertilization occurs, pairs of chromosomes are reunited once again, and the resulting offspring are either *AA, Aa,* or *aa,* depending on which factors happened to be located on the particular chromosome pairs that were reunited.

You will recall that Mendel followed the patterns of inheritance of seven pairs of alternate traits of the common garden pea. It turns out that pea plants have seven pairs of chromosomes, which correlates with Mendel's observation that all seven of the traits he studied behaved independently. That is, Mendel's factors showed *independent assortment* because virtually all of the traits he studied are determined by factors located on different pairs of chromosomes. (Mendel was thus not only careful, but lucky, too.)

It is now known that the number of chromosome pairs normally present in the nucleus can vary widely from species to species. (Remember that only one member of each pair is present in an animal's sex cells.) It is also known that each chromosome pair usually carries factors responsible for many different traits.

How does all of this relate to people? Our discussion of Mendel's work provides a background for discussing human chromosomes and for relating them to the patterns of inheritance shown by some alternate traits in human families. But before we go any further we should first introduce some terms that describe the genetic make-up of an individual, human or otherwise. Familiarize yourself with these words now, for they will be used repeatedly in the discussion that follows.

Some Definitions

Mendel's inherited factors, the units of heredity, are now called *genes.* We will discuss the biochemical basis of genes in following chapters. The two (or more) forms of genes responsible for alternate traits (*A* and *a* in our example) are called *alleles.*

Individuals in whom the two alleles of a given pair are the same (*AA* or *aa*) are called *homozygotes,* whereas *heterozygotes* are individuals in whom the two alleles of a given pair are different (*Aa*).

Recall that you cannot distinguish *Aa* heterozygotes from *AA* homozygotes merely by looking at them. (Both have round seeds.) But the two can be told apart if they are crossed with known heterozygotes (*Aa*). Thus, if wrinkled seeds (*aa*) turn up in the offspring, we can conclude that both parents must have been *Aa.* If only round seeds are produced, then the offspring are either *Aa* or *AA,* and the parent crossed with the known heterozygote must have been *AA.* To distinguish individuals that look alike but nonetheless have different genetic constitutions geneticists use the terms *phenotype* and *genotype.* Phenotype is a description of what an individual looks like, and genotype describes the individual's genetic constitution. In our example, individuals of phenotype "round" may be either of two genotypes, *Aa* and *AA.*

With these terms in mind, let us discuss the chromosomes of the human species and relate them to some fairly obvious patterns of inheritance in human families.

Human Chromosomes

Although the existence of chromosomes has been known for over a hundred years, the exact number of chromosomes that characterizes the human species was not discovered until 1956. This seems extraordinary, but in fact, the chromosomes of most mammals are not only numerous but difficult to prepare for detailed microscopic study; only relatively recently have satisfactory and consistent methods for visualizing them been developed. Human cells contain a total of 46 chromosomes in 23 pairs. That is, body cells, or *somatic cells,* contain 46 chromosomes in their nuclei, and sex cells contain only a single member of each pair, or a total of 23.

Human chromosomes can be made visible for detailed study in the following way. First, a few cubic centimeters of blood are collected in a syringe and then inoculated into a culture dish containing chemicals that stimulate some white blood cells in the sample to divide, thereby making their chromosomes visible. When the white cells have started to divide, other chemicals are added to make the chromosomes swell and to stop the process of cell division at a stage when the chromosomes are most easily distinguished from one another. The cells are then broken open and examined under the microscope. The result, as shown in Figure 16-5, for a blood sample obtained from a normal human male,

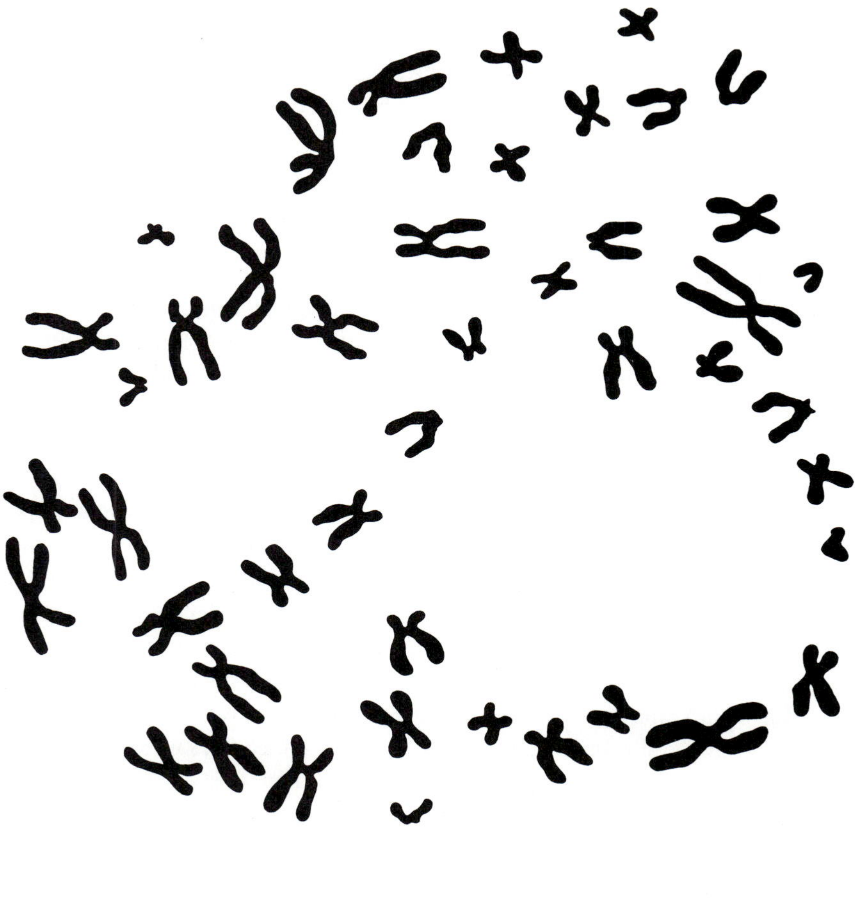

16–5
A chaotic array of chromosomes of a normal human male.

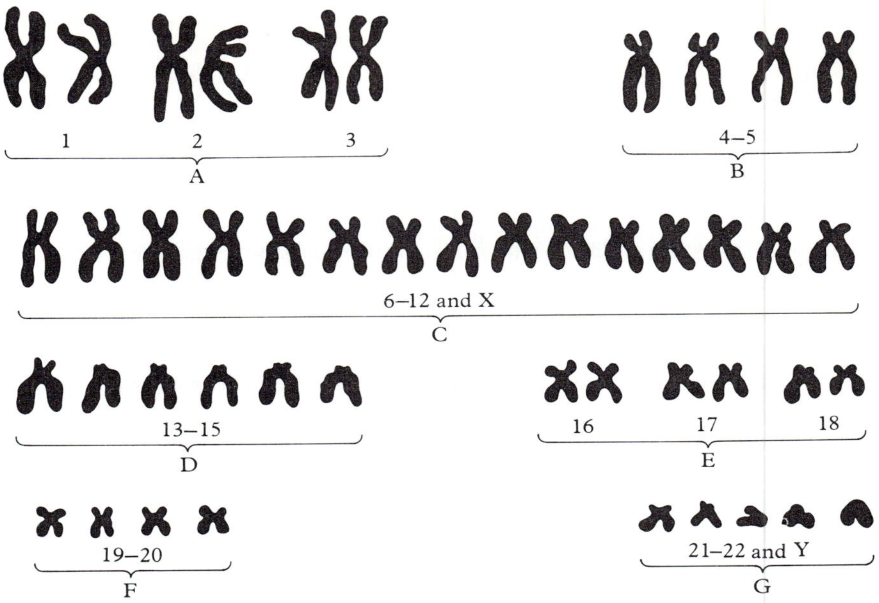

16–6
The karyotype of a normal human male. The chromosomes have been photographed, cut out, and arranged in groups according to size and shape.

is a distinct but chaotic array of human chromosomes. These are then photographed and individual chromosomes are cut out of the photo like paper dolls and lined up with one another in matching pairs to form a *karyotype,* as shown in Figure 16-6. (More recently, computers have been used for the matching process.)

The 23 pairs of human chromosomes are each placed in one of seven groups designated by the letters A through G (Figure 16-6). Of the 23 pairs, 22 are for all practical purposes perfectly matched in both sexes and are called *autosomes.* The remaining pair are called *sex chromosomes,* and though the members of this pair are apparently identical in women, they are not identical in men. With regard to genotype, women are said to be of sex chromosome constitution *XX,* whereas men are said to be *XY.* We will discuss sex chromosomes further in the following chapter. For now, we turn our attention to the inheritance of some human traits determined by genes located on autosomes. But first, it is helpful to elaborate a little on exactly how the terms dominant and recessive apply to human characteristics.

Most of what is known of human genetics has been learned by studying various kinds of diseases or abnormalities that obviously run in families. In the final analysis, all gene-dependent differences among human beings are differences between physiological processes that occur inside cells. Sometimes the way in which genetic differences between cells can result in phenotypic differences between individuals is obvious. For example, albinos are lightly pigmented because their cells are unable to properly synthesize the dark-colored pigment melanin. But oftentimes it is not at all obvious how phenotypic abnormalities relate to gene-dependent abnormalities in cellular physiology and in biochemistry. For example, it is not obvious how the gene for six-fingered dwarfism (see the following discussion) brings about its effect. Moreover, the exact biochemical or physiological defect that is associated with a genetically determined abnormality is known for only about one trait in five.

Many heritable human disorders, most of them individually rare, are the result of a single abnormal allele for which an affected person may be either homozygous or heterozygous. These rare disorders therefore have simple Mendelian patterns of inheritance, as we are about to discuss. An abnormal allele is dominant or recessive depending on whether its effects are evident in a single dose (in heterozygotes) or whether the allele must be present in a double dose (in homozygotes) to produce its effects.

On the other hand, many common disorders, such as high blood pressure and some relatively frequent congenital malformations such as cleft lip and palate, do not have simple Mendelian patterns of inheritance. This is because these abnormalities, like many others, result from the interaction of many genes and many nongenetic environmental factors. As we will discuss in Chapter 20, this is also true for many "normal" human characteristics such as height and intelligence. In the discussion of the patterns of inheritance shown by autosomal dominant and recessive traits that follows, it should be borne in mind that the existence of rare abnormal alleles in affected persons implies the presence of normal alleles in normal people. The study of genetic abnormalities can thus help us to get an idea of how extensive the normal human genetic program really is.

Autosomal Dominant Inheritance

About 450 human traits are known to have their genetic basis in dominant genes located on autosomes. As you know, Mendel discovered that dominant traits are manifested both by heterozygotes (*Aa*) and by homozygotes (*AA*). But almost all human beings who manifest documented autosomal dominant traits turn out to be heterozygotes. This is because dominant genes for the most part produce undesirable effects. That is, persons manifesting dominant traits are usually at some kind of disadvantage compared to their normal peers. Apparently, homozygotes for dominant traits are at such a disadvantage that most of them do not survive life before birth and die as embryos. Besides, autosomal dominant traits are rare to begin with, so it is unlikely that two affected persons would come together to produce homozygous offspring. Thus, we can usually assume that persons manifesting autosomal dominant traits are heterozygous.

A good example of autosomal dominant inheritance is provided by the rather benign trait known as "wooly hair," whose distribution has been well documented in Norwegian families. Affected persons have hair that is tightly kinked and very brittle, so that it breaks off before growing very long (Figure 16-7). As usual, people manifesting this dominant trait are heterozygous, and

16–7
A Norwegian family, some of whose members have woolly hair. (From Mohr, Journal of Heredity, 23, *1932.)*

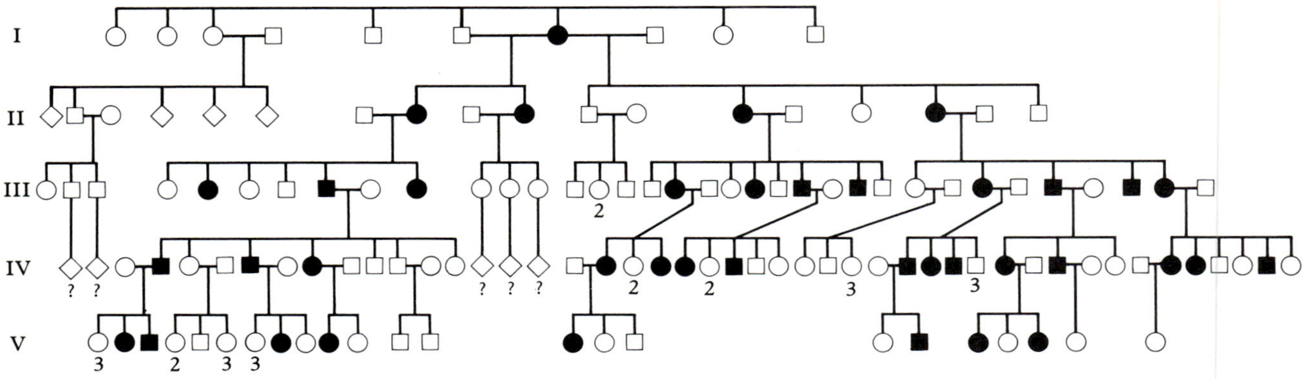

16–8
A pedigree showing the transmission of woolly hair through five generations. If more than one offspring are represented by a symbol, the number represented is given below the symbol. (After Mohr, Journal of Heredity, 23, *1932.)*

their genotype can be symbolized *Ww* (*W* for wooly). If an affected person (*Ww*) and an unaffected, "normal" person (*ww*) produce offspring, we would expect about half of them to have normal hair and half to have wooly hair. Moreover, if the gene that determines the trait is located on an autosome, then the affected offspring should include roughly equal numbers of males and females, and the trait should be transmitted from either parent to both sons and daughters. That this is true is shown in Figure 16-8, which is a human pedigree outlining the transmission of wooly hair through several generations. The symbols used in outlining pedigrees are these: women are represented by circles and men by squares. Matings occur between individuals directly connected by horizontal lines, and their offspring are indicated at the end of short vertical lines. Affected individuals are indicated by blacking in the symbols. For example, suppose a normal man marries a wooly-haired woman and that they produce four offspring, two boys and two girls, one affected and one normal child of each sex. This is symbolized as follows:

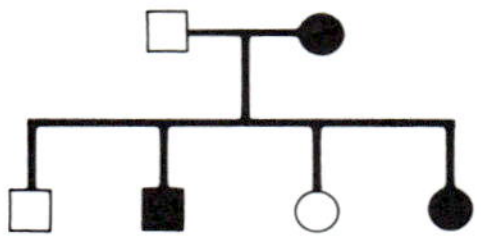

If the affected daughter then marries a normal man and produces five offspring, including three normal daughters and two affected sons, the diagram is extended to:

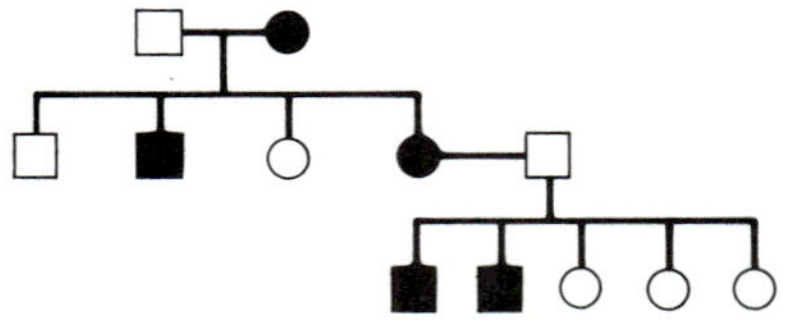

Pedigrees of families including wooly-haired individuals have also been recorded in Holland and the United States. If we pool all the data concerning the offspring from marriages between one affected and one normal person, we find, as we would expect, that within the limits of chance half of the sons and half of the daughters inherit the trait.

A fairly constant feature of most autosomal dominant traits in human families is that they can vary widely in severity from one individual to another.

The degree of severity is referred to as the *expressivity* of a particular trait. For example, the condition known as Marfan's syndrome is an autosomal dominant disorder of connective tissue that is manifested by abnormalities of the position of the lens of the eye, by excessively long bones in the hands, feet, and extremities, and by defects in the wall of the aorta as it comes off the heart. People who have Marfan's syndrome can be anywhere from severely affected with all of these abnormalities to slightly affected with only abnormally long fingers to reveal the presence of the disorder. (In fact, because of his distinctive appearance and family background, it has been suggested that Abraham Lincoln may have been mildly afflicted with Marfan's syndrome. See Figure 16-9.)

By and large, whether or not a particular autosomal dominant gene is fully expressed depends on the rest of the person's genes. In other words, the presence of certain other genes or combinations of genes can markedly influence the expressivity of a genetically determined trait, particularly of autosomal dominant ones.

Another feature of autosomal dominant traits is that they sometimes appear unexpectedly among the offspring of unaffected parents. In Marfan's syndrome, this occurs about 15 percent of the time. Those who are affected thereafter pass the trait on to about half of their offspring, as expected. In general, the sudden, unexpected appearance of an autosomal dominant trait in a lineage from which it was not previously known is the result of a *mutation.* In our example, a heritable change occurs spontaneously within the genetic material and thereby transforms a normal gene into the one responsible for Marfan's syndrome. (We will discuss mutations further in following chapters.)

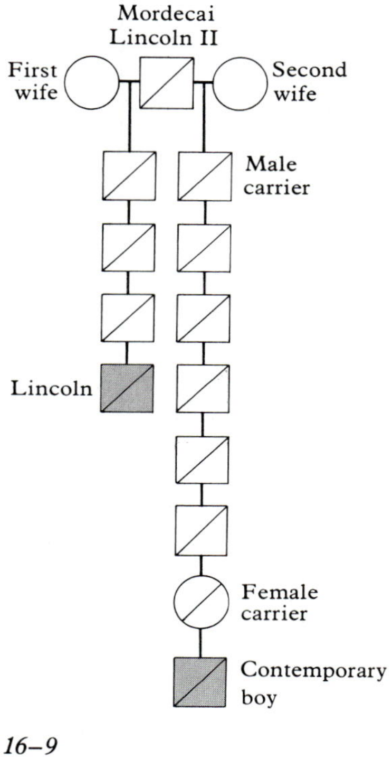

16–9
Abraham Lincoln's extremely long legs and the very unequal lengths of his thumbs may indicate that he was mildly afflicted by Marfan's syndrome. As shown in the pedigree, this disease was recently diagnosed in a boy who is a distant relative of Lincoln's. The boy and the famous president are both descendants of Mordecai Lincoln, II, but they had different mothers. Mordecai almost certainly had the gene for Marfan's syndrome, but he showed no obvious symptoms because in him the gene had a low expressivity.

Codominance, The ABO Blood Group

So far we have been discussing autosomal dominant traits that depend on a single pair of alleles, which is to say that the genes determining such traits come in only two forms, *A* and *a.* As we have seen, most people who manifest autosomal dominant traits are heterozygous for the allele in question. However, *multiple alleles* are also known to influence the inheritance of some well-known and important human characteristics, including the determination of a person's *blood group.* (There are many blood groups, most of which depend on different sets of alleles.) Perhaps the best known set of multiple alleles is that determining the ABO blood group, which is important enough to be worth discussing further.

Three alleles determine to which ABO blood group a person belongs. These can be symbolized: I^A, I^B, and I^O. These alleles are always found at a particular place, or *locus,* on a particular pair of autosomes, and any one person has two out of the three alleles. I^AI^A individuals are of blood group A, as are I^AI^O individuals. Similarly, persons who are I^BI^B or I^BI^O are group B. To complete the possibilities, I^AI^B individuals are of blood group AB, and I^OI^Os are of group O.

Both of the alleles I^A and I^B are dominant to I^O. Moreover, when I^A and I^B occur together, both have an effect on the surface of red blood cells (and both therefore stimulate the production of antibodies; see Chapter 13). Thus I^A and I^B are said to show *codominance.*

ABO blood grouping has been carried out nearly world wide for several reasons. First, ABO compatibility is essential in performing blood transfusions, as discussed in Chapter 13. And second, ABO groups vary enough from one group of people to another that their patterns of variation have been useful to physical anthropologists in the study of human races (see Chapter 19).

An unintended but useful side effect of widespread ABO typing has been its application to cases of disputed paternity. Consider the following example. During the course of divorce proceedings a woman of Type A (genotype $I^A I^A$ or $I^A I^O$) seeks child support from a man of Type O (genotype $I^O I^O$), claiming that he is the father of her recently born child. If it should turn out that the child is of Type AB (genotype $I^A I^B$), the case would be thrown out of court, because the supposed father is capable of contributing only the allele I^O to his offspring. The mother can contribute only I^A or I^O, so the allele I^B must have been contributed by someone else. (The likelihood of a mutation having occurred in one of the parents' sex cells is negligible.) However, if the presumed father were of Type B (genotype $I^B I^B$ or $I^B I^O$) then overall there would be a 50 percent chance that he could in fact be the father and the trial would continue.

We now turn our attention to the inheritance of some autosomal traits whose expression implies that an affected person, unlike most people who are afflicted by a dominant trait, is a homozygote.

16–10
Two albino parents and their albino daughter. (After Davenport, Journal of Heredity.*)*

Autosomal Recessive Inheritance

At least 500 human traits are definitely known to show an autosomal recessive pattern of inheritance. As first discovered by Mendel, heterozygous individuals (*Aa*) do not manifest autosomal recessive traits because of the masking effect of the dominant allele. Thus, people who manifest autosomal recessive traits are generally *homozygous recessive;* that is, their genotype is *aa.*

As compared with their dominant counterparts, autosomal recessive traits are likewise usually associated with diseases or abnormalities, and they are not quite so rare. This is because most people who are heterozygous for an autosomal recessive trait (of genotype *Aa,* also called carriers of the trait) are not at much of a disadvantage compared to normal (*AA*) individuals, so the trait may become widely disseminated, even if affected homozygous people choose not to reproduce. Also, those affected by autosomal recessive traits tend to be less variable than are those affected by autosomal dominant ones. That is, the expressivity of autosomal recessive traits is about the same for all those who are affected.

A good example of autosomal recessive inheritance is the condition known as *albinism.* Albinism is one of the most common and widespread of genetic disorders. Affected individuals include not only human beings of all races, but also other mammals, insects, fish, reptiles, amphibians, and birds. Albinism results from the body's inability to properly synthesize the dark-colored pigment *melanin.* Melanin is the principal pigment that imparts color to human skin, hair, and eyes, so human albinos generally have white hair and pink or only lightly colored irises (Figure 16-10). Because of their lack of pigment, the skins and eyes of albinos are abnormally sensitive to the effects of sunlight, and because of their unusual appearance human albinos may receive special treatment from other members of their species. For example, the Aztec emperor Montezuma is said to have included many albinos among the members of his "museum" of living human "curiosities," and albinos among the present-day San Blaz Indians of Panama (known as "moon children" because they avoid bright sunlight) are not permitted to marry.

Until quite recently it was thought that albinism was the result of a single, specific defect in the synthesis of melanin from the amino acid tyrosine. (In later chapters we will discuss the biochemistry of melanin synthesis as it relates to inborn errors of metabolism.) But then a well-documented pedigree of albinism in England showed that two albino parents produced four children, none of whom were albinos. As shown in Figure 16-11, the mating of parents manifesting the same autosomal recessive trait can produce only affected offspring, because each of the parents contributes an *a*. How then can the pattern observed in this unusual English family be explained?

It turns out that there are at least two, and perhaps as many as six, genes that result in albinism, depending on where the biochemical block in the synthesis of melanin occurs. And all of these defects in melanin synthesis are inherited as autosomal recessive traits. Therefore, it is possible for two albinos who are homozygous recessive for different defects in melanin synthesis (and who are therefore albinos for different reasons) to produce normal offspring, as shown in Figure 16-11.

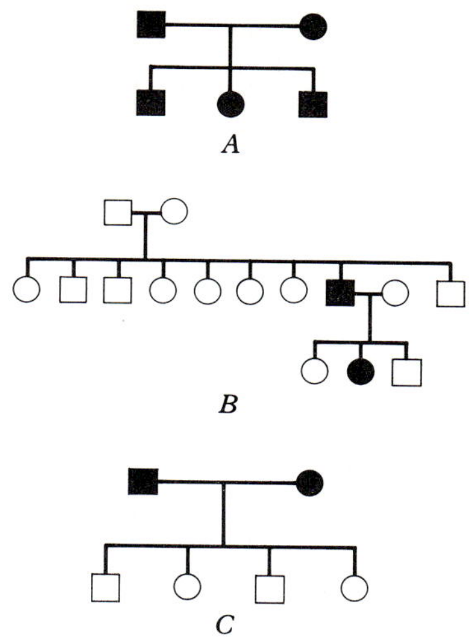

16–11
A, *The mating of albino parents almost always results in all albino offspring.* B, *A pedigree showing albinism among the offspring of unaffected carrier parents and among the offspring of an albino father and an unaffected carrier mother.* C, *A pedigree of albinism in which albino parents produced four nonalbino offspring.*

In the United States, about one white person in 38,000 and one black person in 22,000 are albinos. But circumstances are known under which the percentage of albino offspring produced is higher than in the population at large. The best known example is that of marriages between relatives, or *consanguineous* marriages. Although only about 0.1 percent of marriages in the United States are between first cousins, about 8 percent of albino children result from first-cousin marriages. (First cousins are the offspring of brothers and sisters who married unrelated spouses. See Figure 16-12.) How does the incidence of albinism, or any other autosomal recessive trait, relate to the degree to which an affected person's parents are related?

Most autosomal recessive traits are rare. Nonetheless, the brothers and sisters of someone who carries an allele for a rare autosomal recessive trait are very likely to be carriers too. (If a person carries an allele for albinism, the person's normal brothers and sisters have a 50 percent chance of also carrying the allele, because at least one of their parents must be a heterozygote. Work this out for yourself.) Similarly, people descended from a common ancestor known to have manifested or carried a particular trait are also more likely to be carriers. Thus, people manifesting rare autosomal recessive alleles tend to cluster in certain family lines because the mating of rather closely related individuals is likely to bring two rare autosomal recessive alleles together to produce an affected individual.

One example is the autosomal recessive condition known as six-fingered dwarfism, which is unusually frequent among the Old Order Amish of Lancaster County, Pennsylvania (Figure 16-13). Amish people usually choose to marry other Amish people, and because their numbers are relatively small to begin with, this means that marriages between individuals who have common ancestors occur frequently. Accordingly, whereas six-fingered dwarfs are very rare elsewhere, they exist in at least 33 Amish families. Apparently, one of the original founders of the sect in eastern Pennsylvania was a heterozygous carrier of the gene for six-fingered dwarfism: the gene has since become widespread in the population and is frequently found in the homozygous state because of consanguineous marriages.

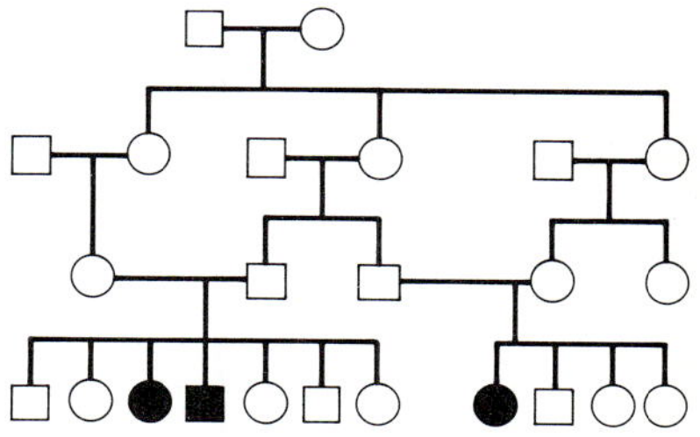

16–12
A pedigree showing cousin marriages that resulted in albino offspring. The consanguineous matings occurred in the third generation.

In general, the more closely related two people are, the greater the chance that their offspring will manifest some (usually detrimental) autosomal recessive trait. This is best illustrated by data about the offspring of incestuous unions of fathers and daughters and brothers and sisters. Information on the offspring of

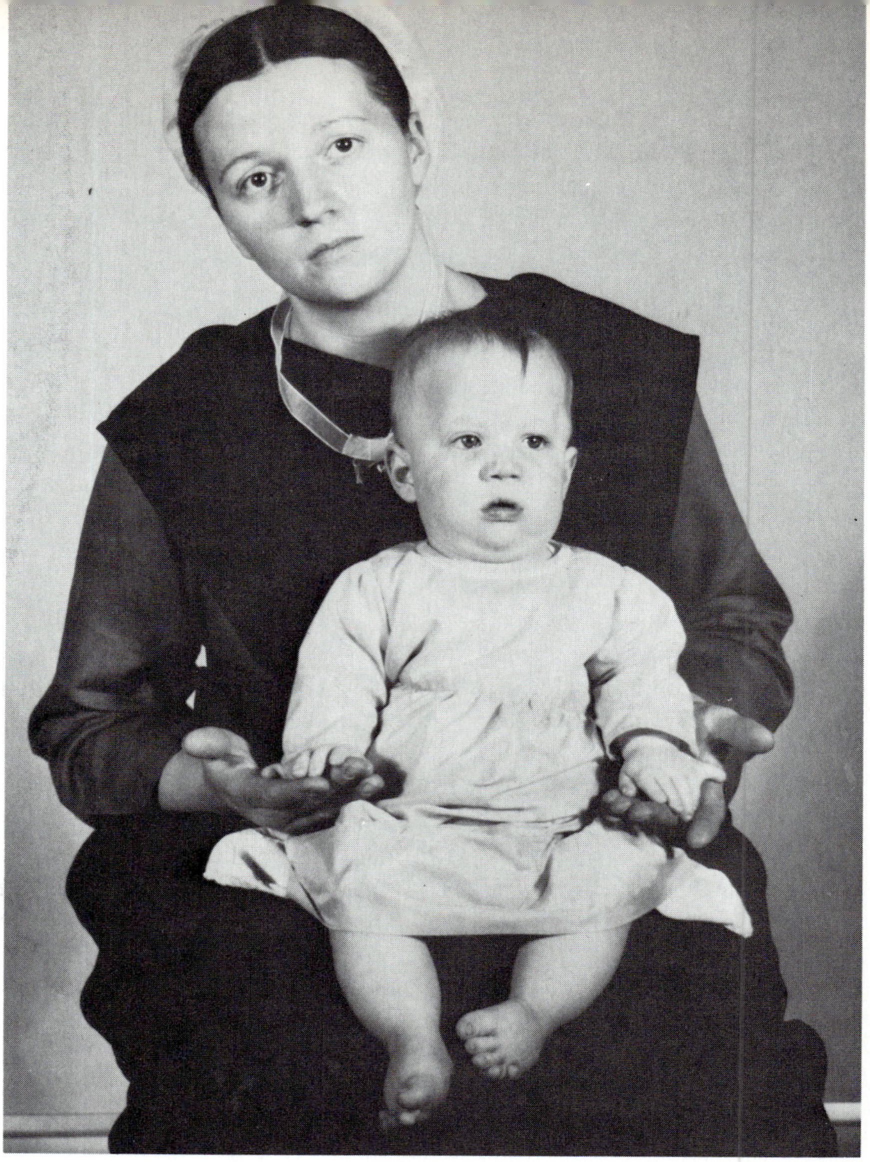

16–13
An Amish mother and her child, who has six-fingered dwarfism. (Photo courtesy of Dr. Victor A. McKusick.)

31 such unions is available from England and the United States. Six of the offspring died early in life and twelve were severely affected physically or severely retarded mentally. Only 42 percent of the children were apparently normal. (Almost all human societies have strict cultural prohibitions against incestuous unions. Is this because the deleterious genetic consequences of such unions are so obvious? Almost certainly not. Rather, it appears that incest taboos are outgrowths of social and cultural factors, and not the result of a primitive form of applied genetics.)

Autosomal recessive traits are of particular interest in genetic counseling. Most often the relatives of an affected person seek advice about whether they are carriers, or about the chances of their having an affected child. It is sometimes possible to infer a person's genotype directly from pedigree studies, and the person can thereby be told his or her status as a carrier. But even when a person's genotype cannot be deduced by pedigree analysis, it is often possible to determine it by performing various physiologic and biochemical tests. For some heterozygotes for autosomal recessive traits do indeed manifest the fact that they harbor both dominant and recessive alleles.

We have already discussed the fact that homozygous individuals who manifest autosomal recessive traits usually do so about equally. The effects of a dominant allele generally mask the effects of a recessive one in heterozygotes, but in some instances it is possible to infer the presence of the recessive allele by subjecting the suspected carrier to some kind of stress. For example, heterozygous carriers for sickle-cell anemia (whose biochemistry and genetics we will discuss further in Chapter 18) can be detected if their blood is subjected to lower-than-normal concentrations of oxygen. Under this stress, some of the carrier's red blood cells take on the characteristic sickle shape, and thus reveal heterozygosity.

In general, it is possible to devise some sort of measurement that will reveal heterozygous carriers of autosomal recessive traits, as long as the physiological or biochemical defect in question is known. But the exact defect is known for only about one genetically determined abnormality in five. Besides, the inheritance of many traits of normal human beings does not fit into any of Mendel's patterns, in spite of the fact that the traits obviously have some sort of genetic basis. This is especially true of traits that are not clearly alternate, but rather are continuously distributed throughout the population. Such traits do not manifest themselves as sharply defined pairs of phenotypes like "round" or "wrinkled," but rather as continuous gradations. For example, normal body height varies widely but in general is continuously distributed in a given population (Figure 16-14). Even though there is a definite tendency toward tallness or shortness in families, there may be widespread differences in the heights of parents and offspring. Also, as with most continuously varying traits, height is known to be affected, not only by a person's genes, but by the environmental conditions in which the person grows up and lives.

It was Mendel himself who first proposed an explanation for the inheritance of continuously graded traits. Based on some experiments he performed on bean plants with colored and white flowers, he suggested that *more than one pair of genes* transmitted such traits. For the most part, Mendel's explanation has stood the test of time, though we now are much more aware of the effects of the environment on the expression of continuously varying traits than Mendel was. We will return to the inheritance of continuously varying traits and to the effects of the environment on the expression of a person's genotype in later chapters. But for now, let us return to our discussion of autosomal dominant and recessive traits and ask a deceptively simple question. Which genes are located on which chromosomes?

1	0	0	1	5	7	7	22	25	26	27	17	11	17	4	4	1
4:10	4:11	5:0	5:1	5:2	5:3	5:4	5:5	5:6	5:7	5:8	5:9	5:10	5:11	6:0	6:1	6:2

16–14
A company of 175 soldiers arranged in groups according to height. The lower row of numbers indicates height in feet and inches, the upper row the number of men in each group. (From Blakeslee Journal of Heredity, 5, *1914.)*

Mapping Human Chromosomes

About 1200 human genes are definitely known to exist, and new ones responsible for normal or abnormal traits are discovered each year. It has been estimated that the total number of genes per average human being will probably eventually be measured in thousands or perhaps tens or even hundreds of thousands, but nobody knows for sure. At any rate, there are clearly thousands more genes than chromosomes, which means that in general each chromosome contains a large number of genes. Parents contribute entire chromosomes to their offspring, so we would expect all of the genes on a given chromosome to be inherited *en masse,* and such collections of genes make up a chromosomal *linkage group.* Thus, for human beings we would expect 23 groups of linked genes corresponding to the 23 pairs of chromosomes. This will probably turn out to be true, but our present knowledge of human linkage groups is meager. Only one human chromosomal linkage group, that associated with sex chromosomes, has been described in any detail, as will be discussed in the following chapter. One of the major reasons linkage groups are so difficult to figure out is that most of the time the linkage of genes on any chromosome is not complete. What this means is well illustrated by the test cross that first demonstrated the existence of linkage in the early 1900s.

You will recall that the seven traits of the common garden pea studied by Mendel showed independent assortment because the genes responsible for the traits are located on different pairs of chromosomes. But some experiments on sweet peas by later investigators suggested that not all traits are inherited independently. True-breeding sweet peas that had red flowers and spherical pollen grains (both dominant) were crossed with plants that bore purple flowers and had cylindrical pollen grains (both recessive). If these traits were determined by factors that showed independent assortment, the offspring of crosses of members of the F_1 generation with individuals who were homozygous recessive for both traits were expected to be of four different phenotypes in the ratio 1:1:1:1. But to the investigators' surprise the actual ratios turned out to be 7:7:1:1. Furthermore, the two smaller categories of offspring manifested combinations of traits that were not found in the parents. The explanation offered was that combinations of traits that occurred together almost all of the time did so because the factors responsible for them were located on the same chromosome. But what about the unexpected *recombinations* of traits observed in the offspring? Further experiments revealed a remarkable fact: the members of each pair of chromosomes sometimes physically exchange sections with one another, thus unlinking traits that are found together on the same chromosome and allowing for the generation of new combinations. This exchange occurs when sex cells are produced during meiosis, and the exchange of sections between chromosomes of a given pair is known as *crossing over.*

Meiosis is a process in which cells divide in such a way as to reduce the number of chromosomes in the nucleus by half. Meiosis is thus often referred to as *reduction division,* and in general it applies only to the production of sex cells. As you may recall from our discussion in Chapter 5, when a somatic cell divides, the two cells that are produced have the same number of chromosomes as the original cell before it divided. This is because somatic cells duplicate their entire set of chromosomes before they divide and then distribute a complete set to each of the two cells produced by division. (This type of cell division is called *mitosis.*) But in meiosis, one cell divides twice and produces four cells, each of which has only half of a complete set of chromosomes, one member of each chromosome pair. How this occurs is shown in Figure 16-15.

The most important features of meiosis as it relates to crossing over and linkage can be summed up in the following way. Like a somatic cell, a cell undergoing meiosis duplicates its entire set of chromosomes before it divides the first time. But before the duplicated sets of chromosomes separate in the first division of meiosis they do something generally unheard of among the chromosomes of somatic cells. Before the first division most chromosomes

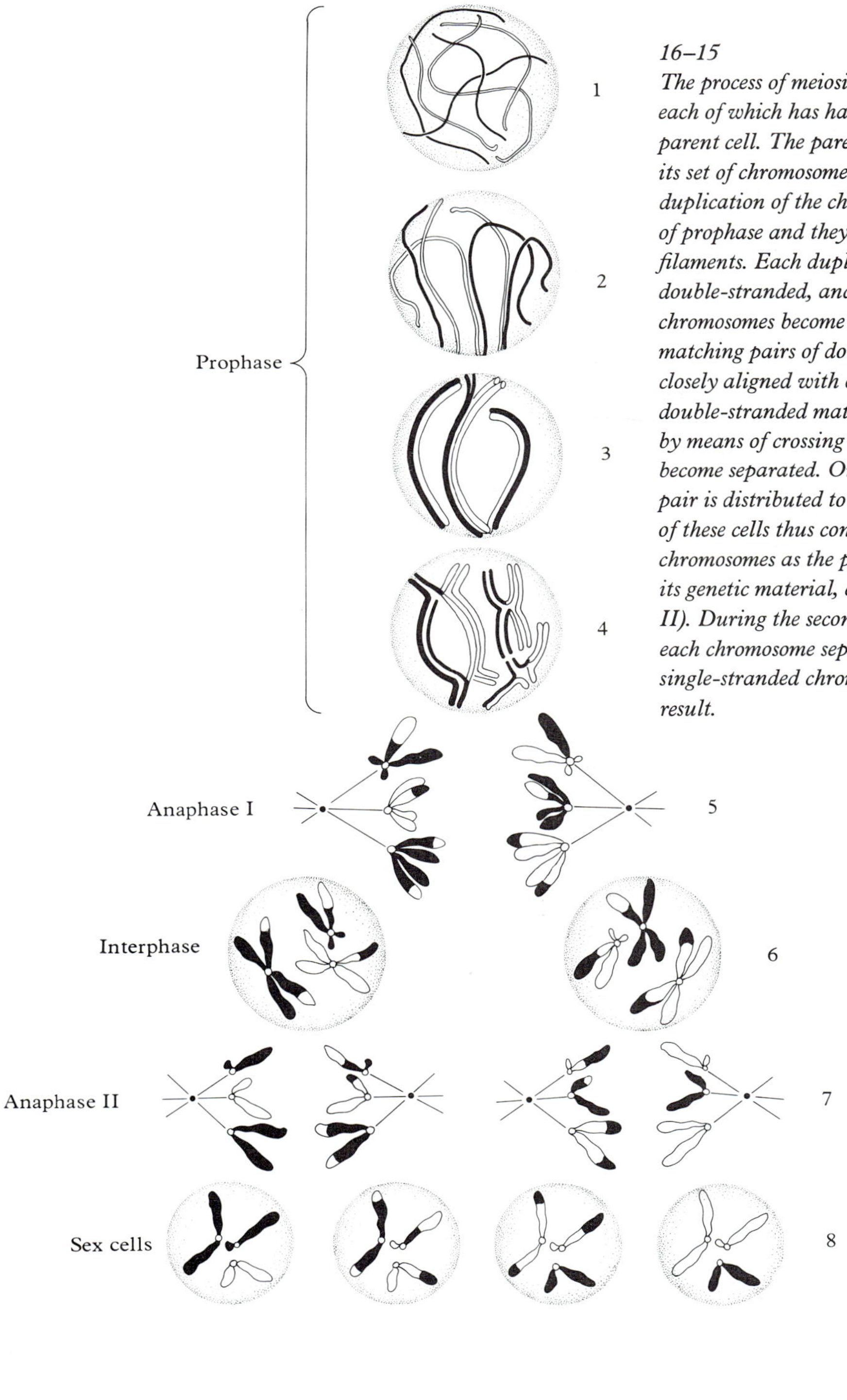

16–15
The process of meiosis results in the formation of four cells, each of which has half the number of chromosomes as the parent cell. The parent cell accomplishes this by duplicating its set of chromosomes once and then dividing twice. The duplication of the chromosomes occurs just before the onset of prophase and they first become visible as thread-like filaments. Each duplicated chromosome is actually double-stranded, and this becomes more apparent as the chromosomes become shorter and thicker. During prophase matching pairs of double-stranded chromosomes become closely aligned with one another and segments of the double-stranded matching pairs may be physically exchanged by means of crossing over. During anaphase I the pairs become separated. One double-stranded member of each pair is distributed to each of the two cells that result. Each of these cells thus contains half as many double-stranded chromosomes as the parent cell. Then, without duplicating its genetic material, each of these cells divides again (anaphase II). During the second division the two strands that make up each chromosome separate and each strand itself becomes a single-stranded chromosome in one of the four cells that result.

undergoing meiosis become closely aligned with one another in matching pairs along their entire lengths, as shown in Figure 16-15. And it is during this time of close alignment that the physical exchange of sections of matching pairs of chromosomes becomes apparent. As they separate during the first division, paired meiotic chromosomes are connected to one another at X-shaped areas called *chiasmata* (singular, *chiasm*), which presumably are points at which crossing over has occurred. Then, after the duplicate pairs have fully separated and formed two cells, each newly formed cell divides again to produce four cells, each of which has half of a complete set of chromosomes.

By about 1910 or so it was known that genes are strung out on a chromosome like beads on a string. But the exact nature of the relationship between genes and chromosomes and of the physical and biochemical events of crossing over between duplicated sets of chromosomes remain largely unknown. Whatever its physical basis, crossing over is of enormous importance in helping to produce the genetic variability that is the raw material of evolution. This is because crossing over generates an enormous number of combinations of genes both in eggs and in sperm cells, and thus helps to provide the raw material on which natural selection can operate (see Chapter 1). (The production of human egg and sperm cells is discussed more fully in Chapter 14.)

In general, how much crossing over occurs between genes on the same pair of chromosomes depends on the location of the genes with respect to one another. Genes farther away from one another cross over more often than those closer together. This observation provides a basis for constructing *chromosome maps.* Thus, if crossing over between two genes occurs about 1 percent of the time, then the two are separated by "1 map unit" of distance. The further away the two genes are, the greater the number of crossovers. In fact, if two genes are separated from one another by more than 50 map units, crossovers may be so frequent as to suggest that the genes are not linked, but located on different chromosomes (Figure 16-16).

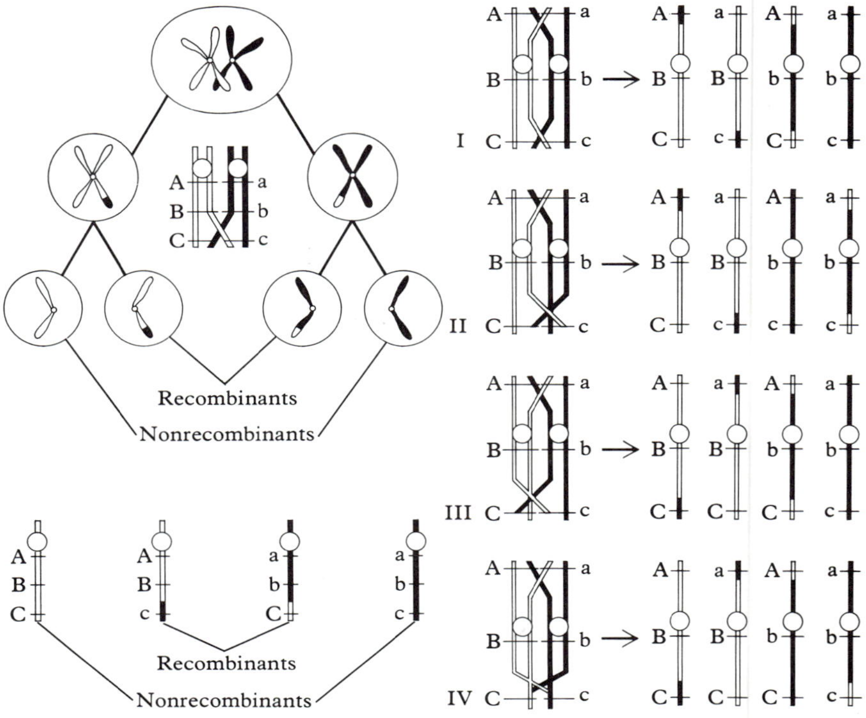

16–16
Segments of matching pairs of chromosomes are physically exchanged by crossing over. Each member of the pair of chromosomes shown here carries three alleles. In crossing over, corresponding segments of the arms of the chromosomes can be physically exchanged (left.). This results in the four chromosomes shown at the bottom left, two of which have a different combination of alleles than was present in either of the original chromosomes. Right, the four kinds of recombinations for three dominant and three recessive alleles.

Extensive chromosome maps have been worked out for a few experimental animals, including mice and fruit flies. (Nearly complete maps are also known of the single, circular chromosome of some prokaryotes, particularly bacteria. See Chapter 18.) But our ignorance of the human map is truly profound. It is estimated that less than 5 percent of the human chromosome map is known. This amounts to a total of some 200 or so human genes whose chromosomal locations have been assigned. And of the genes mapped so far more than half are located on a single chromosome, the sex chromosome known as the X-chromosome. The entire X-chromosome probably has about as many genes as an autosome of similar length. But as we shall discuss in the following chapter, it is relatively easy to tell that a trait is on the X-chromosome because of the characteristic pattern of inheritance of X-linked traits. This is clearly not true for autosomes. Although it is sometimes easy to determine whether a given trait is inherited as an autosomal dominant or recessive, it is usually impossible to say on which pair of autosomes the trait is located. Even more rarely can we determine exactly where on a particular autosome the gene for a given trait is.

The mapping of human autosomes is indeed a formidable undertaking, even with the assistance of some recently developed experimental techniques that have made the task somewhat more approachable. Until recently the construction of human chromosome maps depended first on the demonstration that two traits were linked and therefore on the same chromosome. This is not easily demonstrated in human pedigrees because not often are the data concerning two traits clear-cut enough and common enough to allow determination of whether the combinations observed in a given lineage are best explained by assuming that the traits are linked and that they undergo recombination by crossing over a certain percentage of the time. But even if linkage can be established this tells nothing about the particular autosome on which the genes that determine the traits are located. That can be determined if the frequencies of linked genes correspond with the presence of a particular autosomal abnormality but they very rarely do.

More recently, the technique of *somatic cell hybridization* has been applied to the mapping of human chromosomes. This technique is the experimental fusion of body cells (not sex cells) from other mammals, especially mice, with human cells (Figure 16-17). As the hybrid cells divide, the human chromosomes are selectively and progressively lost, and *clones* of cells, which contain a full complement of mouse chromosomes and a few or a single human chromosome, are produced. Under such circumstances it is sometimes possible to detect the presence of enzymes known to be produced by humans but not by mice. One can then conclude that the gene responsible for the enzyme is located on the particular human chromosome present in the hybrid cell. And, as shown in Figure 16-18, recently developed staining techniques make it possible to distinguish individual autosomes even if they are isolated from one another. By using these techniques is has been possible to assign at least 50 human genes to 18 different autosomes.

Why go to all the trouble of mapping human chromosomes? First of all, detailed maps would be of enormous use in genetic counseling and in prenatal diagnosis. In fact, what meager knowlege we have of the map has already been applied in identifying the presence of hard-to-detect detrimental traits in unborn fetuses. For example, the gene for myotonic dystrophy (characterized by wasting of muscles and by other abnormalities) is closely linked to an easily detected "marker trait" that can be detected in fetal cells obtained by amniocentesis (see Chapter 20). A second reason for mapping chromosomes is

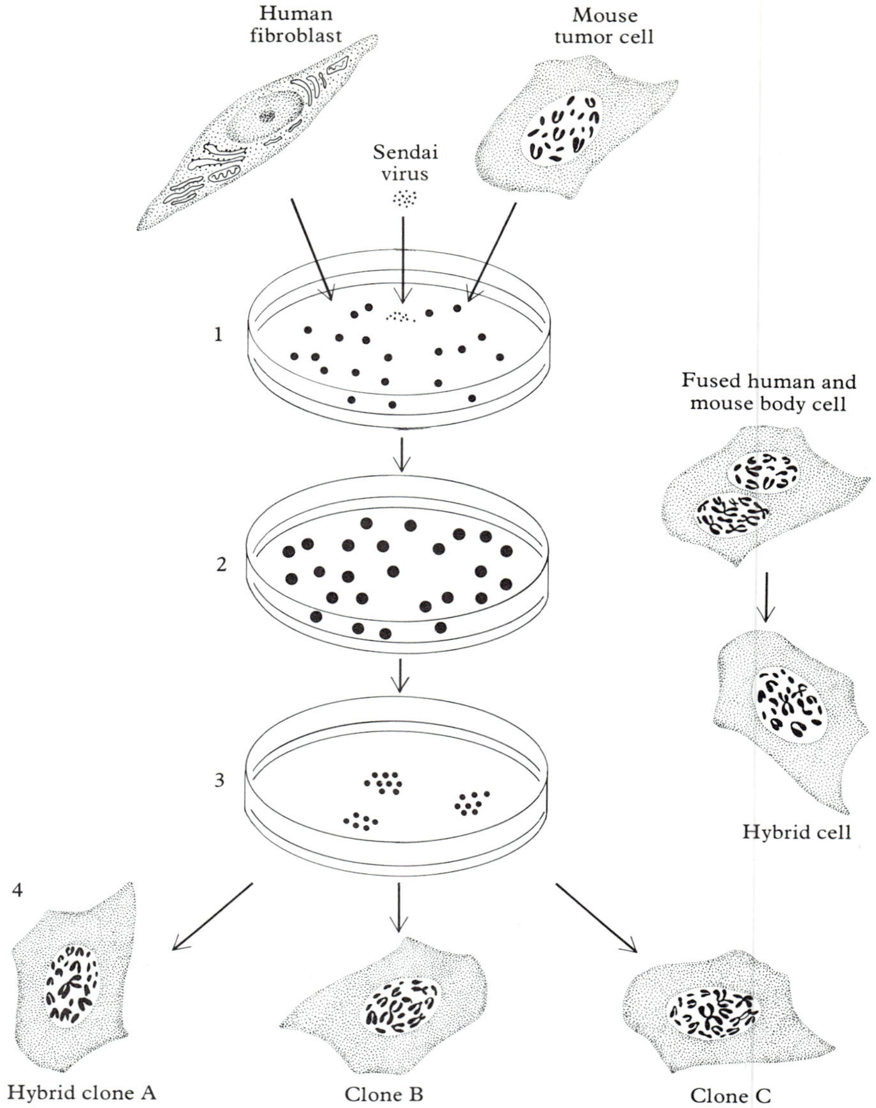

16–17
The techniques of somatic cell hydridization. Human fibroblasts are mixed with mouse tumor cells and incubated with various chemicals and a virus that enhances the fusion of the cells. Hybrid cells result and they can be separated into clones, each of which has special properties, depending on which chromosomes it contains. (From "Hybrid cells and Human Genes," by Frank H. Ruddle and Raju S. Kucherlapati. Copyright © 1974 by Scientific American, Inc. All rights reserved.)

that the attempt to do so has unexpectedly provided information both on the regulation of genes and on their expression in cells. And third, as a leading investigator has put it, the mapping of human chromosomes is rather like climbing Mount Everest. It is exciting because it is there to be accomplished.

Before we leave the subject of autosomes and turn our attention to sex determination and sex chromosomes, we should first consider some human disorders that result from abnormalities in the total number of autosomes within the nuclei of an individual's somatic cells. The most important of these disorders is the presence of an extra copy of chromosome-21 in addition to the usual pair. This condition is known as *trisomy-21,* or *Down's syndrome.*

Down's Syndrome and Other Abnormalities in the Number of Autosomes

People who have Down's syndrome have surely existed since ancient times, but the condition was not described in detail until 1866. Those who are affected with

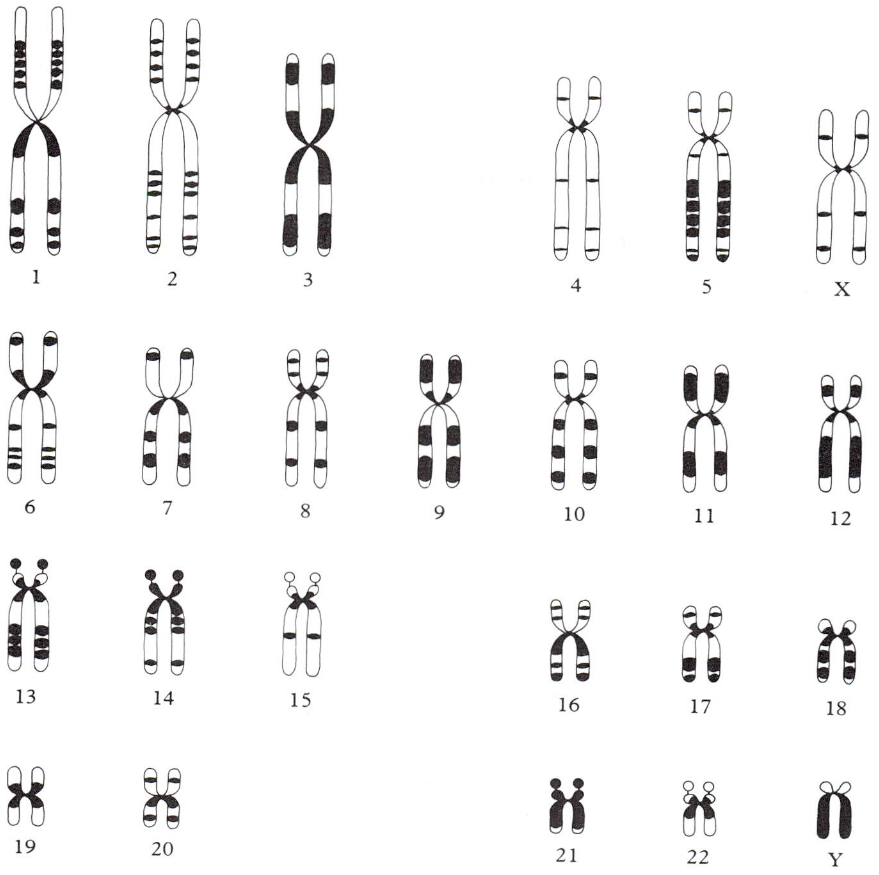

16–18
A simplified diagram of the banding patterns observed among specially stained human chromosomes. (After Drets and Shaw, Proceedings of the National Academy of Sciences, 68, *1971.)*

Down's syndrome are short in stature, frequently have serious malformations of the heart, and usually have characteristically shaped heads with distinctive eyelids and faces. More important, people who have Down's syndrome are almost without exception severely mentally retarded. (To the Europeans who first described the condition, the characteristic appearance of the eyelids suggested the facial features of Mongoloid peoples, and the condition was referred to as "mongolism" or "mongoloid idiocy." In fact, the eyelids of people who have Down's syndrome are quite different from those of people belonging to Mongoloid races, and the earlier, inaccurate terminology has therefore been dropped.)

Down's syndrome occurs sporadically. That is, affected individuals are usually the offspring of normal parents. After it was known that the incidence of the disease varies directly with the age of the mother at the time of birth (as discussed later, older mothers produce more affected offspring), it was assumed that the syndrome was produced by an unfavorable interaction between mother and fetus during the course of pregnancy. Then, in 1959 (shortly after techniques for seeing human chromosomes had been perfected) some French investigators discovered that the somatic cells of those who had Down's syndrome contained 47 chromosomes, one more than usual. As shown in Figure 16-19, the extra chromosome is a rather small autosome and by convention it is designated chromosome-21.

How do the somatic cells of people who have Down's syndrome end up with an extra chromosome-21? In brief, trisomy-21 is usually the result of an accident that occurs either during meiosis or during the first few cell divisions that take

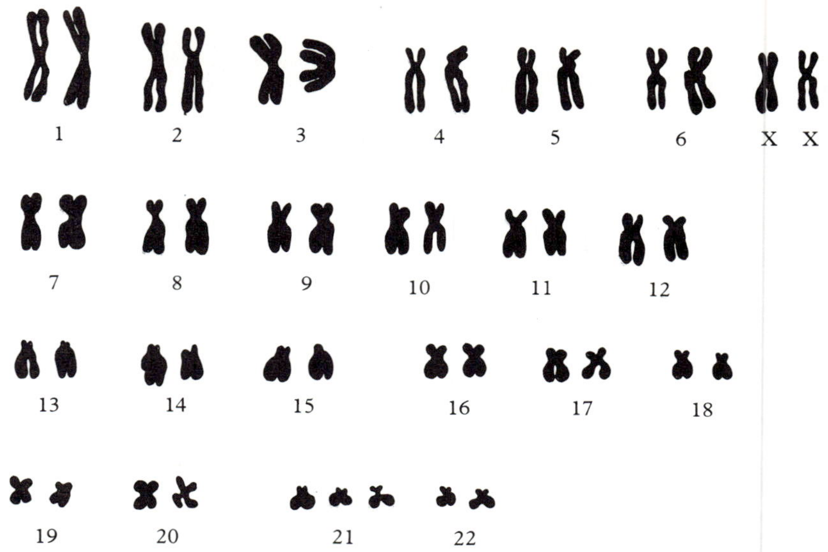

16–19
Down's syndrome most often results because of the presence of an extra copy of chromosome-21, as in the karyotype shown here.

place after fertilization. What happens is this: either duplicated pairs of chromosomes fail to separate during the first division of meiosis or duplicated chromosomes fail to sort out equally in the second division. Either way some sex cells receive two chromosomes-21 and some receive none. This failure of chromosomes to sort out properly during cell division is called *nondisjunction* (see Figure 16-20).

What happens if nondisjunction occurs during the production of a human egg that is then fertilized by a normal sperm? With regard to Down's syndrome, there are two possibilities. If the abnormal egg contains two chromosomes-21 to

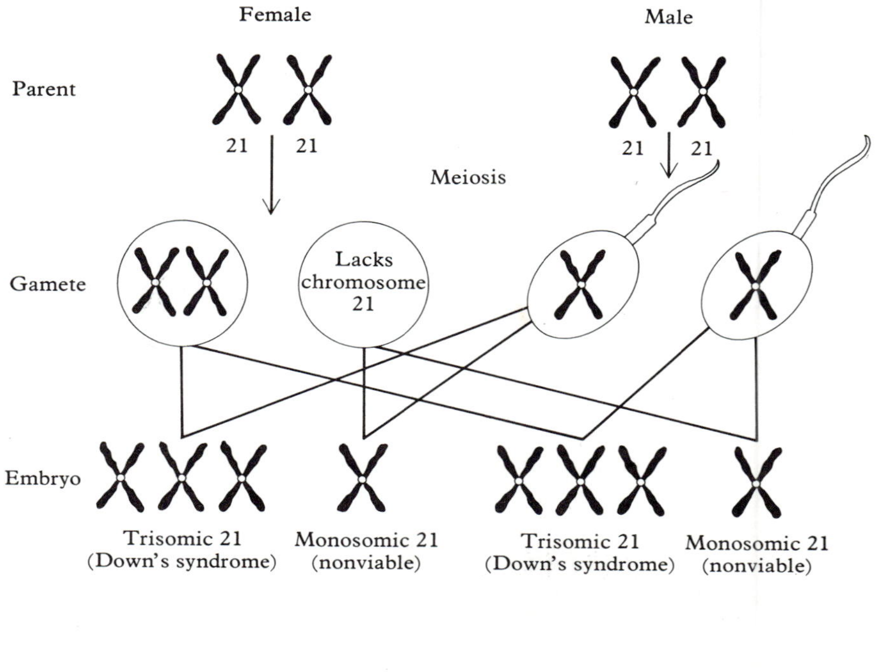

16–20
Nondisjunction, a failure of chromosomes to separate properly during meiosis, can result in Down's syndrome. When an egg with an extra chromosome is fertilized, the resulting person has three chromosomes-21. From "Prenatal Diagnosis of Genetic Disease," by Theodore Friedmann. Copyright © 1971 by Scientific American, Inc. All rights reserved.)

begin with, then on fertilization it becomes trisomic, and if the pregnancy goes to completion a child with Down's syndrome will result. (As discussed in Chapter 20, it is now possible to detect the presence of Down's syndrome while a fetus is only a few months old.) If the egg had no chromosome-21 to begin with it has only one copy after fertilization; this apparently leads to early spontaneous abortion of the developing fetus because no living person who has a single copy of chromosome-21 has been reported. The result is the same if a normal egg is fertilized by an abnormal sperm produced by nondisjunction in the male. Also, nondisjunction can occur in the first few cell divisions following fertilization of a normal egg by a normal sperm. In this case, the cells with a single chromosome-21 die off, and development of the embryo proceeds by further division of those cells trisomic for chromosome-21.

As we shall soon see, variations in the number of human sex chromosomes generally produce abnormalities that are less severe than those produced by the presence of abnormal numbers of autosomes. Trisomy for chromosomes 13 and 18 is known to occur, but those in whom it occurs usually die as infants. Trisomy for other autosomes is apparently lethal before birth because it is not observed among living human beings. Also, deletions of entire autosomes are very rarely seen. In fact, the deletion of even part of an autosome often produces serious, even fatal, abnormalities. A large proportion of the 10 to 15 percent of all pregnancies that terminate in spontaneous abortions before 20 weeks of gestation do so because of chromosomal abnormalities that are incompatible with normal development.

The fact that trisomy-21 is compatible with life no doubt relates at least in part to the small size of the chromosome present in triplicate. Down's syndrome occurs with surprising frequency. It is observed in one out of 500 or 600 births, and has been detected in up to one in 40 fetuses aborted before 20 weeks of gestation. And the chances for the occurrence of trisomy-21 increase with the age of the mother (Figure 16-21). Why is this so?

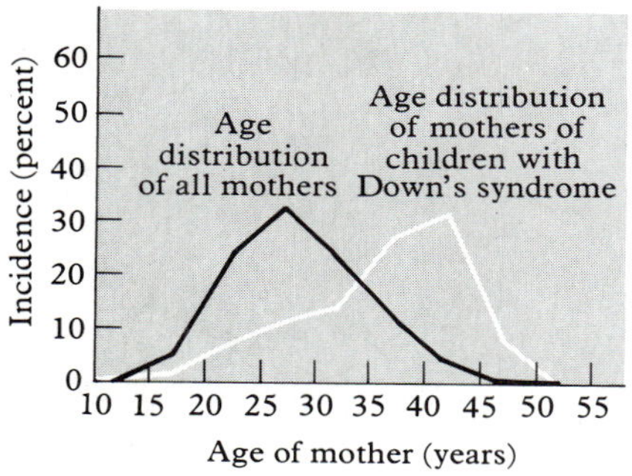

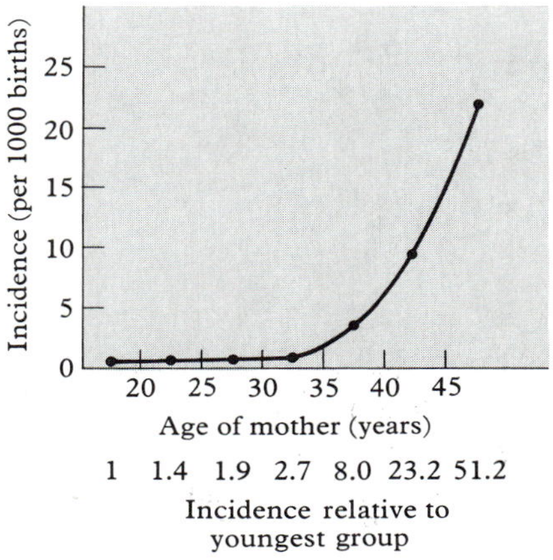

16–21
Left, the age distribution of mothers of children who have Down's syndrome compared to that of all mothers. Right, data about 1,119 cases of Down's syndrome recorded in Victoria, Australia. (Left, after Dr. Victor McKusick; right, after Collman and Stoller, Amer. J. Public Health, 52, *1962.)*

First, it is generally believed that the nondisjunction that leads to trisomy-21 occurs most often in the egg and not in the sperm cell. Perhaps this is because egg cells can sit in the ovary for decades before being ovulated and thus undergo some kind of metabolic or physical damage that later leads to nondisjunction in meiosis (see Chapter 14). At any rate, the incidence of Down's syndrome increases with maternal age even if the age of the fathers is constant. And, Down's syndrome occurs with a rather characteristic frequency among the offspring of women of the same age, no matter how old their husbands are.

Although maternal age is generally associated with the incidence of Down's syndrome, certain circumstances greatly increase the chance that a particular pair of parents will produce an affected child at any age. These cases of Down's syndrome are not due to trisomy for chromosome-21, but rather are the result of an abnormality of meiosis known as *translocation,* which results in the fusion of two normal chromosomes, or at least parts of chromosomes. Most often the long arms of chromosomes 21 and 14 become fused to form a larger, composite chromosome known as *translocation chromosome.* A person who has one chromosome-21, one chromosome-14, and one translocation chromosome is phenotypically normal even though possessed of only 45 chromosomes (Figure 16-22). This is because all of the genetic material is represented and none is in excess. Such a person is called a carrier of the translocation chromosome. (The loss of the short arms of chromosomes 21 and 14 apparently does not have much effect on a person's phenotype.)

Those who are carriers of a translocation chromosome can produce sex cells of two types—those containing the translocation chromosome and those without

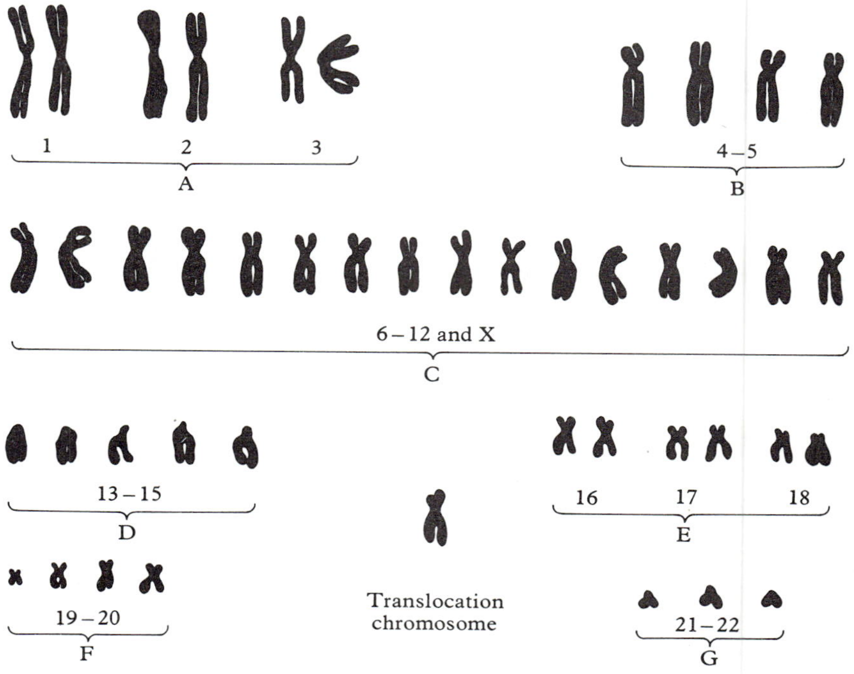

16–22
The 45 chromosomes of a woman who is a translocation carrier of Down's syndrome.

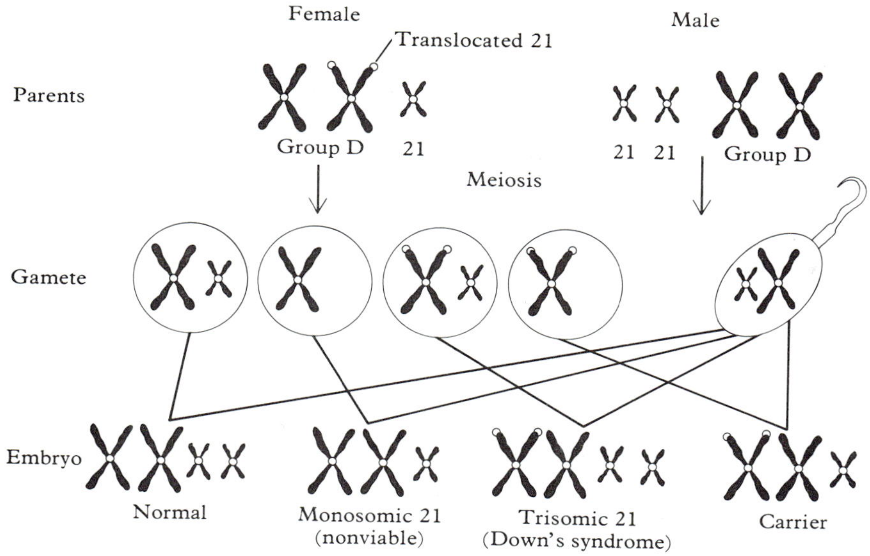

16–23
The kinds of sex cells and offspring that can be produced by a translocation carrier of Down's syndrome. (From "Prenatal Diagnosis of Genetic Disease," by Theodore Friedmann. Copyright © 1971 by Scientific American, Inc. All rights reserved.)

it (Figure 16-23). Either sex cells without the translocation chromosome are normal or they die before or shortly after fertilization because they contain chromosome-21 in but a single copy. On the other hand, after they are fertilized, sex cells containing the translocation chromosome result in the chromosome combinations shown in Figure 16-23. Overall we would expect Down's syndrome to occur regularly in the offspring of people who are translocation carriers, and this is what happens. For this reason, if a child with Down's syndrome is born to relatively young parents, they are advised to have their chromosomes studied to determine whether either parent is a translocation carrier for Down's syndrome and therefore likely to produce another affected child.

With this discussion of autosomes and their abnormalities as background, we are now prepared to take up the subjects of sex determination and sex chromosomes. We do so in the following chapter.

Summary

Species retain their identity in nature because they are isolated from one another behaviorally and genetically. Variation between and within species is due to differences in genetic programs, and variation within a species is not randomly distributed. Some traits show definite patterns in their inheritance, as Mendel discovered.

Mendel proposed that alternate traits were passed from parents to offspring as independent units that are now called genes. He found that the factors responsible for some traits are dominant whereas others are recessive and that he could predict what would happen in crosses involving several traits.

The discovery of the behavior of chromosomes during cell division provided a physical basis for Mendel's patterns. Human beings usually have 46 chromosomes in 23 pairs. Of these, 22 pairs are autosomes and the remaining pair are sex chromosomes.

In human pedigrees, autosomal dominant and recessive traits are usually associated with diseases or abnormalities. Dominant traits vary more in expressivity than recessive ones. Recessive traits can be manifested in a lineage by means of consanguineous marriages. Many human traits are determined by more than one pair of genes, and the expression of most traits depends in part on an individual's environment.

Human chromosomes can be mapped by determining the frequency with which pairs of genes undergo recombination during meiosis or by experiments on somatic cell hybridization. Meiosis results in sex cells that contain one half of a complete set of chromosomes, and recombination helps to provide the raw material for evolution. The mapping of human chromosomes has barely begun, but our knowledge is increasing rapidly.

Abnormalities in the number of autosomes usually produce severe phenotypic abnormalities. The most common of these is Down's syndrome, which occurs more frequently among the offspring of older mothers and which can result from either nondisjunction or translocation.

Suggested Readings

The first five references are for those who would like to learn more about genetics in general. Each of these books is highly recommended. Some are more difficult than others, but all should be understandable to those who can make it through the present volume.

1. *Principles of Human Genetics,* 3d Ed., by Curt Stern. W. H. Freeman, 1973.
2. *Human Genetics,* 2d Ed., by Victor A. McKusick. Prentice-Hall, 1969.
3. *Genetics, Evolution, and Man,* by W. F. Bodmer and L. L. Cavalli-Sforza. W. H. Freeman, 1976.
4. *Heredity Evolution and Society,* by I. Michael Lerner and William J. Libby. W. H. Freeman, 1976.
5. *An Introduction to Genetic Analysis,* by David T. Suzuki and Anthony T. F. Griffiths. W. H. Freeman, 1976.
6. "Chromosomes and Disease," by A. G. Bearn and James L. German, III. *Scientific American,* Nov. 1961, Offprint 150. How advances in the visualization of human chromosomes opened up a new frontier in the study of human heredity.

7. "The Mapping of Human Chromosomes," by Victor A. McKusick. *Scientific American,* April 1971. How human chromosomes can be mapped by applying statistical techniques to data from pedigrees.

8. "Hybrid Cells and Human Genes," by Frank H. Ruddle and Raju S. Kucherlapati. *Scientific American,* July 1974, Offprint 1300. How the fusion of human somatic cells with the cells of other mammals can yield information that can be used in mapping human chromosomes.

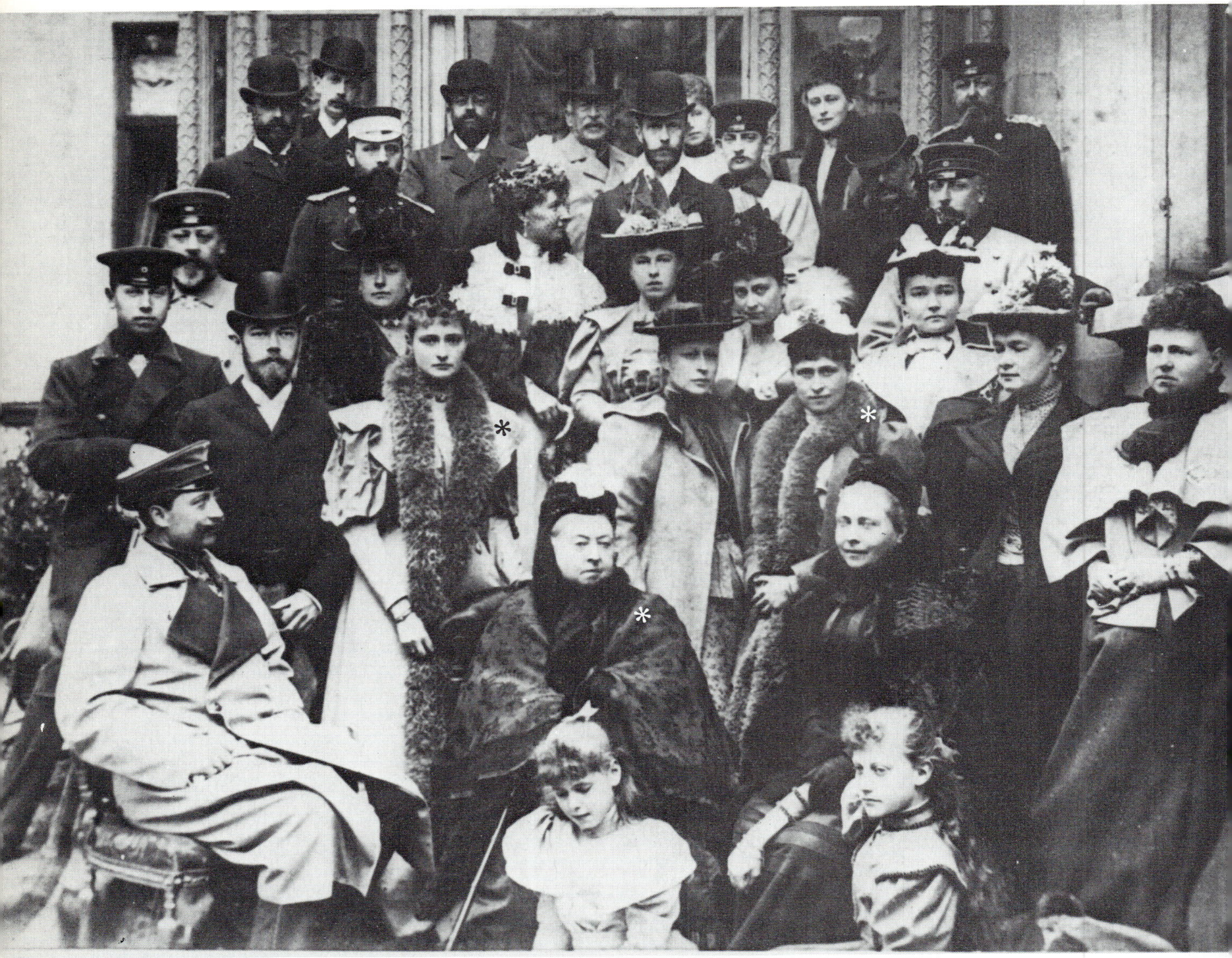

Seventeen of the people in this 1894 photo are descendants of Queen Victoria (seated, center). She and the other two women indicated by an asterisk were carriers of hemophilia, a disease that is inherited as an X-linked recessive trait and that is characterized by a prolonged blood clotting time. The other two women are Princess Irene (Henry) of Prussia (right), and Princess Alix (Alexandra) of Hesse (left) who later married Nicholas II, the last tsar of Russia. (The future tsar is standing beside Alexandra.) A pedigree of hemophilia in royal families of Europe is found later in this chapter. (Courtesy of the Gernsheim Collection, Humanities Research Center, University of Texas, Austin.)

CHAPTER 17

Sex Determination and Sex-Linked Traits

In the shallow offshore waters of several continents live some rather drab-looking but remarkable snails known as slipper shells (Genus *Crepidula*). What makes these animals remarkable is the way in which their sex is determined. To simplify a little, the sex of a slipper shell depends on where it happens to land when it settles down to become an adult.

All young slipper shells are males, and they propel themselves through the water by means of winglike structures, as shown in Figure 17-1. Nonetheless, when they lose their wings and settle to the bottom to begin their adult lives, the young males are transformed into females. That is, the young males are transformed into females if they do not land on a female. Young male slipper shells that land on females remain males throughout their lives, unless they become detached. If a mature male slipper shell becomes detached from a female, the male automatically changes into a female, as it would have done had it not landed on a female in the first place.

What does all of this accomplish for the slipper shell? In brief, this unusual method of sex determination makes it likely that males and females will live in the same area and that they will therefore mate with one another.

The slipper shell's method of sex determination is intriguing because it is so unusual. The sex of most animals, especially those with which we are more familiar, is determined at the time of fertilization and remains the same throughout life. Unlike slipper shells, most animals cannot adjust their sex according to circumstances. Rather, sexual reproduction among animals that have separate sexes usually depends on individuals of opposite sex finding and mating with one another. And under such circumstances it is usually advantageous for a species to have about as many males as females.

Most of us would probably agree that the human species has roughly equal numbers of males and females, as do most other species that reproduce sexually. But the biological mechanism underlying the maintenance of this familiar, nearly equal distribution of the sexes eluded biologists until early in this century. At that time, investigators first turned their attention to male–female differences in chromosomes.

In 1902 it was discovered that the body cells of female grasshoppers contain one more chromosome than those of males. Shortly thereafter, the female's extra chromosome was rather romantically dubbed the *X chromosome* (X for unknown) and it was suggested that the presence or absence of the X chromosome determined whether a grasshopper was a female or a male. Since then it has been learned that chromosomes do have a role in sex determination for the great majority of living things that have separate sexes, including most animals. But it turns out that male and female grasshoppers (and some of their close relatives)—with their unequal numbers of chromosomes—are the exceptions rather than the rule.

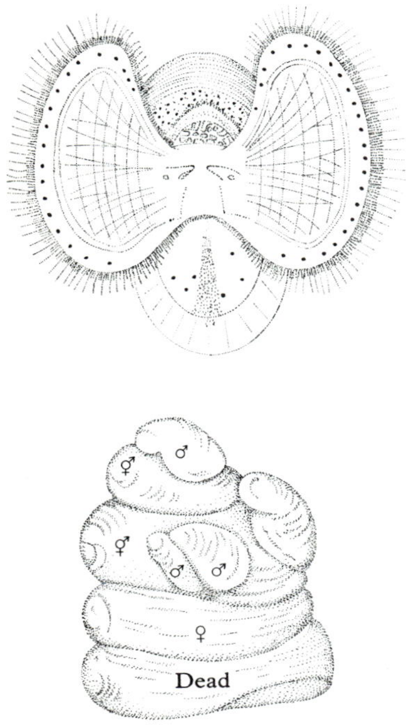

17–1
Top, an immature male slipper shell that has not yet shed his wings, as seen from below. Bottom, a cluster of slipper shells (♀ = female, ♂ = male, ⚥ = individual of intermediate sex).

Male and female animals of the same species generally have the same number of chromosome pairs. But although all of the pairs are matched in females, males have one pair that does not match. As discussed in Chapter 16, the chromosome pairs that match in both sexes are called autosomes, and the members of the remaining pair, which match in females but not in males, are called *sex chromosomes.* For most animals, including all mammals, females are said to be of sex chromosome constitution *XX* and males of *XY.* (In birds the sex chromosomes match in males but not females.)

Figure 17-2 compares the chromosomes of normal males and females. Notice that the X and Y chromosomes are easily distinguished from one another, as, generally, are men and women. In this chapter we discuss the chromosomal basis of maleness and femaleness in human beings and then consider the patterns of inheritance and special properties of traits determined by genes located on sex chromosomes. We begin by discussing the observed human sex ratio and how it relates to the XY mechanism of sex determination.

The Human Sex Ratio

The XY chromosome mechanism of sex determination yields a reliable and nearly equal distribution of the sexes because of the sorting out of chromosomes during meiosis. You will recall that the body cells of women contain 22 pairs of autosomes and two X chromosomes. So when meiosis reduces the number by half during the production of sex cells, eggs that contain 22 unpaired autosomes and a single X chromosome are produced. Thus all normal eggs have an X chromosome. But not so sperm. The body cells of males contain 22 pairs of autosomes, one X chromosome, and one Y chromosome. So when men produce sex cells, meiosis results in two kinds of sperm with regard to sex chromosomes. All human sperm normally contain 22 unpaired autosomes, and on the average

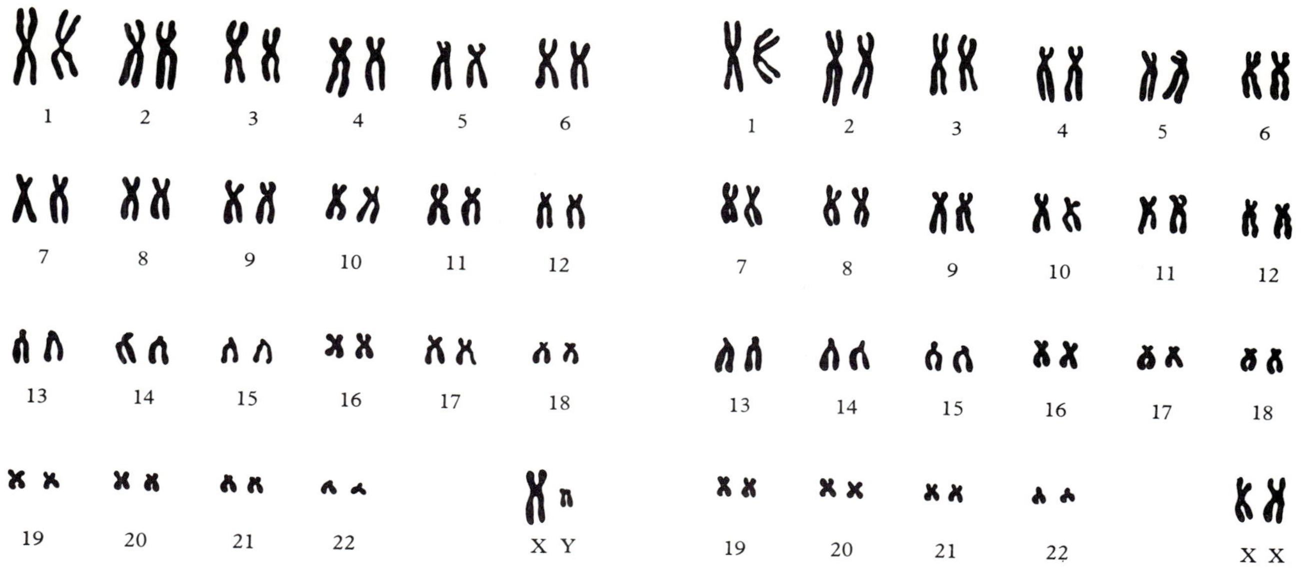

17–2
The chromosomes of normal males and females. Left, males have 22 pairs of autosomes, one X chromosome, and one Y chromosome. Right, females have 22 pairs of autosomes, and two X chromosomes.

half of the sperm have an X chromosome and half have a Y chromosome. Thus, if X-bearing and Y-bearing sperm fertilize normal eggs and result in normal development about equally often, then we would expect that the sexes ought to be about equally distributed, as shown in Figure 17-3. But are they?

Relevant data concerning the number of males and females born in recent decades throughout the world exist and they reveal some rather surprising facts. By convention, the sex ratio is usually reported as the number of males per one hundred females. Among Caucasians in the United States the data show that approximately 106 boys are born for every 100 girls, so the sex ratio at birth is 106. The sex ratio at birth varies somewhat from country to country and it can vary from one racial group to another. For example, the ratio is 113 in Korea, whereas it is about 102.6 among American blacks. Nonetheless, the worldwide data show that on the average more males than females are born in every time interval for which reliable data exist.

The sex ratio at the time of fertilization, known as the *primary sex ratio,* is not necessarily the same as the ratio at the time of birth, the *secondary sex ratio.* Human males have at least a slight numerical edge on females at the time of birth. Is this because more females than males die as embryos? Apparently not, for statistical studies on unborn fetuses have revealed that the primary sex ratio is even higher than the secondary—perhaps as high as 130. Of course, there are many sources of error in determining the sex of unborn fetuses, and the task of assigning a definite sex to the youngest embryos is most difficult. The possibility remains that the observed primary sex ratio results from a large number of female deaths at *very* early stages of development, a period for which we have very little data. But overall, it appears likely that human males do have a numerical advantage over human females both at the time of fertilization and at the time of birth. How can this rather unexpected observation be explained?

If there really are more males than females at the time of fertilization, then several explanations are possible. For example, it may be that Y-bearing sperm win over X-bearing sperm in the race to the waiting egg. It has been suggested

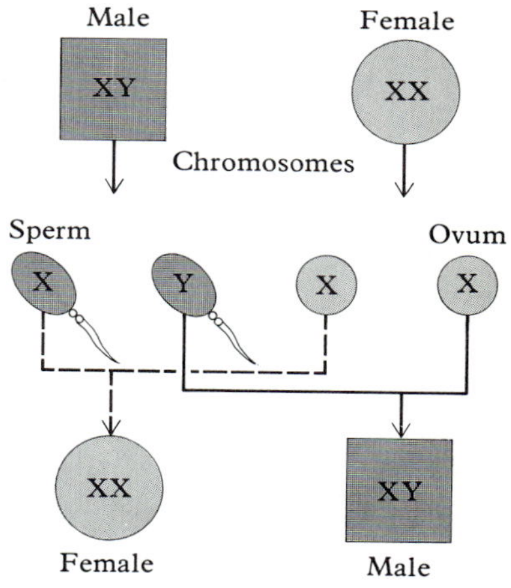

17–3
The chromosomal basis for the existence of nearly equal number of men and women. (From "Sex Differences in Cells" by Ursula Mittwoch. Copyright © 1963 by Scientific American, Inc. All rights reserved.)

that because the Y chromosome is smaller than the X, Y-bearing sperm are lighter and therefore able to swim faster than X-bearing sperm. But, swimming is not the primary means by which human sperm reach the egg. Rather, as discussed in Chapter 15, muscular contractions and ciliary currents within the female reproductive tract are primarily responsible for transporting human sperm to the oviduct (Fallopian tube), where fertilization occurs. Perhaps Y-bearing sperm really do have some as yet unidentified advantage over X-bearing sperm that accounts for the greater numbers of male conceptions and births. (The two types of sperm are produced in approximately equal numbers.) One explanation for the preponderance of male conceptions could be that the physiological environment of the female reproductive tract favors the survival of Y-bearing over X-bearing sperm. Still another possibility is that the surface of the egg attracts Y- more strongly than X-bearing sperm, and this by no means exhausts the possible explanations. In the end, we really do not know what accounts for the greater numbers of males at conception and birth; all we know is that they exist.

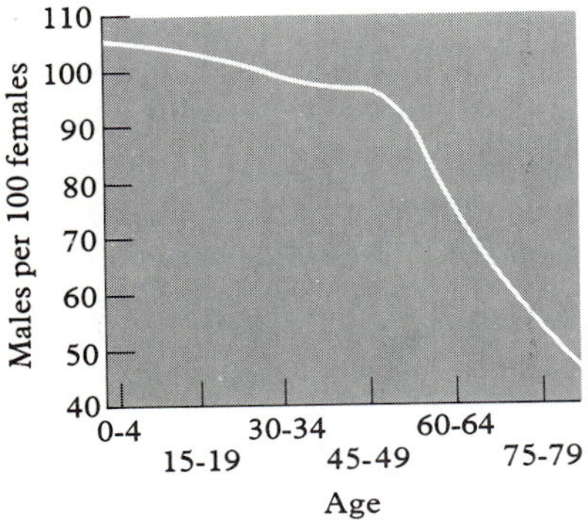

17–4
Data concerning the secondary sex ratio in England and Wales in 1960. (From A. S. Parkes.)

Human males start life with a numerical advantage over females, but they finish a weak second. This is because more males than females die at every stage of life from conception to old age. Thus the numerical advantage of males at birth becomes progressively smaller until at a certain age the sexes exist in equal numbers. The exact age at which this occurs varies from population to population. As shown in Figure 17-4, in some countries males and females are in equal numbers by the time they reach 30 years of age; in others this happens as early as age 18 or as late as age 55. At whatever age it occurs, the equal number of males and females is not maintained. Females soon become the clear numerical majority.

It has been suggested that the male's greater susceptibility to death at every age may be related to the fact that men have only one X chromosome. (As we shall soon discuss, the Y chromosome is very nearly a genetic blank for inherited human characteristics.) It is argued that males, with their single X chromosome, are more vulnerable to the effects of deleterious genes located on it. Even if they do have a deleterious gene on one X chromosome, women are likely to have a corresponding normal allele on their other X chromosome and thus are less likely to be severely affected. Once again, we do not know whether the human male's possession of a single X chromosome is really related to his greater constitutional weakness at every stage. But arguments from population genetics and recent advances in our understanding of the genetics of sex chromosomes have made it seem unlikely that men are at much of a disadvantage simply because of possessing a single X chromosome. As we shall mention later in this chapter, in normal women only one X chromosome is genetically active in each cell; the other X chromosome is nonfunctional.

Voluntary, predictable changes in the primary and secondary sex ratios may soon be a reality. Technological means of altering the sex ratio of the human population already exist. For example, the secondary ratio could be altered by selective early abortion of fetuses of unwanted sex, but this means of controlling the sex of offspring is not likely to become widely accepted or widely available. A more appealing approach to voluntary sexual preselection of offspring is separation of X- and Y-bearing sperm outside the body followed by artificial insemination of either X- or Y-bearing sperm. This technique has been successfully employed in animal husbandry and has been made more reliable by the recent development of a staining method that can identify the Y chromosome in living human cells, including sperm. Other less complicated techniques, such as the use of chemicals or prophylactics designed to block the entry of X- or Y-bearing sperm into the female reproductive tract, would probably be more widely accepted, but they are only theoretical possibilities at the present time.

What would happen to the sex ratio at birth if sexual selection were freely available in the United States? Of course, we can only speculate, but there is good reason to believe that the present ratio of about 106 male to 100 female births would probably not change much. Based on the data from sociological surveys it has been predicted that what would change would be the probability that the first born child would be a male and the second a female. What effect, if any, this would have on our society remains to be seen, as do any other long-term effects of sexual preselection.

We now return our attention to a discussion of the sex chromosomes themselves and in particular to the relationship between femaleness and the X chromosome and maleness and the Y chromosome.

The Roles of the X and Y Chromosomes in Sex Determination— Abnormalities in the Number of Sex Chromosomes

What is the critical genetic difference between normal men and women? From our discussion so far, the answer would appear to be: women are of sex chromosome constitution XX and men are XY. But this tells us nothing of the exact relationship between the X and Y chromosomes and human maleness and femaleness. After all, the sexes are distinguished chromosomally not only by the male's having a Y chromosome, but also by his having only one X chromosome instead of two. Is it the presence of two X chromosomes that determines femaleness and the presence of a single X that results in maleness? Or does the Y chromosome determine maleness? These questions can now be answered. As so often happens in biology, our understanding of the normal process came about through the study of those persons in whom the normal mechanism of chromosomal sex determination had gone awry.

In 1949 it was discovered that two well-known but puzzling human afflictions, *Turner's syndrome* and *Klinefelter's syndrome,* are the result of abnormal sex-chromosome constitutions. (A syndrome is a group of signs and symptoms that occur together and characterize a particular abnormality.) Those who have Turner's syndrome are phenotypic females, which is to say that their genitalia are recognizably female. But these women are generally sterile because most have underdeveloped uteruses and no functional ovarian tissue. Other features of Turner's syndrome are short stature, rather distinctive facial features, a broad shieldlike chest, and a peculiar webbing of the neck. Those who are afflicted with Turner's syndrome have a single, unpaired X chromosome and are thus of sex chromosome constitution XO.

This discovery was made still more interesting when in the same year Klinefelter's syndrome was found to be associated with the sex chromosome constitution XXY. Those who have Klinefelter's syndrome are phenotypic males, but they usually have very small testes and are sterile. Most are also very long-legged and have breast development resembling that of a female.

How does an individual come to have an XO or XXY sex chromosome constitution? Usually by nondisjunction, a term you will recall from our discussion of Down's syndrome in the preceding chapter. Either the sex

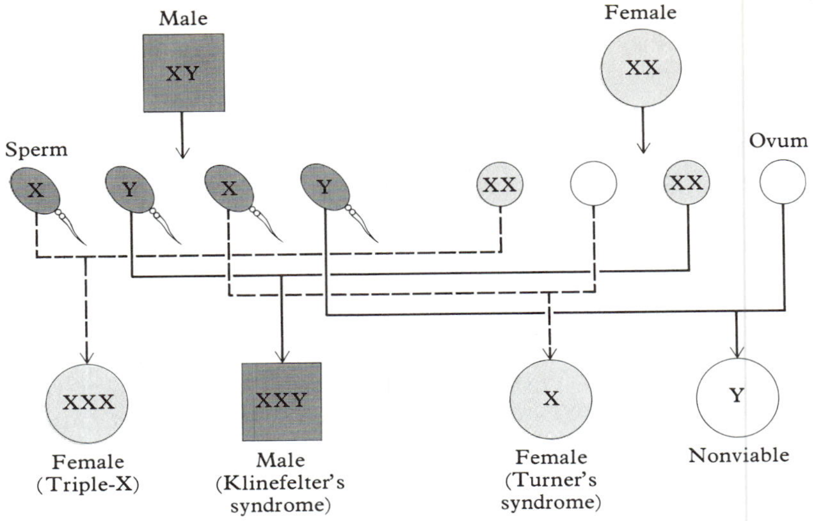

17–5
How nondisjunction can result in certain abnormalities in the number of sex chromosomes. (From "Sex Differences in Cells," by Ursula Mittwoch. Copyright © 1963 by Scientific American, Inc. All rights reserved.)

chromosomes fail to sort out properly during the production of sex cells by meiosis, or they fail to sort out properly in the first few cell divisions following fertilization (see Figure 17-5).

The discovery of the chromosomal basis of Turner's and Klinefelter's syndromes was exciting, not only because the existence of the abnormal sex chromosomes could be explained as the results of nondisjunction, but also because of the insights it provided into the workings of the normal XY mechanism. That those who have XO chromosomes are phenotypic females, whereas people who have a sex chromosome constitution of XXY are phenotypic males suggests that the Y chromosome determines maleness and that the reason those who have XO chromosomes appear female is that in the absence of the Y chromosome even one X is enough to result in a female phenotype.

Klinefelter's syndrome occurs once in every 400 to 600 male births, whereas Turner's syndrome occurs once in about every 3,500 female births. (But Turner's syndrome is observed in many fetuses that undergo spontaneous abortion before twenty weeks of gestation.) In the years since 1949 a rather startling number of cases of human abnormalities caused by unusual numbers of sex chromosomes have been reported. Most of these conditions occur less frequently than Turner's and Klinefelter's syndromes. Many of those who are affected are mentally retarded, and most have physical abnormalities. Sex chromosome constitutions of XXX, XXXX, and XXXXX have been reported, and those who have them are all phenotypic females. On the other hand, sex chromosome constitutions of XXXY, XXXXY, and XXXXXY are also known (Table 17-1) and all belong to phenotypic males. Thus it seems well established that in human beings the Y chromosome determines maleness and its absence results in a phenotypic female, as long as at least one and up to five X chromosomes are present.

Further evidence of the male-determining role of the human Y chromosome comes from the study of persons known as *genetic mosaics.* These persons are remarkable in that their bodies consist of two or more cell lines (that is, cells with different numbers of chromosomes) side by side. Mosaics are the result of accidents that occur during cell division. Most often the accident is either nondisjunction or the accidental loss of a particular chromosome, and the error usually occurs in the first few cell divisions following fertilization. Mosaics *can*

TABLE 17-1

How sex chromosomes relate to sex phenotype. (Barr bodies are discussed later in this chapter.)

SEX CHROMOSOME CONSTITUTION	SEX PHENOTYPE	NUMBER OF BARR BODIES
XX (normal woman)	Female	1
XY (normal man)	Male	0
XO (Turner's syndrome)	Female	0
XXY (Klinefelter's syndrome)	Male	1
XYY (see chapter 20)	Male	0
XXX	Female	2
XXXY	Male	2
XXXX	Female	3
XXXXY	Male	3
XXXXX	Female	4
XXXXXY	Male	4

result from double fertilizations or from the fusion of two embryos very early in development, but this rarely happens.

Mosaics for sex chromosomes are encountered more frequently than those for autosomes (although the latter do occur). In fact, the first person with Klinefelter's syndrome whose chromosomes were studied was a sexual mosaic. Some cells in his bone marrow were XXY, whereas others were XX. This sex chromosome constitution is thus designated XX/XXY.

More recent studies have shown that mosaics in whom only X chromosomes are different (for example, XO/XX, XX/XXX, and XXX/XXXX, all of which have been reported) are phenotypic females (Table 17-2). Sexual mosaics in whom every cell has a Y chromosome, including XY/XXY, XY/XXXY, and XXXY/XXXXY, are, as you might expect, phenotypic males. (Nonmosaic persons of genotype XYY are also phenotypic males. The XYY genotype is further discussed in Chapter 20.)

Of special interest are sexual mosaics whose bodies are made up of cells of different sex. As summarized in Table 17-2, such individuals are usually phenotypic males if one Y chromosome is present. Nonetheless, some of those who have both male and female cell lines exhibit some of the secondary sexual

TABLE 17-2

Human sex-chromosomal mosaics. The mosaics may combine two or three chromosomal constitutions. Phenotypically the mosaics may be female, male, or mixed.

FEMALE	MALE	MIXED
XO/XX	XY/XXY	XO/XY
XO/XXX	XY/XXXY	XO/XYY
XX/XXX	XXXY/XXXXY	XO/XXY
XXX/XXXX	XY/XXY/XXYY	XX/XY
XO/XX/XXX	XXXY/XXXXY/XXXXXY	XX/XXY
XX/XXX/XXXX		XX/XXYY
		XO/XX/XY
		XO/XY/XXY
		XX/XXY/XXYYY

Source: From *The Principles of Human Genetics*, 3d ed., by Curt Stern. W. H. Freeman and Co. Copyright © 1973.

characteristics of the two sexes simultaneously. For example, XX/XXY mosaics may have one male and one female breast, bearded and unbearded facial areas, and, more important, both testicular and ovarian tissue side by side in the same gonad. People who have both kinds of sex tissue generally have a mixture of male and female features in their external genitalia and are known as *hermaphrodites.* (The word is derived from the names of the Greek deities Hermes and Aphrodite.)

Although most hermaphrodites are mosaics for cells of different sex, nonmosaic hermaphrodites of sex chromosome constitution XX (normal female) and XY (normal male) have been reported. As you know, the human Y chromosome is strongly male determining. How then can we account for the presence of male characteristics in these rare XX individuals and female characteristics in these rare XY individuals?

The most widely accepted explanation is this: the mixture of male and female features that exists in exceptional XX and XY hermaphrodites results from the effects of genes that determine sex but that are located on autosomes. (This assumes that the presence of tumors that secrete sex hormones has been excluded. Such tumors may make the secondary sexual characteristics of normal XX females and XY males resemble those of the opposite sex.)

The existence of autosomal genes that influence sex is further supported by well-documented reports of normal males, who have fathered normal-looking sons, but who are nonetheless apparently of sex chromosome constitution XX! Furthermore, it is known that at least one gene, that for testicular feminization, can override the normal XY mechanism of sex determination, because those who are affected by it are phenotypic females, though their sex chromosomes are XY. The available data concerning the transmission of this rare gene do not allow us to decide whether the gene is located on an autosome or on the X chromosome. Nonetheless, the existence of autosomal genes that play a part in sex determination has been demonstrated many times among nonhuman animals whose XY mechanism functions in the same way as that of humans. Overall, there is little doubt that the 22 pairs of human autosomes do indeed carry genes that play a part in determining sex.

Most of the time the sex-determining effects of autosomal genes are balanced by the normal XY mechanism. Thus normal men and women both have genes for maleness and femaleness, both on sex chromosomes and on autosomes. In men, male-determining genes on the Y chromosome and on autosomes outweigh the female-determining effect of genes on the X chromosome and on autosomes. In women, female-determining genes on the X chromosome and on autosomes outweigh autosomal male-determining genes. (The latter are generally much weaker in their male-determining effects than is the presence of a Y chromosome.) But in rare instances the effects of autosomal male-determining or female-determining genes may override the effects of sex-determining genes located on sex chromosomes, thus resulting in persons whose phenotypes and sex chromosomes are at odds with one another.

This concludes our discussion of human sex chromosomes as they relate to sex determination. We now turn our attention to the genetics of genes that do not influence sex but are nonetheless located on the X and Y chromosomes. Such genes are said to be *sex linked* and they have characteristic patterns of inheritance, as we are about to discuss. The X chromosome probably has about as many genes as an autosome of similar length. But the Y chromosome, in spite of its strong male-determining effect is, as far as we know, nearly completely lacking in other genes.

Y-Linked Inheritance

The pattern of inheritance of genes located on the Y chromosome is very simple. Only men have a Y chromosome and all sons but no daughters receive this chromosome from their fathers. So a man who manifests a trait that is determined by a gene on the Y chromosome will pass the trait on to all of his sons and to none of his daughters.

The restriction of a trait to males is not sufficient evidence to prove the existence of a Y-linked gene. This is because some autosomal traits are expressed only in males (or only in females). But such "sex-influenced" autosomal traits can generally be distinguished from those determined by genes on the Y chromosome because traits that depend on autosomal genes are transmitted by both parents, whereas Y-linked traits neither appear in women nor are transmitted by them.

Of the nearly 2000 human traits known to have a genetic basis, only a few are known to be Y linked. The only human trait known to be definitely determined by a gene on the Y chromosome is the presence on the surface of all male cells of a certain protein, called a histocompatibility antigen (H-Y antigen), that is not found on the surface of the cells of females. H-Y antigen was discovered when it was observed that female mice reject skin grafts from males of the same inbred line. There is evidence that the gene for H-Y antigen and the gene that determines the development of testes may be one and the same. At any rate, the presence of H-Y antigen on the cell surface is now considered a reliable criterion of maleness and it can be used to ascertain the sex of infants whose sex is ambiguous at birth.

The only other trait that has stood the test of time as a probable example of Y linkage is a rather unromantic but harmless one known as hairy ear rims. Affected men have long, stiff hairs on the rims of their ears, as shown by the three Muslim brothers from South India in Figure 17-6. The trait also occurs among Caucasians, Australian aborigines, and, more rarely, among Japanese and Nigerian men. Not all instances of hairy ear rims can be attributed to the effects of Y-linked genes. In some instances genes located on autosomes are clearly reponsible for the trait. Nonetheless, in some groups, especially those from India, Y-linkage appears to be established beyond all reasonable doubt.

The human Y chromosome is not unique in its apparent genetic inertness (aside from its role in sex determination). In some insects the Y chromosome is also known to be nearly devoid of genes other than those that determine sex. On the other hand, some fish have numerous Y-linked genes, as do some mice. In all, the Y chromosome remains rather poorly understood, but we can expect our knowledge of it to increase rapidly. Chromosomes are once again the objects of intensive study, just as they were at the turn of the century. As we learn more about the structure of chromosomes, the role of the Y chromosome in sex determination and in Y-linked inheritance will become more clear.

Although the Y chromosome is a near blank for inherited traits, the X chromosome is far from it. We now turn our attention to the distinctive patterns of inheritance of traits determined by genes located on the X chromosome.

X-Linked Inheritance

About 100 abnormal traits are known to be determined by genes located on the X chromosome. (This means, of course, that the corresponding normal traits are

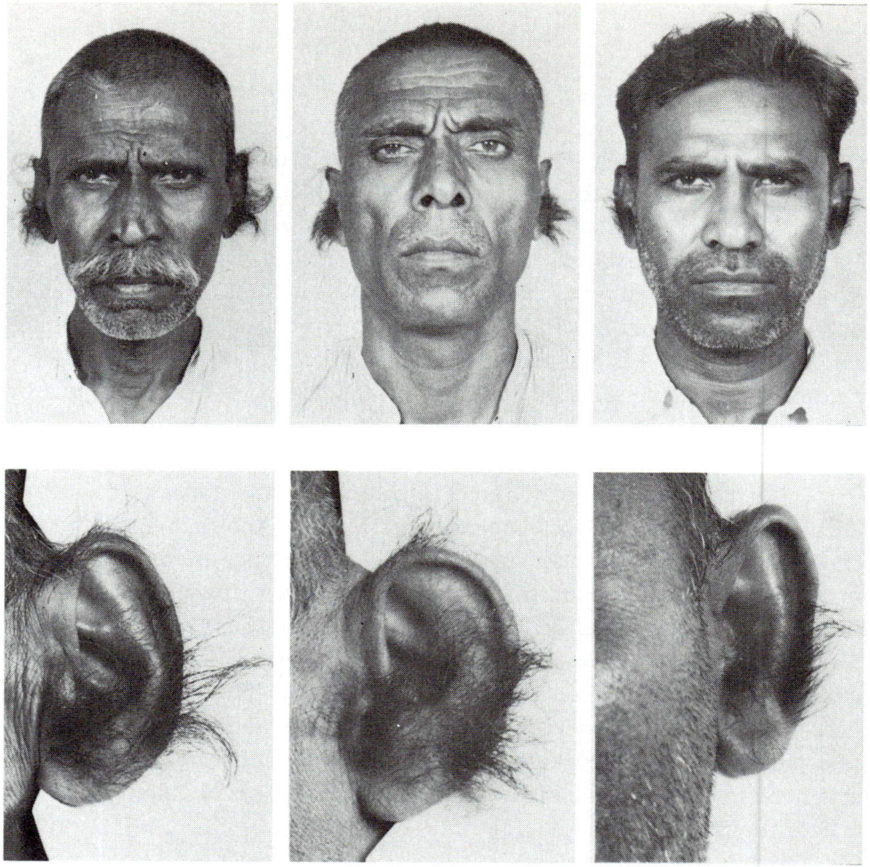

17–6
The strikingly hairy ear rims of three Muslim brothers from South India. This trait is probably determined by a Y-linked gene. (Photo by S. D. Sigamoni, Photography Dept., Christian Medical College Hospital, Vellore. From Stern, Centerwall, and Sarkar, American Journal of Human Genetics, 16. *Copyright © 1964.)*

determined by normal X-linked genes.) The X chromosome is by far the best known of human chromosomes for two reasons. First of all, the patterns of inheritance of X-linked traits are very distinctive, and whenever we observe these patterns in a pedigree we can usually conclude that the gene responsible for the trait is on the X chromosome. (As you know from our discussion of autosomal genes, pedigree analysis usually allows us to decide whether a trait is transmitted as an autosomal dominant or recessive, but tells us nothing about the particular pair of autosomes on which the gene is located.) Second, assigning so many genes to the X chromosome allows us to at least begin to construct a genetic map of this chromosome. As discussed in the preceding chapter, genes located close to each other on a particular chromosome undergo recombination by crossing over during meiosis less frequently than genes that are farther apart. By studying the rates of occurrence of crossovers for two or more X-linked traits we can get an idea of the relative distances between the responsible genes on the X chromosome. Not surprisingly, it is both hard to get and hard to interpret data about recombinations between X-linked genes. Nonetheless, the actual construction of a detailed map of the X chromosome has already begun, as shown in Figure 17-7. This map may seem primitive, but it is by far the most detailed and accurate map of any human chromosome.

Like autosomal traits, those traits determined by X-linked genes may be either dominant or recessive. Nonetheless, X-linked traits have some peculiarities in their patterns of inheritance because of the presence of two X chromosomes in females and only one in males. Females, with their two X chromosomes, may be either heterozygous or homozygous for an abnormal X-linked allele. If the abnormal allele is a dominant one, then a woman heterozygous for it will manifest the trait. But not so if the abnormal allele is recessive. In that case, heterozygous women usually do not manifest the trait, but instead are carriers who appear normal.

On the other hand, males, with their single X chromosome, will always show the effects of an abnormal allele on their X chromosome regardless of whether the trait is inherited as a dominant or a recessive, and in general all males who have abnormal alleles on their X chromosomes manifest X-linked traits about equally. (In other words, males who have X-linked traits are always affected; they cannot be unaffected carriers.)

Of the 100 or so X-linked traits we now know, only a few appear to be inherited as dominants. Included among them are brown discoloration of the teeth, the presence or absence of a particular antigen on the surface of red blood cells, and rickets resistant to vitamin D (a syndrome that is usually characterized by skeletal deformities and by low concentrations of phosphate in the blood).

What are the characteristic features of X-linked dominant inheritance? As shown in Figure 17-8, most women affected by X-linked dominant traits are heterozygous. (Recall that both men and women who manifest autosomal dominant traits are also usually heterozygous.) Women heterozygous for an X-linked dominant trait are affected, and they transmit the trait equally to their sons and daughters. Affected men transmit the trait to all of their daughters (all daughters receive their father's X chromosome), but to none of their sons (no sons receive their X chromosome from their father). In a given pedigree the fact that an affected father does not produce affected sons allows us to distinguish X-linked dominant traits from autosomal dominant ones. In autosomal dominant traits, both affected women and affected men pass the trait on to half of their daughters and to half of their sons. (Work this out for yourself.)

As you may gather from this discussion, the great majority of human abnormalities known to be determined by genes on the X chromosome are inherited as recessive traits. As is also true of X-linked dominant traits, a critical characteristic of X-linked recessive inheritance is the absence of father-to-son transmission. The following are some other characteristics of X-linked recessive inheritance, all of which are summarized in the pedigrees shown in Figure 17-9.

First, virtually all people affected by X-linked recessive traits are men. This is because X-linked traits are rare, and because in order for a woman to be affected, she has to have an abnormal allele on each of her X chromosomes. In

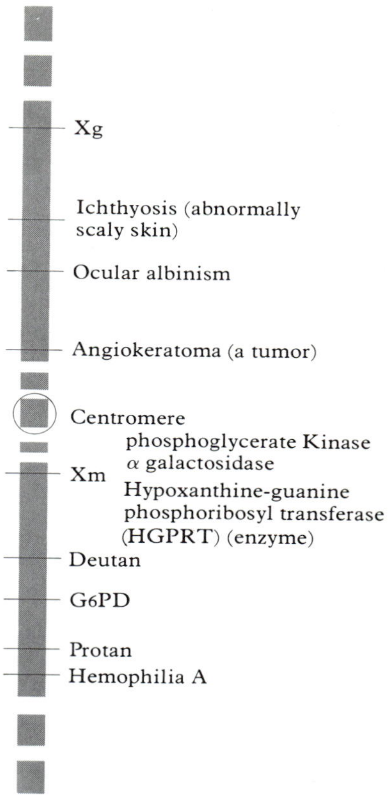

17–7
A tentative map of the human X chromosome. Distances are proportional to the amount of crossing over that takes place between duplicated pairs of X chromosomes during meiosis. Deutan and protan are two forms of color blindness. Xg is a blood group protein, and Xm is a protein found in the blood. There are at least 93 loci known to belong to the X chromosome and almost as many others that are believed to be X-linked. (From Genetics, Evolution, and Man *by W. F. Bodmer and L. L. Cavalli-Sforza. W. H. Freeman and Company. Copyright © 1976.)*

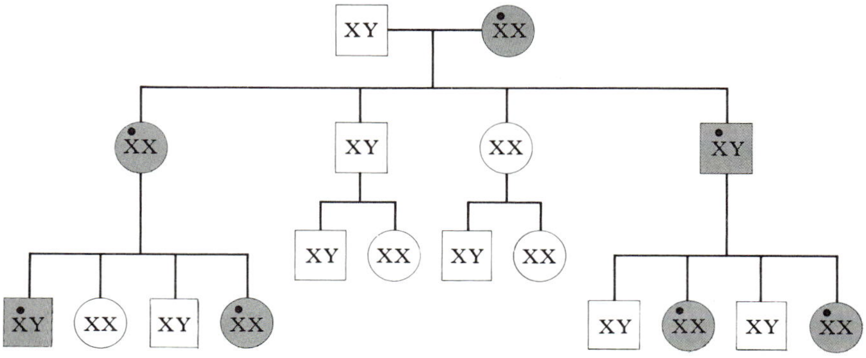

17–8
A pedigree showing X-linked dominant inheritance. Affected individuals are indicated by shaded symbols. The X chromosome bearing the abnormal allele is indicated by a black dot.

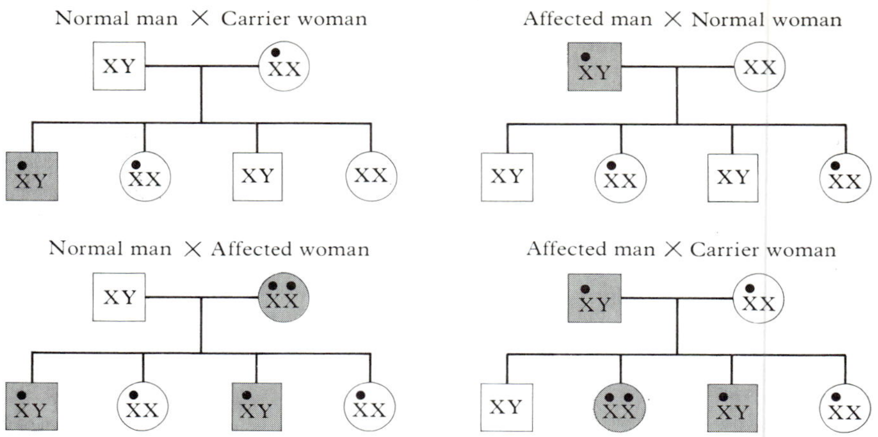

17–9
Pedigrees showing X-linked recessive inheritance. Affected individuals are indicated by shaded symbols. The X chromosome bearing the abnormal allele is indicated by a black dot.

the absence of a spontaneous mutation, this could only occur if her mother were affected or a carrier and her father were affected (Figure 17-9). Such rare, affected women have been recorded in human pedigrees. Second, all of the sons of affected men married to normal women are normal, whereas all of their daughters are carriers. Third, on the average, half of the sons of heterozygous (carrier) women married to normal men are normal, and half are affected (see Figure 17-9).

The most widely publicized of all human pedigrees are probably those in which the X-linked recessive trait known as *classical hemophilia* was transmitted among the royal families of Europe, particularly England. People who have hemophilia are sometimes called "bleeders." Their blood does not clot properly because of a deficiency of one of the many factors that participate in the normal clotting mechanism. (Normal clotting is discussed in Chapter 7. In classical hemophilia the deficiency is in Factor VIII; other types of hemophilia result from deficiencies of different factors.) Those who are affected are usually males (though rare female hemophiliacs have been reported) and they tend to bruise easily and to bleed heavily either into their joints, from their gums, or through open lacerations, often as a result of relatively minor injuries.

The pedigree of Queen Victoria and her descendants is shown in Figure 17-10. Analysis of it reveals that Queen Victoria must have been a carrier for classical hemophilia. One of her sons, Leopold, Duke of Albany, died of hemophilia at age 31. None of Victoria's forebears and neither her husband nor any of her then-living relatives had hemophilia, so the trait probably first appeared as a spontaneous mutation in one of the X chromosomes before Victoria inherited it from one of her parents. Or a mutation could have occurred in one of Victoria's X chromosomes during the queen's early embryonic life. Either way, the pedigree indicates that Victoria must have been a carrier. (The present-day royal family of England is completely free from the gene because the ruling Queen Elizabeth traces her descent through Edward VII, one of Victoria's sons who did not have hemophilia.)

X-linked recessive genes are also responsible for some other rather familiar traits. Of these perhaps the most common are pattern baldness, the common kind of red-green color blindness, and one form of muscular dystrophy, a disease in which the muscles of young males waste away in spite of the presence of an apparently normal nervous system.

For some traits determined by genes located on the X chromosome the distinction between "normal" and "abnormal" is not always clear. One example is the trait known as *G6PD deficiency,* which results from the relative lack of the enzyme glucose-6-phosphate dehydrogenase, an enzyme that participates in carbohydrate metabolism. Affected persons are completely normal under most circumstances, but if they come into contact with certain environmental substances—ranging from the inhalation of the pollen of fava beans to the ingestion of primaquine, a drug used in treating malaria—there may be disastrous results. In the presence of these materials, among others, the red blood cells of affected individuals tend to break open; thus, severe anemia may result.

You will recall that heterozygous carriers of traits determined by genes located on autosomes can usually be identified by some kind of measurement, provided that the underlying biochemical defect is known. It turns out that people who are heterozygous for autosomal traits usually have about half of the normal product of the gene for which they have an abnormal allele. (Thus men and women who are carriers of an allele that results in albinism in homozygotes have about half of the normal concentration of products that have a role in the manufacture of the pigment melanin.) This suggests that autosomal alleles contribute about equally to the total concentration of their normal biochemical product.

This raises an important question. Normal women have two X chromosomes and normal men have only one. Therefore, to return to the example of G6PD, normal women have two normal alleles for the production of the enzyme G6PD, whereas men have only one. Is the concentration of the enzyme in the blood of men therefore only one half of what it is in women? No, it is about the same in both sexes. And this is true not only of the enzyme G6PD, but of the biochemical products of the X-linked genes in general. How can we account for this?

Dosage Compensation in X-Linked Genes—Lyon's Hypothesis

There are at least two possible explanations for the fact that women, with two X chromosomes, have about the same amount of any product determined by an X-linked gene as men, with their single X chromosome. First, the male's single X chromosome could work twice as hard—that is, produce twice the amount of gene product—as each of the female's X chromosomes. (This is true for the fruit fly.) Or second, the activity of one or both of the female's X chromosomes could be less than that of the male's. There is little doubt that for the human species the second explanation is the correct one. Some of the earliest proof for this came from the study of persons who had abnormal sex chromosome constitutions.

In the late 1940s it was discovered that the nondividing cells of female cats contain a small but distinct and stainable blob within their nuclei that is absent from the nuclei of males. This rather mysterious object became known as the *Barr body,* named after one of the first perons to describe it in detail. Barr bodies are found within the nuclei of most female cells from many kinds of animals, including humans (Figure 17-11). Nonetheless, Barr bodies are never observed within the nuclei of normal males.

Not long after Barr bodies were first demonstrated in the tissues of normal women, it was reported that women who had Turner's syndrome (XO) did *not* have a Barr body, whereas men who had Klinefelter's syndrome (XXY) *did!* This led to the hypothesis that the Barr body is actually an X chromosome that is

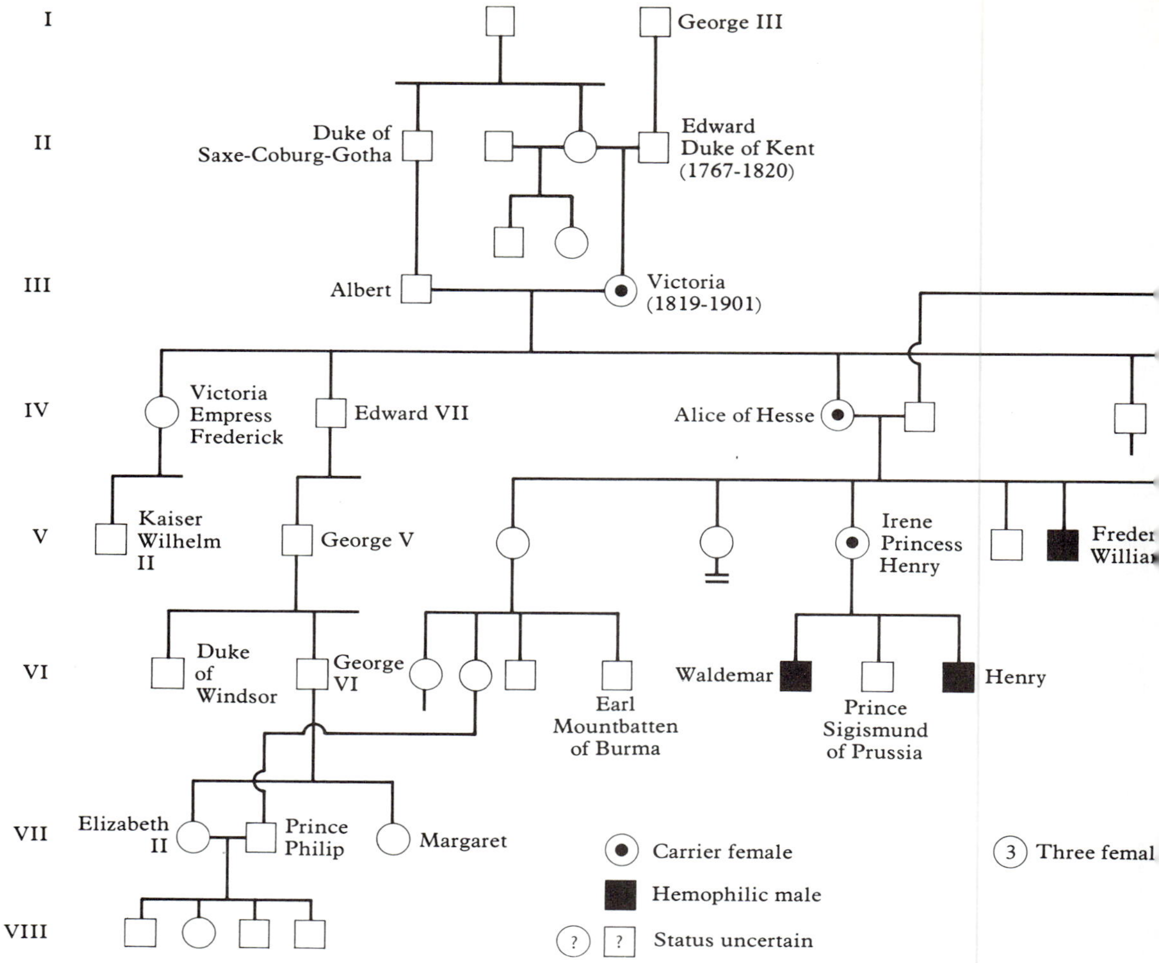

17–10

The pedigree of Queen Victoria and her descendants illustrates the transmission of the X-linked recessive trait hemophilia. (After Dr. Victor McKusick.)

tightly coiled in a dark-staining nuclear blob that is genetically inert. Thus, an obvious explanation for the occurrence of dosage compensation seemed to be this: women and men have the same concentration of gene products determined by X-linked genes because in normal women one X chromosome is genetically inactive.

Further support for the idea that the Barr body is an inactive X chromosome came from the study of the concentration of G6PD in persons who had abnormal sex chromosome constitutions. Thus XO individuals (with no Barr body) and XXY individuals (with one Barr body) both were shown to have roughly normal concentrations of G6PD. Later, it was discovered that individuals whose sex chromosomes are XXX have normal G6PD levels and *two* Barr bodies. Similarly, people who are XXXX have normal G6PD levels and *three* Barr bodies. The number of Barr bodies is thus one less than the total number of X chromosomes, which means that one X chromosome is always available to function normally.

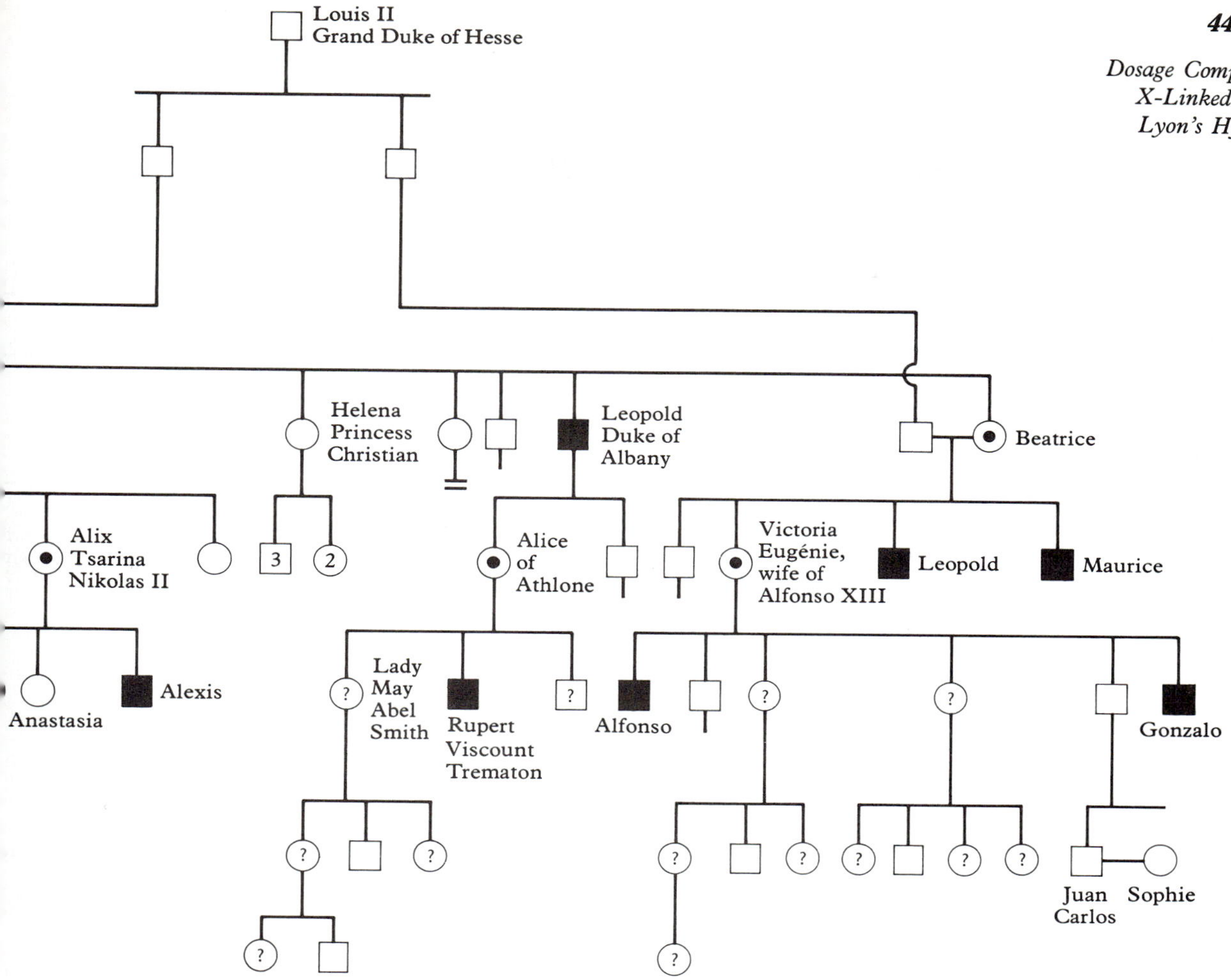

Our present understanding of how the inactivation of one X chromosome relates to dosage compensation is best stated by what has become known as *Lyon's hypothesis,* after the woman who was among the first to propose the idea. (The idea occurred almost simultaneously to several other investigators, and it has been refined over the years as more data have become available.)

The present-day version of Lyon's hypothesis is this. First, very early in embryonic life, when the number of cells in the body of a human female is relatively small, one of the X chromosomes becomes genetically inactive and forms a Barr body. Second, in some cells it is the X chromosome from the mother that is turned off, and in others it is that from the father. In other words, X chromosomes are turned off at random. Third, once the paternal or maternal X chromosome has been turned off in a given cell the same X chromosome is turned off in all of the descendents from that cell during the later development of the embryo. We now turn our attention to some of the evidence that one of the female's X chromosomes really does behave in this rather remarkable way.

If one or the other of the human female's X chromosomes is turned off at random during early fetal development, then one would anticipate that normal women would be mosaics for the X chromosome in that different X chromosomes may be turned on in different cell lines. That this may be true is supported by studies that show that the red blood cells of women who are heterozygous

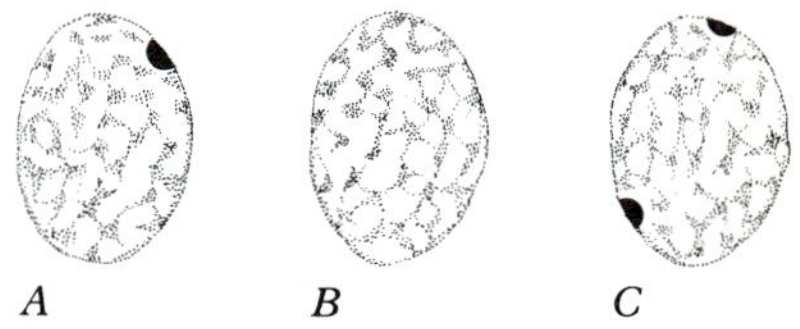

17–11
A, the nondividing nuclei of cells of normal females contain a single Barr body. B, the nuclei of cells of normal males lack a Barr body. C, the nuclei of cells of persons who have sex chromosomes XXX or XXXY have two Barr Bodies. Also see Table 17-1.

carriers for G6PD deficiency clearly fall into two types. They have either normal enzyme activity or almost none. Presumably, the two different populations are the descendants of early embryonic cells in which opposite X chromosomes were inactivated.

Further evidence supporting the inactivation of one or the other X chromosome in different female cell lines comes from determinations of the concentration of G6PD in the red cells of women who are affected by G6PD deficiency. As we would expect, most women who are affected by G6PD deficiency are homozygous recessives, which is to say that they have an abnormal allele on each X chromosome. Nonetheless, some G6PD-deficient women turn out to be heterozygous, and although the concentration of G6PD inside their red blood cells is usually intermediate, it may vary from as low as that of homozygous recessive women to as high as that of normal men and women. This makes sense if we assume that most of the red blood cells of heterozygous women with very low concentrations of G6PD are the descendants of embryonic cells in which the X chromosome that had the abnormal allele was the active one. Similarly, the red cells of heterozygous women who have very high concentrations of G6PD are presumably the descendants of embryonic cells that happened to have the X chromosome whose abnormal allele was turned off. (Women have manifested X-linked recessive traits in spite of the fact that they are heterozygous for hemophilia and some other traits. Such women are known as *manifesting heterozygotes;* presumably, most of their normal X chromosomes are inactivated by chance in early embryonic life.)

Although the evidence in favor of the Lyon hypothesis is convincing overall, there are still some unanswered questions. Perhaps the most pressing of them is this. Normal XX females and normal XY males both have only one active X chromosome. Why then is it that those who have Turner's syndrome (XO) and Klinefelter's syndrome (XXY), both of whom also have only one active X chromosome, are not only sterile but distinctly abnormal in several other ways?

It has been suggested that the abnormal phenotypes of XO and XXY individuals may result because inactivity affects most, but not all, of the X chromosome. If the proposed portion of the mostly-turned-off-X chromosome that remains active carries genes that determine the phenotypic differences between XO and XX individuals, then the phenotypic abnormalities of the XO genotype could be accounted for. Although XO individuals have one complete X chromosome, they lack the supposedly active portion of a normal woman's other, mostly inactivated, X chromosome. Similarly, individuals of genotype XXY would be abnormal because they have a normal X, a normal Y, *plus* the active portion of the other, mostly inactive, X chromosome.

Overall, Lyon's hypothesis seems fairly well established, and evidence is accumulating fast enough that it should be possible to accept or reject the hypothesis in the near future.

Summary

For most sexually reproducing animals it is advantageous to have about as many males as females, and these equal numbers are usually maintained by male-female differences in chromosomes.

All normal human beings have 22 pairs of autosomes; in addition, normal women have two X chromosomes, whereas normal men have one X chromosome and one Y chromosome. The sex of an individual is determined at fertilization and depends on whether the egg is fertilized by an X-bearing or a Y-bearing sperm.

The sex ratio of the human population varies with age. More males than females are probably conceived and born, but at every stage of life males are more likely to die than females. Voluntary selection of the sex of offspring may soon be a reality for some populations.

The discovery of the chromosomal basis of Turner's syndrome (XO) and Klinefelter's syndrome (XXY) suggested a male-determining role for the human Y chromosome, and this was borne out by studies of human sexual mosaics. In the absence of the Y chromosome the phenotype is female as long as at least one X chromosome is present. Autosomal genes must also influence sex determination, and both sexes have genes for "maleness" and "femaleness" on their autosomes.

The Y chromosome is nearly devoid of genes not involved in sex determination, but the X chromosome is known to carry at least 100 genes other than those that determine sex. Most X-linked abnormalities are inherited as recessive traits. X-linked traits, such as classical hemophilia, have distinctive patterns of inheritance.

Dosage compensation occurs for X-linked traits because although women have two X chromosomes, they do not have twice the concentration of products of X-linked genes that men, with one X chromosome, have. Lyon's hypothesis suggests that one or the other of the female's X chromosomes is randomly inactivated very early in development. Inactivated X chromosomes can be observed within the nuclei of female cells, and they are called Barr bodies. The number of Barr bodies is always one less than the number of X chromosomes, and in normal women opposite X chromosomes may be active in different cell lines. Normal women are therefore mosaics for traits determined by X-linked genes.

Suggested Readings

1. "Sex Differences in Cells," by Ursula Mittwoch. *Scientific American,* July 1963, Offprint 161. A review of the major chromosomal differences between men and women.
2. *Genetic Mosaics and Other Essays,* by Curt Stern. Harvard University Press, 1968. A short, rather technical work for those especially interested in genetic mosaics.
3. "Sex Preselection in the United States: Some Implications," by Charles F. Westoff and Ronald R. Rindfuss. *Science,* vol. 184, 10 May 1974. What would happen to the secondary sex ratio if sex preselection were freely available in the United States?

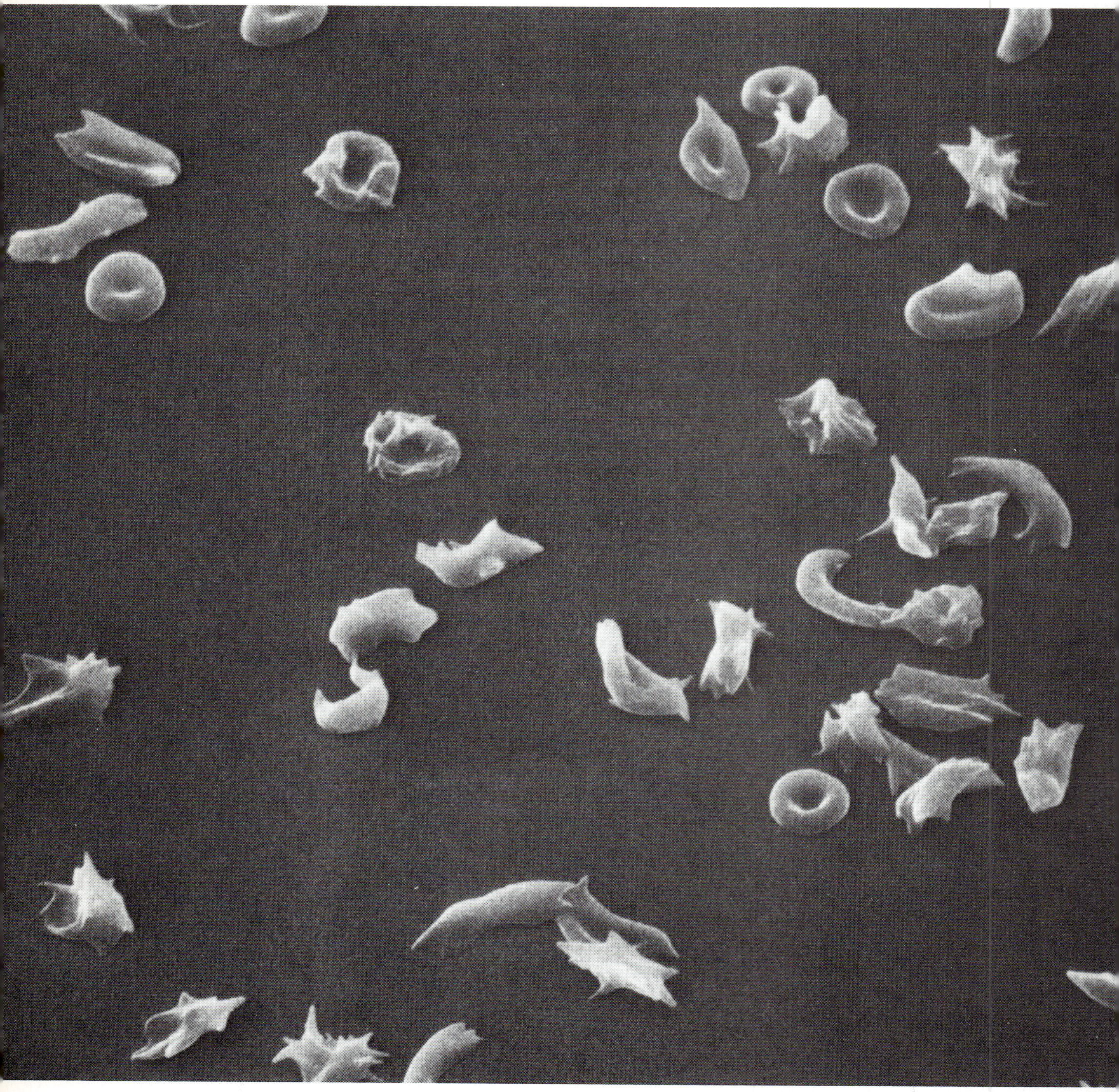

Upon exposure to unusually low concentrations of oxygen, the red blood cells of persons who have sickle-cell trait may become distorted, as shown in this scanning electron micrograph. (Courtesy of Patricia Farnsworth.)

CHAPTER 18

Genes and the Individual

In 1869, only three years after Mendel had published his manuscript, a Swiss biochemist treated some cells with the enzyme pepsin and discovered that although the nuclei of the treated cells shrank in size they were not completely dissolved. Because pepsin is an enzyme that digests protein molecules, it was concluded that nuclei consist, not only of protein, but of other substances, too. Further testing of the undigested nuclear material showed that it was rich in phosphorus and that it could be purified into a white powder. This purified nuclear material eventually became known as *nucleic acid,* and, as was also true of Mendel's manuscript, the scientists of the day took little note of it. The white powder was to spend many years on dusty laboratory shelves before it was realized that it was the hereditary material itself, deoxyribonucleic acid (DNA).

By 1924 specific staining techniques had been devised so that it was possible to see the exact location of DNA within the nucleus. It was discovered that in dividing cells nuclear DNA is always found in chromosomes. It also became known that all of the body cells of a given organism contain the same amount of DNA, whereas the sex cells (eggs and sperm) contain only half as much. This clearly suggested that DNA might be the genetic material, but for some reason the idea was not widely accepted by biologists until the early 1950s.

Before then, most biologists were convinced that the genetic material was not DNA, but protein, which is abundant in both nondividing nuclei and in chromosomes. Proteins were known to be very complicated molecules and what little was then known of the structure of DNA suggested that, although its molecules were very long and threadlike, they were probably simpler than protein molecules. Most biologists felt that protein was more likely to be the genetic material than DNA because complex protein molecules seemed more appropriate bearers of genetic information than the apparently simpler molecules of DNA.

Then in 1944 it was reported that DNA alone, and not protein, could account for a phenomenon known as *transformation* that had been discovered some years earlier. Transformation can be thought of as a lasting change in the genetic program of certain bacteria brought about by DNA from other bacteria. Thus, as shown in Figure 18-1, when purified DNA from dead bacteria of Type 1 is added to living cultures of organisms of Type 2, some *living* organisms that have characteristics of Type 1 may form. This is because some cells of Type 2 take up Type 1 DNA from the culture medium and thus become transformed. Once they are transformed, the organisms breed true to their new type and it is possible to recover more Type 1 DNA from the transformed bacteria than was originally added to the growing culture. Thus not only does Type 1 DNA transform some Type 2 cells into cells that have some of the characteristics of Type 1, but the resulting Type 1 organisms reproduce and manufacture new Type 1 DNA as they do so.

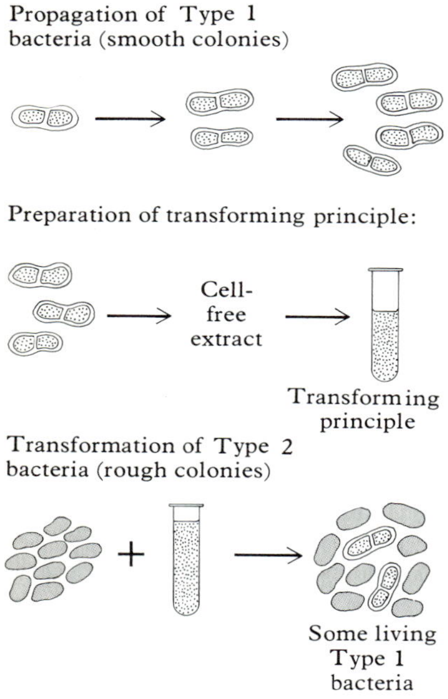

18–1
The "transforming principle" was found to be DNA, not protein. (From Gunther S. Stent, Molecular Genetics. *W. H. Freeman and Company. Copyright © 1971.)*

Further support for the idea that DNA is the genetic material came from the study of certain kinds of *viruses.* In general, viruses are composed of a protein coat surrounding a core of nucleic acid. (The nucleic acid in viruses is usually, but not always, DNA.) In 1952 it was reported that certain kinds of viruses that attack and usually destroy the bacterium *Escherichia coli (E. coli),* a normal inhabitant of the human digestive tract, are able to do so because they inject DNA into the bacterium like tiny parasitic syringes. Such a virus is known as a *bacteriophage,* or *phage* (see Figure 18-2). The phage's protein coat attaches the virus to the bacterium, but, unlike the DNA, the protein coat does not get inside the cell. The DNA injected into the bacterium contains all of the information necessary to manufacture entire new phage particles, including the protein coat.

Thus, by 1952 there was little doubt that DNA was the hereditary material. As we will discuss further in this chapter, it is now known that the protein in the nondividing nucleus and in chromosomes provides a physical support, or scaffolding, for DNA molecules, and plays a part in determining which genes are expressed when. But it is in the enormously long, double-stranded, self-replicating molecule of DNA that we find the final, biochemical basis of heredity.

The unraveling of the biochemical basis of heredity is one of the greatest intellectual achievements of this or any other century. But a detailed description of how our knowledge of biochemical genetics came about, no matter how engaging it may be, is beyond our present purposes. In this chapter, our main concern is with identifying the biochemical basis of some rather well-understood genetic abnormalities of human beings and with how genes bring about their effects in individuals. Our present understanding of human biochemical genetics, indeed of biochemical genetics in general, is in large part an outgrowth of a proposal for the detailed structure of DNA first put forth in 1953. The structure of DNA is so important, and so revealing, that we must discuss it further.

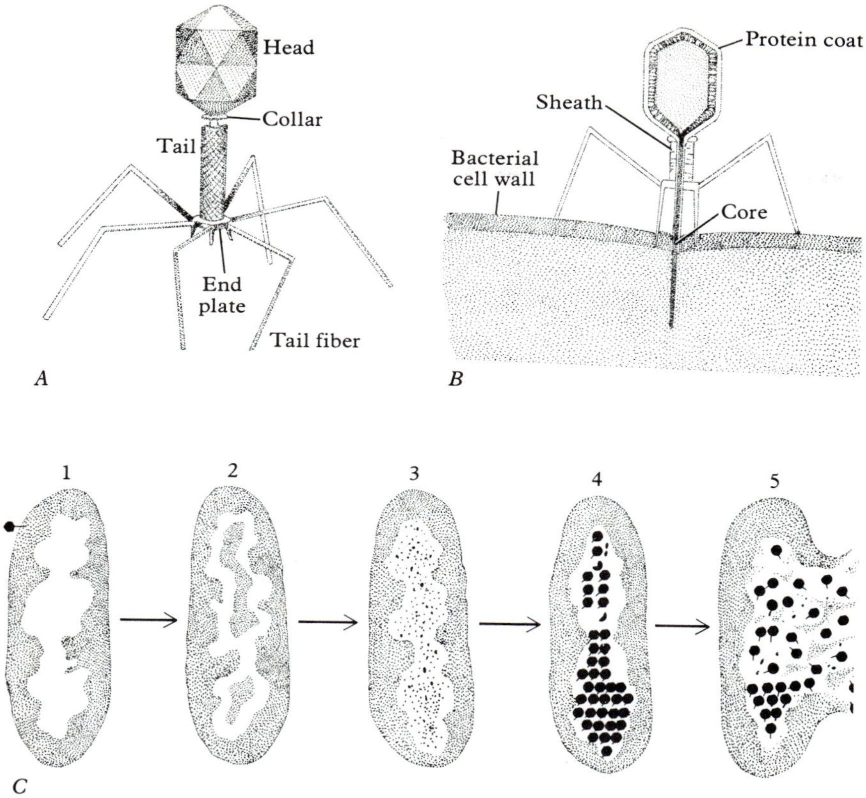

18–2
Phage viruses are parasites of bacterial cells. A, the head of the virus is a protein shell that contains DNA. B, the virus attaches to a bacterial cell and injects viral DNA into it. C, following the injection of viral DNA the bacterial cell is converted into a virus-making factory. The bacterium eventually bursts, releasing the newly formed phage. (From "Building a Bacterial Virus," by William B. Wood and R. S. Edgar. Copyright © 1967 by Scientific American, Inc. All rights reserved.)

The Watson-Crick Model for the Structure of DNA

Like most other very large, naturally occurring molecules, DNA is made up of a few relatively simple chemical building blocks that are joined to one another in sequence by means of chemical bonds. In DNA, these building block compounds are *nucleotides.* Each nucleotide is made up of three parts: a phosphate group, a sugar that contains five carbon atoms and is known as deoxyribose, and a nitrogen-containing base (Figure 18-3).

There are four kinds of nucleotides in DNA. All of them contain the phosphate and the sugar, but they differ in their nitrogen-containing bases. The four bases fall naturally into two categories according to their structure. Two of the bases, *adenine* and *guanine,* have a double-ring structure, and the remaining two bases, *cytosine* and *thymine,* are made up of a single ring, as shown in Figure 18-4.

Within a DNA molecule the nucleotides are bonded in such a way that the sugar of one nucleotide is always attached to the phosphate group of the next nucleotide in line. This arrangement results in a long chain consisting of alternating sugar and phosphate groups, and from this backbone the bases stick out to the side. (Figure 18-5). This was well known before Watson and Crick published their model for the structure of DNA in 1953. Their great accomplishment was not an explanation of the *composition* of DNA, but rather of the detailed three-dimensional architecture of the molecule.

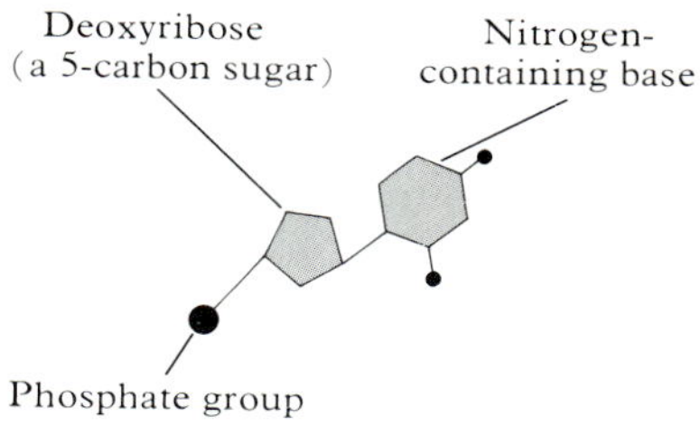

18–3
The structure of a nucleotide, one of the building blocks of DNA.

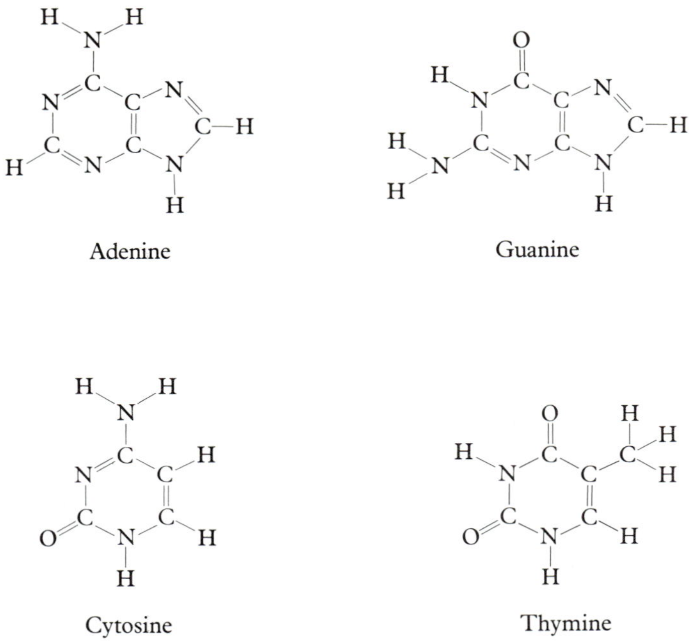

18–4
The structural formulas of the four nitrogen-containing bases that are found in DNA.

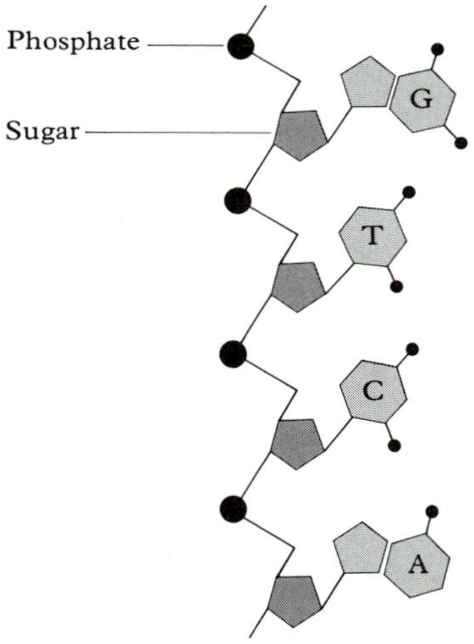

18–5
The backbone of the DNA molecule is made up of alternating sugar and phosphate groups. The nitrogen-containing bases stick out from the backbone. Adenine (A) and guanine (G) are the double-ring bases; thymine (T) and cytosine (C), the single-ring bases.

The model was based on information from several sources. It had been reported a few years earlier that all DNA molecules, no matter what their source, have something in common in the number of bases they possess. Although the amounts of the four bases may vary widely from species to species, in all DNA molecules the number of adenine bases is exactly equal to the number of thymine bases, and the number of guanine bases is exactly equal to the number of cytosine bases.

Watson and Crick used this information and results from X-ray diffraction studies and from other experiments aimed at determining the exact distances between atoms in the DNA molecule and devised a model of DNA structure that is simple and elegant. The overall model can be summed up in the following way.

First, the DNA molecule is a double helix. To visualize a helix, think of the backbone of alternating sugar and phosphate groups as wrapped around a long thin cylinder. In the double helix of DNA, two backbones (two strands) are present, and they are held together because the bases stick out into the interior of the molecule in a way that enables them to form weak chemical bonds with one another. (See Figure 18-6.)

Second, the amount of adenine is equal to that of thymine because an adenine on one strand is always bonded to a thymine on the opposite strand. Similarly, the amount of guanine is equal to that of cytosine because in DNA these two bases are always bonded to one another across the double helix.

One of the most convincing points of this model was that it immediately suggested how DNA might replicate itself. Because adenine always pairs with thymine and guanine always pairs with cytosine, if the two strands of a DNA molecule are separated by breaking the bonds between the bases, then each chain provides all the information necessary to synthesize a new partner. It was soon discovered that DNA is indeed capable of self-replication, as any molecule reputed to be genetic material must be. As shown in Figure 18-7, DNA

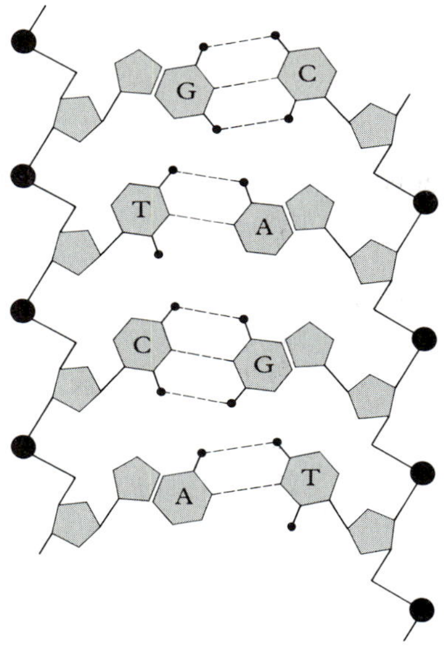

18–6
The structure of DNA. Top, a schematic, flattened view of the molecule. A pairs with T, and G pairs with C. The paired bases A and T are held together by two hydrogen bonds. C and G are held together by three hydrogen bonds. Bottom, three-dimensional representations of the double helix of DNA. (Top, from "The Synthesis of DNA," by Arthur Kornberg. *Bottom, from "The Repair of DNA," by Philip C. Hanawalt and Robert H. Haynes.*

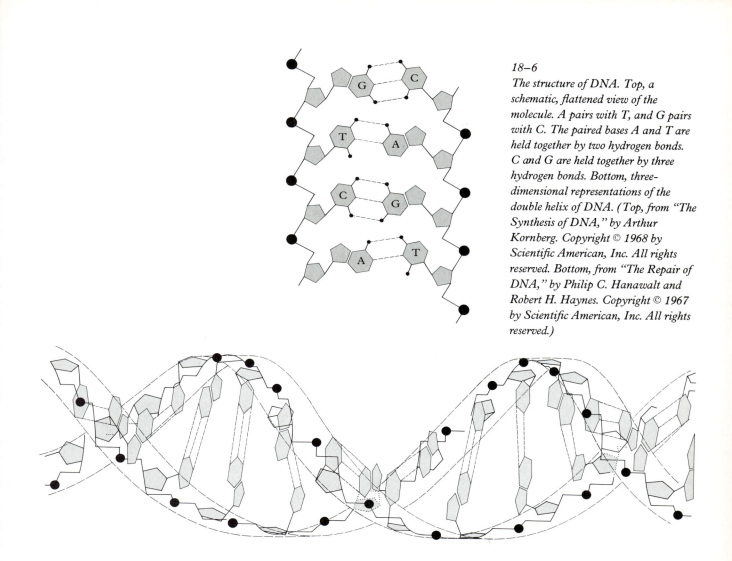

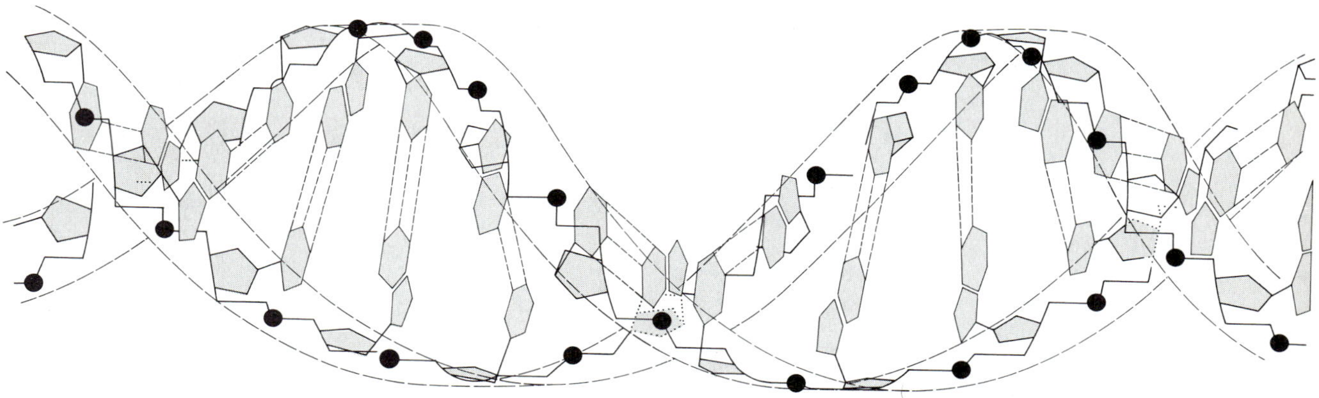

replication is accomplished by the separation of the two strands of the double helix. Each strand then serves as a template for the manufacture of a new complementary strand.

But there remained an obvious question about the Watson and Crick model. DNA had been proved to be the hereditary material, but nobody yet knew how the double helix brings about its biochemical effects. Nonetheless, within a few years laboratories from all over the world had contributed to an overall biochemical scheme of gene action whose details are still being worked out today. In sum, at the biochemical level genes usually participate in *protein synthesis,* and within the double helix is encoded all of the information necessary for a cell to synthesize the proteins it needs to survive and reproduce.

The biochemistry of protein synthesis is also discussed in Chapter 5, which should be consulted if the following outline of protein synthesis is not clear. For our purposes, as shown in Figure 18-8, the relationship between DNA and protein synthesis can be summed up in this way. Proteins are made up of building blocks called *amino acids,* of which there are at least 20 kinds. The number of amino acids in a protein molecule may vary from about 50 to 50,000 or more, and in general the properties of a particular protein depend above all on the sequence of amino acids that makes up the molecule (Figure 18-9).

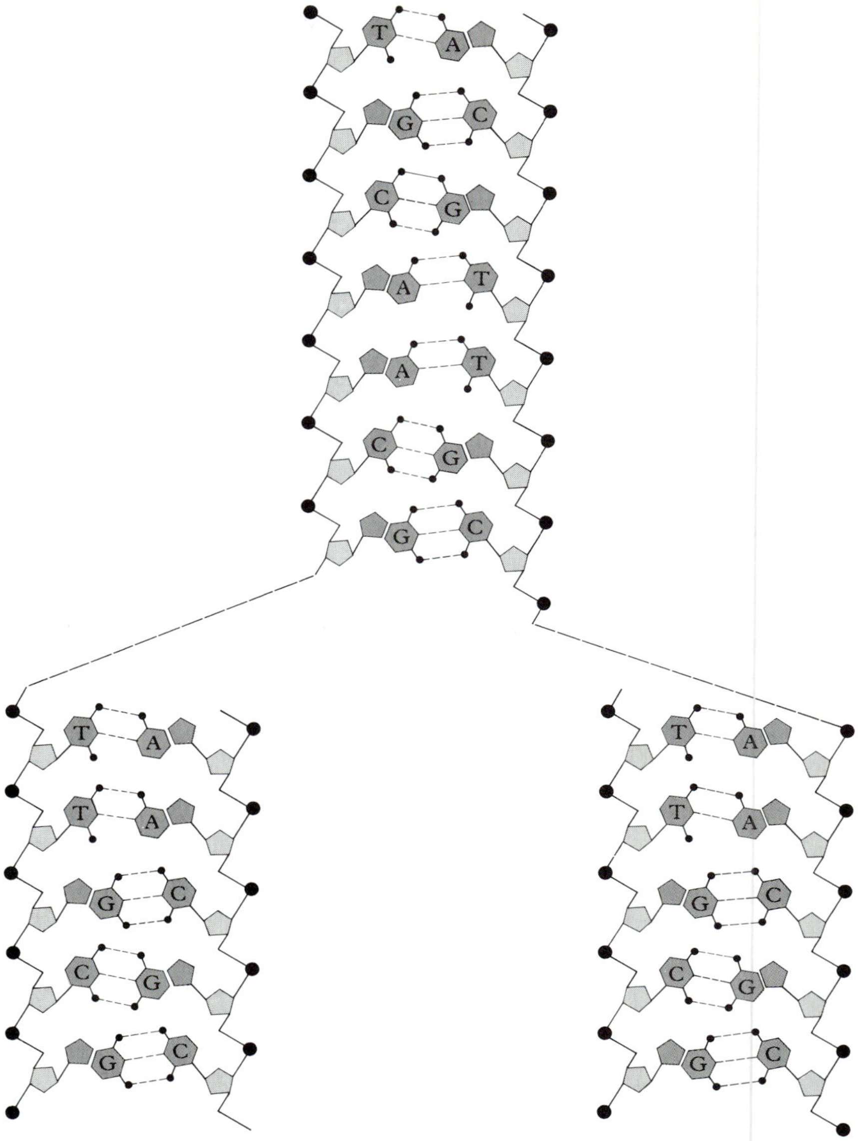

18–7
During DNA replication the two strands of the double helix separate and each of them serves as a template for the manufacture of a new complementary strand. (From "The Synthesis of DNA," by Arthur Kornberg. Copyright © 1968 by Scientific American, Inc. All rights reserved.)

What determines the sequence of amino acids in a given protein is the sequence of the bases in a particular segment of the molecule of DNA. As shown in Figure 18-8, *three* bases in the DNA molecule ususally code for *one* amino acid.

Protein synthesis also requires a second kind of nucleic acid known as ribonucleic acid (RNA). In RNA the five-carbon sugar is ribose, and RNA does not contain thymine but rather a different single-ring base that pairs with adenine, *uracil.* Also, unlike DNA, most RNA is single stranded. That is, the backbone of RNA is usually a single chain of alternating sugar and phosphate groups.

Protein synthesis begins when a segment of DNA within the nucleus becomes unwound and one of the strands provides the code for the synthesis of *messenger RNA* (mRNA). mRNA then diffuses out of the nucleus into the cytoplasm of the cell where it becomes attached to structures called ribosomes.

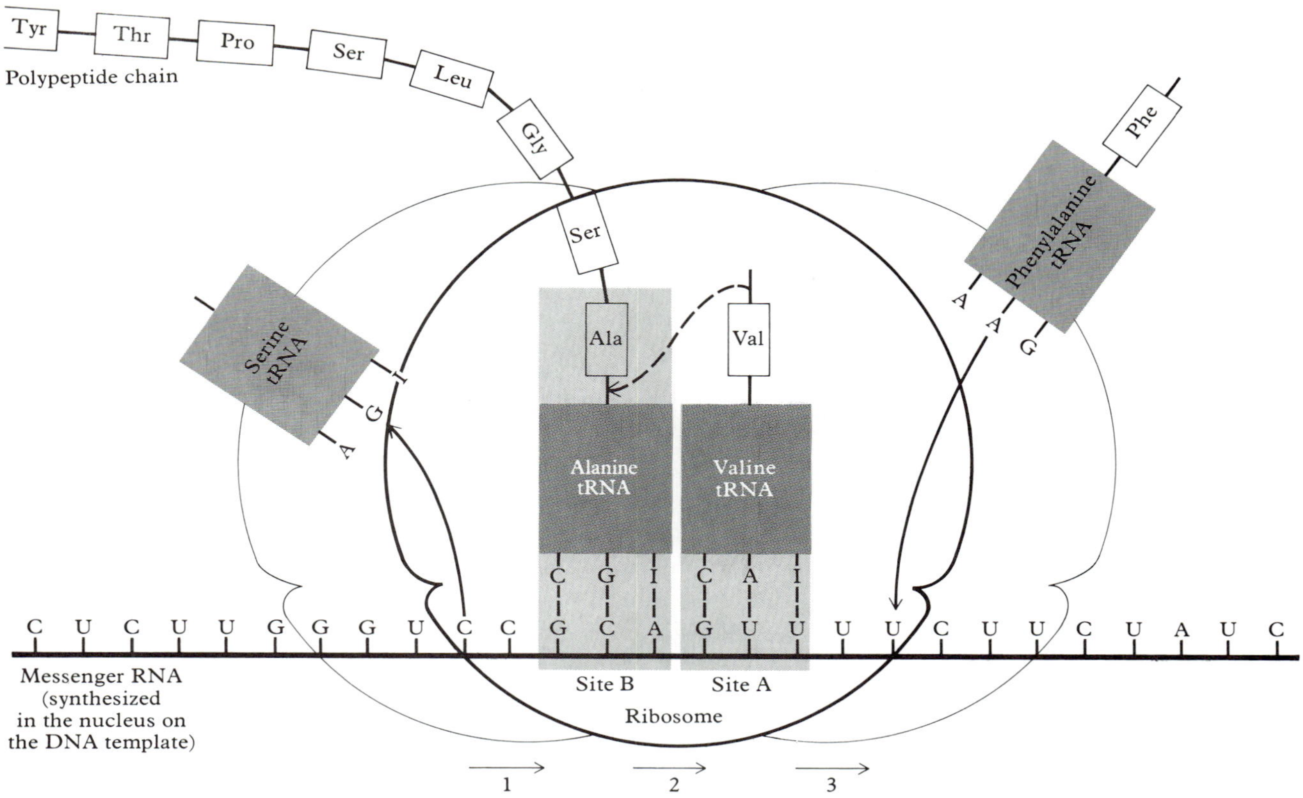

18–8
Protein synthesis occurs on the surfaces of tiny intracellular particles called ribosomes. A ribosome moves along a molecule of mRNA that is complementary to a portion of one of the strands of DNA in the nucleus. On the surface of the ribosome the sequence of bases in the mRNA is translated into a chain of amino acids. Each type of amino acid is carried to the ribosome by a specific tRNA molecule, one portion of which contains three exposed bases. The three exposed bases of tRNA form temporary bonds with a complementary three-base sequence in the mRNA molecule. When this happens, one amino acid is added to the growing chain and the tRNA is released to once again pick up another molecule of its specific amino acid. (Some tRNA molecules contain bases other than those we have already discussed, but each still pairs with a specific base in the mRNA molecule.) (From "The Genetic Code: III," by F. H. C. Crick. Copyright © 1966 by Scientific American, Inc. All rights reserved.)

Ribosomes are the site of the stringing together of specially activated amino acids to form the long chains of protein molecules. On the surface of ribosomes, the base sequence of mRNA, which is complementary to the copied strand of DNA in the nucleus, becomes *translated* into a chain of amino acids. Amino acids are first activated by being attached to another kind of RNA, *transfer RNA* (tRNA). Each amino acid has its own tRNA, and each tRNA contains three exposed bases that can pair with the complementary three bases on the molecule of mRNA. When they do so, tRNA molecules release their particular activated amino acid to the growing amino acid chain. (See Figure 18-8.)

Thus, at the biochemical level, most genes turn out to be specific regions along enormously long DNA molecules that code for the manufacture of a

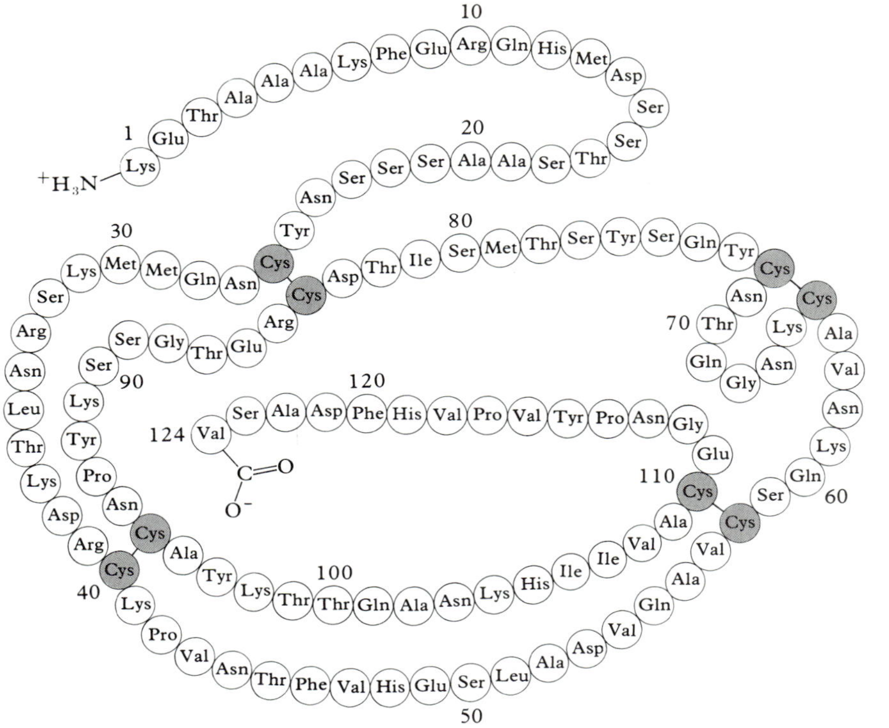

Amino acid	Three-letter abbreviation
Alanine	Ala
Arginine	Arg
Asparagine	Asn
Aspartic acid	Asp
Asparagine or aspartic acid	Asx
Cysteine	Cys
Glutamine	Gln
Glutamic Acid	Glu
Glutamine or glutamic acid	Glx
Glycine	Gly
Histidine	His
Isoleucine	Ile
Leucine	Leu
Lysine	Lys
Methionine	Met
Phenylalanine	Phe
Proline	Pro
Serine	Ser
Threonine	Thr
Tryptophan	Trp
Tyrosine	Tyr
Valine	Val

18–9
The properties of a particular protein depend above all on the sequence of its component amino acids. This is the molecule of bovine ribonuclease, an enzyme that digests RNA. The chemical bonds that form between adjacent cysteine help to determine the shape of the enzyme molecule.

particular protein by the mechanism just described. How does all of this relate to people? To begin with, we can use this information concerning the biochemical bases of protein syntheses to form a rough estimate of what the total number of human genes may be.

DNA and the Structure of Chromosomes

Because most genes can be thought of as sections of DNA molecules that contain coded messages for the manufacture of specific proteins, we can use the total amount of DNA inside the cells of various organisms to estimate how many genes the organisms have. As we have seen, three successive bases in the DNA molecule code for one amino acid. If we assume that an average protein molecule consists of about 200 amino acids (a reasonable estimate), this means that about 600 successive bases in DNA are required to code for an average protein molecule. So if we know the total number of base pairs in the DNA of a particular species, we can estimate the number of genes that code for proteins by dividing the total number of base pairs by 600.

When these calculations are carried out the results are often surprising. For example, ordinary toads are estimated to have about 7 million genes, whereas frogs have 14 million. The same method gives an estimate for human beings of about 5 million genes, and the estimate for a certain species of salamander is a whopping 125 million genes! It is not at all obvious why a frog should have twice as many genes as a toad, or why a salamander should have 25 times as many genes as a human being. On the contrary, most biologists agree that the number of different kinds of protein molecules that have structural roles or that influence metabolism probably does not vary much, perhaps tenfold, from the lowliest multicellular creatures to the most complicated. How then can we account for the widespread variation in the amount of DNA observed from one multicellular creature to another?

At least part of the answer is found in the recent discovery that in most multicellular organisms some segments of the DNA molecule are present in multiple copies. In fact, a significant proportion of the DNA in most multicellular organisms consists of segments that have similar or even identical base sequences, some of them repeated thousands, or as many as a million, times. The genetic role of these widespread duplicated segments of DNA is far from clear, but they can account, at least in part, for the rather extreme variation in the amount of DNA from one species of multicellular organism to another.

In the end, we do not know very much about the exact relation between the amount of DNA and the number of genes in any multicellular creature. According to some experts, a good guess of the "actual" number of human genes, taking duplications and other factors into account, may be somewhere between 25,000 and 100,000 genes. In large part, the reason for our relative ignorance of the workings of DNA in multicellular organisms is that most of what has been learned of biochemical genetics so far has come from the study of viruses and bacteria.

As you may recall, viruses are composed of a protein coat surrounding a nucleic acid core. The core usually consists of a short DNA molecule, though some exceptional viruses contain double-stranded RNA instead. Bacteria are far more complicated than viruses and they contain more DNA. Nonetheless, as discussed more fully in Chapter 1, bacteria are composed of *prokaryotic* cells that are much simpler than the *eukaryotic* cells of higher organisms that we have been discussing. The bacterial cell lacks a distinct nucleus; its chromosome (which is never visible in the light microscope) consists of a single DNA molecule and is in the form of a circle (see Figure 18-10). In general, the bacterial chromosome is present in a single copy.

It is a very large step indeed from understanding the organization and workings of the bacterial chromosome to understanding how the genetic material is organized and how it functions inside the nuclei and chromosomes of eukaryotic cells. However, it is known that the mechanism of protein synthesis is more or less the same in both prokaryotic and eukaryotic cells. It has also been shown that genes known as *regulator genes,* which were first reported from bacterial cells, are also present in the DNA of eukaryotes. These genes do not code for a particular structural protein, but rather play a role in *regulating* protein synthesis. We will discuss them in more detail later in this chapter. Overall, most of what has been learned from bacterial genetics can be applied to the genetics of people or any other eukaryotic organisms, but it can tell us little about how the genetic material in eukaryotes is organized. Nonetheless, there have been some recent advances in our understanding of how DNA is related to the complicated chromosomes of eukaryotic cells. These discoveries are worth discussing further.

By weight, eukaryotic chromosomes contain about equal amounts of protein and DNA. Recent advances in our understanding of eukaryotic chromosomes have for the most part centered on the relationship between DNA and chromosomal proteins. The nuclear proteins known as *histones* have at least two important functions. First, they provide a scaffold for the DNA molecules that is most visible when the nuclear contents condense into chromosomes

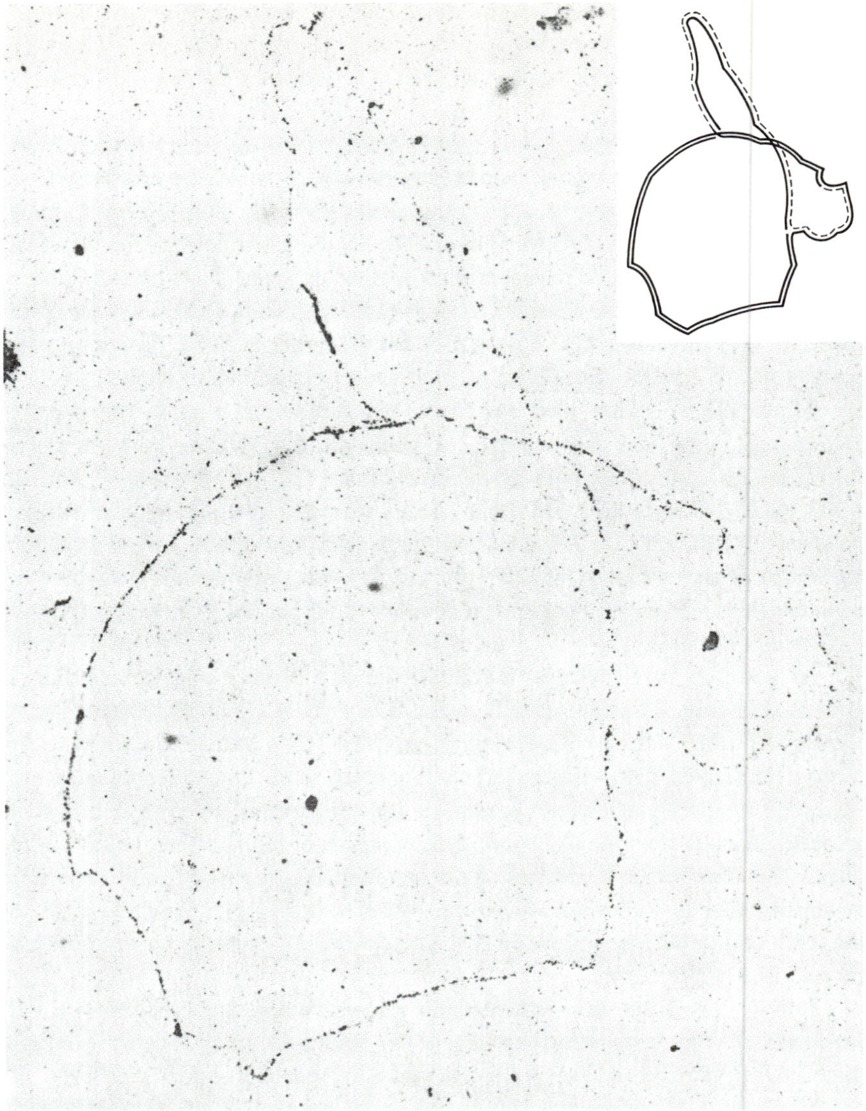

18–10
A bacterial DNA molecule in the act of replicating itself. In the small inset the unduplicated portion of the molecule is indicated by a dotted line. (Courtesy John Cairns.)

during the process of cell division. The central scaffolding may rather resemble beads on a string, with the DNA wrapped around the outside of the beaded structure. Second, histones, along with other proteins in the nucleus, play a major role in regulating the activity of different segments of the DNA molecule. It has been estimated that 10 percent or less of the total genetic material of any eukaryotic cell is biochemically active at any one time, and the nuclear proteins determine which genes are expressed when, though how they do this remains far from clear.

We know very little about the fine structure of the chromosomes of the human species. For example, we do not even know how the total amount of DNA inside human cells is divided among the individual chromosomes. It is possible, though unproved, that the DNA in human chromosomes exists as a very long unbroken thread. For the longest human chromosome, this thread, when completely uncoiled, would be about six centimeters (2.4 inches) long. Obviously, even when not completely unwound, this enormously long molecule would have to be highly folded or coiled within the microscopic nucleus. When eukaryotic cells are not undergoing division, their chromosomes are stretched and loosely entangled with one another inside the nucleus, but the chromosomes are still intact. It is in this relaxed and stretched condition that DNA engages in protein synthesis. On the other hand, the molecular events of crossing over apparently do not occur until the early stages of meiosis, at which time the chromosomes are just beginning to condense into easily distributable packets of genetic information.

With this discussion of biochemical genetics and the structure of eukaryotic chromosomes as background, we now turn our attention once again to the genetics of people. In particular, we can now discuss a famous and frustrating human disease known as *sickle-cell anemia,* which is the result of an abnormality in the genetic code.

Structural Genes—Sickle-Cell Anemia

The clinical syndrome known as sickle-cell disease was first described in 1910. At that time, few facts were known about it. First, those who were affected were blacks of both sexes, and many of them died of it in early childhood. Second, the disease was characterized by the occurrence of sometimes fatal crises that usually lasted for a few days at a time. During a crisis an affected person would develop fever and experience intense and incapacitating pain in the bones, large joints, abdomen, and elsewhere. Third, some of the red blood cells of affected persons were shaped like crescents, or sickles, and all those who were affected had severe anemia. That is, the total number of red blood cells in their circulation was much less than normal (Figure 18-11).

A few years later it was discovered that the sickling of the red cells is related to the state of oxygenation of the iron-containing pigment, *hemoglobin.* As discussed in Chapter 8, hemoglobin molecules inside red blood cells pick up oxygen in the lungs and release it to the tissues. Red blood cells from persons who have sickle cell disease look normal when their hemoglobin molecules are saturated with oxygen. But when the saturation of hemoglobin by oxygen falls to lower than normal, sickling occurs; and the process can be observed through the

18–11

As compared with normal hemoglobin, the abnormal hemoglobin molecules of people who have sickle-cell disease and sickle-cell trait (to be discussed later) show distinctive patterns of movement in an electric field. The areas under the curves correspond to the position and amounts of hemoglobin after an electric current has passed through the liquid in which the hemoglobins are dissolved. The black arrow indicates the location of the specimen before the electric field was applied.

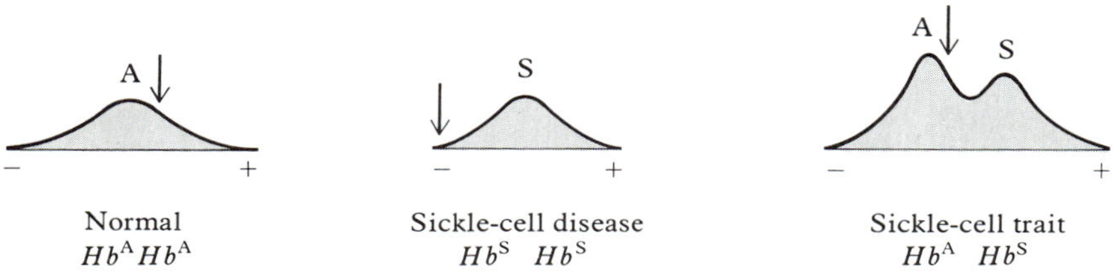

microscope. As the concentration of oxygen falls, more and more of the "normal" cells turn into sickle cells right before one's eyes.

What is it about the red blood cells of persons who have sickle-cell disease that results in their unusual change in shape in the presence of low concentrations of oxygen? The first clue to this question came in 1949 when it was discovered that the hemoglobin of those who have sickle-cell disease differs from normal hemoglobin. As shown in Figure 18-11, sickle-cell hemoglobin molecules were found to have a different pattern of movement than normal hemoglobin molecules in the presence of an electric field. This means that the two molecules cannot be identical. How are they different?

A refinement of the use of an electric field has provided the answer. Hemoglobin can be partially digested by enzymes that split the protein part of the molecule into relatively short amino acid chains of varying length. If the resulting mixture of short amino acid chains is allowed to move under the influence of an electric field, a pattern, or "fingerprint," of the hemoglobin molecule is formed. Each spot of the fingerprint represents a different short chain of amino acids. When this is carried out for normal and sickle-cell hemoglobin, the patterns shown in Figure 18-12 result. Notice that protein digestion results in the formation of 26 short amino acid chains for both kinds of hemoglobin. In sickle-cell hemoglobin, shown on the right in Figure 18-12, 25 of the 26 chains are identical to those in normal hemoglobin. Sickle-cell hemoglobin and normal hemoglobin differ in only one short amino acid chain, that labeled "4" in Figure 18-12.

The exact nature of the difference between sickle-cell and normal hemoglobin is now known. As shown in Figure 18-13, the hemoglobin molecule contains two protein building blocks, the *alpha* chain and the *beta* chain. Each alpha chain consists of 141 amino acids and each beta chain contains 146 amino acids. A normal hemoglobin molecule consists of two alpha chains, two beta chains and four nonprotein, iron-containing portions known as *heme.* As we have just discussed, when the hemoglobin molecule is partially digested by

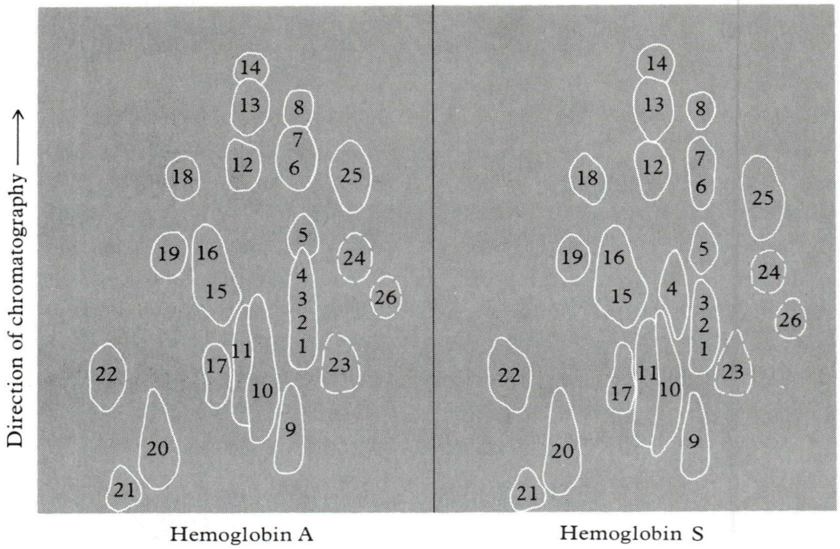

18–12
A "fingerprint" of normal and sickle-cell hemoglobins. Each spot on the fingerprint represents a short chain of amino acids.

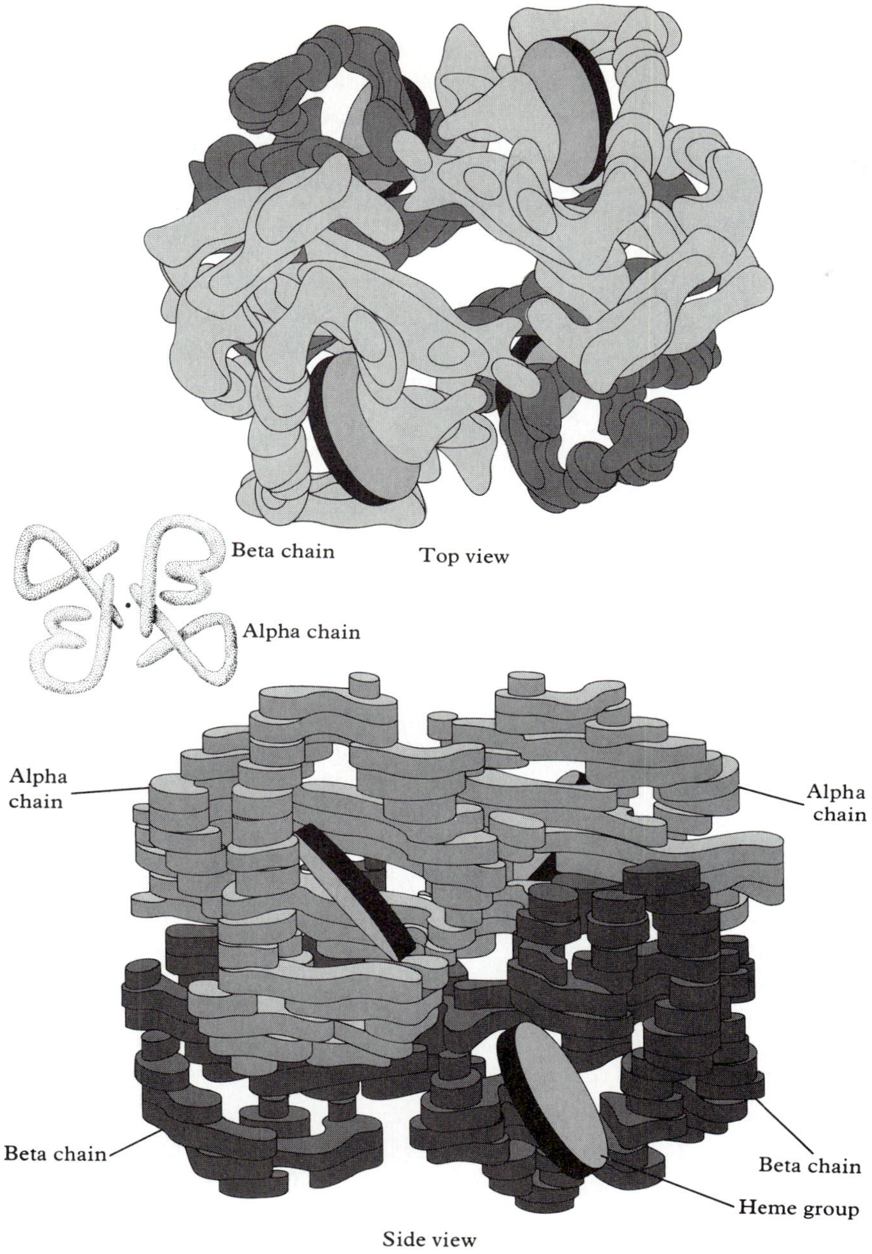

18–13
The hemoglobin molecule consists of two alpha chains, two beta chains, and four heme groups. (From "The Hemoglobin Molecule," by M. F. Perutz. Copyright © 1964 by Scientific American, Inc. All rights reserved.)

protein-splitting enzymes, 26 shorter amino acid chains are left and all but one of these shorter chains (chain 4) are identical in normal and in sickle-cell hemoglobin. That chain on the fingerprint is a short chain consisting of eight amino acids that makes up one of the ends of the beta chain of hemoglobin. Of these eight amino acids, seven are identical in both kinds of hemoglobin. *Only one amino acid* out of a total of 287 is different in sickle-cell hemoglobin. Nonetheless, for persons who have sickle-cell disease, this seemingly trivial molecular difference can mean the difference between life and death.

As you may recall, three base pairs in DNA usually code for one amino acid in a protein molecule. In normal hemoglobin, the three-base sequence CTT (cytosine, thymine, thymine) codes for the amino acid *glutamic acid.* But in sickle-cell hemoglobin, the corresponding three-base sequence is CAT (cytosine, adenine, thymine), which codes for the amino acid *valine.* As shown in Figure 18-14, the amino acid substitution occurs sixth in line from one of the ends of the beta chain.

How does the presence of only one different amino acid out of 287 result in sickle-cell disease? It has recently been shown that the single amino acid difference has no effect on the stability of individual hemoglobin molecules; nor does it alter the molecule's oxygen-carrying ability. Rather, the single amino acid difference results in a unique reaction *between* individual molecules of sickle-cell hemoglobin. Deoxygenated molecules of sickle-cell hemoglobin spontaneously come together to form spiral, rigid, fiberlike structures that distort the red cell and result in sickling. (Most of the red cells quickly resume their normal shape when the hemoglobin is reoxygenated.) The rigid sickle cells tend to become trapped and broken when they circulate through capillaries, and the membrane of the red blood cell may be damaged in the sickling process. Both of these factors contribute to the breakdown of sickle cells that then leads to the anemia that characterizes the disease.

At the present time, many of those who are affected by sickle-cell disease survive beyond the age of fifty. But the improvement in outlook is not due to advances in our understanding of the exact biochemical nature of the disease. For all of our knowledge, we can at present do nothing to alter the composition of the DNA of those who are affected or to alter the translation of DNA into molecules of sickle-cell hemoglobin. Rather, the increased length of survival of affected individuals is largely the result of better nutrition and the prevention of infections that may reduce the amount of oxygen available to the tissues and thus induce widespread sickling.

Shortly after the condition was first described in 1910 it was realized that sickle-cell disease is transmitted as an autosomal recessive trait. This means that affected individuals are homozygous for the gene that results in the production of the abnormal beta chain in the hemoglobin molecule. The genotype of persons who have sickle-cell disease can be represented as $Hb^{S}Hb^{S}$. On the other hand, unaffected individuals are homozygous for normal hemoglobin, or Hemoglobin A, and their genotype is $Hb^{A}Hb^{A}$. People who are *heterozygous* for sickle-cell hemoglobin are thus usually of genotype $Hb^{A}Hb^{S}$. Individuals of genotype $Hb^{A}Hb^{S}$ are said to have *sickle-cell trait* (not sickle-cell disease). Those who have sickle-cell trait do not have anemia and are perfectly normal under most circumstances. Nonetheless, some of the red cells of persons who have sickle-cell trait can be made to sickle in the laboratory by subjecting the cells to lower-than-normal concentrations of oxygen. As shown in Figure 18-11, persons who have sickle-cell trait can also be identified if samples of their hemoglobin molecules are allowed to move in the presence of an electric field. As we might expect, those persons who have sickle-cell trait usually have about equal amounts of normal and sickle-cell hemoglobins throughout their red cells.

Sickle-cell trait occurs in about 8 percent of blacks in the United States, most of whom are of African ancestry. Sickle-cell disease, on the other hand, is becoming less common in the United States, but it still occurs. At least some of the decrease in the number of new cases reported per year is probably related to increased public awareness of the disease and to the detection of normal-appearing individuals who have sickle-cell trait (see Chapter 20). But black

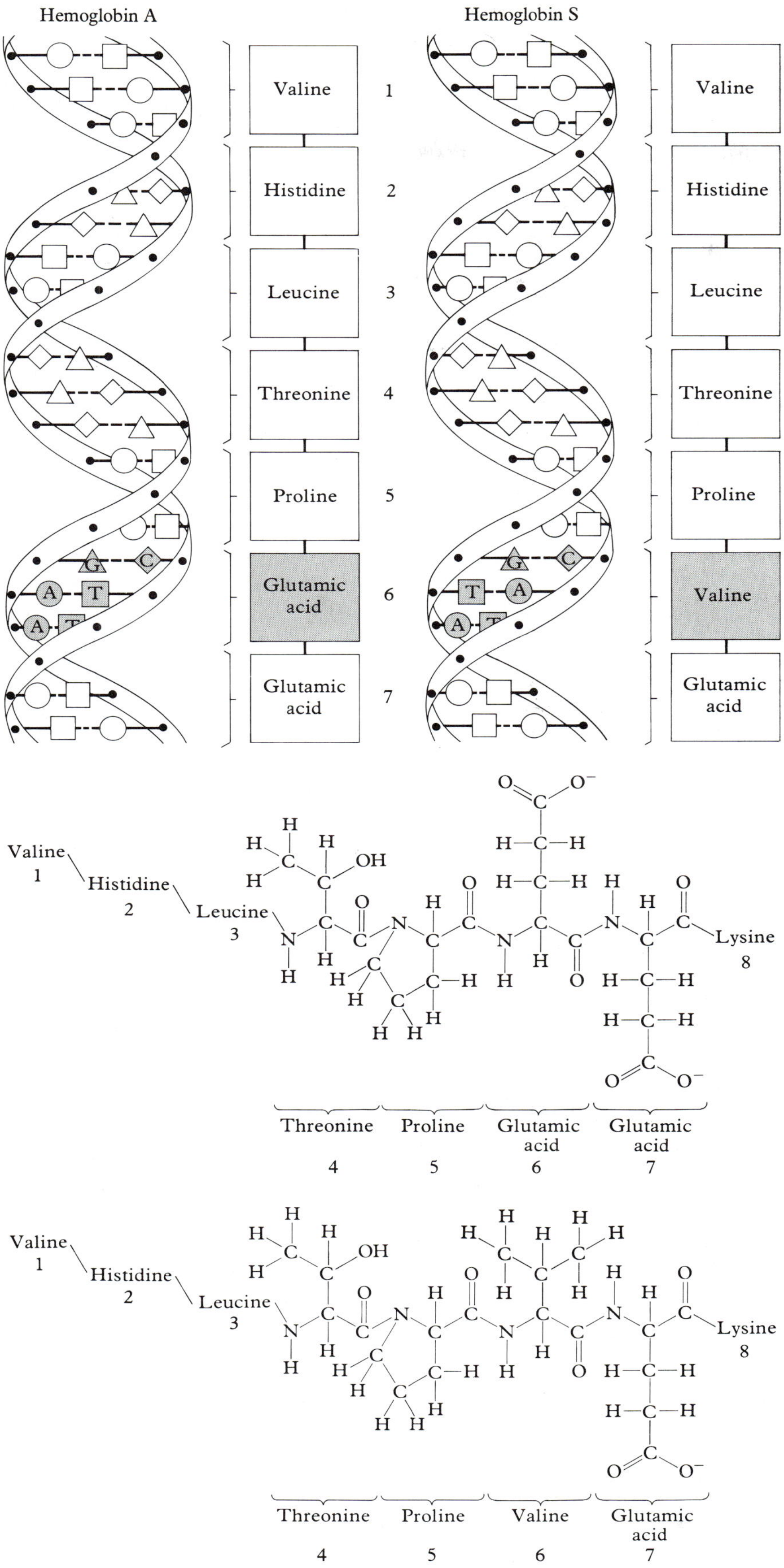

18–14
Only one amino acid out of a total of 287 differs in sickle-cell hemoglobin from normal hemoglobin. As shown here, the amino acid valine replaces glutamic acid. The substitution occurs sixth in line from one of the ends of the beta chain. (From "The Genetics of Human Population," by L. L. Cavalli-Sforza. Copyright © 1974 by Scientific American, Inc. All rights reserved.)

people are not the only human beings afflicted by sickle-cell disease. The abnormal gene also occurs relatively commonly in people inhabiting the Mediterranean area, Arabia, and India. Clearly, people who have sickle-cell disease are at a disadvantage compared to those who have normal hemoglobin. Why, then, is the gene that in the homozygous condition results in sickle-cell disease so widespread in different areas of the world, particularly Africa?

Figure 18-15 is a map showing the distribution of the sickle-cell gene in various parts of the Old World. The highest frequencies are in a rather broad belt across the African continent in which a severe form of *malaria* is a common cause of death. It has been found that the red blood cells of those who have sickle-cell trait ($Hb^{A}Hb^{S}$), which contain a mixture of normal and sickle-cell hemoglobin, are *more resistant to malaria* than normal red blood cells. This advantage probably accounts in large part for the persistent widespread distribution of a gene that in the homozygous state results in reduced reproductive fitness and may lead to early death.

In the past two decades the number of different types of abnormal human hemoglobins reported has increased enormously, and is still doing so. In fact, the deciphering of the biochemical basis of sickle-cell disease and the research that has followed it have brought a mushrooming of our knowledge of molecular disease, which we will discuss at greater length later in this chapter. But the investigation of abnormal hemoglobin molecules has done more than provide us with an understanding of how molecular changes result in disease. It has also provided a clue about how at least some human genes may be regulated. In particular, we have learned that the hemoglobin of normal human beings is different at different stages of development. That is, the kind of hemoglobin present in a normal fetus changes several times during its intrauterine existence, and it is only after birth that Hemoglobin A ($Hb^{A}Hb^{A}$) comes to predominate. Let us discuss this important discovery further.

Fetal Hemoglobins and Regulator Genes

As you know, the protein portion of a normal hemoglobin molecule usually consists of two alpha chains and two beta chains. Each of these amino acid chains is coded for by a different region of nuclear DNA, that is, by a different gene. The gene that codes for the alpha chain becomes active very early in fetal development and remains so throughout adult life. But the gene that codes for the beta chain behaves differently. Although two beta chains are present in almost all of the hemoglobin molecules of normal adults, they are not present in the hemoglobin molecules of a developing fetus or of a newborn infant. The gene that codes for the beta chain becomes active during the second month of embryonic life, and some hemoglobin molecules made up of two alpha chains and two beta chains are present from then on. (As discussed in following paragraphs, in normal adults such molecules of Hemoglobin A compose about 97 percent of the total hemoglobin.) Like the molecules of normal adults, the hemoglobin molecules of developing fetuses and of infants up to six months old contain two alpha chains, but most do not contain two beta chains. Instead of beta chains, most of the hemoglobin molecules of fetuses and newborn infants contain a protein known as the *gamma chain.* This type of hemoglobin is known as *fetal hemoglobin.*

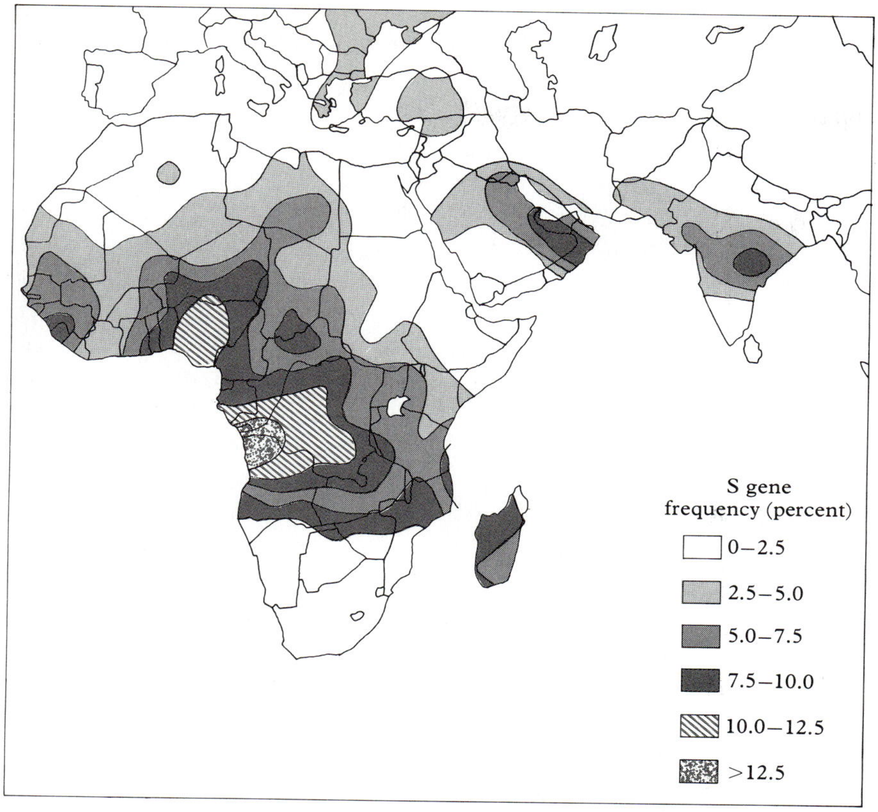

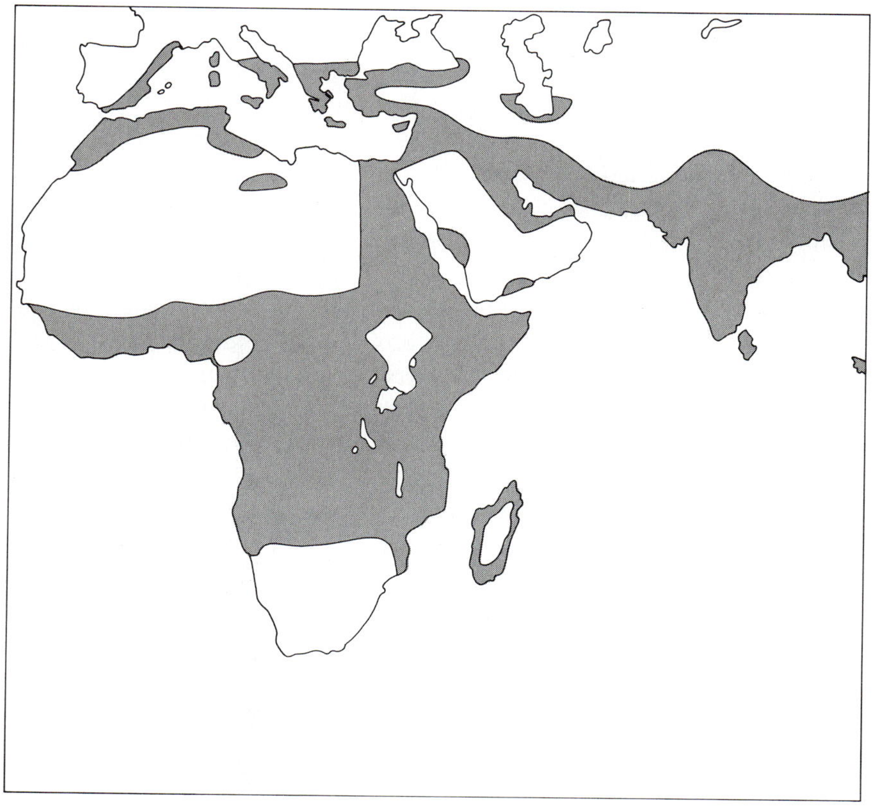

18–15
The distribution of the sickle-cell gene in various parts of the Old World, top, correlates with the distribution of malaria (dark areas), bottom, (From "The Genetics of Human Population," by L. L. Cavalli-Sforza. Copyright © 1974 by Scientific American, Inc. All rights reserved.)

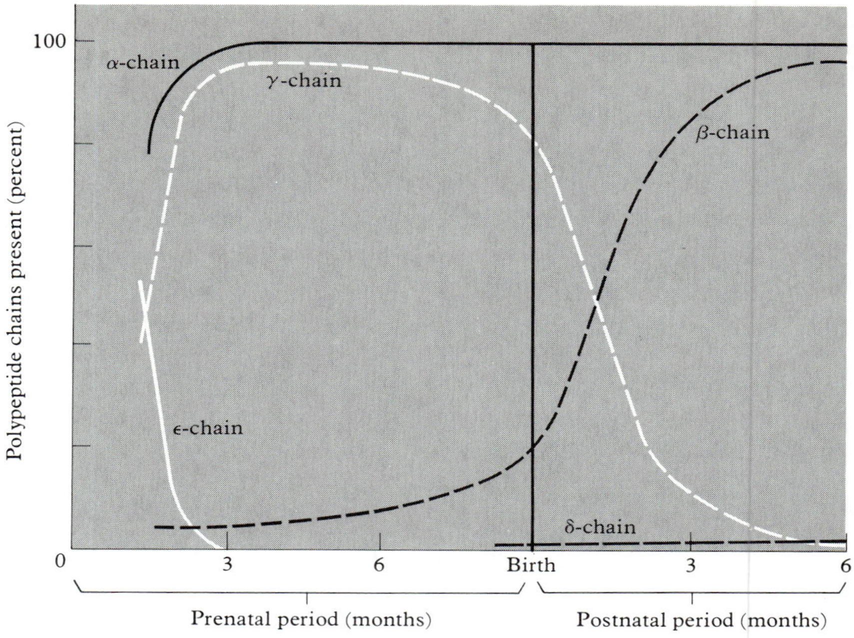

18–16
The percentages of some of the amino acid chains in normal hemoglobin molecules change markedly during development. (After Huehns, et al., "Human Embryonic Hemoglobins," Cold Spring Harbor Symposia on Quantitative Biology, 29, *1964.)*

As shown in Figure 18-16, the number of hemoglobin molecules containing two gamma, rather than two beta chains, changes dramatically shortly after birth. The number of molecules containing gamma chains falls off in proportion to the increase in the number of molecules containing beta chains. By the age of one year, gamma chains are normally found in only about 1 percent of hemoglobin molecules, and this percentage persists throughout adult life. Thus the gene coding for the gamma chain is very active before birth, but becomes almost inactive by the end of the first year of life.

Very early in development some of the hemoglobin molecules of fetuses contain still another chain known as the *epsilon chain.* The gene coding for the epsilon chain is active only during early embryonic life. As shown in Figure 18-16, hemoglobin molecules containing two alpha chains and two epsilon chains (known as *embryonic hemoglobin*) are present only during the first three months of development. Thus by the end of the third month, the gene coding for the epsilon chain is shut off and it never becomes active again.

Finally, shortly before birth the *delta chain* makes its first appearance. Hemoglobin molecules containing two alpha chains and two delta chains are called *Hemoglobin* A_2, and they are never very numerous. In normal adults Hemoglobin A_2 accounts for only about 2 percent of the total hemoglobin (see Figure 18-16). The components of normal human hemoglobins are summarized in Table 18-1.

The genes that code for all of these amino acid chains are examples of *structural genes,* so called because they bring about their effects by coding for the manufacture of specific protein molecules. But not all genes are structural genes. From the study of the genetics of bacteria we know that some genes, rather than coding for large protein molecules, have a controlling or *regulatory* role in protein synthesis. That this is probably also true of some human genes is suggested by certain genetic abnormalities of the kinds of hemoglobin we have just discussed.

In particular, some adults have been found to have abnormally high concentrations of fetal hemoglobin (two gamma chains). It has been suggested that a switch gene, or *regulatory gene,* may cause the rapid changeover from the synthesis of gamma chains to the synthesis of beta chains shortly after birth. Thus those persons who have abnormally high concentrations of fetal hemoglobin can be thought of as having a defective regulatory gene that failed to drastically slow the production of gamma chains.

Adults who have *only* fetal hemoglobin inside their red cells have been reported, and rather surprisingly this condition results in no apparent ill effects. This trait is autosomal recessive, and persons who are homozygous for it not only have no Hemoglobin A (two beta chains), but also lack the small amounts of Hemoglobin A_2 (two delta chains) that are normally present. It is known that the genes that code for the beta and delta chains are closely linked, which is to say that they are found close to one another on a strand of DNA. It has been suggested that persons whose red cells contain only fetal hemoglobin have a defect in the regulatory gene that normally controls the function of the two closely linked structural genes for the beta and gamma chains. One reason for favoring this explanation is that regulatory genes commonly control the function of two or more sequential structural genes in bacteria (see Figure 18-17). Overall, it seems that human regulatory genes do exist, but we have much to learn concerning how they bring about their effects.

From poorly understood regulatory genes we now turn our attention to some rather well-characterized structural genes that code for enzymes. As we discussed in Chapter 5, enzymes are protein molecules that speed the rate at which chemical reactions take place inside living cells, and thousands of them have been isolated from human cells. Human abnormalities resulting from defects in structural genes that code for enzymes are called *inborn errors of metabolism,* and they were first described in detail by the British physician Archibald Garrod in 1908 (Figure 18-18).

TABLE 18-1
Normal human hemoglobins.

CHAIN	COMBINES WITH TWO ALPHA CHAINS AND FOUR HEME GROUPS TO FORM	WHEN PRESENT	PERCENT OF TOTAL ADULT HEMOGLOBIN
Epsilon	Embryonic hemoglobin	Second and third month of fetal life	0
Gamma	Fetal hemoglobin	High fetal levels, drops to adult levels shortly after birth	1%
Beta	Hemoglobin A	From the second fetal month throughout life	97%
Delta	Hemoglobin A_2	From just before birth throughout life	2%

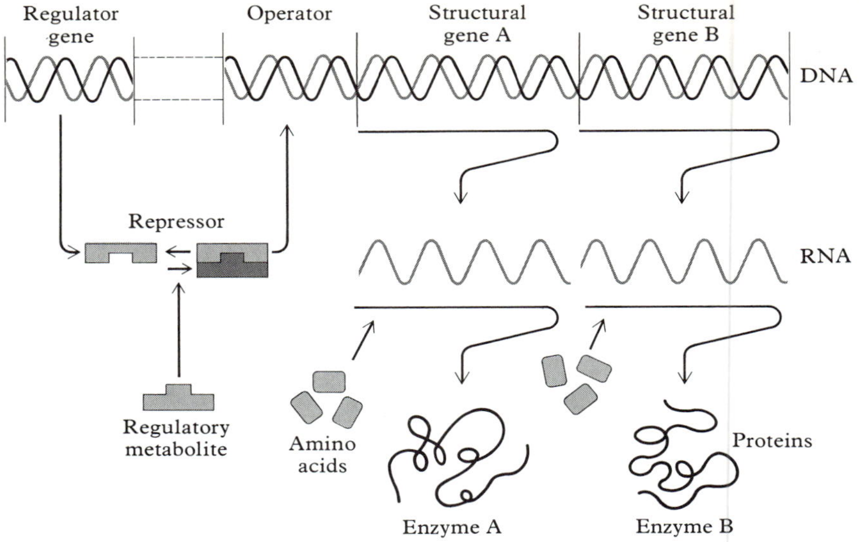

18–17
A model, based on bacterial genetics, of how regulatory genes control the synthesis of enzymes. The regulator gene directs the synthesis of a repressor substance that combines with a regulatory metabolite. This combination then influences the operator, which in turn shuts off the structural genes coding for the enzymes A and B.

Inborn Errors of Metabolism

18–18
Sir Archibald Garrod coined the phrase "inborn errors of metabolism" early in the twentieth century. (Courtesy of the Royal Society.)

One of the first abnormalities studied by Garrod was *alcaptonuria.* Persons who have alcaptonuria appear normal during childhood, but as adults they develop a blue-black discoloration of the ears, the whites of the eyes, the tip of the nose, and other areas of the body where cartilage lies just beneath the skin. Also, on exposure to several hours of sunlight the urine of those who have alcaptonuria turns jet black. Both the black urine and the discoloration of cartilage result from the buildup of a substance called *homogentisic acid.* This compound is also deposited in the cartilage of large joints, and those who have alcaptonuria may therefore develop severe arthritis.

In 1901 Garrod described 11 cases of this rare disorder and noted that at least three of the persons he studied were the offspring of parents who were rather closely related to one another, that is, three were the offspring of consanguineous matings. Garrod used this observation, along with his understanding of Mendel's ratios (which had been rediscovered only the year before), as the basis for a bold and insightful explanation of the nature of the defect that causes alcaptonuria. Garrod suggested that the condition was due to the effects of a single recessive gene that results in the manufacture of a defective enzyme. The disease was thus what Garrod called an "inborn error of metabolism."

Garrod's explanation proved to be correct. The exact nature of the biochemical defect in alcaptonuria is now known. As just mentioned, persons who have alcaptonuria have abnormally high concentrations of homogentisic acid in their urine and cartilage. The excess homogentisic acid results because they have very low, or nonexistent, concentrations of an enzyme that is usually responsible for the further metabolism of homogentisic acid.

Homogentisic acid is one of the metabolites of the amino acid *tyrosine.* In normal individuals the following metabolic pathway operates:

$$\text{Tyrosine} \xrightarrow{\text{enzyme}} \text{homogentisic acid} \xrightarrow{\text{enzymes}} \text{further metabolic products}$$

But in people who have alcaptonuria the lack of a specific enzyme results in a metabolic block to the further processing of homogentisic acid, which therefore accumulates in their tissues, where it may produce its undesirable effects. We can represent this block as follows:

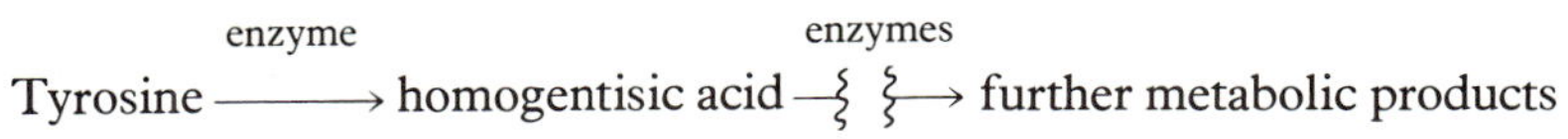
enzyme
enzymes
Tyrosine
homogentisic acid
further metabolic products

Garrod later showed that *albinism,* a condition discussed in Chapter 16, is also the result of a metabolic block that affects the amino acid tyrosine. Tyrosine is not only the precursor of homogentisic acid, but also the basic building block of the pigment melanin.

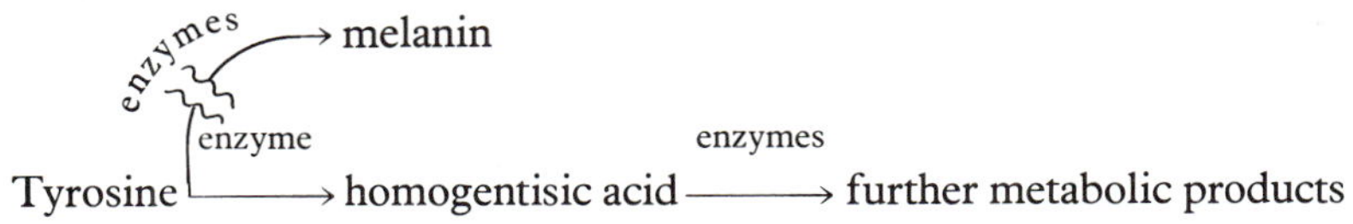
enzymes
melanin
enzyme
enzymes
Tyrosine
homogentisic acid
further metabolic products

The most common kind of albinism results from the lack of an enzyme that participates in the manufacture of melanin from tyrosine. (But albinism can also result because of a different metabolic defect that greatly reduces the amount of tyrosine available to form melanin. In this second kind of albinism affected persons have a normal concentration of the enzyme that facilitates the conversion of tyrosine to melanin.)

Since Garrod's time we have become aware of yet another inborn error of metabolism in the metabolic pathway we have been discussing. This defect is known as *phenylketonuria* (PKU). PKU is a serious disease in that if it is untreated it can result in severe mental retardation. PKU results because of a defect in the enzyme that converts the amino acid *phenylalanine* into tyrosine.

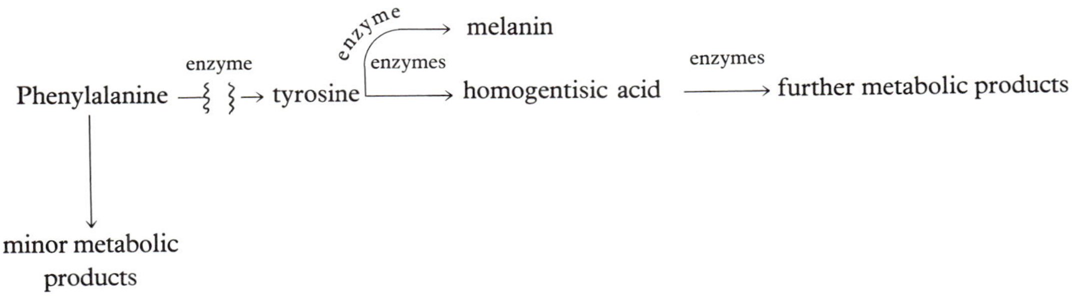

The lack of the appropriate enzyme for converting phenylalanine into tyrosine has two major effects. First, the concentration of phenylalanine in the tissues is greatly elevated. And second, minor metabolic products of phenylalanine that are normally present in only small quantities also accumulate. (It is probably the buildup of these minor metabolic products during the first few months and years of life that accounts for the mental retardation.) Because they cannot convert phenylalanine into tyrosine, persons affected by PKU are lightly pigmented. Nonetheless, enough tyrosine is directly available in the diet that they can still manufacture considerable quantities of melanin (and of homogentisic acid).

PKU can be effectively treated by greatly reducing the amount of phenylalanine present in the diet of an affected infant. Luckily for those who suffer from PKU, phenylalanine is one of the eight essential amino acids that the human body cannot manufacture for itself—all of the body's phenylalanine comes directly from the diet. By restricting dietary intake the buildup of high concentrations of phenylalanine and its minor metabolic products, and thus the mental retardation associated with the disease, can be prevented. At the same time, enough phenylalanine must be provided to allow for normal growth. Affected children are usually kept on their special diet until they are at least five years old, at which time the growth of the human brain is more or less completed.

Like most other inborn errors of metabolism, PKU is inherited as a recessive trait and is manifested only by homozygotes. But it is possible to identify carriers of the abnormal gene by subjecting them to the phenylalanine tolerance test. In this test a large dose of phenylalanine is administered orally and the rate at which it disappears from the bloodstream is measured. As shown in Figure 18-19, when heterozygous carriers of PKU are fed a standard dose of phenylalanine they show higher and more prolonged elevations of phenylalanine in their blood than do normal people. This is because heterozygous carriers of PKU, although they appear normal, have only about half of the normal concentration of the enzyme that aids in converting phenylalanine to tyrosine.

In recent years the number of human diseases known to be due to inborn errors of metabolism has skyrocketed. Hundreds of diseases that result from abnormally low concentrations of critical enzymes have already been described, and the list continues to grow. But the ability to *treat* diseases caused by inborn errors of metabolism has not generally kept pace with the ability both to diagnose them and to identify normal-looking carriers by means of tolerance tests. It is sometimes possible to prevent the development of undesired effects by limiting dietary intake, as with phenylalanine and PKU. But more often than not we can do little or nothing to affect the course of diseases that result from inborn errors of metabolism. We are a very long way indeed from being able to correct the underlying defect in the genetic code of DNA that is ultimately responsible for the production of a faulty enzyme, or of no enzyme at all. Also, it is not now possible to treat these diseases simply by replacing the missing or defective enzymes. This is because enzymes generally do their work inside cells, and even if we are able to prepare concentrated extracts of the missing or defective enzyme, we have no way of getting them inside the cells that need them, where they speed up the rates of crucial biochemical reactions.

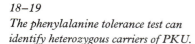

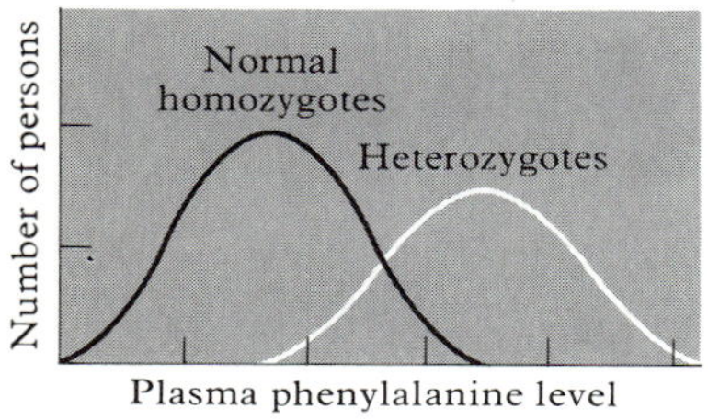

18–19
The phenylalanine tolerance test can identify heterozygous carriers of PKU.

At the present time, and probably for some time to come, the best way to deal with diseases due to inborn errors of metabolism is to prevent them. There are two main ways of doing this. First, genetic counseling, sometimes in combination with tolerance tests, can help identify apparently normal carriers among the relatives of an affected person, or among other people who might be carriers. Second, it is now possible to detect many inborn errors of metabolism in cells obtained from human embryos during the first few months of development. Thus, parents might choose to voluntarily terminate a pregnancy if the fetus were proven to be affected by a serious inborn error of metabolism. (We will have more to say about how inborn errors of metabolism can be detected by prenatal diagnosis in Chapter 20.)

Summary

By the early 1950s experiments with bacteria and viruses had made it clear that the genetic material is not protein, but DNA. The Watson-Crick model of DNA structure was proposed in 1953. The DNA molecule consists of four kinds of nucleotides joined to one another in a double-stranded helix in such a way that the sequence of nucleotides in one strand automatically determines the sequence of nucleotides in the other strand. DNA replication is accomplished by the separation of the two strands; each strand then serves as a template for the manufacture of a new, complementary strand.

Biochemically, genes are specific regions along DNA molecules. Most genes have a role in protein synthesis. Proteins are made up of long chains of amino acids, and three successive nucleotides in the DNA molecule code for one amino acid in a particular protein. Protein synthesis requires a second nucleic acid, RNA, and the actual stringing together of amino acids into proteins takes place outside the nucleus on structures called ribosomes.

We have much to learn about how DNA is packaged inside eukaryotic chromosomes. We do know that some DNA segments are present in multiple copies and that DNA molecules are tightly coiled within the nucleus. Nuclear proteins provide a scaffold for DNA molecules and determine which genes are expressed when.

Persons affected by sickle-cell disease are homozygous for a gene that results in a single difference in the amino acid in the beta chain of their hemoglobin molecules compared with normal. But hemoglobin of persons who have sickle-cell trait (heterozygotes) is more resistant to malaria than is normal hemoglobin. And this probably accounts for the widespread distribution of the allele for the abnormal beta chain among people who live in areas where malaria is common—in spite of the reduced reproductive fitness of homozygotes.

Several different kinds of fetal hemoglobins have been described, and there is evidence that regulatory genes may bring about the rapid changeover in the composition of hemoglobin molecules that occurs during normal development. Regulatory genes commonly control two or more sequential structural genes in bacteria.

Garrod coined the phrase "inborn errors of metabolism" for human abnormalities resulting from defects in genes that code for enzymes. Alcaptonuria, albinism, and phenylketonuria are caused by defects in enzymes that speed up rates of metabolism of the amino acids phenylalanine and tyrosine. The best way to deal with inborn errors of metabolism is to prevent them. This can be accomplished by identifying carriers by means of pedigree analysis or tolerance tests, and identifying affected fetuses at early stages of development by means of prenatal diagnosis.

Suggested Readings

1. *Molecular Biology of the Gene,* 3d ed. by J. D. Watson. W. A. Benjamin, 1976. An authoritative overview of molecular genetics written to be understandable even to those who have little knowledge of chemistry.

2. "The Genetic Code: III," by F. H. C. Crick. *Scientific American,* Oct. 1966, Offprint 1052. A summary of how the sequence of base pairs in DNA provides information for the synthesis of specific protein molecules.

3. "The Visualization of Genes in Action," by O. L. Miller, Jr. *Scientific American,* March 1973, Offprint 1267. With the aid of the electron microscope one can "see" genes being transcribed into RNA and "watch" RNA being translated into protein.

4. "Chromosomal Proteins and Gene Regulation," by Gary S. Stein, Janet Swinehart Stein, and Lewis J. Kleinsmith. *Scientific American,* Feb. 1975, Offprint 1315. A discussion of the genetic role of histones and other proteins associated with chromosomal DNA.

5. "Repeated Segments of DNA," by Roy J. Britten and David E. Kohne. *Scientific American,* April 1970, Offprint 1173. The function of the repeated segments of DNA regularly found in eukaryotic cells has yet to be discovered.

This portrait of people of various human races is based on original photographs taken by Professor Carleton Coon, who kindly granted permission to have this drawing rendered. The groups represented are (right to left): front row, Armenian, Formosan, Bavarian, Veddoid; middle row, Dinaric, Singhalese, Arab; back row, Negrito, Korean, Swedish, Moroccan. See also pp. 486–489.

CHAPTER 19

Genes in the Human Population

A very distinguished group of British scientists, including Sir Ronald Fisher and Sir Julian Huxley, once visited the London Zoo to carry out a peculiar experiment. Their purpose was to find out whether the zoo's chimpanzees were capable of tasting a chemical known as phenylthiocarbamide (PTC). Among human beings, the ability to taste this unpleasantly bitter substance was known to be transmitted as an autosomal dominant trait. One of the scientist's main concerns was that they might not be able to tell whether individual chimpanzees could taste the chemical. But when Fisher gave a sip of a weak solution of PTC to the first chimpanzee to be tested, the creature's reaction left little room for doubt. The chimp was so disgusted by the taste that it spit in Fisher's face!

What prompted these talented scientists to perform such a seemingly foolish experiment? Some years earlier Fisher (Figure 19-1), most of whose formal education was in mathematics and statistics, had worked out the mathematical equations for describing how natural selection can influence the genetic composition of populations in which individuals mate with one another at random. In his book *The Genetical Theory of Natural Selection,* first published in 1930, Fisher showed that the diversity of genetically determined traits can be directly related to fitness for survival. In other words, Fisher's equations indicated that the genetic diversity observed in most populations exists in large part because a high degree of variability allows more possibilities for adapting to the environment and therefore for reproducing successfully. But at that time little was known of how extensive the variety of genetically determined traits is. By offering PTC to chimps at the London Zoo, Fisher and his distinguished colleagues discovered that chimpanzees, like their human relatives, vary in their abilities to taste the chemical PTC. They thus contributed (albeit little) to our knowledge of the range of variability among different individuals of the same species.

In recent years it has been discovered that individual animals of the same species are much more variable than Fisher or anyone else realized only a few decades ago. For the most part, our heightened awareness of how unique individual animals are has been the result of the study of genetically determined differences in protein molecules.

19–1
Sir Ronald Fisher was one of the first persons to formulate equations that describe how natural selection can influence a population's genetic composition. (Courtesy of Godfrey Argent.)

Genetic Variability in Protein Molecules

As discussed in the preceding chapter, protein molecules consist of long chains of amino acids linked in tandem by chemical bonds. There are about twenty kinds of amino acids, and the properties of a particular protein depend above all on the sequence of amino acids in its chain. What determines the sequence of amino acids in a protein is the sequence of the four bases along the particular stretch of the DNA molecule that codes for the protein in question.

A relatively easy and sensitive way to determine whether protein molecules differ from one another is to observe their patterns of movement in an electric field. Most protein molecules, or regions of protein molecules, have either a positive or a negative charge. The distribution of the charge depends on which amino acids are present in the protein and on how they are arranged. Proteins that differ from one another in their amino acid sequences usually have different electrical properties and therefore show different patterns of movement in an electric field. In general, these patterns reflect biochemical differences that are genetically determined. As you may recall, a difference of only one amino acid out of a total of 287 in the protein portion of the hemoglobin molecule can be easily detected by this method. (See Chapter 18.)

The variability of protein molecules, especially enzymes, has been the subject of intensive research in recent years. This systematic research has revealed a rather remarkable fact about the protein molecules of human beings and most other animals studied so far. At the biochemical level organisms of the same species are astoundingly diverse. This is because within a particular species, a given protein molecule is by no means identical in each individual.

Slightly different forms of almost all known human proteins have already been reported. Populations in which several alternate forms of a gene are regularly encountered are said to be *polymorphic* (many-formed) for that gene. As we have mentioned, the presence of some protein polymorphisms in the human population can be readily explained by natural selection. Thus the polymorphism of the beta chain of the hemoglobin molecule that results in sickle-cell disease in homozygotes is maintained in some geographic areas by natural selection because those who are heterozygous for normal and sickle-cell hemoglobin are more resistant to malaria than they would be if they had only normal hemoglobin inside their red cells.

But generally there is little or no evidence that a person is at an advantage or a disadvantage in having a slightly different form of a particular protein molecule. When the affected proteins are enzymes, it is usually possible to detect subtle differences in the activity of each form in the laboratory, but slight differences in protein molecules usually do not produce detectable ill or advantageous effects. Although it is possible to relate the persistence of some genetic polymorphisms in the human population to the effects of natural selection, this is generally *not* true at the biochemical level. Most often we have little or no evidence of how, or whether, natural selection maintains the extraordinary biochemical diversity of the human species. It may be true that many polymorphisms, especially those that do not produce obvious ill effects, are not affected by natural selection. For the present, we simply do not know.

In the rest of this chapter our main concern is with identifying some well-known polymorphisms in the human population and with explaining how the frequencies and distributions of these traits in the present-day population may have come about. We shall discuss the concept of *race* and attempt to explain how the frequencies of various genes observed in different segments of the human population can be affected, not only by natural selection, but also by isolation, chance, and other factors. Ultimately, all polymorphisms are the result of heritable changes in the sequence of the four bases in a particular region of DNA. Such changes in the DNA molecule are known as *mutations*, and in the pages that follow we will discuss the role of mutations in maintaining human diversity. But first, we turn our attention to a well-documented (though incompletely understood) genetic polymorphism about which reliable data are available from around the world. Perhaps the most extensively studied of any human polymorphism is the ABO blood group.

The Worldwide Frequencies and Distribution of the ABO Blood Group

The genes that determine to which ABO blood group an individual belongs come in three alternative forms, or alleles, that can be symbolized I^A, I^B, and I^O Each human being has one of the three alleles at the same location on each member of a particular pair of autosomes. The patterns of inheritance shown by these three alleles and how they determine to which ABO blood group a person belongs are discussed in Chapter 16. The worldwide frequencies of these three alleles in the human population are known with considerable accuracy, and, as determined by medical and anthropological surveys, the worldwide frequencies are: I^O, 62 percent; I^A, 22 percent; and I^B, 16 percent.

But these alleles are not distributed equally throughout the world. Figure 19-2 shows the distribution of the allele I^O in aboriginal populations around the world. Notice that North and South American Indians have very high frequencies of the allele I^O. In fact, in some areas of these two continents, the aboriginal peoples have the allele I^O almost to the exclusion of the other two. On the other hand, the distribution of the allele I^B is almost opposite to that of I^O, as shown in Figure 19-3. I^B is very common in central Asia, but rare among native North and South Americans.

How can we account for these definite patterns in the worldwide distribution of the alleles that determine to which ABO blood group a person belongs? Perhaps the most likely explanation is that the distribution of ABO blood groups is one effect of human migrations, most of which took place in prehistoric times. As discussed in Chapter 2, human beings as we know them today probably

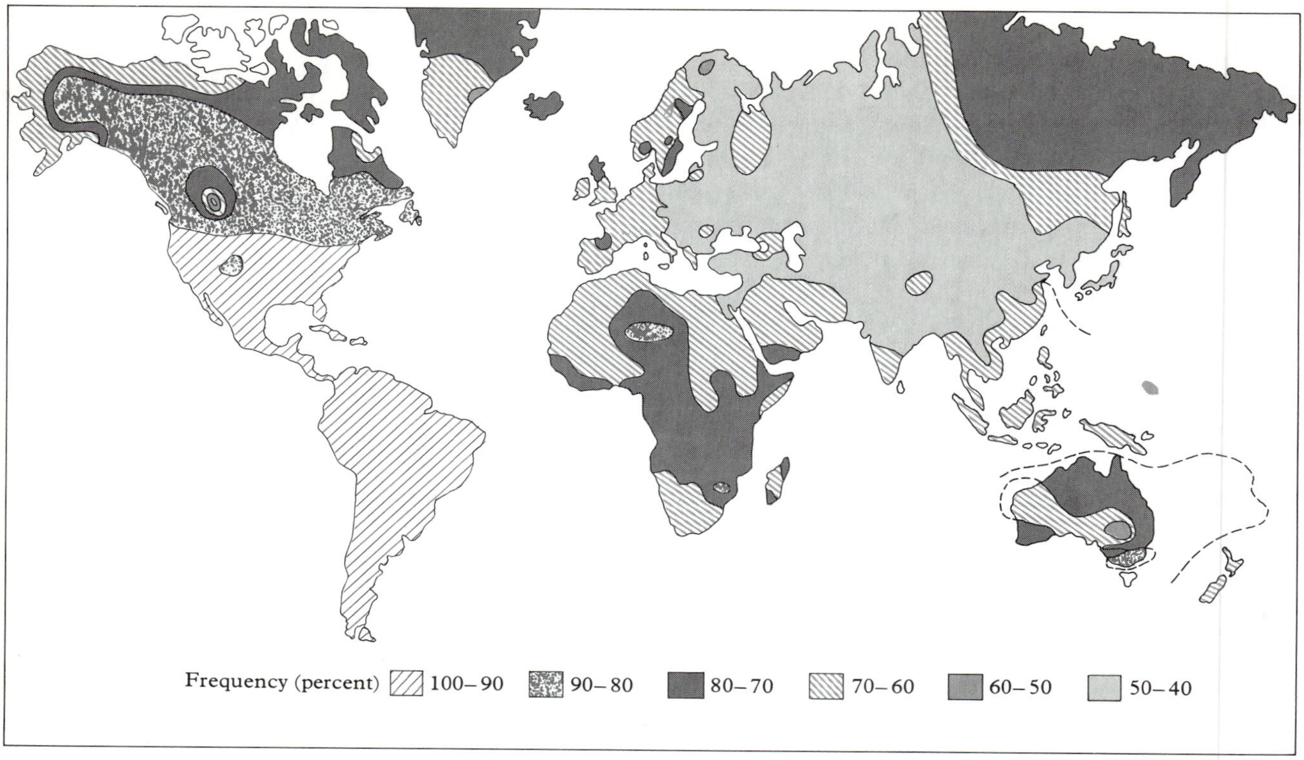

19–2
The distribution of the allele I^{O} *in aboriginal populations around the world. (After Mourant et al.,* The ABO Blood Groups. *Copyright © 1958 by Blackwell Scientific Publications.)*

evolved first in Africa or Asia. Initially human populations must have been rather small, but with their characteristic resourcefulness people soon invented food raising and then increased their numbers and extended their range, not only to the entire African and Asian continents, but to the rest of the Old World as well. Eventually, people made their ways to the Americas, the islands of the Pacific, and Australia. These migrations could easily result in the gradual emergence of the present-day distribution of the ABO blood groups.

One example of how the distribution of ABO blood groups can be influenced by human migrations is shown in Figure 19-4, which is a close-up map of the distribution of the allele I^{B} in Eurasia. In general, the frequency of I^{B} shows a steady decrease from Central Asia toward Western Europe. This pattern has been attributed to the effects of Mongol invasions of Europe, which continued for about 1,000 years and ended about 500 years ago. The Mongols from the East may have had proportionately higher concentrations of I^{B} than their Western counterparts, and as the former moved westward they undoubtedly spread not only their culture but their genes as well. The distribution of Group B in Western Europe could be explained by assuming that before the Mongol invasions most West Europeans did not have the allele I^{B}. A comparison of Figures 19-2 and 19-4 suggests that early West Europeans probably had a preponderance of the allele I^{O} instead, as evidenced by the higher frequency of I^{O} in extreme Western Europe today. (The very low concentrations of I^{B} that are found today in the area of the Pyrenees and Caucasus mountains may have resulted because some of the original Europeans fled to the mountains to evade the Mongol hordes.)

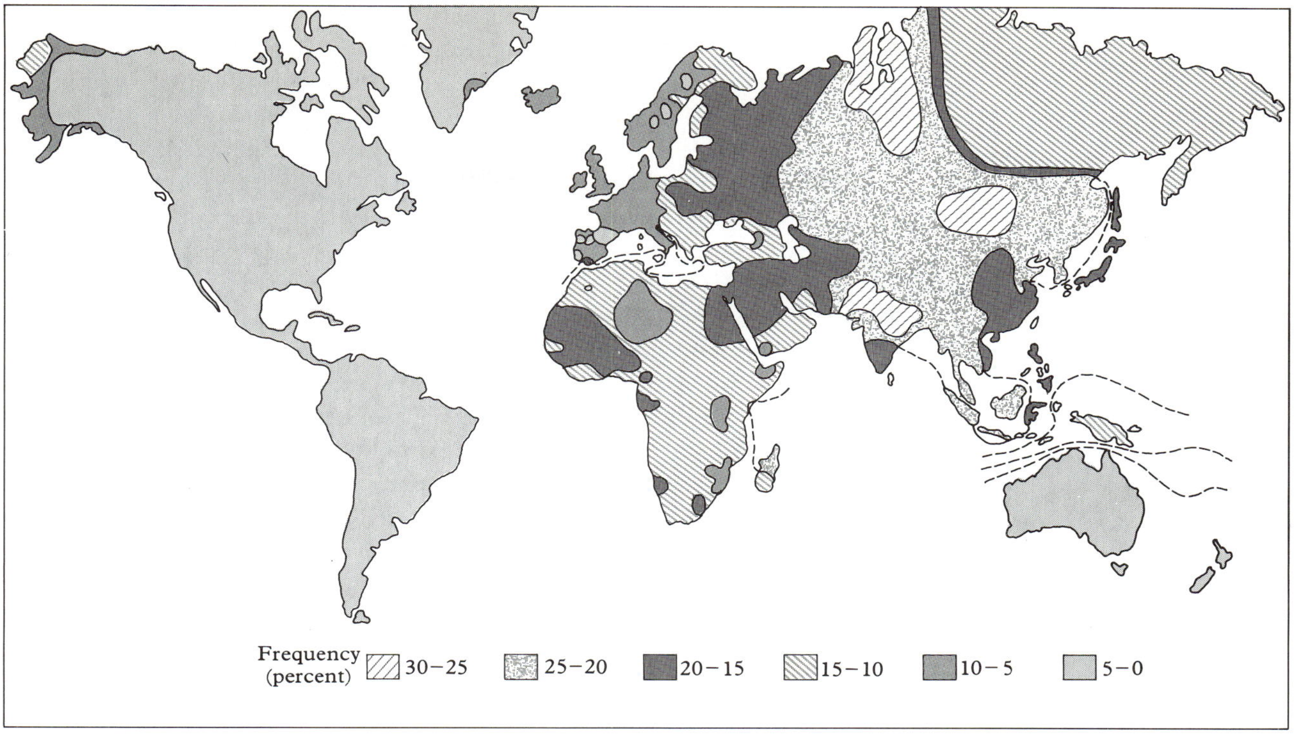

19–3
The distribution of the ellele I^B *in aboriginal populations around the world. (After Mourant et al.,* The ABO Blood Groups. *Copyright © 1958 by Blackwell Scientific Publications.)*

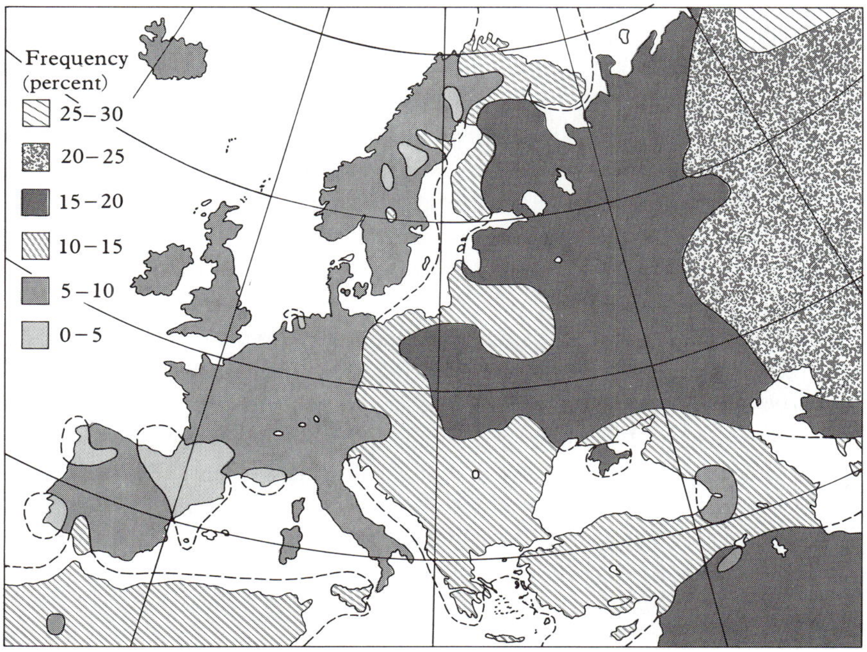

19–4
The distribution of the allele I^B *in Eurasia. Note the relatively low frequencies of* I^B *near the Pyrenees and Caucasus mountains. (After Mourant et al.,* The ABO Blood Groups. *Copyright © 1958 by Blackwell Scientific Publications.)*

Overall, there is little doubt that human migrations have strongly influenced the present distribution of the ABO blood groups. But what about the effects of natural selection? The difficulty in trying to assess its importance in determining the frequencies and distribution of the ABO blood group is that we have no decisive evidence about whether any one of the three alleles is directly influenced by natural selection or not. That is, human beings of one ABO blood group enjoy no obvious reproductive advantage over those of any other ABO group.

But this is not to say that statistical correlations between ABO blood groups and various diseases do not exist. In fact, based on data from Great Britain in the 1950s there is evidence that people of Group O are about 40 percent more likely to develop a duodenal ulcer than members of the other two blood groups. What is not clear, however, is how or whether duodenal ulcers are related to natural selection. If the two are related, then we would expect people who have ulcers to produce either more or fewer offspring than people who do not, but there is no evidence that this happens. It seems unlikely that natural selection has influenced the worldwide distribution and frequencies of the ABO blood groups simply because of the statistical correlation between Group O and ulcers. Besides, the various ABO groups have been statistically correlated, not only with duodenal ulcers, but with many other traits, such as cancer of the stomach and the hard-to-define characteristic of "tender-mindedness" versus "tough-mindedness."

On the other hand, it has been suggested that natural selection may have influenced the present-day distribution and frequencies of the ABO blood groups because people of different groups may be more resistant to certain infectious diseases. For example, people of Groups B and O have natural antibodies against Group A in their bloodstreams, as is discussed more fully in Chapter 13. Antibody against Group A may also be effective against the virus that causes smallpox. Thus, members of Groups B and O may be more resistant to smallpox than those of Group A, and the present-day frequencies and distribution of the ABO groups could reflect devastating epidemics of smallpox or other infectious diseases in the past.

Natural selection could also affect the ABO groups because of antibody-dependent incompatibilities between a mother and her developing fetus. For example, if a woman is Type O and her fetus is Type A, there may be difficulty because the mother has natural antibodies against the red cells of the fetus. But data about the actual outcome of ABO-incompatible pregnancies—although abundant—are inconclusive and some of them are contradictory. For now, we don't know whether natural selection affects the frequencies of ABO blood groups because of antibody-dependent incompatibilities between mother and fetus.

To sum up, natural selection has undoubtedly played a role in determining the frequencies and distribution of the ABO blood groups, but it has probably not been the most important factor. Rather, the present-day frequencies and distribution of the ABO blood groups in the human population can perhaps best be explained by prehistoric migrations and the effects of chance. That chance can play a decisive role in determining the genetic composition of certain human populations is well known. One example of how this occurs is provided by the Dunkers, a group of people who migrated from Germany to Pennsylvania in the early eighteenth century.

In populations made up of large numbers of people who mate with one another at random, the frequencies of those genes not obviously influenced by natural selection tend to remain about the same from one generation to the next. This is because the frequencies of such genes are determined mainly by chance, and in large populations chance fluctuations in gene frequencies tend to balance one another. But in relatively small populations this does not necessarily happen, and significant changes in the genetic composition of the population can occur by chance alone. This chance variation in gene frequencies from one generation to another is known as *genetic drift,* and in general the smaller the population the greater the genetic drift can be.

An ideal population in which to examine the effects of genetic drift is the devout Protestant religious sect known as Dunkers (Figure 19-5). Between 1719 and 1729 fifty families of Dunkers emigrated from the German Rhineland to Pennsylvania and thereby completely transplanted the sect to the New World.

19–5
The Old Order Dunkers of Franklin County, Pennsylvania. Although they seldom marry outside the sect and their attire differs from that of their neighbors, their customs are not otherwise unusual. (From "The Genetics of the Dunkers," by H. Bentley Glass. Copyright © 1953 by Scientific American, Inc. All rights reserved.)

To marry outside the church is considered a grave offense, and a Dunker who does so must either withdraw voluntarily from the community or be expelled from it.

During their first hundred years in Pennsylvania the Dunkers doubled in number, and almost all of them could trace their ancestry to the original fifty families. Then in 1882 the Dunker church underwent schism and a progressive group separated from the old order. At that time the Old Order Dunkers numbered about 3,000, and this number has not changed much to the present day. One of the original Dunker communities, in Franklin County, Pennsylvania, remained with the old order, and in the years since 1882 its size has also changed very little. When the Franklin County group of Dunkers was studied in the early 1950s, the population was about 300 and its size had been nearly the same for several generations. The Franklin County Dunkers were thus an almost ideal population in which to look for the effects of genetic drift.

The effects of genetic drift on the Franklin County Dunkers can be revealed by a comparison of the frequencies of various genes among the Dunkers, their West German forebears, and their present-day American neighbors. For example, these are the frequencies of the ABO blood group alleles I^O, I^A, and I^B among the Dunkers, West Germans, and Americans:

	I^O	I^A	I^B
Dunkers	60%	38%	2%
West Germans	64%	29%	7%
Americans	70%	26%	4%

Notice that the frequencies of these alleles among the Dunkers are not the same as those of West Germans or Americans; nor do they lie between the two. This is to be expected if genetic drift is at work and the frequencies of the alleles are determined largely by chance.

Blood groups are not the only characteristic of the Dunkers that shows the effects of genetic drift. There are clear-cut differences between the Dunkers and surrounding American communities in the frequencies of several other apparently nonadaptive traits. Thus, as compared with their neighbors, fewer Dunkers have hair on the middle segment of one or more fingers, fewer are able to bend the end of the thumb backwards to form an angle of more than 50°, and fewer have earlobes attached to the side of their heads rather than hanging free (Figure 19-6). The best explanation for these observations is that the frequencies among the Dunkers are the result of genetic drift.

The Dunkers are not the only human population in which genetic drift has been detected. Marked variations in the frequencies of certain alleles from those of neighboring populations have also been reported in a number of populations, including aboriginal Australians, Eskimos, Italians in isolated villages, North American Indians, and religious sects in Montana. All of these groups have in common two important features that enable genetic drift to strongly influence their genetic constitutions. First, the populations are relatively small, and second, the groups are isolated from neighboring populations either by physical barriers or by cultural rules.

Although the effects of genetic drift are most pronounced in the smallest populations, genetic drift can also influence the genetic composition of larger populations. It has been estimated that genetic drift can strongly influence the

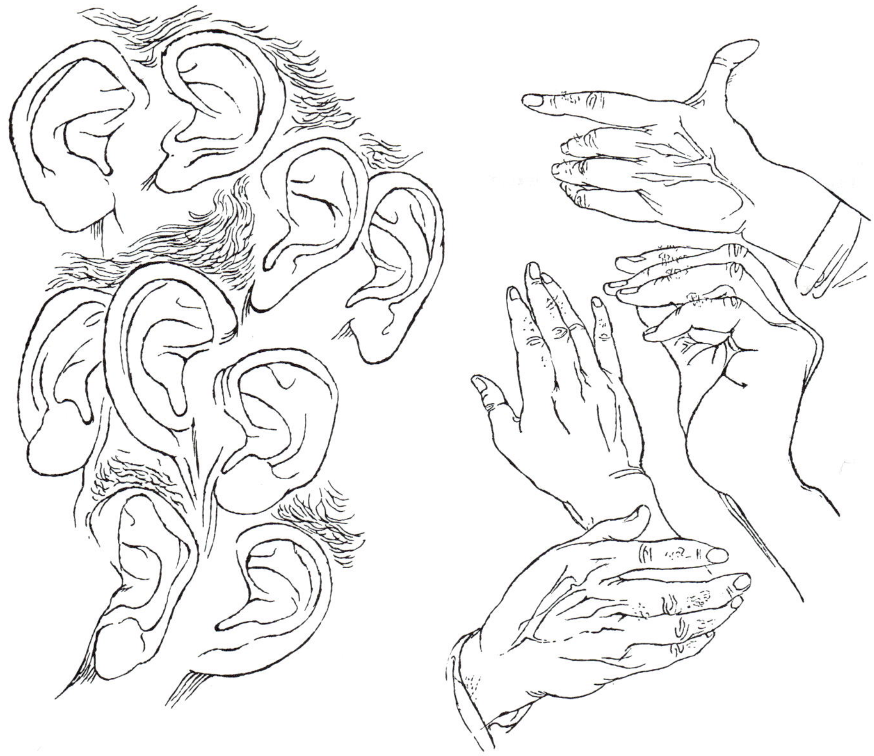

19–6
Some characteristics that show the effects of genetic drift in the Dunkers. Earlobes can be attached or they can hang free; those of most Dunkers hang free. Dunkers have a high incidence of hitchhiker's thumb (the ability to bend the thumb backwards at an angle of more than 50 degrees); and, as compared with the general population, fewer Dunkers have hair on the middle segment of one or more fingers. (From "The Genetics of the Dunkers," by H. Bentley Glass. Copyright © 1953 by Scientific American, Inc. All rights reserved.)

frequencies of apparently nonadaptive traits in human populations if the parents in any generation number a few hundred individuals or fewer. This is important because before the invention of food raising 10,000 years ago most human populations were probably within this size range and were therefore small enough to be strongly influenced by genetic drift. In fact, many inherited differences between human beings belonging to different races may have become established thousands of years ago when people lived in small groups that were physically and reproductively isolated from one another. Let us discuss the concept of *biological race* and its relation to genetic drift and to natural selection.

The Biology of Human Races

As discussed in previous chapters, the species to which all living people belong, known as *Homo sapiens,* originated in either Africa or Asia and then spread outward. From the start our species has been endowed with dextrous, toolmaking hands that are controlled by the most complicated organ that evolution has produced so far—the human brain. The combination of human hand and human brain can be thought of as an adaptation that has allowed our species to virtually cover the surface of the land. No other species is as widely distributed as the human species, with the possible exception of species such as houseflies, body lice, and mice, which directly benefit from human activities and which have therefore followed people in their migrations over the continents.

When widespread species are examined over their full geographic range, it is often found that populations in different places look slightly different. For example, song sparrows from New York and Oregon can easily be told apart, as can zebras from different parts of Africa (Figure 19-7). As we all know, the human species is no exception when it comes to geographic variation. Thus, people from Tokyo, Copenhagen, Bombay, and Nairobi can be told apart as easily as the song sparrow and zebras from different locations. What accounts for the differences in appearance between populations of the same species?

These variations are the result of the interaction of the genes of a population with the environment. The environment determines which genes from among the total range of genes in the species' DNA will be expressed and to what degree they will be expressed. Nonetheless, most populations of widespread species look slightly different from one another because of genetic differences between the groups. These genetic differences arise because populations that are separated from one another by long distances or other barriers cannot interbreed. Whenever local populations of a given species are isolated from one another by a barrier, genetic differences accumulate because of the effects of natural selection, mutation, and, if the populations are small enough, genetic drift. Distinct local breeding groups may thus arise in different areas of a widespread species' range. These distinct local breeding groups within a particular species are known as *races.*

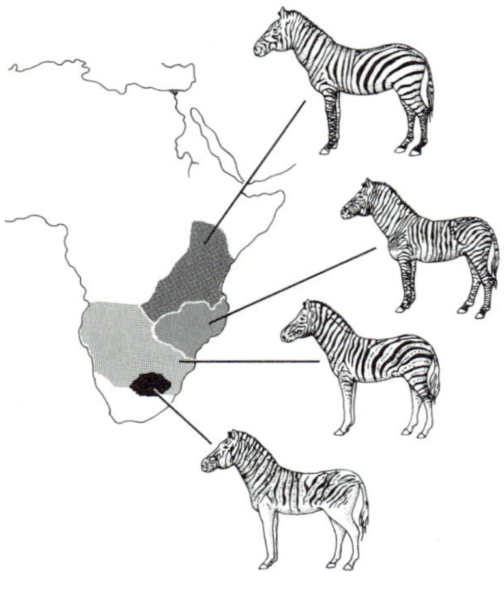

19–7
Variation of stripes of African zebras in different geographic regions. (After Cabrera, Journal of Mammology 17, *1936.)*

There is no doubt that human races originated in the same way as other animal races. About 40,000 years ago most human beings probably lived in small tribes that were relatively isolated from one another by distance and custom. This isolation resulted in chance differences in genetic constitution from one group to another. During this time of relative isolation among early human groups some of the genetic traits that characterize modern races probably became established by genetic drift.

But chance was not the only factor favoring the development of genetic differences between isolated groups of ancient people. Surely natural selection must also have played a role in the evolution of human races. As our ancestors increased in number and extended their range, different groups found themselves in very different environments. You will recall that because of natural selection organisms tend to acquire traits that allow them to closely adapt to the local environment. Thus it seems likely that at least some racial characteristics became established because they were advantageous under certain environmental conditions.

Perhaps the most apparent human racial characteristic is the color of a person's skin. At least 36 shades of human skin color, ranging from jet black to almost white, have been described. As shown in Figure 19-8, there is undoubtedly a worldwide connection between skin color and the intensity of sunlight (especially its ultraviolet component). In general, the most darkly pigmented people live closest to the equator and are exposed to the greatest concentration of sunlight. In some parts of the world, especially Africa, skin color shows a steady gradation from darker to lighter the further away from the equator the population lives. In other locations changes in pigment with latitude are not so clear cut, and the original pattern has been blurred and distorted by human mobility and racial mixing in recent years.

It has been suggested that natural selection influences the worldwide distribution of human skin color because the amount of pigmentation is related to the synthesis of Vitamin D. As discussed in Chapter 9, Vitamin D plays a rather unusual role in human nutrition for several reasons. First, Vitamin D does not exist naturally except in the livers of certain kinds of animals. Instead, some

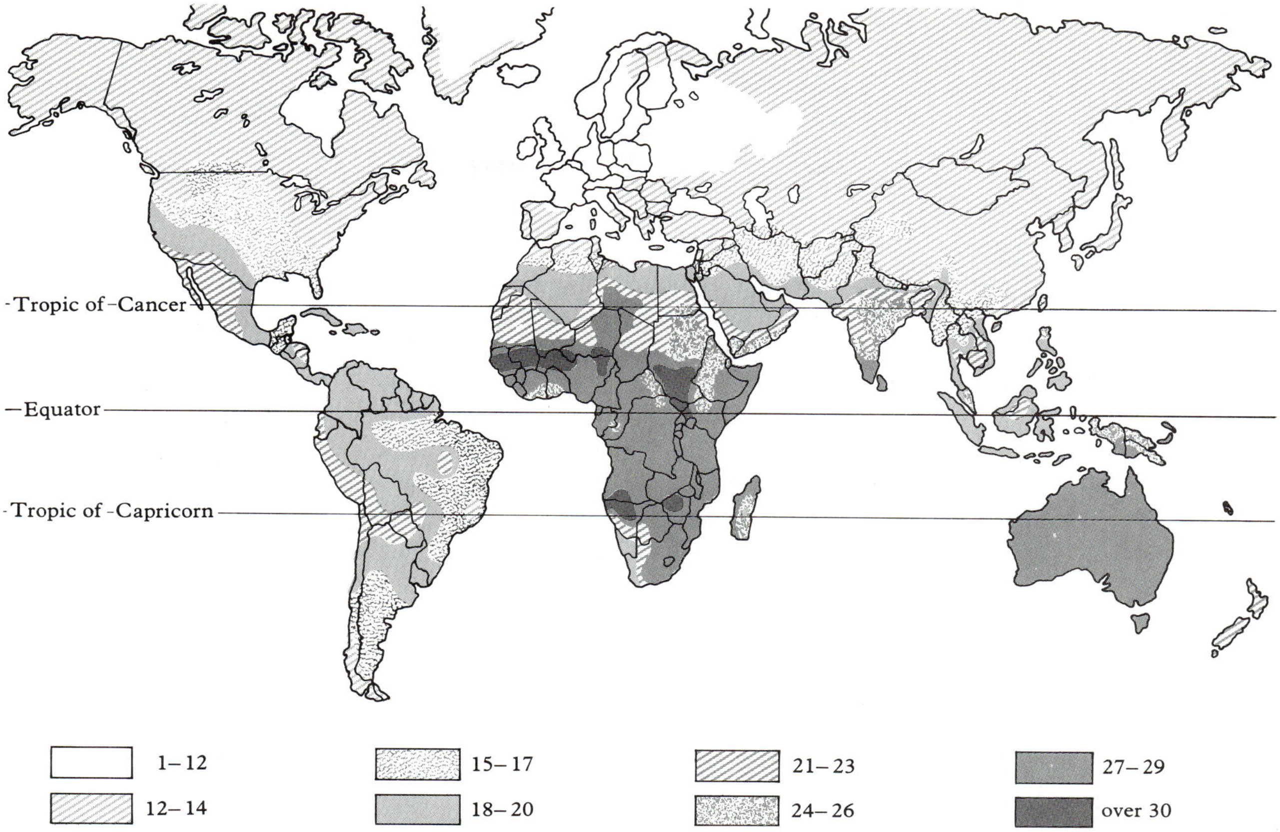

19–8
The distribution of human skin color before Columbus's first voyage to the New World in 1492 A.D. The values increase with darker skin color. (After R. Baisutti, Razze e Popoli della Terra, *Torino: UTET, 1951.)*

foods contain precursors of Vitamin D, and from these precursors Vitamin D is synthesized by the human skin as long as energy from ultraviolet radiation (sunlight) is available. Second, unlike most human vitamins, Vitamin D is toxic in large doses. Vitamin D in normal quantities is essential for proper calcium metabolism, but in excess it can cause kidney stones and can lead to the production of deposits of calcium in areas of the body that are not usually calcified. (For example, calcium deposits may build up in the walls of large arteries or in the cornea of the eye.)

How does the worldwide distribution of human skin color relate to Vitamin D? A possible, though speculative, explanation is as follows. People who live closest to the equator receive the most sunlight and their skins therefore synthesize much more Vitamin D than those of people who live further away from the equator. People near the equator generally have darker skins because the accumulation of pigment in the deeper layers of skin absorbs untraviolet radiation and thus prevents the synthesis of excess Vitamin D. On the other hand, people in northerly latitudes, where ultraviolet radiation is much less intense, tend to have lighter skin so that their bodies can use the available radiation to synthesize sufficient Vitamin D to maintain good health. (Too much Vitamin D does not accumulate in light-skinned people during summertime because their skins become tanned and thus screen out the seasonal excess of ultraviolet radiation. In fact, the tanning reaction is initiated by light of the same wavelength needed for Vitamin D synthesis.)

Because our species almost certainly evolved first in the tropics, it is likely that all human beings were at first darkly pigmented. During the early stages of human evolution natural selection may have favored dark skin, not only because of the relation between Vitamin D and pigmentation, but also because a dark color may have provided better camouflage for our ancestors, who must have been prey for some of the large carnivores of the day. As people migrated from the tropics to more northerly or southerly regions they probably lost most of their skin pigmentation because in doing so they were able to synthesize sufficient quantities of Vitamin D. (Interestingly, the Eskimos are the only northerly people who have dark skins, and they have abundant supplies of Vitamin D in their diets because they regularly eat fish livers.)

You may be wondering why people possess some genetically determined traits, such as skin color, in seemingly endless varieties that differ slightly whereas they either have or do not have others, such as the ability to taste the chemical PTC. The main reason is that the former are under the influence of several pairs of alleles, whereas the latter depend on a single pair. Thus the regular variation in skin color among human populations around the world is the result of the interaction of at least four alleles, as shown in Figure 19-9. That

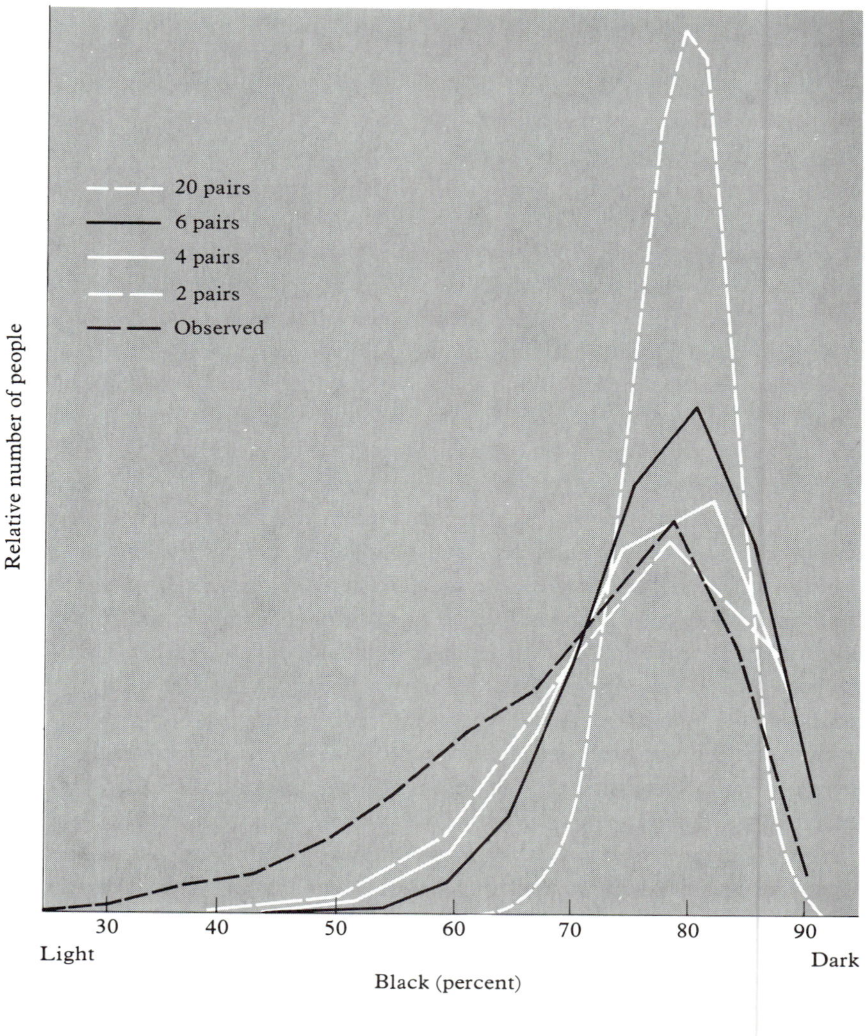

19–9
The distribution of the observed skin color of American blacks and the distribution that would be expected if two, four, six, and 20 pairs of alleles were acting together. (After Curt Stern.)

these alleles are (or were) retained in different frequencies in different human populations can probably be attributed to the effects of natural selection.

Another characteristic that can show gradual variation with latitude, that depends on several pairs of alleles, and that is influenced by natural selection is body build. In general, warm-blooded animals that live in hot equatorial climates are smaller, have longer arms and legs, and have larger ratios of surface area to body weight than warm-blooded animals that live farther north or south. Most of these tendencies are not very clear-cut in the human population, and there are a lot of exceptions to the rule. Nonetheless, as shown in Figure 19-10, human adaptation to climate probably accounts for the differences in body build that allow tall, slender Africans to dissipate more unneeded body heat than Eskimos, who because of where they live must conserve as much body heat as possible.

It is likely that other body features, such as the size of the nose, the color and form of the hair, and the shape of the eyefolds also became established in certain populations because of natural selection. This is because all of these characteristics are external, as are skin color and body build. We would expect the superficial, readily visible characterisitcs to experience the effects of natural selection because our body surfaces are the interface between our bodies and the environment. It is therefore no wonder that human body surfaces have been altered by natural selection to best fit the varied environments into which our ancestors migrated (Figure 19-11).

Biochemical Differences Between Human Races

But differences in traits that are neither superficial nor readily visible parallel the superficial differences of human races to only a limited degree. You will recall that in all species protein molecules, especially enzymes, usually exist in several alternate forms. In recent years it has been possible to compare differences in protein molecules within and between the various races of people. For those protein polymorphisms not obviously influenced by natural selection (and this includes the great majority), the differences in protein molecules between living human races are slight. And protein differences are as slight between Caucasians and African blacks as between Caucasians and Orientals. Moreover, the protein molecules of two Caucasians from opposite ends of Europe differ more than the molecules of two Caucasians from the same isolated European village, but in both differences are about the same as those between the molecules of Caucasians and African blacks or Orientals. Thus at the biochemical level, the differences between human races are generally much less pronounced than the superficial differences that are apparent on body surfaces. In fact, there is so much biochemical overlap that people cannot be assigned with certainty to a given race solely on the particular alleles contained with their genetic programs.

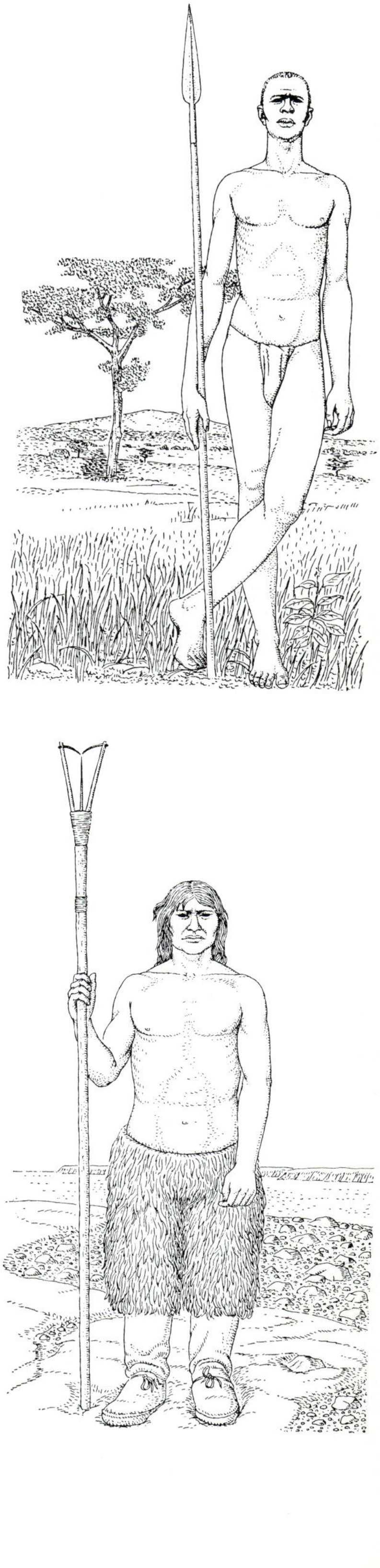

19–10

The greater body surface of the Nilotic Negro from the Sudan (top) allows him to dissipate unneeded body heat, whereas the proportionately greater bulk of the Eskimo (bottom) conserves body heat. (From "The Distribution of Man", by William W. Howells. Copyright © 1960 by Scientific American, Inc. All rights reserved.)

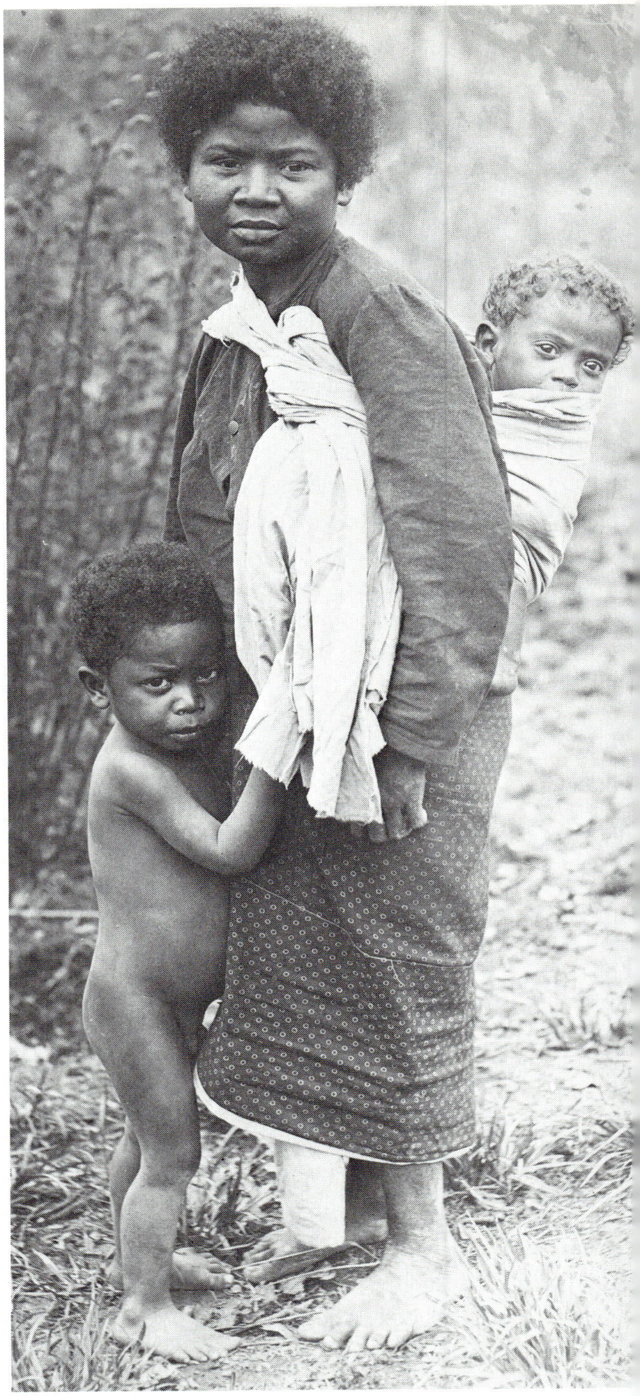

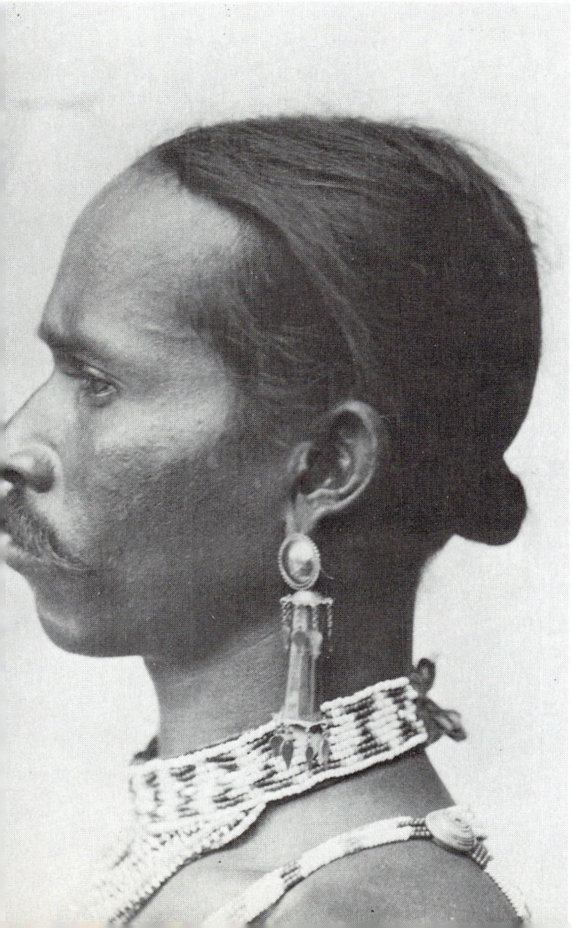

19–11
As shown in these photos, members of the human species are wonderfully diverse in the contours of their body surfaces. The superficial differences between human races probably resulted from the effects of natural selection on our distant ancestors in various parts of the world. Opposite page, top left, Kalahari bushman with two handfuls of stork; top right, young Negrito woman and her children; bottom, Blackheart, a Blackfoot Indian. This page, top, a young Polynesian woman, probably Tahitian; bottom, a Singhalese man (continued on the next two pages).

Opposite page, top left, Melanesian man from Admiralty Islands; top right, Singhalese woman dancer; bottom, young woman from Inner Mongolia. This page, top, three young Swedish women; bottom, Mamayauk Eskimo woman. (Photographs on pp. 486–489 courtesy of American Museum of Natural History.)

The Number of Human Races

How many human races are there? That depends on how the term is defined. The best definition is that races are local breeding groups within a particular species. Races are thus defined by genetic relations between populations and not by differences between individuals. But most local breeding groups in any species tend to blend with one another at their geographic boundaries because adjacent races interchange genes when they reproduce. Thus, races are never clear-cut, precisely defined entities. There are undoubtedly thousands of local breeding groups within the human population today that could legitimately be defined as races. The Dunkers are one example, as are relatively isolated populations in Alpine villages, New Guinea, and Australia.

But it is neither practical nor particularly informative to designate every isolated group of people as a separate race. Rather, most of the time the term *race* is used for any relatively isolated local breeding group that is convenient to distinguish for purposes of a given study. In other words, the number of human races described depends in large part on the purposes of the describer.

For the most part, the description of human races has been undertaken, appropriately, by anthropologists, but some of the criteria used to distinguish individual races are at best vague. The number of human races currently recognized by different anthropologists using different criteria ranges from zero up, but most estimates fall somewhere in between three and thirty human races (see Table 19-1). The criteria used to distinguish human races include not only features of the body surface, but also less obvious physiological and biochemical differences. In general, the genetic differences between currently recognized human races reflect some degree of reproductive isolation, but there is no denying that because of technologic and sociologic changes human populations are much less isolated from one another than ever before. Although human races do exist, it is questionable whether they have much biologic relevance for our species at the present time. Let us discuss this important point further.

The Importance of Culture in Human Adaptation

Among the million and a half species of living things that have been described and named, the human species is unique in that its members adapt to the environment primarily by means of a complicated form of learned behavior called *culture,* which is transmitted from generation to generation by means of the symbol system of language. For all practical purposes geographic variation in our species is now irrelevant because people adapt to the environment primarily by means of behavior, whose biological basis is in the brain and is not reflected in superficial differences in body surfaces. When they first evolved, people who had dark-colored skins were better adapted to climatic conditions near the equator than those who had light-colored skins, but few human beings now live under "natural" conditions. Also, technological advances have assured that at the present time human beings of various races are as likely to reproduce in one environment as another. The superficial differences between existing

TABLE 19-1

Various human "races" that can be identified by means of statistical correlations of the structures of a variety of protein molecules. This analysis was made by the population geneticist R. C. Lewontin and was based on data about 17 proteins. Seven races and many distinct populations can be recognized.

Caucasians

Arabs, Armenians, Austrians, Basques, Belgians, Bulgarians, Czechs, Danes, Dutch, Egyptians, English, Estonians, Finns, French, Georgians, Germans, Greeks, Gypsies, Hungarians, Icelanders, Indians (Hindi speaking), Italians, Irani, Norwegians, Oriental Jews, Pakistani (Urdu-speakers), Poles, Portugese, Russians, Spaniards, Swedes, Swiss, Syrians, Tristan de Cunhans, Welsh

Black Africans

Abyssinians (Amharas), Bantu, Barundi, Batutsi, Bushmen, Congolese, Ewe, Fulani, Gambians, Ghanaians, Hobe, Hottentot, Hututu, Ibo, Iraqi, Kenyans, Kikuyu, Liberians, Luo, Madagascans, Mozambiquans, Msutu, Nigerians, Pygmies, Sengalese, Shona, Somalis, Sudanese, Tanganyikans, Tutsi, Ugandans, U.S. Blacks, "West Africans," Xosa, Zulu

Mongoloids

Ainu, Bhutanese, Bogobos, Bruneians, Buriats, Chinese, Dyaks, Filipinos, Ghashgai, Indonesians, Japanese, Javanese, Kirghiz, Koreans, Lapps, Malayans, Senoy, Siamese, Taiwanese, Tatars, Thais, Turks

South Asian Aborigines

Andamanese, Badagas, Chenchu, Irula, Marathas, Naiars, Oraons, Onge, Tamils, Todas

Amerinds

Alacaluf, Aleuts, Apache, Atacameños, "Athabascans," Ayamara, Bororo, Blackfeet, Bloods, "Brazilian Indians," Chippewa, Caingang, Choco, Caushatta, Cuna, Diegueños, Eskimo, Flathead, Huasteco, Huichol, Ica, Kwakiutl, Labradors, Lacandon, Mapuche, Maya, "Mexican Indians," Navaho, Nez Percé, Paez, Pehuenches, Pueblo, Quechua, Seminole, Shoshone, Toba, Utes, "Venezuelan Indians," Zavante, Yanomama

Oceanians

Admiralty Islanders, Caroline Islanders, Easter Islanders, Ellice Islanders, Fijians, Gilbertese, Guamians, Hawaiians, Kapingas, Maori, Marshallese, Melanauans, "Melanesians," "Micronesians," New Britons, New Caledonians, New Hebrideans, Palauans, Papuans, "Polynesians," Saipanese, Samoans, Solomon Islanders, Tongians, Trukese, Yapese

Australian Aborigines

Source: From R. C. Lewontin in *Evolutionary Biology,* Vol. 6. T. Dobzhansky et al. (eds.). Plenum Publishing Corp. 1972.

human races are thus relics of the past in large part and are not of much functional significance today.

In the evolutionary sense, the significance of races is that under special circumstances some may evolve into new species. The term *subspecies* is used to describe a race that is sufficiently distinct to merit (in the opinion of the person who is classifying) a Latin name in a formal classification. Given enough time and the presence of significant changes in the environment, some isolated groups may become so genetically different from other groups that they can no longer reproduce with one another successfully. When that happens, a new species may evolve, as is discussed more fully in Chapter 1.

The amount of variation observed between populations of living human beings probably does not warrant classifying any of them as official subspecies, especially because racial differences have been blurred as recent years have brought people increased mobility and social changes. Nor is there any evidence that groups of people are becoming reproductively isolated from one another and thus may be in the early stages of evolving into a new species. Even the recently abolished castes of India, which were reproductively isolated from one another for at least 3000 years, showed no signs whatsoever that individuals of particular castes were incapable of successfully reproducing with members of any other caste or a member of any other human population.

But subspecies are officially recognized in extinct members of the human species. Thus Neanderthalers are officially classified as a distinct subspecies, *Homo sapiens neanderthalensis.* This emphasizes that some Neanderthalers probably interbred with fully modern people, known as *Homo sapiens sapiens.* In fact, modern people may well have evolved directly from isolated groups of Progressive Neanderthalers. (The evolution of the human species is discussed at some length in Chapter 2.)

In the end, the wondrous diversity of living things, including our own species, has probably arisen in large part because of the accumulation of genetic differences between isolated populations. The original source of genetic differences between two individuals is found in the phenomenon known as mutation. We now turn our attention to how heritable changes originate within the genetic material and thus produce genetic diversity in any population.

Mutations and Human Diversity

All of the differences between any two living things ultimately originate because of changes that take place within DNA molecules. Heritable changes that arise within existing DNA molecules are known as *mutations,* and in general mutations are the results of accidents. One of the main ways in which mutations arise is by mistakes that occur during DNA replication. As you know, the sequence of the four bases in one strand of a DNA molecule is complementary to the base sequence of the other strand. Thus, when the two strands separate during DNA replication, each strand provides the base sequence necessary to code for a new complementary partner and the two molecules that result are identical, as long as errors in the copying process do not occur. Most of the time the copying process is completely accurate, but rare mistakes do happen, and these unlikely errors help to furnish the raw material for evolution.

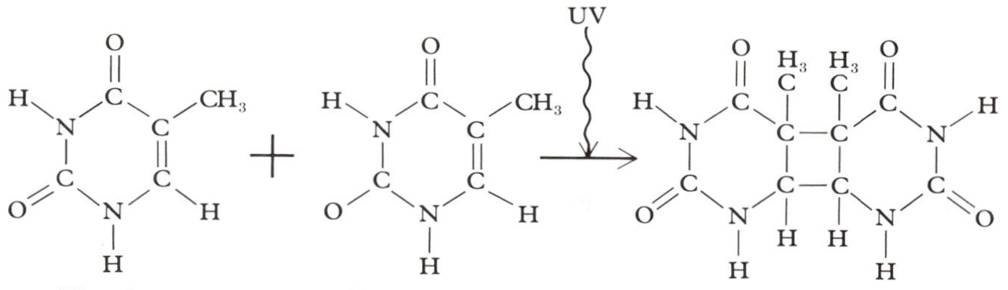

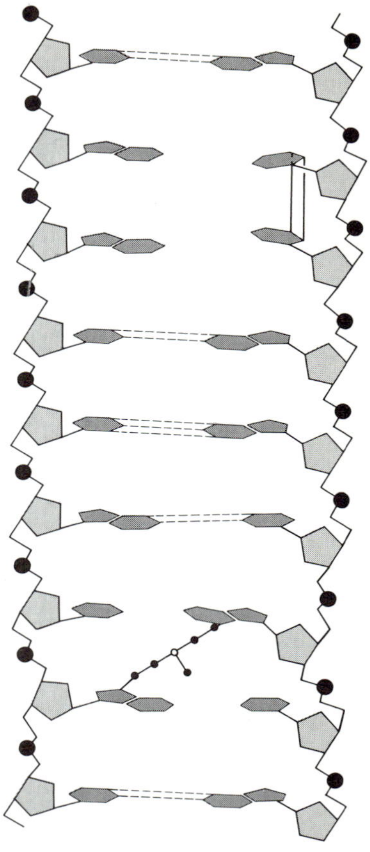

19–12
Thymine dimers can be formed when ultraviolet light interacts with thymine bases in DNA molecules, left. Thymine dimers distort DNA molecules so that they cannot replicate properly, right. (Right-hand portion from "The Repair of DNA," by Philip C. Hanawalt and Robert H. Haynes. Copyright © 1967 by Scientific American, Inc. All rights reserved.)

Several sorts of errors can occur during DNA replication. In brief, most replication errors are caused by the substitution of one base pair for another. This may occur because of the accidental mismatching of two bases that do not usually pair with one another or because of the insertion or excision of one or several base pairs along a particular stretch of DNA. (But there are many other reasons.) In any case, the end result is that a portion of the genetic code is altered. If the alteration in the genetic code occurs in a structural gene (that is, in a stretch of DNA that codes for a particular protein molecule), then we would expect the protein whose amino acid sequence is coded for by the altered gene to be altered too. It turns out that most alterations in protein structure by the kinds of mutations we have been discussing are slight. They usually consist of a single amino acid substitution. For example, it is possible to account for a large part of the naturally occurring variability among the chains of the human hemoglobin molecule by assuming that changes in the amino acid sequence reflect a change in a single DNA base pair in the structural gene in question. Thus, as compared to the corresponding "normal" chains, each of 34 different alpha chains, 56 beta chains, 4 gamma chains and 4 delta chains, all of which are known to exist, can be accounted for by single base pair changes.

Mutations that are produced by more extensive changes in DNA molecules are also known to occur. For example, altered DNA molecules can result from unequal crossing over (crossing over usually occurs shortly before cells undergo meiosis, a form of cell division discussed in Chapter 16). The result of unequal crossing over is that hybrid protein molecules can be produced. Thus persons whose hemoglobin molecules contain a protein chain with an amino acid sequence that begins like that of a normal delta chain and ends like that of a normal beta chain have been reported. These unusual molecules probably result because of unequal crossing over between the structural genes for the human beta and delta chains. (Very recently, it has been suggested that unequal crossing over may also have played a part in generating the repeated segments of DNA that have been reported within the genetic material of most higher organisms.)

Mutations can also result from physical damage to DNA molecules. Localized regions of DNA molecules can be damaged by energy in the form of radiation of various kinds, including ultraviolet light, X-rays, cosmic rays, and various types of natural radiation from radioactive substances. In order to compensate for the damaging effects of certain kinds of radiation, organisms have evolved enzymes that repair radiation-damaged segments of DNA. For example, as shown in Figure 19-12, the main way in which ultraviolet light damages DNA is by causing adjacent thymine bases to become tightly bonded to one another, thus distorting the molecule and interfering with its ability to replicate itself. Human cells usually contain an enzyme that has the rather remarkable ability to excise the abnormal regions of DNA, after which other enzymes then repair the excised segment. Rare persons are deficient in the enzyme that does the excising of the abnormal radiation-damage segments. As expected, their skin is highly sensitive to ultraviolet light and may be severely damaged on exposure to sunlight. (This rare trait is autosomal recessive.)

DNA molecules can also be damaged by heat and by various kinds of chemicals. But before we discuss the relative importance of these and other factors in the production of human mutations, we must first mention what happens to mutated genes in the human population once they have arisen.

A person whose DNA contains a mutated gene may be affected positively, negatively, or to no detectable degree. If the influence is negative, the affected person is usually less likely to reproduce than an unaffected person, or may even die before reaching reproductive age. In either instance the mutation will be lost from the population unless it recurs spontaneously. If the mutation has no detectable effect on the person, then its persistence in the population is a matter of chance. As we have seen, this is probably true of the great majority of polymorphisms involving protein molecules, such as that for the protein chains of the hemoglobin molecule discussed in preceding paragraphs. Finally, if the mutation has a positive effect it will tend to become more prevalent in a population with the passage of time, mostly because of natural selection, though chance can still play a role. The allele for sickle-cell hemoglobin and its relation to malaria provide a good example.

But almost all mutations are either frankly detrimental or have no detectable effect. This is not surprising. Organisms are already so precisely adapted to survival and reproduction in the environments in which they live that any change is much more likely to be a disadvantage than an advantage. In protein polymorphisms, a high degree of diversity is probably maintained in part by chance and in part by natural selection, though the relative roles of each of these factors in most structural differences of human proteins have yet to be explained.

Estimating the Mutation Rates of Human Genes

Mutations are rare events, and it is particularly difficult to estimate the frequency with which they occur in human populations. There are two main ways, known as the *direct* and *indirect* methods, of estimating the rate at which human genes mutate. Both methods are subject to many sources of error and can yield at best, approximate estimates. (Both methods apply only to traits that are obvious and detrimental.)

The direct method of determining the rate at which human genes mutate requires finding out about all occurrences of a particular dominant disorder and determining how many of them appear sporadically among the offspring of unaffected parents. As you may recall, about 15 percent of the cases of Marfan's syndrome occur in this way (see Chapter 16), and data from an obstetrical hospital in Copenhagen indicate that mutation may account for a much larger

19–13
Brother and sister achondroplastic dwarfs of the Owitch family as they appeared in 1949 when they arrived in Israel after having spent several years in Auschwitz concentration camp. Their lives were spared because they were used for medical experiments. The autosomal dominant gene (or genes) responsible for this trait does not impair fertility. (Courtesy United Press International.)

percentage of instances of achondroplastic dwarfism (Figure 19-13). Achondroplastic dwarfism is discussed further in Chapter 15.

The more obvious sources of error in the direct method are these. First, it is hard to be sure that *all* of the instances of any disorder have been identified, no matter how apparent the disorder may be. Second, it is possible that the trait being studied may result from any of several different mutations, all of which produce the same end result. (This is probably true for achondroplastic dwarfism.) Third, a dominant gene may be incompletely expressed in a particular person because of its interaction with other genes. Thus a particular trait may suddenly appear in the offspring of parents who, although they appeared unaffected, actually carried the gene for the disorder. Of course, it would then not be a mutation.

Although the direct method applies only to dominant traits, the indirect method of estimating the mutation rate of human genes can be applied to both dominant and recessive traits. The indirect method is based on the assumption that the rate at which mutant genes are added to a population is balanced by the rate at which they are removed by natural selection. This implies that affected persons are less likely to reproduce than those who are unaffected, which tends to

TABLE 19-2

Estimates of mutation rates of certain human genes from normal to abnormal.

TRAIT	MUTANT GENE PER 100,000 SEX CELLS
Autosomal dominants	
Huntington's chorea	<0.1
Nail-patella syndrome	0.2
Epiloia (type of brain tumor)	0.4–0.8
Aniridia (absence of iris)	0.5
Retinoblastoma (tumor of retina)	0.6–1.8
Multiple polyposis of the large intestine	1–3
Achondroplasia (dwarfness)	4–12
Neurofibromatosis (tumors of nervous tissue)	13–25
X-Linked recessives	
Hemophilia A	2–4
Hemophilia B	0.5–1
Duchenne-type muscular dystrophy	4–10

reduce the frequency of the mutant gene and to numerically cancel out the number of new mutant genes added by spontaneous mutation during any time interval. The indirect method depends on estimating *reproductive fitness,* which is a measure of the likelihood that a person affected by a particular trait will reproduce. The indirect method is subject to the same sources of error as the direct method, and the estimation of reproductive fitness thus adds still another source of error, which makes the indirect method of estimating human mutation rates even less reliable than the direct method.

When all of these factors are taken into account, the most reliable estimate is that a mutation for any particular human gene, or at least for those genes that result in rare, detrimental traits in their mutated forms, is encountered in about one out of every 100,000 sex cells. For the most part, mutations seem to occur about equally often among the sex cells of both sexes. Table 19-2 shows estimates of the mutation rates of certain human genes.

From the total number of autosomal recessive traits that are known and from the rate at which at least some human genes responsible for autosomal recessive traits are known to mutate, it can be calculated that all people probably carry at least several highly detrimental autosomal recessive genes. This estimate is borne out by the rate at which autosomal recessive traits appear among the offspring of consanguineous matings. Overall, it is estimated that the average person probably carries about five recessive genes that in the homozygous condition could result in death before the person could reproduce.

Although the mutation rates for individual genes are somewhat variable, higher mutation rates for almost all genes can be produced in several ways. For example, increased amounts of radiation, higher temperatures, and the effects of various chemicals all tend to increase the mutation rate, sometimes markedly so. (These factors directly alter DNA molecules, but their effects oftentimes show up as damage to entire chromosomes, or to parts of chromosomes. Phenotypic changes that occur because of altered chromosomes are also considered mutations. See Chapter 16.)

It has been estimated that technologically advanced populations, because of exposure to diagnostic X-rays, radioactive fallout, and natural sources of radiation, have tripled the amount of radiation to which their sex cells are exposed during the reproductive years. But there is evidence that overall, radiation from X-rays, fallout, and other sources does not play a major role in the production of human mutations, at least those with apparent effects.

On the other hand, it is known that for various experimental animals the spontaneous mutation rate increases directly with temperature, as is expected of any process that can be explained in molecular terms. This may have some relevance for human genetics because it has been found that the temperature of the gonads of men wearing pants, especially tight-fitting ones, is higher than that of men wearing kilts or wearing nothing at all.

Various chemicals are known to increase the mutation rate among experimental animals, including fruit flies and mice. Nitrogen mustard is a good example, and because of its biochemical effects on rapidly dividing cells, this toxic substance has been of some benefit in the treatment of certain human cancers. Many other chemicals that many of us encounter in our everyday lives—caffeine, for example—are known to increase the mutation rate in bacteria and in some insects, but evidence that they do so in human beings is not conclusive. Nonetheless, in the end most human mutations probably result from short-lived, highly reactive chemicals that are produced inside normal living cells. These highly reactive chemicals are usually formed as the byproducts of normal metabolism, or, more rarely, may result from the effects of radiation.

In recent years it has been asked whether manufactured chemicals, which are so readily available in the environments of industrialized populations, are increasing the rate at which our species' genes are mutating. While environmental chemicals may have had some influence on the mutation rate of some human genes, it is impossible to estimate their overall effect at this time. Mutation rates comparable to those estimated for human beings have also been observed among fruit flies, mice, and other creatures raised under controlled laboratory conditions. It may be true that mutation rates in the human population are not influenced to any presently measurable degree by manufactured chemicals, but rather depend in large part on spontaneous chemical reactions that occur inside normal cells. Nonetheless, it seems wise to continue to question whether or not the benefit of introducing a manufactured chemical into a particular environment warrants the risk that the substance will damage the genetic material of the human and nonhuman populations living there.

Summary

The protein molecules of individuals of the same species are extremely variable. Slightly different forms of almost all human proteins are known to exist, and most of the time slight differences in protein molecules from one person to another do not have ill effects. For most human protein polymorphisms there is no evidence that natural selection influences the frequencies of the slightly different genes that code for the slightly different proteins. Nonetheless, the gene for the abnormal beta chain of hemoglobin that is characteristic of sickle-cell anemia is maintained at a rather high frequency because natural selection favors the reproduction of heterozygotes in malarious areas.

The worldwide distribution of the ABO blood groups is probably best explained as the result of human migrations. Although various blood groups have been correlated with ulcers, with resistance to certain infectious diseases, and with maternal-fetal incompatibilities, the evidence that natural selection plays a major role in how the ABO blood groups are distributed is not convincing.

In large populations random fluctuations in the frequencies of those genes not obviously influenced by natural selection tend to cancel each other. But in small populations, significant changes in genetic constitution often occur by chance alone. Thus the Dunkers, who have been reproductively isolated for many generations, have gene frequencies that are different from those of either their West German forebears or their American neighbors.

Most widespread species look slightly different in different geographic areas, and this reflects some degree of genetic difference between populations. Distinct local breeding groups are known as races. In practice most of these local populations blend with one another because many individuals in neighboring populations interbreed.

External features of the human body, such as skin color and body build, are generally more variable from one local human population to another than are less apparent traits, such as slight differences in protein molecules. Racial differences in body surfaces probably became established in ancient, isolated human groups in large part because of natural selection. The number of human races recognized depends on the purposes and opinions of the person who classifies them.

Although human races surely exist, they are for the most part biologically irrelevant today. This is because most people no longer adapt to the environment primarily by means of body surfaces, but rather by means of language and culture. Human racial differences have been blurred in recent years because of increased mobility and social changes.

Mutations are heritable changes that arise within existing DNA molecules, most of them because of some kind of accident. Mutations that arise during DNA replication usually are produced by the substitution of a single base pair. Unequal crossing-over, radiation of various kinds, heat, and numerous chemicals can either produce mutations directly or speed up the rate at which they occur.

Mutations in the human population are particularly hard to detect. Direct and indirect methods of estimating the mutation rate of certain human genes have been devised, but they are subject to many sources of error, and the estimates are very approximate.

Exposure to X-rays, environmental chemicals, natural radiation, and heat surely have some effect on the rate at which human genes mutate, but they are probably not the most important factors. Rather, intrinsic errors in the replication of DNA as well as the presence of short-lived by-products of normal metabolism produced inside normal cells probably account for the occurrence of most human mutations.

Suggested Readings

1. "The Genetics of Human Populations," by L. L. Cavalli-Sforza. *Scientific American,* Sept. 1974. Discusses how the molecular differences within human populations are greater than those between populations.
2. "Sickle Cells and Evolution," by Anthony C. Allison. *Scientific American,* Aug. 1956, Offprint 1065. How natural selection can maintain an allele that seems to be frankly detrimental.
3. "The Genetics of the Dunkers," by H. Bently Glass. *Scientific American,* Aug. 1953, Offprint 1062. Describes the evidence for genetic drift among the members of a religious sect.
4. "Genetic Drift in an Italian Population," by L. L. Cavalli-Sforza. *Scientific American,* Aug. 1969. Discusses the genetic effects of physical isolation and of consanguineous marriages among people who live in Italy's Parma Valley.
5. "Ionizing Radiation and Evolution," by James F. Crow. *Scientific American,* Sept. 1959, Offprint 55. Discusses the role of X-rays and other kinds of ionizing radiation in the evolutionary process.

This painting by Pablo Picasso ("Girl before a Mirror," 1932, March 14. Oil on canvas, 64 × 51¼") symbolizes the fragmented, emotional consideration that the human species sometimes gives to its own genetic future. (Collection, The Museum of Modern Art, New York. Gift of Mrs. Simon Guggenheim.)

CHAPTER 20

Genes and Human Intervention

Although our understanding of genetics began with the publication of Mendel's manuscript in 1866, people have known how to use it to their advantage since ancient times. Consider the case of "man's best friend," the domestic dog *(Canis familiaris).* Archaeological evidence suggests that the ancestors of domestic dogs were frequent visitors to the camps of prehistoric peoples 30,000 years ago and that the dog had been fully domesticated by 12,000 years ago: its remains are regularly found close to those of human beings from that time on. Our prehistoric forebears must have noticed that some of their canine companions had more desirable characteristics than others and that parents sometimes passed on desirable (and undesirable) physical and behavioral traits to their offspring. People took advantage of this observation by selective breeding—that is, by the deliberate mating, generation after generation of those animals that had the most desirable traits. By means of selective breeding it was possible to produce different *breeds* of dogs that not only looked different from one another but also varied in behavioral traits, such as the degree of tameness, the tendency to bite or to bark, and the willingness to obey commands.

At the present time there are about 110 officially recognized breeds of dogs, almost all of which were developed by means of controlled matings (Figure 20-1). All of these breeds are members of the same species because all dogs can interbreed, or at least can exchange genes through intermediates. (Physical

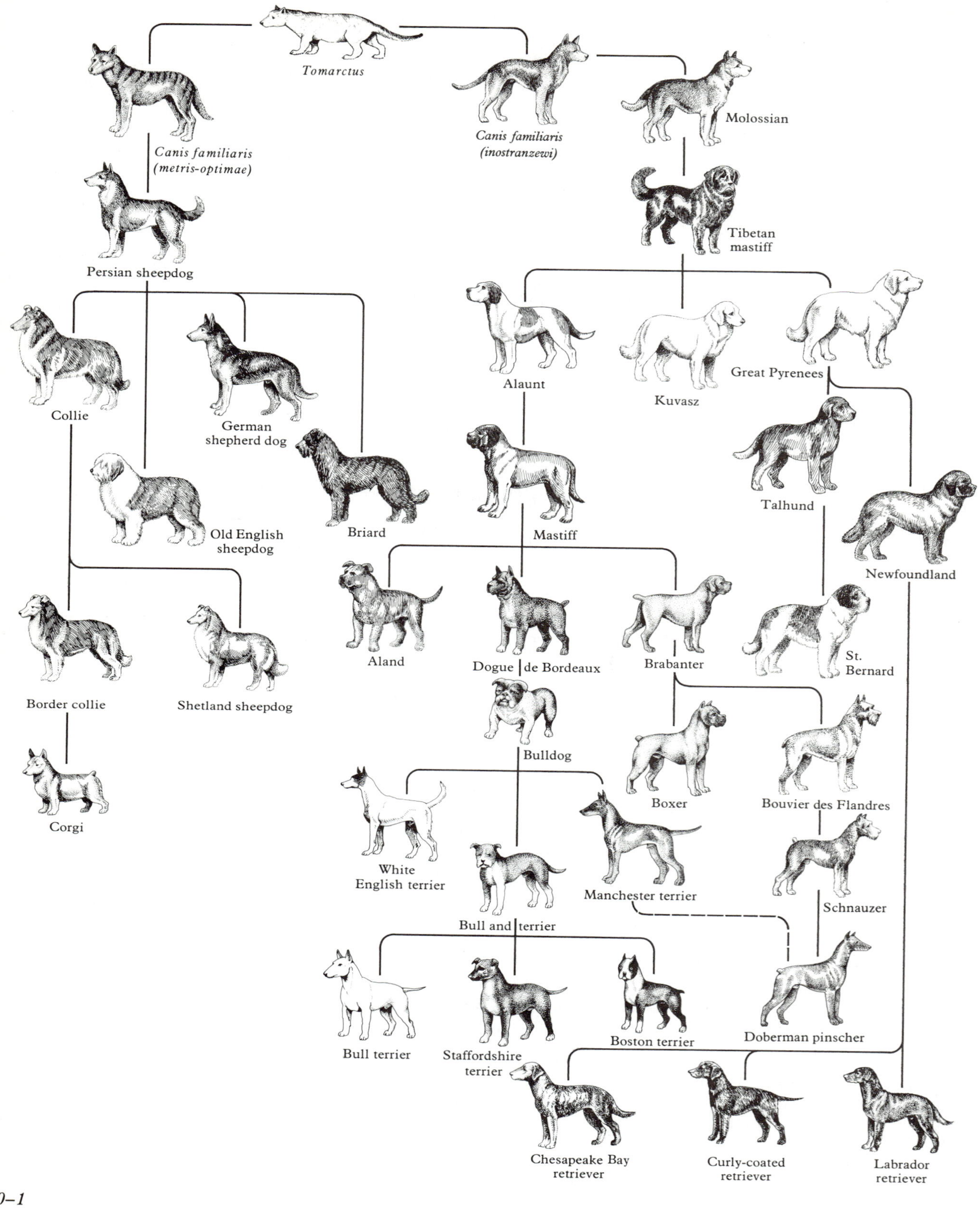
Tomarctus
Canis familiaris
(metris-optimae)
Canis familiaris
(inostranzewi)
Molossian
Persian sheepdog
Tibetan
mastiff
Collie
German
shepherd dog
Old English
sheepdog
Briard
Alaunt
Kuvasz
Great Pyrenees
Talhund
Mastiff
Newfoundland
Border collie
Shetland sheepdog
Corgi
Aland
Dogue de Bordeaux
Brabanter
St.
Bernard
Bulldog
Boxer
Bouvier des Flandres
White
English terrier
Manchester terrier
Schnauzer
Bull and terrier
Bull terrier
Staffordshire
terrier
Boston terrier
Doberman pinscher
Chesapeake Bay
retriever
Curly-coated
retriever
Labrador
retriever

20–1
Some of the major lineages of modern dog.

differences obviously prevent Saint Bernards from directly mating with Chihuahuas, but the two can exchange genes indirectly by mating with breeds of intermediate size.) Dogs are much more variable than most other species of mammals, and this variability is clearly a product of human intervention. The fact that selective breeding has produced the remarkable diversity of domestic dogs (and of other domestic plants and animals) has raised an important thought in human minds from at least Plato's time, and probably from long before. About 2,500 years ago Plato suggested that it might be possible to improve the human species by controlling human variability by means of selective breeding. Plato was probably not the first person to make this suggestion, and he was certainly not the last.

The term *eugenics* was first applied to programs or proposals for the improvement of the human species by means of selective breeding, but the meaning of the term has broadened with technological advances in recent years. To simplify a little, *negative* eugenics has as its goal decreasing the propagation of "undesirable" traits within the human population, and *positive* eugenics is concerned with increasing the propagation of "desirable" traits, not only by selective breeding, but by other means as well.

The obvious problem of eugenics in any form is that in practice it depends on the definition of "desirable" and "undesirable" that is applied to human traits. Sometimes traits are obviously undesirable in that they cause those who have them to be at a severe disadvantage, perhaps to die before reaching reproductive age. But for many variable human characteristics, including most protein polymorphisms, there are no obvious ill or advantageous effects and the terms "desirable" and "undesirable" are meaningless. And, as we shall see, it is often far from clear what is desirable or undesirable in the genetic aspects of human behavior.

A further difficulty of eugenics in any form is that the genetic component of many human traits is complex and incompletely understood. This is especially true of some of the most interesting human traits, including certain features of behavior that are known to have a complicated genetic basis. As we shall discuss in following paragraphs, it is impossible to assess the effects of the environment on the expression of many traits, especially behavioral ones, that are known to be genetically influenced.

In this chapter our main concern is with proposals and programs that have been suggested for genetically improving the human species. We shall first discuss the effects of prenatal diagnosis and genetic counseling and then consider the prospects for, and some of the possible genetic consequences of, what has become known as *genetic engineering.* Then we shall turn our attention to the genetic basis of certain aspects of human behavior. Specifically, we shall discuss the genetics of IQ scores in human populations and the behavior of males who are of sex chromosome constitution XYY. You will see how the environment can strongly influence human behavior. Finally, we shall discuss ways in which natural selection is at work in human populations today and make some cautious speculations about how our species may evolve in the future.

But a word of caution is in order. These are delicate and heady subjects. At best, the application of eugenic measures to human populations could result in the elimination of some frankly detrimental traits or in the widespread distribution of advantageous ones. But at worst, the application of supposed eugenic measures could result in the irrational horrors of genocidal wars or in other kinds of deplorable human activities. The subject of eugenics cannot be discussed with complete objectivity—the issues are too important and too emotion-charged. In particular, anyone who writes about the topic bears the burdens (and the blessings) of a certain background, genes, and political opinions, and this is bound to color any presentation, however careful, of the facts about eugenics. The practice of eugenics arouses opinions, emotions, and sometimes political action. The controversies surrounding the subject will undoubtedly continue for a long time to come.

Let us begin our discussion of eugenics by turning our attention to some traits that show simple Mendelian patterns of inheritance and that could therefore readily lend themselves to eugenic measures.

Genetic Counseling and Prenatal Diagnosis

In earlier chapters we discussed the patterns of inheritance of some human diseases and abnormalities whose genetic basis lies in a single pair of alleles. You will recall that such disorders may be transmitted in either a dominant or a recessive fashion and that their alleles may be located on autosomes or on sex chromosomes. Genetic abnormalities of this sort have simple Mendelian patterns, so reliable predictions can therefore be made about the likelihood that two parents of known genotype will produce an affected child. It is then up to the parents to decide whether the risk is substantial enough to influence their decision about whether to reproduce (see Table 20-1).

In itself genetic counseling is not a form of eugenics, but it becomes one if people who have, or are carriers of, a particular Mendelian trait choose not to reproduce because of the genetic counseling they receive. If they do, genetic

TABLE 20-1

"Recurrence risks" are based on statistical data and on genetic theory. They have been calculated for over 500 known or suspected genetic conditions. The examples show the risk of the birth of an additional affected child to parents who already have one affected offering.

MAGNITUDE OF RISK (%)		GENETIC BASIS
Total	100	Both parents are homozygous recessives
High	75	Both parents are heterozygous for an autosomal dominant with full penetrance
	50	One parent is heterozygous for an autosomal dominant
	50	For sons, a sex-linked gene carried by the mother
Moderate	30	Down's syndrome due to translocation of part of a 21 to another autosome in one parent
	25	Recessives with full penetrance; both parents heterozygous
Low	5 or less	Down's syndrome due to trisomy 21, arising from meiotic nondisjunction in one parent, most likely the mother

Source: After A. G. Motulsky and F. Hecht.

counseling is a form of negative eugenics, because it has the effect of reducing the frequency of a particular abnormal allele in the human population. But although genetic counseling can and does help to alleviate human misery by sparing some parents the financial and emotional expenses of caring for an affected child, genetic counseling has had little effect on our species' overall genetic make-up, at least so far.

To see why this is so, consider the example of *galactosemia*, an autosomal recessive defect of carbohydrate metabolism whose most serious clinical manifestation, if untreated, is severe mental retardation. This serious inborn error of metabolism resembles phenylketonuria (PKU) in two ways: it, too, is caused by an enzyme defect and the mental retardation can be prevented by dietary restriction, in this case, of milk sugar. Galactosemia occurs in about 25 out of every million children born in the United States, and it is estimated that about 10,000 people per million (1 in 100) in the U.S. are actually heterozygous carriers of the disease. (The observed frequency of the disease is less than the high carrier rate might lead us to expect. In large part this is probably because many fetuses affected by galactosemia die before birth.) Biochemical screening tests to detect heterozygotes are available, so it would be possible to identify all carriers of galactosemia in the United States by a mass screening program.

The fact that heterozygotes can be identified makes it theoretically possible to eliminate the abnormal allele for galactosemia from the U.S. population by means of genetic counseling alone. But consider what this would entail. First, assuming a U.S. population of 200 million, there are 2 million carriers of galactosemia in this country, and all those of reproductive age would have to be identified by means of costly screening tests. Second, all carriers of reproductive age and younger would have to agree never to reproduce simply because they were carriers of galactosemia. (When considering the likelihood that a heterozygous carrier would choose not to reproduce, bear in mind that if you live in the United States your chances of being a carrier are one in 100.)

Most people would agree that the benefits of preventing the birth of 25 galactosemic children per million does not warrant either the expense of nationwide screening or the sacrifice of the reproductive potentials of the 10,000 people per million who happen to be carriers. Besides, you will recall that based on the mutation rates of various human alleles and on the frequency with which severe genetic abnormalities are observed among the offspring of consanguineous matings, it is estimated that each of us probably carries several abnormal recessive alleles that would be lethal in the homozygous state. In the extreme, if we could devise screening tests for and then detect all carriers of serious autosomal recessive disorders, and if all of the carriers chose not to reproduce because they were carriers, the end result would be that nobody would reproduce, and our species would become extinct.

The main effect of genetic counseling so far, and probably in the future, too, is that some parents (and other people in affected family lines) are spared the financial and emotional strains of having a child who is severely affected by a predictably inherited trait. But at this time genetic counseling has little overall effect on the genetic constitution of the human species, and it is not likely to have an effect for a long time to come. (We will discuss some of the undesirable effects of the medical treatment of genetic diseases later in this chapter.)

In recent years the effectiveness of genetic counseling has been greatly increased because of the availability of the medical procedure known as *amniocentesis*. This diagnostic technique consists of collecting a sample, by means of a needle carefully inserted through the abdomen of a pregnant woman, of the amniotic fluid that surrounds the developing fetus (Figure 20-2). Cells from the sample are then grown in tissue culture and subsequently analyzed for biochemical (or chromosomal) defects. Also, the fluid itself may be subjected to various biochemical tests. Table 20-2 is a list of some of the inborn errors of metabolism that can be detected by amniocentesis. At the present time the procedure is usually performed during the middle three months of pregnancy; the technique is generally safe and is becoming increasingly accurate. Amniocentesis is very effective in detecting chromosomal abnormalities such as Down's syndrome. In fact, the detection of abnormal chromosome constitutions is at present the most common reason for which amniocentesis is performed. As discussed in Chapter 16, greater maternal age is strongly associated with the increased incidence of Down's syndrome. Pregnant women who are 40 years old or older are usually advised to have the procedure performed, provided that they are willing to abort the development of a fetus that has an abnormal number of chromosomes.)

TABLE 20-2
Some of the inborn errors of metabolism that can be detected by amniocentesis. Note that most of these errors are associated with mental retardation.

DISORDER	DEFFECTIVE ENZYME OR METABOLIC DERANGEMENT
ASSOCIATED WITH MENTAL RETARDATION	
Chromosomal abnormalities (Down's syndrome, Turner's syndrome, XYY, etc.)	Excess or deficiency of total genetic information
Arginosuccinic aciduria	Arginosuccinase
Citrullinemia	Arginosuccinate synthetase
Fucosidosis	Alpha-fucosidase
Galactosemia	Galactose-1-phosphate uridyl transferase
Gaucher's disease	
Infantile type	Absent cerebrosidase
Adult type	Deficient cerebrosidase
Generalized gangliosidosis	Absent beta galactosidase
Juvenile GM_1 gangliosidosis	Deficient beta galactosidase
Juvenile GM_2 gangliosidosis	Deficiency of Hexosaminidase A
Glycogen storage disease type 2	Alpha-1, 4-glucosidase
Hunter's disease	Increased amniotic fluid Heparitin sulfate
Hurler's disease	Increased amniotic fluid Heparitin sulfate

TABLE 20-2 (continued)
Some of the inborn errors of metabolism that can be detected by amniocentesis. Note that most of these errors are associated with mental retardation.

DISORDER	DEFFECTIVE ENZYME OR METABOLIC DERANGEMENT
ASSOCIATED WITH MENTAL RETARDATION	
I-cell disease	Multiple lysomal hydrolases
Isovaleric acidemia	Isovaleryl CoA dehydrogenase
Lesch-Nyhan syndrome	Hypoxanthine-guanine-phosphoribose transferase
Maple syrup urine disease	Alpha-keto isocaproate decarboxylase
Metachromatic leucodystrophy	
Late infantile type	Absent arylsulfatase A
Juvenile and adult types	Deficient arylsulfatase A
Methylmalonic acidemia	Methylmalonyl CoA carbonyl mutase
Niemann-Pick disease	Sphingomyelinase
Refsum's disease	Phytanic acid alpha-oxidase
Sandhoff's disease	Hexosaminidase A and B
Sanfilippo disease	Increased amniotic fluid Heparitin sulfate
Tay-Sachs disease	Hexosaminidase A
Wolman's disease	Acid lipase
POSSIBLY ASSOCIATED WITH MENTAL RETARDATION	
Cystathioninuria	Cystathionase
Homocystinuria	Cystathionine synthase
NOT ASSOCIATED WITH MENTAL RETARDATION	
Adrenogenital syndrome	Increased amniotic fluid corticosteroids
Cystinosis	Increased cellular cystine
Fabry's disease	Alpha-galactosidase
Hypervalinemia	Valine transaminase
Orotic aciduria	Orotidylic pyrophosphorylase and orotidylic decarboxylase

Source: From "Prenatal Diagnosis of Genetic Disease," by T. Friedmann.

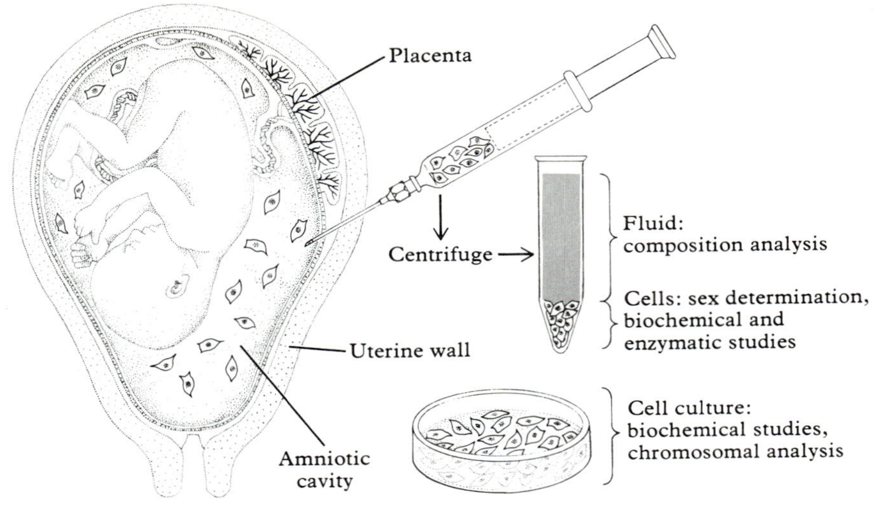

20–2
Cells and amniotic fluid can be obtained for biochemical and chromosomal analysis by means of amniocentesis. (From "Prenatal Diagnosis of Genetic Disease," by Theodore Friedmann. Copyright © 1971 by Scientific American, Inc. All rights reserved.)

Like all other human undertakings that can be considered eugenic in the broad sense of the term, amniocentesis is vigorously and emphatically opposed by some people. At the present time, if a fetus is proved to be affected by a trait detectable by means of amniocentesis, the only recourse other than allowing the birth of a severely affected child is abortion. The practice of aborting fetuses for any reason is repugnant and utterly unacceptable to some people, who therefore oppose amniocentesis because they consider it the first step toward abortion, which it sometimes is. It is hoped that abortion will not always be the only alternative. It will probably be possible someday to treat some genetically determined abnormalities of developing fetuses (such as inborn errors of metabolism) early in pregnancy by correcting the defective DNA that ultimately gives rise to it. Although we are at present a very long way from being able to directly manipulate the genetic material of developing human fetuses, the mid-1970s saw some spectacular (and to some people disturbing) advances in our ability to directly alter the genetic programs of various living things. The direct human manipulation of an organism's genetic program usually is called genetic engineering. It is no longer a futuristic notion, but rather a controversial reality that is worth discussing further.

Genetic Engineering

There are three main ways in which an organism's genetic program could be manipulated: by changing the DNA already present or by either adding to or subtracting from it. At the present time almost all genetic engineering is accomplished by adding new DNA fragments to an existing genetic program, and so far most of the research has been concerned with the introduction of foreign genes into bacterial cells. One technique for doing this depends on the existence of structures called *plasmids,* which are short double-stranded circular pieces of DNA that are found inside certain bacterial cells and that replicate independently of the bacterial chromosome (Figure 20-3). By making use of appropriate enzymes, it is possible to generate fragments of purified DNA molecules from almost any species and then to splice some of the fragments into

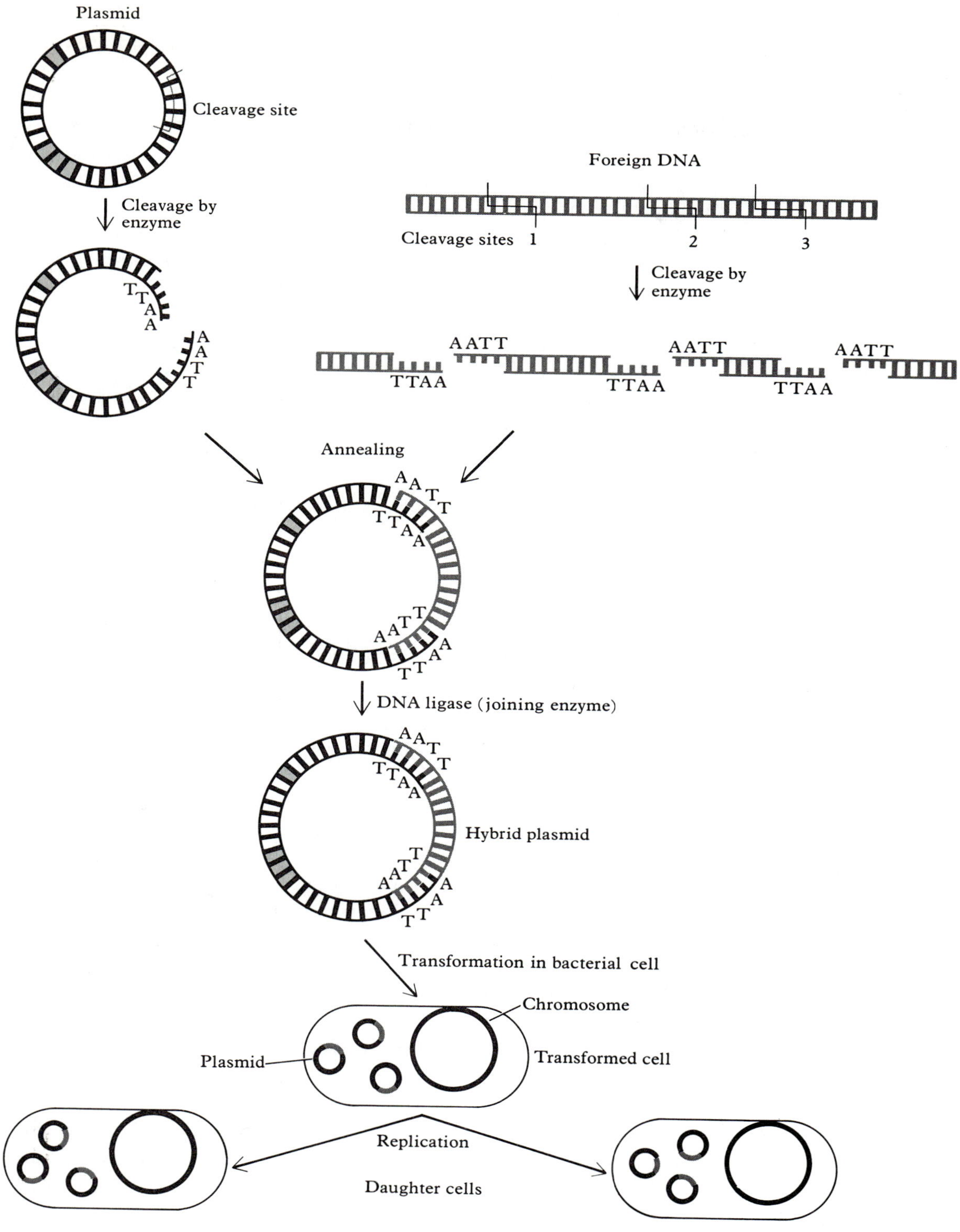

20–3
Foreign DNA can be spliced into a plasmid and introduced with the plasmid into a bacterium. The foreign DNA is replicated along with the bacterial chromosome shortly before cell division occurs. (From "Manipulation of Genes," by Stanley N. Cohen. Copyright © 1975 by Scientific American, Inc. All rights reserved.)

a plasmid that can then be put back inside a living bacterium. (The most commonly used bacterium is *Escherichia coli (E. coli),* a widespread organism that is regularly found in great abundance in the human large intestine.) When a hybrid plasmid replicates, the fragments of foreign DNA may be replicated too, and if these fragments happen to contain a gene not normally found in *E. coli,* it is sometimes possible to detect the corresponding gene product. In other words, it is sometimes possible to splice a foreign gene into a bacterium and to recover from the bacterium the biochemical product of that gene.

At present, the state of the art of genetic engineering is limited almost exclusively to introducing foreign genes into bacterial cells. But there is reason to expect that in the not-too-distant future techniques for incorporating certain genes into human and other mammalian cells will be developed. In fact, this may have already been accomplished, not once, but twice—one time spontaneously and the other intentionally. The spontaneous incident concerned a form of the enzyme arginase that is not usually present in human cells. The blood of scientists who had worked with a certain virus for many years was recently reported to contain some unusual arginase activity. Presumably, this is because some of the scientists' body cells became infected with the virus, whose genetic program is known to contain the gene for the form of arginase in question. (It is known that rabbits intentionally infected by the same virus acquire the capacity to synthesize this enzyme.) It has also been recently reported that cells from a person who has galactosemia can acquire the capacity to synthesize the enzyme they usually lack (galactose-1-phosphate uridyl transferase, or G1PUT) if the cells are infected by a virus carrying a gene that codes for normal G1PUT. The implications of these reports are clear: it may be possible before too long to treat some of those human diseases that result from defective enzymes (or other proteins) by directly introducing into an affected person's cells a particular virus whose DNA normally contains, or has had spliced into it, the normal version of the gene that is defective.

If and when geneticists overcome the technological difficulties of regularly bringing about the manufacture of particular gene products inside the cells of an affected person, what will the impact on human genetics be? To simplify a little, there would probably be little or no positive impact on future generations. This is because at the present time genetic engineers are directing their attention to correcting genetic defects in body cells (somatic cells), not sex cells. To continue with our example, suppose it becomes possible to regularly bring about the manufacture of G1PUT inside the cells of people who have galactosemia. That is surely to be desired, but it is of no consequence to the genetics of our species unless the gene for G1PUT is somehow incorporated into the genetic program of *sex* cells of those who are affected. Technologically, it would be extremely difficult to first detect and then correct, by genetic engineering or any other means, human sex cells that are defective for a particular allele. Therefore, the undeniable good effects of relieving a person of the symptoms of galactosemia by means of genetic engineering would have no positive or negative eugenic effects, because the sex cells would still contain the defective allele.

On the other hand, genetic engineering could have an *adverse* effect on human genetics in that affected individuals would be made healthier and perhaps would be more inclined to pass on their abnormal alleles to future generations. Of course, the modern medical treatment of genetic diseases could have the same detrimental effect. But the accumulation of unfavorable alleles in a large and widespread population is a very slow process, and most geneticists agree that, although human activities may already have increased the frequency of some abnormal alleles in some segments of the human population, any effects on the worldwide population have probably been negligible, at least so far. Moreover, it is possible that the potential adverse eugenic effects of genetic engineering and of modern medical treatment could be offset by the potential favorable effects of genetic counseling. This would occur if the affected people, whose lot is made easier by some kind of human intervention, felt obligated not to pass on their abnormal alleles and therefore chose not to reproduce, or to adopt children. But there is no guarantee that all affected people would elect this course of action, and in the end we may be doing our species a genetic injustice by devising oftentimes elaborate treatments for genetically determined disorders. Whether or not these concerns are grounds for not rendering effective treatment, either through medical practices or through genetic engineering, is left for you to consider.

Although its overall eugenic effects would probably not be of much consequence, the social and ethical problems that could arise from the practice of genetic engineering are formidable and unprecedented. Many geneticists recently showed their concern over a related issue by imposing upon themselves a moratorium on some aspects of research into genetic engineering that could potentially have devastating, though largely unpredictable, effects on people and on other living things. Their concern was that the manipulation of certain genes would give rise to new kinds of bacteria whose properties might allow them to infect people and other creatures with serious, perhaps fatal, effects. The moratorium has now been lifted, but only after serious, sometimes heated, debate in international meetings convened by geneticists in order to collectively discuss how to proceed with their provocative—yet potentially dangerous—research. The result of these international meetings was a set of guidelines for carrying out potentially dangerous experiments in genetic engineering that could result in the release of novel infectious organisms. Importantly, the guidelines were devised, not only by geneticists and other scientists, but also by persons trained in sociology and ethics. It is to be hoped that scientists will continue to acknowledge publicly that they lack expertise in nonscientific matters and that guidelines concerning research on genetic matters that have possible eugenic overtones will result from decisions arrived at not only by professional scientists and professional humanists, but by all of us who are concerned with the genetic future of the human species.

The scientific, ethical and political aspects of traits that show simple Mendelian patterns of inheritance are sometimes in conflict, and traits whose genetic basis is more complex are the subject of even greater controversy. The problem is that some of our species' most familiar and engaging traits do not depend on a single pair of alleles, but rather on many that interact with one another and with the environment in ways that are complicated and poorly understood. Included among such traits are certain aspects of human behavior. But before we discuss the genetic component of human behavior and its important relationship to environment, we must first consider some of the technical difficulties of trying to sort out the genetic and environmental factors that cooperate to make each one of us unique. In other words, we must now discuss the effects of nature (genetic factors) and nurture (environmental factors) as they relate to human uniqueness.

Nature and Nurture—The Concept of Heritability

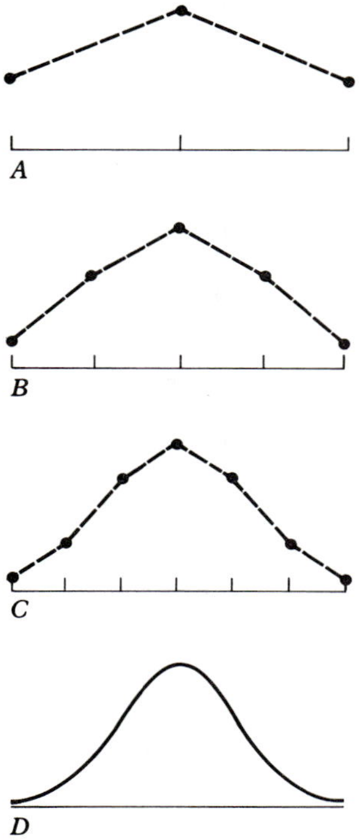

20–4
How the frequency distributions of phenotypes are related to the number of pairs of alleles involved. A, one pair of alleles distributed over three phenotypes. B, two pairs of alleles distributed over five phenotypes. C, three pairs of alleles distributed over seven phenotypes. D, an infinite number of pairs of alleles distributed over a continuous array of phenotypes. (This curve is the normal distribution curve *that characterizes most biological populations. Compare with Figure 16-14.) (From Curt Stern,* Principles of Human Genetics, 3rd Ed. *W. H. Freeman and Company. Copyright © 1973.)*

As discussed in preceding chapters, the inheritance of those traits that show continuous and gradual variation within a population usually depends, not on a single pair of alleles, but on many pairs that in each person interact both with the rest of the individual's genes and with the environment. Consider, for example, the multitude of genetic and environmental factors that interact to determine how tall a person is. Height is influenced by many genes, such as those that code for growth hormone, for intestinal enzymes that digest food and thus provide the body with building blocks for growth, and for the rate at which calcium is deposited in the long bones of the legs. But environmental factors also make an important contribution to a person's stature. For example, the absence of sunlight can result in inadequate synthesis of Vitamin D, which may result in the slowing down of proper bone growth. Chronic poor nutrition in childhood can also influence adult height. The end result of all of these factors affecting height, and of the many others that must also play a part, is that in the worldwide human population, and in any random sample of it we wish to single out, height is "normally" distributed. That is, most individuals in any human population are not far away from being of some "average" height, and although there are very tall and very short people, there are fewer of them than people of average stature (Figure 20-4).

Plant and animal breeders have known for centuries that continuously varying traits are influenced by the environment to different degrees. Because the breeders are concerned with establishing true-breeding lines of plants and animals that have desirable characteristics, it is important for them to be able to assess the relative effects of genes and environment on characteristics they consider desirable, such as copious milk production from cows, large eggs from hens, and long coats on woolly sheep. All of these characteristics are influenced by many genes and by many environmental factors, and breeders have found the statistical concept of *heritability* of use in assessing the relative role of the genetic factors that they would like to "breed into," and thereby genetically improve, the breeds that they have already developed. Heritability is that proportion of the phenotypic variation in any population that can be attributed to genetic factors.

How does one go about assessing the relative effects of genes and environment on a continuously varying trait that is desirable? In practice it is very difficult to estimate heritability, but there are several ways of going about it. For example, heritability can be estimated by mating individuals from a given position in the normal distribution curve for a particular trait, and then examining the distribution of the trait in their offspring, who must be raised under strictly controlled environmental conditions. If all of the variation in a particular trait in a particular population is due to genetic factors alone, the heritability is assigned the number 1 (or 100 percent), and the average value of the trait in the offspring is equal to the value of the position on the curve from which the parents were selected. On the other hand, if all of the variation is due to environmental factors, the heritability is 0 (0 percent), and the average value of the trait in the offspring is the same as the average value in the population from which the parents were selected. Most continuously varying traits have heritabilities between 0 and 1 that can be estimated by determining the difference between the average values in selected parents and their offspring (Figure 20-5).

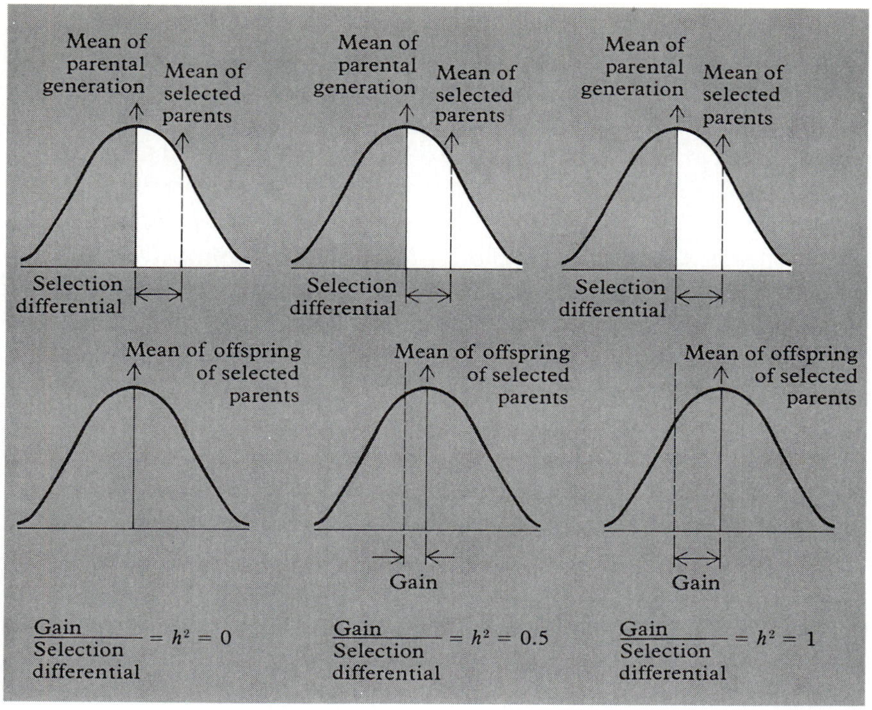

20–5
Heritability can be estimated by breeding experiments that allow one to calculate the ratio of the "gain" to the "selection differential." In the three top curves the difference between the mean (average) of the selected parents and the mean of the population they were selected from (the selection differential*) is the same. In each case the offspring that are produced have characteristics that form a normal distribution curve, and the change in mean between parents and offspring (the* gain*) can be used to estimate the heritability of a particular trait, which is represented by* h^2. *(From I. Michael Lerner and William J. Libby,* Heredity, Evolution, and Society, *2nd Ed. W. H. Freeman and Company. Copyright © 1976.)*

Heritability can also be estimated by experiments in inbreeding, over at least several generations and under controlled environmental conditions, of animals that are related. This method of estimating heritability depends on the fact that different kinds of relatives (brothers and sisters, cousins, etc.) share to different degrees the genes that influence the trait under consideration. Figure 20-6 shows the range of heritabilities, as determined largely by experiments in inbreeding, for various economically important traits of the domestic chicken.

But heritability is a statistical concept that applies to populations, not to individuals. For example, as shown in Figure 20-6, the average heritability for the weight of hens' eggs is about 0.75 (75 percent). A heritability of 0.75 does not mean that for each hen egg weight is determined three-fourths by heredity and one-fourth by environment. What it does mean is that overall three-fourths of the total variation in the weight of hens' eggs is associated with genetic differences, and the remaining one-fourth of the total variation is associated with differences in environmental factors.

Although heritability is never easy to measure, it is particularly difficult to estimate in human populations. There are several reasons for this, the most obvious of which is that people are not experimental animals that can be selectively mated, highly inbred, or raised under strictly controlled conditions. A less obvious reason why heritability is difficult (in fact, impossible) to determine accurately in human populations is that it depends, not only on genes and environment, but on the *interaction* between the two. In human populations, information concerning the exact contribution that the interaction of genes and environment make to the heritability of human traits is in many instances meager, if not nonexistent.

Nonetheless, the heritability of certain human abnormalities that depend on many genes can be crudely estimated from the rate at which the abnormality

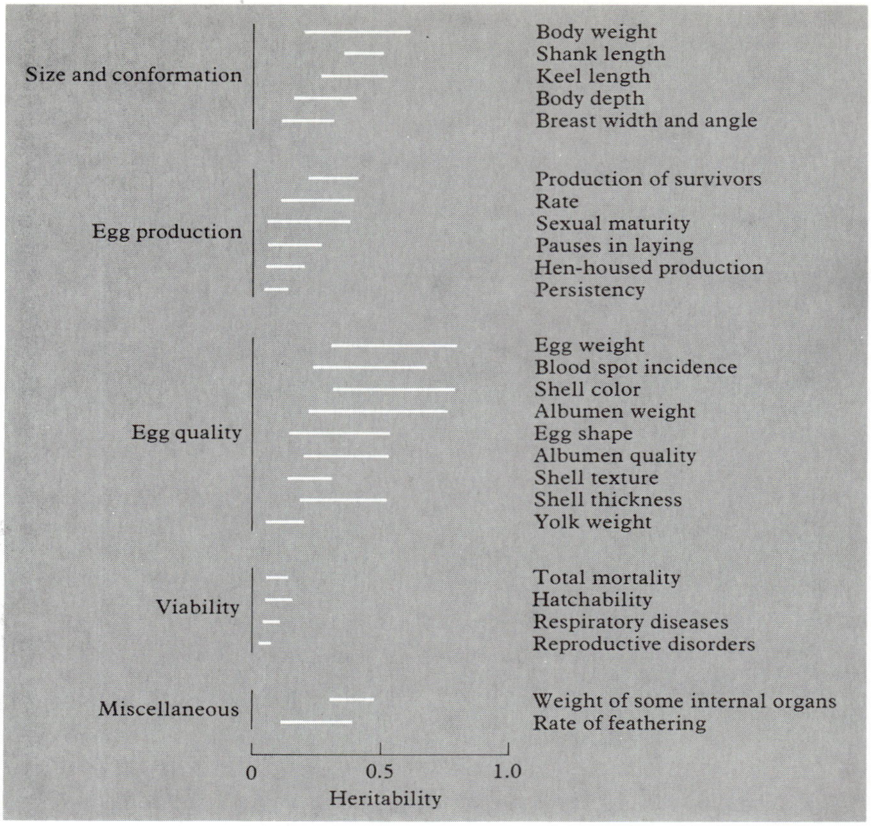

20–6
The range of heritabilities reported for various economically important characteristics of the domestic chicken. (From I. Michael Lerner and William J. Libby, Heredity, Evolution, and Society, *2nd Ed. W. H. Freeman and Company. Copyright ©1976.)*

occurs among close relatives of an affected person, compared with the population at large. Thus it is roughly estimated that the heritability of hydrocephalus (water on the brain) is about 0.4 (40 percent), whereas that for certain kinds of epilepsy is about 0.5 (50 percent). Similarly, the heritability of clubfoot is estimated to be about 0.8 (80 percent), and that of hairlip (with or without cleft palate) is roughly 0.7 (70 percent).

Some biologists and statisticians have recently suggested that the concept of heritability, because of its built-in limitations, should not be applied to human populations at all. There is probably some virtue to this argument, especially for estimates of poorly defined or poorly understood traits such as "intelligence" (see the following discussion). But fortunately, nature has provided us with at least one fairly accurate way of estimating the effects of genes and environment in human populations, whether one has much faith in the concept of heritability or not. Human beings sometimes are produced in almost duplicate copies known as identical twins, who for all practical purposes are genetically identical.

Nature and Nurture—Twin Studies

There are two kinds of human twins: one-egg, or *monozygotic,* twins and two-egg, or *dizygotic,* twins. Dizygotic twins are produced when two eggs, rather than the usual one, are released at ovulation and both are subsequently

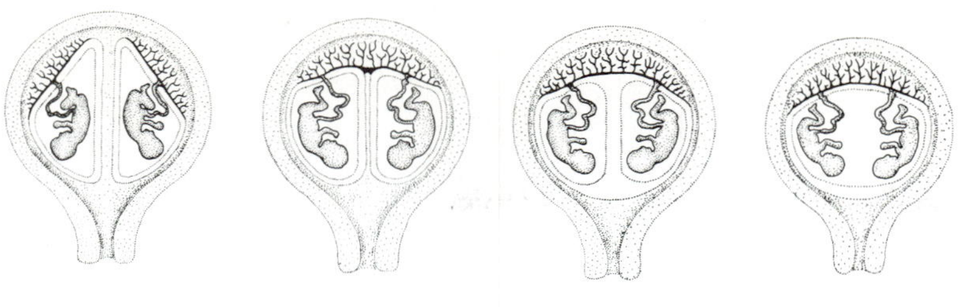

20–7
Monozygotic and dizygotic twins may share the prenatal environment to different degrees because of the arrangement of the membranes surrounding the fetuses.

fertilized by different sperm. The genetic relationship between dizygotic twins is thus the same as between brothers and sisters who are not twins. On the other hand, monozygotic twins orginate from a single fertilized egg that, after having been fertilized by a single sperm, splits into two at a very early stage of development and thus results in two individuals who are, barring mutations during development and in later life, genetically identical (Figure 20-7).

Identical (monozygotic) twins have all of their genes in common, so any variation between two identical twins is in large part due to the effects of the environment. Most of the time identical twins are raised in the same environment, and under such circumstances they tend to strongly resemble each other both physically and in at least some aspects of behavior (Figure 20-8). But sometimes monozygotic twins are raised apart from one another in different environments, and when this happens a study of the differences between the two individuals can provide a fairly accurate measure of the genetic and environmental components of certain human characteristics.

Statistical studies of the differences between identical twins sometimes focus on complex characteristics that depend on many genes and many environmental factors but that nonetheless do not show continuous and gradual variation. The relative role of genetic factors in the expression of such all-or-none traits can be estimated by the degree to which the twins are concordant or discordant. When both twins have a particular trait they are concordant, and when only one does, they are discordant. Table 20-3 shows the degree to which monozygotic and dizygotic twins are concordant for various abnormal conditions that depend on the interaction of many genes and many environmental factors. The percentage of concordance provides an estimate of the degree to which a particular condition is genetically influenced.

Of special interest in studies of the differences between monozygotic twins reared apart are differences in behavior and the degree to which these differences are directly influenced by genes. But before we turn our attention to the genetics of human behavior, we should first discuss some of the difficulties of trying to figure out the genetic basis of some of those traits that impart to our species its complexity and its uniqueness.

The Genetics of Human Behavior—Some Difficulties

The most variable thing about the human species is the endless variety of ways in which people behave. As discussed at some length in Chapter 3, our species is unique among all living things in that its members adapt to the environment primarily by means of a complicated form of learned behavior called culture, which is transmitted from generation to generation by means of the symbol system of language. As you may recall, the biological basis of these complicated

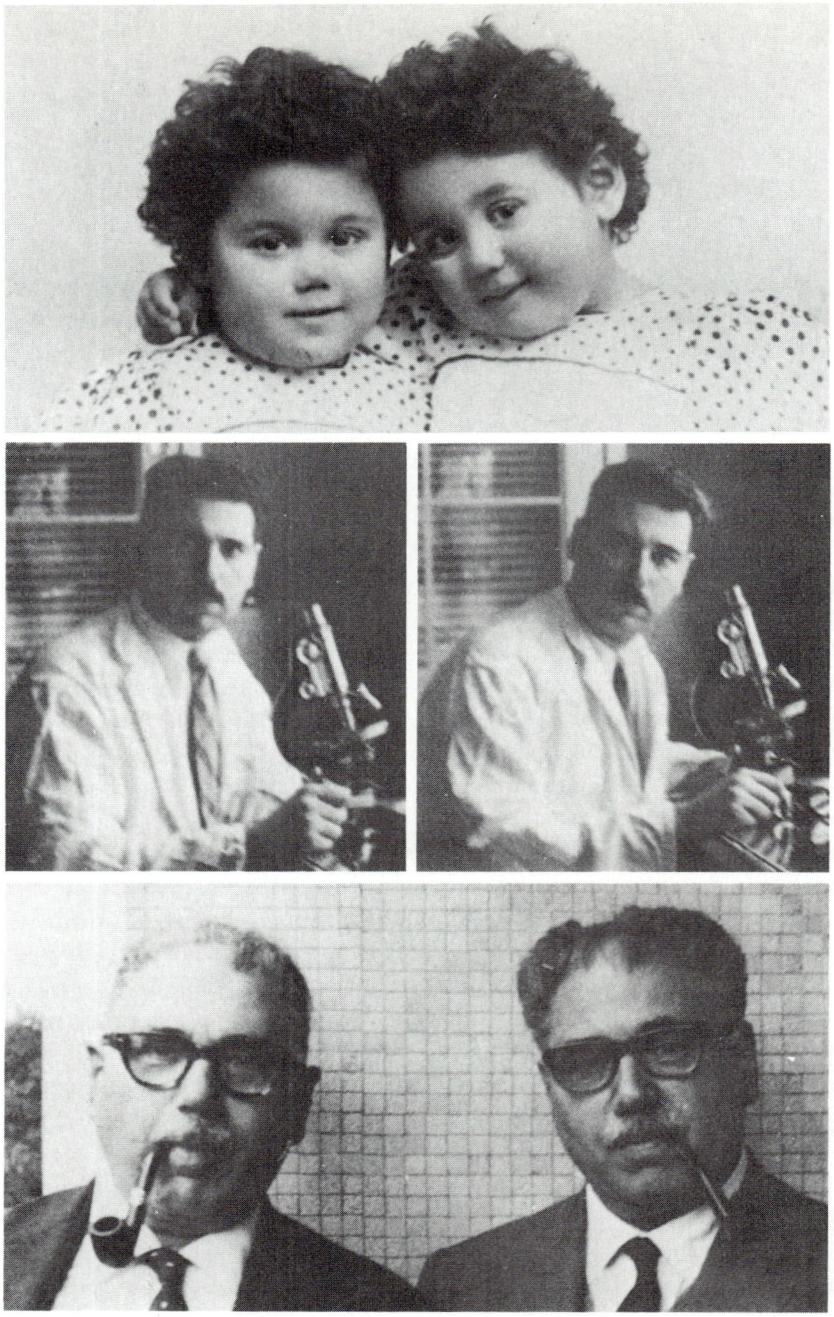

20–8
Top, identical twins Bruno and Giorgio Schreiber at about two years of age. Even the adult twins cannot tell who is who in this photo. Center, as adults, the twins continued to look very much alike. Bottom, a recent photo. Bruno (left) and Giorgio (right) are now professors of zoology at the Universities of Parma (Italy) and Belo Horizonte (Brazil), respectively. (Courtesy of B. Schreiber.)

learned behaviors lies in the human brain, which is the most complicated organ that evolution has produced so far. The capacities for language and culture are shared by all normal human beings, and insofar as they depend on the organization of the human brain the capacities for language and culture may be considered genetically determined.

But very little is known of how the behavior of individuals relates to the brain. How the contributions of genetic and environmental factors interact throughout the course of a person's life to result in behavior is largely unknown to us. Nonetheless, recent years have brought some advances in our understanding of the genetic basis of certain abnormalities in behavior, and these are

TABLE 20-3
The percentage of concordance among twins for some traits that depend on many genes and many environmental factors.

OBSERVED DISEASE OR BEHAVIOR	PERCENTAGE CONCORDANCE	
	MZ TWINS	DZ TWINS
Tuberculosis	54	16
Cancer at the same site	7	3
Clubfoot	32	3
Measles	95	87
Scarlet fever	64	47
Rickets	88	22
Arterial hypertension	25	7
Manic-depressive syndrome	67	5
Death from infection	8	9
Rheumatoid arthritis	34	7
Schizophrenia (1930s)	68	11
Criminality (1930s)	72	34
Feeble-mindedness (1930s)	94	50

Source: From *Heredity Evolution and Society,* 2nd ed., by I. Michael Lerner and William J. Libby. W. H. Freeman and Co. Copyright © 1976.

worth discussing further. Let us now consider some behavioral abnormalities known to result from the presence of a single defective gene.

You will recall that both PKU and galactosemia result from homozygosity for autosomal alleles that code for defective enzymes and that both of these conditions produce mental retardation if they are untreated. (Many genetic defects are known to result in some degree of mental retardation, and this bears witness to the intricacy of the human brain's architecture and biochemistry. Nonetheless, mental retardation can also result from wholly environmental factors, such as accidents during birth and chronically inadequate nutrition of the fetus. See Figure 20-9.) In both PKU and galactosemia the abnormal enzyme presumably leads to a biochemical defect in the developing brains of affected persons, and this defect results in mental retardation, though exactly how this occurs is poorly understood. PKU and galactosemia would thus seem to be examples of a behavioral abnormality that is wholly genetic and independent of environmental factors. Yet environmental factors, in the form of special diets, can prevent the mental retardation associated with both of these conditions. Thus, environmental factors sometimes strongly influence a person's behavior even when genetic factors appear to be of overwhelming importance.

An intriguing human ailment that depends on the presence of a single defective allele and that results in abnormal behavior is the *Lesch-Nyhan syndrome,* first described in 1965. The abnormal allele for this fatal condition is located on the X chromosome (sex-linked) and its presence results in a deficiency for the enzyme hypoxanthine-guanine phosphoribosyl transferase (HGPRT). Only males are affected; affected females die as embryos. A rather shocking fact about this fortunately rare disease is that affected boys have abnormal posture and spastic, involuntary movements (among other abnor-

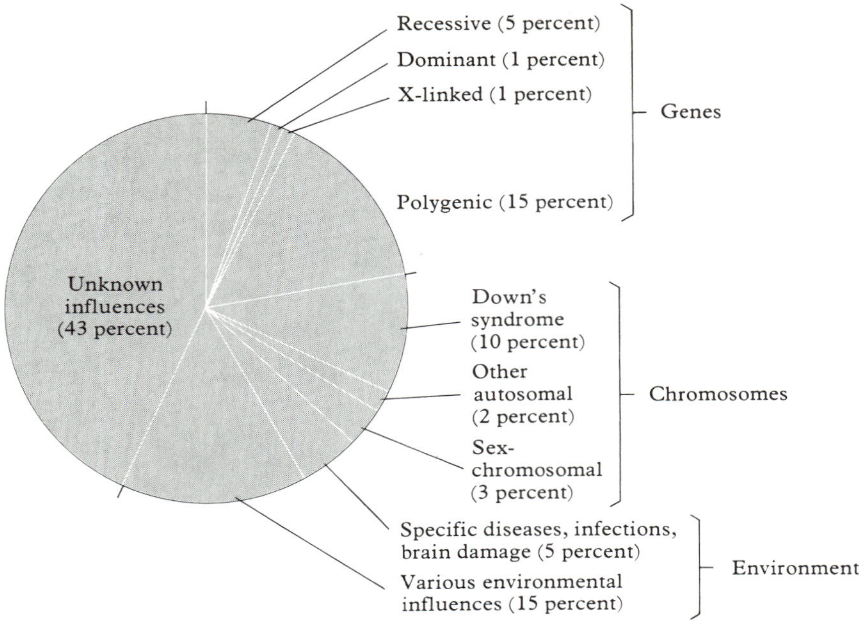

20–9
Estimates of the relative incidence of types of mental retardation that result from genetic and environmental factors. (After Penrose, in Wendt (ed.), Genetik und Gesellschaft, *Wiss. Verlagsges., Stuttgart, 1971.)*

malities), and they regularly engage in a bizarre form of self-destructive behavior. Unless their teeth are removed, they invariably bite off the tissue of their lips, and they frequently use their teeth to tear at the flesh of their hands, sometimes mutilating themselves severely. (The self-mutilation is thought to be involuntary, because they appear to be terrified by their self-destructive activities.) Although the exact relationship between the absence of the enzyme HGPRT in certain parts of the brain and the presence of this unfortunate behavior remain to be explained, the Lesch-Nyhan syndrome is an example of abnormal behavior that as far as we know is determined by simple genetic factors alone.

Genetic factors are also of obvious importance in producing the abnormal behaviors manifested by many (but not all) people who have abnormal numbers of chromosomes. The mental retardation associated with Down's syndrome (trisomy-21) is one example, and, as shown in Figure 20-9, this relatively common disorder accounts for about one-tenth of the total of the people in the United States (about 3 percent of the population) who have some form of mental retardation. On the other hand, the behavior of many people who have abnormal sex chromosome constitutions, such as XO, XXX, and XXY, is apparently normal in spite of their chromosomal abnormalities. (We will discuss the possible behavioral effects of the chromosome constitution XYY later in this chapter.) Overall, the study of the behavioral effects of abnormal numbers of human chromosomes suggests that normal mental activities depend on the effects of many genes, located on many chromosomes, and this is borne out by the rate at which severe mental retardation occurs among the offspring of matings between close relatives.

A good example of an abnormality of human behavior that probably depends on the interaction of many genes and a multitude of environmental factors is *schizophrenia*. Contrary to popular usage of the word, this serious behavioral abnormality, of which several types are officially recognized, is not manifested by "split personality," but rather by a split between thoughts and feelings, and by a loss of contact with the environment (there are many other features). Schizophrenia is by no means rare. It is estimated that in the 1970s at least 2 million people in the United States alone either have schizophrenia or will develop symptoms of the disease sometime during their lives.

Although the issue is far from settled, most geneticists are convinced that genetic factors influence the development of certain types of schizophrenia. Part of the evidence for this comes from studies of monozygotic twins raised apart. As shown in Table 20-4 both develop schizophrenia much more commonly than do both dizygotic twins who are raised in different environments.

But what about the genetics of human behavioral traits that are neither so obvious nor so extreme as severe mental retardation, self-destructive activities, or schizophrenia, but that could nonetheless be of some importance to the future evolution of the human species? For example, what is the genetic basis of the tendency for compassion, for altruism, for perseverence, for leadership, for realism, for wisdom, for curiosity, or for any of the seemingly endless qualities of human behavior that can be singled out, labeled, and measured by some kind of psychological test? As of the late 1970s it has yet to be determined whether most measurable behavioral differences between human beings result primarily from genetic or environmental factors. Nonetheless, there is every reason to expect that such complex and sometimes ill-defined behavioral traits result from the interaction of many genes and a great variety of environmental factors.

A further difficulty in assessing the genetic basis of many aspects of human behavior is that a person's behavior rarely remains the same for any extended period of time. Most of us do not behave the same way we did 10 years ago (or 20 or 30 years ago), and few of us will be behaving as we do now 10 or 20 years hence. Overall, the genetic basis of any human behavior is not only complex, hard to define, and difficult to measure, but changeable as well.

In the end, we have little evidence about the genetic basis of the behavior of individual people, and it is most often impossible to accurately assess the influence that the environment has on it. Nonetheless, the relative contributions of genes and environment to human behavior are sometimes of considerable social interest, in large part because of eugenic proposals that bear on human behavior. We now turn our attention to a human characteristic that is socially important in spite of the fact that its genetic basis is poorly understood and that it is subject to a great number of environmental influences: intelligence.

TABLE 20-4
Concordance of schizophrenia in twins.

COUNTRY	YEAR	MONOZYGOTIC TWINS		DIZYGOTIC TWINS	
		NUMBER OF PAIRS STUDIED	PERCENT CONCORDANCE	NUMBER OF PAIRS STUDIED	PERCENT CONCORDANCE
Denmark	1965	7	29	31	6
Germany	1928	19	58	13	0
Great Britain	1953	26	65	35	11
Japan	1961	55	60	11	18
Norway	1964	8	25	12	17
United States	1946	174	69	296	11

Source: From *Heredity, Evolution and Society,* 2d ed., by I. Michael Lerner and William J. Libby. W. H. Freeman and Co. Copyright © 1976.

The Genetics of IQ Scores

Intelligence quotient, or IQ, is best defined, not as a measure of a person's "intelligence," but as a measure of a person's ability to perform well on IQ tests. This definition is appropriate because it is impossible to define intelligence to everyone's satisfaction, and even if we could define the term, we could not measure it by the same yardstick in all human populations. (To get some idea of the difficulties, how would you measure the intelligence of European college students as compared with that of people of the same age group who live in the Amazonian rain forest, where they must survive without the benefits of the technological achievements of the culture of the former group?) IQ tests were developed to measure some of the mental aptitudes of white middle-class people who live in the United States and in Europe. IQ test results are informative in that they provide a reliable way of predicting success in those environments, as judged by standards considered appropriate to those particular cultures. Thus the degree to which a person is able to score high on IQ tests is important only insofar as it is important to achieve success as measured by academic, occupational, social, and other standards within a given cultural framework. Judgements of what constitutes success are always open to question and subject to change. Whatever its ultimate importance, within white middle-class populations the ability to score high on IQ tests appears to be strongly influenced by genetic factors.

As usual, the main problem in trying to figure out the genetic basis of IQ scores is that it is impossible to accurately sort out the genetic and environmental influences that interact to determine how a person performs on IQ tests. Nonetheless, there are several ways of roughly estimating the genetic component of IQ scores. Monozygotic twins that are raised apart are one source of information, and studies of them indicate a high degree of heritability. (For traits that show continuous and gradual variation, whether between twins or among other relatives, the degree of similarity is best measured by the value of the statistical quantity known as the *correlation coefficient,* whose numerical value can vary between 0 and 1. In general, a high value of the correlation coefficient suggests a high heritability. See Figure 20-10.)

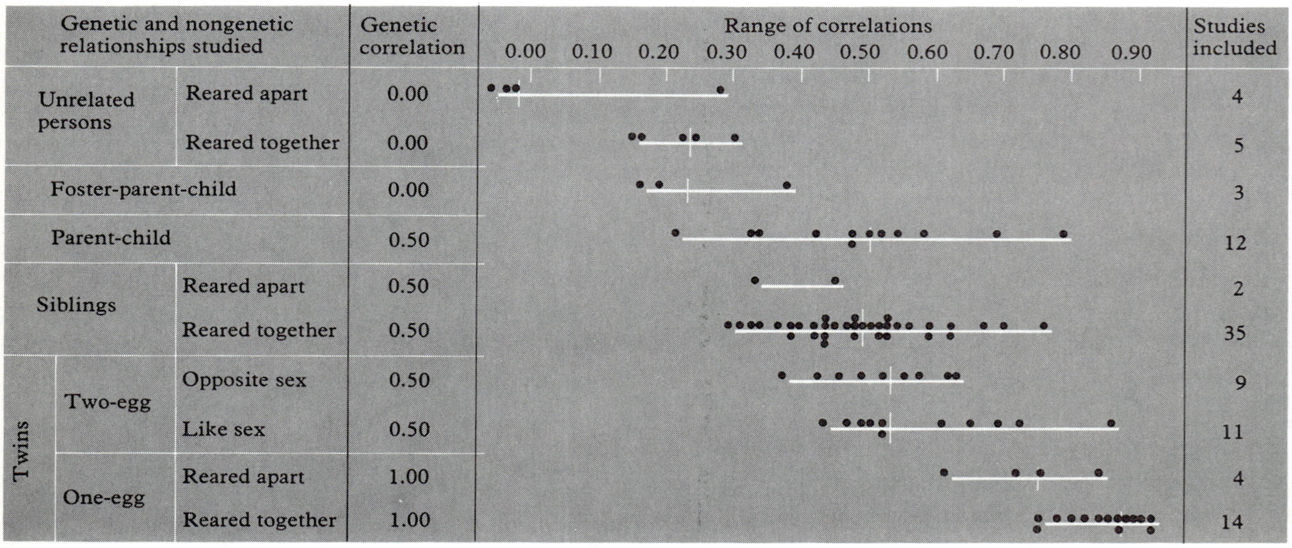
Genetic and nongenetic relationships studied
Genetic correlation
Range of correlations
0.00 0.10 0.20 0.30 0.40 0.50 0.60 0.70 0.80 0.90
Studies included
Unrelated persons
Reared apart
0.00
4
Reared together
0.00
5
Foster-parent-child
0.00
3
Parent-child
0.50
12
Siblings
Reared apart
0.50
2
Reared together
0.50
35
Twins
Two-egg
Opposite sex
0.50
9
Like sex
0.50
11
One-egg
Reared apart
1.00
4
Reared together
1.00
14

20–10
A summary of correlation coefficients concerning IQ scores compiled by L. Erlenmeyer-Kimling and L. F. Jarvik from various sources. The horizontal lines show the range of correlation coefficients for "intelligence" between persons who are related to various degrees either by genes or by environment. (From I. Michael Lerner and William J. Libby, Heredity, Evolution, and Society, *2nd Ed. W. H. Freeman and Company. Copyright © 1976.)*

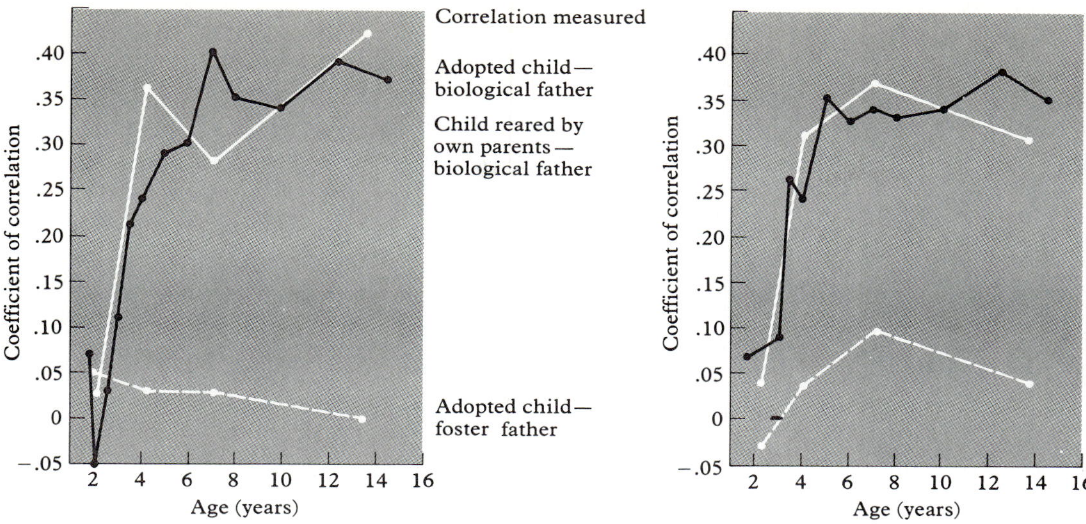

20–11
Correlations between the IQ of children and the education of biological and foster parents. (From M. Honzik, and M. Skodak and B. M. Skeels.)

The genetic component of IQ scores can also be estimated by studies of adopted children. As shown in Figure 20-11, the IQs of parents and their offspring are directly related, whether the offspring are raised by their biological parents or not. On the other hand, the IQs of adopted children are not affected by those of their foster parents.

The heritability of IQ scores as estimated in different populations and by various methods ranges from about 0.3 (30 percent) to 0.9 (90 percent) and this wide range of estimates serves as a reminder that heritability is impossible to accurately measure in human populations. Nonetheless, most studies give values of about 0.5 or 0.6 (50 to 60 percent): this implies that about 40 or 50 percent of the total variation in IQ scores within the population being studied can be attributed to the interaction of genes and environment, or to environmental factors alone.

We have already discussed the role of nutrition in the development of normal mental capacities, and many other environmental factors are known to affect, not only normal mental development, but IQ scores as well. Among the most important environmental components are psychological and social factors, and several are known to influence IQ scores. First, as shown in Figure 20-12, there is evidence that offspring of larger families tend, overall, to have lower IQ scores than those from smaller families. This has been interpreted (though not without criticism) as evidence that large families may provide an environment that is less satisfactory to the development of higher IQ scores. Second, some incomplete but ongoing studies of the effects of the environment within a given social class indicate that the presence of an "enriched," as opposed to a "deprived," environment may favor the development of higher IQ scores. This refers to the fact that most children who have lower IQ scores come from homes in which the kinds of mental activities measured by IQ scores are not emphasized. And third, the children of people who have higher occupational status tend to have higher IQs than children whose parents are of lower occupational status. As shown in Table 20-5, this is true of the populations studied in the United States, Russia, and England.

Although the reasons for the variation observed between the IQ scores of whites of different socioeconomic classes are poorly understood and sometimes hotly disputed, the controversy surrounding them is dwarfed by the ignorance

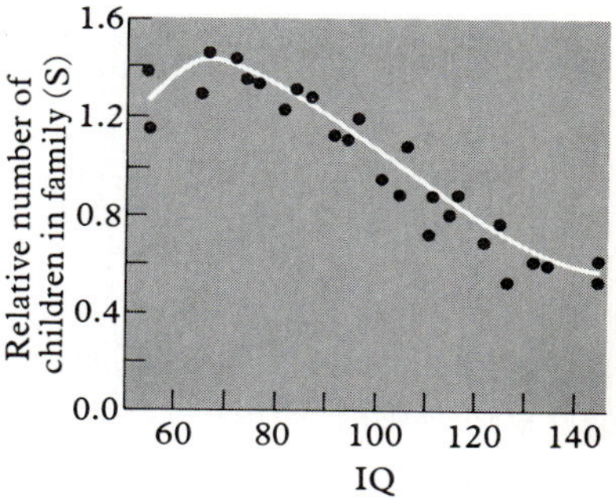

20–12
Data concerning the relation of family size and average IQ. The relative number of children in each family is the actual number of children divided by the average number of children per family in the whole sample. (From Human Diversity, *by Kenneth Mather, Free Press, 1964.)*

TABLE 20-5

Data about the relation between occupational status of parents and the average IQ of biological and adopted children.

OCCUPATIONAL STATUS OF PARENTS	AVERAGE IQ OF CHILDREN					AVERAGE SIZE OF ENGLISH FAMILY
	U.S.	USSR	CHICAGO		ENGLAND	
			ADOPTED	BIOLOGICAL		
Professional	116	117	113	119	115	1.73
Semiprofessional	112	109	112	118	113	1.60
Clerical and retail business	107	105	111 (clerical and retail business, skilled)	107 (clerical and retail business, skilled)	106	1.54
Skilled	105	101			102	1.85
Semiskilled	98	97	109	101	97	2.03
Unskilled	96	92	108	102	95	2.12

Source: From *Heredity Evolution and Society,* 2d ed., by I. Michael Lerner and William J. Libby. W. H. Freeman and Co. Copyright © 1976.

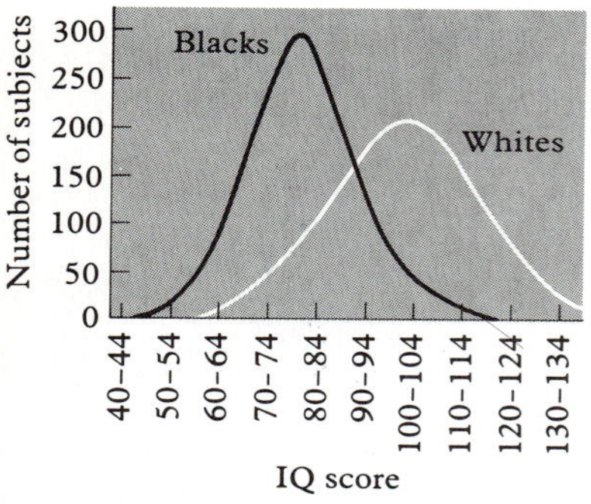

20–13
The distribution of IQ test scores of white school children from all social classes as sampled in 1960 across the United States, as compared with the scores of black school children from the schools of Alabama, Georgia, Florida, Tennessee, and South Carolina in 1963. (From "Intelligence and Race," by W. F. Bodmer and L. L. Cavalli-Sforza. Copyright © 1970 by Scientific American, Inc. All rights reserved.)

and confusion surrounding the average IQ scores of various human races, particularly those of blacks and whites in the United States. As shown in Figure 20-13, the difference between the average IQ scores of blacks and whites in the U.S. population is about 15 IQ points. (The graph shows the distribution of the IQ test scores of white school children of all social classes as sampled in 1960 from across the United States, compared with the scores of black school children from Alabama, Florida, Georgia, Tennessee, and South Carolina in 1963.) A glance at the figure reveals that there is a difference between the average IQ scores of blacks and whites raised in the United States. The question is: what does this difference mean, and how important is it?

Some white academicians, among whom Arthur Jensen and William Shockley are perhaps the most vocal, have asserted that most, if not all, of the differences in IQ scores between black children and white children can be attributed to genetic factors. But it is in fact impossible to sort out the exact contributions of genetic and environmental factors to IQ scores *within* either the black or the white American populations, so it is clearly impossible to make meaningful comparisons *between* the two. No matter how high the heritability of IQ scores actually is, the fact remains that, given the obvious cultural differences between populations of middle-class whites and ghetto-dwelling blacks, we would *expect* to find a difference in their performances on IQ tests. In the face of vast environmental differences, it seems premature at best to attribute the expected IQ score difference entirely to systematic genetic differences between the two populations.

In the opinion of many geneticists, the black–white differences in IQ scores most likely reflect, not genetic, but rather cultural and other environmental factors, such as nutrition, family size, and the psychological impact of racism. In this regard it should be noted that differences in IQ between whites of upper and lower socioeconomic groups are somewhat larger than the IQ differences between American blacks and whites. Also, environmental factors probably vary more between than within these two groups, and IQ tests are culturally biased toward middle-class whites. But this is not to say that genetic factors may not influence IQ differences between races. The point is that most of the differences in IQ scores between American blacks and whites could probably be eliminated by changing the social environments, including the schools, so that they become the same for both groups. Finally, IQ tests only predict a person's success in a middle-class academically oriented environment. They are limited, as is any human endeavor, by the knowledge and experiences of those persons who devise them. In the end, IQ scores measure an aspect of human behavior that, like most other features of the ways in which people behave, is desirable in the opinion of some people and undesirable in the opinion of others.

We now turn our attention to an aspect of human behavior that is clearly undesirable to most people, and that has been said by some to be associated with the presence of an extra Y chromosome. We will now discuss the relationship between criminal behavior and the chromosome constitution XYY.

The Behavior of Men Whose Sex-Chromosomes are XYY

We have already discussed the way in which males whose sex chromosomes are XYY may be produced from various accidents that occur during cell division. XYY males are born with surprising frequency. It is estimated that about one in every 1000 newborn males has XYY sex chromosomes. Most XYY men are taller than those whose chromosomes are XY and some of them have severe acne, but most of them appear otherwise normal.

The controversy that has surrounded the XYY genotype concerns criminal behavior. In 1965 it was reported that the genotype XYY was encountered among men in a certain wing of a "mental-penal institution" (namely, the Carstairs maximum security hospital in Scotland) at a much higher rate than among the general population. (Seven out of 197 men were XYYs, which is about 36 times their frequency in the population at large.) Since that time, other well-documented studies have been published from around the world. The conclusion, though hotly contested by some, seems inescapable to others. It is this: Men of sex-chromosome constitution XYY are somewhat more likely to be incarcerated in a "mental-penal institution" than men whose sex chromosomes are XY. (In general, the offenses of XYY men are similar to those of XYs, and contrary to some earlier reports, XYY men do not appear to be concentrated among the most dangerous, aggressive, or violent inmates.)

Nonetheless, the great majority (at least 98 percent) of men in mental-penal institutions have XY sex chromosomes, and only a small proportion of the total number of men who have sex-chromosome constitution XYY engage in criminal behavior. Based on the percent of XYY men in mental or penal institutions, compared with the incidence of XYY males in the general population, it is estimated that at least 96 percent do not behave in ways that result in their being institutionalized.

Criminal behavior is so widespread and it affects the lives of so many noncriminals that the relation between a slightly increased tendency for criminal behavior and the presence of an extra Y chromosome may be of more than academic interest. But at the same time the issue is so surrounded by prejudice and concerns such important social considerations that there is good reason to argue that it ought not be investigated further. As usual, the problem is that it is impossible to accurately assess the effects of the genetic versus the environmental factors. Environmental factors, such as growing up in a ghetto or associating with people who commit criminal acts, are of great importance in the development of criminal behavior.

Now consider some of the moral and ethical difficulties that would arise in the course of a genetic screening program designed to detect XYY males in the general population. Suppose you have just identified an XYY male infant. What would you advise his parents? What would you expect the parents to do because of your advice? Would you tell the child that he is an XYY and thereby implant in his mind the notion that he may grow up to be criminal? Would you mind if the chromosomes of your own male offspring were routinely screened for the presence of an extra Y chromosome without your knowing about it?

As of May 1976 you need not give much urgent thought to these matters, at least if you live in the United States. The only XYY screening program in the United States has, at least for the present, been shut down because of unrelenting pressure from people opposed to XYY screening. In the spring of 1975 the faculty of Harvard Medical School, brought to caucus by geneticists and psychiatrists interested in continuing a screening program that had been in existence since 1968, voted 200 to 30 to continue the screening project. Nonetheless, a few months later the screening project was shut down because those in charge of the project said they were worn out by the pressures of some activist groups that opposed XYY screening.

How should we proceed in this delicate area that may or may not have some genetic basis and that may or may not be subject to potential eugenic measures? Clearly, that depends on the judgements of individuals, and opinions concerning the matter are abundant and oftentimes strongly felt. Perhaps it would be appropriate to follow the example of molecular geneticists who—when recently confronted by the potentially disastrous effects of their research involving hybrid DNA molecules—convened a series of international conferences to discuss the problem. Any such conference on XYY chromosomes should probably include not only geneticists, sociologists, criminologists, social philosophers, and other humanists, but also people of sex chromosome constitution XYY and representatives of well-informed groups of people that have strong interest in or opposition to XYY screening.

As if all of the uncertainty surrounding the genetics of IQ scores and the behavior of XYY males were not enough, we now turn our attention to a brief consideration of how natural selection may affect human populations in the future, as it does today. (In the discussion of future human evolution that follows, you should remember that, although speculation is sometimes fascinating, sometimes dangerous, and sometimes rewarding, in the end it is just speculation.)

How Natural Selection at Work Today May Affect Human Evolution in the Future

Like all other living things, human beings have evolved and are evolving. But understanding how evolution works does not allow us to make predictions about the future course of the evolution of our own species, or of any other. Nonetheless, we can do some cautious, perhaps meaningful, speculating about the future evolution of the human species, provided that we base our speculations on the assumption that natural selection is at work in the human population today and will be in the future.

You will recall that natural selection is at the core of the theory of evolution, and that in essence it consists of differential reproduction within populations in which individuals differ from one another genetically. Because of natural selection, individuals best suited to survival in a given environment leave more descendants than those who are not so well suited, and because of the effects of natural selection, living things are precisely adapted to the environment. It has been argued that natural selection is no longer important in the evolution of the human species because few human beings now live in "a state of nature," where natural selection can result in the maintenance of human adaptations of benefit in local environments. Moreover, as you will recall from our discussion of human races, differences in body surfaces (which are obviously influenced by natural selection) are now largely irrelevant to our species because people adapt to the environment primarily by means of behavior, not by means of their body surfaces. But this does not mean that natural selection is no longer a major factor in human evolution. In fact, there are at least three major ways in which natural selection operates in human populations today—prenatal selection, postnatal selection, and fecundity selection. Let us now discuss briefly each of these ways in which natural selection is at work in human populations today and speculate as to how they may be affected by some existing and by some proposed eugenic measures.

Prenatal selection refers to any genetic and environmental factors that result in death sometime between fertilization and birth. As you know, prenatal death is frequently the result of genetic factors, such as abnormal chromosome constitutions or homozygosity for disadvantageous alleles; as such, it is beyond our present eugenic reach. Although it is true that genetic counseling and prenatal diagnosis must have some influence on prenatal selection in human populations, any eugenic effect so far has been, and will probably continue to be, meager.

Postnatal selection occurs when infants born alive fail to survive to reproductive age. In large part postnatal selection can be attributed to genetic factors. Included in this category are many of the serious inborn errors of metabolism and other genetic diseases that we have discussed before. Although the effects of modern medical treatment have been spectacular and of enormous benefit to some people and to their families, most geneticists agree that medical treatment has so far had little effect on the average gene frequencies in the global human population. Nonetheless, it must be admitted that the human species could be affected in a very adverse way by the "genetic load" of detrimental alleles that has already begun to build up because of human intervention and that will surely continue to do so in the future.

Fecundity selection, which may take place both within and between populations, occurs when some genetically distinct members of the population leave relatively more descendants than others. Fecundity selection *between* human populations at the present time favors people who live in the less highly developed regions of the world, especially in Latin America, Africa, and Asia, but this pattern may change before too long as effective means of birth control become more widely available. Fecundity selection *within* human populations also occurs, but is hard to measure and subject to frequent changes. Nonetheless, some elitist prophets of doom have made fecundity selection within the United States population a social issue. Their assertion is that the average intelligence of the U.S. population is decreasing because people of lower socioeconomic groups, who have, on the average, lower IQ scores, also tend to have the most offspring. Implicit in this argument is the contention that genetic, not environmental factors, are most important in determining how individuals perform on IQ tests, and, as we have discussed, that question is far from settled. Also, the biological significance, if any, to the human species of a decrease in IQ scores would be far from clear. Overall, fecundity selection, although it does occur both within and between human populations, provides us with little basis from which to predict the future course of human evolution.

Fecundity selection is also largely beyond the reach of existing eugenic measures, and for good reason. Eugenic measures applied to fecundity selection would necessarily require judgements about what constitutes "superior" genotypes and the selective breeding (or artificial insemination) of people who have traits judged to be desirable. Given the social climate of the day, such undertakings are not at all likely to materialize. (Recall that social pressures recently shut down the only XYY screening program in the U.S., in spite of a resounding vote of confidence from professors at the Harvard Medical School to continue the program.) One can only expect that genetic questions that have social overtones will continue to attract the interest of and to elicit reactions from various activist groups whose presence, for better or for worse, will surely strongly influence eugenic proposals of any kind.

In the end, the overall effect of natural selection within the human population today thus appears to be maintenance of the human species as it is at present. That is not surprising, because natural selection is usually a stabilizing influence that tends to put a brake on rapid or extreme evolutionary changes. This is because the extreme phenotypic variants within any population are much less likely to leave descendants than other members of the population.

Certain evidence in the fossil record also suggests a stabilizing influence of natural selection. The human species in its present form has been in existence for about 40,000 years, which though long by human standards is a mere flash in the pan of geological time. The fossil record shows that at least some trends in physical evolution, once they become established, continue for long periods of time, even as measured by geological standards. In general, they continue because they are directly or indirectly influenced by natural selection. One such trend among humans and other primates as well has been (within broad limits) the evolution of a more complicated and more capable brain, which is reflected in solid bone by an increase in cranial capacity—that is, by an increase in the volume of the brain. Yet in the last 40,000 years (a period of time in which somewhat more than 2,500 human generations have come and gone), the volume of the human brain has not changed at all. It is probably safe to assume that science fiction writers are misleading us when they conjure up images of our descendants hundreds of centuries hence, with overgrown brains encased in enormous globular skulls carefully balanced on pale, slender necks, or, worse yet, deposited in receptacles that fall far short of having the form and presence of a human body.

The fossil record also shows a slight tendency for an increase in height in the more recent stages of human evolution, but this is probably due mostly to environmental, rather than genetic, changes. Minor changes in human teeth, with a slight tendency for reduction in the number of wisdom teeth may also have occurred in the past 30,000 years or so, though this is less well documented than the change in stature.

Other than that, as judged by the fossil record, the physical evolution of human beings appears to be at a standstill, or at least proceeding at a rate that is so slow as to be unnoticeable. Of course, only time will tell, but most of us will probably be long dead before any noticeable physical changes in the human species occur. Nonetheless, we have all experienced social and other cultural changes, some of which occur so fast that they bewilder us. Whatever its final basis, whether genetic or environmental, whether measured by IQ scores or by height, whether judged by cranial capacity or by other measures, the fact is that human beings of all races are wonderfully variable. And this enormous stockpile of human differences is our species' greatest asset toward the future. *Variability*—whether physical, behavioral, biochemical, or otherwise—is what is important in the future evolution of any species, including our own. It is to be hoped and expected that people will always be at least as variable as they are today, and that by some means, be it natural or eugenic, people will continue to evolve, and will exist for a very long time to come.

Summary

Eugenics includes all programs and proposals whose aim is genetically improving the human species either by increasing the frequencies of desirable traits (positive eugenics), or by decreasing the frequencies of undesirable traits (negative eugenics). What is desirable or undesirable always depends on judgements of individuals, and differences of opinion concerning eugenic measures often lead to heated social controversy.

Genetic counseling and prenatal diagnosis have greatly benefited some individuals, but have had little effect on human genetics. Genetic engineering, perhaps by means of viruses containing "spliced-in" genes, will probably soon be a reality, but it is not likely to have any noticeable eugenic effects in the human population.

Heritability is a statistical concept that estimates how much of the phenotypic variation in any population is due to genetic factors (nature) as opposed to environmental factors (nurture). Although it is impossible to accurately assess the heritability of any human trait, the study of one-egg twins raised apart provides a good estimate of the relative importance of genetic and environmental factors.

Human behavior is so varied and so changeable that it is most often impossible to sort out the genetic and environmental factors that influence how a person behaves. Mental retardation, schizophrenia, and the ability to score high on IQ tests are behavioral traits known to be strongly influenced both by genetic and by environmental factors. Eugenic measures relating to human behavior engender the most heated controversies of all, as evidenced by the recent dispute over the significance of racial differences in IQ scores, and by the recent shutdown of a screening program for detecting XYY males.

Three types of natural selection are influencing the worldwide human population: prenatal selection, postnatal selection, and fecundity selection. The effects of natural selection on human body surfaces are no longer very important in human evolution, because people now adapt to the environment primarily by means of behavior, not body surfaces. Overall, natural selection acts as a stabilizing influence in human evolution, and our species has not undergone any noticeable physical change in the past 40,000 years at least. Yet social and other cultural changes in human behavior occur frequently and may take place very rapidly, as is well known to all of us. As long as people remain variable they will continue to evolve, and it is to be hoped and expected that they will do so for a very long time to come.

Suggested Readings

1. "Prenatal Diagnosis of Genetic Disease," by Theodore Friedmann. *Scientific American,* Nov. 1971, Offprint 1234. A discussion of amniocentesis and other diagnostic techniques that have both genetic and social aspects.
2. "The Manipulation of Genes," by Stanley N. Cohen. *Scientific American,* July 1975, Offprint 1324. Discusses some of the molecular details and some of the social implications of the newly devised technique of genetic engineering.
3. "Genetic Load," by Christopher Wills. *Scientific American,* March 1970, Offprint 1172. How the accumulated mutations of any species are usually detrimental but at the same time may be a priceless genetic resource.
4. "Intelligence and Race," by W. F. Bodmer and L. L. Cavalli-Sforza. *Scientific American,* October 1970, Offprint 1199. This recap of the differences in IQ scores between American blacks and whites concludes that the heritability of IQ scores cannot be accurately determined from the data presently available.
5. "Behavioral Implications of the Human XYY Genotype," by Ernest B. Hook. *Science,* Vol. 179, 12 Jan. 1973. A review of some of the data on this complicated subject.

APPENDIX

Some Genetics Problems Concerning Human Pedigrees

You may want to test your understanding of the patterns of inheritance discussed in Chapters 16 and 17 by working the following problems. Answers are found at the end of the problem set.

1. Huntington's chorea is a rare, fatal disease of the nervous system whose symptoms are not exhibited until middle age. It is inherited as an autosomal dominant trait. Suppose that an apparently normal man in his early twenties learns that his father has just been diagnosed as having Huntington's chorea.
   a. What are the chances that the son will eventually develop the disease himself?
   b. If the son does not develop the disease, what are the chances that his offspring will have Huntington's chorea?
2. Two bald parents have four children, two bald and two with normal hair. Assuming that it is governed by a single pair of alleles, is this kind of baldness best explained as an example of dominant or of recessive inheritance?

3. Maple syrup urine disease is a rare inborn error of metabolism that derives its name from the odor of the urine of affected people. Those in whom the disease is untreated are severely mentally retarded, and they usually die as infants. Once it has appeared in a given family the disease tends to recur, but the parents of affected children are always normal. Assuming that a single pair of alleles governs the occurrence of this disease, what does this suggest about the inheritance of maple syrup urine disease?

4. About 7 percent of Caucasians in the United States cannot smell the odor of musk. If both parents cannot smell musk, all of their children will be unable to smell it. On the other hand, two parents who can smell musk generally have children who can also smell it; only a few of their offspring will be unable to smell musk. Assuming that a single pair of alleles governs this trait, what does this suggest about the inheritance of the ability to smell the odor of musk?[1]

5. Total color blindness is a rare condition that is inherited as an autosomal recessive trait. Affected people see the world only in shades of gray and can see best in dim light or in the dark. A woman whose father is totally color-blind intends to marry a man whose mother was totally color-blind. What are the chances that they will produce affected offspring?

6. As discussed in Chapter 16, albinism is inherited as an autosomal recessive trait. An albino man marries a normally pigmented woman and they have nine children, all normally pigmented. What are the genotypes of the parents and of the children?

7. A normally pigmented man whose father was an albino marries an albino woman whose parents were both normally pigmented. They have three children, two normally pigmented and one albino. List the genotypes of all these people.

8. A man who has sickle-cell trait marries a woman whose father had sickle-cell disease.
   a. Can they produce offspring who have neither sickle-cell disease nor sickle-cell trait?
   b. What proportion of their offspring will have sickle-cell disease?
   c. What proportion of their offspring will have sickle-cell trait?

9. A man of blood group O marries a woman of blood group A. The woman's father was of blood group O. What are the chances that their children will belong to blood group O?

10. In the following case of disputed paternity, which of the possible fathers can be excluded as the real father? The mother is of blood group B, the child is of blood group O, one possible father is of blood group A, and the other is of blood group AB.

11. Four babies were born in a hospital on a night in which an electrical blackout occurred. In the confusion that followed, their identification bracelets were mixed up. Conveniently, the babies are of four different blood groups:

---

1. From *Genetics, Evolution and Man*, by W. F. Bodmer and L. L. Cavalli-Sforza. W. H. Freeman and Company, 1976.

O, A, B, and AB. The four pairs of parents have the following blood groups: O and O, AB and O, A and B, B and B. Which baby belongs to which parents?[2]

12. A woman who has unusually short fingers marries a man who has fingers of normal length, and they have four children, two of each sex. One of their sons and one of their daughters have unusually short fingers.
    a. Draw the pedigree.
    b. If finger length is governed by a single pair of alleles, could the pedigree be explained by autosomal dominant inheritance?
    c. Could the pedigree be an example of autosomal recessive inheritance?
    d. Could the allele responsible for this trait be located on the X chromosome?

13. Assume that the following pedigree is for a trait that is rare in a particular population.

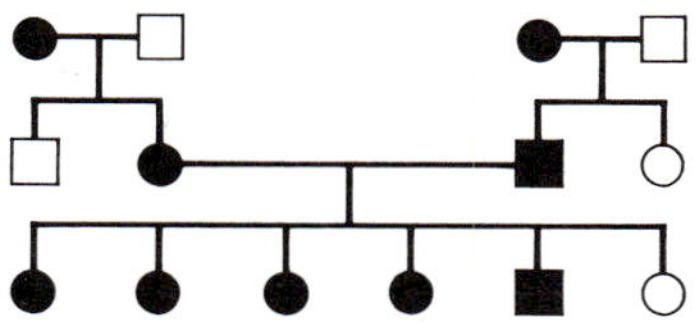

    Indicate whether each of the following patterns of inheritance is consistent with or excluded by this pedigree.[3]
    a. Autosomal recessive.
    b. Autosomal dominant.
    c. X-linked recessive.
    d. X-linked dominant.
    e. Y-linkage.

14. Red–green color blindness is a relatively common condition that is inherited as an X-linked recessive. A normal woman whose father was red–green color-blind marries a man who has normal vision.
    a. What proportion of her sons would you expect to be red–green color-blind?
    b. If she married a man who was red–green color-blind, what proportion of their sons would you expect to have normal vision?
    c. If she married a man who was red–green color-blind, what proportion of their daughters would be carriers?

15. What pattern of inheritance *best* accounts for the following pedigree?

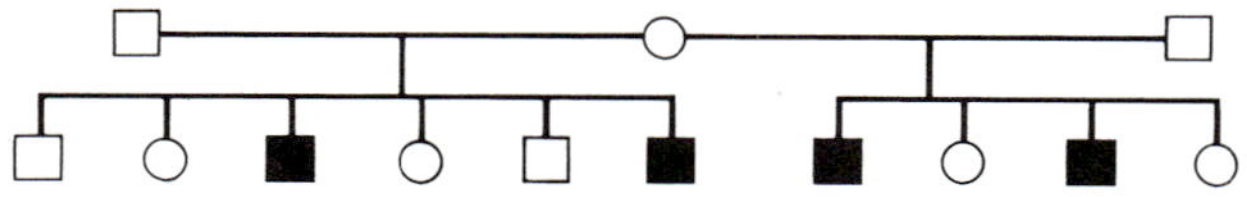

2. From *Heredity Evolution and Society,* by I. Michael Lerner and William Libby. W. H. Freeman and Company, 1976.

3. From *General Genetics,* by Adrian Srb, Ray D. Owen, and Robert Edgar. W. H. Freeman and Company, 1965.

16. As discussed in Chapter 17, hemophilia is inherited as an X-linked recessive trait. An apparently normal woman whose father was a hemophiliac marries a normal man.
    a. What proportion of their sons will have hemophilia?
    b. What proportion of their daughters will have hemophilia?
    c. What proportion of their daughters will be carriers?

17. Suggest three patterns of heredity that could account for the following pedigree. Which is least likely?

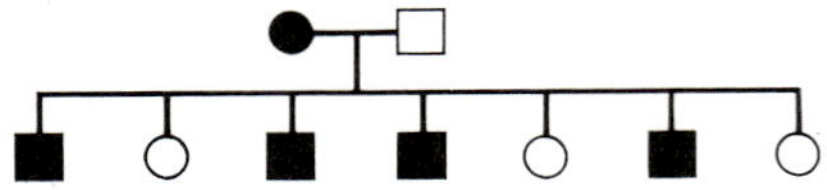

18. A man who is red–green color-blind (X-linked recessive) has four normal sons by his first wife and a color-blind daughter by his second wife. Both wives have normal vision. What is the genotype of each wife?

19. A short-fingered man (autosomal dominant) marries a woman who has normal hands and is red–green color-blind (X-linked recessive). List the possible phenotypes of their sons and daughters.

20. In the following pedigree a dot represents the presence of an extra finger and a shaded area represents the occurrence of an eye disease.[4]

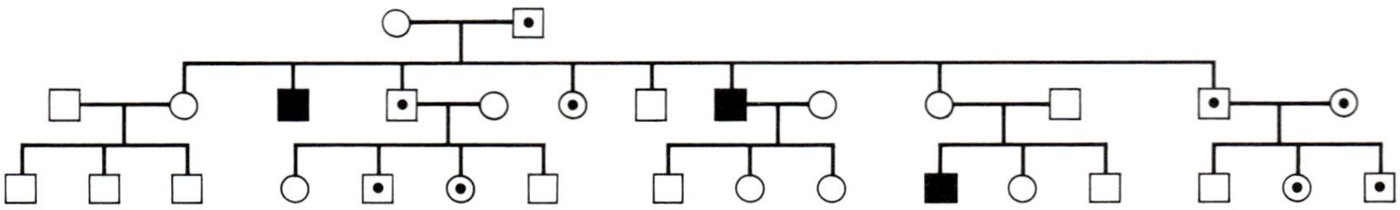

    a. What can you figure out about the inheritance of the extra finger?
    b. What two patterns of heredity might explain the inheritance of the eye disease?

21. Congenital deafness results when a person is homozygous for either or both of two recessive alleles, *d* and *e*. Both of the corresponding dominant alleles *D* and *E* are required for normal hearing, and the two pairs of alleles are inherited independently. A deaf man of genotype ddEE marries a woman with normal hearing who is of genotype *DdEe*. What proportion of their offspring will be deaf?

22. Complete the following table.

CONDITION	SEX CHROMOSOMES	NUMBER OF BARR BODIES
Normal woman		
Normal man		
	XO	
	XXY	
Down's syndrome		

4. From *An Introduction to Genetic Analysis*, by David T. Suzuki and Anthony J. F. Griffiths, W. H. Freeman and Company, 1976.

23. Would you expect a person of sex chromosome constitution XXXX to have higher, lower, or approximately equal concentrations of G6PD as compared to a person whose sex chromosomes are XXXY?

24. A woman who has Turner's syndrome is found to have hemophilia, yet neither of her parents have the disease. How is this possible?

25. How many Barr bodies would be found in a person who is a sexual mosaic with sex chromosomes XX/XXXY?

Answers

1. a. The chances are 50:50. Because the disease is rare, the affected father is most likely heterozygous (of genotype *Hh*, *H* for Huntington's chorea). All of his offspring therefore have a 50:50 chance of inheriting the dominant allele.
   b. Virtually zero, assuming that he marries an unaffected woman. Huntington's chorea can arise by spontaneous mutation, but this is a very rare event.

2. Dominant. If some of the children are not bald, the parents must be heterozygous for the trait (of genotype *Bb*.) If this kind of baldness were recessive, both parents would have to be homozygous recessive (of genotype *bb*); thus they could produce bald offspring only.

3. This pattern suggests that maple syrup urine disease is inherited as a recessive trait.

4. This pattern suggests that the ability to smell the odor of musk is dominant and that the inability to do so is recessive.

5. The chances are one in four. The man and woman must both be heterozygous (of genotype *Cc*,) so these are the kinds of offspring they can produce:

	C	*c*
C	*CC*	*Cc*
c	*Cc*	*cc*

Thus, on the average, one-fourth of their offspring would be normal (*CC*), one-half would be carriers (*Cc*), and one-fourth would be totally colorblind (*cc*).

6. The man is an albino, so he must be homozygous recessive (of genotype *aa*). None of the nine children is an albino, but each must be a carrier (of genotype *Aa*). Since there are no albinos among the nine children, the mother is apparently of genotype *AA*.

7. The man must be of genotype *Aa*, and his father of genotype *aa*. Similarly, the woman must be *aa* and her parents must both be *Aa*. Two of the children are of genotype *Aa*, and the other is *aa*.

8. a. Yes. Both parents must be heterozygous, so they can produce the following kinds of offspring:

	Hb^S	Hb^A
Hb^S	Hb^SHb^S	Hb^SHb^A
Hb^A	Hb^SHb^A	Hb^AHb^A

(Hb^AHb^A = normal, Hb^SHb^A = sickle-cell trait, Hb^SHb^S = sickle-cell disease.)

b. one-fourth (genotype Hb^SHb^S).

c. one-half (genotype Hb^SHb^A).

9. The chances are 50:50. The woman must be of genotype I^AI^O, so these parents can produce the following offspring:

	I^A	I^O
I^O	I^AI^O	I^OI^O
I^O	I^AI^O	I^OI^O

Thus about one-half of their offspring will be of genotype I^OI^O and will therefore belong to blood group O.

10. The man of blood group AB cannot be the real father. The mother is apparently of genotype I^BI^O. Her child is of group O and therefore must have inherited the I^O allele from each parent. The man of group AB can contribute only I^A or I^B and therefore cannot be the father. The man of group A could be of genotype I^AI^O, but he could also be of genotype I^AI^A. Thus we can neither exclude the man of group A nor conclude that he is the real father.

11. a. Baby O could only belong to parents O and O, because the parents must be of genotype I^OI^O.

b. Baby AB must belong to parents A and B because only they could produce an offspring of genotype I^AI^B.

c. Of the remaining two, baby A cannot belong to parents B and B, because they must be of genotype I^BI^B ro I^BI^O. Baby A must therefore belong to parents AB and O.

d. Hence, baby B must belong to parents B and B.

12. a. The pedigree is this:

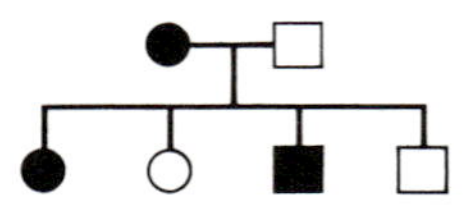

b. Yes. In fact, this trait (officially known as *brachydactyly*) was the first autosomal dominant trait described among human families.

c. Yes. But this is not likely to be an example of autosomal recessive inheritance because the father would have to be a carrier, which is unlikely.

d. Yes. The trait cannot be inherited as an X-linked recessive because one of the sons does not manifest the trait. But the pedigree is consistent with X-linked dominant inheritance.

13. a. Autosomal recessive inheritance is excluded because the mating of two affected people produced a normal daughter.
    b. The pedigree is consistent with autosomal dominant inheritance.
    c. X-linked recessive inheritance is excluded because an affected woman produced a normal son.
    d. X-linked dominant inheritance is also excluded because an affected man produced a normal daughter.
    e. Y-linkage is excluded because of the presence of affected women.

14. a. One-half. The woman must be a carrier. All sons receive one or the other of their mother's X chromosomes, so the chances are 50:50 that any son would receive the chromosome with the abnormal allele.
    b. One-half. No sons receive their father's X chromosome, so the fact that the father is affected does not alter the chances that his sons will be red–green color-blind.
    c. One-half. All daughters receive thier father's X chromosome and one or the other of their mother's X chromosomes. Thus this marriage would produce color-blind and carrier daughters in roughly equal proportions.

15. X-linked recessive inheritance. The pedigree could also be explained by autosomal recessive inheritance, but that would require all three parents to be carriers of the same allele.

16. a. One half. The woman must be a carrier of hemophilia, and on the average half of her sons get the X chromosome that has the abnormal allele.
    b. None of them. An affected daughter could be produced only if the father were affected.
    c. One-half. On the average half of the daughters receive the abnormal X chromosome from their mother and therefore are carriers.

17. The pedigree is probably an example of X-linked inheritance, either dominant or recessive. Autosomal dominant inheritance cannot be excluded, but is least likely because all sons and no daughters are affected.

18. The first wife is most likely normal, and the second wife must be a carrier of red–green color-blindness.

19. All sons will be red–green color-blind and about half of them will have short fingers. All daughters will be carriers of red–green color blindness and about half of them will have short fingers.

20. a. The extra finger is probably an example of autosomal dominant inheritance.
    b. The eye disease is probably inherited as an X-linked recessive trait, but it could be an example of autosomal recessive inheritance.

21. One-half. All of the man's sperm cells must contain the factors *d* and *E*. But the woman can produce four different kinds of eggs, namely: *DE*, *De*, *dE*, and *de*. So we can construct the following table:

	dE	
DE	*DdEE*	(normal)
De	*DdEe*	(normal)
dE	*ddEE*	(deaf)
de	*ddEe*	(deaf)

22. The completed table:

CONDITION	SEX CHROMOSOMES	NUMBER OF BARR BODIES
Normal woman	XX	1
Normal man	XY	0
Turner's syndrome	XO	0
Klinefelter's syndrome	XXY	1
Down's syndrome	XX or XY	1 or 0

23. The concentrations of G6PD in these two persons would be approximately equal. The XXXX individual has one active X chromosome and three Barr bodies, and the XXXY individual has one active X chromosome and two Barr bodies.

24. This could happen if the woman's mother were a carrier and nondisjunction resulted in the daughter's receiving no X chromosome from her father. If the daughter received only her mother's abnormal X chromosome, she would show symptoms of hemophilia.

25. One or two Barr bodies would be found, depending on which cell line is examined. Some of this person's cells are XX and therefore have 1 Barr body, whereas other cells are XXXY and therefore have 2 Barr bodies.

Index

References to illustrations are given in italic type.